The Sally SERIES

Pure and Applied
UNDERGRADUATE TEXTS · 35

Number Theory and Geometry

An Introduction to Arithmetic Geometry

Álvaro Lozano-Robledo

AMERICAN MATHEMATICAL SOCIETY
Providence, Rhode Island USA

2010 *Mathematics Subject Classification.* Primary 11-XX, 11Gxx, 14Gxx.

For additional information and updates on this book, visit
www.ams.org/bookpages/amstext-35

Library of Congress Cataloging-in-Publication Data

Names: Lozano-Robledo, Álvaro, 1978- author.

Title: Number theory and geometry : an introduction to arithmetic geometry / Álvaro Lozano-Robledo.

Description: Providence, Rhode Island : American Mathematical Society, [2019] | Series: Pure and applied undergraduate texts ; volume 35 | Includes bibliographical references and index.

Identifiers: LCCN 2018056079 | ISBN 9781470450168 (alk. paper)

Subjects: LCSH: Arithmetical algebraic geometry–Textbooks. | Number theory–Textbooks. | AMS: Number theory. msc | Number theory – Arithmetic algebraic geometry (Diophantine geometry) – Arithmetic algebraic geometry (Diophantine geometry). msc | Algebraic geometry – Arithmetic problems. Diophantine geometry – Arithmetic problems. Diophantine geometry. msc

Classification: LCC QA242.5 .L6945 2019 | DDC 512.7/4–dc23

LC record available at https://lccn.loc.gov/2018056079

∞ The paper used in this book is acid-free and falls within the guidelines
established to ensure permanence and durability.
Visit the AMS home page at `https://www.ams.org/`

10 9 8 7 6 5 4 3 2 1   24 23 22 21 20 19

For Julia and Natalie,
the twin primes of my life

Knowing, my most beloved daughters, Julia and Natalie, that you are anxious to learn how to investigate problems in numbers, I have tried, beginning from the foundations on which the science is built up, to set forth to you the nature and power subsisting in numbers.

Adapted from the dedication
in Diophantus's *Arithmetica*

CONTENTS

PREFACE

> *Why are numbers beautiful? It's like asking why is Beethoven's Ninth Symphony beautiful. If you don't see why, someone can't tell you. I know numbers are beautiful. If they aren't beautiful, nothing is.*
>
> Paul Erdős

Geometry and the theory of numbers are as old as some of the oldest historical records of humanity. Since Euclid's *Elements* and Diophantus's *Arithmetica*, many excellent geometry and number theory texts have been written, including timeless classics such as [**HW38**]. As we shall lay out in more detail in Chapter 1, the approach of this book is slightly different from more traditional sources, in that the emphasis is in the interactions of number theory with geometry. The field of *arithmetic geometry*, which appears in the subtitle of this book, is indeed the study of the intersection of number theory (arithmetic) and algebraic geometry. The author's reason for this more geometric point of view is the following. Some of the traditional number theory textbooks may seem (to the student) a list of topics, each of which may be of important historical value but that do not readily appear to form a coherent set of topics, well integrated with each other (e.g., prime numbers, congruences, perfect numbers, quadratic reciprocity, and continued fractions). Of course, number theorists understand that these topics are deeply interconnected, and one way to highlight the interwoven nature of number theory is through geometry. In this text, the goal is to use geometry as the motivation to prove the main theorems in the book. For example, the fundamental theorem of arithmetic (the fact that every integer $n \geq 2$ has a unique factorization as a product of prime numbers) is a consequence of the tools we develop in order to find all the integral points on a line in the plane (i.e., the points (x_0, y_0) on a line $L : ax + by = c$ with integer coordinates x_0 and y_0). Similarly, Gauss's law of quadratic reciprocity and the theory of continued fractions naturally arise when we attempt to determine the integral points on a curve in the plane given by a quadratic polynomial equation.

In Chapter 1 we give a brief overview of the types of diophantine equations (i.e., systems of equations given by polynomials) that are the objects of study. The rest of the book is structured in three acts that correspond to linear, quadratic, and cubic curves, respectively.

(I) In Part 1 we introduce the basic tools of number theory. In particular, we discuss the integers and prime numbers and develop the theory of (linear) congruences. We also introduce some basic concepts of abstract algebra (groups, rings, fields) using congruence classes as a motivating example. These tools are applied to determine rational solutions of polynomials in one variable and the integral and rational points on lines in the plane.

(II) In Part 2 we study quadratic equations in one and two variables. We develop the theory of quadratic congruences, we describe the theorem of Hasse and Minkoswki (without a proof), and we also introduce continued fractions. The material is then used to find the integral and rational points on conics: parabolas, ellipses, and hyperbolas.

(III) Part 3 is a brief introduction to the theory of cubic curves. After discussing the projective line and projective space and learning how to work with singular cubic curves, we concentrate on non-singular cubics, and we give a summary of the theory of elliptic curves (projective non-singular cubic curves with at least one rational point).

A number of chapters end with applications of the theory to other topics and, in particular, we highlight the cryptographic applications in Sections 4.6.4, 7.5.3, 8.9.1, 10.7.2, and 16.9.

The book contains much more material than can be covered in a one-semester undergraduate course. For a first course in number theory or arithmetic geometry, we recommend covering Chapters 1 through 10 (Chapter 6 on finite fields is optional). For a second course in arithmetic (or diophantine) geometry, the instructor can cover Chapters 9 through 16 (Chapter 11 on the Hasse–Minkowski theorem is optional). The text assumes that the student has had a sequence of courses in calculus, up to multivariable calculus (a familiarity with matrices is assumed in some exercises). It is recommended that the student has seen an introduction to proofs before reading this book. However, the first few chapters have the secondary goal of providing practice in proof-writing, and they include a review of proofs by induction, in particular.

The material in this text ends where [**Loz11**] begins. There are, of course, many other undergraduate sources on number theory that are highly recommended: [**AC95**], [**Bur10**], [**Chi95**], [**Con1**], [**Gou97**], [**HW38**], [**HPS14**], [**Ros10**], [**ST92**], [**Sil12**], [**Ste08**], [**Was08**], and [**Wei17**], among many others. At the graduate level, the volumes [**DF03**], [**IR98**], [**Lor96**], [**Mil06**], [**Ser73**], and [**Sil86**] are excellent introductions to various aspects of algebra, number theory, and arithmetic geometry.

I started writing my own notes when I taught elementary number theory courses at Cornell University (in the fall of 2006 and 2007) and at the University of Connecticut (in the fall of 2008 and 2011 and the spring of 2014). This book grew out of these notes and the lectures of a special topics course (on diophantine geometry) that I taught at UConn in the fall of 2012. I would like to thank Keith Conrad for many suggestions and corrections. Also, I would like to thank the UConn undergraduate students in my class "MATH 3240Q: Introduction to Number Theory" for carefully reading my notes and providing useful feedback and criticism. In particular, I would like to thank Lia Bonacci, Heather Clinton, Jeremy Driscoll, Randolph Forsyth, Carly Gaccione, Taylor Garboski, Tom Jones, Gregory Knight, David Khondkaryan, Pravesh Mallik, Nicole Raymond, Heather Risley, Antonio Rossini, and Rachel Tangard for their comments, and special thanks go to Michael Lau and Byron Sitaras for their many and very detailed comments. Finally, I would like to thank Jason Dorfman (CSAIL/MIT), the Wikimedia Commons, and the Archives of the Mathematisches Forschungsinstitut Oberwolfach for their permission to use the images from their collections that appear in this book.

CHAPTER 1

INTRODUCTION

As long as algebra and geometry have been separated, their progress have been slow and their uses limited; but when these two sciences have been united, they have lent each mutual forces, and have marched together towards perfection.

Joseph-Louis Lagrange, 1795

The main goal of this book is to study $\mathbb{N}$, $\mathbb{Z}$, and $\mathbb{Q}$, i.e., the natural numbers, the integers, and the rational numbers:

$$\begin{aligned}\mathbb{N} &= \{1, 2, 3, \ldots\},\\ \mathbb{Z} &= \{\ldots, -3, -2, -1, 0, 1, 2, 3, \ldots\},\\ \mathbb{Q} &= \left\{\frac{m}{n} : m, n \in \mathbb{Z},\ n \neq 0\right\}.\end{aligned}$$

In the next chapter, we will be much more careful defining these sets using *axioms*, but, for now, we appeal to our intuition of the properties that these numbers satisfy. One can study these sets of numbers from their intrinsic properties, and much can be gained from such an endeavor, but in this book we study these sets from the point of view of their interaction with geometric objects (graphs of polynomials, lines in the plane, conics, elliptic curves, etc.).

Our generic approach will be as follows: we will define a geometric object G and then we will try to find all the points in the geometric object with coordinates in $\mathbb{N}$, $\mathbb{Z}$, or $\mathbb{Q}$, which we will denote by $G(\mathbb{N})$, $G(\mathbb{Z})$, and $G(\mathbb{Q})$, respectively. As we attempt to find the natural, integral, or rational points, we will develop the theory that is usually called "elementary number theory". Our approach will use the problem of finding arithmetic points on a geometric object as the motivation for the definitions and techniques of elementary number theory. Let us begin with our first example.

1.1. Roots of Polynomials

We begin this section with a discussion about polynomials and, in particular, which polynomials have roots in a given number system. Roots of polynomials will be treated in more detail in Part 1, and in particular in Section 2.8. We will also discuss polynomials (as a *ring*) in Section 5.5.

A polynomial $p(x)$ is an expression of the form

$$p(x) = a_n x^n + a_{n-1} x^{n-1} + \cdots + a_1 x + a_0,$$

where $n \geq 0$ is a non-negative integer and $a_0, a_1, \ldots, a_n$ are constants (in $\mathbb{Z}$ or $\mathbb{Q}$ or $\mathbb{R}$, for example). By a *polynomial equation*, we mean an equation that can be expressed in the form $p(x) = 0$, for some polynomial $p(x)$. A *root* of the polynomial equation $p(x) = 0$ is a number α such that $p(\alpha) = 0$.

For humans, it is natural to work with the natural numbers $\mathbb{N} = \{1, 2, 3, \ldots\}$ as we often need to count things in our daily routine. However, as soon as we try to solve the simplest linear polynomial equations using only natural numbers, we run into problems. An equation of the form

$$3 + x = 5 \tag{1.1}$$

has a (unique!) solution in $\mathbb{N}$, namely $x = 2$. But the similar equation

$$5 + x = 3 \tag{1.2}$$

has no solutions in $\mathbb{N}$, since $5 + x > 5 > 3$, for any $x \in \mathbb{N}$. Indeed, if a and b are natural numbers, then an equation $a + x = b$ has a solution in $\mathbb{N}$ if and only if $a < b$. Thus, in order to solve (1.2), we need to augment $\mathbb{N}$ to include all numbers of the form $-n$, where $n \in \mathbb{N}$. Notice that we also need to include 0 to be able to solve an equation of the form $5 + x = 5$. Thus, we define $\mathbb{Z} = \{\ldots, -3, -2, -1, 0, 1, 2, 3, \ldots\}$ and every equation of the form

$$a + x = b, \tag{1.3}$$

where $a, b \in \mathbb{Z}$, has a (unique!) solution $x = b - a$ in $\mathbb{Z}$. The integers, however, are not enough to solve an equation of the form

$$5x = 3 \tag{1.4}$$

as there is no integer n such that $5n = 3$ (indeed, the number 3 is prime and its only positive divisors are 1 and 3). More generally, an equation of the form $ax + b = 0$, with $a, b \in \mathbb{Z}$ and $a \neq 0$, has solutions in $\mathbb{Z}$ if and only if a is a divisor of b. In order to be able to solve all equations of the form $ax + b = 0$, we need to augment $\mathbb{Z}$ to be a number system such that every non-zero number has a *multiplicative inverse*. And so, we define $\mathbb{Q} = \{\frac{m}{n} : m, n \in \mathbb{Z}, n \neq 0\}$. Now every linear equation $ax + b = c$, with $a, b, c \in \mathbb{Q}$ and $a \neq 0$, has a unique solution $x = \frac{c-b}{a} \in \mathbb{Q}$.

How about quadratic polynomials? Do they all have roots in $\mathbb{Q}$? Of course not. For instance, the polynomial $x^2 - 2 = 0$ does not have any rational roots because $\sqrt{2}$ is not a rational number. (In order to rigorously prove that $\sqrt{2} \notin \mathbb{Q}$ we will first need to prove the fundamental theorem of arithmetic! See Theorems 2.10.2 and 2.10.6 and Section 2.10.1.) We usually represent numbers such as $\sqrt{2}$ by their decimal expansion, i.e., $\sqrt{2} = 1.41421356237309\ldots$. The decimal expansion of a rational number has a period, i.e., the expansion eventually repeats a given pattern of finitely many digits (why?). For example,

$$\frac{13}{17} = 0.76470588235294117647058823529411\ldots7647058823529411\ldots.$$

Conversely, any decimal expansion that has a period represents a rational number (see Section 8.9.2). The expansion of $\sqrt{2}$ has no period, as we have already mentioned that $\sqrt{2} \notin \mathbb{Q}$. In order to be able to solve quadratic equations (and other higher-degree polynomial equations), one can augment $\mathbb{Q}$ to include all decimal expansions and not only those that are periodic. This leads to an informal definition of the real numbers:

$$\mathbb{R} = \{\text{set of all decimal expansions}\},$$

with the usual identification of decimals with "trailing nines"; e.g., the expansion $0.9999\ldots$, with infinitely many nines, is equal to the decimal expansion $1 = 1.0000\ldots$ (see Exercise 1.8.1).

Unfortunately, not all quadratic polynomial equations $ax^2 + bx + c = 0$, with $a, b, c \in \mathbb{R}$ and $a \neq 0$, have a solution in $\mathbb{R}$. In fact, $ax^2 + bx + c = 0$, with $a, b, c \in \mathbb{R}$ and $a \neq 0$, has a solution in $\mathbb{R}$ if and only if $b^2 - 4ac \geq 0$. Similarly, there are higher-degree polynomials with no roots in $\mathbb{R}$. For instance, the polynomial equation $x^4 + x^3 + x^2 + x + 1 = 0$ has no real roots.

In order to ameliorate the "shortcomings" of $\mathbb{R}$, we would like to augment $\mathbb{R}$ so that, at least, all quadratic polynomials have a root. In order to accomplish this, it is sufficient to add a square root of -1 to $\mathbb{R}$, which we shall denote by i, an *imaginary number* such that $i^2 = -1$. Indeed, a polynomial $p(x) = ax^2+bx+c = 0$, with $a, b, c \in \mathbb{R}$ and $a \neq 0$, with $b^2 - 4ac \geq 0$ has real roots

$$x = \frac{-b \pm \sqrt{b^2 - 4ac}}{2a},$$

and if $b^2 - 4ac < 0$, then $p(x) = 0$ has roots

$$x = \frac{-b \pm i\sqrt{|b^2 - 4ac|}}{2a}.$$

Therefore, if we define the complex numbers as

$$\mathbb{C} = \{a + bi : a, b \in \mathbb{R},\ i^2 = -1\},$$

then all linear and quadratic polynomials with coefficients in $\mathbb{C}$ have roots in $\mathbb{C}$ (the reader needs to verify that every complex number $\alpha \in \mathbb{C}$ has a square root $\sqrt{\alpha}$ also in $\mathbb{C}$; see Exercise 1.8.6). Perhaps one of the most surprising and beautiful theorems in algebra is that, in fact, *every* non-constant polynomial (of arbitrary degree ≥ 1) with coefficients in $\mathbb{C}$ has a root in $\mathbb{C}$. This is known as the fundamental theorem of algebra.

Theorem 1.1.1 (Fundamental theorem of algebra). *Let $p(x)$ be a polynomial of degree ≥ 1 with coefficients in $\mathbb{C}$. Then, there is $\alpha \in \mathbb{C}$ such that $p(\alpha) = 0$.*

For example, let $p(x) = x^4 + x^3 + x^2 + x + 1$. As we mentioned above, $p(x)$ is a polynomial that has no real roots. The number $\alpha = \cos(\frac{2\pi}{5}) + i\sin(\frac{2\pi}{5})$ is a complex root of $p(x)$. Indeed, by Euler's formula

$$e^{ix} = \cos x + i \cdot \sin x,$$

we have that $\alpha = e^{2\pi i/5}$. Moreover,

$$x^4 + x^3 + x^2 + x + 1 = \frac{x^5 - 1}{x - 1},$$

and $\alpha^5 - 1 = (e^{2\pi i/5})^5 - 1 = e^{2\pi i} - 1 = 1 - 1 = 0$. Thus, $p(\alpha) = 0$ as well.

Complex numbers are fascinating in their own right, and there is a whole area of mathematics dedicated to the study of $\mathbb{C}$ and complex-valued functions, namely the area known as complex analysis. Here, however, we are (mostly) interested in, and shall concentrate on, the study of $\mathbb{N}$, $\mathbb{Z}$, and $\mathbb{Q}$. Let us try to find out when a polynomial with integer coefficients has a rational root.

Example 1.1.2. Let $p(x) = 3x^3 - 44x^2 - 257x + 190$ be a polynomial. We would like to find the natural ($\mathbb{N}$), integral ($\mathbb{Z}$), or rational ($\mathbb{Q}$) roots of $p(x)$; i.e., we want to find those natural, integral, or rational numbers x that satisfy $p(x) = 0$. Suppose that the natural number $n \in \mathbb{N}$ is a root of $p(x)$. Then,

$$p(n) = 3n^3 - 44n^2 - 257n + 190 = 0,$$

and we may rewrite this expression as $n(3n^2 - 44n - 257) = -190$. Since n is a natural number, the number $3n^2 - 44n - 257$ is an integer (not necessarily in $\mathbb{N}$) and we may conclude that n would necessarily be a divisor of -190. The list of natural divisors of -190 is $L = \{1, 2, 5, 10, 19, 38, 95, 190\}$. Thus, we can try to see whether any of these numbers $n \in L$ is a root of $p(x)$ by calculating $p(n)$. After carrying this out, we find that the only natural number that is a root of $p(x)$ is $n = 19$.

Are there any integral roots of $p(x)$ that are not natural numbers? If $n \in \mathbb{Z}$ and $p(n) = 0$, the expression $n(3n^2 - 44n - 257) = -190$ is still valid, and we may also conclude that n must be a divisor of -190. The *integer* divisors of -190 are those in the list $L' = \{\pm 1, \pm 2, \pm 5, \pm 10, \pm 19, \pm 38, \pm 95, \pm 190\}$. Since we have already checked that the only natural root is 19, we only need to check whether any of the negative divisors is a root. In this manner, we find that $n = 19$ and $n = -5$ are the only integral roots of $p(x)$.

Finally, we wish to find out whether $p(x)$ has any rational roots. Since we know that 19 and -5 are roots, we deduce that $f(x) = (x + 5)(x - 19)$ is a factor of $p(x)$ as polynomials (here we are using the so-called root theorem, Corollary 5.5.15). We may divide $p(x)$ by $f(x)$ to find a third linear factor, and therefore the value of the third root of $p(x)$. Instead, we shall approach this using a divisibility method that works more generally. Suppose $\frac{m}{n} \in \mathbb{Q}$ is a reduced fraction (i.e., m and n share no common divisors) and it is a root of $p(x)$. Then,

$$p\left(\frac{m}{n}\right) = 3\left(\frac{m}{n}\right)^3 - 44\left(\frac{m}{n}\right)^2 - 257\frac{m}{n} + 190 = 0.$$

If we multiply this expression by n^3, we obtain

$$3m^3 - 44m^2n - 257mn^2 + 190n^3 = 0.$$

This expression can be rewritten as $m(3m^2 - 44mn - 257n^2) = -190n^3$. This is an equality of integer numbers and we may deduce that m is a divisor of $-190n^3$.

Since m and n share no common divisors, it follows that m is a divisor of -190; i.e., $m \in L'$ with L' as defined above. The same displayed expression can be rewritten as $3m^3 = n(44m^2 + 257mn - 190n^2)$ and, once again, we may deduce a divisibility property. In this case, we deduce that n is a divisor of $3m^3$. Since m and n share no common divisors, we conclude that n is an integer divisor of 3 and so $n \in \{\pm 1, \pm 3\}$. Therefore, if $m/n \in \mathbb{Q}$ is a rational root of $p(x)$, we have shown that $m \in L'$ and $n \in \{\pm 1, \pm 3\}$. Now it is a matter of checking whether any of these rational numbers are actually roots, and we find that $\frac{m}{n} = \frac{2}{3}$ is indeed the third root we were looking for. Hence, the roots of $p(x)$ are $19 \in \mathbb{N}$, $-5 \in \mathbb{Z}$, and $\frac{2}{3} \in \mathbb{Q}$.

The previous example motivates some of our first definitions and theorems in the book (in Part 1). In the course of finding the roots of a polynomial, we have relied heavily upon the theory of divisibility of natural and integer numbers (and we alluded to divisibility of polynomials too). It is likely that the reader is perfectly comfortable with many of the steps in the example, but one needs to carefully prove some of them. For instance, at some point we used the following fact:

- If m, n, a, b are integers such that $ma = nb$ and m and n share no common factors (i.e., $\gcd(m, n) = 1$), then m is a divisor of b and n is a divisor of a.

Although this fact may be intuitively true, we need a proof! In order to provide a proof, we will need to establish first a number of basic facts about divisibility (see Corollary 2.7.6). But, for now, let us see how our next two examples motivate the study of the greatest common divisor of two integers.

1.2. Lines

In this section we discuss examples of the most basic 1-dimensional object: a line in the plane. We will come back to studying points on a line in detail in Section 2.9.

Example 1.2.1. Let $L : 5x + 17y = 1$ be a line in the plane. See Figure 1.1. There are infinitely many rational points in this line, and they can be found by solving for one of the variables. For example, we may write

$$y = \frac{1 - 5x}{17},$$

and it follows that $Q = \left(x_0, \frac{1-5x_0}{17}\right)$ is a point in L with rational coordinates, for each rational number x_0. In fact, every rational point Q in L is of this form. For instance, the points $(0, 1/17)$ and $(1, -4/17)$ are in L. Are there any points $(x_0, y_0) \in L$ with integer coefficients, with $x_0, y_0 \in \mathbb{Z}$? A quick search for points (using trial and error) reveals at least one point: $(7, -2)$.

Are there more? Yes, in fact, there are infinitely many integral points of the form $P_k = (7 + 17k, -2 - 5k)$ where $k \in \mathbb{Z}$. Let us check that these points belong to L:

$$5(7 + 17k) + 17(-2 - 5k) = 35 + 5 \cdot 17k - 34 - 5 \cdot 17k = 35 - 34 = 1.$$

Interestingly, the points $\{P_k : k \in \mathbb{Z}\}$ are *all* the integral points on L, but this is not so easy to prove (try!). This will be shown in Theorem 2.9.4.

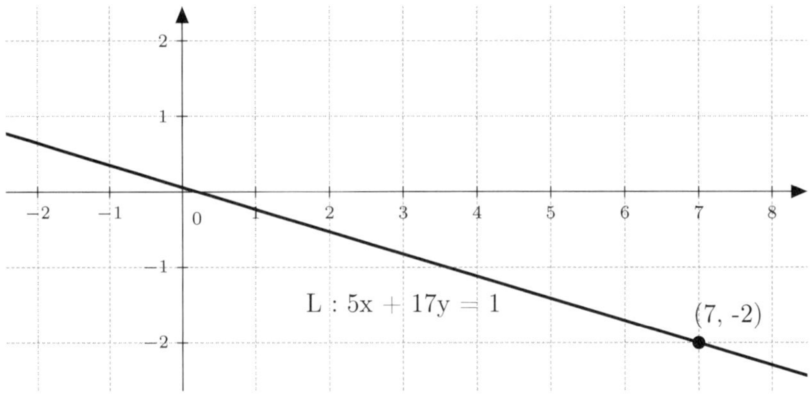

Figure 1.1. The line $5x + 17y = 1$ passes through infinitely many integral points.

Example 1.2.2. Let L' be the line in the plane with equation $5x + 15y = 1$ (see Figure 1.2).

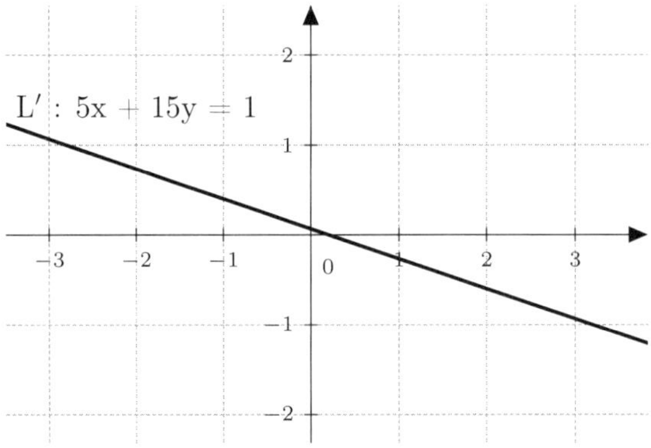

Figure 1.2. The line $5x + 15y = 1$ does not pass through any integral point.

As in our previous example, there are infinitely many rational points on L' given by $(x_0, \frac{1-5x_0}{15})$ for any $x_0 \in \mathbb{Q}$. Are there any integral points on L'? It turns out that there are none. Suppose m and n are integers with $5m + 15n = 1$. Then, $5(m + 3n) = 1$ and we have reached a contradiction because this equation implies that 1 has a non-trivial factorization into integers (other than $1 = 1 \cdot 1 = (-1)(-1)$). Another way to see this is that, in the integers, 5 is not a divisor of 1 (however, the number 5 is a divisor of 1 in the rational numbers: $1 = 5 \cdot \frac{1}{5}$).

Examples 1.2.1 and 1.2.2 show two lines L and L' that behave very differently when we look for integral points on them. Why is their behavior so different? The reason, as we shall see, is that $\gcd(5, 17) = 1$ while $\gcd(5, 15) = 5$. Using an argument similar to that in Example 1.2.2, one can show that a line $L'' : ax+by = c$, with $\gcd(a, b) = d$ and d not a divisor of c, will have no integral points. Indeed, if $m, n \in \mathbb{Z}$ satisfy $am + bn = c$, then d would be a divisor of c and that is a contradiction to one of our assumptions. However, if $\gcd(a, b) = 1$, why should there be integral points on L''? For example, consider $L'' : 1234x + 5007y = 1$. Are there integral points on L''? The greatest common divisor of 1234 and 5007 is 1 and, as we shall see, this implies the existence of integral points and, moreover, we will describe an efficient algorithm to find these points (see Sections 2.6 and 2.7). Here is one such point $P = (-1481, 365)$:

$$1234(-1481) + 365(5007) = -1827554 + 1827555 = 1.$$

All other integral points are of the form $(-1481 + 5007k, 365 - 1234k)$ for any $k \in \mathbb{Z}$.

1.3. Quadratic Equations and Conic Sections

In this section we discuss several examples of rational and integral points on quadratic equations in two variables, i.e., equations of the form

$$ax^2 + bxy + cy^2 + dx + ey + f = 0,$$

where a, b, c, d, e, f are integers and a, b, or c is non-zero. When the graph of a quadratic equation is *smooth* (non-singular; see Section 15.1.5), we call them conic sections (because they arise as sections of cones; see Figure 9.5). We will discuss quadratic equations and conic sections at length in Part 2.

Figure 1.3. Muhammad ibn Musa al-Khwarizmi (c. 780 – c. 850) was a Persian mathematician, astronomer, and geographer. His treaty on algebra contained the first systematic treatment of linear and quadratic equations, including the first demonstration of the "completing the square" method. Image source: Wikimedia Commons.

In our next example, we find rational points on a conic section (a hyperbola, in this case).

Example 1.3.1. Let $C : x^2 - 7y^2 = 1$ be a hyperbola in the plane. Can we find all the rational points on C? Yes, and we will do so using a little bit of geometry. First, notice that there are two (integral) points which are easily found, namely $(\pm 1, 0)$. If we trace a line L that goes through $P = (1, 0)$, it will intersect the hyperbola at exactly two points: the point P and a second point Q (since C is given by a quadratic equation, the intersection with a line is formed by either no points or two points). Let us find the second point of intersection, Q, in terms of the slope of L. The equation of L is given by

$$L : y - 0 = m(x - 1),$$

where m is the slope of L.

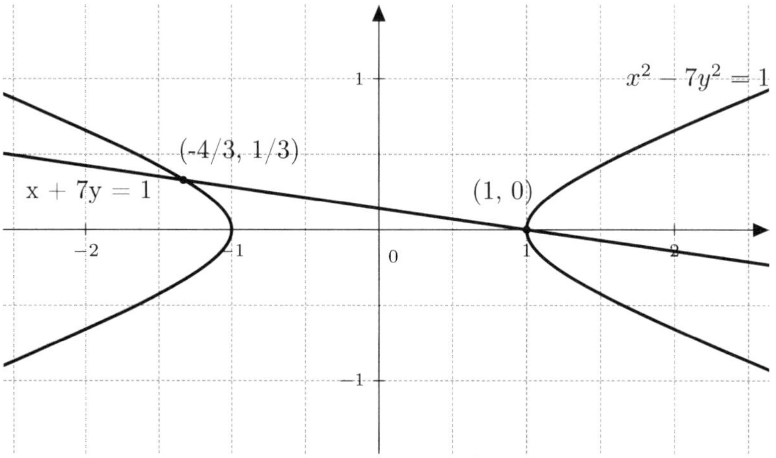

Figure 1.4. The hyperbola $x^2 - 7y^2 = 1$ passes through infinitely many rational points and also through infinitely many integral points.

Notice that L passes through $P = (1, 0)$, as desired. Now we may find the intersection of L and C by solving the system:

$$\begin{cases} x^2 - 7y^2 = 1, \\ y = m(x - 1). \end{cases}$$

Plugging the equation of L into the equation for C, we obtain

$$1 = x^2 - 7(m(x - 1))^2 = (1 - 7m^2)x^2 + 14m^2x - 7m^2,$$

or, equivalently, $(1-7m^2)x^2+14m^2x-(1+7m^2) = 0$. Now we can use the quadratic formula to solve for x:

$$\begin{aligned} x &= \frac{-14m^2 \pm \sqrt{14^2m^4 - 4(1 - 7m^2)(-(1 + 7m^2))}}{2(1 - 7m^2)} \\ &= \frac{-14m^2 \pm \sqrt{14^2m^4 + 4(1 - 7^2m^4)}}{2(1 - 7m^2)} \\ &= \frac{-14m^2 \pm \sqrt{4}}{2(1 - 7m^2)} = \frac{-14m^2 \pm 2}{2(1 - 7m^2)} \\ &= \frac{-7m^2 \pm 1}{1 - 7m^2} = \begin{cases} 1 \text{ or} \\ \frac{7m^2+1}{7m^2-1}. \end{cases} \end{aligned}$$

As we expected, $x = 1$ is a solution (since L passes through $P = (1, 0)$). The second point of intersection, Q_m, has x-coordinate $x = (7m^2 + 1)/(7m^2 - 1)$. The y-coordinate of Q_m is given by

$$y = m(x-1) = m\left(\frac{7m^2+1}{7m^2-1} - 1\right) = \frac{2m}{7m^2-1}.$$

Thus, the point $Q_m = \left(\frac{7m^2+1}{7m^2-1}, \frac{2m}{7m^2-1}\right)$ is a rational point on C for every rational slope $m \in \mathbb{Q}$. For instance, when $m = -1/7$ (see Figure 1.4), the point $Q_1 = (-\frac{4}{3}, \frac{1}{3})$ is in C:

$$\left(-\frac{4}{3}\right)^2 - 7\left(\frac{1}{3}\right)^2 = \frac{16-7}{9} = \frac{9}{9} = 1.$$

It is not difficult to see that this construction yields *all* the rational points on C; i.e., $C(\mathbb{Q}) = \{Q_m : m \in \mathbb{Q}\}$. Indeed, if Q' is a rational point on C and the line $\overline{PQ'}$ has slope m, then $Q' = Q_m$ (notice that the slope cannot be infinite, as there is only one point on C with $x = 1$, namely P).

We have found all the rational points on $C : x^2 - 7y^2 = 1$. Are there integral points on C? If so, how many? It turns out that, in this particular case, there are infinitely many integral points $(m, n) \in C(\mathbb{Z})$ and these points are intimately related with the rational approximations of $\sqrt{7}$. More concretely, if (m, n) is an integral point on C, then $\frac{m}{n}$ is a (very) good rational approximation of $\sqrt{7}$. Indeed,

$$m^2 - 7n^2 = 1$$

implies that

$$7 = \frac{m^2}{n^2} - \frac{1}{n^2} = \left(\frac{m}{n}\right)^2 - \frac{1}{n^2},$$

so that $|7 - (\frac{m}{n})^2| = \frac{1}{n^2}$. For instance, $(m, n) = (8, 3)$ is an integral point on C, and $\frac{8}{3} = 2.666\ldots$ while $\sqrt{7} = 2.645751\ldots$. We will explain later on that, once we have one rational solution (m, n), there is a method to find infinitely many solutions, by squaring the number $m + n\sqrt{7}$ (see Section 14.3.1). More concretely, if (m, n) is an integral point on C and $(m + n\sqrt{7})^2 = a + b\sqrt{7}$, then (a, b) is another integral point on C. In our case,

$$(8 + 3\sqrt{7})^2 = 64 + 48\sqrt{7} + 63 = 127 + 48\sqrt{7},$$

and we can verify that $(127, 48)$ is another point on C. Also, $\frac{127}{48} = 2.6458333\ldots$ is another approximation of $\sqrt{7}$ (see Chapter 13 and Theorem 14.2.3).

Example 1.3.2. Let us now consider the hyperbola $C' : x^2 - 7y^2 = 3$. Are there any integral points? We will show that, in fact, this hyperbola does not have any integral points. Let us assume, for a contradiction, that (m, n) is an integral point on C. It follows that $m^2 = 3 + 7n^2$ and, in particular, the remainder when we divide m^2 by 7 is 3. This is impossible, as the only remainders that occur when we divide a *perfect square* by 7 are 0, 1, 2, or 4, and 3, 5 and 6 never occur as remainders. Let us prove this last claim.

Indeed, every number $m \in \mathbb{Z}$ has a remainder of $0, 1, 2, 3, 4, 5$, or 6 when we divide by 7. In other words, we can always write $m = 7k + r$, where $k \in \mathbb{Z}$ and

$r = 0, 1, 2, 3, 4, 5$, or 6. Let us see what happens when we square $m = 7k + r$, for each possible remainder r:

$$(7k + 0)^2 = 49k^2 = 7(7k^2) + 0,$$
$$(7k + 1)^2 = 49k^2 + 14k + 1 = 7(7k^2 + 2k) + 1,$$
$$(7k + 2)^2 = 49k^2 + 28k + 4 = 7(7k^2 + 4k) + 4,$$
$$(7k + 3)^2 = 49k^2 + 42k + 9 = 7(7k^2 + 6k) + 9 = 7(7k^2 + 6k + 1) + 2,$$
$$(7k + 4)^2 = 49k^2 + 56k + 16 = 7(7k^2 + 8k) + 16 = 7(7k^2 + 8k + 2) + 2,$$
$$(7k + 5)^2 = 49k^2 + 70k + 25 = 7(7k^2 + 10k) + 25 = 7(7k^2 + 10k + 3) + 4,$$
$$(7k + 6)^2 = 49k^2 + 84k + 36 = 7(7k^2 + 12k) + 36 = 7(7k^2 + 12k + 5) + 1.$$

Thus, we have just shown that the remainder of $m^2 = (7k + r)^2$ when we divide by 7 is 0, 1, 2, or 4, and never 3, 5, or 6. Hence, $m^2 = 3 + 7n^2$ is impossible and C does not have any integral points. Similarly, C does not have any rational points either. Suppose $(\frac{m}{a}, \frac{n}{b})$ is a rational point. Then $(\frac{m}{a})^2 - 7(\frac{n}{b})^2 = 3$ and it follows that $(mb)^2 - 7(na)^2 = 3(ab)^2$, or, equivalently, $(mb)^2 = 3(ab)^2 + 7(na)^2$. Suppose that the remainder of dividing $(ab)^2$ by 7 is r; i.e., there is a $k \in \mathbb{Z}$ such that $(ab)^2 = 7k + r$. Then, as before, $r = 0, 1, 2$, or 4 and

$$(mb)^2 = 3(7k + r) + 7(na)^2 = 3r + 7((na)^2 + 3k).$$

In particular, $3r = 0, 3, 6$, or 12. If $3r = 12$, then we may write $(mb)^2 = 5 + 7((na)^2 + 3k + 1)$. Hence, the remainder of dividing $(mb)^2$ by 7 is 0, 3, 6, or 5. We have shown above that it cannot be 3, 5, or 6, so it must be 0 (so that $3r = 0$. Hence, $(mb)^2 = 7((na)^2 + 3k)$, or $A^2 = 7B^2$, where $A = mb$ and $B = (na)^2 + 3k$ are integers. However, the equation $A^2 = 7B^2$ has no solutions in the integers as the left-hand side is a perfect square but the right-hand side is not a square (a consequence of the fundamental theorem of arithmetic, Theorem 2.10.6; see two paragraphs below).

In Example 1.3.1 we have seen that a little bit of geometry can go a long way. The trick of intersecting a curve with a line passing through a known point is very useful. We will see similar tricks in the examples that follow. Another theme that Example 1.3.1 has introduced is that of the approximation of irrational numbers by rationals (e.g., $\sqrt{7} \approx \frac{127}{48}$), usually referred to as diophantine approximation.

In Example 1.3.2 we have claimed that $A^2 = 7B^2$ is impossible, for $A, B \in \mathbb{N}$. This fact relies on the so-called fundamental theorem of arithmetic: every natural number has a unique factorization as a product of prime numbers. Also in Example 1.3.2, we have seen for the first time that working with the remainders of long division can be a very effective technique. This tool will lead us to the study of *congruences* modulo an integer (see Chapter 4). In particular, we were interested in the remainder left out when dividing a perfect square n^2 by a fixed number m. This will lead us to the study of *quadratic residues* and Gauss's law of quadratic reciprocity—one of the theorems in all of mathematics with the largest number of known distinct proofs (Gauss alone published six different proofs; there are now over 200 published proofs). Quadratic congruences and quadratic reciprocity will be dealt with in Chapter 10.

1.4. Cubic Equations and Elliptic Curves

In this section we discuss our first examples of cubic equations in the plane, i.e., equations in two variables of the form

$$ax^3 + bx^2y + cxy^2 + dy^3 + ex^2 + fxy + gy^2 + hx + jy + k = 0,$$

for some integers $a, b, c, \ldots, k \in \mathbb{Z}$ such that not all of a, b, c, d are zero. Cubic equations will be studied in Part 3.

An *elliptic curve* is a smooth cubic curve in the plane, which is also smooth "at infinity", and such that it contains at least one rational point (we will discuss elliptic curves in Chapter 16). Smooth means that the curve has a well-defined tangent line at every point. We will not describe in detail here what is the meaning of the smoothness at infinity condition (see Section 15.1.5 instead). It is sufficient to say that, after an appropriate change of variables, every elliptic curve (defined over $\mathbb{Q}$) can be written in the simpler form

$$y^2 = x^3 + Ax + B$$

where $A, B \in \mathbb{Z}$ and the polynomial $x^3 + Ax + B$ has no repeated roots (which, in turn, is equivalent to $4A^3 + 27B^2 \neq 0$). For example, we have drawn the graph of the elliptic curve $y^2 = x^3 + 1$ in Figure 1.5.

Example 1.4.1. Let E be the elliptic curve given by the equation $y^2 = x^3 + 1$. A quick inspection for points reveals two integral points $P = (-1, 0)$ and $Q = (0, 1)$. By symmetry, there is one additional point $Q' = (0, -1)$. Now, in order to find new points, we may use a trick we have already seen. Let L be the line that goes through P and Q. With a little bit of basic plane geometry, we find an equation for $L : y = x - 1$ (see Exercise 1.8.8). In order to find the intersection points of E and L, we need to solve the system

$$\begin{cases} y^2 = x^3 + 1, \\ y = x + 1. \end{cases}$$

Thus, we plug the equation for L into the equation for E and obtain a polynomial of degree 3 whose roots are the x-coordinates of the points of intersection of E and L:

$$(x + 1)^2 = x^3 + 1, \quad \text{or} \quad x^3 - x^2 - 2x = x(x^2 - x - 2) = 0.$$

The roots are $x = -1, 0, 2$, and the corresponding y-coordinates are $0, 1, 3$, respectively. Hence, we have found a new point $R = (2, 3)$ on E, with natural coordinates. By symmetry, there is an additional point $(2, -3)$ on E.

So far, we have found one natural point, $(2, 3)$, and four additional integral points, $(-1, 0)$, $(0, \pm 1)$, and $(2, -3)$. It turns out that these are all the *rational* points on E, but this is fairly hard to prove.

Example 1.4.2. Let E' be the elliptic curve given by the equation $y^2 = x^3 - 2$. A quick inspection reveals one integral point, $P = (3, 5)$, but no other integral point is easily found. We can modify our previous geometric trick by finding the line L that is *tangent* to E' at P. A little bit of calculus (e.g., implicit differentiation; see

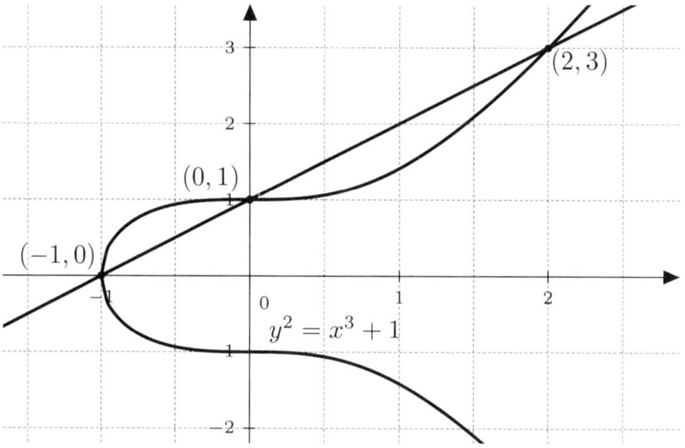

Figure 1.5. The elliptic curve $y^2 = x^3 + 1$.

Exercise 1.8.20) yields that the slope of a tangent line to E at a point $(x, y) \in E$ is given by

$$\frac{dy}{dx} = \frac{3x^2}{2y}.$$

In our case, L has slope $27/10$ and passes through $P = (3, 5)$, so it is given by the equation $L : y = \frac{27}{10}(x-3)+5$. Now we can find the intersection points of L and E' by solving the system

$$\begin{cases} y^2 = x^3 - 2, \\ y = \frac{27}{10}(x-3)+5. \end{cases}$$

Plugging the equation for L into the equation for E' yields a polynomial equation

$$\left(\frac{27}{10}(x-3)+5\right)^2 = x^3 - 2,$$

or, equivalently,

$$x^3 - \frac{729}{100}x^2 + \frac{837}{50}x - \frac{1161}{100} = 0.$$

We know that $x = 3$ is a root of this polynomial (and, in fact, it must be a double-root, because L is *tangent* to E' at P). Thus, this polynomial factors as $(x-3)^2(x-\alpha) = 0$. Hence, we can find the value of α and this turns out to be $\alpha = \frac{129}{100}$. In particular, the x-coordinates of the points of intersection of L and E' are 3 and $\frac{129}{100}$, and their y-coordinates are 5 and $-\frac{383}{100}$, respectively. Hence, we have found a new *rational* point on E', namely $Q = (\frac{129}{100}, -\frac{383}{100})$. By symmetry of the graph of E' with respect to the y-axis, there is an additional point $Q' = (\frac{129}{100}, \frac{383}{100})$.

This construction of a rational point can be repeated to find other points. For instance, we can trace the line L' that goes through P and Q'. This line will intersect E' at a third rational point. We leave it to the reader to verify that the points of intersection of E' and L' are P, Q' and

$$Q'' = \left(\frac{164323}{29241}, \frac{66234835}{5000211}\right).$$

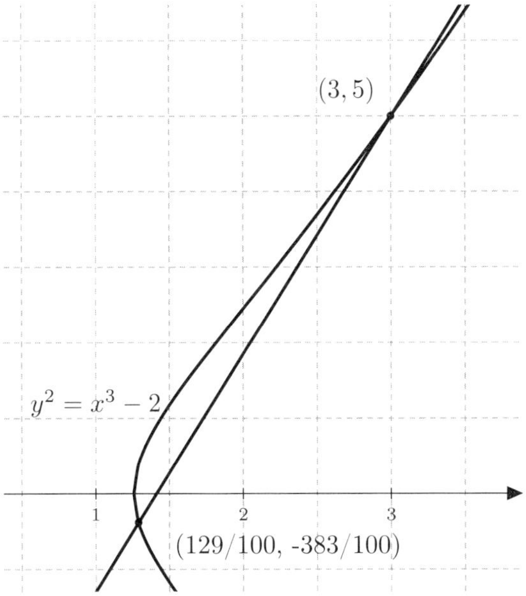

Figure 1.6. The elliptic curve $y^2 = x^3 - 2$.

The previous examples illustrate the method of *chords and tangents* that can be used on cubic curves to find new rational points (see Section 16.3). The most important theorem in the theory of elliptic curves is the following result, proved by Louis Mordell, and vastly generalized by André Weil (see Figure 1.7). The so-called Mordell–Weil theorem says that there is a finite set of rational points S such that every other rational point can be obtained from the points in S, using the method of chords and tangents.

Theorem 1.4.3 (Mordell–Weil theorem). *Let E be an elliptic curve (a smooth cubic curve, together with a given rational point $\mathcal{O}$). Then, there is a set formed by finitely many rational points $P_1, \ldots, P_n$ on E such that if R is any other rational point on E, then R can be obtained from $P_1, \ldots, P_n$ using the method of chords and tangents.*

Example 1.4.4. All the rational points on $E : y^2 = x^3 + 1$ can be generated from the point $P_1 = (2, 3)$ using chords and tangents. In this case, there are only five rational points on E (plus a point at "infinity").

The rational points on $E' : y^2 = x^3 - 2$ are generated from the point $P_1 = (3, 5)$ using chords and tangents. In this case, however, the curve has infinitely many distinct rational points.

The rational points on the curve $E'' : y^2 + y = x^3 - 7x + 6$ are generated using three points $P_1 = (1, 0)$, $P_2 = (2, 0)$, and $P_3 = (0, -3)$. These three points generate infinitely many distinct rational points on E''.

Figure 1.7. Louis Mordell (1888–1972) and André Weil (1906–1998). Images author: Konrad Jacobs (Erlangen). Source: Archives of the Mathematisches Forschungsinstitut Oberwolfach.

1.5. Curves of Higher Degree

By a curve of *higher* degree we refer to equations of the form

$$f(x, y) = 0,$$

where $f(x, y)$ is a polynomial in two variables, with integer coefficients, and such that the largest degree of a monomial is greater than or equal to 4 (here we define the degree of a monomial $x^a y^b$ as $a + b$). For instance, the curves of the form

$$y^2 = p(x),$$

with $p(x)$ a polynomial of degree ≥ 4, are called hyperelliptic curves.

Example 1.5.1. The following is problem 17 in Book VI of Diophantus's *Arithmetica*:

> *Find three squares which when added give a square, and such that the first one is the side of the second, and the second is the side of the third.*

Let A, B, C be integers, and let A^2, B^2, C^2 be the squares mentioned in the problem. Then, $Y^2 = A^2 + B^2 + C^2$, for some $Y \in \mathbb{Z}$, and the first one is the side of the second (so $A^2 = B$) and the second one is the side of the third (so $B^2 = C$). It follows that if $A = x$, then $B = x^2$ and $C = x^4$. Therefore, we are trying to find x and Y integers such that

$$Y^2 = x^2 + x^4 + x^8.$$

If we exclude the unique solution with $x = 0$, i.e., $(0, 0)$, then we can write $Y = xy$, and therefore we are looking for a rational point on the curve

$$x^2y^2 = x^2 + x^4 + x^8,$$

with $x \neq 0$. Thus, we can divide through by x and simplify the equation to

$$\mathcal{C} : y^2 = 1 + x^2 + x^6,$$

and we are looking for all the rational solutions with $x \neq 0$. In his work, Diophantus finds one rational solution of $\mathcal{C}$, namely $(x, y) = (1/2, 9/8)$, which corresponds to $(x, Y) = (1/2, 9/16)$, and therefore

$$A = \frac{1}{2},\ B = \frac{1}{4},\ \text{and}\ C = \frac{1}{16},$$

so that

$$A^2 + B^2 + C^2 = \frac{1}{4} + \frac{1}{16} + \frac{1}{256} = \frac{81}{256} = \left(\frac{9}{16}\right)^2.$$

A natural question arises: are there any other solutions to the problem? In other words, are there any other rational points in $\mathcal{C}$? In 1998, Joseph Wetherell showed in his Ph.D. thesis [**Wet98**] that the only rational points on $\mathcal{C}$ are precisely $(0, \pm 1)$, $(1/2, \pm 9/8)$, and $(-1/2, \pm 9/8)$. Hence, the solution (A, B, C) equal to $(1/2, 1/4, 1/16)$ is the only solution with positive rational numbers to the original problem posed by Diophantus.

Example 1.5.2. The curve $C : y^2 = x^6 - 6x^5 + 11x^4 - 8x^3 + 11x^2 - 6x + 1$ has exactly four rational points, namely $(0, \pm 1)$ and $(1, \pm 2)$. The proof of this fact was the subject of a Ph.D. thesis by Harris Daniels [**Dan13**]. The rational points on C classify elliptic curves that satisfy a certain property, and the fact that $C(\mathbb{Q})$ is finite implies that only finitely many such elliptic curves exist.

In general, when studying rational points, the degree is not the most relevant invariant to classify curves. Instead, we classify curves according to their *genus* (from now on we assume that every curve is smooth). A curve defined over $\mathbb{Q}$ may be regarded as a curve defined over $\mathbb{C}$ and the graph of C in $\mathbb{C} \times \mathbb{C}$ is a 1-complex-dimensional curve (a Riemann surface), which can be viewed as a 2-real-dimensional surface (compact and orientable). Loosely speaking, the genus of C is the number of "holes" in this surface.

Figure 1.8. Curves of genus 1, 2, and 3, defined over $\mathbb{C}$. Images source: Wikipedia Commons.

If $C : f(x, y) = 0$ is a smooth equation for the curve C defined over $\mathbb{Q}$ and the highest degree of a monomial in the polynomial $f(x, y)$ is d, then the genus of C is given by the formula

$$\text{genus}(C) = \frac{(d-1)(d-2)}{2}.$$

For instance, a smooth curve C given by a quadratic equation (a conic, such as an ellipse or a hyperbola) has genus 0, because

$$\text{genus}(C) = \frac{(2-1)(2-2)}{2} = 0.$$

Thus, conics correspond to compact orientable surfaces with no holes, such as a sphere.

An elliptic curve E (as in Section 1.4) is given by an equation $y^2 = x^3 + Ax + B$, with $4A^3 + 27B^2 \neq 0$ to ensure smoothness, so

$$\text{genus}(E) = \frac{(3-1)(3-2)}{2} = 1.$$

Thus, every elliptic curve is a curve of genus 1; i.e., it corresponds to a compact orientable surface with one hole (a torus).

A curve of genus 0 or 1 may have infinitely many rational points (see Examples 1.3.1 and 1.4.4). In contrast, in 1922, Louis Mordell conjectured that any curve with genus > 1 can only have finitely many rational points. This was proved by Gerd Faltings in 1983 (see Figure 1.9).

Figure 1.9. Gerd Faltings (born 1954) is a German mathematician known for his work in arithmetic geometry. Image source: Archives of the Mathematisches Forschungsinstitut Oberwolfach.

Theorem 1.5.3 (Faltings's theorem)**.** *Let C be a smooth curve defined over $\mathbb{Q}$ of genus $g > 1$. Then, C has only finitely many rational points.*

There is some progress on methods to find the rational points on curves of genus 2, but very little is known about how to find the rational points on a curve of genus ≥ 3.

1.6. Diophantine Equations

... his boyhood lasted $\frac{1}{6}$th of his life; he married after $\frac{1}{7}$th more; his beard grew after $\frac{1}{12}$th more, and his son was born 5 years later; the son lived to half his father's age, and the father died 4 years after the son.

Metrodorus ($\sim$ 600 AD), from the *Greek Anthology*, in reference to Diophantus of Alexandria's life

In previous sections, we have discussed examples of finding integral and rational points on polynomials, and curves. More generally, we may ask ourselves how to find integral and rational points on a surface, or on a higher-dimensional algebro-geometric object V (called a *variety*), which, in general will be given by a set of equations

$$V : \begin{cases} f_1(x_1, x_2, \ldots, x_n) = 0, \\ f_2(x_1, x_2, \ldots, x_n) = 0, \\ \vdots \\ f_r(x_1, x_2, \ldots, x_n) = 0, \end{cases}$$

where, for each $1 \leq i \leq r$, the polynomial f_i has n variables $x_1, \ldots, x_n$, and integer coefficients. In this case, we are interested in the integral and rational points of V, namely,

$$V(\mathbb{Z}) = \{(a_1, \ldots, a_n) \in \mathbb{Z}^n : f_1(a_1, \ldots, a_n) = \cdots = f_r(a_1, \ldots, a_n) = 0\} \text{ and}$$
$$V(\mathbb{Q}) = \{(t_1, \ldots, t_n) \in \mathbb{Q}^n : f_1(t_1, \ldots, t_n) = \cdots = f_r(t_1, \ldots, t_n) = 0\}.$$

Each of the equations above is usually called a diophantine equation.

Definition 1.6.1. A polynomial equation of the form $C : f(x_1, \ldots, x_n) = 0$, where f is a polynomial in n variables with integer coefficients, is called a *diophantine equation*. A *rational (resp. integral, resp. natural) solution* of C is an n-tuple $(a_1, \ldots, a_n)$ of rational numbers $a_i \in \mathbb{Q}$ (resp. integers $a_i \in \mathbb{Z}$, resp. natural numbers $a_i \in \mathbb{N}$) such that

$$f(a_1, \ldots, a_n) = 0.$$

The term "diophantine" was coined in honor of Diophantus of Alexandria, whose treaties are the first records of a systematic approach to the study of the rational solutions of algebraic equations. We will write more about Diophantus in Section 1.6.1 below.

Example 1.6.2. Problem 28 in Book II of Diophantus's *Arithmetica* reads as follows:

> *To find two square numbers such that their product added to either gives a square.*

If we write x^2 and y^2 for the squares, then the problem is equivalent to finding rational solutions of the system of equations

$$\begin{cases} x^2y^2 + x^2 = u^2, \\ x^2y^2 + y^2 = v^2. \end{cases}$$

Diophantus finds one rational solution, namely $(x, y) = (3/4, 7/24)$. Let us find all the integral solutions first. From the first equation we see that $x^2(y^2 + 1) = u^2$. Therefore, either $x = u = 0$ or y^2+1 itself is a square. Since the only two consecutive squares are 0 and 1 (see Exercise 1.8.14), it follows that $(x, y, u, v) = (n, 0, \pm n, 0)$ and $(0, m, 0, \pm m)$, for some integers m, n, are the only integral solutions of the problem.

Now, let us find the rational solutions. As before, $x = u = 0$ or $y^2 + 1$ is a square. Thus, there is $t \in \mathbb{Q}$, with $t \neq \pm 1$, such that $y = \frac{2t}{1-t^2}$ (this follows from parametrizing $y^2 + 1 = w^2$; see Exercise 1.8.14). Now the second equation says $y^2(x^2+1) = v^2$, so either $y = 0$ or $x^2 + 1$ is a square. We similarly conclude that $x = \frac{2s}{1-s^2}$ for some $s \in \mathbb{Q}$ with $s \neq \pm 1$. Hence, the rational solutions of the problem are given by

$$(x_s, y_t) = \left(\frac{2s}{1-s^2}, \frac{2t}{1-t^2}\right),$$

and there is one solution for each s and t in $\mathbb{Q}$, other than ± 1. Indeed,

$$\begin{aligned} x_s^2 y_t^2 + x_s^2 &= x_s^2(y_t^2 + 1) \\ &= \left(\frac{2s}{1-s^2}\right)^2 \cdot \left(\frac{1+t^2}{1-t^2}\right)^2 \\ &= \left(\frac{2s(1+t^2)}{(1-s^2)(1-t^2)}\right)^2, \end{aligned}$$

and, similarly, $x_s^2 y_t^2 + y_t^2 = (2t(1+s^2))^2/((1-s^2)(1-t^2))^2$, for any $s \in \mathbb{Q}$ and any $t \in \mathbb{Q}$ not equal to ± 1.

1.6.1. About Diophantus of Alexandria. Diophantus of Alexandria (born between AD 201 and 215 and died between 285 and 299 at, apparently, age 84) is sometimes called "the father of algebra". He was an Alexandrian Greek mathematician and the author of a series of books called *Arithmetica* (see Figure 1.10), a tract *On Polygonal Numbers*, and a collection of results under the title of *Porisms*. Of the original 13 books that formed *Arithmetica*, only six were thought to have survived and it was also thought that the others must have been lost quite soon after they were written. However, in 1968, F. Sezgin made a remarkable discovery of an Arabic manuscript in the library Astan-i Quds in Meshed (The Holy Shrine library of Iran). The book seems to be a translation by Qusta ibn Luqa, who died in 912, of Books IV to VII of the *Arithmetica* by Diophantus of Alexandria.

The *Arithmetica* is not only the major work of Diophantus, but also the most prominent work on algebra in Greek mathematics. The books form a collection of about 130 problems giving numerical solutions of algebraic equations. Here is the dedication at the beginning of *Arithmetica*:

> *Knowing, my most esteemed friend Dionysius, that you are anxious to learn how to investigate problems in numbers, I have tried, beginning from the foundations on which the science is built up, to set forth to you the nature and power subsisting in numbers.*
>
> *Perhaps the subject will appear rather difficult, inasmuch as it is not yet familiar (beginners are, as a rule, too ready to despair of success); but you, with the impulse of your enthusiasm and the benefit of my teaching, will find it easy to master; for eagerness to learn, when seconded by instruction, ensures rapid progress.*

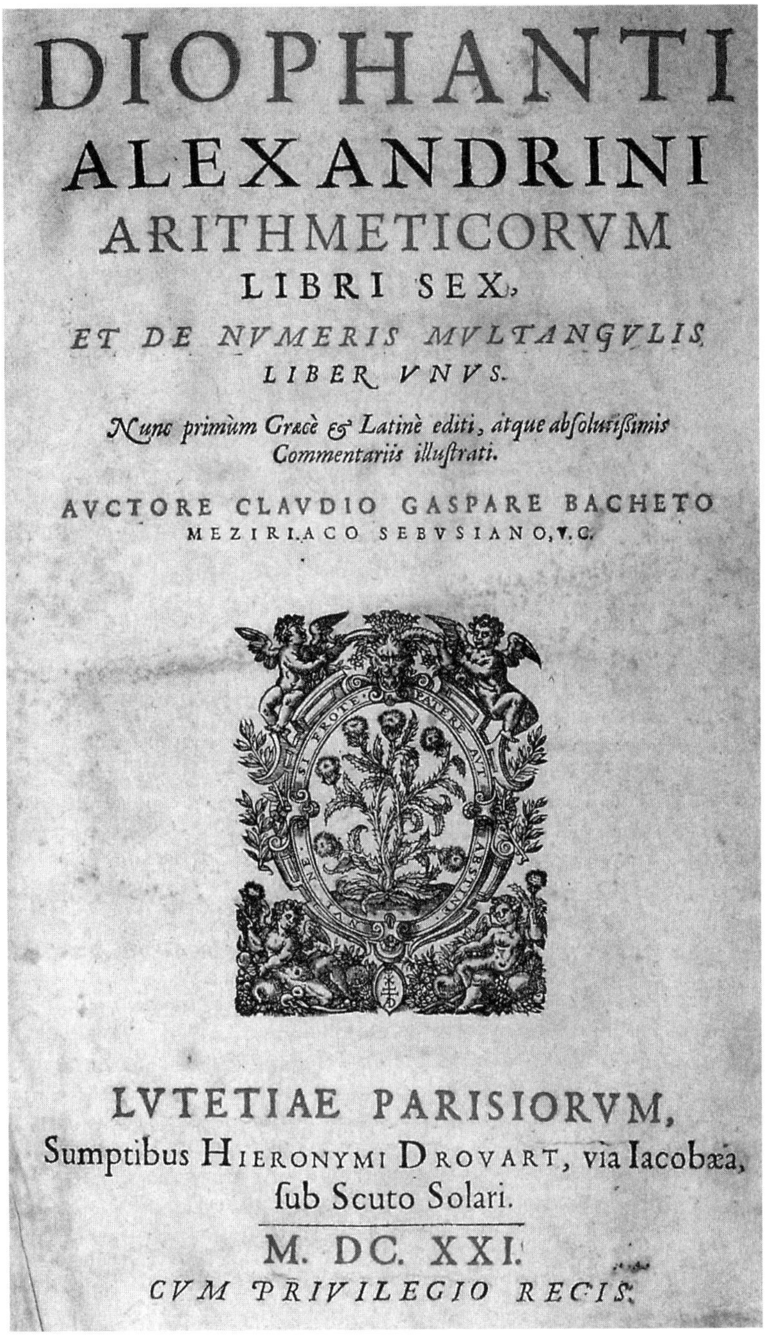

DIOPHANTI
ALEXANDRINI
ARITHMETICORVM
LIBRI SEX,
ET DE NVMERIS MVLTANGVLIS
LIBER VNVS.

Nunc primùm Græcè & Latinè editi, atque absolutissimis
Commentariis illustrati.

AVCTORE CLAVDIO GASPARE BACHETO
MEZIRIACO SEBVSIANO, V.C.

LVTETIAE PARISIORVM,
Sumptibus HIERONYMI DROVART, via Iacobæa,
sub Scuto Solari.

M. DC. XXI.

CVM PRIVILEGIO REGIS.

Figure 1.10. Title page of the 1621 edition of Diophantus's *Arithmetica*, translated from Greek into Latin by Claude Gaspard Bachet de Méziriac. Image source: Wikipedia Commons.

What follows is an example of the exposition in the *Arithmetica*, quoted from **[Hea10]**.

Example 1.6.3 (Diophantus's *Arithmetica*, Book I, Problem 1). *To divide a given number into two having a given difference.*

> *Given number* 100, *given difference* 40.
> *Lesser number required* x. *Therefore*
>
> $$2x + 40 = 100,$$
> $$x = 30.$$
>
> *The required numbers are* 70, 30.

In more modern terminology, the problem is as follows: given natural numbers N and n, find integers x and y, with $x < y$ such that $x + y = N$ and $y - x = n$. Diophantus solves the problem by subtracting both equations, to obtain $2x + n = N$, and therefore $2x = N - n$. If $N - n$ is even, then $x = (N - n)/2$ and $y = x + n = (N + n)/2$. For instance, if $N = 100$ and $n = 40$, then $x = (100 - 40)/2 = 30$ and $y = x + n = 70$.

Perhaps the most famous of all problems proposed by Diophantus in his *Arithmetica* is Problem 8 in Book II, which says

> *8. To divide a given square number into two squares.*

It is next to this proposition that, hundreds of years later, Fermat scribbled his famous note in which he enunciates what is known as "Fermat's last theorem". Pierre de Fermat (1601–1665) was a French lawyer at the Parlement of Toulouse and an amateur mathematician who is given credit for early developments that led to infinitesimal calculus and also for notable contributions to analytic geometry, probability, and optics. Nonetheless, he is particularly famous for his contributions to number theory.

During his lifetime Fermat proposed many challenges to other mathematicians, some of them quite difficult to solve. One by one, his challenges were resolved, except for one claim that took over 350 years to solve (it was proved by Andrew Wiles in 1995). Fermat's original claim was made in 1637, in an intriguing note in the margin of a copy of Diophantus's *Arithmetica*:

> *Cubum autem in duos cubos, aut quadratoquadratum in duos quadratoquadratos, et generaliter nullam in infinitum ultra quadratum potestatem in duos eiusdem nominis fas est dividere cuius rei demonstrationem mirabilem sane detexi. Hanc marginis exiguitas non caperet.*

Or, in English:

> *It is impossible to separate a cube into two cubes, or a fourth power into two fourth powers, or in general, any power higher than the second, into two like powers. I have discovered a truly marvelous proof of this, which this margin is too narrow to contain.*

In more modern notation, Fermat's last theorem can be stated as follows.

Theorem 1.6.4 (Fermat's last theorem). *The equation $x^n + y^n = z^n$ does not have any solutions $x, y, z \in \mathbb{Z}$ with $xyz \neq 0$, if $n \geq 3$.*

Figure 1.11. Pierre de Fermat (1601–1665). Image source: Wikimedia Commons.

The diophantine equation $x^n + y^n = z^n$ is, perhaps, the most studied in the theory of numbers, and it has generated thousands of pages of research articles. Notice that a non-trivial integral solution of $x^n + y^n = z^n$ corresponds to a non-trivial rational point $(\frac{x}{z}, \frac{y}{z})$ on the curve $F_n : X^n + Y^n = 1$ and, conversely, a rational point on F_n provides an integral solution of $x^n + y^n = z^n$. The curve F_n is known as the *nth Fermat curve*.

The curve $F_2 : X^2 + Y^2 = 1$ corresponds to the circle of radius 1 (which is a genus 0 curve), and it has infinitely many rational points. When $n = 3$, the Fermat curve $F_3 : X^3 + Y^3 = 1$ is an elliptic curve (a curve of genus 1), with no rational points other than $(1, 0)$ and $(0, 1)$ (and one point at "infinity"). For $n \geq 4$, the Fermat curve F_n has genus ≥ 2. Thus, by Faltings's theorem (Theorem 1.5.3) for each $n \geq 4$, the curve F_n can have at most finitely many rational points. The proof of the fact that F_n for all $n \geq 3$ has no non-trivial rational points had to wait until 1995, when Andrew Wiles announced the first complete proof of Fermat's last theorem and published it in [**Wil95**]. See [**Loz11**] for an introduction to the concepts that go into Wiles's proof.

1.7. Hilbert's Tenth Problem

Suppose $C : f(x_1, \ldots, x_n) = 0$ is a diophantine equation, as in Definition 1.6.1. The goal of the field of arithmetic geometry is to systematically study the integer and rational solutions of diophantine equations, so we ask ourselves three basic questions:

(a) Can we determine if C has any integral solutions, or rational solutions?

(b) If so, can we find *any* of the integral or rational solutions of C?

(c) Finally, can we find *all* solutions and prove that we have found all of them?

The first question was formalized by David Hilbert (see Figure 1.12): *to devise a process according to which it can be determined in a finite number of operations whether the equation is solvable in rational integers.* This was Hilbert's tenth problem out of 23 fundamental questions that he proposed to the mathematical community during the Second International Congress of Mathematicians in Paris in the year 1900.

Figure 1.12. David Hilbert (1862–1943) was one of the most influential mathematicians of the 19th and early 20th centuries. Image source: Wikimedia Commons.

Julia Robinson's work in the late 1940s on Hilbert's tenth problem (using Pell's equation, a type of equation that we will discuss in Chapter 14) was central to the formulation of a mathematical-logic approach to the problem (see Figure 1.13). Further collaboration among Davis, Matiyasevich, Putnam, and Robinson led to the surprising discovery and proof that, in fact, *there is no such general algorithm* that decides whether a diophantine equation has integer solutions (see [**Mat93**]).

However, if we restrict our attention to solving diophantine equations of certain types, e.g., lines, conics, elliptic curves, then we can answer questions (a), (b), and (c) posed above, and this book is dedicated to describing the techniques that are known in these simpler (but fundamental) cases.

Figure 1.13. Julia Hall Bowman Robinson (1919–1985) was an American mathematician renowned for her contributions to computability theory and computational complexity theory. Her work on Hilbert's tenth problem played a crucial role in its ultimate resolution. Image author: George M. Bergman (Berkeley). Source: Archives of the Mathematisches Forschungsinstitut Oberwolfach.

1.8. Exercises

Exercise 1.8.1. Show that the decimal expansion $0.9999\ldots$, with infinitely many nines, is equal to the decimal expansion $1 = 1.0000\ldots$. In other words, show that the infinite series $\sum_{k=1}^{\infty} 9/10^k$ converges and the sum equals 1. (Hint: use the geometric series test.)

Exercise 1.8.2. Find all the natural, integral, and rational roots of the following polynomial equations:

(1) $x^5 - 9x^4 - 5x^3 + 45x^2 + 4x - 36 = 0.$

(2) $3x^4 + 5x^3 - 3x^2 - 5x = 0.$

(3) $x^4 + 5x^3 - 16x^2 - 17x - 21 = 0.$

(4) $x^4 + x^3 + 21 = 0.$

Exercise 1.8.3. Find $k \in \mathbb{Z}$ such that $x = 5$ is a root of $x^3 + kx^2 + 23x + 285 = 0$.

Exercise 1.8.4. Find integers m and n such that $x = -2$ and $x = 3$ are roots of the polynomial equation $x^3 + 10x^2 + mx + n = 0$.

Exercise 1.8.5. Let $p(x)$ and $q(x)$ be polynomials, and let $\alpha \in \mathbb{Q}$ be a root of $p(x) = 0$ and $\beta \in \mathbb{Q}$ a root of $q(x) = 0$. Show the following statements:

(1) The numbers α and β are roots of the polynomial equation $p(x) \cdot q(x) = 0$.

(2) If $\alpha = \beta$, then α is a root of the polynomial equation $p(x) + q(x) = 0$.

Exercise 1.8.6. The goal of this exercise is to show that every complex number $\alpha \in \mathbb{C}$ has a square root within the complex numbers; i.e., there is some $\sqrt{\alpha} \in \mathbb{C}$. Recall the definition of the complex numbers:

$$\mathbb{C} = \{a + bi : a, b \in \mathbb{R} \text{ and } i^2 = -1\}.$$

(1) Find a square root of $\alpha = 1+i$, within $\mathbb{C}$; i.e., find $\beta \in \mathbb{C}$ such that $\beta^2 = 1+i$. (Hint: write $\beta = c + di$ and find a similar expression for its square β^2.)

(2) Find a square root of $\alpha = a + bi$, within $\mathbb{C}$; i.e., find $\beta = c + di$ such that $\beta^2 = \alpha$. In fact, show that

$$\beta = \pm\left(\sqrt{\frac{\sqrt{a^2+b^2}+a}{2}} + \left(\frac{b}{|b|}\sqrt{\frac{\sqrt{a^2+b^2}-a}{2}}\right)\cdot i\right).$$

(3) For any real number θ, we define

$$e^{i\theta} = \cos(\theta) + \sin(\theta) \cdot i.$$

Show that any complex number α can be written uniquely as $\alpha = re^{i\theta}$, for some $r \geq 0$ and some $\theta \in [0, 2\pi)$. (Hint: find a geometric interpretation of r and θ in the complex plane.)

(4) Show that if $\alpha = re^{i\theta}$, then the square roots of α are $\beta = \sqrt{r}e^{i\theta/2}$ and $-\beta = \sqrt{r}e^{i(\theta/2+\pi)}$.

Exercise 1.8.7. Find all natural numbers n such that its cube minus its square plus itself equals 1.

Exercise 1.8.8. Let $P = (x_0, y_0)$ and $Q = (x_1, y_1)$ be two points in the plane, with $x_0 \neq x_1$, and let $m = (y_1 - y_0)/(x_1 - x_0)$. Show that the line L that passes through P and Q is given by

$$y - y_0 = m \cdot (x - x_0).$$

Exercise 1.8.9. Let $P = (1, 4)$ and $Q = (4, -2)$ be points on the plane.

(1) Find the equation $y = ax + b$ of a line L that passes through P and Q.

(2) Find a formula for all the rational points on L.

(3) Find a formula for all the integral points on L.

(4) How many points on L have natural coordinates; i.e., how many points $R = (x_0, y_0)$ on L are there with $x_0, y_0 \in \mathbb{N}$?

Exercise 1.8.10. Find all the rational points on the circle $x^2 + y^2 = 2$.

Exercise 1.8.11. Let C be the ellipse given by $x^2 + 3y^2 = 784$.

(1) Find all the integral points on C.

(2) Find a parametrization of all the rational points on C.

Exercise 1.8.12. Let C be the hyperbola given by the equation $x^2 - 7y^2 = 2$.

(1) Find all the rational points on the hyperbola $x^2 - 7y^2 = 2$.

(2) Find 3 distinct integral points with positive x-coordinate.

Exercise 1.8.13. Show that the hyperbola $C' : x^2 - 5y^2 = 3$ has no integral points.

Exercise 1.8.14. (1) Are there two perfect squares (i.e., integers of the form n^2, where n itself is an integer) that differ by 1? Write the problem in terms of a diophantine equation, find all integral solutions to the equation, and prove that you have found them all. (Hint: write one square as n^2 and the other square as $(n+m)^2$.)

(2) Find a parametrization of all the rational squares (i.e., rational numbers of the form t^2 for some $t \in \mathbb{Q}$) that differ by 1.

(3) Are there two consecutive integers such that their product is a perfect square? If so, find all such integers.

(4) Are there three consecutive integers such that their product is a perfect square? If so, find all such integers. (*This is hard! Here it suffices to find one diophantine equation in two variables that represents this problem.*)

(5) Are there three integers $u < v < w$ that differ by 5 (i.e., $u + 5 = v$ and $v + 5 = w$) and such that their product is a perfect square? If so, find all such integers. (*There are some .... Can you find any? Finding all solutions is hard! Again, here it suffices to find one diophantine equation in two variables that represents this problem.*)

Exercise 1.8.15. We say that a natural number $n \geq 1$ is a *congruent number* if there is a right triangle with rational sides and area equal to n. Is $n = 5$ a congruent number? If so, find a right triangle with rational sides and area equal to 5.

Exercise 1.8.16. A triple (a, b, c) of natural numbers $a, b, c \in \mathbb{N}$ is said to be *pythagorean* if they satisfy $a^2 + b^2 = c^2$.

(1) Show that $(a, b, c) = (n^2 - m^2, 2nm, n^2 + m^2)$ is a pythagorean triple for any two non-zero distinct integers $n > m > 0$.

(2) Show that if n and m satisfy (i) one of n and m is even and the other one is odd and (ii) n and m are relatively prime, then $(a, b, c) = (n^2 - m^2, 2nm, n^2 + m^2)$ is a *primitive* pythagorean triple; i.e., a, b, and c are pairwise relatively prime.

(3) Use (b) to find five distinct primitive pythagorean triples.

Exercise 1.8.17. An *Euler brick* is just a rectangular box in which all of the edges (length, depth, and height) have integer dimensions and in which the diagonals on all three sides are also integers.

(1) Find the dimensions of two distinct Euler bricks.

(2) A *perfect cuboid* is an Euler brick in which the space diagonal, that is, the distance from any corner to its opposite corner, is also an integer. Can you find a perfect cuboid? (*This is an* **open problem**. *Here, it suffices to find a system of diophantine equations that represents this problem.*)

Exercise 1.8.18 (Diophantus's *Arithmetica*, Book II, Problem 30)**.** Find two numbers such that their product plus or minus their sum gives a square; i.e., find a pair of rational numbers x and y such that there are $u, v \in \mathbb{Q}$ with

$$\begin{cases} xy + x + y = u^2, \\ xy - (x + y) = v^2. \end{cases}$$

Can you find *all* such rational numbers x and y?

Exercise 1.8.19. Find a copy of Book II of the *Arithmetica* by Diophantus of Alexandria and quote two problems (other than numbers 8, 29, or 30). Reproduce Diophantus's solution, and then rewrite it in a more modern language and notation.

Exercise 1.8.20. Let E be the elliptic curve given by $y^2 = x^3 + Ax + B$, for some $A, B \in \mathbb{Z}$.

(1) Use implicit differentiation to show that
$$y' = \frac{dy}{dx} = \frac{3x^2 + A}{2y}.$$

(2) Let E be $y^2 = x^3 - 2$ and $P = (3, 5)$. Find $\frac{dy}{dx}(P)$, i.e., the slope of the tangent line to E at the point P.

(3) Let E be $y^2 = x^3 - x$ and $P = (0, 0)$. What is the slope of the tangent line to E at the point P?

Exercise 1.8.21. Let $f(a, b, c, d) = ad - bc$ and let $C : ad - bc = 1$.

(1) Let $\mathrm{SL}(2, \mathbb{Z})$ be the set of 2×2 matrices with integer coefficients and determinant 1. Show that the set $C(\mathbb{Z})$ of integral points on the diophantine equation C is in bijection with $\mathrm{SL}(2, \mathbb{Z})$.

(2) Show that $S = \begin{pmatrix} 0 & -1 \\ 1 & 0 \end{pmatrix}$ and $T = \begin{pmatrix} 1 & 1 \\ 0 & 1 \end{pmatrix}$ belong to $\mathrm{SL}(2, \mathbb{Z})$.

(3) Show that if A and B are matrices in $\mathrm{SL}(2, \mathbb{Z})$, then $A \cdot B$ is also in $\mathrm{SL}(2, \mathbb{Z})$, where $\cdot$ here denotes matrix multiplication (see Example 5.2.5).

(4) Show that $Q_n = (S \cdot T^2)^n = (S \cdot T^2) \cdots (S \cdot T^2)$ is a matrix in $\mathrm{SL}(2, \mathbb{Z})$ for all $n \geq 1$. Describe the points on C that correspond to the matrices Q_n for $1 \leq n \leq 6$.

(5) Show that there are infinitely many integral points (a, b, c, d) in $C(\mathbb{Z})$ with all non-zero coordinates.

Note: this problem continues in Exercises 2.11.12 and 5.6.4.

Part 1

Integers, Polynomials, Lines, and Congruences

CHAPTER 2

THE INTEGERS

God made the integers, all else is the work of man.
(Die ganzen Zahlen hat der liebe Gott gemacht,
alles andere ist Menschenwerk.)

Leopold Kronecker

In order to build a solid mathematical theory of the integers that does not rely on our intuition, we need to establish a basic set of axioms that define the integers (and the natural numbers).

2.1. The Axioms of $\mathbb{Z}$

The integers, denoted by $\mathbb{Z}$, are a set with the following properties:

(1) There are two operations on elements of $\mathbb{Z}$, namely addition $+$ and multiplication $\times$ (also denoted by $\cdot$), and $\mathbb{Z}$ is closed under these operations; that is, if $a, b \in \mathbb{Z}$, then $a + b$ and $a \cdot b$ are also in $\mathbb{Z}$.

(2) Properties of $+$ and $\times$. For all $a, b, c \in \mathbb{Z}$, we have:
   (2.a) (Commutativity): $a + b = b + a$ and $a \cdot b = b \cdot a$.
   (2.b) (Associativity): $(a + b) + c = a + (b + c)$ and $(a \cdot b) \cdot c = a \cdot (b \cdot c)$.
   (2.c) (Distributivity): $c \cdot (a + b) = c \cdot a + c \cdot b$ and $(a + b) \cdot c = a \cdot c + b \cdot c$.

(3) Existence axioms:
   (3.a) (Additive identity) There exists $0 \in \mathbb{Z}$ such that $a + 0 = 0 + a = a$, for all $a \in \mathbb{Z}$.
   (3.b) (Additive inverses) For all $a \in \mathbb{Z}$ there is $-a \in \mathbb{Z}$ such that $a + (-a) = 0 = (-a) + a$.
   (3.c) (Multiplicative identity) There exists $1 \in \mathbb{Z}$ such that $a \cdot 1 = 1 \cdot a = a$, for all $a \in \mathbb{Z}$.

(4) There exists a subset $\mathbb{N} \subseteq \mathbb{Z}$ satisfying the following properties:
   (4.a) (Non-triviality) $\mathbb{N}$ is non-empty.
   (4.b) (Closure) $\mathbb{N}$ is closed under $+$ and $\cdot$; that is, if $a, b \in \mathbb{N}$, then $a+b$ and $a \cdot b \in \mathbb{N}$.
   (4.c) (Trichotomy) For all $a \in \mathbb{Z}$, precisely one and only one of the following statements is true: $a \in \mathbb{N}$ or $a = 0$ or $-a \in \mathbb{N}$.
   (4.d) (Well-ordering principle) Every non-empty subset of $\mathbb{N}$ has a least element with respect to the ordering defined by $a > b$ if $a + (-b) \in \mathbb{N}$.

Remark 2.1.1. The integers $\mathbb{Z}$ satisfy axioms (1) through (3), but many other number systems satisfy these axioms. For instance, these axioms are satisfied by $\mathbb{Q}$, $\mathbb{R}$, and $\mathbb{C}$ (in general, any *commutative ring* satisfies axioms (1)–(3)). However, the rationals, reals, and complex numbers do not satisfy the axioms in (4), as there is no subset of $\mathbb{Q}$, $\mathbb{R}$, or $\mathbb{C}$ that could replace the subset $\mathbb{N}$ as in the case of $\mathbb{Z}$.

Note, however, that there are other number systems that come close to satisfying axioms (1)–(4). For instance, let $\mathbf{Z} = \{0\}$, with operations $+$ and $\cdot$ defined by $0 + 0 = 0$ and $0 \cdot 0 = 0$. Clearly, $\mathbf{Z}$ satisfies axiom (1) and the group of axioms in (2). Moreover, putting $1 = 0$, the reader can check that $\mathbf{Z}$ also satisfies (3.a), (3.b), and (3.c). By setting $\mathbb{N} = \{0\} = \mathbf{Z}$, we see that axioms (4.a), (4.b), and (4.d) are trivially satisfied. However, $\mathbf{Z}$ does not satisfy the trichotomy axiom (4.c), because it follows from trichotomy that 0 is not a natural number. Indeed, trichotomy implies that a number a is in $\mathbb{N}$ or $a = 0$ or $-a \in \mathbb{N}$, but only one of the three statements occurs for each a. Thus, if $a = 0$, then neither a nor $-a$ is in $\mathbb{N}$.

Remark 2.1.2. The axioms (3.a) and (3.c) declare the existence of at least two elements in $\mathbb{Z}$, namely 0 and 1. The reader, though, should be aware that the axioms do not explicitly say that 1 is a natural number! However, one can deduce this fact from the axioms. Indeed, suppose for a contradiction that 1 is *not* a natural number. Then, trichotomy (axiom (4.c)) says that $-1 \in \mathbb{N}$ (since our previous remark shows that $0 = 1$ is impossible). In Lemma 2.2.2 (together with Exercise 2.11.1) we will show that $(-1) \cdot (-1) = 1$. Since $\mathbb{N}$ is closed under multiplication by axiom (4.b), it follows that $1 \in \mathbb{N}$ as well. But this is a contradiction with trichotomy since a and $-a$ cannot be simultaneously in $\mathbb{N}$. Hence, we have reached a contradiction, and we must have $1 \in \mathbb{N}$. We will prove below in Theorem 2.2.4 that 1 is, in fact, the smallest natural number.

Since $1 \in \mathbb{N}$ and $\mathbb{N}$ is closed under addition, there is a number $1+1 \in \mathbb{N} \subseteq \mathbb{Z}$ that we denote by 2; the number $(1+1)+1$ is called 3; etc. Note that $3 = (1+1)+1 = 1+(1+1)$ by (2.b).

Let us introduce some more notation to ease the discussions about the ordering in the integers and natural numbers.

Definition 2.1.3. Let a, b be integers. The symbol $a - b$ stands for $a + (-b)$, where $-b$ is the additive inverse of b. We say that a is *greater than* b, or $a > b$ if $a - b \in \mathbb{N}$. Similarly, we say that a is *greater than or equal to* b, or $a \geq b$, if $a > b$ or $a = b$. Conversely, we say that a is *less than* b if $b > a$ and $a \leq b$ if $b \geq a$.

Example 2.1.4. The number 5 is greater than 4 because $5 - 4 = 1$ is in $\mathbb{N}$.

In this book, we will pay close attention to divisibility properties of integers. Here is the formal definition of divisibility.

Definition 2.1.5. Let a, b be integers. If there is an integer m such that $b = a \cdot m$, then we say that a is a *divisor* of b, or *a divides b*, or, equivalently, b is *divisible by a*. We also write $a \mid b$.

Example 2.1.6. Let $a = 5$ and $b = 15$. Then, $15 = 5 \cdot 3$ and therefore $a = 5$ is a divisor of $b = 15$. We write $5 \mid 15$.

Definition 2.1.7. A natural number $n \geq 2$ is called a *prime* if it has exactly two positive divisors, namely 1 and n. A number $n \geq 2$ that is not prime is called a *composite number*.

The first few prime numbers are 2, 3, 5, 7, 11, 13, ... Later we will show that there are infinitely many prime numbers (Theorem 3.2.1). As of January 2018, the largest known prime number is $p = 2^{77232917} - 1$, a number with 23,249,425 digits (p is the 50th Mersenne prime; see Exercise 3.5.23).

2.2. Consequences of the Axioms

There are a number of statements that may seem axiom-like but, in fact, they are consequences of the axioms, i.e., theorems. For instance, the fact that if $a, b \in \mathbb{Z}$ and $ab = 0$, then $a = 0$ or $b = 0$, is a direct consequence of the axioms. Before we prove this, first we need two lemmas that show other basic facts that follow from the axioms.

Lemma 2.2.1. *For all $a \in \mathbb{Z}$, we have that $a \cdot 0 = 0 \cdot a = 0$.*

Proof. Let us show that $a \cdot 0 = 0$ for any $a \in \mathbb{Z}$. The fact that $0 \cdot a = 0$ will follow from the commutative law. Notice that $0 = 0 + 0$, because 0 is the additive identity (axiom (3.a)). Thus,

$$a \cdot 0 = a \cdot (0 + 0) = a \cdot 0 + a \cdot 0, \tag{2.1}$$

where in the last equality we have used the distributive law (axiom (2.c)). The number $a \cdot 0$ is an integer (by closure, axiom (1)), and therefore it has an additive inverse (axiom (3.b)), which we call $-(a \cdot 0)$. Adding $-(a \cdot 0)$ to both sides of (2.1) we obtain

$$\begin{aligned} 0 &= a \cdot 0 + (-(a \cdot 0)) = (a \cdot 0 + a \cdot 0) + (-(a \cdot 0)) \\ &= a \cdot 0 + (a \cdot 0 + (-(a \cdot 0))) = a \cdot 0 + 0 \\ &= a \cdot 0, \end{aligned}$$

where we have used axioms (2.c), (3.b), and (3.a). Hence, $0 = a \cdot 0$, as claimed. □

Lemma 2.2.2. *Let a, b be integers. Then, $a \cdot (-b) = -(a \cdot b)$. Similarly, $(-a) \cdot b = -(a \cdot b)$.*

Proof. In order to show that $a \cdot (-b) = -(a \cdot b)$, it suffices to show that $a \cdot (-b)$ is the additive inverse of $a \cdot b$. Hence, we need to show that $a \cdot b + a \cdot (-b) = 0$. Indeed,

$$a \cdot b + a \cdot (-b) = a \cdot (b + (-b)) = a \cdot 0 = 0,$$

where we have used the distributive law (axiom (2.c)), the defining property of the additive inverse $-b$ (axiom (3.b)), and Lemma 2.2.1. □

Theorem 2.2.3. *Let a, b be integers such that $a \cdot b = 0$. Then $a = 0$ or $b = 0$.*

Proof. Let $a, b \in \mathbb{Z}$ such that $a \cdot b = 0$. By trichotomy (axiom (4.c)), we have that $a \in \mathbb{N}$, $a = 0$, or $-a \in \mathbb{N}$. Similarly, $b \in \mathbb{N}$, $b = 0$, or $-b \in \mathbb{N}$. We analyze all the possible cases:

- If a and b are in $\mathbb{N}$, by closure (axiom (4.b)), it would follow that $a \cdot b \in \mathbb{N}$ but, by trichotomy, this is impossible, since we know that $a \cdot b = 0$.
- If $a \in \mathbb{N}$ and $-b \in \mathbb{N}$, then $a \cdot (-b) = -(a \cdot b) \in \mathbb{N}$ (by Lemma 2.2.2) but, again, this is a contradiction with trichotomy, since we know that $a \cdot b = 0$. Similarly, if $-a \in \mathbb{N}$ and $b \in \mathbb{N}$, we have that $(-a) \cdot b = -(a \cdot b) \in \mathbb{N}$ and we reach the same contradiction.
- If $-a$ and $-b$ are in $\mathbb{N}$, then $(-a) \cdot (-b) = -(a \cdot (-b)) = -(-(a \cdot b)) = a \cdot b \in \mathbb{N}$ by closure (the reader should verify that $-(-c) = c$, for all $c \in \mathbb{Z}$). This is a contradiction with trichotomy, since we have assumed that $a \cdot b = 0$.

Thus, the only possibilities that remain satisfy that a or b is 0 (and by Lemma 2.2.1, we know that $a \cdot 0 = 0$ or $0 \cdot b = 0$). □

Finally, we conclude this section by showing that 1 is the smallest natural number.

Theorem 2.2.4. *The number 1 is the smallest natural number; i.e., if n is a natural number, then either $n = 1$ or $1 < n$.*

Proof. In Remark 2.1.2 we have shown that 1 is a natural number, so it remains to show that it is the smallest natural number. By the well-ordering principle, the set of all natural numbers has a least element $a \in \mathbb{N}$. Suppose for a contradiction that $a < 1$. Since a is natural, then $0 < a$ because $a - 0 = a \in \mathbb{N}$. Hence, we have $0 < a < 1$.

Now, consider $b = a^2$. Since $b = a^2 = a \cdot a$, then $b \in \mathbb{N}$ by axiom (1). Moreover, a is the smallest natural number, so either $a = b$ or $a < b$. If $a < b$, then $b - a \in \mathbb{N}$, but

$$b - a = a^2 - a = a \cdot (a - 1)$$

and since $a < 1$, the number $a - 1$ is not natural. In particular, $a \cdot (a - 1)$ cannot be natural (by Exercise 2.11.2). Thus, we reach a contradiction to the fact that $a(a - 1) = b - a \in \mathbb{N}$.

Finally, if $a = b$, then $a = a^2$, and therefore $a(a - 1) = 0$. Hence, by Theorem 2.2.3, we have $a = 0$ or $a = 1$, both of which would be contradictory since we have shown that $0 < a < 1$.

Thus, we have reached a contradiction in all cases, and we must have that the smallest natural number is $a = 1$. □

2.3. The Principle of Mathematical Induction

In this section, we shall prove the principle of mathematical induction using the axioms of the integers that we established in the previous sections. In particular, induction is a consequence of the well-ordering principle.

Theorem 2.3.1 (Mathematical induction). *Let $P(n)$ be a statement such that*

(1) *the statement $P(n_0)$ is true, for some $n_0 \in \mathbb{N}$, and*

(2) *if $P(k)$ is true for some $k \geq n_0$, then $P(k+1)$ is also true.*

Then, $P(n)$ is true for all $n \geq n_0$.

Proof. Let S be the set of all natural numbers $n \geq n_0$ such that $P(n)$ is *false*; that is,

$$S = \{n \in \mathbb{N} : n \geq n_0 \text{ and } P(n) \text{ is false}\}.$$

We would like to show that S is empty, so let us assume, for a contradiction, that there is some natural number n greater than or equal to n_0 such that $P(n)$ is false. Thus, S is non-empty. By the well-ordering principle, there is a minimum element of S which we shall call m. Since $m \in S$, it follows that $m \geq n_0$. However $P(n_0)$ is true, so $m > n_0$. Notice that $m > m-1 \geq n_0$ so $P(m-1)$ must be true, since m is the minimum of S. But our assumptions on the statement of the theorem imply that if $P(m-1)$ is true and $m-1 \geq n_0$, then $P((m-1)+1) = P(m)$ must be true as well. This is a contradiction, for m is an element of S and therefore $P(m)$ is false. Hence, S must be empty and $P(n)$ is true for all $n \geq n_0$, as claimed. □

Remark 2.3.2. In order to prove a result using the principle of mathematical induction, by Theorem 2.3.1, one needs to check two things: (1) is called the *base case*, and one needs to check by hand that the statement $P(n_0)$ is true; (2) is called the *induction step*, and one needs to prove $P(k+1)$ while assuming that $P(k)$ is true, for some $k \geq n_0$ (the assumption of $P(k)$ is usually called the *induction hypothesis*).

Let us see some examples of proofs that use the principle of mathematical induction.

Example 2.3.3. Let us show that, for all $n \geq 1$, one has

$$1 + 2 + 3 + \cdots + n = \frac{n(n+1)}{2}. \tag{2.2}$$

The formula in (2.2) is what in Theorem 2.3.1 we referred to as the statement $P(n)$. We shall prove this formula using the principle of mathematical induction (Theorem 2.3.1). Let us first show that the statement is true for the base case $n_0 = 1$. This is clear since

$$1 = \frac{1(1+1)}{2} = \frac{2}{2}.$$

Next, we need to prove the induction step, i.e., $P(k)$ implies $P(k+1)$. Let us assume that the statement is true for k; i.e., $1 + 2 + 3 + \cdots + k = \frac{k(k+1)}{2}$. We need to show the statement in the $k+1$ case, which reads:

$$1 + 2 + 3 + \cdots + k + (k+1) =^? \frac{(k+1)((k+1)+1)}{2} = \frac{(k+1)(k+2)}{2}.$$

Let us begin working with the left-hand side of the equation to be shown:

$$\begin{aligned}1+2+3+\cdots+k+(k+1) &= (1+2+3+\cdots+k)+(k+1)\\ &= \frac{k(k+1)}{2}+(k+1).\end{aligned}$$

In the last step, we have used the induction hypothesis $P(k)$; i.e., we have used the fact that $1+2+\cdots+k = k(k+1)/2$. Now we may continue, as follows,

$$\begin{aligned}1+2+3+\cdots+k+(k+1) &= \frac{k(k+1)}{2}+(k+1)\\ &= \frac{k(k+1)+2(k+1)}{2}\\ &= \frac{(k+1)(k+2)}{2},\end{aligned}$$

as desired. This shows that the truth of $P(k)$ implies the truth of $P(k+1)$. Hence, we have shown the base case and the induction step, and by the principle of mathematical induction, the formula in (2.2) is true for all $n \geq 1$.

Example 2.3.4. Let us show the following inequality, using mathematical induction:

$$2^n \geq n+1, \qquad \text{for all } n \geq 1.$$

Let us begin with the base case, $n_0 = 1$. Clearly, we have $2^1 = 2 \geq 1+1$, so the base case is true. Next, we show the induction step; i.e., if we assume that $2^k \geq k+1$, then we need to show that $2^{k+1} \geq (k+1)+1$, as well. Let us begin with the left-hand side of the $(k+1)$th equation:

$$2^{k+1} = 2\cdot 2^k \geq 2(k+1) = 2k+2 \geq k+2 = (k+1)+1,$$

where, in the second inequality, we have used our induction hypothesis, $2^k \geq k+1$. Hence, the induction step is shown, and by the principle of mathematical induction, we have that $2^n \geq n+1$, for all $n \geq 1$.

Example 2.3.5. In this example we show that $n! \geq 2^n$, for all $n \geq 4$, using induction. The base case is $n_0 = 4$, and $4! = 24 \geq 16 = 2^4$, so the base case is true. Now, let us assume that the inequality is true for k; i.e., we assume that $k! \geq 2^k$. Then, for any $k \geq 4$,

$$(k+1)! = (k+1)k! \geq (k+1)2^k \geq 2\cdot 2^k \geq 2^{k+1}.$$

Hence, we have shown the induction step, and by the principle of mathematical induction, $n! \geq 2^n$, for all $n \geq 4$.

Example 2.3.6. If $n \in \mathbb{N}$, then n^3+2n is divisible by 3. Clearly, the base case $n_0 = 1$ is true, since $1^3+2\cdot 1 = 3$ is divisible by 3. Let us assume as our induction hypothesis that k^3+2k is divisible by 3 for some $k \geq 1$; i.e., there is some $m \in \mathbb{Z}$ such that $k^3+2k = 3m$. Thus,

$$\begin{aligned}(k+1)^3+2(k+1) &= k^3+3k^2+3k+1+2k+2\\ &= (k^3+2k)+3(k^2+k+1)\\ &= 3(m+k^2+k+1),\end{aligned}$$

and so, $(k+1)^3+2(k+1)$ is divisible by 3. This settles the induction step, and the principle of mathematical induction implies that n^3+2n is divisible by 3, for all $n \geq 1$.

Example 2.3.7. Let us show that n^2-1 is a natural number divisible by 8, for all **odd** numbers $n \geq 3$. We shall use induction. Let us first check the base case, which, in this instance, would be $n_0 = 3$. We know that

$$3^2 - 1 = 8,$$

so it is clear that $3^2-1=8$ is divisible by 8. Next, we need to verify the induction step. More concretely, we need to show that if the statement is true for an odd number $k \geq 3$, then the statement is also true for the *next odd number*. Note that if k is odd, then the next odd number is $k+2$ (and not $k+1$).

Suppose that $k \geq 3$ is an odd number and k^2-1 is divisible by 8. By the definition of divisibility (Definition 2.1.5), there is an integer m such that $k^2-1 = 8m$. Now we need to show that $(k+2)^2-1$ is also divisible by 8. Notice that

$$(k+2)^2 - 1 = (k^2+4k+4) - 1 = (k^2-1) + 4k + 4 = 8m + 4(k+1).$$

Since k is odd, the number $k+1$ is even, and so there is a number t such that $k+1 = 2t$. Hence,

$$(k+2)^2 - 1 = 8m + 4(k+1) = 8m + 4(2t) = 8m + 8t = 8(m+t).$$

Thus, $(k+2)^2-1 = 8(m+t)$ and, by definition, this shows that $(k+2)^2-1$ is divisible by 8. This proves the induction step. Therefore, by the principle of mathematical induction, the number n^2-1 is divisible by 8 for all odd numbers $n \geq 3$.

Remark 2.3.8. Here is another way to prove that if $n \geq 3$ is odd, then n^2-1 is divisible by 8, without using induction. Note that every odd number n can be written in the form $n = 2m+1$, for some $m \geq 0$. Thus,

$$n^2 - 1 = (2m+1)^2 - 1 = 4m^2 + 4m = 4m(m+1).$$

If m is even, then $m = 2m'$, for some $m' \in \mathbb{N}$. Thus, $n^2-1 = 8m'(m+1)$ is divisible by 8. Otherwise, if m is odd, then $m = 2m'+1$ for some $m' \in \mathbb{N}$ and $m+1 = (2m'+1)+1 = 2(m'+1)$ is even. Thus, $n^2-1 = 8m(m'+1)$ is divisible by 8. Hence, in all cases, n^2-1 is divisible by 8 when $n \geq 3$ is odd.

Example 2.3.9. Let us show that a $2^n \times 2^n$ chessboard with one corner removed from the board can be tiled with L-shaped pieces for all $n \geq 1$, where each L piece has 3 square tiles (2 high, 2 wide). In Figure 2.1, we show the case $n=1$ (this is our base case for induction) and the case $n=2$.

Let us assume that a $2^k \times 2^k$ chessboard can be tiled with L-shaped pieces, and consider a $2^{k+1} \times 2^{k+1}$ chessboard C with the top-right corner removed. Since $2^{k+1} = 2 \cdot 2^k$, the board C can be covered with 4 smaller chessboards of size $2^k \times 2^k$, which we will name clockwise and starting with the top-right chessboard by C_1, C_2, C_3, and C_4. The chessboard C_1 has the top-right corner removed and by our induction hypothesis, C_1 can be tiled with the L-shaped pieces. Moreover, if we remove one tile from each of C_2, C_3, and C_4 (as in the case $n=2$ in the figure), each one of them can be tiled with L-shaped pieces (by our induction hypothesis), and the three removed corner pieces can be filled back in with an L-shaped piece. Hence,

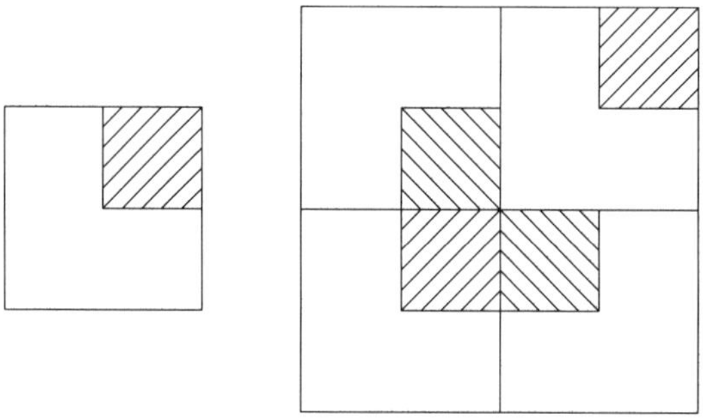

Figure 2.1. A 2×2 and a 4×4 chessboard, with one corner removed, can be tiled with L-shaped pieces.

this shows the induction step, and by the principle of mathematical induction, every $2^n \times 2^n$ chessboard with one corner removed can be tiled with L-shaped pieces, for all $n \geq 1$.

The following theorem is usually referred to as "complete induction". It is another form of induction; however, it is logically equivalent to the regular mathematical induction of Theorem 2.3.1.

Theorem 2.3.10 (Complete (or strong) induction). *Let $P(n)$ be a statement such that*

(1) *the statement $P(n_0)$ is true, for some $n_0 \in \mathbb{N}$, and*
(2) *if the statement $P(t)$ is true for all values of t in the range $n_0 \leq t \leq k$, then $P(k+1)$ is also true.*

Then, $P(n)$ is true for all $n \geq n_0$.

Proof. Let S be the set of all natural numbers $n \geq n_0$ such that $P(n)$ is *false*; that is,

$$S = \{n \in \mathbb{N} : n \geq n_0 \text{ and } P(n) \text{ is false}\}.$$

Let us assume, for a contradiction, that there is some natural number n greater than or equal to n_0 such that $P(n)$ is false. Thus, S is non-empty. By the well-ordering principle, there is a minimum element of S which we shall call m. Since $m \in S$, it follows that $m \geq n_0$. However $P(n_0)$ is true, so $m > n_0$. Notice that if $m - 1 \geq t \geq n_0$, then $P(t)$ must be true, since m is the minimum of S. But our assumptions on the statement of the theorem imply that if $P(t)$ is true for all $m - 1 \geq t \geq n_0$, then $P((m-1)+1) = P(m)$ must be true as well. This is a contradiction, for $m \in S$ and therefore $P(m)$ is false. Hence, S must be empty and $P(n)$ is true for all $n \geq n_0$, as claimed. □

Example 2.3.11. Let us show that every number $n \geq 12$ can be written in the form $4a + 5b$, for some $a, b \geq 0$. For instance,

$$12 = 4 \cdot 3,\ 13 = 4 \cdot 2 + 5,\ 14 = 4 + 5 \cdot 2,\ 15 = 5 \cdot 3,\ 16 = 4 \cdot 4, \ldots.$$

Notice, however, that $n = 11$ cannot be written in the form $4a + 5b$. We shall use complete induction (Theorem 2.3.10). Our base case $n = 12 = 4 \cdot 3 + 5 \cdot 0$ is true.

Now let us fix some $k \geq 12$, and suppose that the statement is true for every t with $12 \leq t \leq k$; i.e., there are some a and $b \geq 1$ such that $t = 4a + 5b$. We would like to show that the result is also true for $k + 1$.

If $k = 12$, 13, or 14, then the result is also true for $k + 1$, as we have seen that, respectively, the result is true for $k + 1 = 13$, 14, and 15 in the examples above. Thus, we may assume that $k \geq 15$, and let $t = k - 3$. Consequently, we have that $12 \leq t \leq k$ and by the induction hypothesis, there are a' and $b' \geq 1$ such that $t = 4a' + 5b'$. Therefore,

$$k + 1 = k + (4 - 3) = (k - 3) + 4 = t + 4 = 4a' + 5b' + 4 = 4(a' + 1) + 5b'.$$

Hence, there exist $a = a' + 1$ and $b = b'$ such that $k + 1 = 4a + 5b$, as desired. This proves the induction step, and the (complete) principle of mathematical induction shows that the statement is true for all $n \geq 12$.

Before we provide our next example, we need to establish one lemma about composite numbers.

Lemma 2.3.12. *If $n \geq 1$ is a composite number, then there are $a, b \in \mathbb{Z}$ such that $n = a \cdot b$ and $1 < a, b < n$.*

Proof. If $n \geq 1$ is a composite number, then, by definition, it is not a prime, so it must have an additional divisor a, other than 1 and n; that is, $n = a \cdot b$, for some $b \in \mathbb{Z}$ and $a \neq 1$ or n. Thus, $b \neq 1$ or n, because if $b = 1$, then $a = n$, and if $b = n$, then $a = 1$. Hence, $n = a \cdot b$, and $1 < a, b < n$. □

Example 2.3.13. Let us show, using complete induction, that each natural number $n \geq 2$ is divisible by at least one prime number. The base case $n_0 = 2$ is prime, and every prime is divisible by a prime, namely itself. Let us assume that each natural number t with $2 \leq t \leq k$ is divisible by a prime number, and consider the natural number $k + 1$.

- If $k + 1$ is prime, then we are done, because $k + 1$ is divisible by itself.
- Otherwise, if $k + 1$ is a composite number, it has a non-trivial factorization $k + 1 = a \cdot b$, with $1 < a, b < k + 1$. In particular, since $1 < a < k + 1$, our induction hypothesis implies that $t = a$ is divisible by a prime number p, i.e., $a = p \cdot a'$ for some $a' \in \mathbb{Z}$. Hence,

  $$k + 1 = a \cdot b = (p \cdot a') \cdot b = p \cdot (a' \cdot b),$$

  and this implies that the prime p is also a divisor of $k + 1$.

In both cases, we have shown that $k+1$ has a prime divisor. Therefore, by complete induction, it follows that all natural numbers $n \geq 2$ are divisible by at least one prime number.

2.3.1. The Pigeonhole Principle. In this section we present an application of induction that is a very useful theorem with surprising consequences. It is usually referred to as the "pigeonhole principle", because in its most colloquial version, it says that if you have more than n pigeons but only n pigeonholes, then there must be at least one pigeonhole with at least two pigeons in it.

Theorem 2.3.14 (Pigeonhole principle). *Let $n \geq 1$ be a natural number. Let S be a set with more than n elements, and define n subsets of S denoted by $S_1, S_2, \ldots, S_n$, such that $\bigcup_{i=1}^{n} S_i = S$; that is, the union of all subsets S_i is equal to S. Then, there is at least one subset S_j, for some $1 \leq j \leq n$, such that S_j contains two or more elements of S.*

Proof. We shall prove the pigeonhole principle using mathematical induction. Let us begin with the base case $n_0 = 1$. Let S be a set with more than 1 element, and let S_1 be a subset of S, such that $S_1 = S$. Then, clearly, $S_1 = S$ has more than 1 element, and the theorem is true for $n_0 = 1$.

Next, let us show the induction step. Let us assume that the theorem is true for sets with more than k elements, and let S be a set with more than $k+1$ elements. Let $S_1, S_2, \ldots, S_k, S_{k+1}$ be $k+1$ subsets of S, such that $\bigcup_{i=1}^{k+1} S_i = S$. If the set S_{k+1} has more than two elements, then we are done. Otherwise, suppose that S_{k+1} contains only one element of S, or $S_{k+1} = \emptyset$. Then, the union $S' = \bigcup_{i=1}^{k} S_i$ contains at least k elements (since S contains $k+1$ elements, and the missing subset S_{k+1} contains ≤ 1 elements, so S' contains $\geq k+1-1 = k$ elements). Hence, using our induction hypothesis on the set S', we conclude that one of $S_1, \ldots, S_k$ contains at least two elements.

Hence, we have shown the base case and the induction step, and by the principle of mathematical induction, the theorem is true for all $n \geq 1$. □

Example 2.3.15. We can demonstrate that there must be at least two people in Boston with the same number of hairs on their heads. A typical head has around 150,000 hairs; therefore it is reasonable to assume that no one has more than 1,000,000 hairs on his or her head. Since there are more than 1,000,000 people in the Greater Boston area (there are about 4.5 million people), we can let S be the set of all people in the Boston area (so S has more than 1,000,000 elements) and we may write S_i for the subset of S formed by people in the Boston area with i hairs on their heads, for each $0 \leq i \leq 999{,}999$. Thus, by the pigeonhole principle (Theorem 2.3.14) there must be at least two people with the same number of hairs on their heads.

2.4. The Division Theorem

Now that we have established the basic building blocks of the integers, i.e., the axioms, we turn to the arithmetic-geometric problems that we are interested in. The goal of this chapter is to prove the following two theorems about roots of polynomials and rational points on lines in the plane.

Theorem. *Let $p(x) = c_n x^n + c_{n-1}x^{n-1} + \cdots + c_2x^2 + c_1x + c_0$ be a polynomial with $c_i \in \mathbb{Q}$, for all $0 \leq i \leq n$. If $\frac{a}{b} \in \mathbb{Q}$ is a rational number in reduced form (i.e., $\gcd(a,b) = 1$) and $\frac{a}{b}$ is a root of $p(x) = 0$ (that is, $p(\frac{a}{b}) = 0$), then c_0 is divisible by a and c_n is divisible by b.*

Theorem. *Let $L : ax + by = c$ be a line in the plane, with $a, b, c \in \mathbb{Z}$ and $ab \neq 0$. Then, $L(\mathbb{Z})$ is non-empty if and only if c is divisible by $\gcd(a,b)$. In other words, the line L has a point with integer coordinates if and only if $\gcd(a,b)|c$. Moreover, $L(\mathbb{Z})$ is non-empty if and only if $L(\mathbb{Z})$ is infinite.*

In order to prove these theorems (in Sections 2.8 and 2.9, respectively), we begin our study of the basic divisibility properties of the integers. Recall that, by Definition 2.1.5, we say that an integer a is a divisor of another integer b if there is a third integer m such that $b = a \cdot m$. Our first theorem is a rigorous proof of the method of "long division" of two integers, also known as division with quotient and remainder. Let us see some examples before we state the theorem.

Example 2.4.1. The number 37 is not divisible by 13. Indeed,

$$13 \cdot 0 = 0, \ 13 \cdot 1 = 13, \ 13 \cdot 2 = 26, \ \text{and} \ 13 \cdot 3 = 39.$$

Thus, there is no integer m such that $13 \cdot m = 37$. The largest multiple of 13 below 37 is $13 \cdot 2 = 26$, and there is a remainder of 11 until 37; i.e., $37 - 13 \cdot 2 = 11$. We write

$$37 = 13 \cdot 2 + 11,$$

and we remark that 11 is the smallest *positive* remainder over all numbers of the form $37 - 13m$ and $0 \leq 11 < 13$. Notice also that $q = 2$ is the only integer such that $37 - 13 \cdot q = 11$, and there is no $m \in \mathbb{Z}$ with $r_0 = 37 - 13 \cdot m$ and $0 \leq r_0 < 11$, since 11 is the smallest positive remainder possible.

Example 2.4.2. The number 127 is not divisible by 4. What is the remainder of division of 127 by 4? In order to find the remainder, we mentally find the largest multiple of 4 below 127 which, in this case, is $4 \cdot 31 = 124$. Thus, the remainder is $127 - 4 \cdot 31 = 3$. We write

$$127 = 4 \cdot 31 + 3.$$

Notice that in order to find the appropriate multiple, we run over all possible numbers of the form $r = 127 - 4 \cdot m$, with $m \in \mathbb{Z}$, until we find the smallest positive $r \geq 0$. Note that we must have a remainder satisfying $0 \leq r < 4$, since we are dividing by 4.

Example 2.4.3. The integer -37 is not divisible by 13. As we did in our previous examples, we can consider all the numbers of the form $r = -37 - 13 \cdot q$ and find the smallest positive r. In this case, such $r \geq 0$ is given by $r = -37 - 13 \cdot (-3) = -37 + 39 = 2$. Hence, we write

$$-37 = 13 \cdot (-3) + 2,$$

and we remark that $0 \leq 2 < 13$. Notice also that $q = -3$ is the only integer such that $-37 - 13 \cdot q = 2$.

We are now ready to state the division theorem.

Theorem 2.4.4 (The division theorem). *Let a, b be integers with $a > 0$. Then, there are unique integers $q, r \in \mathbb{Z}$ such that*

$$b = aq + r,$$

with $0 \leq r < a$.

Proof. Let $a, b \in \mathbb{Z}$ with $a > 0$. Let us first show that there are q and r in $\mathbb{Z}$ with the desired properties, and we will prove the uniqueness later. Let S be the set of integers defined by

$$S = \{b - a \cdot t : \ \text{with } t \in \mathbb{Z} \ \text{and} \ b - a \cdot t \geq 0\}.$$

If $0 \in S$, then there exists $t_0 \in \mathbb{Z}$ such that $b - a \cdot t_0 = 0$, and therefore $b = a \cdot t_0$. In this case a is a divisor of b, and we may pick $q = t_0$ and $r = 0$. Otherwise, if $0 \notin S$, then $S \subseteq \mathbb{N}$. Moreover, S is non-empty (just pick $t \in \mathbb{Z}$ such that $b - at > 0$, i.e., any $t \in \mathbb{Z}$ with $t < b/a$). Therefore, by the well-ordering principle (axiom (4.d) in Section 2.1), the set S has a least element r_1. Since $r \in S$, there must be a $t_1 \in \mathbb{Z}$ such that $b - a \cdot t_1 = r_1$. We claim that $q = t_1$ and $r = r_1$ satisfy the desired property; i.e., $b = at_1 + r_1$ (this follows from the definition of S) and $0 \leq r_1 < a$.

By definition, every element of S is ≥ 0 and so $r_1 \geq 0$. Let us suppose, for a contradiction, that $r_1 \geq a$ (or $r_1 - a \geq 0$). Then,

$$b = at_1 + r_1 = a(t_1 + 1) + (r_1 - a).$$

In particular, $r_1 - a = b - a(t_1 + 1) \in S$, and $0 \leq r_1 - a < r_1$. However, r_1 is the least element of S, but $r_1 - a < r_1$. This is the desired contradiction. Thus, we must have $r_1 < a$.

Finally, let us show that q and r such that $b = aq + r$ and $0 \leq r < a$ are unique. Suppose that q', r' is another pair of integers with $b = aq' + r'$ and $0 \leq r' < a$. We shall show that $q = q'$ and $r = r'$. From the properties of q, r and q', r' we have that

$$b = aq + r = aq' + r',$$

and this implies that $a(q - q') = r' - r$. The left-hand side is a multiple of a, but the right-hand side satisfies $-a < r' - r < a$, since $0 \leq r', r < a$. Thus, $r' - r$ is a multiple of a, strictly between $-a$ and a, and therefore $r' - r = 0 \cdot a = 0$ (in particular, this implies $r' = r$). It follows that $a(q - q') = 0$ as well but, by assumption, $a \neq 0$. Hence, $q = q'$ and $r = r'$. □

Remark 2.4.5. It is worth noting that a is a divisor of b if and only if the remainder when dividing b by a is 0. Indeed, if a is a divisor of b, then there is $k \in \mathbb{Z}$ such that $b = a \cdot k$. Hence, $q = k$ and $r = 0$ are the unique quotient and remainder, as in Theorem 2.4.4. Conversely, if $r = 0$, then $b = aq$, and therefore a divides b.

The following well-known fact, which is used so often in proofs, is a consequence of our previous remark.

Lemma 2.4.6. *Every even integer n (i.e., divisible by 2) is of the form $n = 2s$, for some $s \in \mathbb{Z}$. Similarly, every odd integer m (i.e., not divisible by 2) is of the form $m = 2t + 1$, for some $t \in \mathbb{Z}$. In particular, every integer is either of the form $2k$ or $2k + 1$, for some $k \in \mathbb{Z}$.*

Proof. Suppose that n is even, i.e., divisible by 2. Then, by the definition of divisibility, there is $s \in \mathbb{Z}$ such that $n = 2s$. If m is odd, i.e., not divisible by 2, then our Remark 2.4.5 implies that the remainder when dividing m by 2 is non-zero. But the remainder r satisfies $0 \leq r < 2$, so it must be the case that $r = 1$. Hence, $m = 2q + 1$, and there is $t = q \in \mathbb{Z}$ such that $m = 2t + 1$. Finally, since every number is even or "not even" (i.e., odd), it follows that every number is of the form $2k$ or $2k + 1$, for some $k \in \mathbb{Z}$. □

2.5. The Greatest Common Divisor

Definition 2.5.1. Let a, b be integers, not both zero. The *greatest common divisor* of a and b, denoted by $d = \gcd(a, b)$ or simply $d = (a, b)$, is a natural number $d \geq 1$ with the following properties:

(1) The number d is a common divisor of a and b; i.e., $d|a$ and $d|b$.

(2) The number d is the largest common positive divisor: if n is another common positive divisor of a and b, that is, $n \geq 1$ and $n|a$ and $n|b$, then $n \leq d$.

If $\gcd(a, b) = 1$, then we say that a and b are coprime, or relatively prime.

Example 2.5.2. The numbers 3 and 5 are relatively prime, as they do not share any common positive divisors other than 1. Indeed, the lists of positive divisors of 3 and 5 are, respectively, $\{1, 3\}$ and $\{1, 5\}$. Thus, $\gcd(3, 5) = 1$.

Example 2.5.3. The numbers -8 and 40 are not relatively prime, as they share some divisors ≥ 1. The list of positive divisors of -8 is $\{1, 2, 4, 8\}$, while the list of positive divisors of 40 is given by $\{1, 2, 4, 5, 8, 10, 20, 40\}$. Thus, the greatest common divisor is 8. We write $\gcd(-8, 40) = 8$ or simply $8 = (-8, 40)$.

Example 2.5.4. If $a \neq 0$ is an integer, then $\gcd(a, 0) = |a|$, the absolute value of a, where

$$|a| = \begin{cases} a & \text{if } a \geq 0, \\ -a & \text{if } a < 0. \end{cases}$$

Indeed, every positive divisor n of a is also a divisor of zero (because $0 = n \cdot 0$). This implies that the largest common divisor of a and 0 is the largest divisor of a, and that would be $|a|$.

One can always find the greatest common divisor of two integers a and b by simply listing the positive divisors of each of them and comparing the lists, as in our previous examples. However, this can be extremely time consuming (just try to find $\gcd(7920, 5040)$ this way). Luckily, there is an excellent algorithm to find a gcd that is usually attributed to Euclid of Alexandria (about 325 BC to about 265 BC). Let us first see some examples of usage, and then we will explain why this method works to calculate a greatest common divisor (see Lemma 2.5.7 if you just cannot wait!).

Example 2.5.5. Let us find the greatest common divisor of $b = 7920$ and $a = 5040$, using Euclid's algorithm. The method is based on repeated long division, until we reach a remainder of 0. The first long division is that of b by a (where b is the largest of the two integers):

$$7920 = 5040 \cdot 1 + 2880.$$

In the second step, we perform the long division of $b = 5040$ and the remainder of a divided by b, in this case $r = 2880$:

$$5040 = 2880 \cdot 1 + 2160.$$

In the next step, we divide 2880 by the previous remainder 2160, and so on until we reach a remainder of 0:

$$2880 = 2160 \cdot 1 + 720,$$
$$2160 = 720 \cdot 3 + 0.$$

Then, the greatest common divisor of the numbers 7920 and 5040 is the last positive reminder that we found in our chain of long divisions. In this case, $\gcd(7920, 5040) = 720$.

Example 2.5.6. Let us calculate the GCD of 321 and 123 using Euclid's algorithm:

$$\begin{aligned} 321 &= 123 \cdot 2 + 75, \\ 123 &= 75 \cdot 1 + 48, \\ 75 &= 48 \cdot 1 + 27, \\ 48 &= 27 \cdot 1 + 21, \\ 27 &= 21 \cdot 1 + 6, \\ 21 &= 6 \cdot 3 + 3, \\ 6 &= 3 \cdot 2 + 0. \end{aligned}$$

The GCD is the last non-zero remainder in this chain, so $\gcd(321, 123) = 3$.

The main ingredient for the formal justification of Euclid's algorithm is the following lemma.

Lemma 2.5.7. *Let a, b be integers, and let q, r be integers such that $b = aq + r$. Then, $\gcd(b, a) = \gcd(a, r)$.*

Proof. Let $d = \gcd(a, b)$ and $d' = \gcd(a, r)$. We claim that $d = d'$. First, let n be a common positive divisor for a, b, with $a = nk$ and $b = nj$, for some $k, j \in \mathbb{Z}$. Then, $n|a$, and $r = b - aq = nj - nkq = n(j - kq)$, so $n|r$ as well. Thus, n is a common positive divisor of a and r.

Conversely, suppose that $m \geq 0$ is a common divisor of a and r, with $a = mh$ and $r = ml$, for some $h, l \in \mathbb{Z}$. Then $b = aq + r = mhq + ml = m(hq + l)$, and m is also a divisor of b. Thus, m is also a positive divisor of a and b.

We have shown that the common positive divisors of the pairs of integers a, b and a, r are the same. Thus, the greatest common positive divisor of a, b and a, r must be the same or, in other words, $d = d'$, as desired. □

2.6. Euclid's Algorithm to Calculate a GCD

Let $a, b \in \mathbb{Z}$, not both zero, and let us assume that $b > a$.

(E1) Find $q_1, r_1 \in \mathbb{Z}$ such that $b = aq_1 + r_1$ and $0 \leq r_1 < a$. By Euclid's lemma, Lemma 2.5.7, we have $\gcd(a, b) = \gcd(a, r_1)$.

(E2) Find $q_2, r_2 \in \mathbb{Z}$ such that $a = r_1 q_2 + r_2$ and $0 \leq r_2 < r_1$. As before, $\gcd(a, r_1) = \gcd(r_1, r_2)$.

(E3) Given r_{k-1} and r_k, find $q_{k+1}, r_{k+1} \in \mathbb{Z}$ such that

$$r_{k-1} = r_k q_{k+1} + r_{k+1} \text{ and } 0 \leq r_{k+1} < r_k.$$

By Lemma 2.5.7, we have

$$\gcd(r_k, r_{k+1}) = \gcd(r_{k-1}, r_k) = \cdots = \gcd(a, r_1) = \gcd(a, b).$$

Moreover,

$$r_{k+1} < r_k < r_{k-1} < \cdots < r_1 < r_0 = a.$$

(E4) Since each $r_i \geq 0$ and $r_{k+1} < r_k$, we must have $r_n = 0$ for some $n \geq 1$. In particular,

$$\gcd(a, b) = \gcd(r_{n-1}, r_n) = \gcd(r_{n-1}, 0) = r_{n-1}.$$

Hence, $\gcd(a, b) = r_{n-1}$ or, in other words, the GCD of a and b is the last non-zero remainder in the chain of long divisions.

Example 2.6.1. Let us revisit Example 2.5.6 and see how what we did fits the theoretical description of Euclid's algorithm given above. Let us calculate the GCD of 321 and 123 using Euclid's algorithm. We shall use Lemma 2.5.7 to keep track of the gcds involved:

$$\begin{aligned}
321 &= 123 \cdot 2 + 75; && \text{thus,} \quad (321, 123) = (123, 75),\\
123 &= 75 \cdot 1 + 48; && \text{thus,} \quad (123, 75) = (75, 48),\\
75 &= 48 \cdot 1 + 27; && \text{thus,} \quad (75, 48) = (48, 27),\\
48 &= 27 \cdot 1 + 21; && \text{thus,} \quad (48, 27) = (27, 21),\\
27 &= 21 \cdot 1 + 6; && \text{thus,} \quad (27, 21) = (21, 6),\\
21 &= 6 \cdot 3 + 3; && \text{thus,} \quad (21, 6) = (6, 3),\\
6 &= 3 \cdot 2 + 0; && \text{thus,} \quad (6, 3) = (3, 0) = 3.
\end{aligned}$$

The GCD is the last non-zero remainder in this chain, so $\gcd(321, 123) = 3$.

Example 2.6.2. Let us calculate the GCD of 337 and 271 using Euclid's algorithm.

$$\begin{aligned}
337 &= 271 \cdot 1 + 66; && \text{thus,} \quad (337, 271) = (271, 66),\\
271 &= 66 \cdot 4 + 7; && \text{thus,} \quad (271, 66) = (66, 7),\\
66 &= 7 \cdot 9 + 3; && \text{thus,} \quad (66, 7) = (7, 3),\\
7 &= 3 \cdot 2 + 1; && \text{thus,} \quad (7, 3) = (3, 1),\\
3 &= 1 \cdot 3 + 0; && \text{thus,} \quad (3, 1) = (1, 0) = 1.
\end{aligned}$$

The GCD is the last non-zero remainder in this chain, so $\gcd(337, 271) = 1$.

2.7. Bezout's Identity

Let $a, b \in \mathbb{Z}$, not both zero, and suppose that $\gcd(a, b) = d$. Suppose we have found d using Euclid's algorithm. Then, the steps in the algorithm can be used *backwards* to find $r, s \in \mathbb{Z}$ such that $ar + bs = d$. The identity

$$ar + bs = \gcd(a, b)$$

is known as *Bezout's identity*. Before we write a formal algorithm, let us see one example.

Example 2.7.1. Let $a = 5$ and $b = 7$. Clearly, $\gcd(5,7) = 1$. Let us run Euclid's algorithm to find this gcd:

$$7 = 5 \cdot 1 + 2,$$
$$5 = 2 \cdot 2 + 1,$$
$$2 = 1 \cdot 2 + 0.$$

As we expected, the algorithm tells us that $\gcd(5,7) = 1$. In order to find a solution to $5r + 7s = 1$, we use the steps produced by Euclid's algorithm *backwards*. More concretely, we recursively plug in each equation into the previous one, as follows:

$$\begin{aligned} 1 &= 5 - 2 \cdot 2 \\ &= 5 - 2 \cdot (7 - 5 \cdot 1) \\ &= (1 + 2) \cdot 5 - 2 \cdot 7 \\ &= 3 \cdot 5 - 2 \cdot 7. \end{aligned}$$

Thus, we have found that $1 = 3 \cdot 5 - 2 \cdot 7$. In other words, if $a = 5$, $b = 7$, $r = 3$, and $s = -2$, then Bezout's identity $ar + bs = \gcd(a,b)$ is satisfied.

Figure 2.2. Étienne Bézout (1730–1783) was a French mathematician who was well known for his work on algebraic equations and for his mathematics textbooks. Image source: Wikimedia Commons.

The algorithm to find a solution to Bezout's identity is the following.

Bezout's identity algorithm. Let $a, b \in \mathbb{Z}$, with $\gcd(a,b) = d \geq 1$.

(B1) Run Euclid's algorithm to find $\gcd(a,b) = d$. Label each step in Euclid's algorithm $E_1, E_2, \ldots$. The last three lines E_{k-2}, E_{k-1}, and E_k should look like this:

$$\begin{aligned} E_{k-2}: &\quad r_{k-2} = r_{k-1}q_k + r_k, \\ E_{k-1}: &\quad r_{k-1} = r_k q_{k+1} + r_{k+1}, \\ E_k: &\quad r_k = r_{k+1}q_{k+2} + 0. \end{aligned}$$

Notice that $\gcd(a,b) = r_{k+1}$. We rewrite line E_{k-1} to obtain $\gcd(a,b)$ as a $\mathbb{Z}$-linear combination of r_{k-1} and r_k, as follows:

$$\gcd(a,b) = r_{k+1} = r_{k-1} - r_k q_{k+1}.$$

(B2) Solve for r_k in line E_{k-2} of Euclid's algorithm, to write $\gcd(a,b)$ as a $\mathbb{Z}$-linear combination of r_{k-2} and r_{k-1}, as follows:

$$\begin{aligned} \gcd(a,b) &= r_{k-1} - r_k q_{k+1} \\ &= r_{k-1} - (r_{k-2} - r_{k-1}q_k)q_{k+1} \\ &= (1 + q_k q_{k+1})r_{k-1} - q_{k+1}r_{k-2}. \end{aligned}$$

(B3) At each step, we have an equation of the form $\gcd(a,b) = cr_i + dr_{i-1}$, for some $c, d \in \mathbb{Z}$. Solve for r_i in line E_{i+2}, to write $\gcd(a,b)$ as a $\mathbb{Z}$-linear combination of r_{i-1} and r_{i-2}, as follows:

$$\begin{aligned} \gcd(a,b) &= cr_i + dr_{i-1} \\ &= c(r_{i-2} - r_{i-1}q_i) + dr_{i-1} \\ &= cr_{i-2} + (d - cq_i)r_{i-1}. \end{aligned}$$

(B4) After several repetitions of step (B3) and noticing that $r_{-1} = b$ and $r_0 = a$, we will finish by writing $\gcd(a,b)$ as a $\mathbb{Z}$-linear combination of a and b; i.e., $\gcd(a,b) = ar + bs$, for some $r, s \in \mathbb{Z}$, as desired.

Let us see some examples.

Example 2.7.2. In Example 2.6.2, we have found that the greatest common divisor of 337 and 271 is 1:

$$\begin{aligned} 337 &= 271 \cdot 1 + 66; &\quad \text{thus,} &\quad (337, 271) = (271, 66), \\ 271 &= 66 \cdot 4 + 7; &\quad \text{thus,} &\quad (271, 66) = (66, 7), \\ 66 &= 7 \cdot 9 + 3; &\quad \text{thus,} &\quad (66, 7) = (7, 3), \\ 7 &= 3 \cdot 2 + 1; &\quad \text{thus,} &\quad (7, 3) = (3, 1), \\ 3 &= 1 \cdot 3 + 0; &\quad \text{thus,} &\quad (3, 1) = (1, 0) = 1. \end{aligned}$$

Let us work our way backwards to find a solution to $337r + 271s = 1$:

$$\begin{aligned} 1 &= 7 - 3 \cdot 2 \\ &= 7 - (66 - 7 \cdot 9) \cdot 2 \\ &= 19 \cdot 7 - 2 \cdot 66 \\ &= 19 \cdot (271 - 66 \cdot 4) - 2 \cdot 66 \\ &= 19 \cdot 271 - 78 \cdot 66 \\ &= 19 \cdot 271 - 78 \cdot (337 - 271) \\ &= -78 \cdot 337 + 97 \cdot 271. \end{aligned}$$

Hence, we have achieved our goal and written $1 = -78 \cdot 337 + 97 \cdot 271$. In other words, $r = -78$ and $s = 97$ satisfy Bezout's identity $337r + 271s = 1$.

Example 2.7.3. Let us find $r, s \in \mathbb{Z}$ such that $13 = 91r + 221s$. First, we find the GCD of 221 and 91 using Euclid's algorithm:

$$\begin{aligned} 221 &= 91 \cdot 2 + 39, \\ 91 &= 39 \cdot 2 + 13, \\ 39 &= 13 \cdot 3 + 0. \end{aligned}$$

Therefore, $\gcd(221, 91) = 13$. Now we can run Euclid's algorithm backwards to find r and s:

$$\begin{aligned} 13 &= 91 - 39 \cdot 2 \\ &= 91 - (221 - 91 \cdot 2) \cdot 2 \\ &= 5 \cdot 91 - 2 \cdot 221. \end{aligned}$$

Hence, $r = 5$ and $s = -2$ solve $13 = 91r + 221s$.

Let us write the result of the Bezout's identity algorithm as a theorem, for later use.

Theorem 2.7.4 (Bezout's identity). *Let $a, b \in \mathbb{Z}$, not both zero, such that* $\gcd(a, b) = d$. *Then, there exist r and $s \in \mathbb{Z}$ such that $ar + bs = d$.*

The previous theorem turns out to be extremely useful in the theory of divisibility of integers. Let us see some consequences. First, we note that from the definition of GCD (Definition 2.5.1) it follows that if $\gcd(a, b) = d$ and n divides a and b, then $n \leq d$. However, it is not directly obvious that n *divides* d. Let us show that this is indeed the case.

Corollary 2.7.5. *Let $a, b \in \mathbb{Z}$, not both zero, with* $\gcd(a, b) = d$. *Suppose that n is a common positive divisor of a and b. Then, n is a divisor of d.*

Proof. Let a, b, and $d = \gcd(a, b)$ be as above. Let $n \in \mathbb{Z}$ such that $n|a$ and $n|b$; i.e., there are $h, k \in \mathbb{Z}$ such that $a = nh$ and $b = nk$. By Bezout's identity (Theorem 2.7.4), there exist $r, s \in \mathbb{Z}$ such that $ar + bs = d$. Thus,

$$d = ar + bs = (nh)r + (nk)s = n(hr + ks),$$

and this shows that n is a divisor of d. □

The following corollary will be a key ingredient in the proof of the fundamental theorem of arithmetic and other applications to arithmetic geometry.

Corollary 2.7.6. *Let $a, b, c \in \mathbb{Z}$ such that a divides bc and* $\gcd(a, b) = 1$. *Then, a divides c.*

Proof. Suppose that $\gcd(a, b) = 1$. Then, by Bezout's identity, there are $r, s \in \mathbb{Z}$ such that $ar + bs = 1$. In particular,

$$c = c \cdot 1 = c \cdot (ar + bs) = acr + bcs.$$

If we further assume that a is a divisor of bc, then there is some $m \in \mathbb{Z}$ such that $bc = am$. Hence,

$$c = acr + bcs = acr + ams = a(cr + ms),$$

and this shows that a is a divisor of c, as desired. □

2.8. Integral and Rational Roots of Polynomials

In this section we apply Bezout's identity, and more concretely Corollary 2.7.6, to the problem of finding natural, integral, and rational roots of a polynomial with integer coefficients. The following theorem generalizes and formalizes the results that we already used in Example 1.1.2.

Theorem 2.8.1. *Let $p(x)$ be a polynomial of degree $n \geq 1$ given by*

$$p(x) = c_n x^n + c_{n-1}x^{n-1} + \cdots + c_2 x^2 + c_1 x + c_0,$$

with $c_0, c_1, \ldots, c_n \in \mathbb{Z}$. Suppose that $\frac{a}{b}$ is a rational number written in reduced form; i.e., $a \in \mathbb{Z}$ and $b \in \mathbb{N}$ with $\gcd(a, b) = 1$. Further, suppose that $p(\frac{a}{b}) = 0$ or, in other words, $\frac{a}{b}$ is a root of $p(x)$. Then,

- *the numerator $a \in \mathbb{Z}$ is a divisor of c_0, and*
- *the denominator $b \in \mathbb{N}$ is a divisor of c_n.*

Example 2.8.2. Let $p(x) = 12x^3 + 11x^2 - 208x + 185$. The rational number $\frac{37}{12}$ is a root of $p(x) = 0$ and, indeed, 37 is a divisor of $185 = 5 \cdot 37$ and the coefficient of x^3 is divisible by 12 (it is *equal* to 12 in this case).

Proof of Theorem 2.8.1. Let $p(x) = c_n x^n + c_{n-1}x^{n-1} + \cdots + c_2 x^2 + c_1 x + c_0$ be a polynomial with integer coefficients $c_i \in \mathbb{Z}$, for $0 \leq i \leq n$, and let $\frac{a}{b} \in \mathbb{Q}$ be a root, with $a \in \mathbb{Z}$, $b \in \mathbb{N}$, and $\gcd(a, b) = 1$. Then,

$$c_n \left(\frac{a}{b}\right)^n + c_{n-1}\left(\frac{a}{b}\right)^{n-1} + \cdots + c_2\left(\frac{a}{b}\right)^2 + c_1\left(\frac{a}{b}\right) + c_0 = 0.$$

If we multiply both sides of this expression by b^n, we obtain an equality of integers

$$c_n a^n + c_{n-1}a^{n-1}b + \cdots + c_2 a^2 b^{n-2} + c_1 a b^{n-1} + c_0 b^n = 0. \tag{2.3}$$

It follows that

$$a(c_n a^{n-1} + c_{n-1}a^{n-2}b + \cdots + c_2 a b^{n-2} + c_1 b^{n-1}) = -c_0 b^n \text{ and} \tag{2.4}$$

$$b(c_{n-1}a^{n-1} + \cdots + c_2 a^2 b^{n-3} + c_1 a b^{n-2} + c_0 b^{n-1}) = -c_n a^n. \tag{2.5}$$

Hence, by (2.4) and (2.5), there exist u and $v \in \mathbb{Z}$ such that $au = -c_0 b^n$ and $bv = -c_n a^n$, respectively. In particular, $a|c_0 b^n$ and $b|c_n a^n$. Since $\gcd(a, b) = 1$, this implies that $\gcd(a, b^n) = 1$ and $\gcd(b, a^n) = 1$ (see Exercise 2.11.24). Thus, $a|c_0 b^n$ and $\gcd(a, b^n) = 1$, and this implies that $a|c_0$, by Corollary 2.7.6. Similarly, $b|c_n a^n$ and $\gcd(b, a^n) = 1$ implies $b|c_n$, as desired. □

Example 2.8.3. Let $p(x) = x^2 - 2$. Suppose $x_0 = a/b$ is a rational root of $p(x) = 0$, with $\gcd(a, b) = 1$. Then, by Theorem 2.8.1, a is a divisor of 2 and b is a divisor of 1. Hence, $x_0 = \pm 2$, but $2^2 - 2 = (-2)^2 - 2 = 2 \neq 0$. Hence, $p(x) = 0$ has no rational roots. Since the roots of $p(x) = 0$ are precisely those x_0 with $x_0^2 = 2$, i.e., the square roots of 2, this implies that $x_0 = \pm\sqrt{2}$ are not rational numbers. See also Section 2.10.1.

Remark 2.8.4. A polynomial $p(x)$ may not have rational roots but still factor into polynomials of smaller degrees. For instance, let $p(x) = x^4 + 2x^2 + 1$. Theorem 2.8.1 implies that the only possible roots of $p(x)$ are ± 1, but $1^4 + 2 \cdot 1^2 + 1 = (-1)^4 + 2 \cdot (-1)^2 + 1 = 3 \neq 0$. Hence, $p(x)$ has no rational root. However,

$$p(x) = x^4 + 2x^2 + 1 = (x^2 + 1)^2.$$

Clearly, Theorem 2.8.1 also tells us how to find the integral and natural roots of a given polynomial. We record this in the form of a corollary, for later use.

Corollary 2.8.5. *Let $p(x)$ be a polynomial of degree $n \geq 1$ given by*

$$p(x) = c_n x^n + c_{n-1} x^{n-1} + \cdots + c_2 x^2 + c_1 x + c_0,$$

with $c_0, c_1, \ldots, c_n \in \mathbb{Z}$. If $N \in \mathbb{Z}$ is an integer root of $p(x)$, i.e., $p(N) = 0$, then N is a divisor of c_0.

Example 2.8.6. Is there an integer N such that the difference of its cube and its square equals its square minus 1? In other words, is there N such that

$$N^3 - N^2 = N^2 - 1.$$

If so, N is a root of $p(x) = x^3 - 2x^2 + 1$ and, therefore, by Corollary 2.8.5, the number N must be a divisor of 1. Thus, $N = \pm 1$. We have $p(-1) = -2$ and $p(1) = 0$, and so the only such integer is $N = 1$.

2.9. Integral and Rational Points in a Line

Our second application of Bezout's identity is to the problem of finding all the integral points in a line $L : ax + by = c$. Finding all rational points in L is much easier, and it will be done at the end of the section.

Proposition 2.9.1. *Let $a, b, c \in \mathbb{Z}$ such that a and b are not both zero. The line in the plane $L : ax + by = c$ has an integral point $(r, s) \in L(\mathbb{Z})$; i.e., there are integers $r, s \in \mathbb{Z}$ such that $ar + bs = c$ if and only if $\gcd(a, b)$ is a divisor of c.*

Proof. Let us first assume that there exist $r, s \in \mathbb{Z}$ such that $ar + bs = c$. If n is a common divisor of a and b, then $a = nh$ and $b = nk$, for some $h, k \in \mathbb{Z}$, and therefore $c = ar + bs = n(hr + ks)$ is also divisible by n. Since the GCD of a and b is a common divisor, it follows that $\gcd(a, b)$ is also a divisor of c.

Conversely, suppose that $d = \gcd(a, b)$ is a divisor of c, with $c = dk$ for some $k \in \mathbb{Z}$. By Bezout's identity, there are $r', s' \in \mathbb{Z}$ such that $ar' + bs' = \gcd(a, b) = d$. Hence,

$$c = dk = (ar' + bs')k = a(r'k) + b(s'k).$$

Hence, $r = r'k$ and $s = s'k$ satisfy the desired property $ar + bs = c$. □

Example 2.9.2. The line $L : 15x + 12y = 1$ does not contain any integral points, because $\gcd(15, 12) = 3$ and 3 is not a divisor of 1. Of course, this line contains infinitely many *rational* points (see Proposition 2.9.6 below),

$$L(\mathbb{Q}) = \left\{ \left(t, \frac{1 - 15t}{12} \right) : t \in \mathbb{Q} \right\},$$

but none of them are integral points; i.e., none of the points in $L(\mathbb{Q})$ have simultaneously integer x- and y-coordinates.

Example 2.9.3. Let L be the line in the plane with equation $3x + 11y = 5$. The GCD of 3 and 11 is 1, so the previous proposition implies that L has at least one integral point. Let us find it using Euclid's algorithm and Bezout's identity:

$$\begin{aligned} 11 &= 3 \cdot 3 + 2, \\ 3 &= 2 \cdot 1 + 1, \\ 2 &= 1 \cdot 2 + 0. \end{aligned}$$

Now, we work backwards,

$$1 = 3 - 2 = 3 - (11 - 3 \cdot 3) = 4 \cdot 3 - 1 \cdot 11,$$

to find $1 = 4 \cdot 3 - 1 \cdot 11$. If we multiply both sides of this identity by 5, we obtain

$$5 = 20 \cdot 3 - 5 \cdot 11,$$

and this shows that $(20, -5)$ is an integral point on L. Are there other integral points on L? Yes. Indeed, there are infinitely many points, given by the formula $P_k = (x_k, y_k)$, where $x_k = 20 - 11k$ and $y_k = -5 + 3k$, for any $k \in \mathbb{Z}$. Let us verify that P_k belongs to $L(\mathbb{Z})$. The coordinates are integers, and

$$3x_k + 11y_k = 3(20 - 11k) + 11(-5 + 3k) = 60 - 33k - 55 + 33k = 5,$$

and so $P_k \in L(\mathbb{Z})$. It turns out that all the integral points on L are of the form P_k for some $k \in \mathbb{Z}$. This will be shown in the following theorem.

Theorem 2.9.4. *Let $a, b \in \mathbb{Z}$, not both zero, with $\gcd(a, b) = d$. Let c be an integer divisible by d and let L be the line in the plane with equation $ax + by = c$. Suppose that (x_0, y_0) is an integral point on L. Then, the point $P_k = (x_k, y_k)$ with*

$$x_k = x_0 + \frac{bk}{d}, \quad y_k = y_0 - \frac{ak}{d}$$

is an integral point on L, for any $k \in \mathbb{Z}$. Furthermore, every integral point on L is of the form P_k, for some $k \in \mathbb{Z}$.

Proof. Let us first show that $P_k \in L(\mathbb{Z})$. Clearly, the coordinates of P_k are in $\mathbb{Z}$, so it suffices to verify that $ax_k + by_k = c$:

$$ax_k + by_k = a\left(x_0 + \frac{bk}{d}\right) + b\left(y_0 - \frac{ak}{d}\right) = ax_0 + by_0 + \frac{abk}{d} - \frac{abk}{d} = c,$$

where we have used our assumption that $ax_0 + by_0 = c$.

It remains to show that every integral point on L is of the form P_k for some $k \in \mathbb{Z}$. Let (x', y') be an integral point on L; i.e., $x', y' \in \mathbb{Z}$ and $ax' + by' = c$. Since $ax_0 + by_0 = c$, we may subtract these equations to obtain

$$a(x' - x_0) + b(y' - y_0) = 0,$$

which, in turn, implies that $a(x' - x_0) = -b(y' - y_0)$. The GCD of a and b, the number $d = \gcd(a, b)$, is a common divisor of a and b, so we can divide both sides by d to obtain an equality of *integers* (and not just rational numbers):

$$\frac{a}{d}(x' - x_0) = -\frac{b}{d}(y' - y_0), \tag{2.6}$$

and we remark again that $\frac{a}{d}$ and $\frac{b}{d}$ are integers. Also, by Exercise 2.11.25, we have $\gcd(\frac{a}{d}, \frac{b}{d}) = 1$. Thus, by Corollary 2.7.6, the integer $\frac{b}{d}$ is a divisor of $x' - x_0$ and, similarly, the integer $\frac{a}{d}$ is a divisor of $y' - y_0$. In particular, there is an integer $k \in \mathbb{Z}$ such that

$$x' - x_0 = k \cdot \frac{b}{d}.$$

In other words, $x' = x_0 + \frac{bk}{d}$ and we may use this fact in (2.6) to obtain

$$-\frac{b}{d}(y' - y_0) = \frac{a}{d}(x' - x_0) = \frac{a}{d} \cdot \frac{b}{d} \cdot k,$$

and therefore $\frac{ak}{d} = -(y' - y_0)$. Hence, $y' = y_0 - \frac{ak}{d}$ and $(x', y') = P_k$, as claimed. □

Example 2.9.5. Let us find all the integral points on the line $L : 15x + 12y = 3$. Clearly, $(1, -1)$ is one integral solution. Thus, by our previous theorem, all the points are of the form

$$P_k = \left(1 + \frac{12k}{3}, -1 - \frac{15k}{3}\right) = (1 + 4k, -1 - 5k), \qquad \text{for some } k \in \mathbb{Z}.$$

For instance, $P_0 = (1, -1)$, $P_1 = (5, -6)$, $P_2 = (9, -11)$, and $P_{-1} = (-3, 4)$.

Now that we know how to find all the integral points in a line, we note that a line in the plane always has infinitely many rational points.

Proposition 2.9.6. *Let $a, b, c \in \mathbb{Z}$ such that a and b are not both zero. The rational points in the line in the plane $L : ax + by = c$ are given by*

$$L(\mathbb{Q}) = \left\{\left(s, \frac{c - as}{b}\right) : s \in \mathbb{Q}\right\}$$

if $b \neq 0$ and by

$$L(\mathbb{Q}) = \left\{\left(\frac{c}{a}, t\right) : t \in \mathbb{Q}\right\}$$

if $b = 0$ (and $a \neq 0$).

Proof. Assume first that $b \neq 0$ and $(s, t) \in L(\mathbb{Q})$, for some $s, t \in \mathbb{Q}$. Then, $as + bt = c$ and, therefore, $t = (c - as)/b$, as claimed. Otherwise, if $b = 0$, then $a \neq 0$ and $as = c$, so $s = c/a$ and t can take any value in $\mathbb{Q}$. □

Example 2.9.7. The rational points in the line $L : 15x + 12y = 3$ are given by

$$L(\mathbb{Q}) = \left\{\left(s, \frac{3 - 15s}{12}\right) : s \in \mathbb{Q}\right\} = \left\{\left(s, \frac{1 - 5s}{4}\right) : s \in \mathbb{Q}\right\}.$$

Notice that it is not immediately obvious which of these rational points are actually integral points. Thus, we need Theorem 2.9.4 to find the integral points in L (see Example 2.9.5).

2.10. The Fundamental Theorem of Arithmetic

In this final section we present our third and most important application of Bezout's identity: the fundamental theorem of arithmetic.

Example 2.10.1. The numbers 7919, 7927, 7933, and 7937 are prime numbers (Definition 2.1.7). Is it possible that there is an equality

$$7919 \cdot 7937 = 7927 \cdot 7933?$$

Of course, this is false since

$$7919 \cdot 7937 = 62853103 \neq 62884891 = 7927 \cdot 7933.$$

But this example raises an interesting question. Is it possible to find distinct primes p, q, r, s such that $pq = rs$? The answer is *no*, but in order to prove why there are no such primes, we need to invoke a corollary of Bezout's identity (Corollary 2.7.6). Indeed, if p, q, r, s are distinct primes, then $\gcd(p, r) = 1$. Since we are assuming $pq = rs$, this implies that p divides rs. Hence, by Corollary 2.7.6, p divides s. But p and s are distinct primes, and this is impossible.

We will extend the argument in the previous example to show the fundamental theorem of arithmetic: every natural number $n \geq 2$ has a unique factorization as a product of primes. Let us first show that every natural number can be factored as a product of primes.

Theorem 2.10.2 (Fundamental theorem of arithmetic: existence). *Every natural number $n \geq 2$ has a factorization as a product of primes.*

Proof. We will use complete induction (Theorem 2.3.10) to prove this result. The base case $n = 2$ has a factorization as a product of primes as 2 itself is a prime number.

Suppose as our induction hypothesis that the theorem is true for all natural numbers t with $2 \leq t \leq k$ and consider the number $k + 1$. If $k + 1$ is prime, then we are done as $k + 1$ would have a trivial factorization as a product of primes $k + 1 = k + 1$. Otherwise, assume that $k + 1$ is not prime. Then $k + 1$ is composite and, by Lemma 2.3.12, there are $a, b \in \mathbb{N}$ with $k + 1 = a \cdot b$ and $1 < a, b < k + 1$. In particular, $2 \leq a, b \leq k$ so the induction hypothesis applies for a and b. Thus, a and b have factorizations as products of primes. That is, there are primes $p_1, \ldots, p_i$ and $q_1, \ldots, q_j$ such that

$$a = p_1 \cdots p_i, \quad b = q_1 \cdots q_j.$$

Thus,

$$k + 1 = a \cdot b = (p_1 \cdots p_i) \cdot (q_1 \cdots q_j) = p_1 \cdots p_i \cdot q_1 \cdots q_j.$$

Hence, $k+1$ also has a factorization as a product of primes. This completes the proof of the induction step, and by the principle of (complete) mathematical induction, every natural number $n \geq 2$ satisfies the theorem. □

We say that two factorizations as a product of primes are the same if they are equal up to a reordering of the prime factors. For instance, $2 \cdot 3 \cdot 5^2$ and $2 \cdot 5 \cdot 3 \cdot 5$ are the same factorization as a product of primes. In order to show the uniqueness of the factorization, we first need a key lemma that settles a key property of the prime numbers.

Lemma 2.10.3. *If $p \geq 2$ is a prime and p divides $a \cdot b$, then p divides a or p divides b. In other words, if a prime divides a product, then the prime divides (at least) one of the factors.*

Proof. Let $a, b \in \mathbb{Z}$ and let p be a prime such that $p|ab$. If p divides a, we are done. Otherwise, we claim that $\gcd(p, a) = 1$. Indeed, if e is a positive common divisor of a and p, then $e = 1$ or $e = p$, but p does not divide a, so we must have $e = 1$. Hence, 1 is the only positive common divisor of a and p and $\gcd(a, p) = 1$.

Thus, we have that p divides ab and $\gcd(a, p) = 1$. By Corollary 2.7.6, we conclude that p divides b, as claimed. □

As a corollary of Lemma 2.10.3 one can show that if $p|a_1 \cdots a_n$, then p divides at least one a_i, for some $1 \leq i \leq n$ (see Exercise 2.11.26).

Remark 2.10.4. The conclusion of Lemma 2.10.3 does not hold if p is not prime. For instance, 6 divides $4 \cdot 15 = 60$, but 6 does not divide 4 or 15.

Example 2.10.5. It remains to show that the factorization of a number n as a product of primes is unique. Let us see why this is true in one particular case: let us assume that n is prime. Then, n is written already as a product of primes, as n itself is prime. Let us suppose that we also have

$$n = p_1 p_2 \cdots p_t$$

for some $t \geq 1$ and for some primes p_i for $1 \leq i \leq t$. If $t > 1$, then n would have more than two positive divisors. For instance, 1, p_1, p_2, and $p_1 p_2$ would be distinct positive divisors of n. But n is prime and it should have only two positive divisors. Thus, $t = 1$ and $n = p_1$. It follows that there is only one factorization of n as a product of primes.

Theorem 2.10.6 (Fundamental theorem of arithmetic: uniqueness)**.** *Every natural number $n \geq 2$ has a unique factorization as a product of prime numbers.*

Proof. By Theorem 2.10.2, every natural number $n \geq 2$ has at least one factorization as a product of primes. It only remains to show that this factorization is unique, up to a reordering of the factors. We will show this using complete induction. It is clear that $n = 2$ has a unique prime factorization, since 2 is a prime itself (see Example 2.10.5). Let us assume, as our induction hypothesis, that every number $2 \leq t \leq k$ has a unique factorization as a product of primes, up to a reordering.

Let us consider $k + 1$. If $k + 1$ is prime, then it has a unique factorization into primes, given by $k+1 = k+1$. Otherwise, suppose that $k+1$ has two factorizations into primes:

$$k + 1 = p_1 p_2 \cdots p_r = q_1 q_2 \cdots q_s,$$

for some primes p_i and q_j and $r, s \geq 2$. In particular, p_1 divides $q_1 \cdots q_s$. By Lemma 2.10.3 (and more concretely by Exercise 2.11.26), there is some j, with $1 \leq j \leq s$, such that p_1 divides q_j. After a reordering of the primes q_j, we may assume $q_j = q_1$ and $p_1|q_1$. Since p_1 and q_1 are primes, it follows that $p_1 = q_1$. Therefore,

$$\frac{k+1}{p_1} = p_2 \cdots p_r = q_2 \cdots q_s.$$

Since p_1 is a (prime) divisor of $k+1$, the quotient $\frac{k+1}{p_1} \in \mathbb{N}$ is a natural number. Since $p_1 \geq 2$, we know that $\frac{k+1}{p_1} < k+1$. Finally, $\frac{k+1}{p_1} \neq 1$, because $k+1$ was assumed to be composite. Thus, $2 \leq \frac{k+1}{p_1} < k+1$, and our induction hypothesis implies that $\frac{k+1}{p_1}$ has a unique prime factorization. Hence, the equality

$$p_2 \cdots p_r = q_2 \cdots q_s$$

is only possible if these two factorizations are the same, up to a reordering (i.e., $r = s$ and $p_i = q_i$ for all $2 \leq i \leq r$). Hence, $p_1 p_2 \cdots p_r$ and $q_1 q_2 \cdots q_s$ are also the same factorization. Thus, we have shown that any two prime factorizations of $k+1$ are the same, and therefore $k+1$ has a unique factorization as a product of primes. This proves the induction step and, by the principle of mathematical induction, the theorem is true for all $n \geq 2$. □

Remark 2.10.7. It should now be clear why we do not consider the number 1 a prime number. If 1 was a prime number, then Theorem 2.10.6 would be false! Indeed, if 1 was a prime, then $15 = 3 \cdot 5$ and $15 = 1 \cdot 1 \cdot 3 \cdot 5$ would be two distinct factorizations of 15 as a product of primes (notice that $3 \cdot 5$ is a factorization as a product of 2 primes, while $1 \cdot 1 \cdot 3 \cdot 5$ would be a factorization as a product of 4 primes). Thus, we define a prime p to be a number with precisely two distinct positive divisors, 1 and p, and $1 \neq p$.

The fundamental theorem of arithmetic implies that there is a *canonical* way to write numbers as a product of primes, and this representation is unique.

Corollary 2.10.8. *Let $n \geq 2$ be a natural number. Then, there is a unique factorization of n as a product of prime numbers of the form*

$$n = p_1^{e_1} p_2^{e_2} \cdots p_t^{e_t},$$

where $t \geq 1$, the numbers $p_1 < p_2 < \cdots < p_t$ are the prime divisors of n, and $e_1, e_2, \ldots, e_t \geq 1$ are unique.

Example 2.10.9. The number $7! = 7 \cdot 6 \cdot 5 \cdot 4 \cdot 3 \cdot 2 \cdot 1$ can be expressed uniquely as

$$7! = 2^4 \cdot 3^2 \cdot 5 \cdot 7.$$

2.10.1. Irrational Numbers.

Let us apply the fundamental theorem of arithmetic to prove that $\sqrt{2}$ is an *irrational number*.

Definition 2.10.10. A real number $\alpha \in \mathbb{R}$ is said to be *irrational* if α is not rational. In other words, $\alpha \neq \frac{m}{n}$ for all $m \in \mathbb{Z}$ and $n \in \mathbb{N}$.

Theorem 2.10.11. *The real number $\sqrt{2} = 1.41421356237309\ldots$ is irrational.*

Proof. Suppose for a contradiction that $\sqrt{2}$ is rational; i.e., there are $m \in \mathbb{Z}$ and $n \in \mathbb{N}$ such that $\sqrt{2} = \frac{m}{n}$. Without loss of generality, we may assume that the fraction $\frac{m}{n}$ is in reduced form; i.e., $\gcd(m,n) = 1$. In particular,

$$2 = (\sqrt{2})^2 = \left(\frac{m}{n}\right)^2 = \frac{m^2}{n^2}.$$

Hence, $2n^2 = m^2$. By Lemma 2.10.3 and since 2 is a prime and $2|m^2 = m \cdot m$, it follows that 2 divides m. Let us write $m = 2m'$, for some $m' \in \mathbb{Z}$. Then,

$$2n^2 = (2m')^2 = 4(m')^2,$$

and so, $n^2 = 2(m')^2$. As before, by Lemma 2.10.3, this implies that 2 divides n. Since 2 also divides m, we conclude that $\gcd(m,n) \geq 2$. However, this is a contradiction, as we had assumed that $\gcd(m,n) = 1$. Therefore, $\sqrt{2}$ cannot be rational. □

Here is a similar proof.

Alternative proof of Theorem 2.10.11. Suppose for a contradiction that $\sqrt{2}$ is rational; i.e., there are $m \in \mathbb{Z}$ and $n \in \mathbb{N}$ such that $\sqrt{2} = \frac{m}{n}$. In particular,

$$2 = (\sqrt{2})^2 = \left(\frac{m}{n}\right)^2 = \frac{m^2}{n^2}.$$

Hence, $2n^2 = m^2$. By Corollary 2.10.8, we may write

$$n = p_1^{e_1} p_2^{e_2} \cdots p_t^{e_t} \text{ and } m = q_1^{f_1} q_2^{f_2} \cdots q_s^{f_s}$$

for some distinct primes $p_1 < p_2 < \cdots < p_t$ and $q_1 < q_2 < \cdots < q_s$ and integers $e_i, f_i \geq 1$. Since $2n^2 = m^2$, we obtain

$$2p_1^{2e_1} p_2^{2e_2} \cdots p_t^{2e_t} = q_1^{2f_1} q_2^{2f_2} \cdots q_s^{2f_s}. \tag{2.7}$$

By the fundamental theorem of arithmetic (Theorem 2.10.6), these two factorizations must be the same. In particular, 2 divides the right-hand side and therefore we must have $q_1 = 2$. But this implies that the power of 2 in the right-hand side of (2.7) is $2^{2f_1} \geq 2^2$, so the right-hand side is divisible by at least 4. Thus, 4 also divides the left-hand side, and we must have $p_1 = 2$. Hence,

$$2^{1+2e_1} p_2^{2e_2} \cdots p_t^{2e_t} = 2^{2f_1} q_2^{2f_2} \cdots q_s^{2f_s}.$$

Since these two factorizations must be identical, we must have $2f_1 = 1 + 2e_1$, with $e_1, f_1 \geq 1$, but this is impossible since it implies that $1 = 2f_1 - 2e_1 = 2(f_1 - e_1)$ and 2 is not a divisor of 1. Thus, we have reached a contradiction, and $\sqrt{2}$ cannot be a rational number. □

Remark 2.10.12. Many other well-known constants have been shown to be irrational, but the proofs of these facts can be quite involved. For example:

- The first proof of the irrationality of π is due to Johann Heinrich Lambert in the 18th century. Later, in the 19th century, Charles Hermite found a proof that π^2 (and therefore π) is irrational. In 1945 and 1947, respectively, Dame Mary Cartwright and Ivan Niven published simplified versions of Hermite's proof, which are the proofs that are usually taught nowadays (see Figure 2.3.)
- The irrationality of e was first shown by Euler in 1737. The most well-known proof of this fact was given by Joseph Fourier in 1815.

It is worth pointing out that both Lambert's and Euler's proofs of the irrationality of π and e, respectively, use continued fractions. We will discuss continued fractions and some of their applications to irrationality in Chapter 13. For proofs of the irrationality of π and e, see [**Con4**].

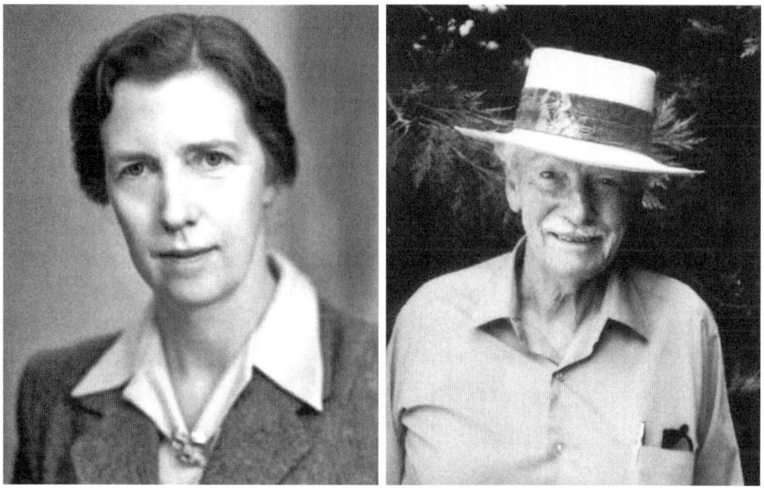

Figure 2.3. Dame Mary Cartwright (1900–1998) and Ivan Niven (1915–1999) simplified Hermite's proof of the irrationality of π to an elementary level. Image author (left): Anitha Maria S, used under Creative Commons Attribution-Share Alike 4.0 International license. Image author (right): Konrad Jacobs (Erlangen). Source: Archives of the Mathematisches Forschungsinstitut Oberwolfach.

2.11. Exercises

Exercise 2.11.1. Show that $(-1)\cdot(-1) = 1$, using the axioms of $\mathbb{Z}$ and Lemma 2.2.2.

Exercise 2.11.2. Let a and b be integers, such that a is a natural number but b is not natural. Show that $a \cdot b$ is not a natural number. (Hint: use Lemmas 2.2.1 and 2.2.2.)

Exercise 2.11.3. In this exercise we show the basic properties of divisibility. Prove the following statements, directly from the axioms of $\mathbb{Z}$ and Definition 2.1.5. Here a, b, and c are arbitrary integers.

(1) For every $a \in \mathbb{Z}$, the number a is a divisor of a.
(2) Every integer a is a divisor of 0.
(3) If a is a divisor of b and c, then a is a divisor of $b + c$ and $b - c$.
(4) More generally, show that if a divides b and c, then a divides $br + cs$, for any integers r and s.
(5) If a divides b and b divides c, then a divides c.

Exercise 2.11.4. Prove that $1^3 + 2^3 + \cdots + n^3 = \left(\frac{n(n+1)}{2}\right)^2$, for all $n \geq 1$.

Exercise 2.11.5. Use induction to prove the following statements:

(1) Prove that, for all $n \geq 1$, we have
$$\frac{1}{1-x} = 1 + x + x^2 + \cdots + x^{n-1} + \frac{x^n}{1-x}.$$

(2) Prove that, for all $n \geq 1$, we have $1 + 2 + 2^2 + \cdots + 2^{n-1} = 2^n - 1$.

Exercise 2.11.6. Prove that 5 divides $3^{4n} - 1$, for all $n \geq 1$.

Exercise 2.11.7. Prove that for any **odd** number $m \geq 1$, the number $4^m + 5^m$ is divisible by 9.

Exercise 2.11.8. The *Tower of Hanoi* is a mathematical puzzle that consists of three rods and a number of disks of different sizes, which can slide onto any rod (see Figure 2.4). The puzzle starts with the disks in a neat stack in ascending order of size on one rod, the smallest at the top, thus making a conical shape. The objective of the puzzle is to move the entire stack to another rod. The player is allowed to move only one disk at a time, and only smaller disks can be on top of a bigger disk.

Find and prove a formula for the least number of moves required to move a Tower of Hanoi with n disks to another rod. (Hint: find the least number of moves for $n = 1$, $n = 2$, and $n = 3$; then conjecture a formula, and prove your formula using induction.)

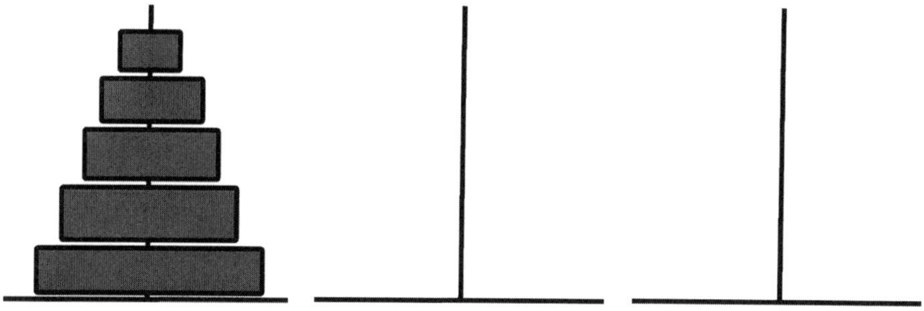

Figure 2.4. The *Tower of Hanoi* puzzle, with $n = 5$ disks.

Exercise 2.11.9. Use induction to prove the following statements.

(1) $n! \leq n^n$ for all $n > 0$.

(2) $(n+1)^{(n-1)} \leq n^n$ for all $n > 0$.

Exercise 2.11.10. What is wrong with the following proof?

Theorem. All babies have the same color eyes.

"Proof". The base case is clear: one baby has the same color eyes as herself or himself. In order to prove the induction step, let us assume as our induction hypothesis that any set of n babies has the same color eyes, and suppose that we have $n+1$ babies, say $\{B_1, \ldots, B_n, B_{n+1}\}$. By the induction hypothesis, the babies in sets $\{B_1, \ldots, B_n\}$ and those in $\{B_2, \ldots, B_{n+1}\}$ have the same color eyes. Since B_2 is in both sets, we conclude that all of the babies $B_1, \ldots, B_{n+1}$ have the same color eyes.

Exercise 2.11.11. Prove that the sum of the interior angles of an n-sided convex polygon is $180(n-2)$ degrees, for any $n \geq 3$.

Exercise 2.11.12. Let $S = \begin{pmatrix} 0 & -1 \\ 1 & 0 \end{pmatrix}$ and $T = \begin{pmatrix} 1 & 1 \\ 0 & 1 \end{pmatrix}$ be the matrices in the set $\mathrm{SL}(2, \mathbb{Z})$ that we previously defined in Exercise 1.8.21. Use induction to show that

$$(S \cdot T^2)^n = (S \cdot T^2) \cdots (S \cdot T^2) = \begin{pmatrix} -(n-1) & -n \\ n & n+1 \end{pmatrix},$$

for all $n \geq 1$, where the operation here is matrix multiplication (see Example 5.2.5).

Exercise 2.11.13. Prove that any natural number $n \geq 2$ is either a prime or factors into a product of primes.

Exercise 2.11.14. Let $0 \leq k \leq n$ be integers, and let

$$\binom{n}{k} = \frac{n!}{k!(n-k)!}$$

be the usual binomial (or combinatorial "n-choose-k") coefficient.

(1) Show that for all $1 \leq k \leq n$, the binomial coefficients satisfy the identity

$$\binom{n}{k} + \binom{n}{k-1} = \binom{n+1}{k}.$$

(2) Use induction to show the binomial theorem; i.e., if $n \geq 1$ and x, y are real numbers, then

$$(x+y)^n = \sum_{k=0}^{n} \binom{n}{k} x^k y^{n-k}.$$

Exercise 2.11.15. Prove that out of any set of 52 integers, two can always be found such that the difference of their squares is divisible by 100. (Hint: use the pigeonhole principle, Theorem 2.3.14. See also Exercise 10.8.9.)

Exercise 2.11.16. Let $n \geq 1$ be an integer.

(1) Show that n and $n+1$ are coprime for all $n \geq 1$.
(2) Prove that if n is odd, then n and $n+2$ are relatively prime.
(3) Prove that if n is even, then the greatest common divisor of n and $n+2$ is 2.

Exercise 2.11.17. Prove that if $k \geq 1$, the integers $6k+5$ and $7k+6$ are relatively prime.

Exercise 2.11.18. Prove that every odd natural number is the difference of two squares.

Exercise 2.11.19. Use Euclid's algorithm to find the following GCDs:

(1) $\gcd(121, 365)$,
(2) $\gcd(89, 144)$,
(3) $\gcd(295, 595)$,
(4) $\gcd(1001, 1309)$.

Exercise 2.11.20. Find the GCD of 17017 and 18900 using Euclid's algorithm.

Exercise 2.11.21. Find d, the GCD of a and b, i.e., $d = (a, b)$, and $r, s \in \mathbb{Z}$ such that $ar + bs = d$:

(1) $a = 267$ and $b = 112$,

(2) $a = 242$ and $b = 1870$.

Exercise 2.11.22. Find all solutions with integer coefficients x and y:

(1) $267x + 112y = 3$,

(2) $376x + 72y = 18$.

Exercise 2.11.23. Find all solutions with integer coefficients x and y:

(1) $203x + 119y = 47$, 48, or 50,

(2) $203x + 119y = 49$.

Exercise 2.11.24. Let $c, d \in \mathbb{Z}$ such that $\gcd(c, d) = 1$. Prove by induction that $\gcd(c, d^n) = 1$, for all $n \geq 1$. (Hint: use Corollary 2.7.6.)

Exercise 2.11.25. Show that if $\gcd(a, b) = d$, then $\gcd(\frac{a}{d}, \frac{b}{d}) = 1$.

Exercise 2.11.26. Let $a_1, \ldots, a_n \in \mathbb{Z}$ and let p be a prime number such that $p | a_1 \cdots a_n$. Show (using induction on n) that p divides at least one a_i, for some $1 \leq i \leq n$.

Exercise 2.11.27. Let $a, b \in \mathbb{N}$. We define the least common multiple of a and b as the number $t \in \mathbb{N}$ such that (i) $a|t$ and $b|t$ and (ii) if $s \in \mathbb{N}$ is another number such that $a|s$ and $b|s$, then $t \leq s$. We write $t = \operatorname{lcm}(a, b)$. Prove the following statements:

(1) If k is a common multiple of a and b, then $\operatorname{lcm}(a, b)$ divides k. In other words, every common multiple of a and b is a multiple of $\operatorname{lcm}(a, b)$.

(2) Prove that $a \cdot b = \gcd(a, b) \cdot \operatorname{lcm}(a, b)$, or, equivalently, $\operatorname{lcm}(a, b) = \frac{ab}{\gcd(a,b)}$.

Exercise 2.11.28. You take a 12-quart jug and a 17-quart jug to a stream and want to bring back 8 quarts of water. How do you do it?

Exercise 2.11.29 (Proposed by the Indian mathematician Bhaskara, c. 600–680 AD)**.** Two men are equally rich. One has 5 rubies, 5 pearls, and 90 gold coins; the other has 8 rubies, 9 pearls, and 48 gold coins. If rubies cost more than pearls, find the price in gold coins of each kind of gem.

Exercise 2.11.30. Find all the natural, integral, and rational roots of the polynomial equation

$$5x^3 + 27x^2 - 153x + 81 = 0.$$

Exercise 2.11.31. Find all the natural, integral, and rational solutions for the equation

$$6x^4 - 23x^3 - 132x^2 - 13x + 42 = 0.$$

Exercise 2.11.32. (1) Is there an integer n such that $(n+1)(n-2)(n+3)$ equals $(n-4)(n+5)(n-6)$? If so, find all such n, or prove there are none.

(2) Is there an integer n such that $(n+1)(n-2)(n+3)(n-4)$ is equal to the quantity $(n+5)(n-6)(n+7)(n-8)$? If so, find all such n, or prove there are none.

Exercise 2.11.33. Is 44497 prime? Why or why not?

Exercise 2.11.34. Prove the following statements.

(1) A natural number is a square if and only if the exponent of each prime factor is even.

(2) If a number n is not a square, then $\sqrt{n}$ is irrational.

Exercise 2.11.35. Show that $100^{(1/3)}$ is irrational.

Exercise 2.11.36. Show that if a, b are natural numbers with $\gcd(a, b) = 1$ and ab is a square, then a and b are also squares.

Exercise 2.11.37. Let n, a, b, c, d be natural numbers.

(1) Show that if d^2 is a divisor of n^2, then d is a divisor of n.

(2) Suppose $a^2 + b^2 = c^2$. Show that if d is a common divisor of any two of a, b, c, then it is also a divisor of the third number. Conclude that

$$\begin{aligned}\gcd(a, b) &= \gcd(b, c)\\ &= \gcd(a, c)\\ &= \gcd(a, b, c).\end{aligned}$$

Exercise 2.11.38. We define the Fibonacci numbers by the following recursive relationship: $F_0 = 1$, $F_1 = 1$, and $F_{n+1} = F_n + F_{n-1}$. Prove that any two consecutive Fibonacci numbers are relatively prime.

Figure 2.5. Leonardo di Pisa (c. 1175 – c. 1250), also known as *Fibonacci*, was an Italian mathematician considered to be "the most talented Western mathematician of the Middle Ages". Fibonacci popularized the Hindu-Arabic numeral system in the Western world through his book *Liber Abaci*. Image source: Wikipedia Commons.

Exercise 2.11.39. Let $\{F_n\}_{n\geq 0}$ be the sequence of Fibonacci numbers, defined in Exercise 2.11.38.

(1) Show that the sequence $\{F_{n+1}/F_n\}_{n\geq 0}$ of ratios of consecutive Fibonacci numbers is convergent. (Hint: use the monotone convergence theorem.)

(2) Show that the limit of $\{F_{n+1}/F_n\}_{n\geq 0}$ is $\varphi = \frac{1+\sqrt{5}}{2}$, the *golden ratio*. (Hint: divide $F_{n+1} = F_n + F_{n-1}$ through by F_n.)

(3) Show that the limit of $\{F_n/F_{n+1}\}_{n\geq 0}$ is $\frac{1}{\varphi} = \frac{\sqrt{5}-1}{2}$.

CHAPTER 3

THE PRIME NUMBERS

Then, just when you are about to surrender, when you no longer have the desire to go on counting, you come across another pair of twin primes, clutching each other tightly.

Paulo Giordano, *The Solitude of Prime Numbers*

The fundamental theorem of arithmetic explains that the prime numbers ought to be regarded as the fundamental building blocks of the natural numbers, when we consider $\mathbb{N}$ from the multiplicative point of view. Recall that we have defined a prime number as a natural number with exactly two positive divisors. The first few prime numbers are

$$2, 3, 5, 7, 11, 13, 17, 19, 23, 29, 31, 37, \ldots.$$

From the definition of prime, we immediately deduce an algorithm to check whether a number n is prime:

> *A number $n > 1$ is prime if and only if n is not divisible by any number between 1 and $n - 1$.*

This algorithm can be improved by noticing that every number n is divisible by at least one prime number (this was shown in Example 2.3.13). Thus, if n is not prime, then n has a positive divisor $d \neq 1, n$, and d itself has a prime divisor p. It follows that n has a prime divisor $p \neq 1, n$. Hence:

> *A number $n > 1$ is prime if and only if n is not divisible by any prime number between 1 and $n - 1$.*

Furthermore, if n is composite, then it has a prime divisor $p \leq \sqrt{n}$ (this is Exercise 3.5.2). Hence, we have shown the following criterion for primality.

Theorem 3.0.1. *A number $n > 1$ is prime if and only if n is not divisible by any prime number between 1 and $\sqrt{n}$.*

Figure 3.1. Eratosthenes of Cyrene (c. 276 BC – c. 194 BC). Image source: Wikimedia Commons.

Example 3.0.2. Let us show that 97 is a prime number. Suppose for a contradiction that 97 is composite. Then, it has a prime divisor $p \leq \sqrt{97} \leq \sqrt{100} = 10$. Hence, 97 is divisible by 2, 3, 5, or 7. However,

$$97 = 2 \cdot 48 + 1 = 3 \cdot 32 + 1 = 5 \cdot 19 + 2 = 7 \cdot 13 + 6.$$

It follows that the remainder when dividing 97 by $2, 3, 5$, or 7, respectively, is $1, 1, 2$, or 6. Hence, 97 is not divisible by $2, 3, 5$, or 7, and it must be prime.

In the next section we discuss a method (due to Eratosthenes of Cyrene; see Figure 3.1) to list all the prime numbers below a bound N.

3.1. The Sieve of Eratosthenes

The following algorithm is named after Eratosthenes, an ancient Greek mathematician (and also a geographer, poet, astronomer, and music theorist). Although none of Eratosthenes's works have survived, the sieve was described and attributed to Eratosthenes in the *Introduction to Arithmetic* by Nicomachus. The goal of the algorithm is to find all primes $p \leq N$, for a fixed $N \in \mathbb{N}$.

The sieve works as follows. We begin with a list all the numbers $2 \leq n \leq N$. The sieve works best if we organize the numbers consecutively, in ten columns. As an example, we will demonstrate the sieve with $N = 30$:

	2	3	4	5	6	7	8	9	10
11	12	13	14	15	16	17	18	19	20
21	22	23	24	25	26	27	28	29	30

In our first step, we cross out all the multiples of 2, except 2 itself (i.e., all the even numbers > 2). In the tables below we have removed the even numbers greater than 2:

	2	3	5	7	9
11		13	15	17	19
21		23	25	27	29

Now, we find the next number in our list that has not been crossed out, in this case 3, and we cross out all the multiples of 3 in the list, except 3 itself. In the tables, the multiples of 3 that were not already crossed out for being a multiple of 2 have been removed:

	2	3	5	7	
11		13		17	19
		23	25		29

As before, we find the next number in our list that has not been crossed out, in this case 5, and we cross out all the multiples of 5 in the list, except 5 itself. The remaining multiples of 5 have been eliminated (i.e., we removed the number 25):

	2	3	5	7	
11		13		17	19
		23			29

We continue in this fashion, until we find a number n that has not been crossed out but $n > \sqrt{N}$. Once we have reached this stage, we claim that all the numbers that remain intact in the list are all the primes $\leq N$. Indeed, suppose that n is a composite number $\leq N$. Then, by Theorem 3.0.1, we know that n has a prime divisor $p \leq \sqrt{N}$; i.e., there is some $k \in \mathbb{N}$ such that $n = pk$. In particular, n is a multiple of p and $p \leq \sqrt{n} \leq \sqrt{N}$ (by Exercise 3.5.2), but all such multiples have been previously crossed out in our list. It follows that all the numbers $\leq N$ that remain intact in the list are prime numbers.

In our example, if $N = 30$, then $\sqrt{30} < 6$. Thus, we need to cross out all the multiples of 2, 3, and 5, but once we reach the number $n = 7$, we can stop and we do not need to cross out any other numbers. The prime numbers ≤ 30 are those that remain intact in our list, namely

$$2,\ 3,\ 5,\ 7,\ 11,\ 13,\ 17,\ 19,\ 23, \text{ and } 29.$$

3.2. The Infinitude of the Primes

The sieve of Eratosthenes is a fairly efficient method to find all the primes below a given bound. However, a question remains. Is it possible that there is some $n_0 \in \mathbb{N}$ such that each number $n \geq n_0$ is a multiple of some prime $p \leq n_0$ and therefore all numbers $n \geq n_0$ have been crossed out using the sieve method? In other words, is it possible that the set of prime numbers is finite?

The following theorem and proof are due to Euclid of Alexandria (born 325 BC, died 265 BC), one of the most prominent mathematicians of ancient Greece.

Theorem 3.2.1. *There exist infinitely many prime numbers.*

Figure 3.2. Euclid of Alexandria (c. 325 BC – c. 265 BC). Image source: Wikimedia Commons.

Proof. Suppose for a contradiction that there exist only finitely many primes $p_1, p_2, \ldots, p_n$, where p_n is the largest of them. Let $\mathcal{N}$ be defined by

$$\mathcal{N} = (p_1 \cdot p_2 \cdot \ldots \cdot p_n) + 1.$$

Since $\mathcal{N} \geq 2$ is a natural number, it must have a prime divisor p (by Example 2.3.13 or by the fundamental theorem of arithmetic, Theorem 2.10.6), and since p is prime, it is one in the list $p_1, \ldots, p_n$. However, if some prime number p_i divides $\mathcal{N}$, it must also divide $\mathcal{N} - (p_1 \cdot \ldots \cdot p_n)$ since both numbers are multiples of p_i. Therefore p_i divides 1, which is impossible. Thus, our first assumption must be false and there exist infinitely many prime numbers. □

The proof of Euclid's theorem also indicates how to find new primes out of a set of known primes.

Corollary 3.2.2. *Let $p_1, p_2, \ldots, p_n$ be prime numbers. Then, $\mathcal{N} = p_1 p_2 \cdots p_n + 1$ is divisible by a new prime number p; i.e., there is a prime number p such that $p \mid \mathcal{N}$ and $p \neq p_i$ for all $1 \leq i \leq n$.*

Example 3.2.3. For instance,

$$\begin{aligned} 2 \cdot 3 + 1 &= 7, \\ 2 \cdot 3 \cdot 5 + 1 &= 31, \\ 2 \cdot 3 \cdot 5 \cdot 7 + 1 &= 211, \\ 2 \cdot 3 \cdot 5 \cdot 7 \cdot 11 + 1 &= 2311, \end{aligned}$$

and 7, 31, 211, and 2311 are prime numbers. It is worth stressing that Euclid's proof *does not* imply that $\mathcal{N} = p_1 p_2 \cdots p_n + 1$ is a prime number. It simply implies that $\mathcal{N}$ is divisible by a new prime number not in the list $\{p_1, p_2, \ldots, p_n\}$. For instance,

$$2 \cdot 3 \cdot 5 \cdot 7 \cdot 11 \cdot 13 + 1 = 59 \cdot 509,$$
$$2 \cdot 3 \cdot 5 \cdot 7 \cdot 11 \cdot 13 \cdot 17 + 1 = 19 \cdot 97 \cdot 277,$$

where 59, 509, 19, 97, and 277 are prime numbers.

Example 3.2.4. Imagine that we only knew of one prime, $p = 2$. Then, Euclid's proof tells us that $2 + 1 = 3$ must be either prime or divisible by new primes. Indeed, 3 is a prime. Now we know two primes $\{2, 3\}$. Once again, Euclid's trick, computing $2 \cdot 3 + 1 = 7$, reveals a new prime, namely 7. We may continue in this fashion, adding to our list the smallest prime that divides the product of those already in our list, plus one:

$$\begin{aligned} 2 \cdot 3 + 1 &= 7, \\ 2 \cdot 3 \cdot 7 + 1 &= 43, \\ 2 \cdot 3 \cdot 7 \cdot 43 + 1 &= 13 \cdot 139, \\ 2 \cdot 3 \cdot 7 \cdot 43 \cdot 13 + 1 &= 53 \cdot 443, \\ 2 \cdot 3 \cdot 7 \cdot 43 \cdot 13 \cdot 53 + 1 &= 5 \cdot 248867, \end{aligned}$$

and so on, so that our list of primes, as they appear in the sequence above, is given by

$$\{2, 3, 7, 43, 13, 53, 5, \ldots\}.$$

In 1963, Albert A. Mullin asked whether every prime appears in this sequence, at some point, and this question remains open to this day.

Finding large primes is an extremely difficult task. A method such as the sieve of Eratosthenes will produce primes as large as we want, but it requires a huge amount of computer time. There are formulas that produce primes, but these formulas either need a lot of initial input (often including the knowledge of a large number of large primes) or the formulas require a large number of calculations to produce one large prime.

Example 3.2.5. Let us define one real number α given by

$$\alpha = \sum_{n=1}^{\infty} \frac{1}{10^{p_n}},$$

where p_n is the nth prime number. Since the infinite sum only contains positive terms and $\alpha \leq \sum_{i=1}^{\infty} \frac{1}{10^i} = \frac{1}{9}$, we conclude that the infinite series that defines α is convergent and $\alpha \in \mathbb{R}$. Notice that the decimal expansion of α has a 1 as the ith digit after the period if and only if i is prime. That is,

$$\alpha = 0.011010100010100010100010000010 1\ldots.$$

A number such as α is no more than a curiosity, since we need prior knowledge of all primes in order to construct α. However, if one found an alternative formula for α, one could find its digits and therefore primes.

Example 3.2.6. Let us define a sequence $\{a_n : n \geq 1\}$ recursively by setting $a_1 = 7$ and

$$a_n = a_{n-1} + \gcd(n, a_{n-1}).$$

For instance,

$$a_1 = 7,\ a_2 = 7 + \gcd(2,7) = 8,\ a_3 = 8 + \gcd(3,8) = 9,$$
$$a_4 = 9 + \gcd(4,9) = 10,\ a_5 = 10 + \gcd(5,10) = 15,$$
$$a_6 = 15 + \gcd(6,15) = 18,\ a_7 = 18 + \gcd(7,18) = 19, \ldots.$$

Now let us define a sequence $\{b_n : n \geq 1\}$ by $b_n = a_{n+1} - a_n$, so that

$$b_1 = 8 - 7 = 1,\ b_2 = 9 - 8 = 1,\ b_3 = 10 - 9 = 1,\ b_4 = 15 - 10 = 5,$$
$$b_5 = 18 - 15 = 3,\ b_6 = 19 - 18 = 1, \ldots.$$

In 2008, Eric Rowland showed that the value of b_n is either 1 or a prime number, for all $n \geq 1$ (see [**Row08**]).

There are also famous cases of formulas that *fail* to produce primes.

Example 3.2.7. In or about the year 1630, Fermat (see Section 1.6.1) claimed that each number

$$F(n) = 2^{2^n} + 1, \text{ for each } n \geq 0,$$

is a prime number. For instance,

$$F(0) = 2 + 1 = 3,\ F(1) = 2^2 + 1 = 5,\ F(2) = 2^4 + 1 = 17,$$
$$F(3) = 257, \quad \text{and} \quad F(4) = 65537$$

are all primes. The numbers $F(n)$ are usually called *Fermat numbers*, and if $F(n)$ is prime, it is called a Fermat prime. However, in 1732, Euler showed that

$$F(5) = 2^{32} + 1 = 4294967297 = 641 \cdot 6700417.$$

And since then, no other Fermat number has been found to be a prime number. The Fermat numbers $F(5)$ through $F(32)$ are known to be composite numbers.

Surprisingly, Fermat primes are relevant in other areas of mathematics. We remind the reader that a straightedge is an idealized ruler, infinite in length, with no markings on it and only one edge.

Theorem 3.2.8 (Gauss–Wantzel theorem; [**DF03**], §14.5, Prop. 29)**.** *A regular polygon with n sides can be constructed with compass and straightedge if and only if n is the product of a power of 2 and any number of distinct Fermat primes; i.e.,*

$$n = 2^k \cdot p_1 \cdot p_2 \cdots p_t,$$

for some $k \geq 0$ and distinct Fermat primes $p_1, p_2, \ldots, p_t$.

For instance, one can construct a regular 65537-sided polygon using a compass and straightedge, but the regular 19-sided polygon cannot be constructed in this manner, because 19 is not a Fermat prime.

One direction of Theorem 3.2.8 was proven by Carl Friedrich Gauss in 1801, and he claimed the other direction to be true but did not provide a proof. In 1837, Pierre Wantzel published a complete proof of the theorem. Although Gauss proved that the regular 17-gon is constructible, he did not actually show how to do it. The first construction is due to Erchinger, a few years after Gauss's work. The first explicit constructions of a regular 257-gon were given by Magnus Georg Paucker (1822) and Friedrich Julius Richelot (1832). A construction for a regular 65537-gon was first given by Johann Gustav Hermes (1894).

3.3. Theorems on the Distribution of Primes

Mathematicians have tried in vain to this day to discover some order in the sequence of prime numbers, and we have reason to believe that it is a mystery into which the human mind will never penetrate.

Leonhard Euler

When we display the prime numbers on the number line, at first glance several patterns seem apparent. Some of these patterns dissipate as we consider larger numbers, and some patterns seem to appear somewhat periodically. In general, it is very difficult to prove any of these patterns. In this section we present some of the most important and famous theorems about the distribution of prime numbers. Unfortunately the proofs of these theorems are beyond the scope of this book.

3.3.1. Bertrand's Postulate. The following statement was first conjectured in 1845 by Joseph Bertrand (1822–1900).

Theorem 3.3.1 (Bertrand's postulate). *For all $n > 1$ there is a prime number p with $n < p < 2n$.*

Bertrand himself verified his statement for all numbers in the interval $[2, 3 \cdot 10^6]$. His postulate (or conjecture) was completely proved by Chebyshev (1821–1894) in 1850 and so Bertrand's postulate is sometimes called the Bertrand–Chebyshev theorem or Chebyshev's theorem. The result is still sometimes referred to as "Bertrand's postulate" for historical reasons, even though we now have a proof.

Example 3.3.2. The prime 3 is between 2 and 4, the prime 5 is between 3 and 6, and the prime 5 is between 4 and 8 (in fact, 5 and 7 are between 4 and 8).

Sylvester (1814–1897) proved the following generalization of Bertrand's postulate. We will leave it to the reader to verify that Sylvester's theorem implies Bertrand's postulate (see Exercise 3.5.13).

Theorem 3.3.3 (Sylvester's theorem). *Let $k \geq 1$ be a natural number. Then, the product of any k consecutive integers greater than k is divisible by a prime number greater than k.*

Example 3.3.4. Let $k = 3$. Then, for instance, $5 \cdot 6 \cdot 7$ is divisible by 5 (a prime greater than 3). Or $12 \cdot 13 \cdot 14$ is divisible by 7.

In 1952, Jitsuro Nagura proved the following strengthened version of the postulate.

Theorem 3.3.5 (Nagura's theorem). *If $n \geq 25$, then there is always a prime p between n and $(1 + 1/5) \cdot n$.*

For instance, if $n = 25$, then Nagura's theorem says that there is a prime between 25 and $(1 + 1/5)25 = 30$ and, indeed, $p = 29$ is such a prime. If $n = 30$, then there is a prime $p = 31$ between 30 and 36, and so on.

Further improvements of Bertrand's postulate have been shown by Schoenfeld, Dusart, Baker, Harman, and Pintz, among others. Legendre proposed the following conjecture, which is still open.

Conjecture 3.3.6 (Legendre's conjecture). *For all $n \geq 1$, there is a prime number p such that $n^2 < p < (n+1)^2$.*

3.3.2. The Prime Number Theorem. In order to study the distribution of prime numbers in more detail, we define two functions.

Definition 3.3.7. Let $\mathcal{P} \subset \mathbb{N}$ be the set of all prime numbers among the natural numbers. We define a function $p : \mathbb{N} \to \mathcal{P}$ such that $p(n)$ is the nth prime number, so that $\mathcal{P} = \{p(1) = 2,\ p(2) = 3,\ p(3) = 5, \ldots\}$. Sometimes we will write $p(n) = p_n$.

We also define the *prime counting function* $\pi(x)$ as the number of primes $p \leq x$, for any given real number $x \geq 0$; i.e.,

$$\pi(x) = \#\{\text{primes } p \leq x\}.$$

For instance, $\pi(2) = 1$, $\pi(5) = 3$, $\pi(6.7) = 3$, and $\pi(10) = 4$. In general, $\pi(p(n)) = n$ as there are n primes less than or equal to the nth prime. The following table provides some values of $\pi(x)$:

n	10^2	10^3	10^4	10^5	10^6	10^7
$\pi(n)$	25	168	1229	9592	78498	664579

The first mathematicians to find a pattern in the values of $\pi(x)$ were Carl Friedrich Gauss (1777–1855) in 1793 and, independently, Adrien-Marie Legendre (1752–1833) in 1798. They conjectured a certain asymptotic behavior for $\pi(x)$, which was proved (simultaneously) in 1896.

Theorem 3.3.8 (Prime number theorem; Hadamard, de la Vallée-Poussin, 1896).

$$\lim_{x \to \infty} \frac{\pi(x)}{\frac{x}{\log x}} = 1.$$

In other words, if we define a function $\pi'(x) = \frac{x}{\log x}$, then the prime number theorem says that the quotient of $\pi(x)$ by $\pi'(x)$ can be made arbitrarily close to 1 as x goes to ∞. Or, in less technical terms, $\pi(x)$ is approximately $\frac{x}{\log x}$ for large values of x. In 1838, Dirichlet (see Figure 3.3), in a letter to Gauss, suggested that the logarithmic integral, or *li* for short, given by

$$\operatorname{li}(x) = \int_2^x \frac{dt}{\log t},$$

would be an even better approximation of $\pi(x)$. In 1899, de la Vallée-Poussin showed that $\pi(x)$ is also asymptotic to $\operatorname{li}(x)$ and gave a precise error estimate. We compare the values of $\pi(x)$, $\pi'(x)$, and $\operatorname{li}(x)$ in Table 3.1.

One can also use the prime number theorem to obtain an asymptotic expression for the nth prime number. Here is a heuristic argument. By definition $\pi(p(n)) = n$. Thus, by the prime number theorem, we obtain

$$n = \pi(p_n) \approx \frac{p_n}{\log p_n}, \tag{3.1}$$

Table 3.1. Comparison of values of the prime counting function $\pi(x)$, the approximating function $\pi'(x) = x/\log x$, and the logarithmic integral $\mathrm{li}(x)$.

n	10^2	10^3	10^4	10^5	10^6	10^7
$\pi(n)$	25	168	1229	9592	78498	664579
$\pi'(n)$	21.7...	144.7...	1085.7...	8685.8...	72382.4...	620420.6...
$\mathrm{li}(n)$	30.1...	177.6...	1246.1...	9629.8...	78627.5...	664918.4...

or $p_n \cong n \log p_n$ for large n. One can also show that

$$\lim_{x\to\infty} \frac{\log x}{\log(\frac{x}{\log x})} = 1.$$

In particular, $\log x \cong \log(\frac{x}{\log x})$ for large x, and therefore

$$\log n \cong \log\left(\frac{p_n}{\log p_n}\right) \cong \log p_n.$$

It follows that $p_n \cong n \log p_n \cong n \log n$. The previous heuristic argument can be made into a formal proof and can show the following theorem, which is an equivalent formulation of the prime number theorem.

Theorem 3.3.9.

$$\lim_{n\to\infty} \frac{p_n}{n \log n} = 1.$$

In Table 3.2 we compare values of $p(n) = p_n$ with values of $n \log n$.

Table 3.2. Comparison of the values of the nth prime and $n \log n$.

n	10^2	10^3	10^4	10^5
p_n	541	7919	104729	1294709
$n \log n$	460.5...	6907.7...	92103.4...	1151292.5...

While the approximate formula $p_n \cong n \log n$ cannot be used to find primes, it does provide a quick way to find out the approximate order of magnitude of the nth prime.

Remark 3.3.10. The prime number theorem can also be used to estimate the average gap between two consecutive primes. Let p_n and p_{n+1} be two consecutive primes, for some (large) number $n > 0$. By Theorem 3.3.9, we have

$$p_n \cong n \log n \quad \text{and} \quad p_{n+1} \cong (n+1)\log(n+1).$$

Moreover, for large n we can approximate $\log n \cong \log(n+1)$. Thus,

$$\begin{aligned} p_{n+1} - p_n &\cong (n+1)\log(n+1) - n \log n \\ &\cong (n+1)\log n - n \log n \cong \log n. \end{aligned}$$

Hence, according to the prime number theorem, the gap between two consecutive primes is typically $p_{n+1} - p_n \cong \log n$ or, in other words, p_{n+1} is typically of size $\cong p_n + \log n$. Notice that Bertrand's postulate (Theorem 3.3.1) shows that p_{n+1} can be found in the interval $(p_n, 2p_n)$.

3.3.3. Primes in Arithmetic Progressions. An arithmetic progression is a sequence of natural numbers such that the difference of any two consecutive numbers in the sequence is constant. In other words, an arithmetic progression is a sequence of numbers of the form

$$\{a,\ a+m,\ a+2m,\ a+3m,\ldots\} = \{a+mk : k \geq 0\},$$

for some fixed $a \geq 0$ and $m > 0$. For instance,

$$\{1,\ 6,\ 11,\ 16,\ 21,\ 26,\ 31,\ldots\} = \{1+5k : k \geq 0\}$$

is an arithmetic progression. We immediately see that this progression contains at least two prime numbers, namely 11 and 31, and, in fact, it contains many prime numbers. If $p = 1+5k$ is prime, then k must be even because otherwise $1+5k$ is even and p is not a prime. Thus, the primes in the progression $\{1+5k\}$ are in fact in the subarithmetic progression $\{1+10k : k \geq 0\}$. Here are all the prime numbers below 300 that form part of the progression $\{1+10k : k \geq 0\}$:

$$11,\ 31,\ 41,\ 61,\ 71,\ 101,\ 131,\ 151,\ 181,\ 191,\ 211,\ 241,\ 251,\ 271,\ 281. \tag{3.2}$$

Suppose that we fix an arbitrary arithmetic progression $S_{a,m} = \{a+mk : k \geq 0\}$. Are there necessarily infinitely many primes in $S_{a,m}$? The answer is *no* because if a and m share a common prime factor, then $S_{a,m}$ can contain at most one prime. For instance, if $m = 10$ and $a = 5$, then

$$S_{5,10} = \{5,\ 15,\ 25,\ 35,\ldots\} = \{5+10k : k \geq 0\},$$

and therefore every number in $S_{5,10}$ is divisible by 5. Thus, the only prime in $S_{5,10}$ is $p = 5$. However, if a and m are relatively prime, then $S_{a,m}$ contains infinitely many prime numbers. This important theorem was first shown in 1837 by Johann Peter Gustav Lejeune Dirichlet (1805–1859).

Figure 3.3. Johann Peter Gustav Lejeune Dirichlet (1805–1859) was a German mathematician who made contributions to number theory and analysis. Image source: Wikimedia Commons.

Theorem 3.3.11 (Dirichlet's theorem on primes in arithmetic progressions). *Let $m > 0$ and $a \geq 0$ be fixed integers such that $\gcd(a, m) = 1$. Then, there are infinitely many primes of the form $a+mk$. That is, there are infinitely many natural numbers $k_1, k_2, \ldots$ such that $p_i = a+mk_i$ is a prime number. Or, equivalently, the arithmetic progression $\{a + mk : k \geq 0\} \subseteq \mathbb{N}$ contains infinitely many prime numbers.*

The proof of Dirichlet's theorem is beyond the scope of this book. The reader can find a proof in Chapter 7 of [**Apo76**].

Example 3.3.12. Let us use Dirichlet's theorem to show that there are infinitely many primes whose decimal representation ends in $\ldots 321$. Here are all such primes below 30000:

$$1321,\ 7321,\ 10321,\ 11321,\ 14321,\ 17321,\ 23321,\ 25321,\ 26321.$$

Notice that those numbers that end in 321 form an arithmetic progression $\{321 + 1000k : k \geq 0\}$ and $\gcd(321, 1000) = 1$. Thus, by Dirichlet's theorem, the arithmetic progression $S_{321,1000}$ contains infinitely many prime numbers.

Example 3.3.13. If q is a prime, then there are primes $p_r \geq q$, one for each $r = 0, 1, \ldots, q-1$, such that the remainder when we divide p_r by q is exactly r.

Indeed, when $r = 0$, the prime $p_0 = q$ has a remainder of $r = 0$. If $1 \leq r \leq q-1$, then $\gcd(q, r) = 1$, and Dirichlet's theorem implies that the arithmetic progression $\{r + qk : k \geq 0\}$ contains infinitely many primes. It follows that there is a prime p in $S_{r,q}$ larger than q. If $p = qk + r$ and $1 \leq r \leq q-1$, then the remainder of p divided by q must be r, so we can pick $p_r = p = qk + r$.

For example, if $q = 5$, we can pick

$$p_0 = 5,\ p_1 = 11,\ p_2 = 7,\ p_3 = 13, \text{ and } p_4 = 19,$$

as the primes that leave a remainder of $0, 1, 2, 3$, and 4 when divided by 5.

The arithmetic progression $\{1 + 10k : k \geq 0\}$ contains infinitely many prime numbers, by Dirichlet's theorem, and all the prime numbers in $S_{1,10}$ below 300 were listed in (3.2). Notice that the subsequence S of primes in $S_{1,10}$ *belongs to* an arithmetic progression, but on their own they do not form an arithmetic progression. For example, $31-11 = 20$ but $41-31 = 10$ and $181-151 = 30$; thus, their difference is not constant and the sequence S is not an arithmetic progression.

One can show that there cannot be infinitely many primes in arithmetic progression; i.e., there is no sequence of primes $\{q_1, q_2, \ldots\}$ such that $q_i - q_{i-1} = m > 0$ is constant, for all $i \geq 2$ (see Exercise 3.5.20). However, we can find finite arithmetic progressions of primes. For instance,

$$3, 5, 7$$

is an arithmetic progression of three primes because $7 - 5 = 5 - 3 = 2$. Also,

$$5, 11, 17, 23, 29$$

is an arithmetic progression of five primes because the difference of consecutive primes in the sequence is constant, equal to 6. Is there an arithmetic progression of k primes, for all $k \geq 2$? In other words, if we fix $k \geq 2$, are there k primes $q_1, q_2, \ldots, q_k$ such that $q_i - q_{i-1} = m > 0$ is constant for all $i = 2, \ldots, k$? The answer to these questions had been conjectured to be *yes* for a long time, until the statements were proved in 2004 by Benjamin Green and Terence Tao.

Theorem 3.3.14 (Green–Tao theorem). *For every $k \geq 2$, we can find a set of k primes in arithmetic progression. In other words, if we fix k, then there exist k primes $q_1, q_2, \ldots, q_k$ such that $q_i - q_{i-1} = m > 0$ is a constant for all $i = 2, \ldots, k$.*

3.4. Famous Conjectures about Prime Numbers

There are many properties of prime numbers that are believed to be true but that no one knows how to prove. Here we present some of them.

3.4.1. The Twin Prime Conjecture.

The numbers 2 and 3 are the only consecutive primes. Indeed, if p and $p+1$ were primes, then one needs to be even, so one of them is 2. Thus, $p = 2$ and $p+1 = 3$. Hence, except for 2 and 3, the difference of any two prime numbers is at least 2. For instance, 3 and 5 differ by 2, and so do the following pairs of prime numbers:

$$(3,5),\ (5,7),\ (11,13),\ (17,19),\ (29,31),\ (41,43),\ (59,61)\ldots.$$

Two primes p and q whose difference is 2, i.e., $q = p+2$, are called *twin primes*. Here is a list of all twin primes below 1000:

$$\begin{gathered}(71,73),(101,103),(107,109),(137,139),(149,151),(179,181),(191,193),\\(197,199),(227,229),(239,241),(269,271),(281,283),(311,313),(347,349),\\(419,421),(431,433),(461,463),(521,523),(569,571),(599,601),(617,619),\\(641,643),(659,661),(809,811),(821,823),(827,829),(857,859),(881,883).\end{gathered}$$

The twin prime conjecture claims that there are infinitely many twin primes.

Conjecture 3.4.1 (Twin prime conjecture). *There are infinitely many primes p such that $p+2$ is also a prime number.*

In fact, there is another conjecture, known as the Hardy–Littlewood conjecture, which predicts an asymptotic estimate for the number of twin primes below a given $x > 0$, just as the prime number theorem gives an asymptotic estimate for the number of primes below x.

Conjecture 3.4.2 (Hardy–Littlewood twin prime conjecture, [**HL23**]). *Let $\pi_2(x)$ denote the number of primes $p \leq x$ such that $p+2$ is also a prime. Then, there is a constant*

$$C = 2\prod_{p\geq 3}\left(1 - \frac{1}{(p-1)^2}\right) = 1.320323631\ldots$$

such that

$$\lim_{x\to\infty} \frac{\pi_2(x)}{\frac{Cx}{(\log x)^2}} = 1.$$

In particular, the Hardy–Littlewood conjecture may be used to give a rough estimate of $\pi_2(x)$ as, approximately, given by $\frac{Cx}{(\log x)^2}$. For instance, there are 35 prime pairs $(p, p+2)$ with $p \leq 1000$, so $\pi_2(1000) = 35$. The asymptotic estimate predicts about 28 pairs of twin primes below 1000:

$$\frac{C \cdot 1000}{(\log 1000)^2} = 27.6698311973002\ldots.$$

The twin prime conjecture can be restated as follows: there are infinitely many primes p such that the *next* prime is in the interval $(p, p+2]$. Thus, one can formulate a weaker conjecture, for any $N \geq 2$.

Conjecture 3.4.3 (Infinitely many prime gaps of size $\leq N$). *Let $N \geq 2$ be fixed. Then, there are infinitely many primes p such that the next prime is in the interval $(p, p+N]$. More precisely,*

$$p_{n+1} - p_n \leq N$$

occurs for infinitely many values of $n \geq 1$, where p_n is the nth prime number.

Figure 3.4. Yitang Zhang is a Chinese-born American mathematician. He was awarded a 2014 MacArthur Award for his work on the least gap between consecutive primes. Image source: Wikimedia Commons.

In 2013, the mathematics community was surprised when Yitang Zhang ([**Zha13**]) announced a proof of Conjecture 3.4.3 for $N = 7 \cdot 10^7$, i.e., a proof of the fact that there are infinitely many pairs of prime numbers that are at most 70 million units apart, or, equivalently,

$$p_{n+1} - p_n \leq 7 \cdot 10^7$$

occurs infinitely often. Following this groundbreaking result, one of the first "crowd-sourced" projects in mathematics (the so-called PolyMath 8a and 8b projects) set as a goal improving Zhang's results in order to prove Conjecture 3.4.3 for the smallest value of N possible (where $N = 2$ would prove the twin prime conjecture). As of this writing, the conjecture has been shown for $N = 246$; i.e., there are infinitely many pairs of prime numbers that differ by at most 246.

Remark 3.4.4. Even though the twin prime conjecture claims that there are infinitely many small gaps between two consecutive primes, i.e., there are infinitely many consecutive primes p_n and p_{n+1} that only differ by $p_{n+1} - p_n = 2$, the gaps between primes can be arbitrarily large; i.e., for any $N \geq 1$, there are consecutive primes p_n and p_{n+1} such that $p_{n+1} - p_n > N$ (see Exercise 3.5.6 for a proof of this fact). For instance, $3 - 2 = 1$, $5 - 3 = 2$, $11 - 7 = 4$, $29 - 23 = 6$, $97 - 89 = 8$, $127 - 113 = 14$, etc.

3.4.2. Prime Constellations and the Bateman–Horn Conjecture. The twin prime conjecture (as in Section 3.4.1) can be restated as follows: both coordinates of the tuple $(n, n+2)$ are prime numbers for infinitely many values of $n \geq 3$. It is natural, then, to consider other possible tuples that may yield simultaneous prime numbers, for instance $(n, n+4)$ yields $(3, 7)$, $(7, 11)$, $(13, 17)$, etc. One can also increase the length of the tuple and ask the same question; for example, is there a natural number n such that $(n, n+2, n+4)$ is a triple of primes? The answer is yes, but this only occurs when $n = 3$ and the triple is $(3, 5, 7)$, because one can show that if n is an integer, then one of n, $n+2$, and $n+4$ is divisible by 3 (see Exercise 3.5.16). However, the triple $(n, n+2, n+6)$ seems to take prime values for infinitely many values of $n \geq 3$; e.g., $n = 5$ yields the triple $(5, 7, 11)$. In other words, we conjecture that the numbers p, $p+2$, and $p+6$ are primes for infinitely many (prime) numbers $p \geq 3$. The tuples $(n, n+2)$ or $(n, n+2, n+6)$ are called *constellations* of primes when all the coordinates are prime numbers.

Definition 3.4.5. Let $k \geq 0$, let $0 = a_0 < a_1 < \cdots < a_{k-1}$, and let $n \geq 2$ be integers. Let $c = (n + a_0, n + a_1, \ldots, n + a_{k-1})$.

(1) If c is a k-tuple of prime numbers, then we say that c is a *prime constellation* of length k. The number a_{k-1} is called the *diameter* of the constellation.

(2) Let p be a prime number, and let r_i be the remainder of division of a_i by p, for $0 \leq i \leq k-1$. We say the k-tuple $T = (0 = a_0, \ldots, a_{k-1})$ of integers is *p-admissible* if the set $\{r_0, \ldots, r_{k-1}\}$ does not contain all possible remainders $\{0, \ldots, p-1\}$.

(3) Finally, we say that the k-tuple $T = (0 = a_0, \ldots, a_{k-1})$ is *admissible* if it is admissible for all primes p.

Example 3.4.6. The triple $(0, 2, 4)$ is 2-admissible, because the set of remainders when dividing by 2 is just $\{0\}$, but it is not 3-admissible, because the remainders of 0, 2, and 4 when dividing by 3 are respectively 0, 2, and 1, which is a complete set.

In **[HL23]**, Hardy and Littlewood conjectured a specific asymptotic formula for the number of twin primes (Conjecture 3.4.2), but they also conjectured the following result about admissible k-tuples and constellations of prime numbers.

Conjecture 3.4.7 (Hardy–Littlewood k-tuple conjecture, **[HL23]**). *Let $k \geq 0$ be an integer, and let $(0 = a_0, \ldots, a_{k-1})$ be an admissible k-tuple of integers. Then, there are infinitely many natural numbers $n \geq 2$ such that $(n, n+a_1, \ldots, n+a_{k-1})$ is a prime constellation. Moreover, a conjectural asymptotic formula for the number of such constellations can be explicitly formulated.*

In 1962, building on previous work and conjectures of Bunyakovsky and Schinzel, mathematicians Paul T. Bateman and Roger A. Horn proposed a generalization of the Hardy–Littlewood conjecture to a much broader context of simultaneous prime values of polynomials. The twin prime conjecture can be rephrased in terms of polynomial values as follows: let $f(x) = x$ and let $g(x) = x + 2$. Then, the twin prime conjecture claims that there are infinitely many integer values $n_1, n_2, \ldots$ such that $f(n_i)$ and $g(n_i)$ are both prime numbers, for all $i \geq 1$. In other words, the conjecture says that there are infinitely many integers $n \in \mathbb{Z}$ such that the polynomials $f(x)$ and $g(x)$ take simultaneously prime numbers as values, when evaluated at $x = n$. It is now natural to replace $f(x)$ and $g(x)$ by other irreducible polynomials $f_1(x), \ldots, f_m(x)$ and ask whether they take prime values at a common argument $x = n$.

Example 3.4.8. Let $f_1(x) = x$, $f_2(x) = x + 2$, and $f_3(x) = x + 6$. Then, these polynomials take simultaneous prime values at $x = n$ if and only if $(n, n+2, n+6)$ is a prime constellation. For instance, $n = 5$ yields the triple $(f_1(5), f_2(5), f_3(5)) = (5, 7, 11)$.

Example 3.4.9. Let $f_1(x) = x$ and $f_2(x) = 2x + 1$. Then, f_1 and f_2 take simultaneous prime values at $x = p$ if p and $q = 2p + 1$ are both primes. Such a prime p is called a *Sophie Germain prime*. See Exercise 3.5.17 and Figure 3.8.

The Bateman–Horn conjecture, which we state next, predicts what sets of polynomials can take simultaneous prime values, and it also gives a conjectural asymptotic density for how often these coincidences can occur.

Conjecture 3.4.10 (Bateman–Horn conjecture, **[BH62]**). *Let $m \geq 1$ be an integer, and let $f_1(x), \ldots, f_m(x)$ be irreducible polynomials with integer coefficients. Also define:*

- *$f(x) = f_1(x) \cdots f_m(x)$, their product,*
- *$D_f = (\deg f_1(x)) \cdot (\deg f_2(x)) \cdots (\deg f_m(x))$,*
- *for a prime p, let $N_f(p)$ be the number of solutions of $f(x) \equiv 0 \bmod p$, and*
- *$C_f = \prod_p \dfrac{(1 - N_f(p)/p)}{(1 - 1/p)^m}$, where the product is over all primes p.*

Let $\pi_f(x)$ be the number of values $n \leq x$ such that $(f_1(n), \ldots, f_m(n))$ is an m-tuple of prime numbers. Then,

$$\lim_{x \to \infty} \frac{\pi_f(x)}{\frac{C_f}{D_f} \cdot \int_2^x \frac{1}{(\log t)^m}\, dt} = 1.$$

In other words, the Bateman–Horn conjecture says that the counting function $\pi_f(x)$ is asymptotic to $\frac{C_f}{D_f} \cdot \left(\int_2^x \frac{1}{(\log t)^m} dt\right)$. Let us see some examples.

Example 3.4.11. Let $m = 1$ and $f_1(x) = f(x) = x$. Then, $D_f = 1$, and $N_f(p) = 1$ for every prime number p. Thus, $C_f = 1$, and $\pi_f(x)$ is the usual prime counting function $\pi(x)$. The Bateman–Horn conjecture then predicts that $\pi(x)$ is asymptotic to $\int_2^x \frac{1}{\log t} dt$, which is in fact the logarithmic integral function $\mathrm{li}(x)$ that Dirichlet defined (see Section 3.3.2). The result $\lim_{x\to\infty} \pi(x)/\operatorname{li}(x) = 1$ is an equivalent formulation of the prime number theorem, and therefore the Bateman–Horn conjecture is known to hold in this case.

Example 3.4.12. Let $m = 2$ and let $f_1(x) = x$ and $f_2(x) = x + 2$. Then, $D_f = 1$. Moreover, $N_f(2) = 1$, and $N_f(p) = 2$ for all $p > 2$. Thus,

$$C_f = \frac{(1-1/2)}{(1-1/2)^2} \prod_{p\geq 3} \frac{(1-2/p)}{(1-1/p)^2} = 2 \prod_{p\geq 3} \frac{(1-2/p)}{(1-1/p)^2} = 2 \prod_{p\geq 3} \left(1 - \frac{1}{(p-1)^2}\right).$$

In this case, $\pi_f(x)$ coincides with $\pi_2(x)$, the counting function for twin primes $(p, p+2)$ with $p \leq x$, and the Bateman–Horn conjecture is the logarithmic integral version of Conjecture 3.4.2 on the distribution of twin primes:

$$\lim_{x\to\infty} \frac{\pi_f(x)}{C_f \cdot \int_2^x \frac{1}{(\log t)^2}\, dt} = 1.$$

This version is equivalent to that of the Hardy–Littlewood conjecture on twin primes. Therefore, the Bateman–Horn conjecture is not known in this case.

The Bateman–Horn conjecture can be used to formulate an asymptotic conjecture on the number of Sophie Germain primes. We will leave this to be worked out by the reader, in Exercise 3.5.18.

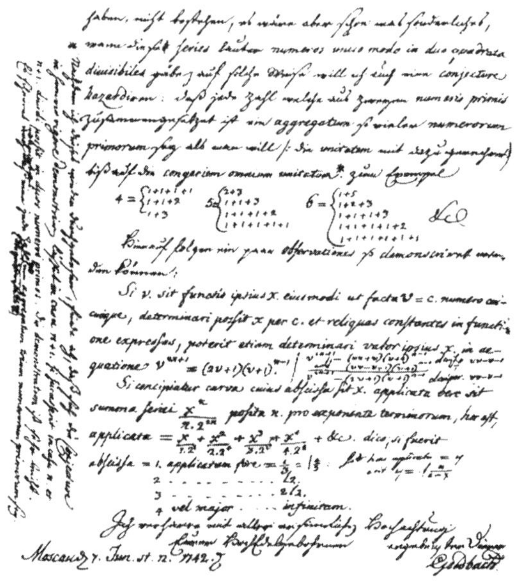

Figure 3.5. A letter from Christian Goldbach to Leonhard Euler on June 7, 1742. Image source: Wikimedia Commons.

3.4.3. Goldbach's Conjecture.

In 1742, the German mathematician Christian Goldbach wrote a letter to Leonhard Euler in which he proposed the following conjecture.

Conjecture 3.4.13. *Every even number $n > 2$ can be written as the sum of two prime numbers.*

For instance, $4 = 2 + 2$ and $6 = 3 + 3$. Also,

$$8 = 3 + 5,\ 10 = 3 + 7 = 5 + 5,\ 12 = 5 + 7,\ 14 = 3 + 11 = 7 + 7, \ldots$$

and, in fact, large even numbers should have many (distinct) representations as the sum of two primes (we consider $p + q = q + p$ the same representation). For example, $n = 100$ has six distinct representations:

$$100 = 3 + 97 = 11 + 89 = 17 + 83 = 29 + 71 = 41 + 59 = 47 + 53,$$

and $n = 1000$ has 28 representations as a sum of two primes.

Several mathematicians (Chudakov, Van der Corput, Estermann, Montgomety, and Vaughan) have shown results that prove an asymptotic version of Goldbach's conjecture or, in other words, that "almost all" even numbers can be written as the sum of two primes. More concretely, it has been shown that

$$\lim_{x\to\infty} \frac{\{2n \in \mathbb{N} : 2n \le x \text{ and } 2n = p + q \text{ for some prime numbers } p, q\}}{\{2n \in \mathbb{N} : 2n \le x\}} = 1.$$

In 2013, Harald Helfgott proved a version of Goldbach's conjeture, the so-called ternary (or odd or weak) Goldbach's conjecture.

Figure 3.6. Harald Helfgott is a Peruvian mathematician who proved the ternary Goldbach conjecture. Image source: Archives of the Mathematisches Forschungsinstitut Oberwolfach.

Theorem 3.4.14 (Helfgott, 2013). *Every odd number greater than 7 is the sum of three odd primes.*

Goldbach's conjecture implies the ternary version proved by Helfgott. Indeed, if $n > 7$ is odd, then $n - 3$ is even, larger than 4, and, therefore, if we assume Goldbach's conjecture, then $n - 3 = p + q$ for some odd primes p and q. Thus, $n = 3 + p + q$.

3.4.4. The Riemann Hypothesis.

Hiervon wäre allerdings ein strenger Beweis zu wünschen; ich habe indess die Aufsuchung desselben nach einigen flüchtigen vergeblichen Versuchen vorläufig bei Seite gelassen, da er für den nächsten Zweck meiner Untersuchung entbehrlich schien.

Bernhard Riemann (a translation follows in the text below)

Perhaps one of the most famous conjectures in number theory (and perhaps in all of mathematics) is the Riemann hypothesis. Bernhard Riemann (Figure 3.7) proposed this conjecture in 1859 while studying the distribution of prime numbers.

Figure 3.7. Georg Friedrich Bernhard Riemann (1826–1866) was a German mathematician who made fundamental contributions to analysis, number theory, and differential geometry. Image source: Wikimedia Commons.

The conjecture is usually stated (as Riemann did) in terms of the zeros of the Riemann zeta function $\zeta(s)$, which is defined as follows:

$$\zeta(s) = \sum_{n\geq 1} \frac{1}{n^s} = \frac{1}{1^s} + \frac{1}{2^s} + \frac{1}{3^s} + \cdots$$

for any complex number $s \in \mathbb{C}$ with real part greater than 1. The zeta function $\zeta(s)$ may be extended and defined for any complex number $s \neq 1$ (the function has a different definition for complex number with real part ≤ 1; we will not do this here). The conjecture then says that the only "non-trivial" zeros of $\zeta(s)$ are in the line $s = \frac{1}{2} + it$ for some $t \in \mathbb{R}$; i.e., the non-trivial zeros have real part equal to $1/2$ (the *trivial* zeroes occur at negative even integer values $s = -2, -4, -6$, etc.). Riemann tried to prove this statement but could not find such a proof:

> *Of course one would wish for a rigorous proof here; I have for the time being, after some fleeting vain attempts, provisionally put aside the search for this, as it appears dispensable for the immediate objective of my investigation.*

The Riemann hypothesis was number 8 of the 23 unsolved problems that David Hilbert listed as fundamental in 1900 (and it continues to be unsolved today). It is also one of the seven Clay Mathematics Institute's Millennium (one million dollar) Prize Problems.

Instead of writing a precise statement for the conjecture in terms of $\zeta(s)$, we will write an equivalent statement to the Riemann hypothesis (due to von Koch in 1901) that makes the connection to prime numbers explicit.

Conjecture 3.4.15 (Riemann hypothesis). *There is a constant c such that*

$$|\pi(x) - \mathrm{li}(x)| < c \cdot \sqrt{x}\log(x)$$

for all sufficiently large x, where $\pi(x)$ is the prime number counting function and $\mathrm{li}(x) = \int_2^x \frac{1}{\log(t)}\,dt$.

In 1976, Schoenfeld gave this explicit version of the Riemann hypothesis:

$$|\pi(x) - \mathrm{li}(x)| < \frac{\sqrt{x}\log(x)}{8\pi} \quad \text{for all } x \geq 2657. \tag{3.3}$$

For example $\pi(10^4) = 1229$ and $\mathrm{li}(10^4) = 1246.1\ldots$, so

$$|\pi(10^4) - \mathrm{li}(10^4)| < 18$$

and the Riemann hypothesis says that the difference is less than

$$\frac{\sqrt{10^4}\log(10^4)}{8\pi} = 36.646\ldots$$

so the bound in (3.3) holds for $x = 10^4$.

3.5. Exercises

Exercise 3.5.1. The author's first daughter was born in January, and the second daughter was born in November of the following year. How many times will their ages be consecutive primes before the first daughter turns 100? For instance, their ages are currently 5 and 7.

Exercise 3.5.2. Show that if n is not prime, then n has a prime divisor $p \leq \sqrt{n}$.

Exercise 3.5.3. Prove that there are infinitely many primes of the form $4n - 1$. (Hint: suppose that there are only finitely many of them, say $p_1, p_2, \ldots, p_t$. Now consider $N = 4p_1p_2\cdots p_t - 1$.)

Exercise 3.5.4. Prove that there are infinitely many primes of the form $6n - 1$.

Exercise 3.5.5. Let $a_1 = 2$ and $a_{n+1} = a_n(a_n - 1) + 1$. Prove that $a_{n+1} = a_1a_2\cdots a_n + 1$. Prove that for all $m \neq n$, the numbers a_m and a_n are relatively prime.

Exercise 3.5.6. Prove that for any $N \geq 1$ there are N consecutive composite numbers. (Hint: consider $(N+1)! + 2$, $(N+1)! + 3, \ldots,$ $(N+1)! + N + 1$.)

Exercise 3.5.7. Prove that for any $n \geq 2$ there is a prime p with $n < p \leq n! + 1$.

Exercise 3.5.8. Use a sieve method to find all the prime numbers between 105 and 115. Explain how you did it.

Exercise 3.5.9. Find all the primes between 200 and 250 by using the sieve of Eratosthenes on the list of numbers in that range.

Exercise 3.5.10. Find all primes p such that $17p + 1$ is a square.

Exercise 3.5.11. Find all the primes p such that $p + 1$ is a cube.

Exercise 3.5.12. Find all prime numbers x and y such that $x - y$ and $x + y$ are also prime.

Exercise 3.5.13. Show that Sylvester's theorem implies Bertrand's postulate. (Hint: let $k = n$, and consider $(n+1)\cdots(n+k)$. Then, use Theorem 3.3.3.)

Exercise 3.5.14. Use Nagura's theorem (Theorem 3.3.5) to show that if $n \geq 6$, then there are two distinct prime numbers between n and $2n$. (Hint: verify this by hand for $6 \leq n \leq 24$ and then use Nagura's theorem for $n \geq 25$.)

Exercise 3.5.15. Use Exercise 3.5.14 to show that if p_n denotes the nth prime number, then $p_{n+2} < p_n + p_{n+1}$, for all $n \geq 1$.

Exercise 3.5.16. Are there infinitely many primes p such that $(p, p+2, p+4)$ are all primes? Why? Are there infinitely many primes p such that $(p, p+2, p+6, p+8, p+12, p+14)$ are all primes? Why? Make a generalization of the twin prime conjecture for 6-tuples; i.e., make an educated conjecture for the existence of 6-tuples of primes. (In other words, make a concrete conjecture for the existence of infinitely many primes in a constellation of size six.)

Exercise 3.5.17. A prime p is called a *Sophie Germain prime* if $q = 2p+1$ is also a prime number. Find the first five Sophie Germain primes.

(Note: prime numbers of this type were introduced by Sophie Germain (see Figure 3.8) in order to prove certain cases of Fermat's last theorem. Moreover, these primes also have cryptographic applications. See also Exercises 4.7.38 and 10.8.32.)

Exercise 3.5.18. Let $\pi_G(x)$ be the number of Sophie Germain primes $p \leq x$ (defined as in Exercise 3.5.17). Use the Bateman–Horn conjecture, Conjecture 3.4.10, to formulate a precise asymptotic formula for $\pi_G(x)$. (Hint: see Examples 3.4.9 and 3.4.12.)

Exercise 3.5.19. Two primes p and q are called *sexy*, or *sexy primes*, if they differ by six (e.g., 5 and 11).

(1) Find the first five pairs of sexy primes.

(2) If p, q, r, and s are primes such that the pairs (p, q), (q, r), and (r, s) are sexy primes, we say that (p, q, r, s) is a sexy prime quadruple. If $p > 5$ and (p, q, r, s) is a sexy prime quadruple, show that the remainder of dividing p by 10 is 1; i.e., p is of the form $1 + 10k$ for some $k \geq 1$.

(3) Find four sexy prime quadruples.

Exercise 3.5.20. Show that there cannot be an infinite sequence of prime numbers in an arithmetic progression; i.e., if $\{q_1, q_2, \ldots\}$ is a sequence of integers in arithmetic progression, then there is some $k \geq 0$ such that q_k is not a prime number.

Exercise 3.5.21. Suppose that $\{q_1, \ldots, q_k\}$ are k primes in arithmetic progression. Show that $q_1 \geq k$.

Figure 3.8. Marie-Sophie Germain (1776–1831) was a French mathematician, physicist, and philosopher. Her brilliant approach to Fermat's last theorem inspired other mathematicians' work for hundreds of years. Image source: Wikimedia Commons.

Exercise 3.5.22. Show that if $a > 0$ and $a^n + 1$ is prime for some $n > 1$, then a is even and n is a power of 2. (Hint: if k is odd, then we can find a factorization of $x^k + 1$.)

Exercise 3.5.23. Show that if $a > 0$ and $a^n - 1$ is prime for some $n > 1$, then $a = 2$ and n is prime. (Hint: find a factorization of $x^k - 1$, for any $k \geq 2$.)

Exercise 3.5.24. Goldbach's conjecture says that every even integer $n \geq 4$ can be expressed as the sum of two primes. Show that there exist infinitely many odd integers that cannot be expressed as the sum of two prime numbers (for instance, $n = 3$).

CHAPTER 4

CONGRUENCES

If the Sun refused to shine,
I don't mind, I don't mind.
If the mountains fell in the sea,
Let it be, it ain't me.
Now, if six turned out to be nine,
Oh I don't mind, I don't mind...

Jimi Hendrix, *If Six Was Nine*, from the album *Axis: Bold as Love*, 1967

Before we give a formal definition of congruence, we will give two examples, one from ordinary life and another one that we have already encountered during our introduction in Chapter 1.

Example 4.0.1. Let us look at a clock. What will be the time after 27 hours? Since it is difficult for the author to predict the time at the reader's end, let us assume that it is 2 pm. After 27 hours, one day and 3 hours will have passed and, therefore, it will be 5 pm.

If, instead, 327 hours go by, what will be the time then? In order to solve this problem, first we need to know how many complete days (24 hours) will pass. Thus, we need to divide 327 by 24, using long division. Since $327 = 24 \cdot 13 + 15$ and the current time is 2 pm, it follows that, after 327 hours, 13 days and 15 hours will have passed. Thus, the time will be 2 pm plus 15 additional hours, and that brings us to 5 am. This result could have also been obtained by finding the remainder of dividing $341 = 327 + 14$ (here 14 represents 2 pm) by 24:

$$341 = 24 \cdot 14 + 5.$$

The remainder 5 tells us that the answer is the 5th hour of the day, i.e., 5 am.

In general, suppose that it is the mth hour of the day (where we are using a 24-hour clock; that is, $m = 0, 1, 2, \ldots, 23$), N hours go by and we want to find out the time. In order to do so, we need to find the remainder of division of $N + m$

by 24; i.e.,

$$N + m = 24q + r,$$

and r, with $0 \le r < 24$, is the new time (in a 24-hour clock).

Example 4.0.2. Let us show that the hyperbola $C : x^2 - 5y^2 = 2$ does not have any integral points. Suppose, for a contradiction, that $C(\mathbb{Z})$ is not empty and $(m, n) \in C(Z)$. In other words, there are integers $m, n \in \mathbb{Z}$ such that $m^2 - 5n^2 = 2$. In particular, $m^2 = 2 + 5n^2$ and the remainder when dividing m^2 by 5 must be 2. This is impossible. Indeed, by the division theorem (Theorem 2.4.4), we know that $m = 5q + r$, for some unique $q, r \in \mathbb{Z}$ such that $0 \le r < 5$. Thus, we can calculate r', the remainder of m^2 modulo 5, according to the value of r.

- If $r = 0$, then $m = 5q$ and $m^2 = 25q^2 = 5(q^2)$, so $r' = 0$.
- If $r = 1$, then $m = 5q + 1$ and $m^2 = 5(5q^2 + 2q) + 1$, so $r' = 1$.
- If $r = 2$, then $m = 5q + 2$ and $m^2 = 5(5q^2 + 4q) + 4$, so $r' = 4$.
- If $r = 3$, then $m = 5q + 3$ and $m^2 = 5(5q^2 + 6q + 1) + 4$, so $r' = 4$.
- If $r = 4$, then $m = 5q + 4$ and $m^2 = 5(5q^2 + 8q + 3) + 1$, so $r' = 1$.

Hence, the remainder of m^2 when divided by 5 is 0, 1, or 4, but never 2. It follows that $m^2 = 2 + 5n^2$ is impossible and $C(\mathbb{Z})$ must be empty.

In both of the previous examples, we were more interested in the remainder of a number modulo some other number, rather than in the number itself. Congruences will help us work with remainders in a much easier manner, and they will allow us to solve problems that, without congruences, would be very difficult to solve.

Example 4.0.3. Is the number $N = 4^{3001} - 1$ divisible by 5? If not, what is the remainder of N when divided by 5? We will return to this problem after we have defined congruences and shown a number of properties. (Can't wait that long? See Example 4.2.4.)

4.1. The Definition of Congruence

> *The invention of the symbol $\equiv$ by Gauss affords a striking example of the advantage which may be derived from an appropriate notation, and marks an epoch in the development of the science of arithmetic.*
>
> G. B. Matthews, in *Theory of Numbers*, 1892

> *Numerorum congruentiam hoc signo, $\equiv$, in posterum denotabimus, modulum ubi opus erit in clausulis adiugentes, $-16 \equiv 9(\text{mod} 5)$, $-7 \equiv 15(\text{mod} 11)$.*
>
> C. F. Gauss, in *Disquisitiones Arithmeticae*, 1801

Definition 4.1.1. Let $m > 1$ be fixed. We say that two integers a and b are *congruent modulo* m, and we write $a \equiv b \bmod m$ if m divides $a - b$. If we fix an integer a and a modulus m, the set of integers congruent to a modulo m is called the *congruence class* of a mod m.

Example 4.1.2. The congruence $8 \equiv 3 \bmod 5$ holds because $8 - 3 = 5$. Similarly, $23 \equiv 3 \bmod 5$ because $23 - 3 = 20 = 5 \cdot 4$.

The number -3 is congruent to 18 mod 7. Indeed, $-3 - 18 = -21 = 7 \cdot (-3)$ and it follows that $-3 \equiv 18 \bmod 7$. On the contrary, $55 \not\equiv 31 \bmod 7$, because $55 - 31 = 24$ is not a multiple of 7.

Remark 4.1.3. If a is equal to b plus a multiple of m, then $a \equiv b \bmod m$. Indeed, if $a = b + km$, for some $k \in \mathbb{Z}$, then $a - b = km$ and, therefore, m divides $a - b$; i.e., $a \equiv b \bmod m$. Hence, if we fix 3 mod 7, then

$$3 + 7 = 10,\ 3 + 2 \cdot 7 = 17,\ 3 + 3 \cdot 7 = 24,\ 3 + 4 \cdot 7 = 31, \ldots$$

are all numbers congruent to 3 modulo 7. Similarly,

$$3 - 7 = -4,\ 3 - 2 \cdot 7 = -11,\ 3 - 3 \cdot 7 = -18,\ 3 - 4 \cdot 7 = -25, \ldots$$

are also numbers congruent to 3 mod 7. The congruence class of 3 mod 7 is the (infinite) set of integers

$$\{\ldots, -25, -18, -11, -4, 3, 10, 17, 24, 31, \ldots\}.$$

Proposition 4.1.4. *Let $m > 1$ be fixed. Every integer is congruent modulo m to exactly one of the numbers in the set $C_m = \{0, 1, 2, \ldots, m-1\}$. In other words, for every N there is a number r with $0 \leq r \leq m - 1$ such that $N \equiv r \bmod m$.*

Proof. Let $m > 1$ be fixed. By the division theorem (Theorem 2.4.4), there are unique $q, r \in \mathbb{Z}$, with $0 \leq r < m$, such that $N = qm + r$. Therefore, $N - r = qm$, and this implies that $N \equiv r \bmod m$. Since $0 \leq r < m$, it follows that $r \in C_m$, and the division theorem guarantees that r is unique with this property. □

Definition 4.1.5. Let $m > 1$ and $N \in \mathbb{Z}$ be fixed. The unique number r in the set $C_m = \{0, 1, \ldots, m - 1\}$ such that $N \equiv r \bmod m$ is called the *least non-negative residue* of N modulo m. We say that r is the *reduction, or residue*, of N modulo m.

Corollary 4.1.6. *Let $m > 1$ be fixed. Then, for any $N \in \mathbb{Z}$, the remainder of division of N by m is precisely the least non-negative residue of N modulo m. In other words, if q and r are the unique integers with $N = qm + r$ and $0 \leq r \leq m - 1$, then r is the least non-negative residue of N mod m.*

Proof. As we have seen in the proof of Proposition 4.1.4, the residue r satisfies $N \equiv r \bmod m$ and $r \in \{0, 1, \ldots, m - 1\}$. Thus, r is the least non-negative residue of N modulo m. □

Example 4.1.7. Suppose that it is 2 pm and 327 hours go by. What is the time? We can deduce the time of the day by finding the least non-negative residue of $14 + 327 = 341$ modulo 24. Using long division, we find that $341 = 24 \cdot 14 + 5$, and therefore $341 \equiv 5 \bmod 24$. Hence, the time will be 5 am.

Definition 4.1.8. Let $m > 1$ be fixed. A *complete residue system* modulo m is a set $S = \{s_1, s_2, \ldots, s_m\}$ with m integers $s_i \in \mathbb{Z}$, for $1 \leq i \leq m$, such that each integer N is congruent to exactly one element of S. In other words, for every $N \in \mathbb{Z}$ there is an $s_i \in S$ such that $N \equiv s_i \bmod m$ and $N \not\equiv s_j \bmod m$ if $i \neq j$.

Example 4.1.9. By Proposition 4.1.4, the set $C_m = \{0, 1, 2, \ldots, m-1\}$ is a complete residue system modulo m. Indeed, if N is an integer and r is the least non-negative residue of N mod m, then $r \in C_m$; also $N \equiv r \bmod m$. Moreover, if $1 \leq s \leq m-1$ and $s \neq r$, then $N \not\equiv s \bmod m$.

For instance, $\{0, 1, 2, 3, 4\}$ is a complete residue system modulo 5. If N is an arbitrary integer, say $N = 1138$, then $N \equiv 3 \bmod 5$, and $N \not\equiv 0, 1, 2, 4 \bmod 5$.

Example 4.1.10. The set $S = \{0, 2, 4, 6, 8\}$ is also a complete residue system modulo 5. Indeed, $6 \equiv 1 \bmod 5$ and $8 \equiv 3 \bmod 5$, so each possible residue modulo 5 is represented uniquely in the set S. Similarly, $S' = \{1, 3, 5, 7, 9\}$ is a complete residue system modulo 5. Notice that S is formed by even integers, while S' is formed by odd integers, but both sets contain one unique representative for each residue modulo 5. One can also write a complete residue system formed only by primes; for instance, $S'' = \{2, 3, 5, 11, 19\}$ is such a system. However, there is no complete residue system modulo 6 formed by 0 and prime numbers. (Why? See Exercises 4.7.3 and 4.7.4.)

4.2. Basic Properties of Congruences

In this section we present several basic (but fundamental) properties of congruences.

Proposition 4.2.1. *For all integers a, b, c, a', b', k and for all $m > 1$, the following properties hold.*

(i) *If $a \equiv b \bmod m$, then $ka \equiv kb \bmod m$.*
(ii) *If $a \equiv b \bmod m$ and $b \equiv c \bmod m$, then $a \equiv c \bmod m$.*
(iii) *If $a \equiv b \bmod m$, then $a^j \equiv b^j \bmod m$, for all $j \geq 1$.*
(iv) *If $a \equiv b \bmod m$ and $a' \equiv b' \bmod m$, then*
  (a) *$a + a' \equiv b + b' \bmod m$, and*
  (b) *$a \cdot a' \equiv b \cdot b' \bmod m$.*

Before we dive into proving these properties, let us illustrate their usefulness with some examples.

Example 4.2.2. Suppose we want to reduce $70001 + 3504$ modulo 7. In other words, we want to calculate the least non-negative residue of 73505 mod 7. By Proposition 4.2.1, in particular part (iv)(a), it suffices to reduce 70001 and 3504 separately modulo 7, then add their least non-negative residues, and reduce again. We have

$$70001 \equiv 1 \bmod 7 \quad \text{and} \quad 3504 \equiv 4 \bmod 7,$$

because 70000 and 3500 are multiples of 7. Thus,

$$73505 = 70001 + 3504 \equiv 1 + 4 \equiv 5 \bmod 7.$$

The properties of congruences prove to be even more useful if we want to calculate $70001 \cdot 3504$ mod 7. Indeed, by Proposition 4.2.1 (iv)(b), it suffices to reduce each factor separately and then multiply the least non-negative residues together. Thus,

$$245283504 = 70001 \cdot 3504 \equiv 1 \cdot 4 \equiv 4 \bmod 7.$$

Example 4.2.3. Let us find the least non-negative residue of $10!$ mod 11. From the definition of congruence, we would first calculate $10! = 3628800$ and then use the division theorem to write

$$3628800 = 329890 \cdot 11 + 10.$$

Thus $10! \equiv 10 \bmod 11$. Let us do this calculation again, using congruences and their properties. We will use Proposition 4.2.1 (iv)(b) repeatedly, but we will also use the fact that $a \equiv a - m \bmod m$. We remark that there are *many* ways to perform this calculation, and this is only one of them:

$$\begin{aligned} 10! &= 10 \cdot 9 \cdot 8 \cdot 7 \cdot 6 \cdot 5 \cdot 4 \cdot 3 \cdot 2 \cdot 1 \\ &\equiv (-1) \cdot (-2) \cdot (-3) \cdot (-4) \cdot (-5) \cdot 5 \cdot 4 \cdot 3 \cdot 2 \cdot 1 \bmod 11 \\ &\equiv -6 \cdot 20 \cdot 20 \cdot 6 \bmod 11 \\ &\equiv -6 \cdot 9 \cdot 9 \cdot 6 \bmod 11 \\ &\equiv -6 \cdot (-2) \cdot (-2) \cdot 6 \bmod 11 \\ &\equiv 12 \cdot (-12) \bmod 11 \\ &\equiv 1 \cdot (-1) \bmod 11 \\ &\equiv -1 \bmod 11 \\ &\equiv 10 \bmod 11. \end{aligned}$$

Example 4.2.4. Is the number $N = 4^{3001} - 1$ divisible by 3? Is it divisible by 5? The number N has more than 1800 digits, so a direct computation is tricky. However, property (iii) of congruences in Proposition 4.2.1 tells us that, since $4 \equiv 1 \bmod 3$, then

$$N = 4^{3001} - 1 \equiv 1^{3001} - 1 \equiv 1 - 1 \equiv 0 \bmod 3.$$

Thus, N is a multiple of 3. Similarly, $4 \equiv -1 \bmod 5$ and so

$$N = 4^{3001} - 1 \equiv (-1)^{3001} - 1 \equiv (-1) - 1 \equiv -2 \equiv 3 \bmod 5,$$

and it follows that the residue of N modulo 5 is 3. Thus, the number N is not divisible by 5.

Let us now tackle the proof of the proposition.

Proof of Proposition 4.2.1. (i) This is left for the reader as Exercise 4.7.6.

(ii) If $a \equiv b \bmod m$ and $b \equiv c \bmod m$, then $a - b$ and $b - c$ are divisible by m; i.e., there are h and k such that $a - b = hm$ and $b - c = km$. Thus, $a - c = a - b + b - c = (h + k)m$ and, therefore, $a \equiv c \bmod m$.

(iv) (a) This is left for the reader as Exercise 4.7.6.

(b) Suppose that $a \equiv b \bmod m$ and $a' \equiv b' \bmod m$; then there are $h, k \in \mathbb{Z}$ such that $a = b + hm$ and $a' = b' + km$. Thus,

$$\begin{aligned} a \cdot a' &= (b + hm)(b' + km) \\ &= bb' + bkm + b'hm + hkm^2 \\ &= bb' + m(bk + b'h + hkm). \end{aligned}$$

This implies that $a \cdot a' \equiv b \cdot b' \bmod m$.

(iii) We will show this by induction on $j \geq 1$. Suppose that $a \equiv b \bmod m$. The base case $j = 1$ is clear. Let us now assume that $a^j \equiv b^j \bmod m$. Thus,

$$a^{j+1} \equiv a^j \cdot a \equiv b^j \cdot b \equiv b^{j+1} \bmod m,$$

as desired, where we have used property (iv)(b) and the induction hypothesis. Hence, by the principle of mathematical induction, $a^j \equiv b^j \bmod m$ for all $j \geq 1$. □

Remark 4.2.5. A congruence is an *equivalence relation*; that is, the congruence relation ($\equiv$) satisfies three properties: reflexivity, symmetry, and transitivity. More concretely, for all $a, b \in \mathbb{Z}$ and $m > 1$:

(1) $\equiv$ is *reflexive*: $a \equiv a \bmod m$, for all $a \in \mathbb{Z}$ and $m > 1$. Indeed, $a - a = 0$ is always divisible by $m > 1$.

(2) $\equiv$ is *symmetric*: $a \equiv b \bmod m$ if and only if $b \equiv a \bmod m$, because $a - b$ is divisible by m if and only if $b - a$ is divisible by m.

(3) $\equiv$ is *transitive*: if $a \equiv b \bmod m$ and $b \equiv c \bmod m$, then $a \equiv c \bmod m$. This is Proposition 4.2.1, part (ii).

This will become useful later when we define congruence classes as the equivalence classes of $\mathbb{Z}$ with respect to the congruence relation.

In the following two examples we illustrate how useful and efficient congruences can be to calculate the remainders of high powers of a number. In both examples we make repeated use of Proposition 4.2.1, particularly part (iii).

Example 4.2.6. What is the remainder of 3^{253} when divided by 7? In other words, what is the least non-negative residue of 3^{253} mod 7? Let us first calculate some small powers of 3 modulo 7:

$$3^2 \equiv 9 \equiv 2 \bmod 7,$$
$$3^3 \equiv 3^2 \cdot 3 \equiv 2 \cdot 3 \equiv 6 \equiv -1 \bmod 7.$$

It follows that $3^6 \equiv (3^3)^2 \equiv (-1)^2 \equiv 1 \bmod 7$. Since the 6th power of 3 is congruent to 1 mod 7, we calculate the long division $253 = 6 \cdot 42 + 1$ and, therefore,

$$3^{253} \equiv 3^{6 \cdot 42+1} \equiv (3^6)^{42} \cdot 3 \equiv 1^{42} \cdot 3 \equiv 3 \bmod 7.$$

Hence, the remainder of 3^{253} when divided by 7 is 3.

Example 4.2.7. What is the least non-negative residue of 5^{22} mod 11? We begin calculating small powers of 5 mod 11 and square consecutively until we have enough powers of 5 to build 5^{22} mod 11, using the fact that $22 = 16 + 4 + 2$:

$$5^2 \equiv 25 \equiv 3 \bmod 11,$$
$$5^4 \equiv (5^2)^2 \equiv 3^2 \equiv 9 \equiv -2 \bmod 11,$$
$$5^8 \equiv (5^4)^2 \equiv (-2)^2 \equiv 4 \bmod 11,$$
$$5^{16} \equiv (5^8)^2 \equiv 4^2 \equiv 16 \equiv 5 \bmod 11.$$

Hence,

$$5^{22} \equiv 5^{16} \cdot 5^4 \cdot 5^2 \equiv 5 \cdot (-2) \cdot 3 \equiv -10 \cdot 3 \equiv 1 \cdot 3 \equiv 3 \bmod 11.$$

Thus, the least non-negative residue of 5^{22} mod 11 is 3.

We finish this section with an additional example that illustrates how congruences are very useful for arguments about remainders of integers.

Example 4.2.8. In Example 4.0.2, we demonstrated that the conic $C : x^2 - 5y^2 = 2$ does not have any integral points. Let us repeat the same proof, but this time we will use the language of congruences.

Suppose, for a contradiction, that $C(\mathbb{Z})$ is not empty and $(m, n) \in C(\mathbb{Z})$. In other words, there are integers $m, n \in \mathbb{Z}$ such that $m^2 - 5n^2 = 2$. In particular, $m^2 = 2 + 5n^2$ and $m^2 \equiv 2 \bmod 5$. This is impossible. Indeed, by Proposition 4.1.4, we know that $\{0, 1, 2, 3, 4\}$ is a complete residue system modulo 5. Thus, $m \equiv 0, 1, 2, 3,$ or $4 \bmod 5$. Thus, we can calculate $m^2 \bmod 5$ according to the value of $m \bmod 5$.

- If $m \equiv 0 \bmod 5$, then $m^2 \equiv 0^2 \equiv 0 \bmod 5$.
- If $m \equiv 1 \bmod 5$, then $m^2 \equiv 1^2 \equiv 1 \bmod 5$.
- If $m \equiv 2 \bmod 5$, then $m^2 \equiv 2^2 \equiv 4 \bmod 5$.
- If $m \equiv 3 \bmod 5$, then $m^2 \equiv 3^2 \equiv 9 \equiv 4 \bmod 5$.
- If $m \equiv 4 \bmod 5$, then $m^2 \equiv 4^2 \equiv (-1)^2 \equiv 1 \bmod 5$.

Hence, the only possible values for $m^2 \bmod 5$ are $0, 1,$ or $4 \bmod 5$, but never 2 or $3 \bmod 5$. It follows that $m^2 = 2 + 5n^2$ is impossible and $C(\mathbb{Z})$ must be empty.

4.3. Cancellation Properties of Congruences

When we have an equality of integers $ab = ac$, for some $a, b, c \in \mathbb{Z}$, such that $a \neq 0$, then it follows that $b = c$. Indeed, the equality $ab = ac$ implies $a(b - c) = 0$, and Theorem 2.2.3 implies that $a = 0$ or $b - c = 0$. Since we assume that $a \neq 0$, it follows that $b - c = 0$, or, equivalently, $b = c$. However, a congruence may not satisfy this cancellation property. For example,

$$3 \cdot 5 \equiv 3 \cdot 7 \bmod 6$$

but $5 \not\equiv 7 \bmod 6$. The problem is that the number being cancelled (3) is not relatively prime with the modulus of the congruence (6). Here is another example where cancellation works as in the case of equalities of integers:

$$3 \cdot 5 \equiv 3 \cdot 12 \bmod 7$$

and, indeed, $5 \equiv 12 \bmod 7$ as well. This time cancellation worked because $\gcd(3, 7) = 1$. Let us see a result that explains when cancellation in congruences works as we would hope.

Proposition 4.3.1. *Let $a, b, k \in \mathbb{Z}$ and $m, n > 1$.*

(1) *If $a \equiv b \bmod m$ and $d > 1$ is a divisor of m, then $a \equiv b \bmod d$.*

(2) *If $ka \equiv kb \bmod m$, then $a \equiv b \bmod \frac{m}{\gcd(k,m)}$. In particular:*

  (a) *If $ka \equiv kb \bmod kn$, then $a \equiv b \bmod n$.*

  (b) *If $ka \equiv kb \bmod m$ and $\gcd(k, m) = 1$, then $a \equiv b \bmod m$.*

Proof. (1) Suppose that $a \equiv b \bmod m$ and $d > 1$ is a divisor of m; then $m = dm'$ for some $m' \in \mathbb{Z}$ and $b - a = hm = hdm'$ for some $h \in \mathbb{Z}$. Hence, d divides $b - a$ and $a \equiv b \bmod d$.

(2) If $ka \equiv kb \bmod m$, then there is some $h \in \mathbb{Z}$ such that $kb - ka = k(b-a) = hm$. Suppose that $\gcd(k, m) = g$. Then, $k = gk'$ and $m = gm'$ and $k(b-a) = hm$ implies $k'(b-a) = hm'$. Hence, m' divides $k'(b-a)$. Since $\gcd(\frac{k}{g}, \frac{m}{g}) = \gcd(k', m') = 1$ (by Exercise 2.11.25), it follows that m' divides $b-a$ by Corollary 2.7.6. Thus, we have shown that $a \equiv b \bmod m'$, where $m' = \frac{m}{g} = \frac{m}{\gcd(k,m)}$.

(a) If $ka \equiv kb \bmod kn$, then $a \equiv b \bmod n$. This is a direct consequence of part (2): put $m = kn$, and then $\gcd(k, kn) = k$.

(b) If $ka \equiv kb \bmod m$ and $\gcd(k, m) = 1$, then $a \equiv b \bmod m$. This is a direct consequence of part (2). □

Example 4.3.2. The congruence $29 \equiv 44 \bmod 15$ implies that $29 \equiv 44 \bmod 3$ and $29 \equiv 44 \bmod 5$, by Proposition 4.3.1, part (1), because 3 and 5 are divisors of 15. Indeed, $29 \equiv 2 \equiv 44 \bmod 3$ and $29 \equiv 4 \equiv 44 \bmod 5$.

Example 4.3.3. The congruence $3 \cdot 5 \equiv 3 \cdot 7 \bmod 6$ holds true, but $5 \not\equiv 7 \bmod 6$. Proposition 4.3.1 says that, instead, we can cancel the 3 on both sides if we consider congruences modulo $\frac{6}{\gcd(3,6)} = \frac{6}{3} = 2$. Indeed, $3 \cdot 5 \equiv 3 \cdot 7 \bmod 6$ implies that $5 \equiv 7 \bmod 2$.

Similarly, the congruence $5 \cdot 2 \equiv 5 \cdot 9 \bmod 35$ implies that $2 \equiv 9 \bmod 7$, because $7 = \frac{35}{\gcd(5,35)} = \frac{35}{5}$.

Example 4.3.4. If $ka \equiv kb \bmod p$, where p is a prime that does not divide k, then $a \equiv b \bmod p$, as a consequence of Proposition 4.3.1, part (2)(b), because $\gcd(p, k) = 1$. For instance, $33 \equiv 11 \cdot 3 \equiv 11 \cdot 16 \equiv 176 \bmod 13$, and since $\gcd(11, 13) = 1$, it follows that we can cancel 11 on both sides of the congruence to obtain $3 \equiv 16 \bmod 13$ as well, without needing to change the modulus.

4.4. Linear Congruences

The goal of this section is to investigate whether a linear congruence of the form

$$ax \equiv b \bmod m$$

has solutions $x \in \mathbb{Z}$ and, if it is solvable, find all such integral solutions. We begin with some examples that illustrate the possible scenarios.

Example 4.4.1. Consider the congruence

$$3x \equiv 5 \bmod 7.$$

The number $x = 4$ is a solution, because $3 \cdot 4 = 12 \equiv 5 \bmod 7$. Since the congruence is modulo 4, if we replace $x = 4$ by any other number x' congruent to $4 \bmod 7$, then x' should also be a solution. For instance, $x' = 11$ is also a solution, for $3 \cdot 11 = 33 \equiv 5 \bmod 7$. Similarly, the numbers

$$\ldots,\ -10,\ -3,\ 4,\ 11,\ 18, \ldots$$

are all solutions. In other words, if x is any integer such that $x \equiv 4 \bmod 7$, then x must be a solution for $3x \equiv 5 \bmod 7$.

Is there any other solution $x \in \mathbb{Z}$ for $3x \equiv 5 \bmod 7$ such that $x \not\equiv 4 \bmod 7$? It turns out that there are no other solutions. We know that $\{0, 1, 2, 3, 4, 5, 6\}$ is a complete residue system modulo 7. Thus, any $x \in \mathbb{Z}$ is congruent to $0, 1, 2, 3, 4, 5$, or 6 mod 7. Now we can make a table of values of $3x$ mod 7, for each possible value of x mod 7:

x	0	1	2	3	4	5	6
$3x$	0	3	6	2	5	1	4

Hence, if $x \in \mathbb{Z}$ and $3x \equiv 5 \bmod 7$, then it follows that $x \equiv 4 \bmod 7$ necessarily.

Example 4.4.2. In stark contrast with the previous example, the congruence

$$3x \equiv 5 \bmod 6$$

has no solutions with $x \in \mathbb{Z}$. As before, the set $\{0, 1, 2, 3, 4, 5\}$ is a complete residue system modulo 6. Thus, any $x \in \mathbb{Z}$ is congruent to $0, 1, 2, 3, 4$, or 5 mod 6. Now we can make a table of values of $3x$ mod 6, for each possible value of x mod 6:

x	0	1	2	3	4	5
$3x$	0	3	0	3	0	3

It follows that $3x \equiv 0$ or $3 \bmod 6$, but never $\equiv 1, 2, 4$, or $5 \bmod 6$. The same table implies that the congruence

$$3x \equiv 3 \bmod 6$$

has solutions $x \equiv 1$, $x \equiv 3$, and $x \equiv 5 \bmod 6$. In other words, the numbers

$$\ldots, -3, -1, 1, 3, 5, 7, 9, 11, 13, \ldots$$

are all solutions for the congruence $3x \equiv 3 \bmod 6$. Notice that these are all the odd numbers. Indeed, by Proposition 4.3.1, the congruence $3x \equiv 3 \bmod 6$ implies the simpler congruence $x \equiv 1 \bmod 2$ (i.e., x is odd).

When we encounter a linear congruence, say $119x \equiv 14 \bmod 203$, we can always determine whether there is a solution by *brute force*; i.e., calculate $119x$ for each x in the complete residue system $\{0, 1, 2, 3, \ldots, 202\}$ modulo 203. Obviously, this method can be very long and tedious. Alas, the readers need not worry, as we already have all the tools to solve a linear congruence efficiently, namely Euclid's algorithm and Bezout's identity.

Theorem 4.4.3. *Let $a, b \in \mathbb{Z}$, with $a \neq 0$, and let $m > 1$. Then, the following statements are equivalent:*

1. *The linear congruence $ax \equiv b \bmod m$ has a solution $x_0 \in \mathbb{Z}$.*
2. *There is a $y_0 \in \mathbb{Z}$ such that (x_0, y_0) is an integral point on the line $L : ax + my = b$.*
3. *The number b is divisible by $\gcd(a, m)$.*

Moreover, if $ax \equiv b \bmod m$ has a solution, then it has exactly $d = \gcd(a, m)$ distinct solutions modulo m, given by $x \equiv x_0 + (m/d) \cdot k \bmod m$, for $k = 0, 1, \ldots, d - 1$, where x_0 is any solution of $(a/d)x \equiv b/d \bmod (m/d)$.

Proof. We need to show that (1), (2), and (3) are equivalent statements. In order to do so, we will show that (1) implies (2), that (2) implies (3), and that (3) implies (1).

Assume first that x_0 is a solution for the congruence $ax \equiv b \bmod m$; that is, we have that $ax_0 \equiv b \bmod m$. By definition, this is equivalent to $b - ax_0$ being divisible by m. In other words, there is some $y_0 \in \mathbb{Z}$ such that $b - ax_0 = my_0$, or $ax_0 + my_0 = b$. Thus, the line $L : ax + my = b$ has an integral point. This shows (2).

Now suppose that (2) is true; that is, the line $L : ax + my = b$ has an integral point. By Proposition 2.9.1, the line $L : ax+my = b$ has integral points if and only if $\gcd(a, m)$ divides b. Thus, (3) is true. (Notice that Proposition 2.9.1 actually says that (2) and (3) are equivalent.)

Finally, suppose that (3) holds and b is a multiple of $\gcd(a, m)$. Then, by Proposition 2.9.1, the line $L : ax + my = b$ has an integral point (x_0, y_0); i.e., $ax_0 + my_0 = b$. Therefore, $b - ax_0 = my_0$ and, by definition, $ax_0 \equiv b \bmod m$.

We have shown that (1), (2), and (3) are equivalent. If $ax \equiv b \bmod m$ has a solution, then, by (3), b is divisible by $d = \gcd(a, m)$. Consider the equation $a'x \equiv b' \bmod m'$, where $a' = a/d$, $b' = b/d$, and $m' = m/d$. By Exercise 2.11.25, $\gcd(a', m') = 1$, and by Theorem 2.9.4, all the solutions to $a'x + m'y = b'$ are given by

$$x_k = x_0 + m'k, \quad y_k = y_0 - a'k$$

for any $k \in \mathbb{Z}$, where (x_0, y_0) is one fixed solution. In particular, the solutions of $a'x \equiv b' \bmod m'$ are all of the form $x_k = x_0 + m'k$, for some $k \in \mathbb{Z}$. If $x_k \equiv x_j \bmod m$, then

$$x_0 + m'k \equiv x_0 + m'j \bmod m$$

and, therefore, $m'k \equiv m'j \bmod m$, and by Proposition 4.3.1, $k \equiv j \bmod d$. Hence, the solutions to $ax \equiv b \bmod m$ that are distinct modulo m are

$$x_k \equiv x_0 + m'k \bmod m$$

for $k = 0, 1, \ldots, d - 1$, as claimed. □

Corollary 4.4.4. *Let $m \geq 2$, and let $\gcd(a, m) = 1$. Then, the congruence $ax \equiv b \bmod m$ has a unique solution modulo m.*

Proof. Since $d = \gcd(a, m) = 1$, any number b is divisible by d. By Theorem 4.4.3, the congruence $ax \equiv b \bmod m$ has a solution, and, in fact, it has only $d = 1$ different solutions modulo m. □

The proof of Theorem 4.4.3 also outlines a method to solve any linear congruence; namely, use Euclid's algorithm and Bezout's identity.

Example 4.4.5. Let us find all solutions to the congruence $11x \equiv 10 \bmod 35$. First, we find the greatest common divisor of 35 and 11, using Euclid's algorithm:

$$35 = 11 \cdot 3 + 2,$$
$$11 = 2 \cdot 5 + 1,$$
$$2 = 1 \cdot 2 + 0.$$

Therefore, $\gcd(35, 11) = 1$. Now, we reverse Euclid's algorithm to find a solution to Bezout's identity $11x + 35y = \gcd(35, 11) = 1$:

$$\begin{aligned} 1 &= 11 - 2 \cdot 5 \\ &= 11 - (35 - 11 \cdot 3) \cdot 5 = 16 \cdot 11 - 5 \cdot 35. \end{aligned}$$

Thus, we have found $1 = 16 \cdot 11 + (-5) \cdot 35$, and if we multiply through by 10, we find

$$10 = 160 \cdot 11 + (-50) \cdot 35.$$

Moreover, by Theorem 2.9.4, all the points in the line $10 = 11x + 35y$ are given by

$$x = 160 + \frac{35}{\gcd(35, 11)} k = 160 + 35k, \quad y = -50 - \frac{11}{\gcd(35, 11)} k = -50 - 11k,$$

for all $k \in \mathbb{Z}$. It follows that all the solutions to $11x \equiv 10 \bmod 35$ are given by $x = 160 + 35k$, or, equivalently,

$$x \equiv 160 \equiv 20 \bmod 35.$$

Therefore, $x \equiv 20 \bmod 35$.

Example 4.4.6. Let us find all solutions to the congruence $119x \equiv 14 \bmod 203$. First, we find the greatest common divisor of 203 and 119, using Euclid's algorithm:

$$\begin{aligned} 203 &= 119 \cdot 1 + 84, \\ 119 &= 84 \cdot 1 + 35, \\ 84 &= 35 \cdot 2 + 14, \\ 35 &= 14 \cdot 2 + 7, \\ 14 &= 7 \cdot 2 + 0. \end{aligned}$$

Therefore, $\gcd(203, 119) = 7$. Now, we reverse Euclid's algorithm to find a solution to Bezout's identity $119x + 203y = \gcd(203, 119) = 7$:

$$\begin{aligned} 7 &= 35 - 14 \cdot 2 \\ &= 35 - (84 - 35 \cdot 2) \cdot 2 = 5 \cdot 35 - 2 \cdot 84 \\ &= 5 \cdot (119 - 84) - 2 \cdot 84 = 5 \cdot 119 - 7 \cdot 84 \\ &= 5 \cdot 119 - 7 \cdot (203 - 119) = 12 \cdot 119 - 7 \cdot 203. \end{aligned}$$

Thus, we have found $7 = 12 \cdot 119 + (-7) \cdot 203$, and if we multiply through by 2, we find

$$14 = 24 \cdot 119 + (-14) \cdot 203.$$

Moreover, by Theorem 2.9.4, all the points in the line $14 = 119x + 203y$ are given by

$$x = 24 + \frac{203}{\gcd(203, 119)} k = 24 + 29k, \quad y = -14 - \frac{119}{\gcd(203, 119)} k = -14 - 17k,$$

for all $k \in \mathbb{Z}$. It follows that all the solutions to $119x \equiv 14 \bmod 203$ are given by $x = 24 + 29k$, or, equivalently, $x \equiv 24 \bmod 29$. Hence, the solutions modulo 203 are

$$x \equiv 24, 53, 82, 111, 140, 169, \text{ and } 198 \bmod 203.$$

4.5. Systems of Linear Congruences

In the previous section we have learned to solve a single linear congruence $ax \equiv b \bmod m$. What if we need to find the solutions of a *system* of linear congruences?

Example 4.5.1. Suppose we are trying to find a solution of the congruence

$$x^2 \equiv 29 \bmod 35.$$

Since $35 = 5 \cdot 7$, it follows that

$$(S_1) \quad \begin{cases} x^2 \equiv 29 \equiv 4 \bmod 5, \text{ and} \\ x^2 \equiv 29 \equiv 1 \bmod 7. \end{cases}$$

Each of these congruences has solutions that we can spot because the modulus is relatively small. Namely, the solutions are $x \equiv \pm 2 \bmod 5$ and $x \equiv \pm 1 \bmod 7$, respectively. Thus, if $x^2 \equiv 29 \bmod 35$, then

$$(S_2) \quad \begin{cases} x \equiv \pm 2 \bmod 5, \\ x \equiv \pm 1 \bmod 7. \end{cases}$$

In this section we will learn how to solve a system of congruences such as (S_2). In particular, if $x \equiv 2 \bmod 5$ and $x \equiv 1 \bmod 7$, we shall find that $x \equiv 22 \bmod 35$. Moreover, $x \equiv 22 \bmod 35$ is a solution of our initial quadratic congruence, because $22^2 = 484 \equiv 29 \bmod 35$. In fact, we will show (in Lemma 4.5.4) that $x^2 \equiv 29 \bmod 35$ if and only if x satisfies the system (S_1), which in turn is equivalent to (S_2).

This example shows that one can trade a congruence with a composite modulus with a system of congruences with smaller modulus (that are often more tractable).

Example 4.5.2. Find the smallest positive integer x that satisfies the following system of linear congruences:

$$(S) \quad \begin{cases} x \equiv 3 \bmod 7, \\ x \equiv 7 \bmod 3. \end{cases}$$

In this section, we will learn two methods to solve systems of congruences. The first method consists of solving each linear congruence separately and then finding common solutions. In this particular example, we begin by solving $x \equiv 3 \bmod 7$, which is equivalent to $x = 3 + 7k$, for some $k \in \mathbb{Z}$. Now, we also need $x \equiv 7 \bmod 3$, so $3 + 7k \equiv 7 \equiv 1 \bmod 3$, or, equivalently, $7k \equiv 1 - 3 \equiv 1 \bmod 3$. The latter implies that there is some $j \in \mathbb{Z}$ such that $1 - 7k = 3j$. In other words, (k, j) is an integral point in the line $1 = 7k + 3j$. By the theory we have developed (Proposition 2.9.1, Theorem 2.9.4), we can find all the integral points in the line $1 = 7k + 3j$. They are given by

$$k = 1 + 3t, \quad j = -2 - 7t, \quad \text{for all } t \in \mathbb{Z}.$$

In particular, $k = 1 + 3t$, and

$$x = 3 + 7k = 3 + 7(1 + 3t) = 10 + 21t,$$

or, equivalently, $x \equiv 10 \bmod 21$. The smallest positive integer that satisfies $x \equiv 10 \bmod 21$ is $x = 10$. Finally, we may check that $x = 10$ is indeed a solution to the system of equations: $10 \equiv 3 \bmod 7$ and $10 \equiv 7 \bmod 3$.

In our next example, we will solve the same system using an alternative method. This new method is much more efficient for systems with more than two equations, as we shall see in other examples to follow later in this section.

Example 4.5.3. We shall solve the system

$$(S)\ \begin{cases} x \equiv 3 \bmod 7, \\ x \equiv 1 \bmod 3 \end{cases}$$

by solving two simpler systems first. (Notice that we have already simplified the system (S) by replacing $x \equiv 7 \bmod 3$ with $x \equiv 1 \bmod 3$.) We will begin by solving the systems

$$(S_1)\ \begin{cases} x_1 \equiv 1 \bmod 7, \\ x_1 \equiv 0 \bmod 3 \end{cases} \quad \text{and} \quad (S_2)\ \begin{cases} x_2 \equiv 0 \bmod 7, \\ x_2 \equiv 1 \bmod 3. \end{cases}$$

(S_1) The congruence $x_1 \equiv 0 \bmod 3$ means that x_1 is a multiple of 3; i.e., $x_1 = 3m$ for some $m \in \mathbb{Z}$. In order to satisfy $x_1 \equiv 1 \bmod 7$ as well, we need $x_1 = 3m \equiv 1 \bmod 7$. Thus, we need to find an n such that $1 - 3m = 7n$; that is, $1 = 7n + 3m$. It follows that

$$n = 1 + 3s, \quad m = -2 - 7s, \quad \text{for all } s \in \mathbb{Z}.$$

Hence, $x_1 = 3m = 3(-2 - 7s) = -6 - 21s$, or, equivalently,

$$x_1 \equiv -6 \equiv 15 \bmod 21.$$

(S_2) In the system (S_2), the first congruence $x_2 \equiv 0 \bmod 7$ means $x_2 = 7u$ for some $u \in \mathbb{Z}$. In order to satisfy $x_2 \equiv 1 \bmod 3$ as well, we need $x_2 = 7u \equiv 1 \bmod 3$, and if we reduce $7 \equiv 1 \bmod 3$, we obtain $u \equiv 1 \bmod 3$. Thus, $u = 1 + 3v$ for any $v \in \mathbb{Z}$, and

$$x_2 = 7u = 7(1 + 3v) = 7 + 21v,$$

or, in other words, $x_2 \equiv 7 \bmod 21$.

Once we have solved systems (S_1) and (S_2), we claim that

$$x \equiv 3 \cdot x_1 + 1 \cdot x_2 \bmod 21$$

is the solutions for the original system (S). Indeed, if we reduce $x \bmod 7$, we obtain

$$x \equiv 3 \cdot x_1 + 1 \cdot x_2 \equiv 3 \cdot 1 + 1 \cdot 0 \equiv 3 \bmod 7$$

because x_1 and x_2 were constructed so that $x_1 \equiv 1$ and $x_2 \equiv 0 \bmod 7$. And if we reduce $x \bmod 3$, we obtain

$$x \equiv 3 \cdot x_1 + 1 \cdot x_2 \equiv 3 \cdot 0 + 1 \cdot 1 \equiv 1 \bmod 3$$

because x_1 and x_2 were constructed so that $x_1 \equiv 0$ and $x_2 \equiv 1 \bmod 3$. Thus, x satisfies (S). We plug in the values of x_1 and $x_2 \bmod 21$, and we obtain

$$x \equiv 3 \cdot 15 + 1 \cdot 7 \equiv 52 \equiv 10 \bmod 21.$$

Therefore, all numbers of the form $x \equiv 10 \bmod 21$ are solutions for (S). The Chinese remainder theorem will explain that these are, indeed, all the solutions for (S). Before we state and prove the theorem, let us see one useful lemma and then one additional example with a system of three linear congruences.

Lemma 4.5.4. *Let $a, b, m, n \in \mathbb{Z}$ such that $\gcd(m, n) = 1$. Then,*

$$\begin{cases} a \equiv b \bmod m, \\ a \equiv b \bmod n, \end{cases}$$

if and only if $a \equiv b \bmod mn$.

Proof. Suppose first that $a \equiv b \bmod mn$. Then, by Proposition 4.3.1, part (1), we also have $a \equiv b \bmod m$ and $a \equiv b \bmod n$.

Conversely, suppose that $a \equiv b \bmod m$ and $a \equiv b \bmod n$. Then, m divides $a - b$, and there is some $k \in \mathbb{Z}$ such that $a - b = mk$. Also, n divides $a - b$, and so n divides mk. Since $\gcd(m, n) = 1$ and $n | mk$, Corollary 2.7.6 implies that n is a divisor of k; i.e., $k = nk'$ for some $k' \in \mathbb{Z}$. Thus, $a - b = mk = mnk'$ and mn is a divisor of $a - b$. This means that $a \equiv b \bmod mn$, as desired. □

Example 4.5.5. The difference $103 - 33 = 70$ implies the congruences $103 \equiv 33 \bmod 5$ and $103 \equiv 33 \bmod 7$. In turn, these congruences and Lemma 4.5.4 imply that $103 \equiv 33 \bmod 35$.

See Exercise 4.7.29 for a generalization of Lemma 4.5.4. Using induction one can show the analogue of Lemma 4.5.4 for a system of congruences.

Corollary 4.5.6. *Suppose that $a \equiv b \bmod m_1, m_2, \ldots, m_r$, for some m_i such that $\gcd(m_i, m_j) = 1$. Then, $a \equiv b \bmod M$, where $M = m_1 \cdot m_2 \cdots m_r$. Conversely, if $a \equiv b \bmod M$, then $a \equiv b \bmod m_i$, for each $i = 1, \ldots, r$.*

The following problem appeared in the 4th century AD, in the work of the Chinese mathematician Sun Zi (see Figure 4.1). This is, essentially, Problem 26 in Chapter 3 of *Sun Zi Suanjing* (this means "mathematical manual of Sun Zi"). This is the earliest known occurrence of this type of problem.

Example 4.5.7. There are a certain number of horses, less than 100, but whose exact number is unknown. When ordered in groups of 3 or in groups of 7, two horses remain. When ordered in groups of 5, three horses remain. Find the number of horses.

The statement of the problem amounts to finding a natural number x such that $x < 100$ and x satisfies a system of congruences

$$(S) \quad \begin{cases} x \equiv 2 \bmod 3, \\ x \equiv 3 \bmod 5, \\ x \equiv 2 \bmod 7. \end{cases}$$

One method to solve this system would be to first solve the system formed by the first two congruences (as in Example 4.5.2), whose solution would be a congruence $x \equiv * \bmod 15$, and then solve the system formed by $x \equiv * \bmod 15$ and $x \equiv 2 \bmod 7$. Instead, we will employ the method of Example 4.5.3. We will first solve three

孫子算經卷上
唐朝議大夫行太史令上輕車都尉臣李淳風等奉 勅注釋
度之所起起於忽欲知其忽蠶吐絲爲忽十忽
爲一絲十絲爲一毫十毫爲一氂十氂爲一分
十分爲一寸十寸爲一尺十尺爲一丈十丈爲
一引五十尺爲一端四十尺爲一疋六尺爲一
步二百四十步爲一畝三百步爲一里
稱之所起起於黍十黍爲一絫十絫爲一銖二
十四銖爲一兩十六兩爲一斤三十斤爲一鈞

Figure 4.1. A page from a Qing dynasty edition of *Sun Zi Suanjing*. Image source: Wikimedia Commons.

simpler systems:

$$(S_1)\begin{cases} x_1 \equiv 1 \bmod 3, \\ x_1 \equiv 0 \bmod 5, \\ x_1 \equiv 0 \bmod 7, \end{cases} \quad (S_2)\begin{cases} x_2 \equiv 0 \bmod 3, \\ x_2 \equiv 1 \bmod 5, \\ x_2 \equiv 0 \bmod 7, \end{cases} \quad \text{and } (S_3)\begin{cases} x_3 \equiv 0 \bmod 3, \\ x_3 \equiv 0 \bmod 5, \\ x_3 \equiv 1 \bmod 7. \end{cases}$$

It will suffice to find one solution $x_i \in \mathbb{Z}$ for each system (S_i), for $i = 1, 2, 3$.

(S_1) The second and third equations imply that x_1 is a multiple of 5 and 7, respectively. Thus, by Lemma 4.5.4, x_1 is a multiple of 35; i.e., $x_1 = 35k$. We also require $x_1 \equiv 1 \bmod 3$. Thus, $35k \equiv 1 \bmod 3$. If we reduce $35 \equiv 2 \bmod 3$, we simply need $2k \equiv 1 \bmod 3$. Since $2 \cdot 2 = 4 \equiv 1 \bmod 3$, it follows that $k = 2$ works, so $x_1 = 35k = 70$ is a solution for (S_1).

(S_2) The solution x_2 is a multiple of 21, so $x_2 = 21j$. Moreover $x_2 = 21j \equiv 1 \bmod 5$, or, equivalently, $j \equiv 1 \bmod 5$. Clearly $j = 1$ works, so $x_2 = 21$ is a solution for (S_2).

(S_3) The number $x_3 = 15h$ for some $h \in \mathbb{Z}$. Moreover, $x_3 = 15h \equiv 1 \bmod 7$, so $h \equiv 1 \bmod 7$ and $h = 1$ works. The number $x_3 = 15$ is a solution for (S_3).

Next, we claim that

$$x \equiv 2 \cdot x_1 + 3 \cdot x_2 + 2 \cdot x_3 \bmod (3 \cdot 5 \cdot 7)$$

is the solution set for (S). Indeed, notice that

$$\begin{aligned} x &\equiv 2 \cdot 1 + 3 \cdot 0 + 2 \cdot 0 \equiv 2 \bmod 3, \\ x &\equiv 2 \cdot 0 + 3 \cdot 1 + 2 \cdot 0 \equiv 3 \bmod 5, \\ x &\equiv 2 \cdot 0 + 3 \cdot 0 + 2 \cdot 1 \equiv 2 \bmod 7, \end{aligned}$$

and so, $x \bmod 105$ is a solution for (S). Simplifying the expression that defined x, we obtain

$$\begin{aligned} x &\equiv 2 \cdot x_1 + 3 \cdot x_2 + 2 \cdot x_3 \bmod 105 \\ &\equiv 2 \cdot 70 + 3 \cdot 21 + 2 \cdot 15 \bmod 105 \\ &\equiv 23 \bmod 105. \end{aligned}$$

Finally, we notice that the only number $x \equiv 23 \bmod 105$ that is a natural number < 100 is $x = 23$, so the number of horses must be 23.

Example 4.5.8. Not every system of congruences has a solution. For instance,

$$\begin{cases} x \equiv 1 \bmod 2, \\ x \equiv 4 \bmod 6 \end{cases}$$

cannot have any solutions. Indeed, the first equation says that $x = 1 + 2k$ for some $k \in \mathbb{Z}$, or, in words, x is odd. However, $x \equiv 4 \bmod 6$ implies that $x = 4 + 6j$, for some $j \in \mathbb{Z}$, and in particular $x \equiv 0 \bmod 2$; i.e., x is even. Hence, the equations in this system are incompatible.

4.5.1. The Chinese Remainder Theorem. We are ready to state and prove the Chinese remainder theorem, which will tell us that congruence systems under certain conditions have solutions. Moreover, the proof of the theorem outlines a method to find the solutions (a method that we have already hinted at in Examples 4.5.3 and 4.5.7).

Theorem 4.5.9 (Chinese remainder theorem). *Let $m_1, m_2, \ldots, m_r$ be natural numbers which are relatively prime in pairs; i.e., $\gcd(m_i, m_j) = 1$ for all $i \neq j$. Then, for any $a_1, \ldots, a_r \in \mathbb{Z}$, the simplified system*

$$(S) \quad \begin{cases} x \equiv a_1 \bmod m_1, \\ x \equiv a_2 \bmod m_2, \\ \vdots \\ x \equiv a_r \bmod m_r \end{cases}$$

has a common solution $x_0 \in \mathbb{Z}$. Moreover, if x_0 and y_0 are two solutions for (S), then $x_0 \equiv y_0 \bmod M$ where $M = m_1 \cdot m_2 \cdots m_r$. In other words, the system (S) has a unique solution modulo M.

Proof. Let $M = m_1 \cdot m_2 \cdots m_r$. We will begin by showing that (S) has a solution. First, let us show that, for each $i = 1, \ldots, r$, the system

$$(S_i) \quad \begin{cases} x_i \equiv 0 \bmod m_1, \\ \vdots \\ x_i \equiv 1 \bmod m_i, \\ \vdots \\ x_i \equiv 0 \bmod m_r \end{cases}$$

has a solution $x_i \in \mathbb{Z}$. By Corollary 4.5.6, a number x_i satisfies (S_i) if and only if it satisfies the more compact version

$$(S_i') \begin{cases} x_i \equiv 0 \bmod M_i, \\ x_i \equiv 1 \bmod m_i, \end{cases}$$

where $M_i = \frac{M}{m_i} = m_1 \cdots m_{i-1} m_{i+1} \cdots m_r$. If x_i satisfies (S_i'), then x_i must be a multiple of M_i; i.e., $x_i = M_i b_i$, for some $b_i \in \mathbb{Z}$, and $x_i = M_i b_i \equiv 1 \bmod m_i$. Since $\gcd(m_i, m_j) = 1$, for all $i \neq j$, it follows that $\gcd(m_i, M_i) = 1$, and therefore, the congruence $M_i x \equiv 1 \bmod m_i$ has a solution $x \equiv c_i \bmod m_i$, by Theorem 4.4.3. Hence, $x_i = M_i c_i \in \mathbb{Z}$ is a solution for (S_i') and also for (S_i), for each $i = 1, \ldots, r$.

Now, consider the congruence class of

$$x_0 \equiv a_1 \cdot x_1 + a_2 \cdot x_2 + \cdots + a_r \cdot x_r \bmod M.$$

We claim that any such $x_0 \bmod M$ satisfies the original system (S). Indeed,

$$\begin{aligned} x_0 &\equiv a_1 \cdot x_1 + \cdots + a_i \cdot x_i + \cdots + a_r \cdot x_r \\ &\equiv a_1 \cdot 1 + \cdots + a_i \cdot 0 + \cdots + a_r \cdot 0 \equiv a_1 \bmod m_1 \\ &\vdots \\ &\equiv a_1 \cdot 0 + \cdots + a_i \cdot 1 + \cdots + a_r \cdot 0 \equiv a_i \bmod m_i \\ &\vdots \\ &\equiv a_1 \cdot 0 + \cdots + a_i \cdot 0 + \cdots + a_r \cdot 1 \equiv a_r \bmod m_r. \end{aligned}$$

Thus, any $x_0 \bmod M$ satisfies (S). Moreover, suppose that x_0 and y_0 are two solutions for (S). Then $x_0 \equiv a_i \equiv y_0 \bmod m_i$, for each $i = 1, \ldots, r$. Thus, by Corollary 4.5.6, it follows that $x_0 \equiv y_0 \bmod m_1 \cdots m_r$, as claimed. □

Example 4.5.10. Let us find all the integer solutions for the system

$$(S) \begin{cases} x \equiv 2 \bmod 3, \\ x \equiv 3 \bmod 4, \\ x \equiv 4 \bmod 5. \end{cases}$$

One way to do this would be to follow the method of the Chinese remainder theorem, as outlined in the proof of the theorem and Examples 4.5.3 and 4.5.7. However, there is a trick that allows for an immediate answer. Notice that (S) is equivalent to

$$(S') \begin{cases} x \equiv -1 \bmod 3, \\ x \equiv -1 \bmod 4, \\ x \equiv -1 \bmod 5. \end{cases}$$

Therefore, $x = -1$ is a solution for the system. The Chinese remainder theorem, Theorem 4.5.9, says that there is a **unique** solution for (S) modulo $3 \cdot 4 \cdot 5 = 60$ and, therefore, $x \equiv -1 \equiv 59 \bmod 60$ must be such a solution (the reader can verify that indeed 59 satisfies all three specified congruences by (S)). Hence, all solutions are of the form $x \equiv 59 \bmod 60$.

One of the conditions in the statement of Theorem 4.5.9 is that the moduli m_i need to be relatively prime. However, the Chinese remainder theorem can be applied, indirectly, to solve systems where the moduli are not relatively prime.

Example 4.5.11. Find all solutions for the system

$$(S)\ \begin{cases} x \equiv 4 \bmod 60, \\ x \equiv 10 \bmod 21. \end{cases}$$

The moduli 60 and 21 are not relatively prime, so we cannot use the Chinese remainder theorem directly. However, the $\gcd(60, 21) = 3$ and we can write $60 = 3 \cdot 20$ and $21 = 3 \cdot 7$, and we may expand each congruence in system (S) into two congruences, using Lemma 4.5.4:

$$x \equiv 4 \bmod 60 \Longleftrightarrow \begin{cases} x \equiv 4 \bmod 3, \\ x \equiv 4 \bmod 20 \end{cases} \quad \text{and} \quad x \equiv 10 \bmod 21 \Longleftrightarrow \begin{cases} x \equiv 10 \bmod 3, \\ x \equiv 10 \bmod 7. \end{cases}$$

Equivalently, if we simplify each congruence, we obtain

$$x \equiv 4 \bmod 60 \Longleftrightarrow \begin{cases} x \equiv 1 \bmod 3, \\ x \equiv 4 \bmod 20 \end{cases} \quad \text{and} \quad x \equiv 10 \bmod 21 \Longleftrightarrow \begin{cases} x \equiv 1 \bmod 3, \\ x \equiv 3 \bmod 7. \end{cases}$$

Hence,

$$(S)\ \begin{cases} x \equiv 4 \bmod 60, \\ x \equiv 10 \bmod 21 \end{cases} \Longleftrightarrow \begin{cases} x \equiv 1 \bmod 3, \\ x \equiv 4 \bmod 20, \\ x \equiv 1 \bmod 3, \\ x \equiv 3 \bmod 7. \end{cases}$$

The larger system contains two congruences modulo 3, which happen to be the same, so this is a redundancy in the system and one congruence can be eliminated. Thus,

$$(S)\ \begin{cases} x \equiv 4 \bmod 60, \\ x \equiv 10 \bmod 21 \end{cases} \Longleftrightarrow \begin{cases} x \equiv 1 \bmod 3, \\ x \equiv 4 \bmod 20, \\ x \equiv 3 \bmod 7. \end{cases}$$

The new system with three equations has moduli $m_1 = 3$, $m_2 = 7$, and $m_3 = 20$, which are relatively prime in pairs. Hence, the Chinese remainder theorem proves the existence of a unique solution modulo $3 \cdot 7 \cdot 20 = 420$. We leave it to the reader to find and verify that the solution is $x \equiv 304 \bmod 420$.

Example 4.5.12. Find all solutions for the system

$$(S)\ \begin{cases} x \equiv 4 \bmod 60, \\ x \equiv 11 \bmod 21. \end{cases}$$

As in the previous example, we first need to modify the system so that the moduli are relatively prime in pairs:

$$x \equiv 4 \bmod 60 \Longleftrightarrow \begin{cases} x \equiv 1 \bmod 3, \\ x \equiv 4 \bmod 20 \end{cases} \quad \text{and} \quad x \equiv 11 \bmod 21 \Longleftrightarrow \begin{cases} x \equiv 2 \bmod 3, \\ x \equiv 3 \bmod 7. \end{cases}$$

Hence,

$$(S) \begin{cases} x \equiv 4 \bmod 60, \\ x \equiv 11 \bmod 21 \end{cases} \iff \begin{cases} x \equiv 1 \bmod 3, \\ x \equiv 4 \bmod 20, \\ x \equiv 2 \bmod 3, \\ x \equiv 3 \bmod 7. \end{cases}$$

Thus, a solution to the system (S) would have to satisfy two incompatible conditions modulo 3, namely $x \equiv 1 \bmod 3$ and $x \equiv 2 \bmod 3$. This is impossible and the original system has no solutions.

In the last example of this section, we use the Chinese remainder theorem to solve a quadratic congruence. We will return to quadratic congruences and study them in detail in Chapter 10.

Example 4.5.13. Is there a solution of the congruence $x^2 \equiv 214 \bmod 1155$? Notice that $1155 = 3 \cdot 5 \cdot 7 \cdot 11$. Therefore, by Corollary 4.5.6, $x^2 \equiv 214 \bmod 1155$ has a solution if and only if the following system has a solution:

$$(S) \begin{cases} x^2 \equiv 214 \bmod 3, \\ x^2 \equiv 214 \bmod 5, \\ x^2 \equiv 214 \bmod 7, \\ x^2 \equiv 214 \bmod 11. \end{cases}$$

The system (S) can be simplified by reducing 214 modulo 3, 5, 7, and 11:

$$(S) \begin{cases} x^2 \equiv 1 \bmod 3, \\ x^2 \equiv 4 \bmod 5, \\ x^2 \equiv 4 \bmod 7, \\ x^2 \equiv 5 \bmod 11. \end{cases}$$

Only the last congruence requires a little bit of work, but one soon realizes that $4^2 \equiv 16 \equiv 5 \bmod 11$. Thus, every single congruence in (S) has a solution. One such solution is given by

$$(S') \begin{cases} x \equiv 1 \bmod 3, \\ x \equiv 2 \bmod 5, \\ x \equiv 2 \bmod 7, \\ x \equiv 4 \bmod 11. \end{cases}$$

Searching among numbers in the congruence class of 4 mod 11, we can easily find that the smallest positive integer that solves (S') is $x = 37$. By the Chinese remainder theorem, the solution of (S') is $x \equiv 37 \bmod 1155$. Therefore, $x \equiv 37 \bmod 1155$ is also a solution of (S) and a solution of $x^2 \equiv 214 \bmod 1155$. Indeed, $37^2 = 1369 \equiv 214 \bmod 1155$.

Notice that every solution of (S) gives a solution of the quadratic congruence $x^2 \equiv 214 \bmod 1155$, and $x \equiv 37 \bmod 1155$ is only one of the possibilities. The system

$$(S'')\begin{cases} x \equiv \pm 1 \bmod 3, \\ x \equiv \pm 2 \bmod 5, \\ x \equiv \pm 2 \bmod 7, \\ x \equiv \pm 4 \bmod 11 \end{cases}$$

offers $2^4 = 16$ distinct possibilities, and each one forms a system that has a unique solution and provides a new congruence class that solves the quadratic congruence. The complete set of solutions of (S) is therefore

$$37, 103, 128, 257, 268, 422, 488, 502, 653, 667, 733, 887, 898, 1027, 1052, \text{ and } 1118$$

modulo 1155.

4.6. Applications

In this section we discuss several applications of congruences to divisibility testing, check digits, and factoring algorithms.

4.6.1. Divisibility Tests. It is easy to recognize even numbers. For instance, 2, 46, and 10201342338 are even numbers because their first digits (2, 6, and 8, respectively) are even; that is, for each, the units digit (the first digit when we start counting from the right-hand side) is even. However, multiples of 11 are not immediately recognizable, unless we know some *trick* or *divisibility test* that allows us to check whether a number n is divisible by 11.

Here is one such trick: a number n is divisible by 11 if the alternating sum of its digits is divisible by 11; e.g., a number $dcba$ (in base 10) is divisible by 11 if and only if $a - b + c - d$ is a multiple of 11. For example, $n = 10201342338$ is divisible by 11 because

$$8 - 3 + 3 - 2 + 4 - 3 + 1 - 0 + 2 - 0 + 1 = 11.$$

And, indeed, $10201342338 = 11 \cdot 927394758$. But here is a *warning!* This divisibility test only works when the number n is expressed *in base* 10. If we express a number in base 2, for instance, this divisibility trick ceases to work. For example $(11)_2 = 1 + 1 \cdot 2^1 = 3$ is not divisible by 11, but the alternating sum of its digits $1 - 1 = 0$ is divisible by 11. In other words, the divisibility tricks are dependent on the basis we are using to express our numbers.

Let us begin with a refresher of what it means to express a number in different bases. Recall that when we write $N = 54321$ in the decimal system, we mean that

$$N = 1 + 2 \cdot 10 + 3 \cdot 10^2 + 4 \cdot 10^3 + 5 \cdot 10^4.$$

The following result explains that every number $N \geq 0$ can be expressed uniquely in base $B > 1$.

Proposition 4.6.1. *Let $B > 1$ be fixed. Then, every number $N \geq 0$ can be expressed uniquely in the form*

$$N = a_0 + a_1 \cdot B + a_2 \cdot B^2 + \cdots + a_t \cdot B^t$$

for some $t \geq 0$ and some $0 \leq a_i \leq B - 1$, for $i = 0, \ldots, t$.

Proof. We will show the following statement by induction on $t \geq 0$: every number $0 \leq N < B^{t+1}$ can be represented uniquely in the form

$$N = a_0 + a_1 \cdot B + a_2 \cdot B^2 + \cdots + a_t \cdot B^t$$

for some $0 \leq a_i \leq B-1$, for $i = 0, \ldots, t$.

The statement is clear for the case of $t = 0$; i.e., every number $0 \leq N \leq B-1$ can be expressed uniquely as $N = a_0$ with $a_0 = N$, since $0 \leq a_0 = N \leq B-1$. Now let us assume that the statement is true for t and suppose that N is a number $< B^{t+1}$. By the division theorem, there are $q \geq 0$ and $0 \leq r \leq B-1$ such that $N = qB + r$. Notice that $q < B^t$, because if $q \geq B^t$, then $N = qB + r \geq B^t \cdot B = B^{t+1}$. Hence, by the induction hypothesis, there is a unique representation of q with some $0 \leq a_i \leq B-1$, for $i = 1, \ldots, t+1$, such that

$$q = a_1 + a_2 \cdot B + a_3 \cdot B^2 + \cdots + a_{t+1} \cdot B^t.$$

Hence,

$$\begin{aligned} N &= qB + r \\ &= (a_1 + a_2 \cdot B + a_3 \cdot B^2 + \cdots + a_{t+1} \cdot B^t)B + r \\ &= r + a_1 \cdot B + a_2 \cdot B^2 + \cdots + a_{t+1} \cdot B^{t+1}. \end{aligned}$$

If we put $r = a_0$ (and we notice that $0 \leq r \leq B-1$), we have shown that the statement is also true for $t+1$. Hence, we have shown the induction step, and by the principle of mathematical induction, the statement is true for all $t \geq 0$. □

Definition 4.6.2. We say that a number $N \geq 1$ is expressed in *base* B if it is written in the (unique) form

$$N = a_0 + a_1 \cdot B + a_2 \cdot B^2 + \cdots + a_t \cdot B^t,$$

where $0 \leq a_i \leq B-1$, for $i = 0, \ldots, t$. The expansion of N in base B is also written as

$$N = (a_t a_{t-1} \cdots a_2 a_1 a_0)_B.$$

If the expansion of N is in the most common base, i.e., the decimal base $B = 10$, we usually drop the parentheses $(\cdot)_{10}$ and simply write the digits $N = a_t a_{t-1} \cdots a_2 a_1 a_0$.

Example 4.6.3. Here are some examples of numbers expressed in different bases:

$$\begin{aligned} (54321)_{10} &= 1 + 2 \cdot 10 + 3 \cdot 10^2 + 4 \cdot 10^3 + 5 \cdot 10^4, \\ (54321)_7 &= 1 + 2 \cdot 7 + 3 \cdot 7^2 + 4 \cdot 7^3 + 5 \cdot 7^4 = (13539)_{10}, \\ (10101)_2 &= 1 + 0 \cdot 2 + 1 \cdot 2^2 + 0 \cdot 2^3 + 1 \cdot 2^4 = (21)_{10}. \end{aligned}$$

We are now ready to state and prove several divisibility tests in base 10 and base 1000.

Proposition 4.6.4. *Let $N \in \mathbb{N}$ be a natural number whose representation in base 10 is given by*

$$N = a_0 + a_1 \cdot 10 + a_2 \cdot 10^2 + \cdots + a_t \cdot 10^t,$$

for some $t \geq 0$ and some $0 \leq a_i \leq 9$, for all $0 \leq i \leq t$. Then:

(a) $N \equiv a_0 \bmod 2$ *and also* mod 5. *In particular, N is divisible by 2 (resp. 5) if and only if a_0 is divisible by 2 (resp. 5).*

(b) $N \equiv a_0 + a_1 + \cdots + a_t \bmod 3$ *and also* $\bmod 9$. *In particular, N is divisible by 3 (resp. 9) if and only if the sum of the digits of N is divisible by 3 (resp. 9).*

(c) $N \equiv a_0 - a_1 + a_2 - \cdots + (-1)^t a_t \bmod 11$. *In particular, N is divisible by 11 if and only if the alternating sum of the digits of N is divisible by 11.*

(d) *Suppose that N has the following representation in base 1000:*

$$N = b_0 + b_1 \cdot 1000 + b_2 \cdot 1000^2 + \cdots + b_s \cdot 1000^s,$$

for some $s \geq 0$ and $0 \leq b_j \leq 999$ for all $0 \leq j \leq s$. Then

$$N \equiv b_0 - b_1 + b_2 - \cdots + (-1)^s b_s$$

modulo 7, modulo 11, and modulo 13. In particular, N is divisible by 7 (resp. 11; resp. 13) if and only if the alternating sum of the digits of N in base 1000 is divisible by 7 (resp. 11; resp. 13).

Proof. (a) Since $10 \equiv 0 \bmod 2$ and also $\equiv 0 \bmod 5$, it follows that

$$\begin{aligned} N &= a_0 + a_1 \cdot 10 + a_2 \cdot 10^2 + \cdots + a_t \cdot 10^t \\ &\equiv a_0 + a_1 \cdot 0 + a_2 \cdot 0^2 + \cdots + a_t \cdot 0^t \\ &\equiv a_0 \end{aligned}$$

modulo 2 and also modulo 5. Thus, $N \equiv a_0 \bmod 2$ and $\bmod 5$. Hence, $a_0 \equiv 0 \bmod 2$ (resp. $\bmod 5$) if and only if $N \equiv 0 \bmod 2$ (resp. $\bmod 5$).

(b) Since $10 \equiv 1 \bmod 3$ and also $\equiv 0 \bmod 9$, it follows that

$$\begin{aligned} N &= a_0 + a_1 \cdot 10 + a_2 \cdot 10^2 + \cdots + a_t \cdot 10^t \\ &\equiv a_0 + a_1 \cdot 1 + a_2 \cdot 1^2 + \cdots + a_t \cdot 1^t \\ &\equiv a_0 + a_1 + a_2 + \cdots + a_t \end{aligned}$$

modulo 3 and also modulo 9. Thus, $N \equiv a_0 + a_1 + \cdots + a_t \bmod 3$ and also mod 9. Hence, $a_0 + a_1 + \cdots + a_t \equiv 0 \bmod 3$ (resp. $\bmod 9$) if and only if $N \equiv 0 \bmod 3$ (resp. $\bmod 9$).

(c) Since $10 \equiv -1 \bmod 11$, it follows that

$$\begin{aligned} N &= a_0 + a_1 \cdot 10 + a_2 \cdot 10^2 + \cdots + a_t \cdot 10^t \\ &\equiv a_0 + a_1 \cdot (-1) + a_2 \cdot (-1)^2 + \cdots + a_t \cdot (-1)^t \bmod 11 \\ &\equiv a_0 - a_1 + a_2 - \cdots + (-1)^t a_t \bmod 11. \end{aligned}$$

Thus, $N \equiv a_0 - a_1 + a_2 - \cdots + (-1)^t a_t \bmod 11$. Hence, $a_0 - a_1 + a_2 - \cdots + (-1)^t a_t \bmod 11$ if and only if $N \equiv 0 \bmod 11$.

(d) Suppose that N has the following representation in base 1000

$$N = b_0 + b_1 \cdot 1000 + b_2 \cdot 1000^2 + \cdots + b_s \cdot 1000^s.$$

Since $1001 = 7 \cdot 11 \cdot 13$, it follows that $1000 \equiv -1 \bmod 7$, 11, and 13. It follows that

$$N \equiv b_0 - b_1 + b_2 - \cdots + (-1)^s b_s$$

modulo 7, modulo 11, and modulo 13. Hence, $b_0 - b_1 + b_2 - \cdots + (-1)^s b_s \equiv 0 \bmod 7$ (resp. $\bmod 11$; resp. $\bmod 13$) if and only if $N \equiv 0 \bmod 7$ (resp. mod 11; resp. mod 13). □

Example 4.6.5. The number $N = 13574 = 4 + 7 \cdot 10 + 5 \cdot 10^2 + 3 \cdot 10^3 + 1 \cdot 10^4$ is even, because the first digit, 4, is divisible by 2. The number N is not divisible by 3, because $N \equiv 1 + 3 + 5 + 7 + 4 \equiv 2 \bmod 3$. However, N is divisible by 11 because $N \equiv 4 - 7 + 5 - 3 + 1 \equiv 0 \bmod 11$. Indeed, $N = 11 \cdot 1234$.

Example 4.6.6. Let $N = 95061659$. We may express N in base 1000 as

$$N = 95061659 = 659 + 61 \cdot 1000 + 95 \cdot 1000^2.$$

Thus, $N \equiv 659 - 61 + 95 \equiv 693 \bmod 7$, 11, and 13. Since $693 = 9 \cdot 7 \cdot 11 \equiv 0 \bmod 7$ and 11, it follows that N is also divisible by 7 and by 11. However, $693 \equiv 4 \bmod 13$ and so N is not divisible by 13. Indeed,

$$N = 95061659 = 7 \cdot 11 \cdot 1234567 = 7 \cdot 11 \cdot 127 \cdot 9721.$$

Example 4.6.7 ("Casting out nines"). The method of *casting out nines* is a quick way to provide some evidence that a large calculation is correct, without having to repeat the whole computation. This method *does not prove* that the calculation is correct but rather checks whether there are small inaccuracies in the calculation. In particular, the method is based on reducing all numbers modulo 9 and checking that the calculation is correct modulo 9.

For instance, suppose that we want some assurance that the multiplication

$$3325 \cdot 182 = 605150$$

has been done correctly. The method of casting out nines amounts to redoing this calculation modulo 9. Remember that a number N in base 10 is congruent to the sum of its digits modulo 9, so

$$3325 \cdot 182 \equiv (3+3+2+5) \cdot (1+8+2) \equiv 13 \cdot 11 \equiv (1+3) \cdot (1+1) \equiv 4 \cdot 2 \equiv 8 \bmod 9$$

and

$$605150 \equiv 6 + 0 + 5 + 1 + 5 + 0 \equiv 17 \equiv 1 + 7 \equiv 8 \bmod 9.$$

Thus, $3325 \cdot 182 \equiv 605150 \bmod 9$, and the method of casting out nines has not found any errors. (The calculation is, in fact, correct!)

Now, suppose we have done a second multiplication

$$12345 \cdot 678 = 8379910.$$

We verify our work modulo 9, as follows:

$$\begin{aligned} 12345 \cdot 678 &\equiv (1+2+3+4+5) \cdot (6+7+8) \\ &\equiv 15 \cdot 21 \\ &\equiv (1+5) \cdot (2+1) \\ &\equiv 6 \cdot 3 \equiv 18 \equiv 0 \bmod 9, \end{aligned}$$

and

$$8379910 \equiv 8 + 3 + 7 + 9 + 9 + 1 + 0 \equiv 37 \equiv 3 + 7 \equiv 10 \equiv 1 \bmod 9.$$

Since $12345 \cdot 678 \equiv 0 \not\equiv 1 \equiv 8379910 \bmod 9$, we conclude that the multiplication is wrong and we have made an error. In fact,

$$12345 \cdot 678 = 8369910.$$

Proposition 4.6.8 (Divisibility tests for powers of 2 and 5). *Let N be a natural number whose expression in base 10^a, for some $a \geq 1$, is given by*

$$N = b_n \cdot (10^a)^n + \cdots + b_2 \cdot (10^a)^2 + b_1 \cdot (10^a) + b_0,$$

for some $0 \leq b_i \leq (10^a - 1)$, for each $i = 0, 1, \ldots, n$. Then, $N \equiv b_0 \bmod 2^a$ and mod 5^a. In particular:

(1) *A number N in base 10 is divisible by 2 (resp. 5) if and only if the first digit is divisible by 2 (resp. 5).*
(2) *A number N in base 10 is divisible by 4 (resp. 25) if and only if the number formed by the first two digits is divisible by 4 (resp. 25).*
(3) *A number N in base 10 is divisible by 2^a (resp. 5^a) if and only if the number formed by the first a digits is divisible by 2^a (resp. 5^a).*

Proof. Suppose that N can be written in base 10^a as in the statement of the proposition. Since $10^a = 2^a \cdot 5^a$, it follows that $10^a \equiv 0 \bmod 2^a$ and also mod 5^a. Thus,

$$N = b_n \cdot (10^a)^n + \cdots + b_2 \cdot (10^a)^2 + b_1 \cdot (10^a) + b_0 \equiv b_0$$

modulo 2^a and also mod 5^a. Since b_0 is the number formed by the first a digits of N in base 10, it follows that N is divisible by 2^a (resp. 5^a) if and only if b_0 is divisible by 2^a (resp. 5^a), and the results of the proposition follow. □

Example 4.6.9. Let $N = 34765296$ (in base 10). Since N ends in 65296, we know that:

- N is even, because $6 = 2 \cdot 3$ is divisible by 2;
- N is divisible by 4, because $96 = 4 \cdot 24$ (or, simply, $96 \equiv 0 \bmod 4$ suffices);
- N is divisible by 8, because $296 = 8 \cdot 37$ (or, $296 \equiv 0 \bmod 8$);
- N is divisible by 16, because $5296 = 16 \cdot 331$.

However, N is not divisible by 32, because $65296 \equiv 16 \bmod 32$.

Example 4.6.10. Let $N = 5276771250$ (in base 10). Since N ends in 71250, we know that:

- N is divisible by 5, because $0 = 5 \cdot 0$ is divisible by 5;
- N is divisible by 25, because $50 = 2 \cdot 25$ (or, simply, $50 \equiv 0 \bmod 25$ suffices);
- N is divisible by $5^3 = 125$, because $250 = 2 \cdot 125$ (or, $250 \equiv 0 \bmod 125$);
- N is divisible by $5^4 = 625$, because $1250 = 2 \cdot 625$.

However, N is not divisible by $5^5 = 3125$, because $71250 \equiv 2500 \bmod 3125$.

4.6.2. Congruences in Real Life: Check Digits. When a code is created, there are several issues to be considered, for instance: versatility of the code, security of the encryption, and reliability in the transmission of the code. In order to ensure reliability, codes are usually built with redundancies so that a machine (or a very perceptive human) can detect whether the code is valid. One way to incorporate error detection into the code, sometimes the code contains a check digit or the code verifies some sort of check sum. For example, credit card numbers, the International Standard Book Number (ISBN), the Universal Product Code (UPC or UPC-A), the National Provider Identifier for the US healthcare industry, the routing transit number in bank codes, and national identification numbers in certain countries, among others, include a check digit to a check sum as part of the code. Let us see some examples.

4.6.2.1. *UPC-A.* The Universal Product Code (UPC) is a barcode symbology (i.e., a specific type of barcode) that is widely used in North America and in countries including the UK, Australia, and New Zealand, for tracking trade items in stores. Its most common form, the UPC-A, consists of 12 numerical digits, which are uniquely assigned to each trade item, plus a check digit. A UPC-A code is a string of numbers

$$C\ a_1a_2a_3a_4a_5a_6\ a_7a_8a_9a_{10}a_{11}a_{12}$$

where each digit is between 0 and 9 and C is a check digit such that

$$C \equiv -3\left(\sum_{n=0}^{5} a_{2n+1}\right) - \sum_{n=1}^{6} a_{2n} \bmod 10.$$

For example, a certain (must-have) product has UPC-A number

$$9\ 780821\ 852422,$$

and the check digit $C = 9$ satisfies

$$\begin{aligned} C &\equiv 9 \\ &\equiv -3(7+0+2+8+2+2) - (8+8+1+5+4+2) \\ &\equiv -3 \cdot 21 - 28 \equiv -3 - 8 \equiv -11 \equiv -1 \equiv 9 \bmod 10. \end{aligned}$$

ISBN 978-0-8218-5242-2

9 780821 852422

Figure 4.2. An example of a UPC-A code (below the barcode) and an ISBN-13 code (above the barcode).

4.6.2.2. *ISBN-10.* The International Standard Book Number, or ISBN, is a 10-digit code used to identify books (here by ISBN we mean ISBN-10; the ISBN-13 has a different check digit system, very similar to that of the UPC-A). The 10-digit ISBN has a check sum built in so that multiplying each digit by its position in the number (counting from the right) and taking the sum of these products modulo 11 is 0. The digit in the first position is chosen so that the check sum works, and it may need to have the value 10, which is represented by the letter X. More concretely, if we have an ISBN

$$a_{10}a_9a_8a_7a_6a_5a_4a_3a_2a_1$$

where $0 \le a_1 \le 10$ (if $a_1 = 10$, then we write $a_1 = X$) and $0 \le a_i \le 9$, for $i = 2, \ldots, 10$, then the code satisfies

$$10a_{10} + 9a_9 + 8a_8 + 7a_7 + 6a_6 + 5a_5 + 4a_4 + 3a_3 + 2a_2 + a_1 \equiv 0 \bmod 11.$$

For instance, a certain (must-read) book has ISBN 0821852426, and

$$10 \cdot 0 + 9 \cdot 8 + 8 \cdot 2 + 7 \cdot 1 + 6 \cdot 8 + 5 \cdot 5 + 4 \cdot 2 + 3 \cdot 4 + 2 \cdot 2 + 6 \equiv 198 \equiv 0 \bmod 11.$$

4.6.3. Factoring Large Numbers. One important application of congruences is to factoring large numbers. Let us describe consecutive improvements on an algorithm to factor a (large) number N. The algorithm succeeds if it finds one divisor d of N (if d is a divisor, then we can repeat the algorithm replacing N by d, and also by N/d, until we find the factorization of N into primes). The first versions of our algorithm are very time consuming, so in each stage we improve the efficiency of the algorithm by reducing the number of possible divisors we test.

(1.0) Check if d divides N, for all $d = 2, 3, \ldots, N-1$. In this method we may have to check up to $N - 1$ possible divisors.

(1.1) Check if d divides N, for all $2 \le d \le \sqrt{N}$. Here we may have to check up to $\sqrt{N}$ possible divisors.

(1.2) Check if p divides N, for all primes $2 \le p \le \sqrt{N}$. Due to the prime number theorem, with this method we may be checking roughly $\frac{\sqrt{N}}{\log \sqrt{N}}$ possible divisors.

If N is large, however, in order to run our version (1.2) of the algorithm, we need to know a large list of primes, which may not be available or which may be very time consuming to produce. Instead, we will restrict our search to possible prime numbers modulo M, for some fixed $M > 1$. For instance, if $M = 6$ and $x \ge 5$ is a prime, then $x \equiv 1$ or $5 \bmod 6$. Indeed, if $x \equiv 0, 2$, or $4 \bmod 6$, then x is even, and if $x \equiv 0$ or $3 \bmod 6$, then x is divisible by 3, so the only possibilities for a prime x are 1 or 5 mod 6.

(2.0) Check if d divides N, for all $2 \le d \le \sqrt{N}$, such that $d = 2$, $d = 3$, or $d \equiv 1$ or $5 \bmod 6$. In this version, we are checking $2/6 = 0.333\ldots$, or about 33%, of the numbers that were checked in version (1.1) of the algorithm.

If $M = 30$ and $x \ge 7$ is a prime, then $\gcd(x, 30) = 1$. Thus, the only congruence classes modulo 30 that are possible for x are those numbers $1 \le n \le 30$ such that $\gcd(n, 30) = 1$. Thus, $x \equiv 1, 7, 11, 13, 17, 19, 23$, or $29 \bmod 30$. These represent 8 out of 30 possibilities mod 30.

(2.1) Check if d divides N, for all $2 \le d \le \sqrt{N}$, such that $d = 2, 3$ or $d = 5$ or $d \equiv 1, 7, 11, 13, 17, 19, 23$, or $29 \bmod 30$. In this version, we are checking about $8\sqrt{N}/30 = (0.2666\ldots)\sqrt{N}$ possible divisors, or about 26% of the numbers that were checked in version (1.1) of the algorithm.

When $M = 2 \cdot 3 \cdot 5 \cdot 7 = 210$, there are 48 numbers $1 \le n \le 210$ such that $\gcd(n, 210) = 1$.

(2.2) Check if d divides N, for all $2 \le d \le \sqrt{N}$, such that $d = 2, 3, 5$ or $d = 7$ or $d \ge 8$ such that $\gcd(d, 210) = 1$. In this version, we are checking $48\sqrt{N}/210 = (0.2285\ldots)\sqrt{N}$ possible divisors, or about 23% of the numbers that were checked in version (1.1) of the algorithm.

Example 4.6.11. Let $N = 2491$. Let us see how different methods of factorization work in this particular case.

(1.0) We would try dividing $N = 2491$, by $d = 2, 3, 4, 5, 6, \ldots, 2490$. In practice, we need to calculate $N \bmod d$, for each $d = 2, 3, \ldots$. The method would stop at $d = 47$, which is the smallest non-trivial positive divisor (after having tested 46 possible divisors). In fact $N = 47 \cdot 53$ is the prime factorization for N.

(1.1) We would check if $N = 2491$ is divisible by $d = 2, 3, 4, \ldots, 49$; i.e., test all the numbers $2 \le d \le \sqrt{N} = \sqrt{2491} = 49.90991\ldots$. The method would stop at $d = 47$ after having tested 46 possible divisors.

(1.2) We would check if $N = 2491$ is divisible by $p = 2$, 3, 5, 7, 11, 13, 17, 19, 23, 29, 31, 37, 41, 43, or 47, which are all the primes below $\sqrt{N} = 49.9\ldots$. The method would stop at $p = 47$, after having tested 15 possible (prime) divisors.

(2.0) We would check if $N = 2491$ is divisible by d in the set

$$\{2, 3, 5, 7, 11, 13, 17, 19, 23, 25, 29, 31, 35, 37, 41, 43, 47, 49\},$$

which are all the numbers d such that $d = 2$ or 3, or $d \equiv 1$ or $5 \bmod 6$, and $d \le 49 < \sqrt{N}$. The method would stop at $d = 47$ after having tested 17 possible divisors.

(2.1) We would check if $N = 2491$ is divisible by d in the set

$$\{2, 3, 5, 7, 11, 13, 17, 19, 23, 29, 31, 37, 41, 43, 47, 49\},$$

which are all the numbers d such that $d = 2, 3$ or 5, or $d \equiv 1$, 7, 11, 13, 17, 19, 23, or $29 \bmod 30$, and $d \le 49 < \sqrt{N}$. The method would stop at $d = 47$ after having tested 15 possible divisors (just as good as (1.2) but without having to check that these numbers are primes!).

(2.2) We would check if $N = 2491$ is divisible by d in the set

$$\{2, 3, 5, 7, 11, 13, 17, 19, 23, 29, 31, 37, 41, 43, 47\},$$

which are all the numbers d such that $d = 2, 3, 5$ or 7, or $\gcd(d, 210) = 1$, and $d \le 49 < \sqrt{N}$. The method would stop at $d = 47$ after having tested 15 possible divisors (just as good as (2.1)).

For any $n \ge 1$, one can test whether N is divisible by $p_1, \ldots, p_n$ or numbers $d \le \sqrt{N}$ with $\gcd(d, M) = 1$, where $M = p_1 p_2 \cdots p_n$. This is a fine method, but it can be improved. Next, we present another method to factor large numbers, of a completely different nature, but that also relies on congruences.

4.6.3.1. *The Pollard's Rho Factorization Algorithm.* Pollard's rho factoring algorithm is based on two facts:

- Euclid's algorithm to find the greatest common divisor of two numbers is efficient and fast.

- If we have a p-sided die, where p is a prime, the expected value of consecutive rolls before there are two repeated values is about $(1.2)\sqrt{p}$.

The idea of the method is the following. Let p be the smallest prime divisor of N. If we can generate a list of integers $x_1, x_2, \ldots$ that behave as if they were random modulo p, then we expect that after $(1.2)\sqrt{p}$ values, there will be a repetition (mod p) in the list. Suppose $x_i \equiv x_j \bmod p$; then $x_i - x_j$ is divisible by p, and $\gcd(x_i - x_j, N)$ should be at least p. Here are the steps in the algorithm:

(1) Generate a list of values modulo N recursively with $x_0 \equiv 6 \bmod N$ and $x_{n+1} \equiv x_n^2 + 1 \bmod N$.

(2) For each $n \geq 1$, calculate $d = \gcd(x_{2n} - x_n, N)$. If $1 < d < N$, then d is a divisor of N, and the algorithm stops.

Remark 4.6.12. The function $x^2 + 1$ is essentially random modulo N and also modulo p for any prime p. Thus, the list of integers $x_0, x_1, \ldots$ is essentially random modulo N and modulo p. The initial value $x_0 \equiv 6 \bmod N$ is called the *seed value* and can be changed to any other value, therefore producing a different list of integers, also random modulo N and p.

Example 4.6.13. Let us factor $N = 2491$ using Pollard's rho algorithm.

(1) We calculate the sequence $x_0 \equiv 6 \bmod 2491$ and $x_{n+1} \equiv x_n^2 + 1 \bmod 2491$:

$$6, 37, 1370, 1178, 198, 1840, 332, 621, 2028, 144, 809, 1840, 332, 621, 2028, \ldots.$$

(2) Each time we calculate one value in an even position (e.g., the 2nd value x_2, the 4th value x_4, the 6th value x_6, etc.) we calculate $\gcd(x_{2n} - x_n, 2491)$:

$$\gcd(x_2 - x_1, N) = \gcd(1370 - 37, 2491) = \gcd(1333, 2491) = 1,$$
$$\gcd(x_4 - x_2, N) = \gcd(198 - 1370, 2491) = \gcd(-1172, 2491) = 1,$$
$$\gcd(x_6 - x_3, N) = \gcd(332 - 1178, 2491) = \gcd(-846, 2491) = 47.$$

The algorithm stops. We have found one divisor of $N = 2491$, namely $d = 47$. Indeed, 47 is prime and $N = 47 \cdot 53$ is its prime factorization.

4.6.4. Substitution and Vigenère Ciphers. In this section we introduce our first applications of congruences to the art of communicating in secret, i.e., cryptography. We will see more sophisticated applications in Sections 7.5.3, 8.9.1, 10.7.2, and 16.9. For an introduction to mathematical cryptography, we highly recommend **[HPS14]**.

One of the simplest methods to encrypt a message is what we call a *substitution cipher*. An example of substitution cipher is the so-called *Caesar cipher*, a type of encryption that was used by Roman Emperor Julius Caesar (100 BC – 44 BC) to secure his private communications. A substitution cipher simply replaces each letter for another letter (or symbol). In other words, if $\mathcal{A} = \{A, B, \ldots, Z\}$ is our alphabet, a substitution cipher is just a bijection $\tau \colon \mathcal{A} \to \mathcal{A}$. In order to work with a substitution cipher more efficiently, we will use congruences. First, we identify each letter with a 2-digit number $0 \leq N \leq 25$; i.e.,

Letter	A	B	C	D	E	F	G	H	I	$\cdots$	X	Y	Z
Number	00	01	02	03	04	05	06	07	08	$\cdots$	23	24	25.

For example, $A = 00$ and $W = 22$. The word "HELLO" would be transcribed as the string of numbers "07 04 11 11 14".

Since the alphabet now corresponds to numbers $0 \leq N \leq 25$, a substitution cipher is a bijective function between the numbers modulo 26.

Definition 4.6.14. Let $C_{26} = \{0, \ldots, 25\}$ be a complete residue system mod 26.

(1) A *substitution cipher* is a bijection $\tau \colon C_{26} \to C_{26}$. The inverse function τ^{-1} is called the *deciphering map*.

(2) Let $k \geq 0$ be an integer and let $\tau_k \colon C_{26} \to C_{26}$ be the function

$$\tau_k(N) \equiv N + k \bmod 26.$$

Then, τ_k is called a *shift cipher*, or *Caesar cipher*. The deciphering map τ_k^{-1} is given by $\tau_k^{-1}(M) \equiv M - k \bmod 26$.

Let us show that a Caesar cipher is indeed a substitution cipher. In other words, we need to show that τ_k is a bijection, for all $k \geq 1$. First, τ_k is injective because $N + k \equiv N' + k \bmod 26$ implies that $N \equiv N' \bmod 26$, by Proposition 4.2.1. Since C_{26} is a complete residue system mod 26, it follows that $N = N'$, as desired. Since C_{26} is finite and τ_k is injective, it must be surjective and therefore a bijection. Alternatively, the reader can check that $\tau_k(\tau_k^{-1}(M)) = M$ and $\tau_k^{-1}(\tau_k(N)) = N$ for all $M, N \in C_{26}$.

Example 4.6.15. Let $k = 1$. Then, τ_1 is the substitution cipher that sends each letter to the next letter of the alphabet, and Z is replaced by A. Indeed, $\tau_1(N) \equiv N + 1 \bmod 26$, so the Nth letter is sent to the $(N+1)$th letter for $0 \leq N \leq 24$, and

$$\tau_1(25) \equiv 25 + 1 \equiv 26 \equiv 0 \bmod 26.$$

Thus, the letter Z is sent to A, as claimed.

We can now use τ_1 to encrypt a message. Suppose our message is "HELLO", which is transcribed as "07 04 11 11 14". Thus, using τ_1 on each pair of numbers, we obtain a number string "08 05 12 12 15" which corresponds to the encrypted message "IFMMP".

A substitution cipher is susceptible to attack by spies, using a method known as (letter) *frequency analysis*. This type of attack is based on the fact that, in a long piece of written language, some letters are more frequent than others. For instance, the most frequent letter in the English language is E, followed by T, and then A. See Figure 4.3 for a complete chart of frequencies. If a text is encrypted using a substitution cipher, then every instance of the letter E will be transcribed as a fixed letter, and this character will appear as often as E appears in the English language.

In particular, if we intercept a long message that has been encrypted using a substitution cipher, it is likely that the most common character corresponds to the letter E. If this guess is correct and the cipher was a Caesar cipher, then a spy can now break the entire code and translate the message (see Exercise 4.7.51). In order to strengthen the encryption, the message can be encoded using a rotation of Caesar ciphers. This method is called a Vigenère cipher, named after Blaise de Vigenère (1523–1596), although the method was first described by Giovan Battista

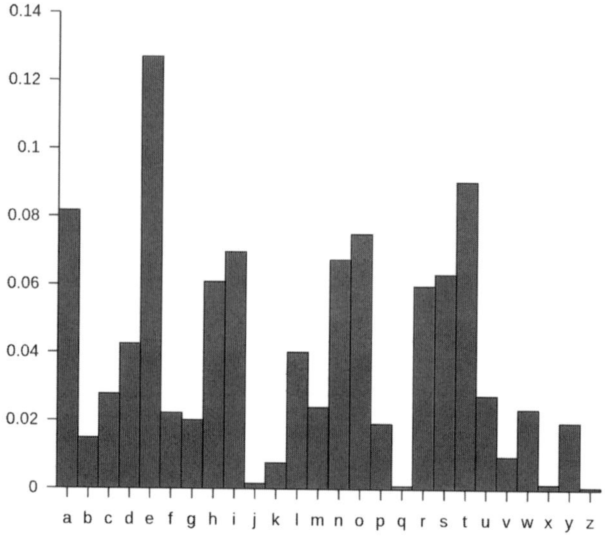

Figure 4.3. Relative frequency of letters in the English language. Image source: Wikimedia Commons.

Bellaso in 1553. Before we write a formal definition of a Vigenère cipher, let us see a simple example.

Example 4.6.16. Let us encrypt the message "HELLO" using a rotation of three Caesar ciphers: τ_1, τ_2, and τ_3. First, the message is written in numeric form as "07 04 11 11 14" as before. Then, the encrypted message will be given by the string

$$\tau_1(07)\ \tau_2(04)\ \tau_3(11)\ \tau_1(11)\ \tau_2(14) = 08\ 06\ 14\ 12\ 16,$$

which in turn corresponds to the letters "IGOMQ". Notice that the letter L has been encrypted in two different ways, because the different positions corresponded to different Caesar ciphers.

Definition 4.6.17. Let $n \geq 0$ and let $K = (k_1, \ldots, k_{n-1}, k_n = k_0)$ be an n-tuple of non-negative integers. Let τ_{k_i} be the Caesar cipher corresponding to the integer k_i. Finally, let (N, i) be a pair formed by a number $N \in C_{26}$ and a natural number i that gives the position of N in a string. Then, a *Vigenère cipher with key* K is a map

$$\tau_K \colon C_{26} \times \mathbb{N} \to C_{26}$$

given by $\tau_K((N,i)) \equiv N + k_{i \bmod n} \bmod 26$ where $i \bmod n$ is the least non-negative residue of i modulo n.

For instance, in Example 4.6.16 we used a Vigenère cipher with key $K = (k_1, k_2, k_3 = k_0) = (1, 2, 3)$. The 5th letter of HELLO is O $= 14$, so the cipher will replace it by

$$\tau_K((14, 5)) \equiv 14 + k_{5 \bmod 3} \equiv 14 + k_2 \equiv 14 + 2 \equiv 16 \bmod 26,$$

since $5 \equiv 2 \bmod 3$, and we have $k_2 = 2$. Thus, the cipher changes O by the letter corresponding to the number 16, which is the letter Q. See Exercises 4.7.53 and 4.7.54 for some other examples of Vigenère-ciphered messages.

4.7. Exercises

Exercise 4.7.1. Find the least non-negative residues in the following congruence classes:

(a) 365 mod 5.

(b) −3122 mod 3.

(c) 3122082546 mod 10.

(d) −2445678 mod 10.

Exercise 4.7.2. Find one integer $a \in \mathbb{Z}$ that satisfies, simultaneously, both congruences $a \equiv 5 \bmod 8$ and $a \equiv 3 \bmod 7$.

Exercise 4.7.3. Let $N > 4$ be a composite number, and suppose $\{0, a_2, \ldots, a_N\}$ is a complete residue system modulo N, for some integers a_i, for $1 \leq i \leq N$, where we fix $a_1 = 0$. Show that at least one of $a_2, \ldots, a_N$ is not a prime number.

Exercise 4.7.4. Let p be a prime number.

(a) Show that there is a complete residue system $\{q_1, q_2, \ldots, q_p\}$ modulo p, such that each q_i is a prime number, for $1 \leq i \leq p$. (Hint: use Dirichlet's theorem on primes in arithmetic progressions, Theorem 3.3.11.)

(b) Find a complete residue system $\{q_1, \ldots, q_{11}\}$ modulo 11 such that each q_i is a prime number.

Exercise 4.7.5. Show that if $n > 4$ is not prime, then $(n-1)! \equiv 0 \bmod n$.

Exercise 4.7.6. Let a, b, a', b', k be arbitrary integers and let $m > 1$. Show that:

(a) If $a \equiv b \bmod m$, then $ka \equiv kb \bmod m$.

(b) If $a \equiv b \bmod m$ and $a' \equiv b' \bmod m$, then $a + a' \equiv b + b' \bmod m$.

Exercise 4.7.7. Use congruences to show that $6 \cdot 4^n \equiv 6 \bmod 9$ for any $n \geq 0$.

Exercise 4.7.8. Find the least non-negative residues in the following congruence classes:

(a) $5^{18} \bmod 7$.

(b) $68^{105} \bmod 13$.

(c) $6^{47} \bmod 12$.

Exercise 4.7.9. Let a and b be odd numbers, and let $m = a + b$ and $n = a - b$. Show that either m or n is congruent to 0 mod 4 and the other one is congruent to 2 mod 4.

Exercise 4.7.10. Let a, b, c be non-zero integers such that $\gcd(a, b, c) = 1$ and $a^2 + b^2 = c^2$ (that is, (a, b, c) is a pythagorean triple). Prove that one of a or b is even and the other one is odd.

Exercise 4.7.11. Are there three consecutive numbers whose sum is divisible by 37? Either find an example or prove that such an example does not exist.

Exercise 4.7.12. Let $n > 1$ be a natural number, and let S be a set formed by $n+1$ distinct integers. Show that there are two numbers of S such that their difference is divisible by n. (Hint: consider the least non-negative residue modulo n of each element in S and the pigeonhole principle, Theorem 2.3.14.)

Exercise 4.7.13. Show that $5^e + 6^e \equiv 0 \bmod 11$ for all odd numbers e.

Exercise 4.7.14. Prove part (a) below, and then find the least non-negative residue modulo 7, 11, and 13 in parts (b), (c), and (d).

(a) A number N is congruent modulo 7, 11, or 13 to the alternating sum of its digits in base 1000. (For example, $123456789 \equiv 789 - 456 + 123 \equiv 456 \bmod 7$, 11, or 13.)

(b) 11233456.

(c) 58473625.

(d) 100,000,000,000,000,001.

Exercise 4.7.15. Find divisibility tests for numbers in base 34 for $2, 3, 5, 7, 11$, and 17. Is $(5368)_{34}$ divisible by 11 or 17?

Exercise 4.7.16. Show that $2^{560} \equiv 1 \bmod 561$.

Exercise 4.7.17. Show that $36^{100} \equiv 16 \bmod 17$.

Exercise 4.7.18. Show that 257 is a divisor of $100 \cdot 2^{25} - 57 = 3355443143$.

Exercise 4.7.19. Show that 42 is a divisor of $n^7 - n$ for all positive n.

Exercise 4.7.20. Show that $5555^{2222} + 2222^{5555}$ is divisible by 7.

Exercise 4.7.21. Prove that for any natural number $n \geq 1$, $3^{6n} - 2^{6n}$ is divisible by 35 (Hint: work modulo 5 and modulo 7, separately).

Exercise 4.7.22. Find the remainder when $14!$ is divided by 17.

Exercise 4.7.23. Find the smallest number ≥ 120120 which is not divisible by any prime $p < 20$, using congruences. (Hint: calculate $120120 \bmod p$, for every prime $p < 20$.)

Exercise 4.7.24. What time does a clock read 100 hours after it reads 2 o'clock? If the time is now 2 pm, after 100 hours, will it be in the pm or in the am?

Exercise 4.7.25. Find all $x \in \mathbb{Z}$ that satisfy the following linear congruence or explain why no integral solution exists (these are individual congruences and not a system!).

(a) $6x \equiv 9 \bmod 11$.

(b) $6x \equiv 11 \bmod 9$.

(c) $6x \equiv 9 \bmod 15$.

Exercise 4.7.26. Let p be a prime, and let $k \in \mathbb{Z}$ such that $1 \leq k \leq p$. Define the p-over-k binomial coefficient (or p-choose-k coefficient) by

$$\binom{p}{k} = \frac{p!}{k!(p-k)!}.$$

Prove that $\binom{p}{k}$ is divisible by p. (You may assume that $\binom{p}{k}$ is an integer.)

Exercise 4.7.27. Let p be a prime, and let a and b be integers. Prove that

$$(a+b)^p \equiv a^p + b^p \bmod p.$$

(Hint: use the binomial theorem $(a+b)^n = \sum_{k=0}^{n} \binom{n}{k} a^k b^{n-k}$, Exercise 2.11.14, and Exercise 4.7.26.)

Exercise 4.7.28. Let p be a prime, and let a, s, t be integers with $s \geq t \geq 1$. Prove that there are p^{s-t} integers congruent to a mod p^t in the interval $[1, p^s]$.

Exercise 4.7.29. Let m, n be two integers greater than 1, and suppose a, b are integers such that $a \equiv b \bmod m$ and $a \equiv b \bmod n$.

(a) Show that $a \equiv b \bmod \operatorname{lcm}(m, n)$.

(b) Find values of m, n, a, and b such that $a \equiv b \bmod m$ and $a \equiv b \bmod n$, but $a \not\equiv b \bmod mn$.

Exercise 4.7.30. Find the smallest positive integer n that satisfies, simultaneously, all three congruences

$$\begin{aligned} n &\equiv 1 \bmod 3, \\ n &\equiv 2 \bmod 4, \\ n &\equiv 3 \bmod 5. \end{aligned}$$

You must use the method that appears in the proof of the Chinese remainder theorem (see also Example 4.5.7).

Exercise 4.7.31. Find the smallest positive integer that leaves remainders of $2, 4, 6$ when divided by $3, 5, 7$, respectively. You must use the Chinese remainder theorem.

Exercise 4.7.32. Find the smallest positive integer n such that

$$n \equiv 7 \bmod 3, \quad n \equiv 5 \bmod 5, \quad n \equiv 3 \bmod 7.$$

Exercise 4.7.33. Find three consecutive integers x, y, and z that are divisible by 3, 5, and 7, respectively (i.e., x is divisible by 3, y by 5, and z by 7).

Exercise 4.7.34. Solve each of the following systems:

$$\begin{cases} x \equiv 2 \bmod 7, \\ x \equiv 4 \bmod 8, \\ x \equiv 3 \bmod 9, \end{cases} \qquad \begin{cases} y \equiv 1 \bmod 7, \\ y \equiv 3 \bmod 8, \\ y \equiv 6 \bmod 9, \end{cases} \qquad \begin{cases} z \equiv 5 \bmod 7, \\ z \equiv 2 \bmod 8, \\ z \equiv 1 \bmod 9. \end{cases}$$

Exercise 4.7.35. Solve each of the following systems:

$$\begin{cases} x \equiv -3 \bmod 11, \\ x \equiv 103 \bmod 13, \\ x \equiv 3 \bmod 15, \end{cases} \qquad \begin{cases} y \equiv 25 \bmod 11, \\ y \equiv 35 \bmod 13, \\ y \equiv 31 \bmod 15. \end{cases}$$

Exercise 4.7.36. Determine if the following system is compatible and, if so, then determine all the solutions in integers $x \in \mathbb{Z}$:

$$\begin{cases} x \equiv 1 \bmod 2, \\ x \equiv 2 \bmod 5, \\ x \equiv 5 \bmod 6, \\ x \equiv 5 \bmod 12. \end{cases}$$

Exercise 4.7.37. A troop of 17 monkeys store their bananas in 11 piles of equal size with a twelfth pile of 6 left over. When they divide the bananas into 17 equal groups, none remain. What is the smallest number of bananas they can possibly have?

Exercise 4.7.38. A prime p is a safe prime if $p = 2q+1$ where q is also prime. The prime q, in turn, is called a Sophie Germain prime. For instance, $p = 5 = 2 \cdot 2 + 1$ and $p = 7 = 2 \cdot 3 + 1$ are the first two safe primes, and $q = 2$ and $q = 3$ are the first two Sophie Germain primes. Suppose that $p > 7$ is a safe prime and prove the following.

(a) Show that $p \equiv 2 \bmod 3$.

(b) Show that $p \equiv 3 \bmod 4$.

(c) Show that if $p > 11$, then $p \not\equiv 1 \bmod 5$.

(d) Use the previous congruences to show that $p \equiv 23, 47$, or $59 \bmod 60$.

(e) Use (d) to find 10 safe primes larger than 1000.

Note: safe primes are important in cryptographic applications because of their use in discrete logarithm-based techniques like the Diffie–Hellman key exchange.

Exercise 4.7.39. Find all the solutions for the following congruences.

(a) Find all solutions for the congruence $x^2 \equiv 1 \bmod 8$.

(b) Find all solutions for $x^2 \equiv 1 \bmod 5$.

(c) Use (a) and (b) and the Chinese remainder theorem to find all solutions for $x^2 \equiv 1 \bmod 40$.

Exercise 4.7.40. Solve the following quadratic congruences.

- Find all solutions of $x^2 \equiv 1 \bmod 133$.
- Prove that there are no solutions: $x^2 \equiv 2 \bmod 133$.
- Find (at least) one solution: $x^2 \equiv 93 \bmod 133$.

Note: Trial and error will yield no points. (Hint: use the Chinese remainder theorem and the fact that $133 = 7 \cdot 19$.)

Exercise 4.7.41. Prove that the equation $x^2 - 7y^3 + 21z^5 = 3$ has no solution with x, y, z in $\mathbb{Z}$ (Hint: calculate all possible squares modulo 7.)

Exercise 4.7.42. Show that $2^{2^n} + 5$ is composite for every positive integer n.

Exercise 4.7.43. The 7-digit number $n = 72x20y2$, where x and y are digits between 0 and 9, is divisible by 72. What are the possibilities for x and y?

Exercise 4.7.44. Show that $n(n-1)(2n-1)$ is divisible by 6 for every $n > 0$.

Exercise 4.7.45. Find 3 primes in each category:

(1) Find 3 primes $p \equiv 1 \bmod 3$ and also 3 primes $p \equiv 2 \bmod 3$.

(2) Find 3 primes $p \equiv 1 \bmod 5$ and also 3 primes $p \equiv 2 \bmod 5$.

(3) Find 3 primes $p \equiv 3 \bmod 5$ and also 3 primes $p \equiv 4 \bmod 5$.

(4) Are there any primes $p \equiv 3 \bmod 21$? Why? Why not?

(5) Are there any primes $p = 3 \bmod 22$? Why? Why not?

(6) Are there infinitely many primes in each category above? How do you know? (Hint: read Chapter 3.)

Exercise 4.7.46. Let $n \geq 0$ and let $F(n) = 2^{2^n} + 1$ be the nth Fermat number (see Example 3.2.7). Show that $F(n) + 2$ is composite for infinitely many values of n. (Hint: reduce modulo 7.)

Exercise 4.7.47. Let $F(n)$ be the nth Fermat number (as in Exercise 4.7.46). Show that the decimal expansion of $F(n)$ ends in 7 for all $n > 1$ (i.e., the unit digit is 7). For instance, $F(2) = 2^{2^2} + 1 = 17$.

Exercise 4.7.48. Let p be a prime number.

(1) Let a and b be integers such that $ab \equiv 0 \bmod p$. Show that either $a \equiv 0 \bmod p$ or $b \equiv 0 \bmod p$.

(2) Let $n > 3$ be a composite number. Show that there are integers a, b, both non-zero when reduced modulo n and such that $ab \equiv 0 \bmod n$.

Exercise 4.7.49. Let $C_{26} = \{0, 1, 2, \ldots, 25\}$ be a complete residue system mod 26. Decide whether each map $\tau\colon C_{26} \to C_{26}$ given below is a substitution cipher, as in Definition 4.6.14.

(1) The map given by $\tau(N) \equiv 2N \bmod 26$.

(2) The map given by $\tau(N) \equiv N^3 \bmod 26$.

(3) The map given by $\tau(N) \equiv 5N + 3 \bmod 26$.

Exercise 4.7.50. Let τ be the substitution cipher given by $\tau(N) \equiv 3N \bmod 26$ for any letter $0 \leq N \leq 25$.

(1) Use the function τ to encrypt the word "PIN" letter by letter.

(2) Describe the decryption function τ^{-1} such that $\tau^{-1}(\tau(N)) \equiv N \bmod 26$ for any $0 \leq N \leq 25$.

(3) Decipher the message "OYN" that was encrypted using τ.

Exercise 4.7.51. The following text was encrypted using a Caesar cipher:

ZNGU VF NJRFBZR OHG AHZORE GURBEL VF GUR ORFG.

Use frequency analysis (Figure 4.3) to break the cipher and decrypt the message.

Exercise 4.7.52. You have intercepted a ciphertext "URYYB" that was encrypted using a Caesar cipher and later learned it corresponded to the word "HELLO". Decipher the next intercepted ciphertext using the same Caesar cipher: "JBEYQ".

Exercise 4.7.53. Use a Vigenère cipher with key $K = (3, 1, 2)$ to encrypt the message "HELLO SILENCE MY OLD FRIEND".

Exercise 4.7.54. The message "OIYFV WFPO NI WII REHV" was encrypted using a Vigenère cipher with key $K = (1, 4, k_3)$. What is k_3? What was the original message?

CHAPTER 5

GROUPS, RINGS, AND FIELDS

Algebra is the offer made by the devil to the mathematician.... All you need to do is give me your soul: give up geometry.

Michael Atiyah

In the previous chapter we defined the concept of congruence,

$$a \equiv b \bmod m \text{ if and only if } m|a-b,$$

and we proved a number of properties satisfied by congruences. We also saw that the set $C_m = \{0, 1, 2, 3, \ldots, m-1\}$ is a complete residue system modulo m. In particular, when we work with congruences modulo m, every equation can be reduced to a statement about the numbers in C_m. In this chapter, we will begin by defining an addition operation and a multiplication operation on C_m, which will make C_m into a number system with only m elements, that resembles the usual number system $\mathbb{Z}$ is some respects, but it is also fundamentally different, as C_m only has m elements, while $\mathbb{Z}$ has infinitely many elements. The set C_m, together with its addition and multiplication operations, will be denoted by $\mathbb{Z}/m\mathbb{Z}$.

5.1. $\mathbb{Z}/m\mathbb{Z}$

We begin by reminding the reader about the definition of a congruence class (which we already saw in Definition 4.1.1).

Definition 5.1.1. Let $m > 1$ be fixed. The *congruence class* of an integer $a \in \mathbb{Z}$ modulo m is the set of all integers $b \in \mathbb{Z}$ such that $b \equiv a \bmod m$. The congruence class of a modulo m will be denoted by $a \bmod m$. In other words,

$$\begin{aligned} a \bmod m &= \{b \in \mathbb{Z} : b \equiv a \bmod m\} \\ &= \{\ldots, a-3m, a-2m, a-m, a, a+m, a+2m, a+3m, \ldots\}. \end{aligned}$$

Example 5.1.2. Let $m = 5$. The congruence class of 2 modulo 5 is the set of all integers that are congruent to 2 modulo 5; i.e.,

$$\begin{aligned} 2 \bmod 5 &= \{b \in \mathbb{Z} : b \equiv 2 \bmod 5\} \\ &= \{\ldots, -13, -8, -3, 2, 7, 12, 17, \ldots\}. \end{aligned}$$

Proposition 5.1.3. *Let $m > 1$ be fixed. Then $\mathbb{Z}$ is the disjoint union of the m congruence classes $0 \bmod m$, $1 \bmod m$, $\ldots$, and $m-1 \bmod m$. In other words,*

$$\mathbb{Z} = (0 \bmod m) \cup (1 \bmod m) \cup \cdots \cup (m-1 \bmod m) = \bigcup_{i=0}^{m-1} (i \bmod m),$$

and $(i \bmod m) \cap (j \bmod m) = \emptyset$ if $i \not\equiv j \bmod m$.

Proof. By Proposition 4.1.4, the set $C_m = \{0, 1, 2, \ldots, m-1\}$ is a complete residue system modulo m. In particular, every integer $a \in \mathbb{Z}$ is congruent to exactly one i in the range $0 \le i \le m-1$. Thus, $a \equiv i \bmod m$ and a belong to the congruence class $i \bmod m$, and it does not belong to any other $j \bmod m$, for $i \not\equiv j \bmod m$. This shows that $\mathbb{Z}$ is the disjoint union of the congruence classes $0 \bmod m$, $1 \bmod m$, $\ldots$, and $m-1 \bmod m$, as claimed. □

Now we are ready to define operations of addition and multiplication on C_m and, therefore, define $\mathbb{Z}/m\mathbb{Z}$ (we read this as "zee modulo m" or "the integers modulo m").

Definition 5.1.4. We define $(\mathbb{Z}/m\mathbb{Z}, +, \cdot)$ as the set of all congruence classes modulo m; i.e.,

$$\mathbb{Z}/m\mathbb{Z} = \{0 \bmod m, 1 \bmod m, \ldots, m-1 \bmod m\}$$

together with two operations $+$ and $\cdot$, defined by

(1) addition: $(a \bmod m) + (b \bmod m) = (a+b) \bmod m$ and
(2) multiplication: $(a \bmod m) \cdot (b \bmod m) = (a \cdot b) \bmod m$,

for any $0 \le a, b \le m-1$.

Remark 5.1.5. Notice that if $a' \equiv a \bmod m$ and $b' \equiv b \bmod m$, then $a' + b' \equiv a + b \bmod m$ by Proposition 4.2.1, so that

$$\begin{aligned} (a \bmod m) + (b \bmod m) &= (a+b) \bmod m \\ &= (a'+b') \bmod m = (a' \bmod m) + (b' \bmod m). \end{aligned}$$

In other words, addition of congruence classes is a well-defined operation and it does not depend on the choice of representative $a \in \mathbb{Z}$ of a congruence class $a \bmod m$. Similarly, multiplication is also well-defined.

Remark 5.1.6. The congruence relation $\equiv$ in $\mathbb{Z}$ is an equivalence relation (see Remark 4.2.5). Each congruence class is an equivalence class for the congruence relation. The set $\mathbb{Z}/m\mathbb{Z}$ is the set of all equivalence classes, or, in other words, $\mathbb{Z}/m\mathbb{Z}$ is the quotient set $\mathbb{Z}/\equiv$.

Example 5.1.7. Let $m = 2$. Then $(\mathbb{Z}/2\mathbb{Z}, +, \cdot)$ has only two elements, namely $0 \bmod 2$ and $1 \bmod 2$. For instance,

$$(1 \bmod 2) + (1 \bmod 2) = (1 + 1 \bmod 2) = (0 \bmod 2),$$

which we will usually abbreviate as $1 + 1 \equiv 2 \equiv 0 \bmod 2$. Here are the addition and multiplication tables in $\mathbb{Z}/2\mathbb{Z}$:

+	0	1
0	0	1
1	1	0

and

·	0	1
0	0	0
1	0	1

.

Example 5.1.8. Let $m = 5$. The set $(\mathbb{Z}/5\mathbb{Z}, +, \cdot)$ has 5 elements, namely $\mathbb{Z}/5\mathbb{Z} = \{0, 1, 2, 3, 4 \bmod 5\}$. Here are some examples of calculations in $\mathbb{Z}/5\mathbb{Z}$:

$$(2 \bmod 5) + (4 \bmod 5) = (6 \bmod 5) = (1 \bmod 5),$$
$$(2 \bmod 5) \cdot (4 \bmod 5) = (8 \bmod 5) = (3 \bmod 5),$$

which we will abbreviate as $2 + 4 \equiv 1 \bmod 5$ and $2 \cdot 4 \equiv 3 \bmod 5$. Here are the full addition and multiplication tables for $\mathbb{Z}/5\mathbb{Z}$:

+	0	1	2	3	4
0	0	1	2	3	4
1	1	2	3	4	0
2	2	3	4	0	1
3	3	4	0	1	2
4	4	0	1	2	3

and

·	0	1	2	3	4
0	0	0	0	0	0
1	0	1	2	3	4
2	0	2	4	1	3
3	0	3	1	4	2
4	0	4	3	2	1

.

Example 5.1.9. Let $m = 6$ and consider $\mathbb{Z}/6\mathbb{Z}$. The addition and multiplication tables for $\mathbb{Z}/6\mathbb{Z}$ are as follows:

+	0	1	2	3	4	5
0	0	1	2	3	4	5
1	1	2	3	4	5	0
2	2	3	4	5	0	1
3	3	4	5	0	1	2
4	4	5	0	1	2	3
5	5	0	1	2	3	4

and

·	0	1	2	3	4	5
0	0	0	0	0	0	0
1	0	1	2	3	4	5
2	0	2	4	0	2	4
3	0	3	0	3	0	3
4	0	4	2	0	4	2
5	0	5	4	3	2	1

.

Some peculiarities occur in $\mathbb{Z}/6\mathbb{Z}$ that are not present in $\mathbb{Z}/2\mathbb{Z}$ and $\mathbb{Z}/5\mathbb{Z}$; for instance,

$$(3 \bmod 6) \cdot (4 \bmod 6) = (12 \bmod 6) = (0 \bmod 6),$$

or $3 \cdot 4 \equiv 0 \bmod 6$. Notice that $3 \not\equiv 0 \bmod 6$ and $4 \not\equiv 0 \bmod 6$ but their product is $0 \bmod 6$. Also, we have

$$3 \cdot 5 \equiv 3 \cdot 7 \bmod 6$$

but $5 \not\equiv 7 \bmod 6$. In particular, the cancellation law (Theorem 2.2.3) does not hold in $\mathbb{Z}/6\mathbb{Z}$. See also Proposition 4.3.1 and Exercise 4.7.48.

In the following proposition, we shall prove properties of $(\mathbb{Z}/m\mathbb{Z}, +, \cdot)$, similar to the defining axioms of $\mathbb{Z}$ that we saw in Section 2.1.

Proposition 5.1.10. *Let $m > 1$ be fixed, and let $(\mathbb{Z}/m\mathbb{Z}, +, \cdot)$ be the set of congruence classes modulo m with operations $+$ and $\cdot$ as in Definition 5.1.4. Then:*

(1) *The set $\mathbb{Z}/m\mathbb{Z}$ is closed under the $+$ and $\cdot$ operations; that is, if $a, b \bmod m \in \mathbb{Z}/m\mathbb{Z}$, then $a + b \bmod m$ and $a \cdot b \bmod m$ are also in $\mathbb{Z}/m\mathbb{Z}$.*

(2) *(Properties of $+$ and $\cdot$) For all $a, b, c \bmod m \in \mathbb{Z}/m\mathbb{Z}$, we have:*
 (2.a) *(Commutativity): $a + b \equiv b + a \bmod m$ and $a \cdot b \equiv b \cdot a \bmod m$.*
 (2.b) *(Associativity): $(a + b) + c \equiv a + (b + c) \bmod m$ and $(a \cdot b) \cdot c \equiv a \cdot (b \cdot c) \bmod m$.*
 (2.c) *(Distributivity): $c \cdot (a + b) \equiv c \cdot a + c \cdot b \bmod m$ and $(a + b) \cdot c \equiv a \cdot c + b \cdot c \bmod m$.*

(3) *(Existence of distinguished elements)*
 (3.a) *(Additive identity) There exists $0 \bmod m \in \mathbb{Z}/m\mathbb{Z}$ such that $a + 0 \equiv 0 + a \equiv a \bmod m$, for all $a \bmod m \in \mathbb{Z}/m\mathbb{Z}$.*
 (3.b) *(Additive inverses) For all $a \bmod m \in \mathbb{Z}/m\mathbb{Z}$ there is $m - a \bmod m \in \mathbb{Z}/m\mathbb{Z}$ such that $a + (m - a) \equiv 0 \equiv (m - a) + a \bmod m$.*
 (3.c) *(Multiplicative identity) There exists $1 \bmod m \in \mathbb{Z}/m\mathbb{Z}$ such that $a \cdot 1 \equiv 1 \cdot a \equiv a \bmod m$, for all $a \bmod m \in \mathbb{Z}/m\mathbb{Z}$.*

Proof. The properties of $(\mathbb{Z}/m\mathbb{Z}, +, \cdot)$ follow directly from the axioms of $\mathbb{Z}$, the definition of congruence $\equiv$, and the definitions of $+$ and $\cdot$ on $\mathbb{Z}/m\mathbb{Z}$. For example, the fact that $+$ and $\cdot$ are closed in $\mathbb{Z}/m\mathbb{Z}$ follows from Proposition 5.1.3 and, in particular, from Proposition 4.1.4. Indeed, if $a, b \in \mathbb{Z}$ are arbitrary, then $a + b$ will belong to a unique congruence class modulo m, which we denote by $(a + b) \bmod m$ and, by definition, $(a \bmod m) + (b \bmod m)$ is the class $(a + b) \bmod m$. Thus, $+$ is closed in $\mathbb{Z}/m\mathbb{Z}$ (see also Remark 5.1.5).

As another example, let us show that $+$ and $\cdot$ are commutative:

$$\begin{aligned}(a \bmod m) + (b \bmod m) &= (a + b) \bmod m \\ &= (b + a) \bmod m = (b \bmod m) + (a \bmod m), \text{ and} \\ (a \bmod m) \cdot (b \bmod m) &= (a \cdot b) \bmod m \\ &= (b \cdot a) \bmod m = (b \bmod m) \cdot (a \bmod m),\end{aligned}$$

where we have used that $a + b = b + a$ and $a \cdot b = b \cdot a$ for any $a, b \in \mathbb{Z}$. □

The previous proposition shows that $\mathbb{Z}$ and $\mathbb{Z}/m\mathbb{Z}$, for any $m > 1$, share a number of properties: namely the axioms (1), (2), and (3) for $\mathbb{Z}$ as in Section 2.1 also hold for $\mathbb{Z}/m\mathbb{Z}$. However, the fourth axiom that defines $\mathbb{Z}$, i.e., the existence of $\mathbb{N}$, with its characteristic properties, is a fundamental difference between $\mathbb{Z}$ and $\mathbb{Z}/m\mathbb{Z}$. The reader might want to play around with the axioms that define $\mathbb{N}$ and try to find a subset $\mathbb{N}'$ of $\mathbb{Z}/m\mathbb{Z}$ that would satisfy said properties... and eventually prove that such a set $\mathbb{N}'$ cannot exist (see Exercise 5.6.1).

5.1.1. $\mathbb{Z}/m\mathbb{Z}$ and Geometry. From the arithmetic geometry point of view, our interest in $\mathbb{Z}/m\mathbb{Z}$ comes from the following theorem, which says that if a diophantine equation has an integral point, then it also has a solution over $\mathbb{Z}/m\mathbb{Z}$, for all $m > 1$.

Theorem 5.1.11. *Let $C : f(x_1, \ldots, x_n) = 0$, where f is a polynomial in n variables with integer coefficients, and suppose that C has at least one integral point. Then, for each $m > 1$, the equation $f_m(x_1, \ldots, x_n) \equiv 0 \bmod m$ has at least one solution defined over $\mathbb{Z}/m\mathbb{Z}$, where f_m is the polynomial obtained from f by reducing each coefficient modulo m.*

Proof. If $v = (a_1, \ldots, a_n) \in \mathbb{Z}^n$ is an integral point on C, then $f(a_1, \ldots, a_n) = 0$, and in particular $f(a_1, \ldots, a_n) \equiv 0 \bmod m$. Now, by the properties of congruences (Proposition 4.2.1), we must have

$$f_m(a_1 \bmod m, \ldots, a_n \bmod m) \equiv f(a_1, \ldots, a_n) \equiv 0 \bmod m,$$

and therefore $(a_1 \bmod m, \ldots, a_n \bmod m)$ is a point defined over $\mathbb{Z}/m\mathbb{Z}$ of the congruence equation $f_m(x_1, \ldots, x_n) \equiv 0 \bmod m$, as claimed. □

Example 5.1.12. The diophantine equation $C : x^2 + y^2 + z^2 = 107$ has a solution $(9, 5, 1)$. Therefore, $x^2 + y^2 + z^2 \equiv 107 \bmod m$ has a solution over $\mathbb{Z}/m\mathbb{Z}$ for all $m > 1$, namely $(9 \bmod m, 5 \bmod m, 1 \bmod m)$. For instance, if $m = 5$, then $(4 \bmod 5, 0 \bmod 5, 1 \bmod 5)$ is a solution of $x^2 + y^2 + z^2 \equiv 2 \bmod 5$.

Example 5.1.13. The curve $C : x^2 - 5y^2 = 2$ has no integral points. Indeed, assume for a contradiction that C has an integral point. Then, Theorem 5.1.11 shows that there must be solutions modulo every $m > 1$. In particular, there would be solutions for $m = 5$, but we showed in Example 4.2.8 that the equation $x^2 - 5y^2 \equiv 2 \bmod 5$ (or, equivalently, $x^2 \equiv 2 \bmod 5$) has no solutions modulo 5. Therefore, we have reached a contradiction, and C cannot have integral points.

Notice, however, that C does have points modulo m, for some values of m (in fact, for *most* values of m!). For instance, let $m = 7$. Then, C reduces to $x^2 - 5y^2 \equiv 2 \bmod 7$, or $x^2 + 2y^2 \equiv 2 \bmod 7$. Thus, $(0 \bmod 7, 1 \bmod 7)$ is a solution of C defined over $\mathbb{Z}/7\mathbb{Z}$.

Remark 5.1.14. It is natural to ask whether the converse of Theorem 5.1.11 is also true. More concretely, suppose that $C : f(x_1, \ldots, x_n) = 0$ is a diophantine equation that has solutions modulo m for every value of $m > 1$. Is it true that C has an integral (or perhaps rational) point? This is a very deep and interesting question, and it will be discussed in depth in Chapter 11.

For now, let us show an example (due to K. Conrad) with solutions modulo m, for all $m > 1$, but no integral solution. Consider the ellipse $C : 2x^2 + 7y^2 = 1$. Clearly, it has no integral solutions because if x or y is non-zero, then $2x^2 + 7y^2 \geq 2$. However, there are rational points on C such as $P = (1/3, 1/3)$ and $Q = (3/5, 1/5)$. Now, if m is relatively prime to 3 (resp. 5), the point P (resp. Q) is well-defined in $\mathbb{Z}/m\mathbb{Z}$ (as $3^{-1} \bmod m$ would make sense), and therefore there is a point in $C(\mathbb{Z}/m\mathbb{Z})$. If m is a multiple of 3 and 5, say $m = s \cdot t$ with $\gcd(s, 5) = 1 = \gcd(t, 3)$, then we can use the Chinese remainder theorem to find a point R modulo m such that

$$R \equiv \begin{cases} P \bmod s, \\ Q \bmod t. \end{cases}$$

Since $P \in C(\mathbb{Z}/s\mathbb{Z})$ and $Q \in C(\mathbb{Z}/t\mathbb{Z})$, it follows that $R \in C(\mathbb{Z}/m\mathbb{Z})$, as desired. We leave it as an exercise for the reader to find a point $R \bmod 15$ on the curve C following the method outlined above.

5.2. Groups

Groups, as men, will be known by their actions.

Guillermo Moreno

In this section we look at other sets with operations that satisfy properties similar to those satisfied by $\mathbb{Z}$ and $\mathbb{Z}/m\mathbb{Z}$. We begin with a definition that captures properties of sets with a binary operation.

Definition 5.2.1. Let G be a set. A *binary operation, or binary operator*, on G is a map $G \times G \to G$.

Definition 5.2.2. A pair $(G, *)$ formed by a set G with a binary operation $*$ is called a *group* if the following properties are satisfied:

(1) The set G is closed under the binary operation $*$; i.e., $g * h \in G$ for any $g, h \in G$.

(2) The operation $*$ is associative: $g * (h * k) = (g * h) * k$, for any $g, h, k \in G$.

(3) There exists an identity element in G for $*$; i.e., there is some $e \in G$ such that $e * g = g * e = g$, for all $g \in G$.

(4) There are inverses in G for $*$; i.e., for each $g \in G$ there is an element $g^{-1} \in G$ such that $g * g^{-1} = e$.

In addition, suppose the operation $*$ satisfies the following additional property:

(5) The operation $*$ is commutative: $g * h = h * g$, for any $g, h \in G$.

Then we say that $(G, *)$ is a commutative group, or abelian group.

Example 5.2.3. The pair $(\mathbb{Z}, +)$ is an abelian group. In this case the identity element is $e = 0 \in \mathbb{Z}$ and the inverses with respect to $+$ are simply the additive inverses. In other words, for each $a \in \mathbb{Z}$ there is $-a \in \mathbb{Z}$ such that $a + (-a) = 0$.

However, the pair $(\mathbb{Z}, \times)$ does not form a group. The identity element with respect to multiplication is $1 \in \mathbb{Z}$, but not every integer $a \in \mathbb{Z}$ has a multiplicative inverse in $\mathbb{Z}$. For instance, there is no $2^{-1} \in \mathbb{Z}$ such that $2 \cdot 2^{-1} = 1$. As we know, $2^{-1} \in \mathbb{Q}$, but it does not belong to $\mathbb{Z}$.

Example 5.2.4. The pairs $(\mathbb{Q}, +)$ and $(\mathbb{Q} - \{0\}, \times)$ are two examples of abelian groups.

Example 5.2.5. Let $(G, \cdot)$ be the set of all 2×2 matrices with real entries and non-zero determinant, and let $\cdot$ be matrix multiplication (we usually denote this group by $\mathrm{GL}(2, \mathbb{R})$ and call it the general linear group). In other words,

$$G = \left\{ \begin{pmatrix} a & b \\ c & d \end{pmatrix} : a, b, c, d \in \mathbb{R},\ ad - bc \neq 0 \right\}$$

and

$$\begin{pmatrix} a & b \\ c & d \end{pmatrix} \cdot \begin{pmatrix} e & f \\ g & h \end{pmatrix} = \begin{pmatrix} ae + bg & af + bh \\ ce + dg & cf + dh \end{pmatrix}.$$

The set $(G, \cdot)$ is a group with identity element $\begin{pmatrix} 1 & 0 \\ 0 & 1 \end{pmatrix}$. The reader can check associativity and the existence of inverses. However, $(G, \cdot)$ is not a commutative group. For instance,

$$\begin{pmatrix} 1 & 2 \\ 0 & 1 \end{pmatrix} \cdot \begin{pmatrix} 0 & -1 \\ 1 & 0 \end{pmatrix} = \begin{pmatrix} 2 & -1 \\ 1 & 0 \end{pmatrix} \neq \begin{pmatrix} 0 & -1 \\ 1 & 2 \end{pmatrix} = \begin{pmatrix} 0 & -1 \\ 1 & 0 \end{pmatrix} \cdot \begin{pmatrix} 1 & 2 \\ 0 & 1 \end{pmatrix}.$$

Example 5.2.6. Let $(C^0(\mathbb{R}), +)$ be the set of all continuous functions $f : \mathbb{R} \to \mathbb{R}$ with addition of functions as an operation; i.e., $(f+g)(x) = f(x) + g(x)$. Then $(C^0(\mathbb{R}), +)$ is a group, because (1) the sum of two continuous function is also continuous (so $C^0(\mathbb{R})$ is closed under addition), (2) addition of functions is associative, (3) there is an identity element, namely $f(x) = 0$ for all $x \in \mathbb{R}$, and (4) for every $f(x) \in C^0(\mathbb{R})$ the function $g(x) = -f(x)$ is also continuous and $f(x) + g(x) = 0$, for all $x \in \mathbb{R}$.

Let us define an alternative operation $*$ on $C^0(\mathbb{R})$ given by composition of functions; i.e., $(f * g)(x) = f(g(x))$. Is $(C^0(\mathbb{R}), *)$ a group?

(1) $C^0(\mathbb{R})$ is closed under composition of functions, because the composition of two continuous functions is also continuous.

(2) Composition of functions is associative:

$$(f * (g * h))(x) = f(g(h(x))) = ((f * g) * h)(x).$$

(3) There is an identity element $e(x) = x$ such that

$$(f * e)(x) = f(e(x)) = f(x) = e(f(x)) = (e * f)(x).$$

However, not every element of $C^0(\mathbb{R})$ has an inverse with respect to composition. For instance, $f(x) = x^2$ does not have an inverse function defined over all of $\mathbb{R}$. Also, the inverse of $f(x) = e^x$ would be $f^{-1}(x) = \log x$, which is not defined anywhere in the interval $(-\infty, 0]$.

Proposition 5.2.7. *Let $m > 1$ be fixed. Then, $(\mathbb{Z}/m\mathbb{Z}, +)$ is an abelian group.*

Proof. This follows from Proposition 5.1.10 and in particular from parts (1), (2.a), (2.b), (3.a) and (3.b). □

Let us consider the pair $(\mathbb{Z}/m\mathbb{Z}, \cdot)$. Is this a group? Obviously not, because $0 \bmod m$ does not have a multiplicative inverse. Let us define *multiplicative inverse* in the particular case of $\mathbb{Z}/m\mathbb{Z}$, for completeness sake.

Definition 5.2.8. Let $m > 1$ be fixed. The congruence class of $b \bmod m$ is a *multiplicative inverse* of the congruence class of $a \bmod m$ if $a \cdot b \equiv 1 \bmod m$. If so, we sometimes write $b \equiv a^{-1} \bmod m$, or $b \equiv \frac{1}{a} \bmod m$. We say that a congruence class $a \bmod m$ is *invertible* if $a \bmod m$ has a multiplicative inverse modulo m.

Example 5.2.9. In $\mathbb{Z}/15\mathbb{Z}$ we have $2 \cdot 8 \equiv 1 \bmod 15$. Thus, $8 \bmod 15$ is a multiplicative inverse for $2 \bmod 15$ and, conversely, the class of 2 is a multiplicative inverse for $8 \bmod 15$. We say that the class of 2 and the class of 8 are invertible.

Alternatively, we write

$$2 \equiv 8^{-1} \bmod 15 \quad \text{and} \quad 8 \equiv 2^{-1} \bmod 15,$$

or simply $2 \equiv \frac{1}{8} \bmod 15$ and $8 \equiv \frac{1}{2} \bmod 15$.

Now, let us consider $(\mathbb{Z}/m\mathbb{Z} - \{0\}, \cdot)$. Is this a group? By Proposition 5.1.10, the set $\mathbb{Z}/m\mathbb{Z}$ is closed under the multiplication operation, and this operation is commutative and associative and there is an identity element $1 \bmod m$. It only remains to check whether every non-zero element has a multiplicative inverse.

Example 5.2.10. Let $m = 2$. The set $\mathbb{Z}/2\mathbb{Z} - \{0\}$ consists of only one congruence class, namely $1 \bmod 2$. Since $1 \cdot 1 \equiv 1 \bmod 2$, we conclude that 1 is its own multiplicative inverse, and therefore $(\mathbb{Z}/2\mathbb{Z} - \{0\}, \cdot)$ is a group.

Example 5.2.11. Let $m = 3$. The set $\mathbb{Z}/3\mathbb{Z} - \{0\}$ consists of two congruence classes: $1 \bmod 3$ and $2 \bmod 3$. Since

$$1 \cdot 1 \equiv 1 \bmod 3 \quad \text{and} \quad 2 \cdot 2 \equiv 1 \bmod 3,$$

we conclude that the classes of 2 and 3 are invertible, and therefore every non-zero class in $\mathbb{Z}/3\mathbb{Z}$ has a multiplicative inverse. In particular, $(\mathbb{Z}/3\mathbb{Z} - \{0\}, \cdot)$ is a group.

Example 5.2.12. Let $m = 4$. The set $\mathbb{Z}/4\mathbb{Z} - \{0\}$ consists of three congruence classes: $1 \bmod 4$, $2 \bmod 4$, and $3 \bmod 4$. We have

$$1 \cdot 1 \equiv 1 \bmod 4 \quad \text{and} \quad 3 \cdot 3 \equiv 1 \bmod 4,$$

so 1 and $3 \bmod 4$ have multiplicative inverses, but $2 \bmod 4$ does not have a multiplicative inverse. Indeed, if $x \bmod 4$ is a multiplicative inverse for $2 \bmod 4$, then $2 \cdot x \equiv 1 \bmod 4$. But this congruence does not have solutions, for any $x \in \mathbb{Z}$, by Theorem 4.4.3, because $\gcd(2, 4) = 2$ and 2 does not divide 1. Hence, $2 \bmod 4$ is not invertible, and $(\mathbb{Z}/4\mathbb{Z} - \{0\}, \cdot)$ is not a group. Notice that, in fact, $\mathbb{Z}/4\mathbb{Z} - \{0\}$ is not closed under multiplication, as $2 \cdot 2 \equiv 0 \bmod 4$.

Proposition 5.2.13. *Let $m > 1$. The congruence class of $a \bmod m$ is invertible in $\mathbb{Z}/m\mathbb{Z}$ if and only if $\gcd(a, m) = 1$.*

Proof. The class of $a \bmod m$ is invertible if and only if the linear congruence $a \cdot x \equiv 1 \bmod m$ has a solution. By Theorem 4.4.3, this linear congruence has a solution if and only if $\gcd(a, m) = 1$. □

Corollary 5.2.14. *Let $m > 1$ be fixed. The pair $(\mathbb{Z}/m\mathbb{Z} - \{0\}, \cdot)$ is a group if and only if m is a prime number.*

Proof. If m is not prime, then there are some $a, b \in \mathbb{Z}$ with $m = ab$ and $1 < a, b < m$. In particular, the class of $a \bmod m$ is non-zero (because $a \not\equiv 0 \bmod m$), but $\gcd(a, m) = a$, and Proposition 5.1.10 implies that a is not invertible modulo m. Thus, not all non-zero congruence classes are invertible, and $(\mathbb{Z}/m\mathbb{Z} - \{0\}, \cdot)$ is not a group.

If m is prime, by Proposition 5.1.10, the set $\mathbb{Z}/m\mathbb{Z}$ is closed under the multiplication operation, and this operation is commutative and associative and there is an identity element $1 \bmod m$. It only remains to check whether every non-zero element has a multiplicative inverse. If a is an integer in the interval $1 \leq a \leq m-1$, then $\gcd(a, m) = 1$ because m is prime and $m > a$. Thus, by Proposition 5.2.13, it follows that a is invertible in $\mathbb{Z}/m\mathbb{Z}$. Hence, every non-zero congruence class in $\mathbb{Z}/m\mathbb{Z}$ is invertible, and $(\mathbb{Z}/m\mathbb{Z} - \{0\}, \cdot)$ is a group. □

Next, we introduce the concept of a subgroup of a group.

Definition 5.2.15. Let $(G, *)$ be a group with identity element e, and let H be a subset of G containing e, such that $(H, *)$ is a group also, where the operation $*$ is the same as in G. Then, H is said to be a *subgroup* of G. If $H \subsetneq G$, then H is called a *proper subgroup* of G.

Example 5.2.16. Let $(G, *) = (\mathbb{Q}, +)$ be the group of rational numbers under addition. Then, $(\mathbb{Z}, +)$ is a subgroup of $\mathbb{Q}$. However, $(\mathbb{N}, +)$ is not a subgroup of $(\mathbb{Z}, +)$ because, for instance, $\mathbb{N}$ does not have an identity element $0 \in \mathbb{N}$.

Example 5.2.17. Let $(G, *) = (\mathbb{Z}/6\mathbb{Z}, +)$ be the group of congruence classes modulo 6, with respect to addition. Let

$$H = \{0, 2, 4 \bmod 6\}.$$

Then, $(H, +)$ is a subgroup of $(\mathbb{Z}/6\mathbb{Z}, +)$. We leave it up to the reader to check that $(H, +)$ is a group.

Example 5.2.18. Let $G = \{1, 2, 4, 5, 7, 8 \bmod 9\} \subseteq \mathbb{Z}/9\mathbb{Z}$ and let $(G, \cdot)$ be the group formed by G under multiplication. Then $(H, \cdot)$, where $H = \{1, 4, 7 \bmod 9\}$, forms a subgroup of G. We again leave it up to the reader to check that $(G, \cdot)$ and $(H, \cdot)$ are indeed groups.

We finish this section citing one of the most important theorems in group theory.

Theorem 5.2.19 (Lagrange's theorem). *Let G be a finite group and let H be a subgroup of G (with respect to a common operation $*$). Then, the size of G (usually denoted by $|G|$) is divisible by the size of H.*

Figure 5.1. Joseph-Louis Lagrange (1736–1813). Image source: Wikimedia Commons.

The proof of Lagrange's theorem has been left as an exercise for the reader (Exercise 5.6.6). Here we simply provide two examples to illustrate the idea behind the proof of the theorem.

Example 5.2.20. Let $(G, \cdot)$ and $(H, \cdot)$ be as in Example 5.2.18, so that

$$G = \{1, 2, 4, 5, 7, 8 \bmod 9\} \quad \text{and} \quad H = \{1, 4, 7 \bmod 9\}.$$

We can define an equivalence relation on G by setting $g_1 \sim_H g_2$ if and only if there is some $h \in H$ such that $g_1 = g_2 \cdot h$ (the reader needs to verify that this, indeed, is an equivalence relation). Since H contains the identity element of G, any two elements of H are equivalent with respect to $\sim_H$. However, 2 is not equivalent to any of the elements of H, so 2 belongs to a different equivalence class, which we denote as

$$2 \cdot H = \{g \in G : g = 2 \cdot h \text{ for some } h \in H\} = \{2, 5, 8 \bmod 9\}.$$

Then, $G = H \cup 2 \cdot H$, and, in fact, G is the disjoint union of H and $2 \cdot H$. Moreover, H and $2 \cdot H$ have the same number of elements, namely 3. Therefore, $|G| = 2 \cdot |H|$ and, in particular, the size of H, namely $|H| = 3$, is a divisor of the size of G, which is $|G| = 6$.

Example 5.2.21. Let $(G, +)$ be the group $\mathbb{Z}/6\mathbb{Z}$ under addition, and let $H = \{0, 3 \bmod 6\}$. Then $(H, +)$ is a subgroup of $(G, +)$. Let us define an equivalence relation on G by $g_1 \sim_H g_2$ if and only if $g_1 \equiv g_2 + h \bmod 6$, for some $h \in H$. Then, there are three equivalence classes; namely,

$$H = \{0, 3 \bmod 6\}, \; 1 + H = \{1, 4 \bmod 6\}, \quad \text{and} \quad 2 + H = \{2, 5 \bmod 6\}.$$

In particular,

$$G = H \cup (1 + H) \cup (2 + H),$$

and each equivalence class is disjoint from the others. Hence, $|G| = |H| + |1 + H| + |2 + H|$. Moreover, all equivalence classes have the same size, namely $|H| = 2$, and therefore

$$|G| = 3 \cdot |H|.$$

In particular, $|H| = 2$ is a divisor of $|G| = 6$.

5.2.1. Group Homomorphisms. We use functions between two groups G and H to study their relationships, or to understand one group in terms of the other. In order to be able to deduce useful group-theoretic information from a map, such a function needs to respect the group structures. Such a map is called a group homomorphism.

Definition 5.2.22. Let $(G, *)$ and $(H, \star)$ be groups. A function $f : G \to H$ is a *group homomorphism* if

$$f(g * g') = f(g) \star f(g')$$

for all g and g' in G.

It is important to remark that the binary operations $*$ on G and $\star$ on H may be quite different (see Example 5.2.25 below) and $g * g'$ is an operation in G, while $f(g) \star f(g')$ occurs in H. Next, we give a few examples of group homomorphisms. We leave it up to the reader as an exercise to verify that each of the maps f below satisfies the required equation $f(g * g') = f(g) \star f(g')$.

Example 5.2.23. The zero map $f_0 : G \to G$ that sends all elements of G to the trivial element $e \in G$, i.e., $f_0(g) = e$ for all $g \in G$, is a group homomorphism. Similarly, the identity map $f_1 : G \to G$ with $f_1(g) = g$ is also a group homomorphism.

Example 5.2.24. Let $n \geq 1$ be an integer, and let $(G, *)$ be a group. The multiplication-by-n map, sometimes denoted by $[n]\colon G \to G$, is a group homomorphism defined by

$$[n](g) = g * g * \cdots * g,$$

so that we have used the group operation $n - 1$ times on n elements of G. For instance, if the group is $(\mathbb{Z}, +)$, then $[n](g) = g + g + \cdots + g = n \cdot g$ for any $g \in \mathbb{Z}$. If the group under consideration is $(\mathbb{R}^\times, \cdot)$, non-zero real numbers under multiplication, then $[n](g) = g \cdot g \cdots g = g^n$ for any $g \in \mathbb{R}$.

Example 5.2.25. Let $(\mathrm{GL}(2, \mathbb{R}), \cdot)$ be the group of all 2×2 matrices with real entries and non-zero determinant (as in Example 5.2.5), and let $(\mathbb{R}^\times, \cdot)$ be the non-zero real numbers under multiplication. Then, the determinant map

$$\det\colon\ \mathrm{GL}(2, \mathbb{R}) \to \mathbb{R}^\times$$

that sends a matrix $M = \begin{pmatrix} a & b \\ c & d \end{pmatrix}$ to its determinant $\det(M) = ad - bc$ is a group homomorphism from $\mathrm{GL}(2, \mathbb{R})$ to $\mathbb{R}^\times$.

Example 5.2.26. Consider the groups $(\mathbb{Z}, +)$ and $(\mathbb{Z}/m\mathbb{Z}, +)$, for some $m \geq 2$. Then, the reduction-by-m map from $\mathbb{Z}$ to $\mathbb{Z}/m\mathbb{Z}$ that sends an integer $g \mapsto g \bmod m$ is a group homomorphism.

Example 5.2.27. The map $f\colon \mathbb{R} \to \mathbb{R}$ that sends $x \mapsto x + 1$ is not a group homomorphism from $(\mathbb{R}, +)$ to itself, because

$$f(x + y) = x + y + 1 \neq (x + 1) + (y + 1) = f(x) + f(y).$$

Next, we state (without proof) a few results about homomorphisms.

Proposition 5.2.28. *Let $(G, *)$ and $(H, \star)$ be groups and suppose that $f\colon G \to H$ is a group homomorphism. Then:*

(1) *$f(e_G) = e_H$, where e_G and e_H are the identity elements of G and H, respectively.*

(2) *The kernel of f, defined by*

$$\mathrm{Ker}(f) = \{g \in G : f(g) = e_H\},$$

is a subgroup of G.

(3) *The image of f, i.e.,*

$$f(G) = \{h \in H : \text{ there is } g \in G \text{ such that } f(g) = h\},$$

is a subgroup of H.

Example 5.2.29. Let $m > 1$, and consider the multiplication-by-m homomorphism $[m]\colon \mathbb{Z} \to \mathbb{Z}$ given by $g \mapsto m \cdot g$. Then, the image of $[m]$ is $m\mathbb{Z}$, which is an additive subgroup of $\mathbb{Z}$.

If instead we consider the reduction-by-m homomorphism $\mathbb{Z} \to \mathbb{Z}/m\mathbb{Z}$, then the kernel is precisely given by $m\mathbb{Z}$.

Example 5.2.30. The determinant map $\det\colon \mathrm{GL}(2,\mathbb{R}) \to \mathbb{R}^\times$ is surjective (why?), so the image is all of $\mathbb{R}^\times$. The kernel of det is given by those 2×2 matrices with determinant 1; i.e.,

$$\mathrm{Ker}(\det) = \left\{ \begin{pmatrix} a & b \\ c & d \end{pmatrix} : a, b, c, d \in \mathbb{R} \text{ and } ad - bc = 1 \right\}.$$

The subgroup of matrices of determinant 1 is the special linear group, and it is usually denoted by $\mathrm{SL}(2,\mathbb{R})$.

Definition 5.2.31. Let $(G, *)$ and $(H, \star)$ be groups. A homomorphism $f\colon G \to H$ is said to be an *isomorphism* if it is also a bijection. If such an isomorphism exists, then we say that G and H are *isomorphic*.

Example 5.2.32. Let $m > 1$ be an integer, and let μ_m be the set of all mth roots of unity in the complex numbers; i.e.,

$$\mu_m = \{z \in \mathbb{C} : z^m = 1\} = \{e^{2n\pi i/m} : 0 \le n < m\}.$$

Then, $(\mu_m, \cdot)$ is a group under multiplication (Exercise 5.6.3). Moreover, there is a homomorphism $\varphi\colon \mathbb{Z}/m\mathbb{Z} \to \mu_m$ given by $a \bmod m \mapsto e^{2a\pi i/m}$. We leave it to the reader to show that this map is well-defined, a homomorphism, and in fact a bijection. Thus, φ is an isomorphism of groups.

We conclude this section with a useful criterion to determine whether a homomorphism is injective.

Proposition 5.2.33. *Let $f\colon G \to H$ be a group homomorphism and let K be its kernel. Then, f is injective if and only if $K = \{e_G\}$; i.e., the kernel is trivial (it only contains the identity element of G).*

Example 5.2.34. Let $f\colon \mathbb{Z} \to \mathrm{GL}(2,\mathbb{R})$ be the map given by

$$n \mapsto \begin{pmatrix} 1 & n \\ 0 & 1 \end{pmatrix}.$$

We leave it up to the reader in the exercises to check that f is a group homomorphism from $(\mathbb{Z}, +)$ to $(\mathrm{GL}(2,\mathbb{R}), \cdot)$. Proposition 5.2.33 says that f is injective because the kernel of f consists only of the identity matrix. Hence, if we define

$$H = \left\{ \begin{pmatrix} 1 & n \\ 0 & 1 \end{pmatrix} : n \in \mathbb{Z} \right\},$$

then H is a subgroup of $\mathrm{GL}(2,\mathbb{R})$, and $f\colon \mathbb{Z} \to H$ is an isomorphism of groups.

5.3. Rings

In the previous section we defined groups, i.e., sets together with one binary operation. The integers and $\mathbb{Z}/m\mathbb{Z}$, however, have two operations, namely addition and multiplication. Next, we define a structure of a set with two operations.

Definition 5.3.1. A *ring* (with identity) is a triple $(R, +, \cdot)$ where R is a set with two binary operations $+$ and $\cdot$ that satisfy the following properties:

(1) The set R is closed under the operations $+$ and $\cdot$; i.e., $r + s, r \cdot s \in R$, for all $r, s \in R$.

(2) The pair $(R, +)$ is an abelian group.

(3) The operation $\cdot$ is associative; i.e., $r \cdot (s \cdot t) = (r \cdot s) \cdot t$, for all $r, s, t \in R$.

(4) The operations $+$ and $\cdot$ satisfy the distributive laws:

$$r \cdot (s + t) = r \cdot s + r \cdot t \quad \text{and} \quad (s + t) \cdot r = s \cdot r + t \cdot r,$$

for all $r, s, t \in R$.

(5) There is an identity with respect to $\cdot$ in R; i.e., there is some $1 \in R$ such that $1 \cdot r = r \cdot 1 = r$, for all $r \in R$.

In addition, suppose $\cdot$ satisfies the following property:

(6) The operation $\cdot$ is commutative; i.e., $r \cdot s = s \cdot r$, for all $r, s \in R$.

Then we say that R is a *commutative ring* (with identity).

Example 5.3.2. The integers $\mathbb{Z}$ are our prototype of a (commutative) ring with identity. The ring axioms of Definition 5.3.1 are precisely the axioms (1) through (3) that defined $\mathbb{Z}$ in Section 2.1.

Example 5.3.3. The rational numbers $\mathbb{Q}$, the real numbers $\mathbb{R}$, and the complex numbers $\mathbb{C}$ are also examples of commutative rings.

Example 5.3.4. Let $(C^0(\mathbb{R}), +, \cdot)$ be the triple given by the set of all continuous functions $f : \mathbb{R} \to \mathbb{R}$ with operations of addition and multiplication defined by

$$(f + g)(x) = f(x) + g(x) \quad \text{and} \quad (f \cdot g)(x) = f(x) \cdot g(x),$$

for any $f, g \in C^0(\mathbb{R})$. Then, $(C^0(\mathbb{R}), +, \cdot)$ is a commutative ring with additive identity given by the zero function $0 : x \mapsto 0$ and multiplicative identity $1 : x \mapsto 1$.

Example 5.3.5. Let $M_{2\times 2}(\mathbb{C})$ be the set of all 2×2 matrices with complex entries. Define a triple $(M_{2\times 2}(\mathbb{C}), +, \cdot)$, where $+$ is defined coordinatewise, i.e.,

$$\begin{pmatrix} a & b \\ c & d \end{pmatrix} + \begin{pmatrix} e & f \\ g & h \end{pmatrix} = \begin{pmatrix} a+e & b+f \\ c+g & d+h \end{pmatrix},$$

and multiplication $\cdot$ is as in Example 5.2.5. Then, $(M_{2\times 2}(\mathbb{C}), +, \cdot)$ is a ring with additive identity given by $\begin{pmatrix} 0 & 0 \\ 0 & 0 \end{pmatrix}$ and multiplicative identity $\begin{pmatrix} 1 & 0 \\ 0 & 1 \end{pmatrix}$. This ring is not commutative, as Example 5.2.5 shows.

Proposition 5.3.6. *Let $m > 1$ be fixed. The triple $(\mathbb{Z}/m\mathbb{Z}, +, \cdot)$ is a commutative ring with identity.*

Proof. The fact that $(\mathbb{Z}/m\mathbb{Z}, +, \cdot)$ satisfies the axioms of a commutative ring was shown in Propositions 5.1.10 and 5.2.7. $\square$

By definition, if $(R,+,\cdot)$ is a ring, then $(R,+)$ is an abelian group. However, $(R-\{0\},\cdot)$ is not necessarily a group. For instance, $(\mathbb{Z}-\{0\},\cdot)$ is not a group, because only 1 and -1 have a multiplicative inverse, and any $n \neq \pm 1$ is not invertible in $\mathbb{Z}$; i.e., there is no $m \in \mathbb{Z}$ such that $n \cdot m = 1$. As another example, the pair

$$\left(M_{2\times 2}(\mathbb{C}) - \left\{\begin{pmatrix} 0 & 0 \\ 0 & 0 \end{pmatrix}\right\}, \cdot\right)$$

is not a group because many matrices do not have multiplicative inverses (indeed, a matrix $\begin{pmatrix} a & b \\ c & d \end{pmatrix}$ is invertible if and only if its determinant $ad - bc \neq 0$).

In order to study whether $(R-\{0\},\cdot)$ is a group, we define two concepts: unit and zero-divisor in a ring.

Definition 5.3.7. Let $(R,+,\cdot)$ be a ring with multiplicative identity $1 \in R$ and additive identity $0 \in R$.

(1) We say that an element $u \in R$ is a *unit* if there is some $v \in R$ such that $u \cdot v = 1$. The set of all units in R is denoted by $R^\times$.

(2) We say that an element $r \in R$ is a *zero-divisor* if there is some $s \in R$ such that $r \cdot s = 0$ and $r \neq 0$ and $s \neq 0$. The set of all zero-divisors in R will be denoted by R^0.

Example 5.3.8. The only units in $\mathbb{Z}$ are 1 and -1; i.e., $\mathbb{Z}^\times = \{\pm 1\}$. Indeed, if $u \cdot v = 1$ with $u, v \in \mathbb{Z}$, then u and v are divisors of 1. Thus, both u and v are 1 or -1. There are no zero-divisors in $\mathbb{Z}$, by Theorem 2.2.3, so $\mathbb{Z}^0 = \emptyset$.

Example 5.3.9. Every non-zero number in $\mathbb{Q}$ is a unit; that is, $\mathbb{Q}^\times = \mathbb{Q} - \{0\}$. For all $u = \frac{p}{q} \in \mathbb{Q}$ there is $v = \frac{q}{p} \in \mathbb{Q}$ such that $u \cdot v = \frac{p}{q} \cdot \frac{q}{p} = 1$. There are no zero-divisors in $\mathbb{Q}$, for if $r \cdot s = \frac{p}{q} \cdot \frac{m}{n} = 0$, then $p \cdot m = 0$. Thus, by Theorem 2.2.3, $p = 0$ or $m = 0$, and this implies that $r = 0$ or $s = 0$.

Example 5.3.10. Let $M_{2\times 2}(\mathbb{C})$ be the set of all matrices with entries in $\mathbb{C}$. Then, the set of all units is formed by all matrices with non-zero determinant

$$(M_{2\times 2}(\mathbb{C}))^\times = \left\{\begin{pmatrix} a & b \\ c & d \end{pmatrix} : ad - bc \neq 0\right\},$$

and the set of all zero-divisors is formed by those matrices with zero determinant

$$(M_{2\times 2}(\mathbb{C}))^0 = \left\{\begin{pmatrix} a & b \\ c & d \end{pmatrix} : ad - bc = 0\right\} \setminus \left\{\begin{pmatrix} 0 & 0 \\ 0 & 0 \end{pmatrix}\right\}.$$

Proposition 5.3.11. *Let $m > 1$ be fixed, and consider the ring $(\mathbb{Z}/m\mathbb{Z},+,\cdot)$.*

(1) *A congruence class $a \bmod m$ is a unit in $\mathbb{Z}/m\mathbb{Z}$ if and only if $a \bmod m$ has a multiplicative inverse in $\mathbb{Z}/m\mathbb{Z}$ if and only if $\gcd(a,m) = 1$.*

(2) *The set of all units in $\mathbb{Z}/m\mathbb{Z}$ is denoted by $U_m = (\mathbb{Z}/m\mathbb{Z})^\times$ and it is equal to*

$$U_m = (\mathbb{Z}/m\mathbb{Z})^\times = \{a \bmod m : 1 \leq a \leq m-1 \text{ and } \gcd(a,m) = 1\}.$$

(3) *The set of all zero-divisors in $\mathbb{Z}/m\mathbb{Z}$ is*

$$(\mathbb{Z}/m\mathbb{Z})^0 = \{a \bmod m : 1 \leq a \leq m-1 \text{ and } \gcd(a,m) > 1\}.$$

Proof. By definition of unit, a congruence class $a \bmod m$ is a unit in $\mathbb{Z}/m\mathbb{Z}$ if and only if there is some $b \bmod m$ such that $a \cdot b \equiv 1 \bmod m$. In other words, $a \bmod m$ is a unit if and only if $ax \equiv 1 \bmod m$ has a solution. This means that $a \bmod m$ is a unit if and only if $a \bmod m$ is an invertible congruence class (as in Definition 5.2.8). By Proposition 5.2.13, this is equivalent to $\gcd(a, m) = 1$. This shows (1).

By part (1), the units are equal to the invertible classes, equal to those $a \bmod m$ with $\gcd(a, m) = 1$. Thus, U_m is the set given in the statement of the proposition.

Finally, suppose that $a \bmod m$ is a zero-divisor. Equivalently, $a \not\equiv 0 \bmod m$ but there is some non-zero $b \bmod m$ such that $ab \equiv 0 \bmod m$. This means that m divides ab but m does not divide a and m does not divide b. Thus, $\gcd(a, m) > 1$ and $\gcd(b, m) > 1$. Conversely, if $\gcd(a, m) = d > 1$ and $1 \leq a \leq m - 1$, then we can write $a = dk$ and $m = dh$, for some $k, h \in \mathbb{Z}$. It follows that

$$a \cdot h = dk \cdot \frac{m}{d} = k \cdot m \equiv 0 \bmod m,$$

and, moreover $a \not\equiv 0 \bmod m$ and $b \not\equiv 0 \bmod m$. Thus, a is a zero-divisor. We conclude that the set of all zero-divisors is the set given in the statement of the proposition. □

When we specialize the previous proposition to the case when m is prime, we obtain the following corollary.

Corollary 5.3.12. *Let p be a prime. Then, the ring $(\mathbb{Z}/p\mathbb{Z}, +, \cdot)$ does not have any zero-divisors, and every non-zero element is a unit; i.e.,*

$$U_p = (\mathbb{Z}/p\mathbb{Z})^\times = \mathbb{Z}/p\mathbb{Z} - \{0 \bmod p\}.$$

Proof. This follows directly from Proposition 5.3.11, once we realize that every number $1 \leq a \leq p - 1$ is relatively prime to p. □

Example 5.3.13. Let $m = 15$. Then, the units in $\mathbb{Z}/15\mathbb{Z}$ form the set

$$U_{15} = \{1, 2, 4, 7, 8, 11, 13, 14 \bmod 15\},$$

and the zero-divisors are the congruence classes in the set

$$(\mathbb{Z}/15\mathbb{Z})^0 = \{3, 5, 6, 9, 10, 12 \bmod 15\}.$$

We remark that $\mathbb{Z}/15\mathbb{Z} = \{0\} \cup U_{15} \cup (\mathbb{Z}/15\mathbb{Z})^0$.

Example 5.3.14. Let $m = 16$. Then, the units in $\mathbb{Z}/16\mathbb{Z}$ form the set

$$U_{16} = \{1, 3, 5, 7, 9, 11, 13, 15 \bmod 16\},$$

and the zero-divisors are the congruence classes in the set

$$(\mathbb{Z}/16\mathbb{Z})^0 = \{2, 4, 6, 8, 10, 12, 14 \bmod 16\}.$$

We remark that $\mathbb{Z}/16\mathbb{Z} = \{0\} \cup U_{16} \cup (\mathbb{Z}/16\mathbb{Z})^0$.

Example 5.3.15. Let $m = 17$. Then, the units in $\mathbb{Z}/17\mathbb{Z}$ form the set

$$U_{17} = \{1, 2, 3, 4, 5, 6, 7, 8, 9, 10, 11, 12, 13, 14, 15, 16 \bmod 17\},$$

and the zero-divisors are the congruence classes in the empty set; i.e., the set $(\mathbb{Z}/17\mathbb{Z})^0 = \emptyset$. We remark that $\mathbb{Z}/17\mathbb{Z} = \{0\} \cup U_{17} = \{0\} \cup U_{17} \cup (\mathbb{Z}/17\mathbb{Z})^0$.

Proposition 5.3.16. *Let $(R, +, \cdot)$ be a commutative ring with identity, and let $R^\times$ be the set of all units in R. Then, the pair $(R^\times, \cdot)$ is an abelian group.*

Proof. We need to verify that the pair $(R^\times, \cdot)$ verifies the axioms of a group in Definition 5.2.2.

(1) The $\cdot$ operation is associative in $R^\times$: since $\cdot$ is associative in R and since R and $R^\times$ share the same operation, it follows that $\cdot$ is associative in $R^\times \subseteq R$.

(2) The set $R^\times$ is closed under the $\cdot$ binary operation: let us show that if u and u' are units in R, then uu' is also a unit in R. By the definition of unit, there are v and $v' \in R$ such that $uv = u'v' = 1 \in R$. Thus, the element vv' is an inverse for uu', since

$$uu' \cdot vv' = (uv) \cdot (u'v') = 1 \cdot 1 = 1,$$

where we have used the fact that $\cdot$ commutes in R, because R is assumed to be a commutative ring. Thus, $uu' \in R^\times$ and $R^\times$ is closed under multiplication.

(3) There exists an identity element in $R^\times$ for the $\cdot$ operation: the ring R is assumed to have a unity $1 \in R$, and clearly $1 \cdot 1 = 1$, so in particular 1 is a unit. Moreover, $1 \cdot u = u \cdot 1 = u$ for all $u \in R^\times$, and so 1 is an identity element for $(R^\times, \cdot)$ with respect to its operation.

(4) There are inverses in $R^\times$ for the $\cdot$ operation: for every unit $u \in R^\times$, by definition of unit, there is a $v \in R$ such that $uv = 1$. Notice that v is also a unit, so $v \in R^\times$ and, moreover, $v = u^{-1}$.

(5) The operation $\cdot$ is commutative in $R^\times$: since R is assumed to be a commutative ring, we have that $r \cdot s = s \cdot r$ for all $r, s \in R$. Since $R^\times \subseteq R$ and they share the same multiplication operation, we conclude that $(R^\times, \cdot)$ is commutative.

Hence, the pair $(R^\times, \cdot)$ is a commutative (or abelian) group, as claimed. □

Corollary 5.3.17. *Let $m > 1$ be fixed and let U_m be the set of all units in $\mathbb{Z}/m\mathbb{Z}$; i.e.,*

$$U_m = \{a \bmod m : 1 \leq a \leq m-1 \text{ and } \gcd(a, m) = 1\}.$$

Then, $(U_m, \cdot)$ is a commutative group.

Example 5.3.18. Let $m = 15$. Then, the units in $\mathbb{Z}/15\mathbb{Z}$ form the set

$$U_{15} = \{1, 2, 4, 7, 8, 11, 13, 14 \bmod 15\}.$$

By our previous proposition, the pair $(U_{15}, \cdot)$ forms a (commutative) group. In particular, whenever we multiply any two units in U_{15}, we see that their product is also a unit in U_{15}. For instance,

$$2 \cdot 4 \equiv 8 \bmod 15, \quad 4 \cdot 7 \equiv 28 \equiv 13 \bmod 15, \quad \text{and} \quad 11 \cdot 13 \equiv 8 \bmod 15.$$

It also follows that the multiplicative inverse of any element of U_{15} is also an element of U_{15}:

$$2^{-1} \equiv 8,\ 4^{-1} \equiv 4,\ 7^{-1} \equiv 13,\ 8^{-1} \equiv 2,\ 11^{-1} \equiv 11,\ 13^{-1} \equiv 7,\ 14^{-1} \equiv 14 \bmod 15.$$

Example 5.3.19. If p is a prime, then we have seen that $U_p = \mathbb{Z}/p\mathbb{Z} - \{0 \bmod p\}$. Thus, $(\mathbb{Z}/p\mathbb{Z} - \{0 \bmod p\}, \cdot)$ is a group.

It is worth noting that if $(G, *)$ is a group, then the inverses with respect to $*$ are unique; i.e., if $g \in G$, then there is a *unique* inverse element g^{-1} such that $g * g^{-1} = e$, where e is the identity element in the group G (this is left as Exercise 5.6.5). In particular, since $(R, +)$ and $(R^\times, \cdot)$ are groups, the additive inverse of an element $r \in R$ and the multiplicative inverse of a unit $u \in R^\times$ are unique. Let us show this directly, in the case of $(\mathbb{Z}/m\mathbb{Z}, +, \cdot)$.

Proposition 5.3.20. *Let $m > 1$ be fixed. Then:*

(1) *Every congruence class $a \bmod m$ has a unique additive inverse; i.e., there is a unique congruence class $b \bmod m$ such that*
$$a + b \equiv 0 \bmod m.$$
(2) *Every unit $u \bmod m$ in $(\mathbb{Z}/m\mathbb{Z})^\times$ has a unique multiplicative inverse; i.e., there is a unique congruence class $v \bmod m$ such that*
$$u \cdot v \equiv 1 \bmod m.$$

Proof. Suppose first that b and $b' \bmod m$ are additive inverses for $a \bmod m$. Then, by definition of additive inverse,
$$a + b \equiv 0 \bmod m \quad \text{and} \quad a + b' \equiv 0 \bmod m.$$
In particular, $a + b \equiv a + b' \bmod m$, or, equivalently, $(a + b) - (a + b') = b - b'$ is divisible by m. This means that $b \equiv b' \bmod m$, so they represent the same congruence class. Hence, the additive inverse of $a \bmod m$ is unique modulo m.

Similarly, suppose that v and $v' \bmod m$ are two multiplicative inverses for a unit $u \bmod m$. Then, by definition of multiplicative inverse,
$$u \cdot v \equiv 1 \bmod m \quad \text{and} \quad u \cdot v' \equiv 1 \bmod m.$$
In particular, $u \cdot v \equiv u \cdot v' \bmod m$. This means that $uv - uv' = u(v - v')$ is divisible by m. Since u is a unit, we have $\gcd(u, m) = 1$, by Corollary 5.3.17. Hence, $\gcd(u, m) = 1$ and m divides $u(v - v')$ imply that m divides $v - v'$, by Corollary 2.7.6. Therefore, $v \equiv v' \bmod m$, and v and v' represent the same congruence class modulo m. We conclude that the multiplicative inverse of $u \bmod m$ is unique modulo m, as claimed. □

We finish this section comparing $\mathbb{Z}/m\mathbb{Z}$ and $\mathbb{Z}/n\mathbb{Z}$, when n is a divisor of m.

Proposition 5.3.21. *Let $m \geq 2$ be an integer and let $n \geq 2$ be a divisor of m. Then, the map*
$$\psi\colon \mathbb{Z}/m\mathbb{Z} \to \mathbb{Z}/n\mathbb{Z}$$
that sends $a \bmod m$ to $a \bmod n$ is well-defined and surjective and it induces a surjection $\psi\colon U_m \to U_n$.

Proof. Let us first verify that ψ is well-defined. Suppose a and b are integer representatives of the same class modulo m; i.e., $a \equiv b \bmod m$. Equivalently, there is some $k \in \mathbb{Z}$ such that $a - b = km$. Since n is a divisor of m, there is some $t \in \mathbb{Z}$ such that $m = tn$ and $a - b = ktn$. Hence, $a \equiv b \bmod n$. In particular,
$$\psi(a \bmod m) \equiv a \equiv b \equiv \psi(b \bmod m) \bmod n.$$
Thus, ψ is well-defined.

Clearly, ψ is surjective: if c is a representative of a class modulo n, then $\psi(c \bmod m) \equiv c \bmod n$.

Let us show that ψ sends U_m to U_n. Let u be a representative of a unit modulo m. Then, $\gcd(u, m) = 1$. If we write $m = tn$ as before, then $\gcd(u, tn) = 1$ and in particular $\gcd(u, n) = 1$. Thus, u is also a representative for a unit modulo n. Hence, $\psi(u \bmod m) \equiv u \bmod n$ is a unit modulo n, and $\psi(U_m) \subseteq U_n$.

Finally, it remains to show that $\psi(U_m) = U_n$. Let v be a unit modulo n (so that $\gcd(v, n) = 1$). Let n' be the largest divisor of m such that $n|n'$ and such that n and n' share all their prime divisors. Then, we have that $\gcd(v, n') = 1$ and $m = n't'$ with $\gcd(t', n') = 1$. In particular, the image of $U_{n'}$ is U_n under $\mathbb{Z}/n'\mathbb{Z} \to \mathbb{Z}/n\mathbb{Z}$ and ψ factors as

$$\psi : \mathbb{Z}/m\mathbb{Z} \to \mathbb{Z}/n'\mathbb{Z} \to \mathbb{Z}/n\mathbb{Z}.$$

So to finish our proof, it suffices to show that the image of U_m is $U_{n'}$ under $\psi' : \mathbb{Z}/m\mathbb{Z} \to \mathbb{Z}/n'\mathbb{Z}$. Let v be a unit modulo n' (so that $\gcd(v, n') = 1$). Then, there is a $j \in \mathbb{Z}$ such that

$$v + jn' \equiv 1 \bmod t'.$$

Indeed, the linear equation

$$v - 1 = t'x + n'y$$

by Proposition 2.9.1 has a solution $(x, y) = (i, j) \in \mathbb{Z}^2$ because $\gcd(n', t') = 1$, by the definition of n'. In particular, with this choice of j, the number $v + jn'$ is relatively prime to t'. In addition, $\gcd(v + jn', n') = \gcd(v, n') = 1$, so $v + jn'$ is also relatively prime to n'. Hence, $v + jn'$ is relatively prime to $n't' = m$, and therefore it is a unit modulo m. Finally,

$$\psi'(v + jn' \bmod m) \equiv v + jn' \equiv v \bmod n'.$$

Hence, $\psi'(U_m) = U_{n'}$, and when reduced modulo n, the image of $U_{n'}$ is U_n. Thus, $\psi(U_m) = U_n$, as we wanted to prove. □

Remark 5.3.22. Let $\psi\colon \mathbb{Z}/m\mathbb{Z} \to \mathbb{Z}/n\mathbb{Z}$ be the reduction map that we discussed in Proposition 5.3.21, where n is a divisor of m. Here is a slick proof of the fact that $\psi(U_m) = U_n$ which uses a *deep* theorem about prime numbers, namely Dirichlet's theorem on primes in arithmetic progressions (Theorem 3.3.11).

We want to show that for any unit $v \bmod n$, there is a unit $u \bmod m$ such that $u \equiv v \bmod n$. Since $\gcd(v, n) = 1$, Dirichlet's theorem implies that there are *infinitely many* primes p_i, $i = 1, 2, 3, \ldots$, such that $p_i \equiv v \bmod n$. Since m has only finitely many prime divisors (by the fundamental theorem of arithmetic, Theorem 2.10.6!), we may choose a prime p_i that is not a divisor of m and such that $p_i \equiv v \bmod n$. Hence, $\gcd(p_i, m) = 1$, so p_i is a unit modulo m, and

$$\psi(p_i \bmod m) \equiv p_i \equiv v \bmod n.$$

Therefore, we may pick $u \equiv p_i \bmod m$ to be our unit in U_m that maps to $v \bmod n$ via ψ, as desired.

5.3.1. Ring Homomorphisms. As in the case of groups, we use maps to compare rings, and in order to reach ring-theoretic conclusions, we need our maps to preserve ring structures. Such maps are called ring homomorphisms.

Definition 5.3.23. Let R and S be rings (with identity). A function $f\colon R \to S$ is a *ring homomorphism* if

(1) $f(r + r') = f(r) + f(r')$ and

(2) $f(r \cdot r') = f(r) \cdot f(r')$

for all r and r' in R. A bijective ring homomorphism is called a *ring isomorphism.*

Example 5.3.24. If R is any ring, the zero map $f_0\colon R \to R$ defined by $f_0(r) = 0$ for all $r \in R$ is a ring homomorphism. Similarly, the identity map $f_1(r) = r$ is also a homomorphism of rings.

Example 5.3.25. Let $m > 1$ be an integer and consider the reduction modulo m map $\mathbb{Z} \to \mathbb{Z}/m\mathbb{Z}$. This map is a ring homomorphism from $\mathbb{Z}$ to $\mathbb{Z}/m\mathbb{Z}$.

Example 5.3.26. Let $m > 1$ and let $n > 1$ be a divisor of m. Then, the map $\psi\colon \mathbb{Z}/m\mathbb{Z} \to \mathbb{Z}/n\mathbb{Z}$ that sends $a \bmod m$ to $a \bmod n$ is a ring homomorphism from $\mathbb{Z}/m\mathbb{Z}$ to $\mathbb{Z}/n\mathbb{Z}$. This map was studied in Proposition 5.3.21.

5.3.2. Ideals. An important concept in the theory of rings is that of *ideals*, which is a special type of subset of a ring. Ideals were first introduced by Richard Dedekind in 1876, who was in turn reformulating ideas of Ernst Kummer. However, it was Emmy Noether's work that showcased the wide range of applications of ideals in algebra, algebraic geometry, and other areas of mathematics.

Figure 5.2. Emmy Noether (1882–1935) was a German mathematician, who is regarded as one of the most important women in the history of mathematics. Image source: Wikipedia Commons.

Definition 5.3.27. Let $(R, +, \cdot)$ be a commutative ring with identity. An *ideal* in R is an additive subgroup I of R such that I is closed under multiplication by R.

Equivalently, $I \subseteq R$ is an ideal if the following conditions are satisfied:

- If $a, b \in I$, then $a + b \in I$.
- If $r \in R$ and $a \in I$, then $r \cdot a \in I$.

Example 5.3.28. Let $(R, +, \cdot)$ be a commutative ring with identity. Then, $\{0\}$ and R are ideals. If R is a field, then $\{0\}$ and R are the only ideals of R (see Exercise 5.6.17).

Example 5.3.29. Let $R = C^0(\mathbb{R})$ be the ring of continuous functions $f \cdot \mathbb{R} \to \mathbb{R}$, as in Example 5.3.4. Then, the set of functions that vanish at $x = 0$, given by

$$I = \{f \in C^0(\mathbb{R}) : f(0) = 0\},$$

is an ideal of $C^0(\mathbb{R})$. Indeed, if $f, g \in I$ and $h \in C^0(\mathbb{R})$, then

- $(f + g)(0) = f(0) + g(0) = 0 + 0 = 0$, so $f + g \in I$, and
- $(h \cdot f)(0) = h(0) \cdot f(0) = h(0) \cdot 0 = 0$, so $h \cdot f \in I$.

These two properties show that I is an ideal in $C^0(\mathbb{R})$.

Example 5.3.30. Let m be an integer. Then, $m\mathbb{Z} = \{m \cdot n : n \in \mathbb{Z}\}$ is an ideal of the ring $\mathbb{Z}$. Moreover, every ideal of $\mathbb{Z}$ is of this form (this is left as an exercise for the reader; see Exercise 5.6.16).

Example 5.3.31. Let R and S be commutative rings with identity, and let $\psi \colon R \to S$ be a ring homomorphism. Then, the kernel of ψ, defined by $\mathrm{Ker}(\psi) = \{r \in R : \psi(r) = 0_S\}$, is an ideal of R. Indeed, suppose that $k, j \in \mathrm{Ker}(\psi)$ and $r \in R$. Then,

- $\psi(k + j) = \psi(k) + \psi(j) = 0_S + 0_S = 0_S$; thus, $k + j$ is also in the kernel, and
- $\psi(r \cdot k) = \psi(r) \cdot \psi(k) = \psi(r) \cdot 0_S = 0_S$, so $r \cdot k \in \mathrm{Ker}(\psi)$,

where, in both cases, we have used the properties of a ring homomorphism and the property of the zero element of a ring.

5.4. Fields

Those commutative rings where every non-zero element is a unit have a special name.

Definition 5.4.1. A commutative ring with identity $(R, +, \cdot)$ is called a *field* if $(R - \{0\}, \cdot)$ is a commutative group. Equivalently, $(R, +, \cdot)$ is a field if $R^\times = R - \{0\}$.

Example 5.4.2. The rings $\mathbb{Q}$, $\mathbb{R}$, and $\mathbb{C}$ are fields, because every non-zero element in each ring has a multiplicative inverse in the ring. For instance,

$$\left(\frac{3}{7}\right)^{-1} = \frac{7}{3} \in \mathbb{Q}, \quad \pi^{-1} = \frac{1}{\pi} \in \mathbb{R}, \quad (1 + i)^{-1} = \frac{1}{2} - \frac{i}{2} \in \mathbb{C}.$$

Notice that $\mathbb{Q}$, $\mathbb{R}$, and $\mathbb{C}$ are infinite fields. Our next theorem describes our first examples of finite fields.

Theorem 5.4.3. *Let $m > 1$ be a fixed integer. The ring $(\mathbb{Z}/m\mathbb{Z}, +, \cdot)$ is a field if and only if m is a prime number.*

Proof. We have seen in Proposition 5.3.6 that $(\mathbb{Z}/m\mathbb{Z}, +, \cdot)$ is a commutative ring with identity, for any $m > 1$. In Proposition 5.3.11, we showed that

$$U_m = (\mathbb{Z}/m\mathbb{Z})^\times = \{a \bmod m : 1 \le a \le m-1 \text{ and } \gcd(a,m) = 1\},$$

and in Corollary 5.3.17 we have seen that $((\mathbb{Z}/m\mathbb{Z})^\times, \cdot)$ is a commutative group. Thus, $(\mathbb{Z}/m\mathbb{Z})^\times = \mathbb{Z}/m\mathbb{Z} - \{0 \bmod m\}$ if and only if m is a prime number. Hence, $(\mathbb{Z}/m\mathbb{Z}, +, \cdot)$ is a field if and only if m is prime, as claimed. □

Example 5.4.4. The previous theorem provides an infinite source of examples of *finite* fields: $(\mathbb{Z}/p\mathbb{Z}, +, \cdot)$ is a field for each prime p. To ease notation, we usually denote a field with p elements by $\mathbb{F}_p$:

$$\mathbb{F}_2 = \mathbb{Z}/2\mathbb{Z}, \quad \mathbb{F}_3 = \mathbb{Z}/3\mathbb{Z}, \quad \mathbb{F}_5 = \mathbb{Z}/5\mathbb{Z}, \ldots.$$

Theorem 5.4.3 shows that for any prime p, there is a field with exactly p elements, namely $\mathbb{Z}/p\mathbb{Z}$. Some follow-up questions naturally come to mind: is there a field with 4 elements? Is there a field with 9 elements? Is there a field with 6 elements? We will come back to these interesting questions in Chapter 6 of this book.

Example 5.4.5. Let d be a square-free integer. Let $\mathbb{Q}(\sqrt{d})$ be the subring of complex numbers $\mathbb{C}$ generated by 1 and $\sqrt{d}$, with the addition and multiplication operations inherited from $\mathbb{C}$. Then:

(1) The ring $\mathbb{Q}(\sqrt{d})$ is precisely

$$\mathbb{Q}(\sqrt{d}) = \left\{a + b\sqrt{d} : a, b \in \mathbb{Q}\right\}.$$

(2) The ring $\mathbb{Q}(\sqrt{d})$ is a field.

These fields (called quadratic fields) will be studied in some detail in Sections 12.5 and 14.3.1. The fact that $\mathbb{Q}(\sqrt{d})$ is indeed a field will be shown in Proposition 12.5.3. We will use quadratic fields in our study of the rational and integral points on ellipses and hyperbolas.

The notion of ring homomorphism (as in Definition 5.3.23) can be extended to fields, as follows.

Definition 5.4.6. Let $(F, +, \cdot)$ and $(K, +, \cdot)$ be fields. A map $\psi\colon F \to K$ is a *field homomorphism* if it is a ring homomorphism. In other words, ψ satisfies the following conditions, for all $f, g \in F$:

(1) $\psi(f+g) = \psi(f) + \psi(g)$, and
(2) $\psi(f \cdot g) = \psi(f) \cdot \psi(g)$.

If ψ is injective and surjective, then we say that ψ is an *isomorphism of fields*.

Example 5.4.7. If F and K are fields, the map $F \to K$ that sends every element of F to 0_K is the zero homomorphism.

Example 5.4.8. Let K be a field and let $F \subseteq K$ be a subfield of K. Then, the inclusion map $F \hookrightarrow K$ that sends $f \mapsto f \in K$ is a field homomorphism. For instance, $\mathbb{Q} \subseteq \mathbb{R}$ or $\mathbb{R} \subseteq \mathbb{C}$ gives rise to a field homomorphism.

Moreover, if $\psi\colon F \to K$ is a field homomorphism, then it is either identically zero (as in Example 5.4.7) or it is injective (Exercise 5.6.18).

Example 5.4.9. Let $\mathbb{C}$ be the field of complex numbers and let $c\colon \mathbb{C} \to \mathbb{C}$ be the map defined by

$$a + bi \mapsto a - bi.$$

Then, c is a field homomorphism that we usually call complex conjugation (and we usually write $c(a+bi) = \overline{a+bi} = a-bi$). Let us show that c is a field homomorphism:

$$\begin{aligned} c((a+bi)+(d+ei)) &= c((a+d)+(b+e)i) = (a+d)-(b+e)i \\ &= (a-bi)+(d-ei) = c(a+bi)+c(d+ei), \end{aligned}$$

and

$$\begin{aligned} c((a+bi)\cdot(d+ei)) &= c((ad-be)+(ae+bd)i) = (ad-be)-(ae+bd)i \\ &= (a-bi)\cdot(d-ei) = c(a+bi)\cdot c(d+ei). \end{aligned}$$

Moreover, c is injective (if $a - bi = d - ei$, then $a = d$ and $b = e$) and surjective ($c^{-1}(d+ei) = d-ei$). Thus, c is an isomorphism of $\mathbb{C}$ to $\mathbb{C}$ (i.e., an automorphism of $\mathbb{C}$). Since $(c \circ c)(\alpha) = \alpha$, for every $\alpha \in \mathbb{C}$, we say that c is an involution.

5.5. Rings of Polynomials

In this section we introduce new examples of rings that will be very useful in later chapters: the rings of polynomials. The operations of addition and multiplication of polynomials are defined in the most natural way one could imagine.

Example 5.5.1. The ring of polynomials with coefficients in $\mathbb{Z}$ will be denoted by $\mathbb{Z}[x]$. The addition and multiplication in $\mathbb{Z}[x]$ will be defined naturally so that $+$ and $\cdot$ are commutative and associative and satisfy the usual distributive laws. For instance,

$$(1+x)+(-2+x+3x^2) = -1+2x+3x^2,$$

and

$$(1+x)\cdot(-2+x+3x^2) = -2-x+4x^2+3x^3.$$

Definition 5.5.2. Let $(R,+,\cdot)$ be a commutative ring with identity $1 \in R$, and define a triple $(R[x],+,\cdot)$ as follows.

(1) The set $R[x]$ is formed by all polynomials in the variable x with coefficients in R; i.e.,

$$R[x] = \{r_0 + r_1x + r_2x^2 + \cdots + r_nx^n : n \geq 0,\ r_i \in R\}.$$

(2) The addition operation $+$ of two polynomials $p(x) = \sum_{i=0}^{n} r_ix^i$ and $q(x) = \sum_{j=0}^{m} s_jx^j$ is defined by the formula

$$p(x)+q(x) = \sum_{i=0}^{n} r_ix^i + \sum_{j=0}^{m} s_jx^j = \sum_{k=0}^{\max\{m,n\}} (r_k+s_k)x^k.$$

(3) The multiplication operation $\cdot$ of two polynomials $p(x)$ and $q(x)$ as above is defined by the formula

$$p(x)\cdot q(x) = \left(\sum_{i=0}^{n} r_ix^i\right)\cdot\left(\sum_{j=0}^{m} s_jx^j\right) = \sum_{k=0}^{m+n} t_kx^k,$$

where

$$t_k = \sum_{\substack{i+j=k \\ i,j\geq 0}} r_i s_j.$$

Theorem 5.5.3. *If $(R, +, \cdot)$ is a commutative ring with identity $1 \in R$, then $R[x]$ is also a commutative ring with identity $1 \in R[x]$.*

We will not provide a proof of this theorem in this book, but it is not deep, only tedious, to check that all the axioms are satisfied. For our purposes, it suffices to say that the additive identity of $R[x]$ is the polynomial identically equal to $0 \in R$ and the multiplicative identity is given by the polynomial identically equal to $1 \in R$.

Example 5.5.4. Thanks to our previous theorem, now we can build many new rings. For instance, we may consider $\mathbb{Z}[x]$, $\mathbb{Q}[x]$, $\mathbb{R}[x]$, and $\mathbb{C}[x]$. Since $R = \mathbb{Z}[x]$ is a ring, we may construct a new ring of polynomials with coefficients in R, i.e., $R[y] = (\mathbb{Z}[x])[y]$ which is formed by polynomials in two variables and that we usually denote by $\mathbb{Z}[x, y]$.

Of course, in this book we are particularly interested in the ring $\mathbb{Z}/m\mathbb{Z}$, for some $m > 1$, and we will also be very interested in the study of the polynomial rings $(\mathbb{Z}/m\mathbb{Z})[x]$. This ring of polynomials behaves just as $\mathbb{Z}[x]$ does, except that we always reduce each coefficient modulo m.

Example 5.5.5. Let $m = 15$ and consider the ring of polynomials $(\mathbb{Z}/15\mathbb{Z})[x]$ and, in particular, consider the product of the two polynomials $x - 3 \bmod 15$ and $x - 5 \bmod 15$:

$$(x-3)(x-5) \equiv x^2 - 3x - 5x + 15 \equiv x^2 - 8x \equiv x(x-8) \bmod 15.$$

We notice that the polynomial $x^2 - 8x$ has two distinct factorizations in $(\mathbb{Z}/15\mathbb{Z})[x]$; namely,

$$x^2 - 8x \equiv x(x-8) \equiv (x-3)(x-5) \bmod 15.$$

This indicates that, unlike $\mathbb{Z}$, the elements in the ring $(\mathbb{Z}/15\mathbb{Z})[x]$ do not have a unique factorization into "primes" (however, we have not defined *prime* in a general ring).

As in the case of $\mathbb{Z}$ we can define a concept of divisibility of polynomials.

Definition 5.5.6. Let R be a ring and let $R[x]$ be the associated ring of polynomials with coefficients in R. Let $p(x), q(x) \in R[x]$ be polynomials. We say that $q(x)$ *divides* the polynomial $p(x)$ if there is another polynomial $t(x) \in R[x]$ such that $p(x) = q(x)t(x)$. In this case, we say that $q(x)$ is a *divisor* of $p(x)$ or that $p(x)$ is a multiple of $q(x)$. We also write $q(x) \mid p(x)$.

Every polynomial has a term of maximum degree, and this degree is called the degree of the polynomial.

Definition 5.5.7. Let $(R, +, \cdot)$ be a commutative ring with identity, and let $R[x]$ be the ring of polynomials with coefficients in R. The *degree* of a non-zero polynomial $p(x) \in R[x]$ is the exponent of the highest power of x in $p(x)$ with a non-zero coefficient; i.e., if $p(x)$ is given by

$$p(x) = a_d x^d + a_{d-1}x^{d-1} + \cdots + a_2x^2 + a_1x + a_0,$$

with $a_i \in R$ and $a_d \neq 0$, then the degree of $p(x)$ is d. We write $\deg p(x) = d$.

As the reader may have noticed, Definition 5.5.7 does not assign a degree to the identically zero polynomial $p(x) = 0$, for if we were to assign a value to the degree of 0, it would create conflicts in the formulas we are about to quote below. There is one way, however, to extend the definition of degree to the zero polynomial, and that is by assigning $\deg 0 = -\infty$.

Lemma 5.5.8. *Let $R[x]$ be a ring of polynomials, and let $p(x)$ and $q(x)$ be arbitrary non-zero polynomials in $R[x]$.*

(1) *If $p(x) + q(x) \neq 0$, then $\deg(p(x) + q(x)) \leq \max\{\deg p(x), \deg q(x)\}$.*
(2) *If $p(x) \cdot q(x) \neq 0$, then $\deg(p(x) \cdot q(x)) \leq \deg p(x) + \deg q(x)$. Moreover, if R is a field, then $\deg(p(x) \cdot q(x)) = \deg p(x) + \deg q(x)$.*

The proof of the previous lemma follows directly from the definition of addition and multiplication of polynomials and the fact that every non-zero element in a field is a unit (therefore, there are no zero-divisors). We leave the proof as an exercise for the reader (Exercises 5.6.13 and 5.6.20).

Exercise 5.5.9. Let $(\mathbb{Z}/6\mathbb{Z})[x]$ be the ring of polynomials with coefficients in $\mathbb{Z}/6\mathbb{Z}$, and let $p(x) \equiv 2x + 5$ and $q(x) \equiv 3x^2 + 1 \bmod 6$. Then,

$$p(x) + q(x) \equiv (2x+5) + (3x^2+1) \equiv 3x^2 + 2x + 6 \equiv 3x^2 + 2x \bmod 6,$$

and

$$p(x) \cdot q(x) \equiv (2x+5) \cdot (3x^2+1) \equiv 6x^3 + 15x^2 + 2x + 5 \equiv 3x^2 + 2x + 5 \bmod 6.$$

Thus, $\deg p(x) = 1$, $\deg q(x) = 2$, and

$$\deg(p+q) = 2 = \max\{1,2\}, \quad \deg(p \cdot q) = 2 \leq 1 + 2 = 3.$$

As in the case of $\mathbb{Z}$ and Theorem 2.4.4, we can also perform "long division" of polynomials in a ring $F[x]$ with coefficients in a field F. Before we prove the division theorem for polynomials, we need to verify that polynomials over a field satisfy a cancellation law.

Lemma 5.5.10. *Let F be a field, let $F[x]$ be the associated polynomial ring, and let $p(x), q(x) \in F[x]$. Then, $p(x) \cdot q(x) = 0$ if and only if $p(x) = 0$ or $q(x) = 0$.*

Proof. Let $p(x) = a_d x^d + \cdots + a_1 x + a_0$ and $q(x) = b_e x^e + \cdots + b_1 x + b_0$, where $d, e \geq 0$ and $a_i, b_j \in F$. It is clear that if $p(x) = 0$ or $q(x) = 0$, then $p(x) \cdot q(x) = 0$.

Conversely, suppose that $p(x) \cdot q(x) = 0$, but $p(x), q(x) \neq 0$. Then, we may assume that $a_d, b_e \neq 0$, where $d, e \geq 0$. Moreover,

$$\begin{aligned} p(x) \cdot q(x) &= (a_d x^d + \cdots + a_1 x + a_0) \cdot (b_e x^e + \cdots + b_1 x + b_0) \\ &= (a_d b_e) x^{d+e} + \cdots + (a_1 b_0 + b_1 a_0) x + a_0 b_0. \end{aligned}$$

It follows that, if $p(x) \cdot q(x) = 0$, then $a_d b_e = 0$. Since F is a field, we conclude that $a_d = 0$ or $b_e = 0$ (see Exercise 5.6.19), which is a contradiction to our assumption that $a_d, b_e \neq 0$. Thus, we must have that $p(x) = 0$ or $q(x) = 0$, as claimed. □

Remark 5.5.11. Notice that the cancellation law does not hold in general if the coefficients of the polynomials are not in a field. For instance,

$$(2x+2)(3x+3) \equiv 0 \bmod 6,$$

but the polynomials $2x+2$ and $3x+3$ are non-zero in $(\mathbb{Z}/6\mathbb{Z})[x]$.

We are now ready to prove the division theorem for polynomials:

Theorem 5.5.12 (Division theorem for polynomials). *Let F be a field, and let $F[x]$ be the ring of polynomials with coefficients in F. Then, for all polynomials $a(x), b(x) \in F[x]$, with $a(x) \neq 0$, there exist unique polynomials $q(x)$ and $r(x)$ in $F[x]$ such that*

$$b(x) = a(x) \cdot q(x) + r(x),$$

such that either $r(x) = 0$ or $0 \leq \deg r(x) < \deg a(x)$.

Proof. Let $a(x), b(x) \in F[x]$, with $a(x) \neq 0$. We will first show the existence of polynomials $q(x)$ and $r(x)$ as in the statement of the theorem, and later we shall prove their uniqueness.

If there exists a polynomial $q(x) \in F[x]$ such that $b(x) = a(x)q(x)$, then we can take $r(x) = 0$. Otherwise, $b(x) - a(x)q(x)$ is never zero, and we may consider the set

$$S = \{d \geq 0 : d = \deg(b(x) - a(x) \cdot q(x)) \text{ for some } q(x) \in F[x]\}.$$

Clearly, S is non-empty (as $\deg b(x) \in S$), so either $d = 0 \in S$ or by the well-ordering principle (Section 2.1), the set $S \subseteq \mathbb{N}$ has a least element. In either case, S has a least element d_0. In particular, there is a polynomial $q(x)$ such that d_0 is the degree of $b(x) - a(x)q(x)$. Define $r(x) = b(x) - a(x)q(x)$, so that $\deg r(x) = d_0$. Furthermore,

$$b(x) = a(x) \cdot q(x) + r(x).$$

Next we show that $0 \leq \deg r(x) = d_0 < \deg a(x)$. Since $d_0 \in S$, it follows that $d_0 \geq 0$. Suppose for a contradiction that $\deg r(x) = d_0 \geq \deg a(x)$. Let us write

$$r(x) = r_{d_0}x^{d_0} + r_{d_0-1}x^{d_0-1} + \cdots + r_1x + r_0$$

and

$$a(x) = a_tx^t + \cdots + a_1x + a_0,$$

for some $r_i, a_j \in F$, with $r_{d_0}, a_t \neq 0$. The condition $\deg r(x) \geq \deg a(x)$ implies that $d_0 \geq t$. Let $s = d_0 - t \geq 0$ and define $\tilde{q}(x) = q(x) + r_{d_0}a_t^{-1}x^s$. Notice that a_t has an inverse in F, because F is a field and $a_t \neq 0$. Then, we have

$$\begin{aligned} b(x) &= a(x)q(x) + r(x) \\ &= a(x) \cdot (q(x) + r_{d_0}a_t^{-1}x^s) + r(x) - a(x) \cdot (r_{d_0}a_t^{-1}x^s) \\ &= a(x) \cdot \tilde{q}(x) + \tilde{r}(x), \end{aligned}$$

where we define $\tilde{r}(x) = r(x) - a(x) \cdot (r_{d_0}a_t^{-1}x^s)$. Notice that

$$\begin{aligned} \tilde{r}(x) &= r(x) - a(x) \cdot (r_{d_0}a_t^{-1}x^s) \\ &= r_{d_0}x^{d_0} + \cdots + r_1x + r_0 - (a_tx^t + \cdots + a_1x + a_0) \cdot (r_{d_0}a_t^{-1}x^s) \\ &= r_{d_0}x^{d_0} + \cdots + r_1x + r_0 - r_{d_0}x^{d_0} - r_{d_0}a_t^{-1}a_{t-1}x^{d_0-1} - \cdots - r_{d_0}a_0a_t^{-1}x^s \\ &= r_{d_0-1}x^{d_0-1} + \cdots + r_1x + r_0 - r_{d_0}a_t^{-1}a_{t-1}x^{d_0-1} - \cdots - r_{d_0}a_0a_t^{-1}x^s. \end{aligned}$$

Thus, $\deg \tilde{r}(x) \leq d_0 - 1 < \deg r(x)$. Since

$$\tilde{r}(x) = b(x) - \tilde{q}(x) \cdot a(x),$$

it follows that $\deg \tilde{r}(x)$ belongs to S. But $\deg \tilde{r}(x) < d_0$ and d_0 was supposed to be the least element in S. This is a contradiction and, therefore, our assumption that $\deg r(x) = d_0 \geq \deg a(x)$ must be wrong, and we must have $d_0 < \deg a(x)$, as desired.

It remains to show that $q(x)$ and $r(x)$ are unique. Suppose that $q'(x)$ and $r'(x) \in F[x]$ are other polynomials such that

$$b(x) = a(x)q'(x) + r'(x)$$

and either $r'(x) = 0$ or we have that $0 \leq \deg r'(x) < \deg a(x)$. Since we also have that $b(x) = a(x)q(x) + r(x)$, we have

$$a(x)q(x) + r(x) = a(x)q'(x) + r'(x),$$

and, as a consequence,

$$a(x)(q(x) - q'(x)) = r'(x) - r(x).$$

If $q(x) - q'(x) \neq 0$, then Lemma 5.5.8 implies that the degree of the left-hand side of the equation is at least $\deg a(x)$, while the degree of the right-hand side is less than $\deg a(x)$. This is impossible, so we must have $q(x) = q'(x)$. Hence, $r'(x) - r(x) = 0$ and $r'(x) = r(x)$. It follows that $q(x)$ and $r(x)$ are unique. This concludes the proof of the theorem. □

Example 5.5.13. Let us find quotients and remainders in $\mathbb{Q}[x]$ and $(\mathbb{Z}/5\mathbb{Z})[x]$.

- For instance, when we divide $x^5 + 1$ by $x + 2$ in $\mathbb{Q}[x]$ we obtain a remainder of -31 and a quotient $x^4 - 2x^3 + 4x^2 - 8x + 16$. In particular, the polynomial $x + 2$ is a divisor of $x^5 + 1$ in $(\mathbb{Z}/31\mathbb{Z})[x]$, because

$$\begin{aligned} x^5 + 1 &= (x+2)(x^4 - 2x^3 + 4x^2 - 8x + 16) - 31 \\ &\equiv (x+2)(x^4 - 2x^3 + 4x^2 - 8x + 16) \bmod 31. \end{aligned}$$

- When we divide $x^5 + 1$ by $x^2 + 2$ in $\mathbb{Q}[x]$ we obtain a quotient of $x^3 - 2x$ and a remainder of $4x + 1$, because

$$x^5 + 1 = (x^2 + 2)(x^3 - 2x) + 4x + 1.$$

Corollary 5.5.14 (Remainder theorem). *Let F be a field, let $p(x) \in F[x]$ be a polynomial, and let $f \in F$. Then, the remainder of the division of $p(x)$ by $x - f$ is precisely $p(f)$.*

Proof. By the division theorem (Theorem 5.5.12), we can divide $p(x)$ by $x - f$ with quotient and remainder; i.e., there is a $q(x)$ and $r(x)$ such that

$$p(x) = (x - f)q(x) + r(x),$$

with either $r(x) = 0$ or we have that $0 \leq \deg r(x) < \deg(x - f) = 1$; i.e., either $r(x) = 0$ or we have $0 \leq \deg r(x) < 1$. Hence, $r(x) = r$ is a constant. Moreover,

$$p(f) = (f - f)q(f) + r = 0 \cdot q(f) + r = r$$

and we have shown that $p(f) = r$, as claimed. □

Corollary 5.5.15 (Root theorem). *Let F be a field, let $p(x) \in F[x]$ be a polynomial, and let f be an element of F. Then, the polynomial $x - f$ divides $p(x)$ if and only if $p(f) = 0$.*

Proof. Suppose first that $p(x)$ is divisible by $x - f$. Then, there is some $q(x)$ such that $p(x) = (x - f)q(x)$. Hence,

$$p(f) = (f - f)q(f) = 0 \cdot q(f) = 0.$$

Conversely, suppose that $p(f) = 0$. By Corollary 5.5.14,

$$p(x) = (x - f)q(x) + p(f),$$

for some $q(x) \in F[x]$. Since $p(f) = 0$, it follows that $p(x) = (x - f)q(x)$ and so, by definition, $(x - f)$ divides $p(x)$, as claimed. □

Example 5.5.16. Let $p(x) \equiv x^4 - 1 \bmod 5$ in $(\mathbb{Z}/5\mathbb{Z})[x]$. Let us find a factorization of $p(x)$. The first thing we notice is that $x \equiv 1 \bmod 5$ is a root. Thus, by the root theorem, $x - 1$ is a factor. Indeed,

$$x^4 - 1 \equiv (x - 1)(x^3 + x^2 + x + 1) \bmod 5.$$

Moreover $2^4 - 1 \equiv 15 \equiv 0 \bmod 5$, so $x \equiv 2 \bmod 5$ is also a root. Since 2 is not a root of $x - 1$, it must be a root of $x^3 + x^2 + x + 1$, and therefore the latter should be divisible by $x - 2$. Indeed,

$$x^3 + x^2 + x + 1 \equiv (x - 2)(x^2 + 3x + 2) \bmod 5.$$

The polynomial $x^2 + 3x + 2$ factors as $(x + 1)(x + 2)$. Therefore,

$$x^4 - 1 \equiv (x - 1)(x - 2)(x + 1)(x + 2) \equiv (x - 1)(x - 2)(x - 3)(x - 4) \bmod 5.$$

Definition 5.5.17. Let $R[x]$ be a ring of polynomials, where R is a ring. Let $p(x)$ be a polynomial. We say that $r \in R$ is a *zero or a root* of $p(x)$ if $p(r) = 0$, or, equivalently, if $x - r$ divides $p(x)$. If there is some $e \geq 1$ such that $(x - r)^e$, then we say that r is a zero of $p(x)$ with *multiplicity* e.

Example 5.5.18. Let $p(x) \equiv x^4 + 4x^3 + 4x + 4$ in $(\mathbb{Z}/5\mathbb{Z})[x]$. Since $p(3) = 205$, it follows that $x \equiv 3 \bmod 5$ is a root of $p(x)$ in $(\mathbb{Z}/5\mathbb{Z})[x]$. Hence, by the root theorem (Corollary 5.5.15), the polynomial $x - 3$ is a divisor of $p(x)$. Indeed, when we divide $p(x)$ by $x - 3$ we find that the remainder is 0 mod 5:

$$p(x) \equiv (x - 3)(x^3 + 2x^2 + x + 2) \bmod 5.$$

In fact, notice that 3 is also a root of $x^3 + 2x^2 + x + 2$, as $3^3 + 2 \cdot 3^2 + 3 + 2 \equiv 50 \equiv 0 \bmod 5$. Thus, $x - 3$ is a factor:

$$x^3 + 2x^2 + x + 2 \equiv (x - 3)(x^2 + 1) \bmod 5.$$

Moreover, $x^2 + 1 \equiv x^2 - 4 \equiv (x - 2)(x + 2) \bmod 5$. Thus,

$$\begin{aligned} p(x) &\equiv x^4 + 4x^3 + 4x + 4 \\ &\equiv (x - 3)^2(x - 2)(x + 2) \\ &\equiv (x - 3)^3(x - 2) \bmod 5. \end{aligned}$$

Hence, 3 is a root with multiplicity three, because $(x-3)^3$ also divides $p(x)$. Notice that the roots of the polynomial $p(x)$ are $\{2,3\}$, or if we count them with multiplicities, the roots are $\{2,3,3,3\}$. In either case, the number of roots is ≤ 4 and $\deg(p(x)) = 4$.

Theorem 5.5.19. *Let F be a field and let $F[x]$ be the associated ring of polynomials with coefficients in F. Let $p(x) \in F[x]$ be a non-zero polynomial of degree n. Then, $p(x)$ has at most n roots in F, even when counted with multiplicities.*

Proof. We shall prove the theorem by induction on n, the degree of the non-zero polynomial $p(x)$. If $n = 0$, then $p(x) = p$ is a constant polynomial, for some $0 \neq p \in F$. Then, $p(x)$ has no zeros, as $p(x) = p = 0$ is impossible. Thus, the number of roots is ≤ 0, as needed.

Now, suppose that the theorem is true for all polynomials in $F[x]$ of degree n, and suppose $p(x)$ has degree $n+1$. If $p(x)$ has no roots in F, then the number of roots of $p(x)$ is $0 \leq n+1$, and the theorem holds for $p(x)$. Otherwise, if $p(x)$ has at least one root $r \in F$, Corollary 5.5.15 implies that $x - r$ is a divisor of $p(x)$; i.e., there is a $q(x) \in F[x]$ such that $p(x) = (x-r)q(x)$. Moreover, by Lemma 5.5.8, we have

$$n + 1 = \deg p(x) = \deg((x-r)q(x)) = \deg(x-r) + \deg q(x) = 1 + \deg q(x).$$

Hence, $\deg q(x) = n+1-1 = n$. By our induction hypothesis, the polynomial $q(x)$ has at most n roots, and any root of $q(x)$ is clearly also a root of $p(x)$, because $q(x)$ divides $p(x)$. Moreover, if s is a root of $p(x)$, then either $r = s$ or s is a root of $q(x)$, because

$$0 = p(s) = (s-r)q(s),$$

and so $s - r = 0$ or $q(s) = 0$. Hence, the number of roots of $p(x)$ is at most the number of roots of $x - r$ (1 root) plus the number of roots of $q(x)$ (n roots by the induction hypothesis). It follows that $p(x)$ has at most $n+1$ roots, as desired. This shows the induction step.

Hence, by the principle of mathematical induction, every polynomial of degree $n \geq 0$ has, at most, n roots. □

Remark 5.5.20. The conclusion of Theorem 5.5.19 is **false** if the coefficient ring of the polynomial is not a field. For instance, consider the polynomial $p(x) = x^2 - 1$ as an element of the ring of polynomials $(\mathbb{Z}/105\mathbb{Z})[x]$. This polynomial has 8 distinct roots, namely

$$x \equiv 1, 29, 34, 41, 64, 71, 76, 104 \bmod 105.$$

Thus, $p(x)$ has 8 distinct roots over $\mathbb{Z}/105\mathbb{Z}$, but the degree of $p(x)$ is only 2. See also Example 4.5.13.

Let us specialize Theorem 5.5.19 to the case of the finite field $F = \mathbb{Z}/p\mathbb{Z}$, for some prime p.

Corollary 5.5.21. *Let p be a prime, and let $p(x)$ be a polynomial in $(\mathbb{Z}/p\mathbb{Z})[x]$ of degree n. Then, $p(x)$ has at most n roots in $\mathbb{Z}/p\mathbb{Z}$, even when counted with multiplicities.*

Equivalently, let $p(x) \in \mathbb{Z}[x]$, of degree n, and such that the coefficient of x^n is not divisible by p. Then $p(x) \equiv 0 \bmod p$ has at most n distinct solutions modulo p, even when counted with multiplicities.

We finish this section with a result that says that if a polynomial is defined over F and all but one root is defined over F, then the remaining root is also defined over F. This result will be particularly useful in Parts 2 and 3 of the book (for instance, see Theorem 9.3.4).

Proposition 5.5.22. *Let $n \geq 2$, let F be a field, and let $p(x) \in F[x]$ be a polynomial with coefficients in F and degree n. Suppose that p has at least $n-1$ roots (counted with multiplicity) defined over F. Then, $p(x)$ has n roots defined over F.*

Proof. Let n, F, and $p(x)$ be as in the statement, and let $\alpha_1, \ldots, \alpha_{n-1} \in F$ be roots of $p(x)$, counted with multiplicity. By Corollary 5.5.15 we have

$$p(x) = (x - \alpha_1)(x - \alpha_2) \cdots (x - \alpha_{n-1})q(x),$$

for some $q(x) \in F[x]$. Since the degree of $p(x)$ is n, it follows that the degree of $q(x)$ is 1; i.e., $q(x)$ is a linear polynomial, say $q(x) = ax + b$, for some $a, b \in F$ and $a \neq 0$. Thus, the roots of $p(x)$ are precisely $\alpha_1, \ldots, \alpha_{n-1}$, and $\alpha_n = -b/a \in F$. So all n roots of $p(x)$ are actually defined over F. □

5.5.1. The Discriminant of a Polynomial. Let $p(x) = ax^2 + bx + c$ be a quadratic polynomial in $F[x]$, for some field F. If $F = \mathbb{C}$, then the well-known quadratic formula

$$x = \frac{-b \pm \sqrt{b^2 - 4ac}}{2a}$$

tells us the values of the two (complex) roots of $p(x)$. It is not immediate, however, whether the quadratic formula works for other fields such as $F = \mathbb{F}_p$, for a prime $p > 2$ (what is the meaning of $\sqrt{a}$ for $a \in \mathbb{F}_p$?). This topic will be revisited in detail in Chapter 10 when we introduce the concept of quadratic residues. In this section, instead, we concentrate on the quantity $b^2 - 4ac$ that appears as part of the quadratic formula and which we will refer to as the discriminant of $p(x)$.

When $F = \mathbb{R}$, the reader is probably familiar with the fact that the sign of $\Delta = b^2 - 4ac$ determines whether $p(x)$ has 0, 1, or 2 distinct real roots, according to whether $\Delta < 0$, $\Delta = 0$, or $\Delta > 0$, respectively. We shall define a discriminant for polynomials of arbitrary degree that generalizes the definition in the quadratic case and enjoys a similar *discriminating* property.

Definition 5.5.23. Let $n \geq 2$, and let $p(x) = a_n x^n + \cdots + a_1 x + a_0$ be a polynomial defined over a field F of degree n (i.e., $a_n \neq 0$), with roots $r_1, \ldots, r_n$ (not necessarily different) defined over a field extension L of F. We define the *discriminant* of $p(x)$ as

$$\Delta_{p(x)} = a_n^{2n-2} \cdot \prod_{1 \leq i < j \leq n} (r_j - r_i)^2.$$

Example 5.5.24. Let $p(x) = a_2 x^2 + a_1 x + a_0$ be a quadratic polynomial (i.e., $a_2 \neq 0$) defined over $\mathbb{C}$. Then, the roots of $p(x)$ are

$$r_1 = \frac{-a_1 - \sqrt{a_1^2 - 4a_0 a_2}}{2a_2} \quad \text{and} \quad r_2 = \frac{-a_1 + \sqrt{a_1^2 - 4a_0 a_2}}{2a_2}.$$

Thus,

$$\Delta_{p(x)} = a_2^2 \cdot (r_2 - r_1) = a_2^2 \cdot \frac{a_1^2 - 4a_0a_2}{a_2^2} = a_1^2 - 4a_0a_2,$$

which is the usual discriminant of a quadratic polynomial.

A direct consequence of the definition of a discriminant, as given in Definition 5.5.23, is that $\Delta_{p(x)} = 0$ if and only if $p(x)$ has a root of multiplicity ≥ 2. However, it is difficult to calculate a discriminant directly from this definition for polynomials of degree ≥ 3, when the roots are not already known. The following theorem expresses the discriminant solely in terms of the coefficients of the polynomial.

Theorem 5.5.25. *Let $p(x) = a_nx^n + \cdots + a_1x + a_0$ be a polynomial defined over a field F of degree n (i.e., $a_n \neq 0$). Then,*

$$\Delta_{p(x)} = (-1)^{n(n-1)/2}\frac{R(p,p')}{a_n},$$

where $R(p,p')$ is the resultant of $p(x)$, which is given by the determinant of the $(2n-1) \times (2n-1)$ matrix

$$\begin{pmatrix}
a_n & a_{n-1} & a_{n-2} & \cdots & a_2 & a_1 & a_0 & 0 & 0 & \cdots & 0 \\
0 & a_n & a_{n-1} & a_{n-2} & \cdots & a_2 & a_1 & a_0 & 0 & \cdots & 0 \\
\vdots & \vdots & \vdots & \vdots & \vdots & \vdots & \vdots & \vdots & \vdots & \vdots & \vdots \\
0 & 0 & \cdots & 0 & a_n & a_{n-1} & a_{n-2} & \cdots & a_2 & a_1 & a_0 \\
b_n & b_{n-1} & b_{n-2} & \cdots & b_2 & b_1 & 0 & 0 & 0 & \cdots & 0 \\
0 & b_n & b_{n-1} & b_{n-2} & \cdots & b_2 & b_1 & 0 & 0 & \cdots & 0 \\
\vdots & \vdots & \vdots & \vdots & \vdots & \vdots & \vdots & \vdots & \vdots & \vdots & \vdots \\
0 & 0 & \cdots & 0 & 0 & b_n & b_{n-1} & b_{n-2} & \cdots & b_2 & b_1
\end{pmatrix}$$

with $p'(x) = b_nx^{n-1} + \cdots + b_2x + b_1$, i.e., $b_k = k \cdot a_k$ for $1 \leq k \leq n$.

The determinant $R(p,p')$ is called the resultant of $p(x)$. We will not prove Theorem 5.5.25 here, which is a standard result about discriminants and resultants. Instead, let us apply the theorem to find formulas of discriminants for quadratic and cubic polynomials.

Example 5.5.26. Let $p(x) = a_2x^2 + a_1x + a_0$ be a quadratic polynomial (i.e., $a_2 \neq 0$) defined over a field F. Then, the resultant $R(p,p')$ of $p(x)$ is the determinant of

$$\begin{pmatrix} a_2 & a_1 & a_0 \\ 2a_2 & a_1 & 0 \\ 0 & 2a_2 & a_1 \end{pmatrix}.$$

Thus,

$$R(p,p') = a_1^2a_2 + 4a_0a_2^2 - 2a_1^2a_2 = 4a_0a_2^2 - a_1^2a_2.$$

Hence,

$$\Delta_{p(x)} = (-1) \cdot \frac{R(p,p')}{a_2} = a_1^2 - 4a_0a_2,$$

as expected.

Example 5.5.27. Let $p(x) = a_3x^3 + a_2x^2 + a_1x + a_0$ be a cubic polynomial (i.e., $a_3 \neq 0$) defined over a field F. Then, the resultant $R(p, p')$ of $p(x)$ is the determinant of

$$\begin{pmatrix} a_3 & a_2 & a_1 & a_0 & 0 \\ 0 & a_3 & a_2 & a_1 & a_0 \\ 3a_3 & 2a_2 & a_1 & 0 & 0 \\ 0 & 3a_3 & 2a_2 & a_1 & 0 \\ 0 & 0 & 3a_3 & 2a_2 & a_1 \end{pmatrix}.$$

We leave it up to the reader to verify that $\Delta_{p(x)}$ is given by

$$\Delta_{p(x)} = a_1^2a_2^2 - 4a_1^3a_3 - 4a_0a_2^3 - 27a_0^2a_3^2 + 18a_0a_1a_2a_3.$$

In particular, the discriminant of a polynomial of the form $p(x) = x^3 + Ax + B$ is given by

$$\Delta_{p(x)} = -(4A^3 + 27B^2).$$

The discriminant $\Delta = -(4A^3 + 27B^2)$ will be particularly relevant when we study elliptic curves in Chapter 16 (see Proposition 16.1.2, for instance).

Remark 5.5.28. We will continue our discussion of polynomials in Section 6.3, when discussing finite fields.

5.6. Exercises

Exercise 5.6.1. Let $m > 1$ be fixed. Suppose that $\mathbb{N}' \subseteq \mathbb{Z}/m\mathbb{Z}$ is a subset of congruence classes that satisfies the following properties:

(1) (Non-triviality) $\mathbb{N}'$ is non-empty.

(2) (Closure) $\mathbb{N}'$ is closed under $+$ and $\cdot$; that is, if $a, b \in \mathbb{N}'$, then $a + b$ and $a \cdot b \in \mathbb{N}'$.

(3) (Trichotomy) For all $a \in \mathbb{Z}/m\mathbb{Z}$, precisely one and only one of the following statements is true: $a \in \mathbb{N}'$ or $a \equiv 0$ or $-a \in \mathbb{N}'$.

Show that $\mathbb{N}'$ cannot exist.

Exercise 5.6.2. Let $(G, \cdot)$ and $(H, \cdot)$ be defined as in Example 5.2.18. Prove that $(G, \cdot)$ and $(H, \cdot)$ are groups.

Exercise 5.6.3. Let $m > 1$ be an integer, and let μ_m be the set of all mth roots of unity in the complex numbers; i.e.,

$$\mu_m = \{z \in \mathbb{C} : z^m = 1\} = \{e^{2n\pi i/m} : 0 \leq n < m\}.$$

Show that $(\mu_m, \cdot)$ is a group under multiplication.

Exercise 5.6.4. Let $\mathrm{SL}(2, \mathbb{Z})$ be the set of 2×2 matrices with integer coefficients and determinant 1 that we defined in Exercise 1.8.21.

(1) Show that $\mathrm{SL}(2, \mathbb{Z})$ is a group with respect to matrix multiplication. Give an explicit formula for the multiplicative inverse of an arbitrary element of $\mathrm{SL}(2, \mathbb{Z})$.

(2) Show that $\mathrm{SL}(2, \mathbb{Z})$ is not abelian.

(3) Show that $\mathrm{SL}(2,\mathbb{Z})$ is generated as a group by the matrices S and T defined in Exercise 1.8.21, together with $-\mathrm{Id}$. (Note: this is a *difficult* problem. A proof can be found in [**Ser73**, Chapter VII, Theorem 2].)

Exercise 5.6.5. Prove that if $(G,*)$ is a group, then the inverses with respect to $*$ are unique; i.e., if $g \in G$, then there is a *unique* inverse element g^{-1} such that $g * g^{-1} = e$, where e is the identity element in the group G.

Exercise 5.6.6. The goal of this exercise is to prove Lagrange's theorem (Theorem 5.2.19). Let $(G,*)$ be a finite group, and let $(H,*)$ be a subgroup of G.

(1) Let g_1, g_2 be elements of G. We say that $g_1 \sim_H g_2$ if and only if there is some $h \in H$ such that $g_1 = g_2 * h$. Prove that $\sim_H$ defines an equivalence relation on the elements of G; i.e., prove that $\sim_H$ is reflexive, symmetric, and transitive.

(2) Let G/H be the set of equivalence classes of G with respect to $\sim_H$, and let $g*H$ denote the equivalence class formed by all the elements of G in the same equivalence class as $g \in G$; i.e.,

$$g * H = \{k \in G : k \sim_H g\}.$$

Let g_1*H and g_2*H be two equivalence classes. Show that either $g_1*H = g_2*H$ or $(g_1 * H) \cap (g_2 * H) = \emptyset$, or, in words, two equivalence classes are either identical or completely disjoint.

(3) Prove that all equivalence classes have the same size. (Hint: if $a * H$ and $b * H$ are two equivalence classes, consider the map $f : g_1 * H \to g_2 * H$ given by $f(x) = g_2 * g_1^{-1} * x$.)

(4) Show that $e * H = H$. Conclude that all equivalence classes are of size $|H|$.

(5) Show that $|G| = |G/H| \cdot |H|$, where $|G/H|$ is the number of equivalence classes of elements of G with respect to $\sim_H$. Hence, $|H|$ is a divisor of $|G|$.

Exercise 5.6.7. Let $(G,*)$ be a finite abelian group, and let $(H,*)$ be a subgroup of G. Let $G/H = \{g * H : g \in G\}$ be the set of equivalence classes defined in Exercise 5.6.6, and define a group operation $\star$ on G/H as follows:

$$(g_1 * H) \star (g_2 * H) = (g_1 * g_2) * H.$$

(a) Show that $(G/H, \star)$ is a group, called the quotient group of G by H.

(b) Conclude that G/H is a group of order $|G|/|H|$, where $|G|$ and $|H|$ are the orders of G and H, respectively.

(c) Let $p > 2$ be a prime, let $G = (\mathbb{Z}/p\mathbb{Z})^\times$ be the group of units in $\mathbb{Z}/p\mathbb{Z}$, and let $H = \{\pm 1 \bmod p\}$.
   (1) Show that the quotient group $(\mathbb{Z}/p\mathbb{Z})^\times/\{\pm 1 \bmod p\}$ has $(p-1)/2$ elements.
   (2) Let $p = 11$. The group $G/H = (\mathbb{Z}/11\mathbb{Z})^\times/\{\pm 1 \bmod 11\}$ has five elements $\{H, g_2H, g_3H, g_4H, g_5H\}$. Find representatives $1, g_2, \ldots, g_5$ for the five

equivalence classes of $(\mathbb{Z}/11\mathbb{Z})^\times/\{\pm 1\}$, and write a multiplication table for the elements of G/H of the form

$\cdot$	H	g_2H	g_3H	g_4H	g_5H
H	H	g_2H	$\cdots$		
g_2H					
g_3H					
g_4H					
g_5H					

such that each entry $(g_iH)\cdot(g_jH)$ in the table is identified to be another element g_kH of G/H.

Exercise 5.6.8. Verify that the maps in Examples 5.2.23, 5.2.24, 5.2.25, and 5.2.26 are group homomorphisms.

Exercise 5.6.9. Prove Proposition 5.2.28.

Exercise 5.6.10. Let $\varphi : \mathbb{Z}/m\mathbb{Z} \to \mu_m$ be the map given by $a \bmod m \mapsto e^{2a\pi i/m}$ in Example 5.2.32. Prove the following statements:

(1) Show that φ is well-defined; that is, $\varphi(a \bmod m) = \varphi(b \bmod m)$ whenever a, b are integers with $a \equiv b \bmod m$.

(2) Show that φ is a group homomorphism.

(3) Show that φ is injective and surjective.

(4) Show that φ is an isomorphism of groups.

Exercise 5.6.11. Prove Proposition 5.2.33.

Exercise 5.6.12. Verify that the map f of Example 5.2.34 is an injective group homomorphism.

Exercise 5.6.13. Let $(R, +, \cdot)$ be a commutative ring with identity, such that $0 \neq 1$. Prove that if $u \in R$ is a unit, then u is not a zero-divisor.

Exercise 5.6.14. Let n and m be positive integers such that n is a divisor of m. Let u be an integer relatively prime to n.

(a) Show that there exists an integer v, relatively prime to m, such that $v \equiv u \bmod n$. (Hint: use Proposition 5.3.21.)

(b) Let $n = 5$, $m = 210$, and $u = 12$. Find $v \in \mathbb{Z}$ as in part (a), i.e., an integer v relatively prime to m such that $v \equiv u \bmod n$.

Exercise 5.6.15. Let $(F, +, \cdot)$ be a field. Show that $0_F \cdot f = 0_F$ for every $f \in F$, where 0_F is the zero element for the addition in the field. (Hint: $0 + 0 = 0$.)

Exercise 5.6.16. Let $(R, +, \cdot)$ be a commutative ring with identity. Prove that all the ideals (see Definition 5.3.27) of $\mathbb{Z}$ are of the form $\{0\}$ or $n\mathbb{Z}$ for some $n \in \mathbb{Z}$; i.e., an ideal in $\mathbb{Z}$ is a set of multiples of a fixed integer n:

$$n\mathbb{Z} = \{n \cdot m : m \in \mathbb{Z}\}.$$

(Hint: find the smallest positive integer $n \in I$, and show that every element of I is divisible by n.)

Exercise 5.6.17. Let $(F, +, \cdot)$ be a field. Show that the only ideals of F are $\{0\}$ and F.

Exercise 5.6.18. Let F and K be fields, and let $\psi\colon F \to K$ be a field homomorphism (Definition 5.4.6). We define the kernel of ψ by

$$\operatorname{Ker}(\psi) = \{f \in F : \psi(f) = 0_K\}.$$

Show that $\operatorname{Ker}(\psi)$ is an ideal of F. Use Exercise 5.6.17 to conclude that $\operatorname{Ker}(\psi) = \{0_F\}$ or F.

Exercise 5.6.19. Let $(F, +, \cdot)$ be a field, and let $a, b \in F$. Show that $a \cdot b = 0$ if and only if $a = 0$ or $b = 0$. (Hint: if $ab = 0$ and $a \neq 0$, then we can multiply both sides of $ab = 0$ by a^{-1}.)

Exercise 5.6.20. Prove Lemma 5.5.8.

Exercise 5.6.21. (1) Find all the congruence classes modulo 35 that are zero-divisors in $\mathbb{Z}/35\mathbb{Z}$.

(2) Find all the congruence classes modulo 35 that are units in $\mathbb{Z}/35\mathbb{Z}$.

(3) For each unit modulo 35, find its multiplicative inverse.

(4) Repeat parts (1), (2) and (3) for the ring $\mathbb{Z}/11\mathbb{Z}$.

Exercise 5.6.22. (1) Find all the congruence classes modulo 16 that are zero-divisors in $\mathbb{Z}/16\mathbb{Z}$.

(2) Find all the congruence classes modulo 16 that are units in $\mathbb{Z}/16\mathbb{Z}$.

(3) For each unit modulo 16, find its multiplicative inverse.

Exercise 5.6.23. Find all the units in $\mathbb{Z}/18\mathbb{Z}$ and all the units in $\mathbb{Z}/19\mathbb{Z}$. Find the multiplicative inverse for every unit in $\mathbb{Z}/18\mathbb{Z}$. Does $3x \equiv 1 \bmod 18$ have a solution? What about $3x \equiv 1 \bmod 19$?

Exercise 5.6.24. Verify that:

(1) The numbers $0, 2, 2^2, 2^3, 2^4, 2^5, 2^6, 2^7, 2^8, 2^9, 2^{10}$ are a complete set of representatives modulo 11.

(2) The numbers $0, 2, 2^2, 2^3, 2^4, 2^5, 2^6$ are not a complete set of representatives modulo 7.

Exercise 5.6.25. Find the quotient and remainder in $\mathbb{Q}[x]$ when dividing $x^3 - 7x - 1$ by $x^2 + 2x - 3$.

Exercise 5.6.26. Find the remainder when $x^4 - 7x^2 + 3$ is divided by $x + 1$ in $\mathbb{Q}[x]$.

Exercise 5.6.27. For which values of k in $\mathbb{Q}$ does $x - k$ divide $f(x) = x^3 - kx^2 - 2x + k + 3$?

Exercise 5.6.28. Using Euclid's algorithm, find a greatest common divisor in $\mathbb{F}_3[x]$ of the following:

(1) $x^5 + 1$ and $x^2 + 1$.

(2) $x^3 + 2x^2 + 3x + 2$ and $x^2 - x + 4$.

Exercise 5.6.29. In $\mathbb{Z}/3\mathbb{Z}[x]$, write if possible the polynomial 1 in the form $f(x)p(x) + g(x)q(x) = 1$ where

$$p(x) = x^3 + 1, \quad q(x) = x^3 + x + 1.$$

Exercise 5.6.30. Prove that any polynomial of degree ≥ 1 in $F[x]$, where F is a field, is either irreducible or it factors into a product of irreducibles (see Definition 6.3.3).

Exercise 5.6.31. Factor $x^5 - x$ into a product of irreducibles in $\mathbb{Z}/5\mathbb{Z}[x]$.

Exercise 5.6.32. Show that for any prime p the polynomial $x^p - x$ factors as $x(x-1)(x-2)\cdots(x-(p-1))$ over $\mathbb{Z}/p\mathbb{Z}[x]$.

Exercise 5.6.33. Let $p(x) = 16x^3 + 8x^2 - 7x + 1$ be defined over $\mathbb{Q}$.

(1) Calculate the discriminant of $p(x)$.
(2) Use (1) to show that $p(x)$ has a double root, but not a triple root.
(3) Calculate the roots of $p(x)$ over $\mathbb{Q}$.
(4) Now consider $p(x)$ as defined over $\mathbb{F}_5$. Show that $p(x)$ has a triple root.

Exercise 5.6.34. Compute the discriminant of a polynomial $p(x) = x^3 + Ax + B$, defined over a field F. (Hint: use Theorem 5.5.25.)

CHAPTER 6

FINITE FIELDS

Taking Three as the subject to reason about–
A convenient number to state–
We add Seven, and Ten, and then multiply out
By One Thousand diminished by Eight.
The result we proceed to divide, as you see,
By Nine Hundred and Ninety Two:
Then subtract Seventeen, and the answer must be
Exactly and perfectly true.
The method employed I would gladly explain,
While I have it so clear in my head,
If I had but the time and you had but the brain–
But much yet remains to be said.

Lewis Carroll, from *The Hunting of the Snark*

In Section 5.4 we introduced the concept of field, and we showed that $\mathbb{Z}/p\mathbb{Z}$ is a field, when p is a prime (Theorem 5.4.3). Here is a natural question that arises: are there fields with any fixed finite number of elements? In this chapter we explore this question. We will construct fields of size p^n for any prime p and any $n \geq 1$ and then show that the size of a finite field is always a prime power.

Remark 6.0.1. Let p be a prime number. The ring $\mathbb{Z}/p^2\mathbb{Z}$ has p^2 elements, but it is *not* a field. Indeed, the congruence class $p \bmod p^2$ is not invertible (see Theorem 5.4.3).

6.1. An Example

Let p be a prime such that $p \equiv 3 \bmod 4$. Later in this book, we will show that (as a consequence of Lemma 10.3.4, part (3)) the number -1 is not a square modulo p. In other words, there is no integer n such that $n^2 \equiv -1 \bmod p$. Let us define a new number "i" such that $i^2 \equiv -1 \bmod p$ and also define

$$\mathbb{F}_p[i] = \{a + bi : a, b \in \mathbb{F}_p\}.$$

The set $\mathbb{F}_p[i]$ inherits addition and multiplication laws from $\mathbb{F}_p$; i.e.,

$$(a + bi) + (c + di) \equiv (a + b) + (c + di) \bmod p$$

and

$$(a + bi) \cdot (c + di) \equiv (ac - bd) + (ad + bc)i \bmod p.$$

One can routinely check that the addition and multiplication laws are commutative and associative and the distributive laws are satisfied. Moreover, $\mathbb{F}_p[i]$ has a zero element, $0 \bmod p$; an identity element, $1 \bmod p$; and every element $a + bi$ has an additive inverse, namely $(-a) + (-b)i$. In order to prove that $\mathbb{F}_p[i]$ is a field, it remains to show that every non-zero element $a + bi$ is a unit; i.e., $a + bi$ has a multiplicative inverse.

Lemma 6.1.1. *Let $p \equiv 3 \bmod 4$ be a prime and suppose $a, b \in \mathbb{Z}$ are non-zero modulo p. Then, $a^2 + b^2 \not\equiv 0 \bmod p$.*

Proof. Let $p \equiv 3 \bmod 4$ be a prime. By Lemma 10.3.4, -1 is not a square mod p. Suppose that $a \not\equiv 0 \not\equiv b \bmod p$ (i.e., a, b are units modulo p) and $a^2 + b^2 \equiv 0 \bmod p$. Then, $a^2 \equiv -b^2 \bmod p$ and, therefore, $(ab^{-1})^2 \equiv a^2b^{-2} \equiv -1 \bmod p$. But this is impossible because we just remarked that -1 is a quadratic non-residue. This is a contradiction and it follows that $a^2 + b^2 \not\equiv 0 \bmod p$, as claimed. □

We claim that if $a+bi \neq 0$, then $\frac{a-bi}{a^2+b^2} \equiv (a-bi)(a^2+b^2)^{-1}$ is the multiplicative inverse of $a + bi$. Notice that we are able to find an inverse for $a^2 + b^2$ because the previous lemma shows that it is non-zero modulo p and it is therefore a unit. Indeed,

$$(a + bi) \cdot \frac{a - bi}{a^2 + b^2} \equiv \frac{a^2 - (bi)^2}{a^2 + b^2} \equiv \frac{a^2 + b^2}{a^2 + b^2} \equiv 1 \bmod p,$$

as claimed. Therefore, every non-zero $a + bi$ is a unit in $\mathbb{F}_p[i]$ and we conclude that $\mathbb{F}_p[i]$ is a field. Notice that this field has p^2 elements.

This example prompts several questions. Are there fields of p^2 elements for those primes $p \equiv 1 \bmod 4$? Are there fields of p^n elements, for all $n \geq 1$? Are there fields of m elements, for all $m > 1$? Before we answer this question, we need to learn a little bit more about congruences of polynomials.

6.2. Polynomial Congruences

Let p be a prime. We will write $\mathbb{F}_p$ for $\mathbb{Z}/p\mathbb{Z}$.

Definition 6.2.1. Let $a(x)$, $b(x)$, and $m(x)$ be polynomials in $\mathbb{F}_p[x]$. We say that $a(x) \equiv b(x) \bmod (m(x))$ if the polynomial $a(x) - b(x)$ is divisible by $m(x)$ in $\mathbb{F}_p[x]$.

We work with congruences of polynomials just as we work with congruences of integers. We quote, without proof, an analogue of Proposition 4.2.1 for polynomials.

Proposition 6.2.2. *Let p be a fixed prime. For all polynomials $a(x)$, $b(x)$, $c(x)$, $a'(x)$, $b'(x)$, $k(x)$, and $m(x)$ in $\mathbb{F}_p[x]$, the following properties hold.*

(i) *If $a(x) \equiv b(x) \bmod m(x)$, then $k(x)a(x) \equiv k(x)b(x) \bmod m(x)$.*

(ii) *If $a(x) \equiv b(x) \bmod m(x)$ and $b(x) \equiv c(x) \bmod m(x)$, then $a(x) \equiv c(x) \bmod m(x)$.*

(iii) *If $a(x) \equiv b(x) \bmod m(x)$, then $a(x)^j \equiv b(x)^j \bmod m(x)$, for all $j \geq 1$.*

(iv) *If $a(x) \equiv b(x) \bmod m(x)$ and $a'(x) \equiv b'(x) \bmod m(x)$, then*
  (a) *$a(x) + a'(x) \equiv b(x) + b'(x) \bmod m(x)$, and*
  (b) *$a(x) \cdot a'(x) \equiv b(x) \cdot b'(x) \bmod m(x)$.*

Example 6.2.3. Let us simplify $(x^2 + x + 1)^3 \bmod (x^2 - 1)$ over $\mathbb{F}_5[x]$. Since

$$x^2 + x + 1 \equiv x^2 - 1 + 1 + x + 1 \equiv 0 + x + 2 \equiv x + 2 \bmod (x^2 - 1),$$

it follows from Proposition 6.2.2, part (iii), that

$$(x^2 + x + 1)^3 \equiv (x + 2)^3 \equiv x^3 + 6x^2 + 12x + 8 \equiv x^3 + x^2 + 2x + 3 \bmod (x^2 - 1)$$

since the coefficients are in $\mathbb{F}_5$. Further, $x^2 \equiv 1$ and so

$$x^3 \equiv x \cdot x^2 \equiv x \cdot 1 \equiv x \bmod (x^2 - 1).$$

Thus,

$$(x^2 + x + 1)^3 \equiv x^3 + x^2 + 2x + 3 \equiv x + 1 + 2x + 3 \equiv 3x + 4 \bmod (x^2 - 1),$$

where we have used Proposition 6.2.2, part (iv)(a), to simplify the expression.

Remark 6.2.4. Let $a(x)$ and $m(x)$ be in $\mathbb{F}_p[x]$ and let $q(x), r(x) \in \mathbb{F}_p[x]$ be, respectively, the quotient and reminder of the long division of $a(x)$ by $m(x)$ as polynomials, with $\deg r(x) < \deg m(x)$. In other words,

$$a(x) = m(x)q(x) + r(x), \quad \text{with } \deg r(x) < \deg m(x).$$

Then, $a(x) \equiv r(x) \bmod (m(x))$. This shows that every polynomial in $\mathbb{F}_p[x]$ is congruent mod $m(x)$ to another polynomial of degree less than the degree of $m(x)$.

Example 6.2.5. Let $p = 7$ and put $m(x) = x^2 + 1 \in \mathbb{F}_7[x]$. Let $a(x) = x^4 + x^2 + x + 1$. Then:

- We may use long division and obtain $x^4 + x^2 + x + 1 = (x^2 + 1)(x^2) + (x + 1)$. Thus,

$$x^4 + x^2 + x + 1 \equiv x + 1 \bmod (x^2 + 1).$$

- $x^2 + 1 \equiv 0 \bmod (x^2 + 1)$; thus $x^2 \equiv -1 \bmod (x^2 + 1)$. Therefore,

$$x^3 \equiv x^2 \cdot x \equiv -x \bmod (x^2 + 1).$$

  Alternatively, $-x \equiv 6x$ in $\mathbb{F}_7[x]$. Similarly, we can calculate other powers of x, for example $x^4 \equiv 1$, and $x^5 \equiv x \bmod (x^2 + 1)$.

- We reduce $x^4 + x^2 + x + 1$ in a different way, using the previous remarks:

$$x^4 + x^2 + x + 1 \equiv 1 + -1 + x + 1 \equiv x + 1 \bmod (x^2 + 1).$$

Definition 6.2.6. Let p be a prime and let $m(x)$ be a polynomial in $\mathbb{F}_p[x]$ of degree $n \geq 1$. We define the set of all *congruence classes* in $\mathbb{F}_p[x]$ modulo $m(x)$ by

$$\mathbb{F}_p[x]/(m(x)) = \{a_0 + a_1x + \cdots + a_{n-1}x^{n-1} \bmod (m(x)) : a_i \in \mathbb{F}_p\}.$$

The set $\mathbb{F}_p[x]/(m(x))$ is equipped with addition and multiplication laws, inherited from $\mathbb{F}_p[x]$:

$$(a(x) \bmod m(x)) + (b(x) \bmod m(x)) \equiv (a(x) + b(x) \bmod (m(x))$$

and

$$(a(x) \bmod m(x)) \cdot (b(x) \bmod m(x)) \equiv (a(x) \cdot b(x) \bmod (m(x)).$$

Example 6.2.7. Let $p = 3$ and $m(x) = x^2 + 1 \in \mathbb{F}_3[x]$. Then, the set of all congruence classes modulo $(x^2 + 1)$ is

$$\begin{aligned}\mathbb{F}_3[x]/(x^2+1) &= \{a + bx \bmod (x^2+1) : a, b \in \mathbb{F}_3\} \\ &= \{0, 1, 2, x, x+1, x+2, 2x, 2x+1, 2x+2 \bmod (x^2+1)\}.\end{aligned}$$

The polynomial $x^5 + 2x^3 + x + 2$ is congruent to one of the representatives above. Which one? Notice that $x^2 \equiv -1 \bmod (x^2 + 1)$; thus $x^3 \equiv -x$, $x^4 \equiv 1$, and $x^5 \equiv x \bmod (x^2 + 1)$. Therefore,

$$x^5 + 2x^3 + x + 2 \equiv x - 2x + x + 2 \equiv 2 \bmod (x^2 + 1).$$

What is the representative of $(2x + 1)(2x + 2)$ of degree ≤ 1 modulo $x^2 + 1$? Let us calculate:

$$(2x+1)(2x+2) \equiv 4x^2 + 6x + 2 \equiv -4 + 2 \equiv -2 \equiv 1 \bmod (x^2 + 1).$$

Therefore $(2x + 2)$ is the multiplicative inverse of $(2x + 1)$ in $\mathbb{F}_3[x]/(x^2 + 1)$.

Example 6.2.8. Let $p = 3$ and $m(x) = x^2 + 2x + 1$. Then, the set of all congruence classes modulo $(x^2 + 2x + 1)$ is

$$\begin{aligned}\mathbb{F}_3[x]/(x^2+2x+1) &= \{a + bx \bmod (x^2+2x+1) : a, b \in \mathbb{F}_3\} \\ &= \{0, 1, 2, x, x+1, x+2, 2x, 2x+1, 2x+2 \bmod (m(x))\}.\end{aligned}$$

Notice that $\mathbb{F}_3[x]/(x^2 + 1)$ (in the previous example) and $\mathbb{F}_3[x]/(x^2 + 2x + 1)$ have the same set of representatives. However, the additive and multiplicative structures are quite different. For example, $x^2 \equiv -1$ and $x^3 \equiv -x \bmod (x^2 + 1)$ but $x^2 \equiv -2x - 1 \equiv x + 2 \bmod (x^2 + 2x + 1)$ and

$$x^3 \equiv x^2 + 2x \equiv x + 2 + 2x \equiv 3x + 2 \equiv 2 \bmod (x^2 + 2x + 1).$$

Also, notice that $\mathbb{F}_3[x]/(x^2 + 2x + 1)$ **is not a field** because it has zero-divisors:

$$(x+1)(x+1) \equiv x^2 + 2x + 1 \equiv 0 \bmod (x^2 + 2x + 1)$$

so $x + 1 \bmod (x^2 + 2x + 1)$ is a zero-divisor.

Proposition 6.2.9. *Let p be a prime and let $m(x) \in \mathbb{F}_p[x]$ be a polynomial. Then, $\mathbb{F}_p[x]/(m(x))$ is a commutative ring with identity.*

Proof. The addition and multiplication laws on the set $\mathbb{F}_p[x]/(m(x))$ are inherited from the laws in $\mathbb{F}_p$ and $\mathbb{F}_p[x]$, which are, respectively, a field and a commutative ring with identity. The reader can now finish the proof by checking that $+$ and $\cdot$ on $\mathbb{F}_p[x]/(m(x))$ satisfy all the properties of a commutative ring with identity. $\square$

When is $\mathbb{F}_p[x]/(m(x))$ a field? In Example 6.2.8 we have seen that $\mathbb{F}_3[x]/(x^2 + 2x + 1)$ is not a field, because it has zero-divisors. The problem is that $x^2 + 2x + 1$ factors as $(x + 1)(x + 1)$ in $\mathbb{F}_3[x]$. We will see that, in fact, if $m(x)$ cannot be factored, then $\mathbb{F}_p[x]/(m(x))$ is a field.

6.3. Irreducible Polynomials

Before we construct other finite fields, we need to discuss the concept of irreducible polynomial. In particular, we need to know (a) how to tell if a polynomial is irreducible and (b) how to find irreducible polynomials of a certain degree. In this section, R is a commutative ring with identity and F is an arbitrary field.

Definition 6.3.1. A polynomial $f(x) \in R[x]$ is a *unit polynomial* if $f(x)$ is a unit in the ring $R[x]$, i.e., if there is another polynomial $g(x) \in R[x]$ such that $f(x) \cdot g(x) = 1 \in R$.

If $f(x)$ is defined over a field F instead, then the only unit polynomials are the constants in F. However, if R is not a field, there may be other units, as the following example shows.

Example 6.3.2. Let $R = \mathbb{Z}/4\mathbb{Z}$ and let $f(x) = 1 + 2x \in \mathbb{Z}/4\mathbb{Z}[x]$. Put $g(x) = 1 + 2x = f(x)$. Then,

$$f(x)g(x) \equiv (1 + 2x)(1 + 2x) \equiv (1 + 2x)(1 - 2x) \equiv 1 - 4x^2 \equiv 1 \bmod 4.$$

Thus, $f(x)$ is a unit.

Definition 6.3.3. Let R be a ring and let $f(x)$ be a polynomial in $R[x]$. We say that $f(x)$ is *irreducible* if $f(x)$ is not a unit and whenever $f(x)$ factors as $f(x) = p(x)q(x)$, then $p(x)$ or $q(x)$ is a unit polynomial. Otherwise, we say that $f(x)$ is *reducible*, i.e., if it factors as a product of irreducible polynomials of lesser degree.

We remind the reader of two important results, the remainder theorem (Corollary 5.5.14) and the root theorem (Corollary 5.5.15).

Example 6.3.4. Let $F = \mathbb{Z}/7\mathbb{Z}$ and let $f(x) = x^2 + 5$. Even though $f(x)$ is irreducible over $\mathbb{R}$, $f(x)$ is not irreducible over $\mathbb{Z}/7\mathbb{Z}$, because

$$f(x) \equiv x^2 + 5 \equiv (x + 4)(x + 3) \equiv x^2 + 7x + 12 \bmod 7.$$

The polynomial $g(x) = x^2 + 4$ is irreducible in $\mathbb{Z}/7\mathbb{Z}[x]$. Indeed, if $g(x)$ was reducible, then it would factor as two polynomials of degree 1 and 1. But a polynomial of degree 1 indicates a root, by the root theorem. Since $x^2 + 4$ has no roots in $\mathbb{Z}/7\mathbb{Z}$ (because $-4 \equiv 3$ is not a square modulo 7), the polynomial $g(x)$ must be irreducible.

The argument that was used in the last example is very common, so we record it as a proposition.

Proposition 6.3.5. *Let F be a field and let $f(x) \in F[x]$ be a polynomial of degree 2 or 3. If $f(x)$ has no roots in F, then $f(x)$ is an irreducible polynomial over F.*

Proof. We begin with a simple remark. Suppose that $g(x)$ is a polynomial of degree 1 that divides f. Then, we claim that $f(x)$ has a root. Indeed, if $g(x) = ax + b$ with $a \neq 0$, then $g(-b/a) = 0$ and, therefore, $f(-b/a) = 0$ because $g(x)$ divides $f(x)$.

Now, suppose that $f(x)$ is a polynomial of degree 2 or 3 that has no roots over F. Suppose, for a contradiction, that $f(x)$ is not irreducible. Then $f(x) = p(x)q(x)$ and neither $p(x)$ nor $q(x)$ is a unit (i.e., $\deg(p(x)), \deg(q(x)) \geq 1$). But $p(x)$ and $q(x)$ cannot be polynomials of degree 1 because, by the remark above, this would imply that $f(x)$ has a root. Therefore $\deg(p(x)), \deg(q(x)) \geq 2$ and

$$\deg(f(x)) = \deg(p(x)) + \deg(q(x)).$$

Notice that the previous equality holds because F is a field (and it is not true in general over a ring). If $\deg(f(x)) = 2$ or 3, then $\deg(p) + \deg(q) \geq 4 > \deg(f)$, and we have reached a contradiction. Thus, $f(x)$ must be irreducible. □

Remark 6.3.6. The previous proposition is **false** for polynomials of degree ≥ 4. For example, consider $f(x) = x^4 + 6x^2 + 1$ over $\mathbb{Z}/7\mathbb{Z}$. Then, $f(x)$ has no roots (try to find one!) but

$$f(x) \equiv (x^2 + 4)(x^2 + 2) \bmod 7.$$

Therefore, $f(x)$ is not irreducible but $f(x)$ has no roots over $\mathbb{Z}/7\mathbb{Z}$.

Example 6.3.7. *Is the polynomial $f(x) = x^3 + 2x + 1$ irreducible over $\mathbb{Z}/7\mathbb{Z}$?* By Proposition 6.3.5, since $\deg(f(x)) = 3$, if $f(x)$ has no roots, then it must be an irreducible polynomial. One calculates

$$f(0) \equiv 1,\ f(1) \equiv 4,\ f(2) \equiv 6,\ f(3) \equiv 6,\ f(4) \equiv 3,\ f(5) \equiv 3,\ f(6) \equiv 5 \bmod 7.$$

Thus, $f(x)$ has no roots over $\mathbb{Z}/7\mathbb{Z}$ and, therefore, it is irreducible.

Example 6.3.8. *Is the polynomial $f(x) = x^3 + x^2 + x + 1$ irreducible over $\mathbb{Z}/7\mathbb{Z}$?* By Proposition 6.3.5, since $\deg(f(x)) = 3$, if $f(x)$ has no roots, then it must be an irreducible polynomial. One calculates

$$f(0) \equiv 1,\ f(1) \equiv 4,\ f(2) \equiv 1,\ f(3) \equiv 5,\ f(4) \equiv 1,\ f(5) \equiv 2,\ f(6) \equiv 0 \bmod 7.$$

Thus, 6 is a root of $f(x)$ and, by the root theorem, $(x - 6) \equiv x + 1$ divides $f(x)$ in $\mathbb{Z}/7\mathbb{Z}[x]$. Indeed,

$$x^3 + x^2 + x + 1 \equiv (x + 1)(x^2 + 1) \bmod 7.$$

Hence, $f(x)$ is not irreducible.

6.4. Fields with p^n Elements

We are ready to construct finite fields with p^n elements.

Theorem 6.4.1. *Let p be a prime and let $m(x)$ be a polynomial in $\mathbb{F}_p[x]$ of degree $n \geq 1$. Then, the ring $\mathbb{F}_p[x]/(m(x))$ has p^n elements, and it is a field if and only if the polynomial $m(x)$ is irreducible.*

Proof. From the definitions (see Definition 6.2.6),

$$\mathbb{F}_p[x]/(m(x)) = \{a_0 + a_1x + \cdots + a_{n-1}x^{n-1} \bmod (m(x)) : a_i \in \mathbb{F}_p\}.$$

There are p choices for each coefficient and n coefficients, so $\mathbb{F}_p[x]/(m(x))$ has exactly p^n elements. Suppose first that $m(x)$ is not irreducible. Then, there are polynomials $a(x), b(x) \in \mathbb{F}_p[x]$ such that $m(x) = a(x)b(x)$ and $\deg a(x)$ and $\deg b(x) \geq 1$. Therefore, $a(x), b(x) \not\equiv 0 \bmod (m(x))$ but $a(x)b(x) \equiv m(x) \equiv 0 \bmod (m(x))$. Hence, $a(x)$ and $b(x)$ are zero-divisors, and so $\mathbb{F}_p[x]/(m(x))$ cannot be a field.

Next, suppose that $m(x)$ is irreducible. By Proposition 6.2.9, $\mathbb{F}_p[x]/(m(x))$ is a commutative ring with identity. Thus, in order to show that it is a field, it suffices to show that every non-zero element is a unit; i.e., it has a multiplicative inverse. Let $a(x)$ be a polynomial such that $a(x) \not\equiv 0 \bmod (m(x))$. Therefore, the GCD of $a(x)$ and $m(x)$ must be a unit, because $m(x)$ is an irreducible (if $d(x)$ is their gcd, then $d(x)$ divides $m(x)$, but the only divisors of $m(x)$ are units). Hence, the Bezout's identity

$$a(x) \cdot X + m(x) \cdot Y = 1$$

has solutions $X = f(x), Y = g(x) \in \mathbb{F}_p[x]$, and so $a(x)f(x) + m(x)g(x) = 1$. Hence, $a(x)f(x) \equiv 1 \bmod (m(x))$ and, therefore, $f(x)$ is the multiplicative inverse of $a(x)$ modulo $m(x)$. Since $a(x)$ was arbitrary, this concludes the proof. □

Example 6.4.2. Let $p = 2$ and write $m(x) = x^2 + x + 1$. Then, $m(x)$ is irreducible because it is of degree 2 and it does not have any roots in $\mathbb{F}_2$:

$$m(0) \equiv 0^2 + 0 + 1 \equiv 1 \bmod 2 \quad \text{and} \quad m(1) \equiv 1^2 + 1 + 1 \equiv 1 \bmod 2.$$

Therefore, $\mathbb{F}_2[x]/(x^2 + x + 1)$ is a field with four elements:

$$\mathbb{F}_2[x]/(x^2 + x + 1) = \{0, 1, x, x + 1 \bmod (x^2 + x + 1)\}.$$

Here are tables of addition in $\mathbb{F}_2[x]/(x^2 + x + 1)$:

$+$	0	1	x	$x+1$
0	0	1	x	$x+1$
1	1	0	$x+1$	x
x	x	$x+1$	0	1
$x+1$	$x+1$	x	1	0

and multiplication:

$\times$	0	1	x	$x+1$
0	0	0	0	0
1	0	1	x	$x+1$
x	0	x	$x+1$	1
$x+1$	0	$x+1$	1	x

Example 6.4.3. Let $p = 7$ and $m(x) = x^3 + 2x + 1$, which is irreducible by Example 6.3.7. Therefore, $\mathbb{F}_7[x]/(x^3 + 2x + 1)$ is a field with $7^3 = 343$ elements.

6.5. Fields with p^2 Elements

In the first section of this chapter, Section 6.1, we saw an example of a field with p^2 elements, namely $\mathbb{F}_p[i]$ for $p \equiv 3 \bmod 4$. In this case, the polynomial $m(x) = x^2 + 1$

is irreducible over $\mathbb{F}_p[x]$, because it is of degree 2 and it does not have any roots (see Proposition 6.3.5): a root of x^2+1 would be a number whose square is congruent to -1 modulo p, but -1 is not a square mod p. Hence, $\mathbb{F}_p[x]/(x^2+1)$ is a field with p^2 elements, by Theorem 6.4.1. This is exactly the same construction that we showed in Section 6.1, where the role of i is played by x here (notice that i was defined by $i^2 \equiv -1$ and $x^2 \equiv -1$ as well). Now we can extend this construction to all primes.

Let $p > 2$ be a prime and let $s \in \mathbb{Z}$ be a *quadratic non-residue modulo* p; that is, $(s, p) = 1$ but s is not congruent to a square modulo p (we will discuss quadratic residues in much more detail in Section 10.2). Then, $m(x) = x^2 - s$ is irreducible over $\mathbb{F}_p[x]$ and $\mathbb{F}_p[x]/(x^2-s)$ is a field with p^2 elements. If $p \equiv 3 \bmod 4$, one may choose $s = -1$ in which case $m(x) = x^2+1$ and we recover the example $\mathbb{F}_p[i]$. In general, we may also write $\sqrt{s}$ for the indeterminate x and write our field of p^2 elements as

$$\mathbb{F}_p[\sqrt{s}\,] = \{a + b\sqrt{s} : a, b \in \mathbb{F}_p\},$$

where $(\sqrt{s}\,)^2 \equiv s \bmod p$ and the addition and multiplication laws in $\mathbb{F}_p[\sqrt{s}]$ are given by

$$(a + b\sqrt{s}) + (c + d\sqrt{s}) \equiv (a+b) + (c + d\sqrt{s}) \bmod p$$

and

$$(a + b\sqrt{s}) \cdot (c + d\sqrt{s}) \equiv (ac + bds) + (ad + bc)\sqrt{s} \bmod p.$$

Let us see a concrete example.

Example 6.5.1. Let $p = 5$ and let $s = 2$. Then, s is not a square modulo 5, and we can form the field

$$\mathbb{F}_5[\sqrt{2}] = \{a + b\sqrt{2} : a, b \in \mathbb{F}_5\}.$$

Let us perform some calculations in $\mathbb{F}_5[\sqrt{2}]$. For instance, let $\alpha \equiv 1 + \sqrt{2}$ and $\beta \equiv 3 + 4\sqrt{2} \bmod 5$. Then,

$$\alpha + \beta \equiv 1 + \sqrt{2} + 3 + 4\sqrt{2} \equiv 4 + 5\sqrt{2} \equiv 4 \bmod 5,$$

and

$$\alpha \cdot \beta \equiv (1+\sqrt{2})(3 + 4\sqrt{2}) \equiv 3 + 4\cdot 2 + (3+4)\sqrt{2} \equiv 1 + 2\sqrt{2} \bmod 5.$$

Notice that, as we pointed out above, here $\sqrt{2}$ is playing the role of x in $\mathbb{F}_5[x]/(x^2-2)$. As such, $\alpha \equiv 1 + x$ and $\beta \equiv 3 + 4x \bmod (x^2-2)$, and then

$$\alpha + \beta \equiv 1 + x + 3 + 4x \equiv 4 + 5x \equiv 4 \bmod (x^2-2),$$

in $\mathbb{F}_5[x]/(x^2-2)$, and

$$\begin{aligned}\alpha \cdot \beta &\equiv (1+x)(3+4x)\\ &\equiv 3 + (4+3)x + 4x^2\\ &\equiv 3 + 7x + 4(x^2-2) + 8\\ &\equiv 11 + 7x \equiv 1 + 2x \bmod (x^2-2),\end{aligned}$$

where we have used the fact that $x^2 - 2 \equiv 0$ in $\mathbb{F}_5[x]/(x^2-2)$. Hence, $\alpha\cdot\beta \equiv 1 + 2x \bmod (x^2-2)$, which corresponds to $1 + 2\sqrt{2}$, as we had already calculated above.

We will see an application of finite fields to the problem of finding "square roots" modulo a prime p, the so-called Cipolla's algorithm, in Section 10.6.

6.6. Fields with s Elements

In this section we prove that if a field has s elements, then s is a power of a prime.

Theorem 6.6.1. *Let F be a finite field with $s > 0$ elements. Then, $s = p^d$ for some prime number p and some $d \geq 1$.*

Let us first discuss some preliminaries and define the characteristic of a field.

Definition 6.6.2. Let $(F, +, \cdot)$ be a field. The *characteristic* of F is the smallest positive integer n such that $n \cdot 1_F = 1_F + \cdots + 1_F = 0_F$. If no such n exists, then we say that the characteristic of F is zero.

Example 6.6.3. The characteristic of $\mathbb{Q}$, $\mathbb{R}$, and $\mathbb{C}$ is 0, because $n \cdot 1 = n$ is non-zero as long as $n \geq 1$. However, the characteristic of $\mathbb{Z}/p\mathbb{Z}$ is p because $p \cdot 1 \equiv 0 \bmod p$, and $n \cdot 1 \not\equiv 0 \bmod p$ for any $1 \leq n \leq p-1$.

Remark 6.6.4. If a field F has characteristic $n > 0$, then $n \cdot f = 0_F$ for every $f \in F$. Indeed,

$$n \cdot f = n \cdot (1_F \cdot f) = (n \cdot 1_F) \cdot f = 0_F \cdot f = 0_F,$$

where we have used the associativity of multiplication in a field and the fact that $0_F \cdot f = 0_F$ for every $f \in F$ (see Exercise 5.6.15).

Lemma 6.6.5. *The characteristic of a field F is 0 or a prime number p.*

We leave the proof of Lemma 6.6.5 as an exercise (Exercise 6.7.10). Next we present a proof of Theorem 6.6.1 that assumes some knowledge of linear algebra.

Proof of Theorem 6.6.1. Let F be a field of positive characteristic. By Lemma 6.6.5, the characteristic is a prime number p. Let $\mathbb{F}_p = \mathbb{Z}/p\mathbb{Z}$ be the field of p-elements. Then, we claim that $(F, +)$ is a vector space over $\mathbb{F}_p$ with respect to the following scalar multiplication:

$$(n \bmod p) \cdot f := n \cdot f$$

where $n \cdot f$, for an integer n and an element $f \in F$, is the field multiplication, i.e., $f + \cdots + f$ added n times. Note that the scalar multiplication is well-defined, because if $m \equiv n \bmod p$, then $m = n + pk$ for some $k \in \mathbb{Z}$, and therefore

$$(m \bmod p) \cdot f = (n+pk \bmod p) \cdot f = (n+pk) \cdot f = n \cdot f + pk \cdot f = n \cdot f + p \cdot (k \cdot f) = n \cdot f$$

where we have used that $k \cdot f \in F$ and $p \cdot f' = 0_F$ for any $f' \in F$ because the characteristic of F is p.

We justify that $(F, +)$, with the scalar multiplication defined above, is a vector space. Since F is a field, $(F, +)$ is an abelian group, and therefore $+$ is associative and commutative and there are 0_F and additive inverses in F. Moreover, since the scalar multiplication is defined through the field multiplication operation $\cdot$, it follows that the scalar multiplication and addition operations satisfy all the necessary compatibility and distributive conditions (e.g., $(1 \bmod p) \cdot f = f$ for all $f \in F$, or $(n \bmod p) \cdot (f+g) = (n \bmod p) \cdot f + (n \bmod p) \cdot g)$. See Remark 6.6.6 below for some details.

Now, since F is a vector space over $\mathbb{F}_p$ and F is finite, we conclude that F must be a finite-dimensional vector space, say $\dim_{\mathbb{F}_p}(F) = d \geq 1$. Then, we can choose an $\mathbb{F}_p$-basis $\{f_1, \ldots, f_d\}$ of F such that $F \cong (\mathbb{F}_p)^d$ are isomorphic as vector spaces, where the isomorphism is given by $\psi \colon F \to (\mathbb{F}_p)^d$, such that $\psi(f_i) = e_i = (0, \ldots, 0, 1, 0, \ldots, 0)$ is an element of a canonical basis of $(\mathbb{F}_p)^d$. In particular, ψ is a bijection, and therefore the size of F is $\#(\mathbb{F}_p)^d = p^d$, a power of p, as desired. $\square$

Remark 6.6.6 (The devil is in the details). A vector space $(V, +, \cdot)$ over a field F is a set V with an addition $+$ and a scalar multiplication $\cdot$ by elements of F that satisfy the following properties:

- $(V, +)$ is an abelian group.
- Scalar multiplication: $f \cdot v$ is an element of V for every $f \in F$ and every $v \in V$, and for all $f, g \in F$ and all $v, w \in V$ we have
  (a) $f \cdot (g \cdot v) = (f \cdot g) \cdot v$,
  (b) $1_F \cdot v = v$, where 1_F is the multiplicative identity element in F,
  (c) $f \cdot (v + w) = f \cdot v + f \cdot w$, and
  (d) $(f + g) \cdot v = f \cdot v + g \cdot v$.

The usual example of a vector space is euclidean space over the real numbers, i.e., $\mathbb{R}^n$ as a vector space over $\mathbb{R}$. As we saw in the proof of Theorem 6.6.1, a finite field F of characteristic p is a vector space over $\mathbb{F}_p$. For instance, let us check that the scalar multiplication of $\mathbb{F}_p$ on F satisfies the distributive property (c) above. Let n be an integer, and let $v, w \in F$. Then,

$$(n \bmod p) \cdot (v + w) = n \cdot (v + w) = n \cdot v + n \cdot w = (n \bmod p) \cdot v + (n \bmod p) \cdot w,$$

as desired, where we have used the fact that $(F, +, \cdot)$ satisfies a distributive law (Definition 5.3.1, part (4)) to prove that $n \cdot (v + w) = n \cdot v + n \cdot w$ and we have used the definition of the scalar multiplication $(n \bmod p) \cdot v = n \cdot v$.

Example 6.6.7. Let $F = \mathbb{F}_3[i] \cong \mathbb{F}_3[x]/(x^2+1)$. Then, F is a vector space over $\mathbb{F}_3$. Indeed,

$$\mathbb{F}_3[i] = \{a + bi : a, b \in \mathbb{F}_3\}$$

is isomorphic to $\mathbb{F}_3 \times \mathbb{F}_3 = (\mathbb{F}_3)^2$, via $\psi \colon \mathbb{F}_3[i] \to (\mathbb{F}_3)^2$ such that $\psi(a + bi) = (a, b)$. That is, $\psi(1) = e_1 = (1, 0)$ and $\psi(i) = e_2 = (0, 1)$. Since $(\mathbb{F}_3)^2$ has nine elements, it follows that $\mathbb{F}_3[i]$ has order 3^2, also.

6.7. Exercises

Exercise 6.7.1. Prove parts (i) and (iii) of Proposition 6.2.2.

Exercise 6.7.2. Write down addition and multiplication tables for the ring $F = \mathbb{F}_3[x]/(x^2+1)$ as in Example 6.4.2. Verify that F is a field by finding the multiplicative inverse of each non-zero congruence class in F.

Exercise 6.7.3. Let $\mathbb{F}_5 = \mathbb{Z}/5\mathbb{Z}$ be a finite field with five elements.

(a) Show that 2 mod 5 and 3 mod 5 are not squares modulo 5; i.e., there is no integer n such that $n^2 \equiv 2$ or $3 \bmod 5$.

(b) Show that $F = \mathbb{F}_5[x]/(x^2+2)$ and $K = \mathbb{F}_5[y]/(y^2+3)$ are fields with 25 elements.

(c) Find elements α and β of F such that $\alpha^2 = 2$ (i.e., $\alpha^2 \equiv 2 \bmod (x^2+2)$) and $\beta^2 = 3$.

(d) Find elements δ and γ of K such that $\delta^2 = 2$ and $\gamma^2 = 3$.

(e) Define a map $\psi\colon F \to K$ by

$$\psi((a+bx) \bmod (x^2+2)) \equiv a + b\gamma \bmod (y^2+3),$$

for each $a, b \in \mathbb{F}_5$, where $\gamma \in K$ is such that $\gamma^2 = 3$. Show that ψ is a field isomorphism:

(1) Show that ψ is well-defined; i.e., if $p(x)$ is a polynomial in $\mathbb{F}_5[x]$ such that $p(x) \equiv a + bx \bmod (x^2+2)$, then $\psi(p(x)) \equiv \psi(a+bx) \bmod (y^2+3)$.

(2) Show that ψ is a field homomorphism, as in Definition 5.4.6, that is injective and surjective.

Exercise 6.7.4. Find all the irreducible polynomials in $\mathbb{F}_2[x]$ of degree 1, 2, and 3.

Exercise 6.7.5. Find all the irreducible polynomials in $\mathbb{F}_3[x]$ of degree 1, 2, and 3.

Exercise 6.7.6. Is the polynomial $f = x^4 + 4$ irreducible in $\mathbb{F}_5[x]$? Prove that f is irreducible or find a factorization into irreducible polynomials over $\mathbb{F}_5$.

Exercise 6.7.7. Is the polynomial $f = 2x^4 + x^3 + x + 3$ irreducible in $\mathbb{F}_5[x]$? Prove that f is irreducible or find a factorization into irreducible polynomials over $\mathbb{F}_5$.

Exercise 6.7.8. Let $f(x) = x^3 + 3x + 1$ and let $g(x) = x^3 + 2x + 1$ be defined over $\mathbb{Z}/5\mathbb{Z}[x]$. We will also write $\mathbb{F}_5$ for $\mathbb{Z}/5\mathbb{Z}$.

(a) Is $f(x)$ or $g(x)$ irreducible over $\mathbb{F}_5[x]$? If it is irreducible, prove it. If it is not irreducible, factor the polynomial into other irreducible polynomials of lesser degree.

(b) How many elements are there in $\mathbb{F}_5[x]/(f(x))$ and $\mathbb{F}_5[x]/(g(x))$?

(c) Are $\mathbb{F}_5[x]/(f(x))$ and $\mathbb{F}_5[x]/(g(x))$ fields? If it is a field, explain why. If it is not a field, find a pair of zero-divisors (i.e., you need to find polynomials a, b in $\mathbb{F}_5[x]$ such that $a \cdot b \equiv 0 \bmod f(x)$ or $g(x)$).

(d) Is the polynomial $x + 3$ invertible in $\mathbb{F}_5[x]/(f(x))$? Is it invertible in $\mathbb{F}_5[x]/(g(x))$? If so, find its multiplicative inverse. If not, explain why.

(e) Is the polynomial $x + 1$ invertible in $\mathbb{F}_5[x]/(f(x))$? Is it invertible in $\mathbb{F}_5[x]/(g(x))$? If so, find its multiplicative inverse. If not, explain why.

Exercise 6.7.9. Let $n > 1$, $m_n(x) = x^n + x + 1$, and let $F_n = \mathbb{F}_2[x]/(m(x))$.

(a) Is $m(x)$ irreducible over $\mathbb{F}_2[x]$? Is F_n a field?

(b) How many elements are there in F_n?

(c) Let $n = 3$, so $m_3(x) = x^3 + x + 1$ and $F_3 = \mathbb{F}_2[x]/(x^3 + x + 1)$. Write down a complete set of representative for F_3. (Hint: pick representatives all with degree ≤ 2.)

(d) Find a polynomial $f(x)$ such that

$$f(x) \equiv (x^5 + 1)(x^4 + x^3 + x^2 + 1) \bmod (x^3 + x + 1)$$

and $\deg(f(x)) \leq 2$.

Exercise 6.7.10. Prove Lemma 6.6.5; i.e., show the characteristic of a field F is either 0 or a prime p. (Hint: if $(pq) \cdot 1_F = 0_F$, then $(p \cdot 1_F) \cdot (q \cdot 1_F) = 0_F$. Now use Exercise 5.6.19.)

Exercise 6.7.11. Let p be a prime number, and let F be a finite field of characteristic p. The Frobenius map of F is defined by $\phi_p\colon F \to F$ such that $\phi_p(f) = f^p$ for all $f \in F$. Show the following properties of ϕ_p.

(a) The map ϕ_p is a field homomorphism; i.e.,
- $\phi_p(f + g) = \phi_p(f) + \phi_p(g)$, and
- $\phi_p(f \cdot g) = \phi_p(f) \cdot \phi_p(g)$,

for all $f, g \in F$. (Hint: Exercise 4.7.27.)

(b) The map ϕ_p is injective.

(c) The map ϕ_p is an automorphism of F; i.e., ϕ_p is a bijective homomorphism from F to F. (Hint: it remains to show that ϕ_p is surjective. Use the fact that F is finite.)

Exercise 6.7.12. Let $F = \mathbb{F}_3[i]$ and consider F as a vector space over $\mathbb{F}_3$ (as in Example 6.6.7), with basis $\{1, i\}$. Let $\psi\colon \mathbb{F}_3[i] \to \mathbb{F}_3[i]$ be defined by $\psi(f) = i \cdot f$ for all $f \in \mathbb{F}_3[i]$.

(a) Show that ψ is a linear map of F to F; i.e., $\psi(f + g) = \psi(f) + \psi(g)$ and $\psi((n \bmod 3) \cdot f) = (n \bmod 3) \cdot \psi(f)$, for all $f, g \in F$ and all $n \in \mathbb{Z}$.

(b) What is the matrix M defined over $\mathbb{F}_3$ that represents ψ in coordinates with respect to the basis $\{1, i\}$ of F? That is, find a matrix M such that $\psi(v) = M \cdot v$ for any vector $v \in (\mathbb{F}_3)^2 \cong \mathbb{F}_3[i]$.

(c) Let $\phi_3\colon F \to F$ be the Frobenius automorphism defined in Exercise 6.7.11. Show that ϕ_3 is a linear map, and find the matrix over $\mathbb{F}_3$ that represents ϕ_3 in coordinates with respect to the basis $\{1, i\}$ of F.

Exercise 6.7.13. Let $F = \mathbb{F}_3[x]/(x^2 + 1)$ and let F^* be all the non-zero elements of F. Show that F^* is cyclic; i.e., find an element $f \in F^*$ such that if $g \in F^*$ is any other non-zero element, then $g = f^n$ for some $n \geq 1$.

CHAPTER 7

THE THEOREMS OF WILSON, FERMAT, AND EULER

Mathematics is the queen of the sciences and number theory is the queen of mathematics. (Die Mathematik ist die Königin der Wissenschaften und die Zahlentheorie ist die Königin der Mathematik.)

Carl Friedrich Gauss

When we experiment with the elements of the ring $\mathbb{Z}/m\mathbb{Z}$, lots of patterns emerge. Some are easy to verify, but some others are either very deep and difficult to prove or they still remain as conjectures. In this chapter we present proofs for several of these phenomena. The theorems are named after the mathematicians who first discovered, proved, or published these results: Leonhard Euler (1707–1783), Pierre de Fermat (1601–1665), and John Wilson (1741–1793).

7.1. Wilson's Theorem

In Chapter 5 we saw that if $(R, +, \cdot)$ is a commutative ring with identity, then its subset of units, $(R^\times, \cdot)$, is a commutative group (see Proposition 5.3.16). In particular, in Corollary 5.3.17, we saw that $U_m = (\mathbb{Z}/m\mathbb{Z})^\times$ is a group with respect to multiplication. In this section we present an application of this fact: Wilson's theorem.

Wilson's theorem provides an answer for the following question: what is the least non-negative residue of $(m-1)!$ mod m? In the language of $\mathbb{Z}/m\mathbb{Z}$, we may rewrite this question as follows. We are interested in the product

$$1 \cdot 2 \cdot 3 \cdots (m-2) \cdot (m-1) \bmod m,$$

so we are trying to calculate the value of the product of all non-zero elements in $\mathbb{Z}/m\mathbb{Z}$. For instance,

$$4! \equiv 4 \bmod 5.$$

Example 7.1.1. Let us calculate $(m-1)! \bmod m$ for some small values of m.

$$\begin{aligned}
(2-1)! &\equiv 1 \bmod 2,\\
(3-1)! &\equiv 1\cdot 2 \equiv 2 \bmod 3,\\
(4-1)! &\equiv 1\cdot 2\cdot 3 \equiv 6 \equiv 2 \bmod 4,\\
(5-1)! &\equiv 1\cdot 2\cdot 3\cdot 4 \equiv 24 \equiv 4 \bmod 5,\\
(6-1)! &\equiv 1\cdot 2\cdot 3\cdot 4\cdot 5 \equiv 6\cdot 20 \equiv 0\cdot 20 \equiv 0 \bmod 6,\\
(7-1)! &\equiv 1\cdot 2\cdot 3\cdot 4\cdot 5\cdot 6 \equiv 6\cdot 20\cdot 6 \equiv (-1)(-1)(-1) \equiv -1 \equiv 6 \bmod 7,\\
(8-1)! &\equiv 1\cdot 2\cdot 3\cdot 4\cdot 5\cdot 6\cdot 7 \equiv 8\cdot 2\cdot 5\cdot 6\cdot 7 \equiv 0\cdot 420 \equiv 0 \bmod 8,\\
(9-1)! &\equiv 1\cdot 2\cdot 3\cdot 4\cdot 5\cdot 6\cdot 7\cdot 8 \equiv 18\cdot 3\cdot 4\cdot 5\cdot 7\cdot 8 \equiv 0\cdot 3360 \equiv 0 \bmod 9,\\
(10-1)! &\equiv 1\cdot 2\cdot 3\cdot 4\cdot 5\cdot 6\cdot 7\cdot 8\cdot 9 \equiv 10\cdot 3\cdot 4\cdot 6\cdot 7\cdot 8\cdot 9 \equiv 0 \bmod 10.
\end{aligned}$$

When $m > 4$ is a composite number, a pattern emerges from the data. It seems $(m-1)! \equiv 0 \bmod m$. Why would that be? In Proposition 5.3.11, we saw that the set of all zero-divisors in $\mathbb{Z}/m\mathbb{Z}$ is given by

$$(\mathbb{Z}/m\mathbb{Z})^0 = \{a \bmod m : 1 \le a \le m-1 \text{ and } \gcd(a,m) > 1\}.$$

Thus, if m is composite and we multiply all the non-zero elements of $\mathbb{Z}/m\mathbb{Z}$, there will be a product of two zero-divisors that make the whole congruence zero. Let us show this formally.

Lemma 7.1.2. *Let $m > 4$ be a composite number. Then, $(m-1)! \equiv 0 \bmod m$.*

Proof. Let $m > 4$ be a composite number. Thus, $m = ab$, with $1 < a, b < m$. Suppose first that $a \neq b$. Without loss of generality, we may assume $b > a$. Then, $a \bmod m$ and $b \bmod m$ are two distinct congruence classes, because $0 < b - a < m$ and therefore m does not divide $b - a$, so $a \not\equiv b \bmod m$. Thus, $m = a \cdot b$ is a divisor of $(m-1)! = 1 \cdot 2 \cdots a \cdots b \cdots (m-1)$ and this implies that $(m-1)! \equiv 0 \bmod m$, as claimed.

It only remains to consider the case when $m = ab$ and $a = b$, i.e., $m = a^2$. Since we are assuming that $m > 4$, we have that $a > 2$. Therefore, a and $2a$ are numbers between 1 and $m - 1 = a^2 - 1 > 2a$, since $a \ge 3$. It follows that $a \cdot 2a$ is a factor of $(m-1)! = 1 \cdot 2 \cdots a \cdots 2a \cdots m-1$. Hence, $(m-1)!$ is divisible by $2a^2$, and in particular it is divisible by $a^2 = m$. It follows that $(m-1)! \equiv 0 \bmod m$. □

Example 7.1.3. We have just settled the value of $(m-1)! \bmod m$ when $m > 4$ is composite ($\equiv 0 \bmod m$) or $m = 4$ ($3! \equiv 2 \bmod 4$). What happens when m is prime? We have seen in Example 7.1.1 that

$$1! \equiv 1 \bmod 2,\ \ 2! \equiv 2 \bmod 3,\ \ 4! \equiv 4 \bmod 5,\ \text{ and }\ 6! \equiv 6 \bmod 7,$$

or, equivalently,

$$1! \equiv -1 \bmod 2,\ \ 2! \equiv -1 \bmod 3,\ \ 4! \equiv -1 \bmod 5,\ \text{ and }\ 6! \equiv -1 \bmod 7.$$

Let us calculate one more example to see more evidence of this pattern. Let us calculate $(p-1)! \bmod p$ for $p = 11$. In order to simplify the calculations, we will use the fact that $(\mathbb{Z}/11\mathbb{Z})^\times$ is a group. In particular, every class $a \bmod 11$, for $a = 1, \ldots, 10$, has a multiplicative inverse, and we will pair up together a and $a^{-1} \bmod 11$ because $a \cdot a^{-1} \equiv 1 \bmod 11$:

$$\begin{aligned}(11-1)! &\equiv 1 \cdot 2 \cdot 3 \cdot 4 \cdot 5 \cdot 6 \cdot 7 \cdot 8 \cdot 9 \cdot 10 \\ &\equiv (2 \cdot 6) \cdot (3 \cdot 4) \cdot (5 \cdot 9) \cdot (7 \cdot 8) \cdot 10 \\ &\equiv 1 \cdot 1 \cdot 1 \cdot 1 \cdot 10 \\ &\equiv 10 \equiv -1 \bmod 11.\end{aligned}$$

Notice that the multiplicative inverse of $1 \bmod 11$ is itself. Similarly, the multiplicative inverse of $10 \bmod 11$ is itself. In other words, $1^{-1} \equiv 1 \bmod 11$ and $10^{-1} \equiv 10 \bmod 11$, so those are the two elements of $(\mathbb{Z}/11\mathbb{Z})^\times$ that we were not able to pair up together with their multiplicative inverse, and we ended up with $(11-1)! \equiv 1 \cdot 10 \equiv 10 \equiv -1 \bmod 11$.

We shall generalize the method used to calculate $(p-1)! \bmod p$ in the previous example in order to show that $(p-1)! \equiv -1 \bmod p$ for every prime p. In order to do this, we first need to know that each unit modulo p has a unique multiplicative inverse, but this was already shown in Proposition 5.3.20. We also need to know what congruence classes modulo p are their own multiplicative inverses.

Lemma 7.1.4. *Let p be a prime. Then, the quadratic congruence $x^2 \equiv 1 \bmod p$ has only two solutions; namely, $x \equiv 1$ and $x \equiv -1 \bmod p$. In particular, if $a \bmod p$ is its own multiplicative inverse, then $a \equiv 1 \bmod p$ or $a \equiv -1 \bmod p$.*

Proof. Suppose p is a prime and x is an integer with $x^2 \equiv 1 \bmod p$. In other words, p is a divisor of $x^2 - 1 = (x+1)(x-1)$. Since p is a prime, then p divides $x+1$ or p divides $x-1$, by Lemma 2.10.3. In terms of congruences, this means that $x+1 \equiv 0$ or $x-1 \equiv 0 \bmod p$, or, equivalently, $x \equiv -1$ or $x \equiv 1 \bmod p$. This proves the first part of the lemma.

Now, suppose that the class of $a \bmod p$ is its own multiplicative inverse. This means that $a \cdot a \equiv 1 \bmod p$, or $a^2 \equiv 1 \bmod p$, and, by our previous result, $a \equiv \pm 1 \bmod p$. □

Example 7.1.5. Let us calculate all the squares in $\mathbb{Z}/11\mathbb{Z}$:

$x \bmod 11$	0	1	2	3	4	5	6	7	8	9	10
$x^2 \bmod 11$	0	1	4	9	5	3	3	5	9	4	1

As the previous lemma predicts, the only numbers x whose square is $1 \bmod p$ are precisely $x \equiv \pm 1 \bmod 11$; i.e., $x \equiv 1, 10 \bmod 11$.

Example 7.1.6. Notice, however, that Lemma 7.1.4 is not true if we consider $x^2 \equiv 1 \bmod m$ and m is not a prime. For instance, let us calculate all the squares in $\mathbb{Z}/12\mathbb{Z}$:

$x \bmod 12$	0	1	2	3	4	5	6	7	8	9	10	11
$x^2 \bmod 12$	0	1	4	9	4	1	0	1	4	9	4	1

Therefore, the equation $x^2 \equiv 1 \bmod 12$ has four solutions, namely $x \equiv 1, 5, 7$, or $11 \bmod 12$.

The following theorem was known to Ibn al-Haytham (also known as Alhazen; about 1000 AD), but it is named after John Wilson (a student of the English mathematician Edward Waring) who rediscovered and stated it in the 18th century.

Theorem 7.1.7 (Wilson's theorem). *Let $m > 1$ be fixed.*

(1) *If $m = 4$, then $(m-1)! \equiv 2 \bmod 4$.*

(2) *If $m \neq 4$ is composite, then $(m-1)! \equiv 0 \bmod m$.*

(3) *If m is prime, then $(m-1)! \equiv -1 \bmod m$.*

In particular, $p \geq 2$ is a prime number if and only if $(p-1)! \equiv -1 \bmod p$.

Proof. Parts (1) and (2) have been shown in Example 7.1.1 and Lemma 7.1.2. There are two key ingredients for the proof of part (3). First, each non-zero congruence modulo p has a unique multiplicative inverse mod p (this was shown in Proposition 5.3.20). The second key ingredient is Lemma 7.1.4. These two pieces can be put together to prove that the congruence classes $\{2, 3, \ldots, p-2 \bmod p\}$ can be organized in $\frac{p-3}{2}$ pairs $\{(a_1, a_1^{-1}), (a_2, a_2^{-1}), \ldots, (a_{(p-3)/2}, a_{(p-3)/2}^{-1})\}$ of distinct congruence classes, so that

$$(p-1)! \equiv 1 \cdot a_1 \cdot a_1^{-1} \cdot a_2 \cdot a_2^{-1} \cdots a_{(p-3)/2} \cdot a_{(p-3)/2}^{-1} \cdot (p-1) \bmod p.$$

We leave it as an exercise for the reader to fill in all the details of the proof (see Exercise 7.6.2). □

A surprising consequence of Wilson's theorem is that it provides a criterion for primality of a number p that does not involve finding a factorization of p. This is in practice not very effective (there are a lot of calculations involved in computing the least non-negative residue of $(p-1)! \bmod p$), but it is certainly interesting from a theoretical perspective.

Example 7.1.8. Is $p = 1001$ a prime number? Wilson's theorem tells us that if p is prime, then $(p-1)! = 1000!$ must be congruent to $-1 \bmod 1001$. One can put a computer (or calculator) to work on this and find out that, in fact, $1000! \equiv 0 \bmod 1001$. Thus, we conclude that 1001 is not a prime number and it must be a composite number. Notice, however, that we have not calculated a single prime factor of 1001. (By the way, $1001 = 7 \cdot 11 \cdot 13$, which one can easily deduce from a divisibility test such as Proposition 4.6.4.)

7.2. Fermat's (Little) Theorem

In Section 1.6.1 we already mentioned the French lawyer and mathematician Pierre de Fermat and his famous *last theorem*. In this section, however, we are concerned with another well-known theorem of Fermat, but one that is not quite as hard as the "last" theorem. In fact, this other theorem is not difficult to prove, and it is usually referred to as Fermat's *little* theorem.

Theorem 7.2.1 (Fermat's little theorem). *Let p be a prime number, and let n be an integer. Then, the number $n^p - n$ is always a multiple of p. In other words,*

$$n^p \equiv n \bmod p,$$

for all $n \in \mathbb{Z}$ and any prime p.

For instance, $3^7 - 3 = 2184 = 2^3 \cdot 3 \cdot 7 \cdot 13$ is divisible by 7. Before we attempt to prove Fermat's little theorem, let us think a little bit about powers of numbers modulo m. We have already seen that congruences are very useful in determining the divisibility (or remainder) of a large number by another number m. Let us see another example.

Example 7.2.2. Is $N = 17^{100} - 16$ divisible by 15? Equivalently, is $N \equiv 0 \bmod 15$? By the properties of congruences (Proposition 4.2.1), we know that

$$N = 17^{100} - 16 \equiv 2^{100} - 1 \bmod 15.$$

Moreover, $2^4 \equiv 16 \equiv 1 \bmod 15$. Thus,

$$N \equiv 2^{100} - 1 \equiv (2^4)^{25} - 1 \equiv 1^{25} - 1 \equiv 1 - 1 \equiv 0 \bmod 15.$$

Therefore, the remainder when we divide N by 15 is 0 and this means that 15 divides N.

In this example, the calculation was fairly simple due to the fact that $2^4 \equiv 1 \bmod 15$. In particular, the powers of 2 mod 15 follow a pattern:

$$2,\ 2^2 \equiv 4,\ 2^3 \equiv 8,\ 2^4 \equiv 1,\ 2^5 \equiv 2,\ 2^6 \equiv 4,\ 2^7 \equiv 8,\ 2^8 \equiv 1 \bmod 15, \ldots.$$

In other words, the powers of 2 mod 15 form a repeating sequence

$$2, 4, 8, 1, 2, 4, 8, 1, 2, 4, 8, 1, 2, 4, 8, 1, \ldots.$$

Is this true for any a mod 15? Will the powers of a mod 15 form a repeating sequence? Here is a table of powers for each congruence class modulo 15:

x mod 15	x^2	x^3	x^4	x^5	x^6	x^7	x^8	x^9	x^{10}	x^{11}	x^{12}	...
0	0	0	0	0	0	0	0	0	0	0	0	...
1	1	1	1	1	1	1	1	1	1	1	1	...
2	4	8	1	2	4	8	1	2	4	8	1	...
3	9	12	6	3	9	12	6	3	9	12	6	...
4	1	4	1	4	1	4	1	4	1	4	1	...
5	10	5	10	5	10	5	10	5	10	5	10	...
6	6	6	6	6	6	6	6	6	6	6	6	...
7	4	13	1	7	4	13	1	7	4	13	1	...
8	4	2	1	8	4	2	1	8	4	2	1	...
9	6	9	6	9	6	9	6	9	6	9	6	...
10	10	10	10	10	10	10	10	10	10	10	10	...
11	1	11	1	11	1	11	1	11	1	11	1	...
12	9	3	6	12	9	3	6	12	9	3	6	...
13	4	7	1	13	4	7	1	13	4	7	1	...
14	1	14	1	14	1	14	1	14	1	14	1	...

There are **lots** of patterns in the previous table waiting to be discovered. The first thing to notice is that the powers of a mod 15 follow a repeating pattern, for each congruence class a mod 15 in $\mathbb{Z}/15\mathbb{Z}$, as we had anticipated. It is also apparent that the units and the zero-divisors of $\mathbb{Z}/15\mathbb{Z}$ behave slightly different. Below, the reader can find the same type of table but only for units in $\mathbb{Z}/15\mathbb{Z}$:

x mod 15	x^2	x^3	x^4	x^5	x^6	x^7	x^8	x^9	x^{10}	x^{11}	x^{12}	...
1	1	1	1	1	1	1	1	1	1	1	1	...
2	4	8	1	2	4	8	1	2	4	8	1	...
4	1	4	1	4	1	4	1	4	1	4	1	...
7	4	13	1	7	4	13	1	7	4	13	1	...
8	4	2	1	8	4	2	1	8	4	2	1	...
11	1	11	1	11	1	11	1	11	1	11	1	...
13	4	7	1	13	4	7	1	13	4	7	1	...
14	1	14	1	14	1	14	1	14	1	14	1	...

And this is a table of powers but only for zero-divisors of $\mathbb{Z}/15\mathbb{Z}$:

x mod 15	x^2	x^3	x^4	x^5	x^6	x^7	x^8	x^9	x^{10}	x^{11}	x^{12}	...
3	9	12	6	3	9	12	6	3	9	12	6	...
5	10	5	10	5	10	5	10	5	10	5	10	...
6	6	6	6	6	6	6	6	6	6	6	6	...
9	6	9	6	9	6	9	6	9	6	9	6	...
10	10	10	10	10	10	10	10	10	10	10	10	...
12	9	3	6	12	9	3	6	12	9	3	6	...

Notice, in particular, that if x is a unit modulo 15, then $x^4 \equiv 1 \bmod 15$, which explains and generalizes the fact that $2^4 \equiv 1 \bmod 15$. However, if x is a zero-divisor, then $x^4 \equiv 6$ or $10 \bmod 15$. Moreover $x^4 \equiv 6 \bmod 15$ if $x \equiv 0 \bmod 3$, and $x^4 \equiv 10 \bmod 15$ if $x \equiv 0 \bmod 5$.

Before we attempt to prove some of the patterns observed in the tables of powers modulo 15, let us see some simpler examples.

Example 7.2.3. Let us calculate consecutive powers for each non-zero congruence class modulo 5 and modulo 7:

x mod 5	x^2	x^3	x^4	x^5	x^6	x^7	x^8	x^9	x^{10}	x^{11}	x^{12}	...
1	1	1	1	1	1	1	1	1	1	1	1	...
2	4	3	1	2	4	3	1	2	4	3	1	...
3	4	2	1	3	4	2	1	3	4	2	1	...
4	1	4	1	4	1	4	1	4	1	4	1	...

x mod 7	x^2	x^3	x^4	x^5	x^6	x^7	x^8	x^9	x^{10}	x^{11}	x^{12}	...
1	1	1	1	1	1	1	1	1	1	1	1	...
2	4	1	2	4	1	2	4	1	2	4	1	...
3	2	6	4	5	1	3	2	6	4	5	1	...
4	2	1	4	2	1	4	2	1	4	2	1	...
5	4	6	2	3	1	5	4	6	2	3	1	...
6	1	6	1	6	1	6	1	6	1	6	1	...

From the tables we see that $a^4 \equiv 1 \bmod 5$, for all $a \in \mathbb{Z}$ such that $\gcd(a, 5) = 1$, and, moreover, this is the smallest exponent that works for all such integers. Similarly, $b^6 \equiv 1 \bmod 7$, for all $b \in \mathbb{Z}$ such that $\gcd(b, 7) = 1$, and 6 is the least exponent with this property. Notice that $4 = 5 - 1$ and $6 = 7 - 1$, which suggests that, perhaps, $n^{p-1} \equiv 1 \bmod p$ for any prime p. Let us compute the value of $c^{10} \bmod 11$ for all integers c relatively prime to 11:

$$\begin{aligned}
1^{10} &\equiv 1 \bmod 11,\ 2^{10} \equiv 1024 \equiv 1 \bmod 11,\ 3^{10} \equiv 59049 \equiv 1 \bmod 11,\\
4^{10} &\equiv 1048576 \equiv 1 \bmod 11,\ 5^{10} \equiv 9765625 \equiv 1 \bmod 11,\\
6^{10} &\equiv 60466176 \equiv 1 \bmod 11,\ 7^{10} \equiv 282475249 \equiv 1 \bmod 11,\\
8^{10} &\equiv 1073741824 \equiv 1 \bmod 11,\ 9^{10} \equiv 3486784401 \equiv 1 \bmod 11,\\
10^{10} &\equiv 10000000000 \equiv 1 \bmod 11.
\end{aligned}$$

Thus, it is true that $c^{10} \equiv 1 \bmod 11$ for all $c \in \mathbb{Z}$ with $\gcd(c, 11) = 1$.

Example 7.2.4. We have just seen by direct calculation that $6^{10} \equiv 1 \bmod 11$. Let us show this in a more elegant way, in a manner that we will be able to generalize for any prime p. Consider the two sets

$$S_1 = \{1, 2, 3, 4, 5, 6, 7, 8, 9, 10\}$$

and

$$S_2 = \{6 \cdot i : i = 1, \ldots, 10\} = \{6, 12, 18, 24, 30, 36, 42, 48, 54, 60\}.$$

Notice that

$$\begin{gathered}
6 \equiv 6,\ 12 \equiv 1,\ 18 \equiv 7,\ 24 \equiv 2,\ 30 \equiv 8,\\
36 \equiv 3,\ 42 \equiv 9,\ 48 \equiv 4,\ 54 \equiv 10,\ 60 \equiv 5 \bmod 11.
\end{gathered}$$

Therefore, both S_1 and S_2 are complete residue systems for all the non-zero classes modulo 11. In particular, the numbers in set S_2 cover all the non-zero congruence classes modulo 11, in a different order than S_1. Thus,

$$1 \cdot 2 \cdot 3 \cdot 4 \cdot 5 \cdot 6 \cdot 7 \cdot 8 \cdot 9 \cdot 10 \equiv 6 \cdot 12 \cdot 18 \cdot 24 \cdot 30 \cdot 36 \cdot 42 \cdot 48 \cdot 54 \cdot 60 \bmod 11.$$

Let us write $N = 1 \cdot 2 \cdot 3 \cdot 4 \cdot 5 \cdot 6 \cdot 7 \cdot 8 \cdot 9 \cdot 10$. Hence,

$$\begin{aligned}
N &= 1 \cdot 2 \cdot 3 \cdot 4 \cdot 5 \cdot 6 \cdot 7 \cdot 8 \cdot 9 \cdot 10\\
&\equiv 6 \cdot 12 \cdot 18 \cdot 24 \cdot 30 \cdot 36 \cdot 42 \cdot 48 \cdot 54 \cdot 60\\
&\equiv 6 \cdot 1 \cdot (6 \cdot 2) \cdot (6 \cdot 3) \cdot (6 \cdot 4) \cdot (6 \cdot 5) \cdot (6 \cdot 6) \cdot (6 \cdot 7) \cdot (6 \cdot 8) \cdot (6 \cdot 9) \cdot (6 \cdot 10)\\
&\equiv 6^{10} \cdot (1 \cdot 2 \cdot 3 \cdot 4 \cdot 5 \cdot 6 \cdot 7 \cdot 8 \cdot 9 \cdot 10)\\
&\equiv 6^{10} \cdot N \bmod 11.
\end{aligned}$$

Thus, we have shown that $6^{10} \cdot N \equiv N \bmod 11$. Since $\gcd(N, 11) = 1$, by Proposition 4.3.1, we can cancel N on both sides of the congruence and reach $6^{10} \equiv 1 \bmod 11$, as desired. In other words, since $\gcd(N, 11) = 1$, the congruence class $N \bmod 11$ is invertible. Moreover, $((\mathbb{Z}/11\mathbb{Z})^\times, \cdot)$ is a group and, therefore, there is a multiplicative inverse $N^{-1} \bmod 11$. Thus, given $6^{10} \cdot N \equiv N \bmod 11$, we can multiply both sides by N^{-1} to obtain $6^{10} \equiv 1 \bmod 11$.

As we mentioned, we are going to generalize the method explained in Example 7.2.4 to prove Fermat's little theorem. We will need the following lemma.

Lemma 7.2.5. *Let p be a prime and let $a \in \mathbb{Z}$ be an integer not divisible by p. Then, the set*

$$S = \{a \cdot i : i = 1, 2, \ldots, p-1\} = \{a,\ 2 \cdot a,\ 3 \cdot a, \ldots,\ (p-1) \cdot a\}$$

is a complete residue system of the non-zero classes modulo p. In other words, for each non-zero residue class $b \bmod p$ there is a unique number i, with $1 \leq i \leq p-1$, such that $a \cdot i \in S$ and $a \cdot i \equiv b \bmod p$.

Proof. Let S be the set defined in the statement of the lemma. Clearly, S has $p-1$ elements. Since there are exactly $p-1$ non-zero congruence classes modulo p, we simply need to show that all the elements of S represent different non-zero congruence classes.

First, let us show that $a \cdot i \not\equiv 0 \bmod p$, for all $i = 1, \ldots, p-1$. Suppose, for a contradiction, that $a \cdot i \equiv 0 \bmod p$. This means that p divides $a \cdot i$. Since p is a prime number, it follows from Lemma 2.10.3 that p divides a or p divides i. However, we have assumed that a is not divisible by p and $i = 1, \ldots, p-1$ cannot be divisible by the prime p, so neither a nor i is divisible by p. This is a contradiction and so $a \cdot i$ must be non-zero modulo p.

Next, suppose that $a \cdot i \equiv a \cdot j \bmod p$, for some $1 \leq i, j \leq p-1$. Then, $a \cdot i - a \cdot j \equiv 0 \bmod p$ and this implies that $a(i-j) \equiv 0 \bmod p$. Thus, again by Lemma 2.10.3, this implies that $a \equiv 0 \bmod p$ or $i - j \equiv 0 \bmod p$. Since a is assumed to be relatively prime to p, we conclude that $i - j \equiv 0 \bmod p$, or $i \equiv j \bmod p$. We have assumed that $1 \leq i, j \leq p-1$ and p is prime, so it follows that $i = j$. Thus, if $1 \leq i, j \leq p-1$ and $i \neq j$, then $a \cdot i \not\equiv a \cdot j \bmod p$, as desired.

We conclude that there are $p-1$ distinct, non-zero congruence classes in S, and this concludes the proof of the lemma. □

We are now ready to prove Fermat's little theorem (Theorem 7.2.1).

Proof of Theorem 7.2.1. Let p be a prime number. We will show that $a^p \equiv a \bmod p$, for all integers $a \in \mathbb{Z}$.

First, suppose that a is divisible by p. Then, $a \equiv 0 \bmod p$ and, therefore, $a^p \equiv 0^p \equiv 0 \equiv a \bmod p$. Hence, the theorem is true. It remains to show that the theorem holds for integers a relatively prime to p. We will show that if $\gcd(a, p) = 1$, then $a^{p-1} \equiv 1 \bmod p$. If so, then we may multiply both sides by a and this would show that $a^p \equiv a \cdot a^{p-1} \equiv a \cdot 1 \equiv a \bmod p$, as desired.

In order to show that $a^{p-1} \equiv 1 \bmod p$ for all $\gcd(a, p) = 1$, consider the sets $S_1 = \{1, 2, 3, \ldots, p-1\}$ and

$$S_2 = \{a \cdot i : i = 1, \ldots, p-1\} = \{a, 2a, \ldots, (p-1)a\}.$$

It follows from Lemma 7.2.5 that both sets S_1 and S_2 are complete residue systems for the non-zero congruence classes modulo p. Thus, each set covers each residue class $1 \bmod p, \ldots, p-1 \bmod p$ exactly once. In particular,

$$1 \cdot 2 \cdot 3 \cdots (p-1) \equiv a \cdot 2a \cdot 3a \cdots (p-1)a \bmod p.$$

Let $N = 1 \cdot 2 \cdot 3 \cdots (p-1)$. Then,

$$\begin{aligned} N &= 1 \cdot 2 \cdot 3 \cdots (p-1) \\ &\equiv a \cdot 2a \cdot 3a \cdots (p-1)a \\ &\equiv a^{p-1} \cdot (1 \cdot 2 \cdot 3 \cdots (p-1)) \\ &\equiv a^{p-1} \cdot N \bmod p. \end{aligned}$$

Hence, $a^{p-1} \cdot N \equiv N \bmod p$. Since $\gcd(N, p) = 1$, we may cancel N on both sides of the congruence, by Proposition 4.3.1, to obtain $a^{p-1} \equiv 1 \bmod p$, as desired. □

Remark 7.2.6. The last step in the proof of Fermat's little theorem, i.e., prove that

$$a^{p-1} \cdot N \equiv N \bmod p \text{ implies that } a^{p-1} \equiv 1 \bmod p,$$

where $N = 1 \cdot 2 \cdot 3 \cdots (p-1)$, can be shown in a few different ways:

(1) By Wilson's theorem, $N = (p-1)! \equiv -1 \bmod p$, since p is a prime. Thus, $a^{p-1} \cdot N \equiv N \bmod p$ implies that $-a^{p-1} \equiv -1 \bmod p$, and therefore $a^{p-1} \equiv 1 \bmod p$.

(2) The congruence class $N \bmod p$ is a product of $p-1$ congruence classes, $1 \bmod p, \ldots, p-1 \bmod p$, and each of these congruence classes is a unit modulo p. Since $(\mathbb{Z}/p\mathbb{Z})^\times$ is a group (Corollary 5.3.17), it follows that a product of units is itself a unit, and so $N \bmod p$ is a unit modulo p. Thus, there is a multiplicative inverse $N^{-1} \bmod p$ in $(\mathbb{Z}/p\mathbb{Z})^\times$. In particular,

$$a^{p-1} \equiv a^{p-1} \cdot 1 \equiv a^{p-1} \cdot (N \cdot N^{-1}) \equiv (a^{p-1} \cdot N) \cdot N^{-1} \equiv N \cdot N^{-1} \equiv 1 \bmod p.$$

(3) Since $\gcd(N, p) = 1$, it follows that N is a unit modulo p (by Corollary 5.3.17) and there is a multiplicative inverse $N^{-1} \bmod p$. Then proceed as in the previous item.

Example 7.2.7. Is the number $N = 2379^{11916} - 1$ divisible by 1987? We shall use Fermat's little theorem to show that the answer is "yes". Indeed, it turns out that 1987 is a prime number and $2379 \equiv 392 \bmod 1987$. Since $392 \not\equiv 0 \bmod 1987$, it follows from Fermat's little theorem that

$$392^{1986} \equiv 1 \bmod 1987.$$

Moreover, $11916 = 6 \cdot 1986$. Therefore,

$$2379^{11916} \equiv 392^{11916} \equiv (392^{1986})^6 \equiv 1^6 \equiv 1 \bmod 1987.$$

Hence, $N \equiv 2379^{11916} - 1 \equiv 1 - 1 \equiv 0 \bmod 1987$ and N is divisible by 1987.

Example 7.2.8. Let us show that $n^{12} \equiv 1 \bmod 35$, for all integers n relatively prime to 35. For instance,

$$2^{12} - 1 = 4095 = 35 \cdot 117 \quad \text{and} \quad 3^{12} - 1 = 531440 = 35 \cdot 15184.$$

Since $35 \equiv 5 \cdot 7$, the Chinese remainder theorem (Theorem 4.5.9; see also Lemma 4.5.4) implies that it suffices to prove that

$$n^{12} \equiv 1 \bmod 5 \quad \text{and} \quad n^{12} \equiv 1 \bmod 7.$$

By Fermat's little theorem, we know that $n^4 \equiv 1 \bmod 5$, as long as $\gcd(n,5) = 1$, and $n^6 \equiv 1 \bmod 7$, as long as $\gcd(n,7) = 1$. Hence,

$$n^{12} \equiv (n^4)^3 \equiv 1^3 \equiv 1 \bmod 5$$

and

$$n^{12} \equiv (n^6)^2 \equiv 1^2 \equiv 1 \bmod 7.$$

Thus, $n^{12} \equiv 1 \bmod 35$, as claimed.

7.3. Euler's Theorem

Fermat's little theorem allows us to calculate (large) powers of a number modulo a prime (see Example 7.2.7). It can also be used to calculate powers of numbers modulo a product of distinct primes (as in Example 7.2.8). However, Fermat's little theorem is of little use when trying to calculate powers modulo a number N that is not square-free (e.g., $N = 9$ or $N = 100 = 4 \cdot 25$). In 1736, Leonhard Euler published a proof of a theorem that generalizes Fermat's little theorem to any modulus, which is now known as Eulers' theorem.

Figure 7.1. Leonhard Euler (1707–1783). Image source: Wikipedia Commons.

Leonhard Euler was a Swiss mathematician and one of the greatest mathematicians of all time. He made important discoveries in fields as diverse as infinitesimal calculus and graph theory, but also mechanics, fluid dynamics, optics, and astronomy. He also introduced much of the modern mathematical terminology and notation, particularly for mathematical analysis, such as the notion of a mathematical function.

Example 7.3.1. In the following table we have calculated the first 12 consecutive powers of each invertible congruence class in $\mathbb{Z}/9\mathbb{Z}$:

$x \bmod 9$	x^2	x^3	x^4	x^5	x^6	x^7	x^8	x^9	x^{10}	x^{11}	x^{12}	$\ldots$
1	1	1	1	1	1	1	1	1	1	1	1	$\ldots$
2	4	8	7	5	1	2	4	8	7	5	1	$\ldots$
4	7	1	4	7	1	4	7	1	4	7	1	$\ldots$
5	7	8	4	2	1	5	7	8	4	2	1	$\ldots$
7	4	1	7	4	1	7	4	1	7	4	1	$\ldots$
8	1	8	1	8	1	8	1	8	1	8	1	$\ldots$

It is plain now that $n^6 \equiv 1 \bmod 9$ for all integers n relatively prime to 9. Fermat's little theorem tells us that $n^2 \equiv 1 \bmod 3$, for each n not divisible by 3, but it says nothing about the congruence class of the powers of n modulo 9. Euler's theorem will show that $n^6 \equiv 1 \bmod 9$ when $\gcd(n, 9) = 1$.

Let us show that $5^6 \equiv 1 \bmod 9$ using a method of proof that will generalize to a proof of Euler's theorem (Theorem 7.3.5 below). Let us consider the two sets

$$S_1 = \{1, 2, 4, 5, 7, 8\} \text{ and } S_2 = \{5, 5 \cdot 2, 5 \cdot 4, 5 \cdot 5, 5 \cdot 7, 5 \cdot 8\} = \{5, 10, 20, 25, 35, 40\}.$$

Notice that

$$5 \equiv 5 \bmod 9, \ 10 \equiv 1 \bmod 9, \ 20 \equiv 2 \bmod 9,$$
$$25 \equiv 7 \bmod 9, \ 35 \equiv 8 \bmod 9, \ 40 \equiv 4 \bmod 9.$$

Therefore, both S_1 and S_2 represent the same six congruence classes modulo 9, namely the six congruence classes of the units modulo 9. In particular, the product of the elements of S_1 and the product of the elements of S_2 are congruent modulo 9; i.e.,

$$1 \cdot 2 \cdot 4 \cdot 5 \cdot 7 \cdot 8 \equiv 5 \cdot 10 \cdot 20 \cdot 25 \cdot 35 \cdot 40 \bmod 9.$$

Let $N = 1 \cdot 2 \cdot 4 \cdot 5 \cdot 7 \cdot 8$. Then,

$$\begin{aligned} N &= 1 \cdot 2 \cdot 4 \cdot 5 \cdot 7 \cdot 8 \\ &\equiv 5 \cdot 10 \cdot 20 \cdot 25 \cdot 35 \cdot 40 \\ &\equiv 5 \cdot (5 \cdot 2)(5 \cdot 4)(5 \cdot 5)(5 \cdot 7)(5 \cdot 8) \\ &\equiv 5^6 \cdot (1 \cdot 2 \cdot 4 \cdot 5 \cdot 7 \cdot 8) \\ &\equiv 5^6 \cdot N \bmod 9. \end{aligned}$$

Thus, we have shown that $5^6 \cdot N \equiv N \bmod 9$. Since $\gcd(N, 9) = 1$, it follows that N is a unit modulo 9 and we may cancel N in both sides of the congruence $5^6 \cdot N \equiv N \bmod 9$ to obtain $5^6 \equiv 1 \bmod 9$, as desired. Notice that here the exponent of 5 is 6 because there are 6 elements in S_1 and also in S_2. In other words, there are precisely 6 units in $\mathbb{Z}/9\mathbb{Z}$.

Before we can state and prove Euler's theorem, first we need to introduce an important function.

Definition 7.3.2. We define the *Euler phi function*, or *Euler's totient function*, $\varphi(m)$ as the number of elements in $(\mathbb{Z}/m\mathbb{Z})^\times$, for any $m \geq 2$.

Remark 7.3.3. Let $m > 2$ be a fixed integer. The value of $\varphi(m)$ can be calculated in the following ways:

(1) $\varphi(m)$ is the size of $(\mathbb{Z}/m\mathbb{Z})^\times$.
(2) $\varphi(m)$ is the number of invertible congruence classes in $\mathbb{Z}/m\mathbb{Z}$.
(3) $\varphi(m)$ is the number of integers a in the range $1 \leq a \leq m-1$ which are relatively prime to m.

Indeed, the set $(\mathbb{Z}/m\mathbb{Z})^\times$ is, by definition, the set of units in $\mathbb{Z}/m\mathbb{Z}$, and $a \bmod m$ is a unit if and only if a is invertible modulo m. Finally, we have shown in Proposition 5.3.11 that

$$(\mathbb{Z}/m\mathbb{Z})^\times = \{a \bmod m : 1 \leq a \leq m-1,\ \gcd(a,m) = 1\}.$$

Thus, $\varphi(m)$ counts the size of the set $\{a : 1 \leq a \leq m-1,\ \gcd(a,m) = 1\}$.

Example 7.3.4. Let us calculate some values of $\varphi(m)$.

- The group $(\mathbb{Z}/2\mathbb{Z})^\times$ only contains one element, namely $1 \bmod 2$. Therefore, $\varphi(2) = 1$.
- We have $(\mathbb{Z}/3\mathbb{Z})^\times = \{1, 2 \bmod 3\}$ so $\varphi(3) = 2$.
- We have $(\mathbb{Z}/4\mathbb{Z})^\times = \{1, 3 \bmod 4\}$ so $\varphi(4) = 2$.
- $(\mathbb{Z}/5\mathbb{Z})^\times = \{1, 2, 3, 4 \bmod 5\}$ so $\varphi(5) = 4$.
- $(\mathbb{Z}/6\mathbb{Z})^\times = \{1, 5 \bmod 6\}$ so $\varphi(6) = 2$.
- $(\mathbb{Z}/7\mathbb{Z})^\times = \{1, 2, 3, 4, 5, 6 \bmod 4\}$ so $\varphi(7) = 6$.
- $(\mathbb{Z}/8\mathbb{Z})^\times = \{1, 3, 5, 7 \bmod 8\}$ so $\varphi(8) = 4$.
- $(\mathbb{Z}/9\mathbb{Z})^\times = \{1, 2, 4, 5, 7, 8 \bmod 9\}$ so $\varphi(9) = 6$.

We are now ready to state Euler's theorem.

Theorem 7.3.5 (Euler's theorem). *Let $m > 1$ be fixed and let a be an integer relatively prime to m. Then*

$$a^{\varphi(m)} \equiv 1 \bmod m,$$

where φ is Euler's phi function.

Example 7.3.6. Let $m = 9$. Then, by Example 7.3.4, we know that $\varphi(9) = 6$. Hence, by Euler's theorem, we have that $a^6 \equiv 1 \bmod 9$, for all integers a relatively prime to 9, in agreement with the table we calculated in Example 7.3.1.

Remark 7.3.7. Suppose m is a prime number. Then, every number $1, 2, \ldots, m-1$ is a unit modulo m, because they are all relatively prime to m. It follows that $(\mathbb{Z}/m\mathbb{Z})^\times$ has $m-1$ elements and $\varphi(m) = m-1$. Thus, by Euler's theorem $a^{m-1} \equiv 1 \bmod m$, for all numbers a relatively prime to m. This is precisely the statement of Fermat's little theorem, so now it is clear that Euler's theorem is a generalization of Fermat's result.

In order to prove Euler's theorem, we shall need to generalize Lemma 7.2.5, which was used to prove Fermat's little theorem.

Lemma 7.3.8. *Let $m > 1$ be fixed and let $a \in \mathbb{Z}$ be an integer relatively prime to m. Let $U_m = (\mathbb{Z}/m\mathbb{Z})^\times$ be the set of all units in $\mathbb{Z}/m\mathbb{Z}$. Thus, U_m has $\varphi(m)$ congruence classes*

$$U_m = \{u_1, u_2, \ldots, u_{\varphi(m)} \bmod m\}.$$

Then, the set

$$S = \{a \cdot u_i \bmod m : i = 1, 2, \ldots, \varphi(m)\} = \{a \cdot u_1,\ a \cdot u_2,\ a \cdot u_3, \ldots,\ a \cdot u_{\varphi(m)} \bmod m\}$$

is a complete residue system of the units modulo m. In other words, for each unit congruence class $b \bmod m$ there is a unique number i, with $1 \leq i \leq \varphi(m)$, such that $a \cdot u_i \in S$ and $a \cdot u_i \equiv b \bmod m$.

Proof. Let S be the set defined in the statement of the lemma. Clearly, S has $\varphi(m)$ elements (by definition of $\varphi(m)$). Since there are exactly $\varphi(m)$ units modulo m, we simply need to show that (a) each element of S is a unit and (b) all the elements of S represent different congruence classes.

First, let us show that $a \cdot u_i \bmod m$ is a unit, for all $i = 1, \ldots, \varphi(m)$. Since $\gcd(a, m) = 1$, it follows that a is a unit modulo m (by Corollary 5.3.17). Moreover, u_i is a unit and $U_m = (\mathbb{Z}/m\mathbb{Z})^\times$ is a group; thus the product of two units, a and u_i, is itself a unit. Hence, $a \cdot u_i \bmod m$ is also a unit. (Here is another way to see this: since u_i is a unit, it follows that $\gcd(u_i, m) = 1$. By assumption, we also have $\gcd(a, m) = 1$. Hence, $\gcd(au_i, m) = 1$ and au_i is a unit modulo m.)

Next, suppose that $a \cdot u_i \equiv a \cdot u_j \bmod m$, for some $1 \leq i, j \leq \varphi(m)$. Then, $a \cdot u_i - a \cdot u_j \equiv 0 \bmod m$ and this implies that $a(u_i - u_j) \equiv 0 \bmod m$. Hence, m is a divisor of $a(u_i - u_j)$. Since $\gcd(a, m) = 1$, it follows from Corollary 2.7.6 that m must divide $u_i - u_j$, or, equivalently, $u_i \equiv u_j \bmod m$. Since $\{u_1, \ldots, u_{\varphi(m)}\}$ form a complete residue system for the units modulo m, we have $u_i \equiv u_j \bmod m$ if and only if $i = j$. Hence, if $i \neq j$, then $au_i \not\equiv au_j \bmod m$.

We conclude that there are $\varphi(m)$ distinct unit congruence classes in S, and this concludes the proof of the lemma. □

Proof of Euler's theorem. Let $m > 1$ be fixed and let a be an integer relatively prime to m. By Lemma 7.3.8, the set

$$U_m = (\mathbb{Z}/m\mathbb{Z})^\times = \{u_1, u_2, \ldots, u_{\varphi(m)} \bmod m\}$$

and the set

$$S = \{a \cdot u_i \bmod m : i = 1, 2, \ldots, p-1\} = \{a \cdot u_1,\ a \cdot u_2,\ a \cdot u_3, \ldots,\ a \cdot u_{\varphi(m)} \bmod m\}$$

are both complete residue systems for the units modulo m. Hence, the congruence classes $au_1, \ldots, au_{\varphi(m)} \bmod m$ are simply a reordering of $u_1, \ldots, u_{\varphi(m)} \bmod m$. In particular,

$$u_1 \cdot u_2 \cdot u_3 \cdots u_{\varphi(m)} \equiv (a \cdot u_1) \cdot (a \cdot u_2) \cdot (a \cdot u_3) \cdots (a \cdot u_{\varphi(m)}) \bmod m.$$

Let us write $N \equiv u_1 \cdot u_2 \cdot u_3 \cdots u_{\varphi(m)} \bmod m$. Then,

$$\begin{aligned} N &\equiv u_1 \cdot u_2 \cdot u_3 \cdots u_{\varphi(m)} \\ &\equiv (a \cdot u_1) \cdot (a \cdot u_2) \cdot (a \cdot u_3) \cdots (a \cdot u_{\varphi(m)}) \\ &\equiv a^{\varphi(m)} \cdot (u_1 \cdot u_2 \cdot u_3 \cdots u_{\varphi(m)}) \\ &\equiv a^{\varphi(m)} \cdot N \bmod m. \end{aligned}$$

Thus, we find that $a^{\varphi(m)} \cdot N \equiv N \bmod m$. Since each u_i is a unit modulo m, we conclude that their product, N, is also a unit modulo m. Hence, there is an inverse $N^{-1} \bmod m$, such that $N \cdot N^{-1} \equiv 1 \bmod m$. Hence,

$$a^{\varphi(m)} \equiv a^{\varphi(m)} \cdot 1 \equiv a^{\varphi(m)} \cdot (N \cdot N^{-1}) \equiv (a^{\varphi(m)} \cdot N) \cdot N^{-1} \equiv N \cdot N^{-1} \equiv 1 \bmod m,$$

as desired. This concludes the proof of Euler's theorem. □

Example 7.3.9. What is the first digit (the units digit) in the decimal expansion of 123^{321}? Notice that the units digit in the decimal expansion of a number $N = a_t \cdot 10^t + \cdots + a_2 \cdot 10^2 + a_1 \cdot 10 + a_0$ is the number a_0. Moreover, $N \equiv a_0 \bmod 10$. Thus, we need to calculate the least non-negative residue of $123^{321} \bmod 10$.

We shall use Euler's theorem with $m = 10$, so we first need to calculate $\varphi(10)$. The units in $\mathbb{Z}/10\mathbb{Z}$ are given by the set $\{1, 3, 7, 9 \bmod 10\}$. Thus, $\varphi(10) = 4$. Since $\gcd(123, 10) = 1$, we can conclude that $123^4 \equiv 1 \bmod 10$, by Euler's theorem. Moreover,

$$321 = 4 \cdot 80 + 1.$$

Therefore,

$$123^{321} \equiv (123^4)^{80} \cdot 123 \equiv 1^{80} \cdot 123 \equiv 123 \equiv 3 \bmod 10.$$

We have shown that the first digit of the decimal expansion is 3.

Example 7.3.10. We have seen in Example 7.2.2 that $n^8 \equiv 1 \bmod 15$, for all n relatively prime to 15. Let us show this using Euler's theorem. First, we need to calculate $\varphi(15)$. By Corollary 5.3.17 (see also Example 5.3.18), we know that

$$\begin{aligned} U_{15} = (\mathbb{Z}/15\mathbb{Z})^\times &= \{a \bmod 15 : 1 \leq a \leq 14,\ \gcd(a, 15) = 1\} \\ &= \{1, 2, 4, 7, 8, 11, 13, 14 \bmod 15\}. \end{aligned}$$

In particular, there are 8 units modulo 15 and $\varphi(15) = 8$. Hence, by Euler's theorem, we have that $n^8 \equiv 1 \bmod 15$, for all n relatively prime to 15.

Notice, however, that the tables in Example 7.2.2 imply that $n^4 \equiv 1 \bmod 15$, for all n with $\gcd(n, 15) = 1$, but this does not follow directly from Euler's theorem. We may prove this smaller exponent using Fermat's little theorem. In order to prove that $n^4 \equiv 1 \bmod 15$ whenever $\gcd(n, 15) = 1$, it suffices to show that $n^4 \equiv 1 \bmod 3$ and also modulo 5 (by Lemma 4.5.4). By Fermat's little theorem, we know that

$$n^2 \equiv 1 \bmod 3 \quad \text{and} \quad n^4 \equiv 1 \bmod 5.$$

Thus,

$$n^4 \equiv (n^2)^2 \equiv 1^2 \equiv 1 \bmod 3,$$

and

$$n^4 \equiv 1 \bmod 5.$$

Hence, $n^4 \equiv 1 \bmod 15$, as we wanted to prove.

7.4. Euler's Phi Function

Euler's phi function is prominently featured in the statement of Euler's theorem and, as a consequence, we need to calculate values of $\varphi(m)$ whenever we want to use Euler's theorem (see Example 7.3.4 for some values calculated directly from the definition). In this section we prove a few results that help in calculating values of φ efficiently.

Proposition 7.4.1. *Let φ be Euler's phi function. Then:*

(1) *If p is a prime, then $\varphi(p) = p - 1$.*

(2) *If p is a prime and $n \geq 1$, then $\varphi(p^n) = (p-1)p^{n-1}$.*

Proof. As a consequence of Corollary 5.3.17, if p is a prime, then $(\mathbb{Z}/p\mathbb{Z})^\times = \{1, \ldots, p-1\}$ has exactly $p - 1$ elements. Thus, $\varphi(p) = p - 1$.

Now, suppose that p is prime and $n \geq 1$. Then,

$$(\mathbb{Z}/p^n\mathbb{Z})^\times = \{a \bmod p^n : 1 \leq a \leq p^n - 1,\ \gcd(a, p^n) = 1\}.$$

Notice that $\gcd(a, p^n) = 1$ if and only if $\gcd(a, p) = 1$. Hence, the non-units in $\mathbb{Z}/p^n\mathbb{Z}$ are the multiples of p; i.e., the non-units are the congruence classes

$$0,\ p,\ 2p, \ldots,\ (p-1)p,\ p \cdot p = p^2,\ (p+1)p, \ldots,\ p^n - p = (p^{n-1} - 1)p,$$

or, in other words, the non-units are the classes in the set

$$\{k \cdot p : k = 0, \ldots, p^{n-1} - 1\}.$$

Hence, there are $p^{n-1} - 1 + 1 = p^{n-1}$ non-units. Since there are p^n congruence classes in $\mathbb{Z}/p^n\mathbb{Z}$, there must be

$$p^n - p^{n-1} = p^{n-1}(p-1)$$

units in $\mathbb{Z}/p^n\mathbb{Z}$. We have shown that $\varphi(p^n) = (p-1)p^{n-1}$, as claimed. $\square$

Example 7.4.2. By the previous proposition, we calculate

$$\varphi(4) = (2-1) \cdot 2 = 2 \quad \text{and} \quad \varphi(25) = (5-1) \cdot 5 = 4 \cdot 5 = 20.$$

What is $\varphi(100)$? Proposition 7.4.1 only tells us the values of φ for powers of primes, so we cannot calculate $\varphi(100)$ just yet. However, we shall see that $\varphi(100) = \varphi(4 \cdot 25) = \varphi(4) \cdot \varphi(25) = 2 \cdot 20 = 40$, and this can be done because $\gcd(4, 25) = 1$.

More generally, we shall prove that $\varphi(m \cdot n) = \varphi(m) \cdot \varphi(n)$ whenever m and n are relatively prime. This equality will follow from an important theorem about the structure of $\mathbb{Z}/mn\mathbb{Z}$, which is equivalent to the Chinese remainder theorem (Theorem 4.5.9).

Theorem 7.4.3. *Let $m, n > 1$ be relatively prime integers; i.e.,* $\gcd(m, n) = 1$. *Then, the map*

$$\psi \colon \mathbb{Z}/mn\mathbb{Z} \to \mathbb{Z}/m\mathbb{Z} \times \mathbb{Z}/n\mathbb{Z}$$

defined by

$$\psi(a \bmod mn) = (a \bmod m, a \bmod n)$$

is well-defined and it is a bijection of the set $\mathbb{Z}/mn\mathbb{Z}$ and the direct product

$$\mathbb{Z}/m\mathbb{Z} \times \mathbb{Z}/n\mathbb{Z} = \{(u \bmod m, v \bmod n) : 0 \leq u \leq m-1,\ 0 \leq v \leq n-1\}.$$

Proof. We need to check three things: (1) ψ is well-defined, (2) ψ is injective, and (3) ψ is surjective.

(1) We first need to check that ψ is a well-defined map; i.e., if $a \equiv b \bmod mn$, then $\psi(a \bmod mn) = \psi(b \bmod mn)$. Indeed, suppose that $a \equiv b \bmod mn$. Then, mn divides $a - b$, and therefore m and n are divisors of $a - b$. This shows that $a \equiv b \bmod m$ and $a \equiv b \bmod n$. Hence,

$$\begin{aligned}\psi(a \bmod mn) &= (a \bmod m, a \bmod n)\\ &= (b \bmod m, b \bmod n) = \psi(b \bmod mn)\end{aligned}$$

as we needed to prove.

(2) Let us show that ψ is injective; i.e., $\psi(a \bmod mn) = \psi(b \bmod mn)$ implies that $a \equiv b \bmod mn$. Indeed, if $\psi(a \bmod mn) = \psi(b \bmod mn)$, then

$$(a \bmod m, a \bmod n) = (b \bmod m, b \bmod n).$$

In particular, a and b are solutions to the system

$$\begin{cases} x \equiv a \bmod m \\ x \equiv a \bmod n. \end{cases}$$

However, since $\gcd(m, n) = 1$, the Chinese remainder theorem (Theorem 4.5.9) guarantees that there is a unique solution to this system modulo mn. Hence, we must have $a \equiv b \bmod mn$, as claimed. This proves ψ is injective.

(3) It only remains to show that ψ is surjective; i.e., for all $(u \bmod m, v \bmod n) \in \mathbb{Z}/m\mathbb{Z} \times \mathbb{Z}/n\mathbb{Z}$ there is some $a \bmod mn$ such that $\psi(a \bmod mn) = (u \bmod m, v \bmod n)$. In order to show this, let $(u \bmod m, v \bmod n)$ be an arbitrary element of the direct product $\mathbb{Z}/m\mathbb{Z} \times \mathbb{Z}/n\mathbb{Z}$. Since $\gcd(m, n) = 1$, the Chinese remainder theorem implies that the system

$$\begin{cases} x \equiv u \bmod m \\ x \equiv v \bmod n \end{cases}$$

has a unique solution $x_0 \bmod mn$ such that $x_0 \equiv u \bmod m$ and $x_0 \equiv v \bmod n$. Hence,

$$\psi(x_0 \bmod mn) = (x_0 \bmod m, x_0 \bmod n) = (u \bmod m, u \bmod n),$$

and this shows that ψ is surjective.

We have shown that ψ is well-defined, it is injective, and it is surjective. Therefore, ψ is a bijection. This ends the proof of the theorem. □

Corollary 7.4.4. *Let $m, n > 1$ be relatively prime integers and define a map, as in the previous theorem, $\psi\colon \mathbb{Z}/mn\mathbb{Z} \to \mathbb{Z}/m\mathbb{Z} \times \mathbb{Z}/n\mathbb{Z}$. Then, the restriction of ψ to $(\mathbb{Z}/mn\mathbb{Z})^\times$ is a bijection*

$$\psi\colon (\mathbb{Z}/mn\mathbb{Z})^\times \to (\mathbb{Z}/m\mathbb{Z})^\times \times (\mathbb{Z}/n\mathbb{Z})^\times.$$

Proof. By Theorem 7.4.3, the map $\psi\colon \mathbb{Z}/mn\mathbb{Z} \to \mathbb{Z}/m\mathbb{Z} \times \mathbb{Z}/n\mathbb{Z}$ is a bijection. In order to establish the corollary, we need to check that ψ sends units modulo mn to units modulo m and units modulo n, and vice versa.

Let $a \bmod mn$ be a unit. Then, $\gcd(a, mn) = 1$ and it follows that $\gcd(a, m) = 1$ and $\gcd(a, n) = 1$. Hence, $a \bmod m$ and $a \bmod n$ are units in $\mathbb{Z}/m\mathbb{Z}$ and $\mathbb{Z}/n\mathbb{Z}$, respectively. This shows that the image of $(\mathbb{Z}/mn\mathbb{Z})^\times$ via ψ is contained in $(\mathbb{Z}/m\mathbb{Z})^\times \times (\mathbb{Z}/n\mathbb{Z})^\times$.

Also, let $(u \bmod m, u \bmod n)$ be an arbitrary pair of units modulo m and n. In particular, $\gcd(u, m) = 1$ and $\gcd(v, n) = 1$. Let $a \bmod mn$ be a class such that $\psi(a \bmod mn) = (u, v)$. Then, $a \equiv u \bmod m$ and $a \equiv u \bmod n$, and it follows that $\gcd(a, m) = 1$ and $\gcd(a, n) = 1$. We conclude that $\gcd(a, mn) = 1$. Hence, a is a unit modulo mn. Thus, this implies that the inverse image via ψ of a pair of units in $\mathbb{Z}/m\mathbb{Z} \times \mathbb{Z}/n\mathbb{Z}$ always comes from a unit modulo mn.

Since ψ is a bijection and ψ sends units to units and since units in the image space come from units in the domain, it follows that the induced map

$$(\mathbb{Z}/mn\mathbb{Z})^\times \to (\mathbb{Z}/m\mathbb{Z})^\times \times (\mathbb{Z}/n\mathbb{Z})^\times$$

is also a bijection, as claimed. □

Since the size of a direct product $S \times T$ of two finite sets is the product of the sizes of the finite sets S and T, we obtain the following important corollary.

Corollary 7.4.5. *Let $m, n > 1$ be relatively prime natural numbers. Then*

$$\varphi(mn) = \varphi(m) \cdot \varphi(n).$$

Proof. Since $\gcd(m, n) = 1$, our Corollary 7.4.4 implies that the set $(\mathbb{Z}/mn\mathbb{Z})^\times$ and the direct product $(\mathbb{Z}/m\mathbb{Z})^\times \times (\mathbb{Z}/n\mathbb{Z})^\times$ are in bijective correspondence, and therefore they have the same number of elements. Thus,

$$\begin{aligned}\varphi(mn) &= |(\mathbb{Z}/mn\mathbb{Z})^\times| \\ &= |(\mathbb{Z}/m\mathbb{Z})^\times \times (\mathbb{Z}/n\mathbb{Z})^\times| \\ &= |(\mathbb{Z}/m\mathbb{Z})^\times| \cdot |(\mathbb{Z}/n\mathbb{Z})^\times| \\ &= \varphi(m) \cdot \varphi(n),\end{aligned}$$

as desired. □

Example 7.4.6. Let us calculate the first two digits in the decimal expansion of 123^{321}. Notice that if N has decimal expansion

$$N = a_t \cdot 10^t + \cdots + a_2 \cdot 10^2 + a_1 \cdot 10 + a_0,$$

then the first two digits are a_0 and a_1. Moreover, $N \equiv a_1 \cdot 10 + a_0 \bmod 100$. Hence, it suffices to calculate $123^{321} \bmod 100$. In order to do this, we shall use Euler's theorem. We first need to calculate $\varphi(100)$:

$$\varphi(100) = \varphi(4 \cdot 25) = \varphi(4) \cdot \varphi(25) = 2 \cdot 20 = 40,$$

where we have used the fact that $\gcd(4, 25) = 1$ and Corollary 7.4.5. Notice also that $\gcd(123, 100) = 1$ and

$$321 = 40 \cdot 8 + 1.$$

Therefore, Euler's theorem applies, $123^{40} \equiv 1 \bmod 100$, and we may calculate

$$123^{321} \equiv (123^{40})^8 \cdot 123 \equiv 1^8 \cdot 123 \equiv 123 \equiv 23 \bmod 100.$$

Hence, the first two digits of 123^{321} in its decimal expansion are 23.

7.5. Applications

In this section we discuss applications of Fermat's little theorem to primality testing and cryptography.

7.5.1. Fermat's Primality Test. Fermat's little theorem (Theorem 7.2.1) says that if p is prime, then $a^{p-1} \equiv 1 \bmod p$, for all integers a relatively prime to p. In particular, $a^{p-1} \equiv 1 \bmod p$ for all $a = 1, \ldots, p-1$. The contrapositive statement of Fermat's theorem is as follows:

Theorem 7.5.1 (Fermat's primality test)**.** *Let $n > 2$ be a fixed integer. If there is some integer a, with $1 \leq a \leq n-1$, such that $a^{n-1} \not\equiv 1 \bmod n$, then n is not prime.*

Hence, Fermat's little theorem can be used for primality testing. Let us see a few examples.

Example 7.5.2. The number $n = 6$ is not prime because

$$2^5 \equiv 2 \bmod 6.$$

Similarly, $n = 10$ is not prime because $2^9 \equiv 2 \bmod 10$. Also, $n = 15$ is not prime because $2^{14} \equiv 4 \bmod 15$.

Example 7.5.3. Is $n = 341$ a prime? If we use $a = 2$ in Theorem 7.5.1, we see that

$$2^{340} \equiv 1 \bmod 341,$$

and we might be led to believe that 341 is indeed prime. However,

$$3^{340} \equiv 56 \bmod 341,$$

and therefore 341 is not prime. A composite natural number n such that $2^{n-1} \equiv 1 \bmod n$ is called a 2-pseudoprime. See Exercises 7.6.17 and 7.6.18.

Example 7.5.4. Is $n = 561$ a prime? If we use $a = 2$ in Theorem 7.5.1, we see that

$$2^{560} \equiv 1 \bmod 561.$$

And, in fact, $a^{560} \equiv 1 \bmod 561$ for $a = 2$, 4, 5, 7, 8, and 10. However, $3^{560} \equiv 375 \bmod 561$, so 561 is not a prime. A composite natural number m such that $b^{m-1} \equiv 1 \bmod m$ for all integers b which are relatively prime to m is called a Carmichael number. See Exercises 7.6.19 and Exercise 7.6.20.

Remark 7.5.5. In practice, how useful is Fermat's primality test? Let us introduce some notation. If n is a composite number, an integer a in the interval $[1, n-1]$ that is relatively prime to n and such that $a^{n-1} \not\equiv 1 \bmod n$ is called a *Fermat witness*. Thus, the question is, how many Fermat witnesses are there for a fixed composite number n? One can show (although we will not do this here) that if n is not a Carmichael number, then at least half of the integers in the interval $[1, n-1]$ that are relatively prime to n are Fermat witnesses. Hence, Fermat's primality test is quite useful in practice.

7.5.2. The AKS Primality Test. In 2002, in the landmark paper [**AKS04**], Manindra Agrawal, Neeraj Kayal, and Nitin Saxena (three computer scientists at the Indian Institute of Technology Kanpur) published the first (deterministic) primality test that runs within polynomial time. This is called the AKS primality test, and in it is based on the following consequence of Fermat's little theorem.

Proposition 7.5.6. *Let $n \geq 2$ and a be coprime integers. Then, n is prime if and only if the following congruence of polynomials holds:*

$$(x+a)^n \equiv x^n + a \bmod n.$$

That is, n is prime if and only if there is a polynomial $f(x) \in \mathbb{Z}[x]$ such that

$$(x+a)^n - (x^n + a) = n \cdot f(x).$$

Proof. Suppose $n = p$ is a prime number. Then, the binomial theorem (Exercise 2.11.14, and more concretely Exercise 4.7.26) implies that

$$(x+a)^p \equiv x^p + a^p \bmod p.$$

Moreover, by Fermat's little theorem, we have $a^p \equiv a \bmod p$, and therefore $(x+a)^p \equiv x^p + a \bmod p$, as desired.

We leave the proof of the converse as an exercise for the reader (the crucial step is Exercise 7.6.23). □

Example 7.5.7. For instance,

$$\begin{aligned}(x+5)^7 &= x^7 + 35x^6 + 525x^5 + 4375x^4 + 21875x^3 + 65625x^2 + 109375x + 78125 \\ &\equiv x^7 + 5 \bmod 7\end{aligned}$$

but

$$\begin{aligned}(x+5)^6 &= x^6 + 30x^5 + 375x^4 + 2500x^3 + 9375x^2 + 18750x + 15625 \\ &\equiv x^6 + 3x^4 + 4x^3 + 3x^2 + 1 \bmod 6.\end{aligned}$$

Suppose we want to know if a number n is prime. Let $1 < a < n$ be arbitrary. If we implemented code in a computer to calculate $(x+a)^n \bmod n$, the running time would be exponential with respect to the size of n. In order to improve efficiency, the AKS algorithm works in a different ring. Instead of working over the polynomial ring $(\mathbb{Z}/n\mathbb{Z})[x]$, we choose an appropriate small positive integer r and work in the quotient ring $R_r = (\mathbb{Z}/n\mathbb{Z})[x]/(x^r - 1)$, i.e., polynomials with coefficients in $\mathbb{Z}/n\mathbb{Z}$ modulo $(x^r - 1)$.

Corollary 7.5.8. *If $n \geq 2$ is prime, $a \in \mathbb{Z}$ is not divisible by n, and $1 < r < n$, then*

$$(x+a)^n \equiv x^n + a \bmod (n, x^r - 1).$$

In other words, there are polynomials $f(x)$ and $g(x)$ in $\mathbb{Z}[x]$ such that

$$(x+a)^n - (x^n + a) = n \cdot f(x) + g(x) \cdot (x^r - 1).$$

Proof. By Proposition 7.5.6, if n is prime, then $(x+a)^n \equiv (x^n + a) \bmod n$. Thus, there is a polynomial $q(x) \in \mathbb{Z}[x]$ such that

$$(x+a)^n - (x^n + a) = n \cdot q(x).$$

Thus, the result is true with $f(x) = q(x)$ and $g(x) = 0$. □

The statement of Corollary 7.5.8 is weaker than that of Proposition 7.5.6 (indeed, the converse of Corollary 7.5.8 is no longer true in general), but the benefit is that the congruence modulo $(n, x^r - 1)$ is easier and quicker to check. Finally, the key step shown by Agrawal, Kayal, and Saxena is a partial converse for Corollary 7.5.8: there exist a small value of r and a (relatively) small set S of values for a such that if

$$(x+a)^n \equiv x^n + a \bmod (n, x^r - 1),$$

for all $a \in S$, then n is a prime power. We refer the reader to their paper **[AKS04]** for the rest of the details, which are beyond the scope of this chapter.

7.5.3. RSA Public Key Cryptography. In this section we discuss an application of Fermat's little theorem to "real life", a cryptosystem known as *RSA*, which is an example of public key cryptography. It is a widely used system, relying for its security on the difficulty of factoring a large number. RSA is an acronym for the last names of Ron Rivest, Adi Shamir, and Leonard Adleman, who first publicly described the algorithm in 1977 (notice that the authors were wise not to list their last names in alphabetical order).

The encoding and decoding of messages works as follows. Suppose that there are two people, Alice and Bob, who want to communicate privately. First, they need a way to convert words into numbers. This can be done in many ways. One simple way is to assign a 2-digit number to each letter:

$$00 = A, \quad 01 = B, \quad 02 = C, \; \ldots, \; 24 = Y, \quad 25 = Z.$$

The spaces between words are erased, and we make groups of two consecutive letters to form 4-digit numbers. For example, the message PUBLIC KEY CRYPTOGRAPHY would become

1520 0111 0802 1004 2402 1724 1519 1406 1700 1507 2423

where we have added a dummy letter $X = 23$ at the end of the passage to fill out the final block. Now, we need a secure way to encrypt the messages that Bob will send to Alice. The RSA setup is as follows.

RSA cryptosystem:

(1) Alice chooses large primes p and q, then form $n = pq$.

(2) Alice chooses an integer $e \geq 1$ relatively prime to $\varphi(n) = (p-1)(q-1)$.

(3) Alice publishes (n, e) as her public key and computes her private key d such that $de \equiv 1 \bmod \varphi(n)$.

(4) Bob encrypts a message M as $C \equiv M^e \bmod n$, and he sends C to Alice.

(5) Alice can decrypt the message by computing $M \equiv C^d \bmod n$.

Example 7.5.9. Alice picks "large" primes $p = 43$, $q = 59$, and $n = 43 \cdot 59 = 2537$ as the modulus and $e = 13$ as the exponent. She publishes $(n, e) = (2537, 13)$ in a public channel. Now Bob can encrypt messages using RSA and send them to Alice. For example, the first block of our previous message $M = 1520$ would get encrypted as

$$C \equiv (M)^e \equiv (1520)^{13} \equiv 95 \bmod 2537$$

which Bob would send over to Alice as 0095. The second block is

$$C \equiv (0111)^{13} \equiv 1648 \bmod 2537$$

so Bob would send 1648, and so on. The complete encrypted message would be

$$0095\ 1648\ 1410\ 1299\ 0811\ 2333\ 2132\ 0370\ 1185\ 1957\ 1084.$$

Now, to decrypt the message, since Alice knows p and q, she also knows $\varphi(n) = \varphi(43 \cdot 59) = 2436$. Using Euclid's algorithm she easily finds that $d = 937$ satisfies $de \equiv 1 \bmod \varphi(n)$. Consequently, to decrypt the first block C sent over by Bob, she only needs to compute

$$M \equiv C^{937} \equiv (0095)^{937} \equiv 1520 \bmod 2537$$

and now Alice knows the first two letters of the message, i.e., P and U.

Why does the system work? It works thanks to Fermat's little theorem, Theorem 7.2.1. Let us first see a proof using Euler's theorem, Theorem 7.3.5, but under one additional assumption: M is relatively prime to n (this is not much of an imposition, because Bob can change M slightly to make sure that $\gcd(M, n) = 1$, for instance, changing the dummy letter X to Z at the end of the message).

Proposition 7.5.10. *Let p and q be distinct primes, let $n = pq$, let e be an integer relatively prime to $(p-1)(q-1)$, and let d be such that $de \equiv 1 \bmod (p-1)(q-1)$. Let M be a number relatively prime to $n = pq$, and put $C \equiv M^d \bmod n$. Then,*

$$C^e \equiv M^{de} \equiv M \bmod n.$$

Proof. Since the number d is chosen so that $de \equiv 1 \bmod \varphi(n)$, there is some $k \in \mathbb{Z}$ such that $de = 1 + k\varphi(n)$. Hence,

$$C^d \equiv (M^e)^d \equiv M^{de} \equiv M^{1+k\varphi(n)} \equiv M \cdot (M^{\varphi(n)})^k \equiv M \cdot 1 \equiv M \bmod n,$$

where we have used Euler's theorem to show that $M^{\varphi(n)} \equiv 1 \bmod n$. □

Let us see now a proof of the fact that RSA works, but using Fermat's little theorem instead, in order to avoid any restrictions on the message M.

Proposition 7.5.11. *Let p and q be distinct primes, let $n = pq$, let e be an integer relatively prime to $(p-1)(q-1)$, and let d be such that $de \equiv 1 \bmod (p-1)(q-1)$. Let M be any integer, and put $C \equiv M^d \bmod n$. Then,*

$$C^e \equiv M^{de} \equiv M \bmod n.$$

Proof. Since $n = pq$ and p and q are distinct primes, by the Chinese remainder theorem (Theorem 4.5.9, or simply Corollary 4.5.6), it suffices to show that $M^{ed} \equiv M \bmod p$ and $M^{ed} \equiv M \bmod q$. As before, write $de = 1 + k\varphi(n)$ for some $k \in \mathbb{Z}$.

If $M \equiv 0 \bmod p$, then $M^{ed} \equiv 0 \equiv M \bmod p$, and we would be done, so let us assume that $M \not\equiv 0 \bmod p$. Then,

$$\begin{aligned} M^{ed} &\equiv M^{ed-1}M \equiv M^{k(p-1)(q-1)}M \\ &\equiv (M^{p-1})^{k(q-1)}M \equiv 1^{k(q-1)} \cdot M \\ &\equiv M \bmod p, \end{aligned}$$

where we used the fact that $M^{p-1} \equiv 1 \bmod p$ for any M relatively prime to p, by Fermat's little theorem. The fact that $M^{ed} \equiv M \bmod q$ is proved similarly. Hence, by the Chinese remainder theorem,

$$M^{ed} \equiv M \bmod pq,$$

for any integer M, as desired. □

Remark 7.5.12. It is important to notice that if a spy knows how to factor n, then it is easy to compute $\varphi(n)$ and also the decrypting exponent d, and therefore the spy would be able to decipher Bob's messages to Alice. The security of the algorithm, thus, relies on the fact that factoring a large integer is computationally expensive (i.e., time- and memory-consuming). Currently, it takes several months of computing time (even on the best computers available!) to factor numbers with 200 digits. In practice, the RSA codes used on the internet make use of values of n with 600 or 1200 digits, and the value of n is changed on a weekly basis.

7.6. Exercises

Exercise 7.6.1. Calculate the least non-negative residue of 20! mod 23. Also, calculate the least non-negative residue of 20! mod 25. (Hint: use Wilson's theorem.)

Exercise 7.6.2. The goal of this exercise is to finish the proof of Wilson's theorem (Theorem 7.1.7).

(1) Justify the following congruence modulo 11:

$$\begin{aligned} 10! &\equiv 1 \cdot 2 \cdot 3 \cdot 4 \cdot 5 \cdot 6 \cdot 7 \cdot 8 \cdot 9 \cdot 10 \\ &\equiv 1 \cdot 2 \cdot 2^{-1} \cdot 3 \cdot 3^{-1} \cdot 5 \cdot 5^{-1} \cdot 7 \cdot 7^{-1} \cdot 10 \\ &\equiv 1 \cdot 10 \equiv -1 \bmod 11. \end{aligned}$$

(2) Generalize the formula in (1) to prove that if p is any prime, then $(p-1)! \equiv -1 \bmod p$. (Hint: use Lemma 7.1.4.)

Exercise 7.6.3. Let p be an odd prime. Show that $N = 1 + 2 + \cdots + (p-1)$ is divisible by p. (Hint: let $a \not\equiv 0, 1 \bmod p$, and consider $a \cdot N$. Then, use Lemma 7.2.5.)

Exercise 7.6.4. Show that if $m > 2$, then $\varphi(m)$ is even.

Exercise 7.6.5. Find the least non-negative residue of 2^{47} mod 23.

Exercise 7.6.6. Show that $n^{13} - n$ is divisible by $2, 3, 5, 7$ and 13 for all $n \geq 1$.

Exercise 7.6.7. Show that $\frac{n^5}{5} + \frac{n^3}{3} + \frac{7n}{15}$ is an integer for all n.

Exercise 7.6.8. Show that $37^{100} \equiv 13 \bmod 17$. (Hint: use Fermat's little theorem.)

Exercise 7.6.9. Show that if p and q are distinct primes, then $p^{q-1} + q^{p-1} \equiv 1 \bmod pq$.

Exercise 7.6.10. Show that if p and q are distinct primes, then $p^q + q^p \equiv p + q \bmod pq$.

Exercise 7.6.11. Prove that for any natural number $n \geq 1$, $3^{6n} - 2^{6n}$ is never prime.

Exercise 7.6.12. Find the following values of Euler's phi function:

$$\varphi(5),\ \varphi(6),\ \varphi(16),\ \varphi(11),\ \varphi(77),\ \varphi(10),\ \varphi(100),\ \text{and } \varphi(100^n) \quad \text{for all } n \geq 1.$$

Exercise 7.6.13. For each pair (a, b) below, calculate separately $\varphi(ab)$, $\varphi(a)$, and $\varphi(b)$, and then verify that $\varphi(ab) = \varphi(a)\varphi(b)$, or explain why the equality does not hold.

$$\text{(i) } a = 3, b = 5,\ \text{(ii) } a = 4, b = 7,\ \text{(iii) } a = 5, b = 6, \quad \text{and} \quad \text{(iv) } a = 4, b = 6.$$

Exercise 7.6.14. The goal of this exercise is to provide an alternative proof of $\varphi(ab) = \varphi(a)\varphi(b)$ if $\gcd(a, b) = 1$.

(1) First, we will prove that $\varphi(30) = \varphi(6)\varphi(5)$ as follows. Write down all the numbers $1 \leq n \leq 30$ in 6 rows of 5 numbers

1	7	13	19	25
2	8	14	20	26
3	9	15	21	27
4	10	16	22	28
5	11	17	23	29
6	12	18	24	30

(a) Show that each row is a complete residue system modulo 5; hence each row has $\varphi(5)$ numbers relatively prime to 5.

(b) Show that each column is a complete residue system modulo 6; hence each column has $\varphi(6)$ numbers relatively prime to 6. Show that all the numbers in each row are congruent modulo 6.

(c) Show that if a number is relatively prime to 30, then there are in total $\varphi(5)$ numbers in the same row that are relatively prime to 30.

(d) Conversely, show that if a number is **not** relatively prime to 6, then none of the numbers in the same row are relatively prime to 30.

(e) Conclude that

$$\begin{aligned}\varphi(30) &= \varphi(6)\varphi(5)\\ &= (\varphi(6) \text{ rows with units modulo } 30)(\varphi(5) \text{ units in each row}).\end{aligned}$$

(2) Generalize the previous argument to prove that $\varphi(ab) = \varphi(a)\varphi(b)$ if a and b are relatively prime.

Exercise 7.6.15. Prove that $n^{101} - n$ is divisible by 33 for all $n \geq 1$.

Exercise 7.6.16. Use Euler's theorem to find the first digit (starting from the right-hand side of the expansion, i.e., the units digit) of the decimal expansion of 7^{1000}.

Exercise 7.6.17. Fermat's little theorem says that if p is prime and $\gcd(2, p) = 1$, then $2^{p-1} \equiv 1 \bmod p$. However, the converse is not true: if m is a number, $\gcd(2, m) = 1$, and $2^{m-1} \equiv 1 \bmod m$, this **does not imply** that m is a prime number. A number m is called a 2-pseudoprime if (a) m is composite and (b) $2^{m-1} \equiv 1 \bmod m$. Show that 341 is a 2-pseudoprime; i.e., show that $2^{340} \equiv 1 \bmod 341$, but 341 is a composite number.

Exercise 7.6.18. Let $n \geq 1$ be a natural number.

(1) Verify that if n is composite; i.e., $n = ab$, then the polynomial $x^n - 1$ factors as
$$x^n - 1 = (x^b - 1)(x^{b(a-1)} + x^{b(a-2)} + \cdots + x^b + 1).$$

(2) Show that if n is composite, then $m = 2^n - 1$ is also composite.

(3) Show that if n is a 2-pseudoprime, then $m = 2^n - 1$ is also a 2-pseudoprime.

(4) Use part (3) to show that there are infinitely many 2-pseudoprimes.

Exercise 7.6.19. A Carmichael number is a composite positive integer m such that $b^{m-1} \equiv 1 \bmod m$ for all integers b which are relatively prime to m.

(1) Show that 561 is a 2-pseudoprime and a 5-pseudoprime; i.e., show that
$$2^{560} \equiv 1 \bmod 561 \quad \text{and} \quad 5^{560} \equiv 1 \bmod 561.$$

(2) Show that $b^{80} \equiv 1 \bmod 561$, for all b relatively prime to 561. (Hint: use Fermat's little theorem.)

(3) Use part (2) to conclude that 561 is a Carmichael number. (In fact, 561 is the smallest Carmichael number.)

(4) Prove that 1105 is also a Carmichael number. (1105 is the second Carmichael number.)

Exercise 7.6.20. Chernick showed in [**Che39**] that the number
$$C_k = (6k+1)(12k+1)(18k+1)$$
is a Carmichael number if the three factors $6k+1$, $12k+1$, and $18k+1$ are primes. Find three distinct Carmichael numbers using Chernick's formula.

Exercise 7.6.21. Find as many prime factors as possible of the number $N = 3^{10!} - 1$.

Exercise 7.6.22. Let $a, n > 0$ be natural numbers. Find as many prime factors as possible of the number $N = a^{n!} - 1$.

Exercise 7.6.23. Show that $n \geq 2$ is a prime number if and only if the binomial coefficients $\binom{n}{k} \equiv 0 \bmod n$ for all $1 \leq k \leq n-1$. (Hint: suppose $n = p^k m$ for some prime p and some integer m relatively prime to p. Show that $\binom{n}{p} \not\equiv 0 \bmod p^k$.)

Exercise 7.6.24. Alice wants to set up an RSA encryption scheme with $N = 1147$ as her modulus, as in Section 7.5.3.

(1) Can Alice choose $e = 3$ as an encryption exponent? If so, find the corresponding decryption exponent d. If not, explain why not.

(2) Can Alice choose $e = 7$ as an encryption exponent? If so, find the corresponding decryption exponent d. If not, explain why not.

Exercise 7.6.25. Alice publishes $(N, e) = (1147, 7)$ for her RSA encryption. Bob wants to send the message "HI" to Alice. Each letter is encoded $A = 01$, $B = 02, \ldots$ and messages are encoded in message blocks of two letters (so "YOYO" would be encoded as 2515 2515).

(1) Encode "HI" into numeric form, and encrypt it using RSA with $(N, e) = (1147, 7)$. (Hint: $1956^2 \equiv 691$ and $(-338)^4 \equiv 329 \bmod 1147$.)

(2) Alice receives the message 793 from Bob. Explain how Alice would decrypt the message.

(3) Decrypt the message 793 into an actual English word. (Note: the plaintext message "0868" makes no sense because "68" does not correspond to a letter. However, $0868 \equiv 2015 \bmod 1147$ and "2015" corresponds to the word "TO".) (Hint: $793^{462} \equiv 1062 \bmod 1147$.)

Exercise 7.6.26. Suppose there is a public RSA key $n = 2911$ and $e = 1867$ and you intercept an encrypted message:

$$0785\ 0976\ 1594\ 0481\ 1560\ 2128\ 0917.$$

(1) Can you crack the code and decipher the message? (Note: here letters are encoded as numbers by $00 = A$, $01 = B$, $02 = C$, $\ldots$, $24 = Y$, $25 = Z$.)

(2) Another message is sent with public key $n = 54298697624741$ and $e = 1234567$. Would you be able to decrypt messages that used this RSA code? How would you do it?

Exercise 7.6.27. Let p be a prime such that $p \equiv 3 \bmod 4$, and let $F = \mathbb{F}_p[i]$ be a finite field with p^2 elements. Let $\phi\colon F \to F$ be the Frobenius automorphism such that $\phi(f) = f^p$ for every $f \in F$, as in Exercise 6.7.11.

(a) Show that $\phi(k) = k$ for every $k \in \mathbb{F}_p$.

(b) Show that $\phi(i) = -i$.

(c) Show that $\phi(a + bi) = a - bi$, for all $a, b \in \mathbb{F}_p$.

Exercise 7.6.28. Let p be a prime, let s be an element of $\mathbb{Z}/p\mathbb{Z}$ that is not congruent to a square mod p, and let $F = \mathbb{F}_p[x]/(x^2 - s)$ be a finite field with p^2 elements. Let $\phi\colon F \to F$ be the Frobenius automorphism such that $\phi(f) = f^p$ for every $f \in F$, as in Exercise 6.7.11.

(a) Show that $\phi(k) = k$ for every $k \in \mathbb{F}_p$.

(b) Show that $\phi(x) \equiv -x \bmod (x^2 - s)$.

(c) Show that $\phi(a + bx) \equiv a - bx \bmod (x^2 - s)$, for all $a, b \in \mathbb{F}_p$.

CHAPTER 8

PRIMITIVE ROOTS

Many who have had an opportunity of knowing any more about mathematics confuse it with arithmetic, and consider it an arid science. In reality, however, it is a science which requires a great amount of imagination.

Sofia Kovalevskaya

In this chapter we will explore the concept of multiplicative order of a congruence class and the concept of primitive roots. We will use these tools to find solutions to higher-degree congruences (see Example 8.6.10), which in turn will help us decide whether certain diophantine equations have integral points (see, for instance, Exercises 8.10.12 and 8.10.13). In order to introduce the concept of multiplicative order, let us see an example about the length of the period of a rational number.

Example 8.0.1. As it is well known, the decimal expansion of a rational number a/b is periodic (we will show this in Theorem 8.9.5). For instance,

$$\frac{3}{11} = 0.27272727272727272727272727272727\ldots = 0.\overline{27}\,,$$
$$\frac{1}{17} = 0.05882352941176470588235294117647 0\ldots = 0.\overline{0588235294117647}\,,$$
$$\frac{5}{37} = 0.135135135135135135135135135135135\ldots = 0.\overline{135}\,.$$

What dictates the length of the period of a fraction? In particular, why is the period of $1/17$ so long, while the periods of $3/11$ and $5/37$ are so short? Gauss discusses this sort of questions in his *Disquisitiones Arithmeticae* (articles 308–318), as an application of congruences and primitive roots.

Let us see how one calculates the decimal expansion of $3/11$:

$$\begin{array}{r@{\,}l} & \underline{\;0.27...} \\ 11 \;) & 3.00... \\ & \underline{22\;\;\;\;} \\ & \;\;80 \\ & \underline{\;\;77} \\ & \;\;\;3 \end{array}$$

Since $11 > 3$, we multiply 3 by 10 and we do long division of 30 by 11:

$$30 = 11 \cdot 2 + 8.$$

Then we multiply the remainder by 10 and do long division by 11 once again:

$$8 \cdot 10 = 11 \cdot 7 + 3.$$

At this step, the remainder (3) coincides with our initial numerator, and the computation enters a cycle. It follows that the expansion is periodic and the length of the period is two (namely, the expansion is $0.\overline{27}$, where the digits 2 and 7 are the quotients in the two long divisions performed above). In terms of congruences, we computed

$$3 \cdot 10 \equiv 8 \bmod 11 \quad \text{and} \quad 8 \cdot 10 \equiv 3 \bmod 11,$$

or, equivalently,

$$3 \cdot 10 \equiv 8 \bmod 11 \quad \text{and} \quad (3 \cdot 10) \cdot 10 \equiv 3 \cdot 10^2 \equiv 3 \bmod 11,$$

which implies that $10^2 \equiv 1 \bmod 11$ (because 3 and 11 are relatively prime). Conversely, the fact that $10^2 \equiv 1 \bmod 11$ implies that at the second iteration of the long division the remainder and the initial numerator will coincide modulo 11 and, therefore, the period of the decimal expansion will be of length two.

Similarly, in order to compute the decimal expansion of $5/37$ we do a few long divisions until we find a repeated remainder:

$$50 = 37 \cdot 1 + 13, \;\; 130 = 37 \cdot 3 + 19, \;\; 190 = 37 \cdot 5 + 5,$$

which in terms of congruences modulo 37 read as

$$5 \cdot 10 \equiv 13 \bmod 37, \;\; 5 \cdot 10^2 \equiv 19 \bmod 37, \quad \text{and} \;\; 5 \cdot 10^3 \equiv 5 \bmod 37,$$

which implies that $10^3 \equiv 1 \bmod 37$. Conversely, the fact that $10^3 \equiv 1 \bmod 37$ means that at the third long division the numerator and the remainder agree modulo 37.

Now, let us consider $1/17$. According to our previous experiments, the length of the period should coincide with the first exponent n such that $10^n \equiv 1 \bmod 17$. Thus, we compute a table of powers of 10 modulo 17:

x	x^2	x^3	x^4	x^5	x^6	x^7	x^8	x^9	x^{10}
10	15	14	4	6	9	5	16	7	2

x^{11}	x^{12}	x^{13}	x^{14}	x^{15}	x^{16}	...
3	13	11	8	12	1	...

It follows that the length of the period must be 16 and, indeed,

$$1/17 = 0.\overline{0588235294117647}$$

has the predicted period length.

The smallest positive integer n such that $10^n \equiv 1 \bmod 17$ is called the *multiplicative order* of 10 modulo 17 (see Definition 8.1.1). Since the multiplicative order of 10 is $16 = 17 - 1$, which turns out to be the largest possible multiplicative order mod 17, we say that 10 is a *primitive root* modulo 17 (see Definition 8.2.1). We will come back to decimal expansions in Section 8.9.2 in much more detail.

8.1. Multiplicative Order

Let us consider the powers of congruence classes modulo 7:

$x \bmod 7$	x^2	x^3	x^4	x^5	x^6	x^7	x^8	x^9	x^{10}	x^{11}	x^{12}	$\dots$
1	1	1	1	1	1	1	1	1	1	1	1	$\dots$
2	4	1	2	4	1	2	4	1	2	4	1	$\dots$
3	2	6	4	5	1	3	2	6	4	5	1	$\dots$
4	2	1	4	2	1	4	2	1	4	2	1	$\dots$
5	4	6	2	3	1	5	4	6	2	3	1	$\dots$
6	1	6	1	6	1	6	1	6	1	6	1	$\dots$

From the previous chapter, we know that the column of 1's under x^6 is explained by Fermat's little theorem. Indeed, since $p = 7$ is prime, Fermat's little theorem says that $a^6 \equiv 1 \bmod 7$, whenever $\gcd(a, 7) = 1$. However, as we can see from the table, when we take consecutive powers of a congruence class, some classes reach $1 \bmod 7$ sooner than the sixth power. For instance, $2^3 \equiv 4^3 \equiv 1 \bmod 7$. But for some classes, the first power that is congruent to 1 modulo 7 is precisely the sixth power predicted by Fermat's little theorem. Let us begin this chapter by giving a name to the first power of a unit that is congruent to 1.

Definition 8.1.1. Let $m > 1$ be fixed, and let a be an integer relatively prime to m. The *multiplicative order* of $a \bmod m$, or simply the order of a modulo m, denoted by $\operatorname{ord}_m(a)$, is the smallest positive number n such that $a^n \equiv 1 \bmod m$.

Example 8.1.2. Let $m = 7$. Then, from the previous table, we see that $\operatorname{ord}_7(2) = 3$, because $2^3 \equiv 1 \bmod 7$, and $2^n \not\equiv 1 \bmod 7$ for any positive number n with $1 \leq n < 3$. Moreover,

$$\operatorname{ord}_7(1) = 1,\ \operatorname{ord}_7(2) = 3,\ \operatorname{ord}_7(3) = 6,\ \operatorname{ord}_7(4) = 3,\ \operatorname{ord}_7(5) = 6,$$

and $\operatorname{ord}_7(6) = 2$. Notice then that the possible orders modulo 7 are $1, 2, 3$, and 6, but no congruence class has order 4 or 5. The attentive reader may notice that $\{1, 2, 3, 6\}$ is precisely the set of all positive divisors of $6 = 7 - 1$ and may wonder if this holds for any prime p; i.e., are the possible orders modulo p precisely the positive divisors of $p - 1$? Let us see another example.

Example 8.1.3. Let us calculate a table of all powers $x^1, \ldots, x^{10}$ of the non-zero congruence classes modulo 11:

$x \bmod 11$	x^2	x^3	x^4	x^5	x^6	x^7	x^8	x^9	x^{10}	...
1	1	1	1	1	1	1	1	1	1	...
2	4	8	5	10	9	7	3	6	1	...
3	9	5	4	1	3	9	5	4	1	...
4	5	9	3	1	4	5	9	3	1	...
5	3	4	9	1	5	3	4	9	1	...
6	3	7	9	10	5	8	4	2	1	...
7	5	2	3	10	4	6	9	8	1	...
8	9	6	4	10	3	2	5	7	1	...
9	4	3	5	1	9	4	3	5	1	...
10	1	10	1	10	1	10	1	10	1	...

Once again, the last column of 1's is precisely the content of Fermat's little theorem: if $\gcd(a, 11) = 1$, then $a^{10} \equiv 1 \bmod 11$. Let us make a table of congruence classes modulo 11 and their respective orders:

$a \bmod 11$	1	2	3	4	5	6	7	8	9	10
$\text{ord}_{11}(a)$	1	10	5	5	5	10	10	10	5	2

In particular, the possible multiplicative orders are $1, 2, 5$, or 10, and no congruence class has order 3, 4, 6, 7, 8, or 9. As in the previous example, we notice that $\{1, 2, 5, 10\}$ is precisely the set of all positive divisors of $10 = 11 - 1$.

Example 8.1.4. What happens if m is composite and we consider consecutive powers modulo m? For instance, take $m = 15$. Euler's theorem predicts that $a^8 \equiv 1 \bmod 15$, for all integers a with $\gcd(a, 15) = 1$, because $\varphi(15) = 8$. Are then the possible orders given by the divisors of 8? Are the possible orders $1, 2, 4$, and 8? We have already seen in Example 7.2.2 a table of consecutive powers of units modulo 15:

$x \bmod 15$	x^2	x^3	x^4	x^5	x^6	x^7	x^8	x^9	x^{10}	x^{11}	x^{12}	...
1	1	1	1	1	1	1	1	1	1	1	1	...
2	4	8	1	2	4	8	1	2	4	8	1	...
4	1	4	1	4	1	4	1	4	1	4	1	...
7	4	13	1	7	4	13	1	7	4	13	1	...
8	4	2	1	8	4	2	1	8	4	2	1	...
11	1	11	1	11	1	11	1	11	1	11	1	...
13	4	7	1	13	4	7	1	13	4	7	1	...
14	1	14	1	14	1	14	1	14	1	14	1	...

Thus, we find that the possible orders are $1, 2$, and 4, but no element has order 8. However, we still find that the possible orders are divisors of $\varphi(15)$, even though not every divisor is an actual order of a congruence class modulo 15.

Proposition 8.1.5. *Let $m > 1$ and let a be an integer relatively prime to m, with multiplicative order $n = \operatorname{ord}_m(a)$. Suppose that there is a number $t \geq 1$ such that $a^t \equiv 1 \bmod m$. Then, n is a divisor of t.*

Proof. By the division theorem (Theorem 2.4.4), there exist unique $q, r \in \mathbb{Z}$ such that

$$t = n \cdot q + r$$

and $0 \leq r < n$. Thus,

$$1 \equiv a^t \equiv a^{n \cdot q + r} \equiv (a^n)^q \cdot a^r \equiv 1 \cdot a^r \equiv a^r \bmod m,$$

where we have used the fact that $a^n \equiv 1 \bmod m$, since $n = \operatorname{ord}_m(a)$. It follows that $a^r \equiv 1 \bmod m$ and $0 \leq r < n$. Since n is the order of a modulo m, by definition n is the smallest *positive* integer such that $a^n \equiv 1 \bmod m$. Since $r < n$, the only possibility is that $r = 0$. Hence, $t = n \cdot q + 0 = n \cdot q$, and n is a divisor of t, as claimed. □

Corollary 8.1.6. *Let $m > 1$ be fixed and let $a \in \mathbb{Z}$ be relatively prime to m. Then,* $\operatorname{ord}_m(a)$ *is a divisor of $\varphi(m)$, where φ is the Euler phi function. In particular, if $m = p$ is prime, then* $\operatorname{ord}_p(a)$ *is a divisor of $p - 1$.*

Proof. If $\gcd(a, m) = 1$, then Euler's theorem, Theorem 7.3.5, says that $a^{\varphi(m)} \equiv 1 \bmod m$. Hence, Proposition 8.1.5 implies that $\operatorname{ord}_m(a)$ is a divisor of $\varphi(m)$. If $m = p$ is a prime, then we know that $\varphi(p) = p - 1$ and therefore $\operatorname{ord}_p(a)$ is a divisor of $p - 1$, as claimed. □

Example 8.1.7. As we saw in Example 8.1.2, the order of any unit modulo 7 is 1, 2, 3, or 6, and we see that each order is a divisor of $\varphi(7) = 7 - 1 = 6$. Similarly, in Example 8.1.2 each order modulo 11 is a divisor of $\varphi(11) = 11 - 1 = 10$. Finally, in Example 8.1.4 we have seen that each order mod 15 is a divisor of $\varphi(15) = 8$. However, there is no unit modulo 15 of exact order 8.

In the following proposition we prove a formula for the order of a power a^d of a unit $a \bmod m$, given the order of $a \bmod m$.

Proposition 8.1.8. *Let $m > 1$ be fixed and let $a \in \mathbb{Z}$ be relatively prime to m. Suppose that* $\operatorname{ord}_m(a) = n$. *Then, for any $d \geq 1$ we have*

$$\operatorname{ord}_m(a^d) = \frac{n}{\gcd(n, d)} = \frac{\operatorname{ord}_m(a)}{\gcd(\operatorname{ord}_m(a), d)}.$$

Proof. Fix $m > 1$ and $a \in \mathbb{Z}$ with $\gcd(a, m) = 1$, and let $d \geq 1$. Let $n = \operatorname{ord}_m(a)$ and put $s = \operatorname{ord}_m(a^d)$. We want to show that $s = \frac{n}{\gcd(n,d)}$.

By Exercise 2.11.27, we have an identity $n \cdot d = \operatorname{lcm}(n, d) \cdot \gcd(n, d)$. In particular,

$$(a^d)^{\frac{n}{\gcd(n,d)}} \equiv a^{\operatorname{lcm}(n,d)} \equiv 1 \bmod m,$$

since $\operatorname{lcm}(n, d)$ is a multiple of $n = \operatorname{ord}_m(a)$. It follows from Proposition 8.1.5 that $s \leq \frac{n}{\gcd(n,d)}$. Suppose for a contradiction that $s < \frac{n}{\gcd(n,d)}$.

Since $s = \operatorname{ord}_m(a^d)$, we know that $(a^d)^s \equiv 1 \bmod m$ and s is the smallest positive number with this property. This implies that n divides sd and s is the smallest positive number such that n is a divisor of sd. In particular, sd is a multiple of d and a multiple of n, and therefore $sd \geq \operatorname{lcm}(n,d)$. However, we have assumed $s < \frac{n}{\gcd(n,d)}$ and so

$$sd < d \cdot \frac{n}{\gcd(n,d)} = \frac{nd}{\gcd(n,d)} = \operatorname{lcm}(n,d).$$

This is a contradiction, and it follows that $\operatorname{ord}_m(a^d) = s = \frac{n}{\gcd(n,d)}$, as desired. □

Example 8.1.9. The order of 3 mod 7 is 6. Therefore we can use the formula of Proposition 8.1.8 to find the order of every non-zero congruence class modulo 7:

$$\operatorname{ord}_7(2) = \operatorname{ord}_7(3^2) = \frac{\operatorname{ord}_7(3)}{\gcd(\operatorname{ord}_7(3),2)} = \frac{6}{\gcd(6,2)} = \frac{6}{2} = 3,$$
$$\operatorname{ord}_7(6) = \operatorname{ord}_7(3^3) = \frac{\operatorname{ord}_7(3)}{\gcd(\operatorname{ord}_7(3),3)} = \frac{6}{\gcd(6,3)} = \frac{6}{3} = 2,$$
$$\operatorname{ord}_7(4) = \operatorname{ord}_7(3^4) = \frac{\operatorname{ord}_7(3)}{\gcd(\operatorname{ord}_7(3),4)} = \frac{6}{\gcd(6,4)} = \frac{6}{2} = 3,$$
$$\operatorname{ord}_7(5) = \operatorname{ord}_7(3^5) = \frac{\operatorname{ord}_7(3)}{\gcd(\operatorname{ord}_7(3),5)} = \frac{6}{\gcd(6,5)} = \frac{6}{1} = 6,$$
$$\operatorname{ord}_7(1) = \operatorname{ord}_7(3^6) = \frac{\operatorname{ord}_7(3)}{\gcd(\operatorname{ord}_7(3),1)} = \frac{6}{\gcd(6,1)} = \frac{6}{6} = 1.$$

Example 8.1.10. The number $p = 4001$ is a prime number. The reader can check that the order of 3 mod 4001 is precisely 4000. We know that the order of each congruence class modulo 4001 will be a divisor of $p - 1 = 4000$. Suppose that e is a divisor of 4000. Is there a congruence class of order exactly e? The answer is "yes", and we will be able to find such a congruence class by considering the powers of 3. For instance, $e = 50$ is a divisor of $4000 = 50 \cdot 80$. Let us find some power of 3 whose order is exactly 50. If $d \geq 1$, the order of 3^d mod 4001 will be given by

$$\operatorname{ord}(3^d) = \frac{\operatorname{ord} 3}{\gcd(\operatorname{ord} 3, d)} = \frac{4000}{\gcd(4000,d)}.$$

Thus, if we want the order of 3^d to be 50, we want $\gcd(4000,d) = 80$. It suffices to take $d = 80$. Indeed, $3^{80} \equiv 636 \bmod 4001$ and

$$\operatorname{ord}(636) = \operatorname{ord}(3^{80}) = \frac{4000}{\gcd(4000,80)} = \frac{4000}{80} = 50.$$

More generally, if e is a divisor of 4000 and $4000 = e \cdot d$, then

$$\operatorname{ord}(3^d) = \frac{4000}{\gcd(4000,d)} = \frac{4000}{d} = e,$$

and so, the congruence class of 3^d mod 4001 has exact order e. Notice also that $\gcd(d, 4000) = 1$ if and only if the congruence class of 3^d mod 4001 has exact order 4000.

In the following proposition we prove a formula that will allow us to find elements of higher order out of elements of smaller (relatively prime) orders.

Proposition 8.1.11. *Let $m > 1$ be fixed and suppose a, b are integers relatively prime to m. Further, suppose that* $\operatorname{ord}_m(a) = h$ *and* $\operatorname{ord}_m(b) = k$ *and* $\gcd(h, k) = 1$. *Then,* $\operatorname{ord}_m(ab) = hk$*; i.e.,*

$$\operatorname{ord}_m(ab) = \operatorname{ord}_m(a) \cdot \operatorname{ord}_m(b).$$

Proof. Let m, a, b, h, k be as in the statement of the proposition, and let $s = \operatorname{ord}_m(ab)$. We want to show that $s = hk$. First note that

$$(ab)^{hk} \equiv a^{hk} \cdot b^{hk} \equiv (a^h)^k \cdot (a^k)^h \equiv 1^k \cdot 1^h \equiv 1 \bmod m,$$

because $h = \operatorname{ord}_m(a)$ and $k = \operatorname{ord}_m(b)$. In particular, Proposition 8.1.5 implies that hk is a multiple of $s = \operatorname{ord}_m(ab)$. Moreover, consider $a^{ks} \bmod m$:

$$a^{ks} \equiv a^{ks} \cdot 1 \equiv a^{ks} \cdot (b^k)^s \equiv (ab)^{ks} \equiv ((ab)^s)^k \equiv 1^k \equiv 1 \bmod m,$$

because $s = \operatorname{ord}_m(ab)$. Using Proposition 8.1.5 again, it follows that $h = \operatorname{ord}_m(a)$ is a divisor of ks. Since $\gcd(h, k) = 1$, our favorite corollary (Corollary 2.7.6) tells us that h is actually a divisor of s. By a similar argument, we have that

$$b^{hs} \equiv (ab)^{hs} \equiv ((ab)^s)^h \equiv 1^h \equiv 1 \bmod m,$$

and it follows that k is a divisor of s. Hence, h and k are divisors of s. Since $\gcd(h, k) = 1$, we conclude that hk divides s. We have also shown above that s divides hk, and since they are both positive integers, the only possibility is that $s = hk$, as claimed. □

Example 8.1.12. The order of $2 \bmod 7$ is 3; i.e., $\operatorname{ord}_7(2) = 3$. The order of $6 \equiv -1 \bmod 7$ is 2; i.e., $\operatorname{ord}_7(6) = 2$. Since $\gcd(2, 3) = 1$, the order of -2 must be $2 \cdot 3 = 6$. Indeed, $-2 \equiv 5 \bmod 7$, and the order of $5 \bmod 7$ is precisely 6.

Example 8.1.13. The order of $1444 \bmod 4001$ is 125, i.e., $\operatorname{ord}_{4001}(1444) = 125$, and the order of $3339 \bmod 4001$ is 32, i.e., $\operatorname{ord}_{4001}(3339) = 32$. Since $\gcd(32, 125) = \gcd(2^5, 5^3) = 1$, it follows from Proposition 8.1.11 that

$$\operatorname{ord}_{4001}(1444 \cdot 3339) = \operatorname{ord}_{4001}(1444) \cdot \operatorname{ord}_{4001}(3339) = 32 \cdot 125 = 4000.$$

Hence, the order of $1444 \cdot 3339 \equiv 311 \bmod 4001$ is 4000, which is the largest possible order modulo 4001.

Remark 8.1.14. Beware: if $\operatorname{ord}_m(a)$ and $\operatorname{ord}_m(b)$ are *not* relatively prime, then it is not necessarily true that $\operatorname{ord}_m(ab) = \operatorname{ord}_m(a) \operatorname{ord}_m(b)$. For instance, let $m = 15$ and put $a = 2$ and $b = 11$. Then $\operatorname{ord}_{15}(2) = 4$ and $\operatorname{ord}_{15}(11) = 2$, but $2 \cdot 11 \equiv 7 \bmod 15$ and

$$\operatorname{ord}_{15}(7) = 4 \neq 8 = \operatorname{ord}_{15}(2) \cdot \operatorname{ord}_{15}(11).$$

In previous examples, we have seen several instances of congruence classes that attain the maximum possible order. For instance, $2 \bmod 5$ has order 4, the class of $3 \bmod 7$ has order 6, and the classes of $3 \bmod 4001$ has order 4000. The class of $311 \bmod 4001$ also has order 4000. An element that has the maximum possible order is called a primitive root. We will study these in detail in the following sections.

8.2. Primitive Roots

Definition 8.2.1. Let $m > 1$ be fixed and let $g \in \mathbb{Z}$ be relatively prime to m. We say that the congruence class $g \bmod m$ is a *primitive root* modulo m, or that $g \bmod m$ is a multiplicative generator modulo m, if $\operatorname{ord}_m(g) = \varphi(m)$. In particular, if $m = p$ is a prime number, we say that $g \bmod p$ is a primitive root if $\operatorname{ord}_p(g) = p-1$.

Example 8.2.2. The following congruence classes are primitive roots for the indicated modulus: 1 mod 2, 2 mod 3, 2 mod 5, 3 mod 7, 2 mod 11, 3 mod 4001, etc. Let us show that 2 is a primitive root modulo 11. We only need to show that $\operatorname{ord}_{11}(2) = 10$. Since the order of 2 must divide $p - 1 = 10$, it suffices to show that the order of 2 is not 1, 2, or 5. Indeed,

$$2^1 \equiv 2 \not\equiv 1, \quad 2^2 \equiv 4 \not\equiv 1, \quad 2^5 \equiv 32 \equiv -1 \not\equiv 1 \bmod 11,$$

and, therefore, the order must be exactly 10. Another (lengthier) way to see this is to compute a table of all the powers of 2 modulo 11:

$x \bmod 11$	x^2	x^3	x^4	x^5	x^6	x^7	x^8	x^9	x^{10}
2	4	8	5	10	9	7	3	6	1

Thus, the first power of 2 that is congruent to 1 mod 11 is the tenth power and so, by definition, $\operatorname{ord}_{11}(2) = 10$.

Example 8.2.3. There is no primitive root modulo 15. As we have seen in Example 8.1.4, every order modulo 15 is 1, 2, or 4. A primitive root would be an element of order $\varphi(15) = 8$, but there are none with such order.

However, there are primitive roots for some composite orders. For instance, consider $\mathbb{Z}/9\mathbb{Z}$. The congruence class of 2 mod 9 is a primitive root modulo 9. Indeed, the order of 2 mod 9 is 6 and $\varphi(9) = 6$:

$x \bmod 9$	x^2	x^3	x^4	x^5	x^6
2	4	8	7	5	1

Similarly, the congruence class of 3 mod 50 is a primitive root. Notice that $\varphi(50) = 20$ and the reader can verify that $\operatorname{ord}_{50}(3) = 20$.

Proposition 8.2.4. *Suppose that* $\operatorname{ord}_m(g) = \varphi(m)$*; i.e.,* $g \bmod m$ *is a primitive root in* $\mathbb{Z}/m\mathbb{Z}$*. Then:*

(1) *The set* $\{g, g^2, g^3, \ldots, g^{\varphi(m)}\}$ *is a complete residue system for the unit classes modulo* m*.*

(2) *For every divisor* e *of* $\varphi(m)$*, there is a congruence class whose order is precisely* e*. In particular, if* $\varphi(m) = e \cdot d$*, then the congruence classes modulo* m *with exact order* e *are precisely given by* $\{g^n : n \geq 1,\ \gcd(n, \varphi(m)) = d\}$*.*

(3) *The primitive roots of* $\mathbb{Z}/m\mathbb{Z}$ *are given by* $\{g^n : n \geq 1,\ \gcd(n, \varphi(m)) = 1\}$*. In particular, if there is at least one primitive root modulo* m*, then there are exactly* $\varphi(\varphi(m))$ *primitive roots modulo* m*.*

Proof. Let us assume that $g \bmod m$ is a primitive root.

(1) Since $g \bmod m$ is a primitive root, the order of $g \bmod m$ is precisely $\varphi(m)$; i.e., $g^{\varphi(m)} \equiv 1 \bmod m$ and $g^t \not\equiv 1 \bmod m$ for any $1 \leq t < \varphi(m)$. Now, suppose that
$$g^i \equiv g^j \bmod m$$
for some $1 \leq i \leq j \leq \varphi(m)$. Since g is a primitive root, it is a unit modulo m, and we can multiply both sides of the previous equation by $g^{-i} \equiv (g^{-1})^i \bmod m$ to obtain
$$1 \equiv g^{j-i} \bmod m.$$
But $j - i < \varphi(m)$ and therefore $j - i = 0$ and $i = j$. Hence, all the congruence classes in $S = \{g, g^2, \ldots, g^{\varphi(m)}\}$ are distinct modulo m. Since there are $\varphi(m)$ of them and since $(\mathbb{Z}/m\mathbb{Z})^\times$ has cardinality $\varphi(m)$, we conclude that S is a complete residue system for all the unit classes modulo m.

(2) Let e be a divisor of $\varphi(m)$ with $\varphi(m) = e \cdot d$. Then, by Proposition 8.1.8, we have
$$\operatorname{ord}_m(g^n) = \frac{\operatorname{ord}_m(g)}{\gcd(\operatorname{ord}_m(g), n)} = \frac{\varphi(m)}{\gcd(\varphi(m), n)}.$$
Hence, $\operatorname{ord}_m(g^n) = e$ if and only if $\gcd(\varphi(m), n) = d$. Moreover, if $a \bmod m$ is any congruence class of order e, by part (1) there is some n such that $g^n \equiv a \bmod m$, and we have shown that we must have $\gcd(\varphi(m), n) = d$.

(3) The primitive roots modulo m are those congruence classes modulo m whose order is precisely $\varphi(m)$. By part (2), those elements are given by powers g^n such that $\gcd(\varphi(m), n) = 1$. In other words, the primitive roots are the elements in the set
$$G = \{g^n : 1 \leq n \leq \varphi(m),\ \gcd(\varphi(m), n) = 1\}.$$
The set G is in bijection with the set $\{n : 1 \leq n \leq \varphi(m),\ \gcd(\varphi(m), n) = 1\}$, which, in turn, is in bijection with the set of units modulo $\varphi(m)$. Since the set $(\mathbb{Z}/\varphi(m)\mathbb{Z})^\times$ has exactly $\varphi(\varphi(m))$ elements, we conclude that there are $\varphi(\varphi(m))$ roots of unity modulo m. □

When we specialize the previous proposition to the case when $m = p$ is prime, we obtain the following immediate corollary.

Corollary 8.2.5. *If p is prime and $\mathbb{Z}/p\mathbb{Z}$ has at least one primitive root, then there are exactly $\varphi(\varphi(p)) = \varphi(p-1)$ primitive roots modulo p.*

Example 8.2.6. Let us consider $\mathbb{Z}/7\mathbb{Z}$. We know that $3 \bmod 7$ is a primitive root. Here is a table of powers of $3 \bmod 7$:

$x \bmod 7$	x^2	x^3	x^4	x^5	x^6
3	2	6	4	5	1

Therefore, each primitive root modulo 7 is given by a power of $3 \bmod 7$ whose exponent is relatively prime to $\varphi(7) = 7 - 1 = 6$. Hence, there are $\varphi(\varphi(7)) = \varphi(6) = 2$ primitive roots and they are
$$\{g^1, g^5 \bmod 7\} = \{3, 5 \bmod 7\}.$$

The elements of exact order 3 are given by those powers 3^n whose exponent n satisfies $\gcd(6,n) = \varphi(7)/3 = 2$. Thus, the elements of order 3 are

$$\{g^2, g^4 \bmod 7\} = \{2, 4 \bmod 7\}.$$

Similarly, the elements of exact order 2 are given by those powers 3^n whose exponent n satisfies $\gcd(6,n) = \varphi(7)/2 = 3$. Thus, there is only one element of order 2:

$$\{g^3 \bmod 7\} = \{6 \bmod 7\}.$$

And there is only one element of order 1, namely 1 mod 7.

Example 8.2.7. Let us find all the primitive roots modulo 11. We already know that 2 mod 11 is a primitive root. Here are the powers of 2 modulo 11:

$x \bmod 11$	x^2	x^3	x^4	x^5	x^6	x^7	x^8	x^9	x^{10}
2	4	8	5	10	9	7	3	6	1

By Proposition 8.2.4, all the primitive roots modulo 11 are given by all those powers of 2 whose exponent is relatively prime to $\varphi(11) = 10$. Thus, there are $\varphi(\varphi(11)) = \varphi(10) = 4$ primitive roots and they are given by

$$\{g^1, g^3, g^7, g^9 \bmod 11\} = \{2, 8, 7, 6 \bmod 11\}.$$

Thus, the four primitive roots modulo 11 are $2, 6, 7$, and 8 mod 11.

Example 8.2.8. Let us show that there is a primitive root modulo $p = 43$ (a prime number), by explicitly finding one. Let $\gcd(a, 43) = 1$; then the multiplicative order of $a \bmod 43$ is a divisor of $\varphi(43) = 42 = 2 \cdot 3 \cdot 7$, so $\mathrm{ord}_{43}(a)$ is one of $\{1, 2, 3, 6, 7, 14, 21, 42\}$. Let us find the order of 2 mod 43:

$$2^2 \equiv 4,\ 2^3 \equiv 8,\ 2^6 \equiv 8^2 \equiv 64 \equiv 21,\ 2^7 \equiv 21 \cdot 2 \equiv 42 \equiv -1 \bmod 43,$$

and so, $2^{14} \equiv 1 \bmod 43$, and $\mathrm{ord}_{43}(2) = 14$. By Proposition 8.1.11, it suffices to find an element of order 3, i.e., $b \bmod 43$ such that $b^3 \equiv 1$, but $b \not\equiv 1 \bmod 43$. Since

$$b^3 - 1 = (b-1)(b^2 + b + 1),$$

we are looking for b such that $b^2 + b + 1 \equiv 0 \bmod 43$. We may naively find solutions using the quadratic formula (we will prove that the quadratic formula does indeed work modulo 43 in Chapter 10, Section 10.1) to find that b must satisfy

$$b \equiv \frac{-1 \pm \sqrt{-3}}{2} \bmod 43.$$

Is -3 a square modulo 43? We will dive into questions of this sort in Section 10.2, but for now it suffices to say that after some calculations, we can see that $-3 \equiv 13^2 \bmod 43$, and so

$$b \equiv \frac{-1 \pm \sqrt{-3}}{2} \equiv \frac{-1 \pm 13}{2} \bmod 43,$$

and so $b \equiv 6$ or $b \equiv -7 \bmod 43$. Indeed, $b \equiv 6 \bmod 43$ works as a root of $b^2 + b + 1$ because $36 + 6 + 1 = 43 \equiv 0 \bmod 43$. Therefore,

$$\mathrm{ord}_{43}(12) = \mathrm{ord}_{43}(2 \cdot 6) = \mathrm{ord}_{43}(2) \cdot \mathrm{ord}_{43}(6) = 14 \cdot 3 = 42,$$

and 12 mod 43 is a primitive root.

Once we have found one primitive root, Proposition 8.2.4 tells us that there are $\varphi(\varphi(43)) = 12$ primitive roots in $\mathbb{Z}/43\mathbb{Z}$ and they are given by

$$\{g^n \bmod 43 : n \geq 1,\ \gcd(n, 42) = 1\},$$

where g is one fixed primitive root, say $g \equiv 12 \bmod 43$. Thus, the primitive roots modulo 43 are

$$12,\ 12^5 \equiv 34,\ 12^{11} \equiv 26,\ 12^{13} \equiv 3,\ 12^{17} \equiv 30,\ 12^{19} \equiv 20,\ 12^{23} \equiv 28,\ 12^{25} \equiv 33,$$

$$12^{29} \equiv 29,\ 12^{31} \equiv 5,\ 12^{37} \equiv 19, \quad \text{and} \quad 12^{41} \equiv 18.$$

In order, the primitive roots are $\{3, 5, 12, 18, 19, 20, 26, 28, 29, 30, 33, 34 \bmod 43\}$.

We have shown that *if there is one primitive root*, then there are $\varphi(\varphi(m))$ of them, but we have not yet proved for what values of m there is at least one primitive root. In the next two sections, we shall prove that there is always at least one primitive root when we work modulo a prime.

8.3. Universal Exponents

Definition 8.3.1. Let $m > 1$ be fixed. A *universal exponent* for $\mathbb{Z}/m\mathbb{Z}$ is a positive integer $u \geq 1$ such that $a^u \equiv 1 \bmod m$, for all $a \in \mathbb{Z}$ with $\gcd(a, m) = 1$. The least universal exponent is the smallest positive universal exponent for $\mathbb{Z}/m\mathbb{Z}$.

Remark 8.3.2. By Euler's theorem, $\varphi(m)$ is a universal exponent, so the least universal exponent u for $\mathbb{Z}/m\mathbb{Z}$ satisfies $u \leq \varphi(m)$. If $m = p$ is a prime, then $u \leq p - 1$.

Example 8.3.3. The least universal exponent modulo 11 is 10. Indeed, by Fermat's little theorem, we know that $a^{10} \equiv 1 \bmod 11$ whenever $\gcd(a, 11) = 1$ and this implies that $u \leq 10$. Moreover, $\operatorname{ord}_{11}(2) = 10$. Hence $u \geq 10$, and it follows that $u = 10$.

Example 8.3.4. The least universal exponent modulo 15 is 4. This follows from the table of Example 8.1.4. Notice that here the least universal exponent is strictly less than $\varphi(15) = 8$.

Proposition 8.3.5. *Let $m > 1$ be fixed and let u be the least universal exponent modulo m. Then, u is the least common multiple of all the orders of all units modulo m; i.e.,*

$$u = \operatorname{lcm}(\{\operatorname{ord}_m(a) : 1 \leq a \leq m,\ \gcd(a, m) = 1\}).$$

Proof. Let u be the least universal exponent modulo m and put

$$s = \operatorname{lcm}(\{\operatorname{ord}_m(a)\}),$$

where the least common multiple runs over all orders of all units $a \bmod m$. Fix $a \bmod m$. Then, s is a multiple of $\operatorname{ord}_m(a)$; i.e., $s = d \cdot \operatorname{ord}_m(a)$ for some $d \in \mathbb{Z}$. Hence,

$$a^s \equiv (a^{\operatorname{ord}_m(a)})^d \equiv 1^d \equiv 1 \bmod m.$$

Since a was an arbitrary unit, we conclude that $a^s \equiv 1 \bmod m$, for all units $a \bmod m$. Hence $u \leq s$ because u is the least universal exponent.

Suppose for a contradiction that $u < s$. Since s is the least common multiple of all orders modulo m, it follows that u is not a multiple of at least one order $\mathrm{ord}_m(a)$; i.e., there is some $a \bmod m$ such that $\mathrm{ord}_m(a)$ does not divide u. However,

$$a^u \equiv 1 \bmod m$$

because u is a universal exponent, but then Proposition 8.1.5 implies that $\mathrm{ord}_m(a)$ must divide u. This is a contradiction, and we conclude that $u < s$ is impossible. Thus, $u = s$, as desired. □

Example 8.3.6. The orders modulo 7 are $\{1, 2, 3, 6\}$. Hence, the (least) universal exponent modulo 7 is

$$u = \mathrm{lcm}(1, 2, 3, 6) = 6.$$

Similarly, every element modulo 11 has one of these orders: $1, 2, 5$, or 10. Thus, the least universal exponent modulo 11 is

$$u = \mathrm{lcm}(1, 2, 5, 10) = 10.$$

As a last example, every element modulo 15 has one of these orders: $1, 2$, or 4. Thus, the least universal exponent modulo 15 is

$$u = \mathrm{lcm}(1, 2, 4) = 4.$$

Lemma 8.3.7. *Let p be a prime and suppose that p^k divides $\mathrm{lcm}(a, b)$, for some $k \geq 1$. Then p^k divides a or b.*

Proof. By the definition of *least common multiple*, if e is the largest number such that p^e divides one of a or b, then the largest power of p that divides $\mathrm{lcm}(a, b)$ is also p^e. Thus, if p^k divides $\mathrm{lcm}(a, b)$, then p^k has to divide a or b. □

Lemma 8.3.8. *Let p be a prime and suppose p^e divides u, the least universal exponent modulo m. Then, there is a congruence class $a \bmod m$ whose order is exactly p^e.*

Proof. Suppose that p^e divides u. By Proposition 8.3.5, u is the least common multiple of all orders modulo m. By the previous lemma, there is some order $\mathrm{ord}_m(a)$ which is divisible by p^e. Let us write $\mathrm{ord}_m(a) = p^e \cdot d$, for some $d \geq 1$. Then, by Proposition 8.1.8

$$\mathrm{ord}_m(a^d) = \frac{\mathrm{ord}_m(a)}{\gcd(\mathrm{ord}_m(a), d)} = \frac{p^e \cdot d}{\gcd(p^e \cdot d, d)} = \frac{p^e \cdot d}{d} = p^e.$$

Hence, $a^d \bmod m$ has exact order p^e, as desired. □

The following theorem will be one out of two key results in proving that $\mathbb{Z}/p\mathbb{Z}$ always has a primitive root when p is prime.

Theorem 8.3.9. *Let $m > 1$ be fixed and let u be the (least) universal exponent modulo m. Then, there is some congruence class $a \bmod m$ such that $\mathrm{ord}_m(a) = u$; i.e., there is a congruence class whose order is the least universal exponent.*

Proof. By the fundamental theorem of arithmetic, we can find a factorization of u into prime powers,

$$u = p_1^{e_1} p_2^{e_2} \cdots p_r^{e_r},$$

where each p_i is a prime, $e_i \geq 1$, and $p_i \neq p_j$ if $i \neq j$. Since $p_i^{e_i}$ divides u, it follows from Lemma 8.3.8 that there is some class $c_i \bmod m$ of exact order $p_i^{e_i}$. Since each p_i is a distinct prime, it follows that $\gcd(\text{ord}_m(c_i), \text{ord}_m(c_j)) = 1$ for all $i \neq j$. Let $c = c_1 \cdot c_2 \cdots c_r$. Then, by Proposition 8.1.11,

$$\begin{aligned}\text{ord}_m(c) &= \text{ord}_m(c_1 c_2 \cdots c_r)\\ &= \text{ord}_m(c_1) \cdot \text{ord}_m(c_2) \cdots \text{ord}_m(c_r)\\ &= p_1^{e_1} p_2^{e_2} \cdots p_r^{e_r}\\ &= u.\end{aligned}$$

Hence, the order of $c \bmod m$ is exactly u, the least universal exponent modulo m. □

Example 8.3.10. The least universal exponent of $\mathbb{Z}/7\mathbb{Z}$ is $u = 6 = 3 \cdot 2$. The order of $c_1 = 2 \bmod 7$ is 3. The order of $c_2 = 6 \equiv -1 \bmod 7$ is 2. Hence, the order of $c = c_1 \cdot c_2 \equiv 2 \cdot (-1) \equiv -2 \equiv 5 \bmod 7$ is $3 \cdot 2 = 6$, and the order of 5 mod 7 is the least universal exponent $u = 6$.

8.4. Existence of Primitive Roots Modulo p

Recall that, in Chapter 5, we proved Theorem 5.5.19 that said that a polynomial over a field has at most as many roots as the degree of the polynomial, even when the roots are counted with multiplicity. We shall use this result and Theorem 8.3.9 to show that $\mathbb{Z}/p\mathbb{Z}$ has a primitive root, for every prime p.

Theorem 8.4.1. *Let p be a prime. Then, there is at least one primitive root modulo p; i.e., there is a congruence class $g \bmod p$ such that $\text{ord}_p(g) = p - 1$.*

Proof. Let u be the least universal exponent modulo p. We shall show first that $u = p - 1$. Let $p(x) = x^u - 1$ be a polynomial in $(\mathbb{Z}/p\mathbb{Z})[x]$. Since p is prime, we know that $\mathbb{Z}/p\mathbb{Z}$ is a field (Theorem 5.4.3). Hence, by Theorem 5.5.19 (see also Corollary 5.5.21), we know that the number of roots of $p(x)$ in $\mathbb{Z}/p\mathbb{Z}$ is $\leq u = \deg(p(x))$. On the other hand, each unit a modulo p is a root of $p(x)$ because $a^u \equiv 1 \bmod p$ whenever $\gcd(a, p) = 1$, since u is a universal exponent modulo p. Hence, $a \equiv 1, 2, \ldots, p-1 \bmod p$ are $p-1$ distinct roots of $p(x)$. It follows that $p(x)$ has at least $p - 1$ roots, and so $p - 1 \leq u = \deg(p(x))$. Moreover, we know that the least universal exponent modulo a prime is always $\leq p - 1$, by Fermat's little theorem, and therefore $u = p - 1$.

Hence, the least universal exponent modulo p is always $p - 1$. Now, Theorem 8.3.9 says that there is a congruence class $g \bmod p$ whose order is precisely the least universal exponent $u = p-1$ and, therefore, $\text{ord}_p(g) = p-1$, as we wanted. Hence, $g \bmod p$ is a primitive root modulo p. This concludes the proof of the theorem. □

Example 8.4.2. In the table below, we list all primes $p \le 50$ and the smallest positive integer g such that $g \bmod p$ is a primitive root mod p:

p	2	3	5	7	11	13	17	19	23	29	31	37	41	43	47
g	1	2	2	3	2	2	3	2	5	2	3	2	6	3	5

8.4.1. Finding a Primitive Root Modulo p. In general, the problem of finding a primitive root modulo p is hard, as there is no known formula that will produce a primitive root given a prime number. The best known method to find primitive roots is an "educated trial-and-error" approach, as was illustrated in Example 8.2.8, where we calculated a primitive root modulo 43. In the most basic trial-and-error approach we simply calculate $\mathrm{ord}_p(u)$ for random units $u \bmod p$, until we find a primitive root. Some remarks are in order:

(1) The probability of finding a primitive root at random is in our favor. By Proposition 8.2.4, there are $\varphi(p-1)$ primitive roots modulo p. Hence, the probability that a random unit is indeed a primitive root is $\varphi(p-1)/(p-1)$. This probability can be as good as 50% in some cases (see Exercise 10.8.32). The lowest value of $\varphi(p-1)/(p-1)$ among the first 100000 primes is $0.1852\ldots$, for $p = 870871$. In Table 8.1 we have computed the probability of finding at random a primitive root modulo p, for primes between 3 and 29.

Table 8.1. The probability of finding at random a primitive root modulo p.

p	3	5	7	11	13	17	19	23	29
$\varphi(p-1)/(p-1)$	0.5	0.5	$0.\overline{3}$	0.4	$0.\overline{3}$	0.5	$0.\overline{3}$	$0.\overline{45}$	$0.\overline{428571}$

Moreover, if we fail to find a primitive root in our first pick, then the probability that we find one in our second pick would be $\varphi(p-1)/(p-2)$, and so on. Equivalently, the probability of *not* finding a primitive root in the first random attempt is $\alpha_1 = 1 - \varphi(p-1)/(p-1)$, and the probability of not finding a primitive root after n attempts is

$$\alpha_n = \left(1 - \frac{\varphi(p-1)}{(p-1)}\right)\left(1 - \frac{\varphi(p-1)}{(p-2)}\right)\cdots\left(1 - \frac{\varphi(p-1)}{(p-n)}\right),$$

which decreases to 0 as n approaches $p - \varphi(p-1)$. For instance, when $p = 43$, the probability of *not* finding a primitive root in the first random attempt is $\alpha_1 = 0.714\ldots$, or 71.4%. The probability of not finding a primitive root in four random attempts is down to about 0.245, or 24.5%.

(2) If $p > 3$, then 1 and $-1 \bmod p$ cannot be primitive roots (they have order 1 and 2, respectively, and $p - 1 > 2$). Thus, we search for primitive roots among $2, \ldots, p-2 \bmod p$, and the probability of finding a primitive root at random in our first pick is now $\varphi(p-1)/(p-3)$. See Table 8.2.

Table 8.2. The probability of finding a primitive root between 2 and $p-2$.

p	5	7	11	13	17	19	23	29
$\varphi(p-1)/(p-1)$	1	0.5	0.5	0.4	$0.\overline{571428}$	0.375	0.5	$0.\overline{461538}$

(3) If we compute the order of a unit u mod p and u is not a primitive root, then we have additional information that we can use to discard a number of units that cannot be primitive roots either. Indeed, we know that $\operatorname{ord}_p(u^n)$ is a divisor of $\operatorname{ord}_p(u)$, by Proposition 8.1.8. Hence, if u is not a primitive root, then no power of u can be a primitive root either.

For instance, in Example 8.2.8 we began by computing the order of 2, which happened to be 14, so 2 is not a primitive root. It follows that no power of 2 is a primitive root. Thus, there are 14 units that cannot be primitive roots. So in our second attempt at finding a primitive root, we have reduced the number of candidates from 42 units down to 28 units, and now the probability of finding a primitive root is

$$\frac{\varphi(p-1)}{p-1-\operatorname{ord}_p(u)} = \frac{12}{42-\operatorname{ord}_{43}(2)} = \frac{12}{28} = \frac{3}{7} = 0.\overline{428571},$$

up from approximately a probability 0.28 of success before we calculated the order of 2.

(4) Even if a primitive root has not been found, it is possible to build one out of partial information. We can calculate orders of units $u_1, u_2, u_3, \ldots, u_n$ until

$$\operatorname{lcm}(\operatorname{ord}_p(u_1), \operatorname{ord}_p(u_2), \ldots, \operatorname{ord}_p(u_n))$$

equals $p-1$. At this point, using the multiplicative property of orders (as in Proposition 8.1.11) and the methods outlined in the proofs of Lemma 8.3.8 and Theorem 8.3.9, we can build a unit with order $p-1$, which is therefore a primitive root.

Moreover,

$$\operatorname{lcm}(\operatorname{ord}_p(u), \operatorname{ord}_p(u^n)) = \operatorname{ord}_p(u)$$

for any $n \geq 1$. Thus, when computing primitive roots, it is futile to compute the order of any power of a unit u for which we have already calculated $\operatorname{ord}_p(u)$, as it will not help in any way.

Example 8.4.3. Let $p = 71$, and let us find a primitive root modulo 71. We begin calculating the order of $u \equiv 2 \bmod 71$. Since $70 = 2\cdot 5\cdot 7$, the order of 2 must be one of $1, 2, 5, 7, 10, 14, 35$, or 70. The reader can check that $\operatorname{ord}_{71}(2) = 35$. Therefore, it suffices to find a unit of order $70/35 = 2$. Since $\operatorname{ord}_{71}(-1) = 2$, we conclude that $2 \cdot (-1) \equiv -2 \equiv 69 \bmod 71$ has order

$$\operatorname{ord}_{71}(69) = \operatorname{ord}_{71}(2\cdot(-1)) = \operatorname{ord}_{71}(2)\cdot \operatorname{ord}_{71}(-1) = 35\cdot 2 = 70,$$

where we have used the fact that $\gcd(35, 2) = 1$. Thus, 69 mod 71 is a primitive root modulo 71. (Note: 7 mod 71 is the smallest primitive root modulo 71.)

Example 8.4.4. Let $p = 97$. The order of 2 mod 97 is 48. Although $96/48 = 2$, we cannot use -1 mod 97 to build a unit of order 96, because the orders 48 and 2 are not relatively prime (in fact, the order of -2 is also 48). Before we move on to calculating the order of another unit, we list the powers of 2 modulo 97. These are

$$2, 4, 8, 16, 32, 64, 31, 62, 27, 54, 11, 22, 44, 88, 79, 61, 25, 50, 3, 6, 12,$$
$$24, 48, 96, 95, 93, 89, 81, 65, 33, 66, 35, 70, 43, 86, 75, 53, 9, 18, 36, 72,$$
$$47, 94, 91, 85, 73, 49, 1,$$

and by our remarks above, none of these powers can be of any use to us (note that $94 \equiv -2 \bmod 97$ is a power of 2). Now we calculate the order of the smallest unit *not listed* among the powers of 2, namely 5 mod 97. It turns out that $\mathrm{ord}_{97}(5) = 96$, and therefore it is a primitive root.

8.4.2. Artin's Conjecture on Primitive Roots.

Example 8.4.5. In Example 8.0.1 we saw that if 10 is a primitive root modulo p, then the length of the period in the decimal expansion of $1/p$ must be $p-1$ (we will prove this rigorously and in greater generality in Section 8.9.2; in particular see Corollary 8.9.7). A few natural questions arise: when is 10 a primitive root modulo p? Is 10 a primitive root mod p infinitely often? Artin's conjecture is precisely aimed at predicting how often a fixed integer a is a primitive root modulo p, when p is a prime number. For now, we include here a list of all primes $p \leq 500$ such that 10 is a primitive root modulo p:

$$7, 17, 19, 23, 29, 47, 59, 61, 97, 109, 113, 131, 149, 167, 179, 181, 193, 223, 229, 233,$$
$$257, 263, 269, 313, 337, 367, 379, 383, 389, 419, 433, 461, 487, 491, 499.$$

In Section 8.4.1 we have described a method to find a primitive root modulo p. Since the method is an "educated trial-and-error" search, it is natural to hope for a *small* primitive root. Here is a list of the smallest primitive root modulo $n \geq 2$ (if there is no primitive root modulo n, see Theorem 8.7.4, we included a 0 in the sequence):

$$1, 2, 3, 2, 5, 3, 0, 2, 3, 2, 0, 2, 3, 0, 0, 3, 5, 2, 0, 0,$$
$$7, 5, 0, 2, 7, 2, 0, 2, 0, 3, 0, 0, 3, 0, 0, 2, 3, 0, 0, 6,$$
$$0, 3, 0, 0, 5, 5, 0, 3, 3, 0, 0, 2, 5, 0, 0, 0, 3, 2, 0, 2,$$
$$3, 0, 0, 0, 0, 2, 0, 0, 0, 7, 0, 5, 5, 0, 0, 0, 0, 3, 0, 2,$$
$$7, 2, 0, 0, 3, 0, 0, 3, 0, 0, 0, 0, 5, 0, 0, 5, 3, 0, 0, \ldots.$$

The smallest primitive root modulo 118 is 11, and the smallest one modulo 191 is 19, so sometimes primitive roots are *not so small*. The sequence above indicates, however, that 2 seems to be a primitive root often... but how often is that? The following sequence is formed by those primes $p \leq 700$ such that 2 is a primitive root modulo p:

$$3, 5, 11, 13, 19, 29, 37, 53, 59, 61, 67, 83, 101, 107, 131,$$
$$139, 149, 163, 173, 179, 181, 197, 211, 227, 269, 293, 317,$$
$$347, 349, 373, 379, 389, 419, 421, 443, 461, 467, 491, 509,$$
$$523, 541, 547, 557, 563, 587, 613, 619, 653, 659, 661, 677.$$

There are 125 primes below 700, and of those, the number 2 is a primitive root for 51 of them. That is, 2 is a primitive root for 40.8% of all primes below 700. The mathematician Emil Artin conjectured in 1927 that, indeed, the number 2 is a primitive root infinitely often and, moreover, claimed that 2 is a primitive root modulo p about 37% of the time. In fact, he conjectured that something similar happens more generally (e.g., the number 3 is also a primitive root for about 37% of all prime numbers).

Conjecture 8.4.6 (Artin's conjecture). *Let a be an integer which is not a perfect square and let $a \neq -1$. Let $x \geq 0$ and define $S_a(x)$ as the set of prime numbers $p \leq x$ such that a is a primitive root modulo p. Then:*

(1) *The size of $S_a(x)$ goes to infinity as $x \to \infty$. In other words, there are infinitely many prime numbers p such that a is a primitive root modulo p.*

(2) *Suppose that a is not a perfect power ($a \neq n^m$ for any $n \geq 1$ and $m \geq 2$) and that the square-free part of a is not congruent to $1 \bmod 4$. Then, there is a constant $C_{Artin} = \prod_{q \text{ prime}} \left(1 - \frac{1}{q(q-1)}\right) = 0.3739558136\ldots$ such that*

$$\lim_{x\to\infty} \frac{|S_a(x)|}{\pi(x)} = C_{Artin},$$

where $\pi(x) = |\{p : \text{primes} \leq x\}|$ is the prime counting function (as in Section 3.3.2).

(3) *If a is a perfect power (not a square or -1) or if the square-free part of a is congruent to $1 \bmod 4$, then there is a positive rational number $v(a) \in \mathbb{Q}$ such that*

$$\lim_{x\to\infty} \frac{|S_a(x)|}{\pi(x)} = v(a) \cdot C_{Artin}.$$

We will not give a formula for the rational number $v(a)$ in its utmost generality, but it suffices to say here that such a formula exists.

Figure 8.1. Emil Artin (1898–1962) was an Austrian-American mathematician. Image author: Konrad Jacobs (Erlangen). Image source: Archives of the Mathematisches Forschungsinstitut Oberwolfach.

Remark 8.4.7. In Artin's conjecture, we assume that a is not a perfect square and $a \neq -1$. We leave it as an exercise for the reader to understand why those restrictions are necessary (see Exercise 8.10.10).

Although Artin's conjecture is yet to be proven unconditionally, Hooley, in 1967, proved the conjecture under the assumption of the generalized Riemann hypothesis (an extension of the Riemann hypothesis that we discussed in Section 3.4.4). Further, a series of results by R. Gupta, K. Murty, M. R. Murty, and D. R. Heath-Brown culminated (in [**HB86**]) in the following result.

Theorem 8.4.8. *One of* 2, 3, *or* 5 *is a primitive root modulo* p *for infinitely many primes* p.

Thus, there is at least a number $a \in \{2, 3, 5\}$ such that Artin's conjecture is true for a.

8.5. Primitive Roots Modulo p^k

In the previous sections, we have discussed the set of primitive roots modulo a prime p. Here we will discuss primitive roots modulo a power of a prime (and in the following section we will treat the case of any $m \geq 2$). We begin with a useful lemma.

Lemma 8.5.1. *Let* $m \geq 2$, *and let* n *be a divisor of* m.

(1) *If* g *is a primitive root modulo* m, *then it is also a primitive root modulo* n.
(2) *If* h *is a primitive root modulo* n, *then the order of* h *modulo* m *is divisible by* $\varphi(n)$.

Proof. Suppose first that $g \bmod m$ is a primitive root modulo m. Then, by Proposition 8.2.4, the set $\{g, g^2, \ldots, g^{\varphi(m)}\}$ is a complete residue system for the unit classes modulo m. Now let u be a representative for a unit class modulo n. By Proposition 5.3.21, we may take $u \in \mathbb{Z}$ such that $\gcd(u, m) = 1$ (see Exercise 5.6.14). Then, there is some $t \in \mathbb{Z}$ such that $g^t \equiv u \bmod m$ and, therefore, $g^t \equiv u \bmod n$ as well. In particular, the order g modulo n must be at least $\varphi(n)$ and it follows that g is a primitive root modulo n as well. This shows (1).

For (2), suppose that h is a primitive root modulo n. Without loss of generality, we may assume that $\gcd(h, m) = 1$ (by Proposition 5.3.21). By definition of order, we have $h^{\operatorname{ord}_m(h)} \equiv 1 \bmod m$, but reducing modulo n we also obtain $h^{\operatorname{ord}_m(h)} \equiv 1 \bmod n$. Hence, by Proposition 8.1.5, $\operatorname{ord}_n(h) = \varphi(n)$ is a divisor of $\operatorname{ord}_m(h)$, as claimed. □

Example 8.5.2. Let $n = 5$ and let $m = 25$. Then, 2 is a primitive root modulo 25, and therefore, by the lemma above, it is also a primitive root modulo 5. In particular, $\operatorname{ord}_5(2) = 4$.

Now let $n = 5$ and $m = 35$. Then, by the lemma, the order of $2 \bmod 35$ must be divisible by 4. Indeed, the order of $2 \bmod 35$ is 12. Note, however, that $\varphi(35) = \varphi(5) \cdot \varphi(7) = 4 \cdot 6 = 24$, so 2 is not a primitive root modulo 35.

In the following theorem, we identify the primitive roots modulo a prime power.

Theorem 8.5.3. *Suppose that p is an odd prime, and let g be a primitive root modulo p.*

(1) *The number g is also a primitive root modulo p^2 if and only if $g^{p-1} \not\equiv 1 \bmod p^2$.*

(2) *If $g^{p-1} \equiv 1 \bmod p^2$, then $g+p$ is a primitive root of p^2.*

(3) *If g is a primitive root modulo p^n, for some $n \geq 2$, then g is also a primitive root modulo p^{n+1}.*

Proof. Let p be an odd prime, and let g be a primitive root modulo p. In particular, the order of g modulo p is exactly $p-1$.

(1) Suppose first that g is a primitive root modulo p^2. Then, the order of g modulo p^2 is precisely $\varphi(p^2) = p(p-1)$. Hence, $g^{p(p-1)} \equiv 1 \bmod p^2$, but $g^n \not\equiv 1 \bmod p^2$ for any $1 \leq n < p(p-1)$. In particular, $g^{p-1} \not\equiv 1 \bmod p^2$.

For the converse, suppose that $g^{p-1} \not\equiv 1 \bmod p^2$. By Corollary 8.1.6, the order of $g \bmod p^2$ is a divisor of $\varphi(p^2) = p(p-1)$, but the order cannot be a divisor d of $p-1$, because $(g^d)^{(p-1)/d} \equiv g^{p-1} \not\equiv 1 \bmod p^2$. Therefore, p divides the order of g. Suppose then that $\operatorname{ord}_{p^2}(g) = dp$, for some divisor d of $p-1$. Then, $1 \equiv g^{dp} \equiv (g^d)^p \equiv g^d \bmod p$, by Fermat's little theorem. Since g is a primitive root modulo p, it follows that $d = p-1$ and $\operatorname{ord}_{p^2}(g) = p(p-1)$. Hence, g is also a primitive root modulo p^2, as claimed.

(2) Suppose that $g^{p-1} \equiv 1 \bmod p^2$. By the binomial theorem (Exercise 2.11.14),

$$\begin{aligned}(g+p)^{p-1} &= g^{p-1} + \binom{p-1}{1} p g^{p-2} + \binom{p-1}{2} p^2 g^{p-3} + \cdots + p^{p-1} \\ &\equiv g^{p-1} + \binom{p-1}{1} p g^{p-2} \bmod p^2 \\ &\equiv g^{p-1} + (p-1) p g^{p-2} \bmod p^2 \\ &\equiv 1 + (p-1) p g^{p-2} \bmod p^2.\end{aligned}$$

Hence, $(g+p)^{p-1} \not\equiv 1 \bmod p^2$, because $(p-1)pg^{p-2} \not\equiv 0 \bmod p^2$ (because $(p-1)g^{p-2}$ is a unit modulo p^2). Since $g+p \equiv g \bmod p$ is a primitive root, we conclude from part (1) that $g+p$ is a primitive root modulo p^2.

(3) Let g be a primitive root modulo p^n, and let $d = \operatorname{ord}_{p^{n+1}}(g)$. Then, Lemma 8.5.1 implies that $\varphi(p^n) = (p-1)p^{n-1}$ is a divisor of d. On the other hand, Euler's theorem (Theorem 7.3.5) shows that d is a divisor of $\varphi(p^{n+1}) = (p-1)p^n$. Hence, d is either $\varphi(p^n)$ or $\varphi(p^{n+1})$. We will show that $d = \varphi(p^{n+1})$.

Since g is a primitive root modulo p^n, it also is a primitive root modulo p^{n-1} (by Lemma 8.5.1). Therefore, $g^{\varphi(p^{n-1})} \equiv 1 \bmod p^{n-1}$, and so $g^{\varphi(p^{n-1})} = 1 + kp^{n-1}$, for some $k \in \mathbb{Z}$. Moreover, $k \not\equiv 0 \bmod p$, because $g^{\varphi(p^{n-1})} \not\equiv 1 \bmod p^n$ as g is a primitive root modulo p^n. Since $n \geq 2$, we have

$\varphi(p^n) = p\varphi(p^{n-1})$ and so

$$\begin{aligned} g^{\varphi(p^n)} &= (g^{\varphi(p^{n-1})})^p = (1+kp^{n-1})^p \\ &= 1 + kp^n + \frac{p(p-1)}{2}k^2p^{2n-2} + \binom{p}{3}k^3p^{3n-3} + \cdots \\ &\equiv 1 + kp^n \bmod p^{n+1}, \end{aligned}$$

because $2n-1 \geq n+1$ for all $n \geq 2$ and $p > 2$. Hence, $g^{\varphi(p^n)} \not\equiv 1 \bmod p^{n+1}$, and so the order of g modulo p^{n+1} cannot be $\varphi(p^n)$. It follows that $\operatorname{ord}_{p^{n+1}}(g) = \varphi(p^{n+1})$ and g is a primitive root modulo p^{n+1}, as desired. □

Example 8.5.4. Let us find a primitive root modulo 49. First, we find a primitive root modulo 7. Since $g = 2$ does not work ($2^3 \equiv 1 \bmod 7$), we try $g = 3$:

$$3^2 \equiv 2,\ 3^3 \equiv 6 \equiv -1,\ 3^6 \equiv (3^3)^2 \equiv (-1)^2 \equiv 1 \bmod 7,$$

and we conclude that 3 mod 7 is a primitive root. Further,

$$3^2 \equiv 9,\ 3^3 \equiv 27,\ 3^6 \equiv 27^2 \equiv 729 \equiv 43 \bmod 49.$$

Thus, $3^6 \not\equiv 1 \bmod 49$ and it follows that 3 is also a primitive root modulo 49. Finally, Theorem 8.5.3 also implies that 3 is a primitive root modulo 7^k for all $k \geq 1$.

If we know all the primitive roots modulo p, then the previous theorem allows us to write a complete description of the primitive roots modulo p^k for every $k \geq 2$.

Corollary 8.5.5. *Let p be an odd prime, and for each $n \geq 1$, let $G_n \subseteq \mathbb{Z}/p^n\mathbb{Z}$ be the set of all primitive roots modulo p^n. Then:*

(1) $G_2 = \{h \bmod p^2 : h \in G_1$ *and* $h^{p-1} \not\equiv 1 \bmod p^2\} \subseteq \mathbb{Z}/p^2\mathbb{Z}$.

(2) *Let* $G_1 = \{g_1, \ldots, g_{\varphi(p-1)}\} \subset \mathbb{Z}/p\mathbb{Z}$, *and let* $H_2 = \{g_i^p \bmod p^2 : 1 \leq i \leq \varphi(p-1)\} \subset \mathbb{Z}/p^2\mathbb{Z}$. *Then,*
$$G_2 = \{h \bmod p^2 : h \in G_1,\ h \notin H_2\} \subset \mathbb{Z}/p^2\mathbb{Z}.$$

(3) *If* $n \geq 2$, *then*
$$G_{n+1} = \{g + tp^n \bmod p^{n+1} : g \in G_n, t \equiv 0, 1, \ldots, p-1 \bmod p\} \subseteq \mathbb{Z}/p^{n+1}\mathbb{Z}.$$

Proof. Part (1) is an immediate consequence of Theorem 8.5.3, part (1).

Before we prove (2), let us show that the set H_2 is well-defined. Let $g, g' \in \mathbb{Z}$, such that $g \equiv g' \bmod p$ so that they are in the same congruence class mod p. Then, we claim that $g^p \equiv (g')^p \bmod p^2$. Indeed, since $g \equiv g' \bmod p$, it follows that $g' = g + kp$ for some $k \in \mathbb{Z}$. Thus,

$$\begin{aligned} (g')^p &\equiv (g+kp)^p \\ &\equiv g^p + \binom{p}{1} \cdot g^{p-1} \cdot (kp) + \cdots + \binom{p}{p-1} \cdot g \cdot (kp)^{p-1} + (kp)^p \\ &\equiv g^p \bmod p^2, \end{aligned}$$

where we have used the binomial theorem and Exercises 2.11.14 and 4.7.26. Hence, if $g \bmod p$ is fixed, the class of $g^p \bmod p^2$ is well-defined. Now we are ready to prove (2).

Notice that $|G_1| = \varphi(p-1)$ while $|G_2| = \varphi(\varphi(p^2)) = (p-1)\varphi(p-1)$, so that $|G_2| = (p-1)\cdot|G_1|$. By Lemma 8.5.1, every primitive mod p^2 is also a primitive root modulo p. The set of all lifts of G_1 to G_2, i.e.,

$$L_2 = \{g + kp : g \in G_1,\ k = 0, 1, \ldots, p-1\} \subset \mathbb{Z}/p^2\mathbb{Z},$$

has size $p \cdot |G_1|$ and $G_2 \subseteq L_2$. Therefore $L_2 - G_2$ has size

$$|L_2 - G_2| = p \cdot |G_1| - (p-1)\cdot|G_1| = |G_1| = \varphi(p-1).$$

Let $H_2 = \{g_i^p \bmod p^2 : 1 \leq i \leq \varphi(p-1)\} \subset \mathbb{Z}/p^2\mathbb{Z}$. By Fermat's little theorem, $g_i^p \equiv g_i \bmod p$ and $g_i \in G_1$, so $H_2 \subset L_2$. Moreover, if $g_i^p \equiv g_j^p \bmod p^2$, then $g_i^p \equiv g_j^p \bmod p$, but $g_i^p \equiv g_i$ and $g_j^p \equiv g_j \bmod p$ by Fermat's little theorem, so $g_i \equiv g_i^p \equiv g_j^p \equiv g_j \bmod p$, which implies that $i = j$ because $\{g_1, \ldots, g_{\varphi(p-1)}\}$ is a complete residue system of primitive roots for $\mathbb{Z}/p\mathbb{Z}$, so $g_i \not\equiv g_j \bmod p$ for $i \neq j$. This shows that $|H_2| = \varphi(p-1)$.

Since $L_2 - G_2$ and H_2 have the same size ($\varphi(p-1)$) and since$H_2 \subset L_2$, it suffices to show that $H_2 \cap G_2 = \emptyset$ in order to conclude that $H_2 = L_2 - G_2$, or, equivalently, $G_2 = L_2 - H_2$ (which is equivalent to the statement to show in part (2)). For this, suppose that $h \in H_2$, so that $h \equiv g_i^p \bmod p^2$ for some $i = 1, \ldots, \varphi(p-1)$. Then,

$$h^{p-1} \equiv (g_i^p)^{p-1} \equiv (g_i)^{p(p-1)} \equiv 1 \bmod p^2,$$

by Euler's theorem (Theorem 7.3.5). By Theorem 8.5.3, $h \bmod p^2$ is not a primitive root modulo p^2. Thus, $H_2 \cap G_2 = \emptyset$, as desired. This concludes the proof of (2).

Finally, we prove part (3): for each $n \geq 2$, let G_n be the set of primitive roots modulo p^n. We know that

$$|G_n| = \varphi(\varphi(p^n)) = \varphi(p^{n-1}(p-1)) = \varphi(p^{n-1})\varphi(p-1) = p^{n-2}(p-1)\varphi(p-1),$$

and therefore $|G_{n+1}| = p^{n-1}(p-1)\varphi(p-1)$ and $|G_{n+1}| = p \cdot |G_n|$. Notice that if h is a primitive root modulo p^{n+1}, then $h \bmod p^n$ is a primitive root modulo p^n by Lemma 8.5.1. Therefore, the candidates for primitive roots modulo p^{n+1} are those $h \bmod p^{n+1}$ such that $h \bmod p^n$ is a primitive root modulo p^n. In other words, the primitive roots modulo p^{n+1} are contained in the set

$$G'_{n+1} = \{g + tp^n : g \in G_n, t \equiv 0, 1, \ldots, p-1 \bmod p\} \subseteq \mathbb{Z}/p^{n+1}\mathbb{Z}.$$

But $|G'_{n+1}| = p \cdot |G_n|$ and

$$p \cdot |G_n| = |G_{n+1}| \leq |G'_{n+1}| = p \cdot |G_n|.$$

Thus, $G'_{n+1} = G_{n+1}$. □

Example 8.5.6. Let us apply Corollary 8.5.5 to describe explicitly all the primitive roots of $\mathbb{Z}/49\mathbb{Z}$ and $\mathbb{Z}/7^n\mathbb{Z}$ for $n \geq 3$. In $\mathbb{Z}/7\mathbb{Z}$ there are $\varphi(\varphi(7)) = \varphi(6) = 2$ primitive roots, namely 3 and 5. In $\mathbb{Z}/49\mathbb{Z}$ there are $\varphi(\varphi(49)) = \varphi(42) = \varphi(6)\varphi(7) = 12$. If h is a primitive root modulo 49, then h is also a primitive root modulo 7, so $h \equiv 3$ or $5 \bmod 7$. There are seven such $h \equiv 3$ and seven such $h \equiv 5 \bmod 7$, so we have 14 candidates for primitive roots, of which 12 are primitive roots. By Theorem 8.5.3, those which are *not* primitive roots must satisfy $h^{p-1} \equiv 1 \bmod p^2$. Clearly

$$(3^7)^6 \equiv 3^{(7\cdot 6)} \equiv 1 \quad \text{and} \quad (5^7)^6 \equiv 5^{(7\cdot 6)} \equiv 1 \bmod 49,$$

by Euler's theorem, because $\varphi(49) = 42 = 7 \cdot 6$. Since $3^7 \equiv 3 \bmod 7$ and $5^7 \equiv 5 \bmod 7$, by Fermat's little theorem, we conclude that 3^7 and 5^7 are the two exceptions:

$$3^7 \equiv 31 \bmod 49 \quad \text{and} \quad 5^7 \equiv 19 \bmod 49.$$

Hence, the set G_2 of primitive roots modulo $49 = 7^2$ is the union of

$$\{3 + 7k : 0 \leq k \leq 6, k \neq 4\} \quad \text{and} \quad \{5 + 7j : 0 \leq j \leq 6, j \neq 2\}.$$

Alternatively, in the notation of Corollary 8.5.5, we have $H_2 = \{19, 31 \bmod 49\}$, so

$$G_2 = \{a \bmod p^2 : a \equiv 3 \text{ or } 5 \bmod 7, \text{ and } a \not\equiv 19 \text{ or } 31 \bmod 49\}.$$

Finally, for each $k \geq 2$, the set G_k of primitive roots modulo 7^k are those elements that reduce to one of the elements in G_2 modulo 49.

Theorem 8.5.7. *Let $m = 2$, 4, p^k, or $2p^k$, for some odd prime p and some $k \geq 1$. Then, m has a primitive root.*

Proof. If $m = 2$, then $g \equiv 1 \bmod 2$ is a primitive root. If $m = 4$, then $g \equiv 3 \bmod 4$ is one. If p is an odd prime, then there exists a primitive root modulo p by Theorem 8.4.1. Corollary 8.5.5 shows that there is a primitive root modulo p^k for every $k \geq 1$.

It remains to show that $m = 2p^k$ has a primitive root. Let $g \in \mathbb{Z}$ be a primitive root modulo p^k. We distinguish two cases:

- If g is odd, then every power of g is odd, so $g^j \equiv 1 \bmod 2$ for all $j \geq 1$. Thus, $g^j \equiv 1 \bmod 2p^k$ if and only if $g^j \equiv 1 \bmod p^k$. Hence, the multiplicative order of $g \bmod 2p^k$ is the same as the order of $g \bmod p^k$ which is $\varphi(p^k) = \varphi(2p^k)$. Hence, g is also a primitive root modulo $2p^k$.
- If g is even, then g is not even a unit in $\mathbb{Z}/2p^k\mathbb{Z}$ so it cannot be a primitive root. Let $g' = g + p^k$. Then g' is odd, and $g' \equiv g \bmod p^k$, so it is a primitive root modulo p^k. Hence, by our previous bullet point, g' is a primitive root modulo $2p^k$.

Thus, in all cases, $m = 2p^k$ has a primitive root, as we claimed. □

Example 8.5.8. Let $p = 7$. In Example 8.5.4 we showed that 3 is a primitive root modulo 7^k, for all $k \geq 1$. Since $g = 3$ is odd, it follows that 3 is also a primitive root modulo $2 \cdot 7^k$, for all $k \geq 1$.

Similarly, Example 8.5.6 shows that $g = 10$ is a primitive root modulo 7^k, for all $k \geq 1$. However, 10 is even, so it is not a unit modulo $2 \cdot 7^k$. However, $10 + 7^k$ is a primitive root modulo $2 \cdot 7^k$, for all $k \geq 1$. For instance, this shows that 59 is a primitive root modulo 98.

The converse of Theorem 8.5.7 is also true; i.e., if $m \geq 2$ has a primitive root, then $m = 2$, 4, p^k, or $2p^k$ for some odd prime p. Before we prove this fact, we will introduce the concept of indices, which is an analogue of the concept of logarithm.

8.6. Indices

The logarithm in base b, denoted by $\log_b(x)$, is the inverse function of exponentiation in base b, i.e., b^x. Logarithms are quite useful when solving equations where the unknown is in the exponent. Let us see two examples.

Example 8.6.1. Let us find x such that $x^5 = 16807$, using logarithms. Let us take logarithms (in base e, the natural logarithm) on both sides of the equation:

$$5\log x = \log(x^5) = \log(16807).$$

Thus, $\log x = \log(16807)/5 = 1.945910149\ldots$. Now we use the inverse function of $\log x$, the exponential e^x, to retrieve x:

$$x = e^{\log x} = e^{1.945910149\ldots} = 7.$$

Example 8.6.2. Let us find x such that $7^{x+3} = 16807$. Notice that $16807 = 7^5$. Let us take logarithms in base 7 of both sides:

$$x + 3 = \log_7(7^{x+3}) = \log_7(16807) = \log_7(7^5) = 5.$$

Thus, $x + 3 = 5$, so $x = 2$.

Here are the key properties of the exponential and logarithm functions that make them so useful in the applications. Let $b > 1$ be fixed. Then:

(a) b^x is a bijection, from $\mathbb{R}$ to $\mathbb{R}^+$, and $\log_b(x)$ is a bijection, from $\mathbb{R}^+$ to $\mathbb{R}$;
(b) $\log_b(x)$ is the inverse function of b^x;
(c) $\log_b(x^n) = n \cdot \log_b(x)$;
(d) $\log_b(xy) = \log_b(x) + \log_b(y)$;
(e) and (perhaps the most important property of all) we can calculate b^x and $\log_b(x)$ efficiently.

In this section, we want to define an analog of the logarithm function for the units modulo m, i.e., $U_m = (\mathbb{Z}/m\mathbb{Z})^\times$. Clearly, if g is a primitive root, then g^x is a bijection;

$$g^x \colon \{1, 2, \ldots, \varphi(m)\} \to U_m.$$

Thus, we can define a "logarithm in base g" (an *index* function for the powers of g) as the inverse function of g^x. This is exactly what we will do, and we will show that our index function satisfies properties (a) through (e) above. The following is a *preliminary* definition of the concept of index, which we will refine below in Definition 8.6.7.

Definition 8.6.3. Let $m \geq 2$ be an integer, such that there exists a primitive root g modulo m. We define the *index function* in base g as the function

$$\text{ind}_g \colon (\mathbb{Z}/m\mathbb{Z})^\times \to \{1, 2, \ldots, \varphi(m)\}$$

such that $n = \text{ind}_g(a \bmod m)$ is the smallest integer $n \geq 1$ with $g^n \equiv a \bmod m$.

Example 8.6.4. In Example 8.2.2 we showed that $g = 2$ is a primitive root modulo 11. We indeed calculated a table of powers of 2 mod 11:

x mod 11	x^2	x^3	x^4	x^5	x^6	x^7	x^8	x^9	x^{10}
2	4	8	5	10	9	7	3	6	1

Using this table, we can calculate values of ind_2, the index in base 2. For instance, $\text{ind}_2(9) = 6$, because $2^9 \equiv 6 \bmod 11$. Similarly, $\text{ind}_2(3) = 8$ because $2^8 \equiv 3 \bmod 11$. We can also build a table of all indices in base 2:

a mod 11	1	2	3	4	5	6	7	8	9	10
$\text{ind}_2(a)$	10	1	8	2	4	9	7	3	6	5

Example 8.6.5. In Example 8.2.8, we showed that $g \equiv 3 \bmod 43$ is a primitive root in $\mathbb{Z}/43\mathbb{Z}$. Let us calculate a table of indices in base 3. First, let us calculate a table of powers of 3 modulo 43:

$x \bmod 43$	x^2	x^3	x^4	x^5	x^6	x^7	x^8	x^9	x^{10}	x^{11}	x^{12}	x^{13}
3	9	27	38	28	41	37	25	32	10	30	4	12
	x^{14}	x^{15}	x^{16}	x^{17}	x^{18}	x^{19}	x^{20}	x^{21}	x^{22}	x^{23}	x^{24}	x^{25}
	36	22	23	26	35	19	14	42	40	34	16	5
	x^{26}	x^{27}	x^{28}	x^{29}	x^{30}	x^{31}	x^{32}	x^{33}	x^{34}	x^{35}	x^{36}	x^{37}
	15	2	6	18	11	33	13	39	31	7	21	20
	x^{38}	x^{39}	x^{40}	x^{41}	x^{42}							
	17	8	24	29	1							

And now we can calculate a table of indices in base 3:

$a \bmod 43$	1	2	3	4	5	6	7	8	9	10	11	12	13	14
$\mathrm{ind}_3(a)$	42	27	1	12	25	28	35	39	2	10	30	13	32	20
$a \bmod 43$	15	16	17	18	19	20	21	22	23	24	25	26	27	28
$\mathrm{ind}_3(a)$	26	24	38	29	19	37	36	15	16	40	8	17	3	5
$a \bmod 43$	29	30	31	32	33	34	35	36	37	38	39	40	41	42
$\mathrm{ind}_3(a)$	41	11	34	9	31	23	18	14	7	4	33	22	6	21

Remark 8.6.6. Let m be a positive integer and suppose that $\gcd(a, m) = 1$. Then, $a^s \equiv a^t \bmod m$ if and only if $s \equiv t \bmod (\mathrm{ord}_m(a))$. Indeed, if $a^s \equiv a^t \bmod m$, then $a^{s-t} \equiv 1 \bmod m$, and $\mathrm{ord}_m(a)$ must be a divisor of $s - t$ (by Proposition 8.1.5). Hence $s \equiv t \bmod (\mathrm{ord}_m(a))$.

Conversely, if $s \equiv t \bmod (\mathrm{ord}_m(a))$, then $s - t = n \cdot \mathrm{ord}_m(a)$ and

$$a^{s-t} \equiv (a^{\mathrm{ord}_m(a)})^n \equiv 1^n \equiv 1 \bmod m,$$

and, therefore, $a^s \equiv a^t \bmod m$.

In particular, if g is a primitive root modulo m and $g^s \equiv b \bmod m$, then $g^t \equiv b \bmod m$, for all $t \equiv s \bmod \varphi(m)$, because $\mathrm{ord}_m(g) = \varphi(m)$. This means that $\mathrm{ind}_g(b)$ can be regarded as the congruence class of $s \bmod \varphi(m)$.

In light of Remark 8.6.6, we redefine the index function as follows.

Definition 8.6.7. Let $m \geq 2$ be an integer, such that there exists a primitive root g modulo m. We define the *index function* in base g as the function

$$\mathrm{ind}_g \colon (\mathbb{Z}/m\mathbb{Z})^\times \to \mathbb{Z}/\varphi(m)\mathbb{Z}$$

such that $n \equiv \mathrm{ind}_g(a \bmod m) \bmod \varphi(m)$ is in the unique congruence class modulo $\varphi(m)$ that satisfies $g^n \equiv a \bmod m$.

With this definition, we are ready to show that the index function satisfies properties very similar to the logarithm.

Proposition 8.6.8. *Let $m \geq 2$ be an integer such that there exists a primitive root g modulo m. Then, the function ind_g satisfies the following properties:*

(a) *g^x is a bijection, from $\mathbb{Z}/\varphi(m)\mathbb{Z}$ to $U_m = (\mathbb{Z}/m\mathbb{Z})^\times$, and ind_g is a bijection, from U_m to $\mathbb{Z}/\varphi(m)\mathbb{Z}$.*

(b) *$\operatorname{ind}_g(x)$ is the inverse function of g^x.*

(c) *$\operatorname{ind}_g(x^t) \equiv t \cdot \operatorname{ind}_g(x) \bmod \varphi(m)$.*

(d) *$\operatorname{ind}_g(xy) \equiv \operatorname{ind}_g(x) + \operatorname{ind}_g(y) \bmod \varphi(m)$.*

Proof. Since g is a primitive root, the map g^x is surjective on $(\mathbb{Z}/m\mathbb{Z})^\times$. By Remark 8.6.6, $g^x \equiv g^y \bmod m$ if and only if $x \equiv y \bmod \varphi(m)$. Thus, g^x is injective with domain $\mathbb{Z}/\varphi(m)\mathbb{Z}$. Hence, g^x is a bijection. The index function ind_g is defined to be the inverse function of g^x, so it is also a bijection. This shows (a) and (b).

Let $n \equiv \operatorname{ind}_g(x \bmod m)$. Then, n is in the unique congruence class modulo $\varphi(m)$ that satisfies $g^n \equiv x \bmod m$. It follows that $g^{tn} \equiv x^t \bmod m$, and so $\operatorname{ind}_g(x^t) \equiv t \cdot n \equiv t \cdot \operatorname{ind}_g(x) \bmod \varphi(m)$. This is (c).

Let $u \equiv \operatorname{ind}_g(x \bmod m)$ and $v \equiv \operatorname{ind}_g(y \bmod m) \bmod \varphi(m)$. Then, $g^u \equiv x$ and $g^v \equiv y \bmod m$. Hence,

$$g^{u+v} \equiv g^u \cdot g^v \equiv x \cdot y \bmod m.$$

This implies that

$$\operatorname{ind}_g(x) + \operatorname{ind}_g(y) \equiv u + v \equiv \operatorname{ind}_g(xy) \bmod \varphi(m),$$

as claimed in (d). □

Remark 8.6.9. Note that property (d) in Proposition 8.6.8 would not be true if the index function was integer-valued (as we had preliminarily defined it in Definition 8.6.3) instead of $\mathbb{Z}/\varphi(m)\mathbb{Z}$-valued.

Traditional exponentials and logarithms can be calculated efficiently (any calculator can do that!). In order to use indices, however, (i) there must be a primitive root modulo m, (ii) we need to be able to find an explicit primitive root g modulo m, and (iii) we need a table of indices in base g.

Example 8.6.10. Let us find all the solutions to the congruence $3x^6 \equiv 4 \bmod 11$, using indices. In Example 8.6.4 we calculated a table of indices in base 2:

$a \bmod 11$	1	2	3	4	5	6	7	8	9	10
$\operatorname{ind}_2(a)$	10	1	8	2	4	9	7	3	6	5

Taking indices on both sides of $3x^6 \equiv 4 \bmod 11$ and using the properties of Proposition 8.6.8, we obtain on one hand $\operatorname{ind}_2(4) \equiv 2 \bmod 10$ and on the other hand

$$2 \equiv \operatorname{ind}_2(4) \equiv \operatorname{ind}_2(3x^6) \equiv \operatorname{ind}_2(3) + \operatorname{ind}_2(x^6) \equiv 8 + 6\operatorname{ind}_2(x) \bmod 10.$$

Therefore, $6\operatorname{ind}_2(x) \equiv 2 - 8 \equiv -6 \equiv 4 \bmod 10$. Solving the congruence $6t \equiv 4 \bmod 10$ is equivalent to finding the solutions of $10s + 6t = 4$, which in turn is equivalent to finding solutions to the diophantine equation $5s + 3t = 2$. Using what we learned in Section 2.9, we find the solution to be

$$s = 1 + 3k, \ t = -1 - 5k$$

for each $k \in \mathbb{Z}$. Hence, $t \equiv -1 \equiv 4 \bmod 5$, which means $t \equiv 4$ or $9 \bmod 10$. It follows that the solutions x to our original equation satisfy

$$\mathrm{ind}_2(x) \equiv 4 \text{ or } 9 \bmod 10$$

and by our table, these indices correspond to $x \equiv 5$ or $6 \bmod 11$. Indeed,

$$3 \cdot 5^6 \equiv 46875 \equiv 4 \bmod 11$$

and since $6 \equiv -5 \bmod 11$, it follows that $3 \cdot 6^6 \equiv 3 \cdot (-5)^6 \equiv 3 \cdot 5^6 \equiv 4 \bmod 11$.

In general, there is a formula for the number of solutions of $x^k \equiv a \bmod m$, which is given in the following theorem, and it is an application of indices.

Theorem 8.6.11. *Let $m \geq 2$ and suppose that $\mathbb{Z}/m\mathbb{Z}$ has a primitive root. Let $\gcd(a, m) = 1$. Then, the congruence $x^k \equiv a \bmod m$ has a solution if and only if*

$$a^{\varphi(m)/\gcd(k,\varphi(m))} \equiv 1 \bmod m.$$

If $x^k \equiv a \bmod m$ is solvable, then it has exactly $\gcd(k, \varphi(m))$ different solutions in $\mathbb{Z}/m\mathbb{Z}$.

Proof. Let g be a primitive root modulo m. Then, the congruence $x^k \equiv a \bmod m$ has a solution $x \bmod m$ if and only if $k \cdot \mathrm{ind}_g(x) \equiv \mathrm{ind}_g(a) \bmod \varphi(m)$. Moreover, by Theorem 4.4.3, the congruence $ky \equiv b \bmod \varphi(m)$ has a solution $y_0 \bmod m$ if and only if $d = \gcd(k, \varphi(m))$ is a divisor of b, and if it has a solution, then it has exactly d different solutions modulo $\varphi(m)$. We need a lemma to finish our proof.

Lemma 8.6.12. *Let $m \geq 2$ and suppose that $\mathbb{Z}/m\mathbb{Z}$ has a primitive root. Let $\gcd(a, m) = 1$ and let d be a divisor of $\varphi(m)$. Then, $\mathrm{ind}_g(a) \equiv 0 \bmod d$ if and only if $a^{\varphi(m)/d} \equiv 1 \bmod m$ if and only if $\mathrm{ord}_m(a)$ is a divisor of $\varphi(m)/d$.*

Proof. Suppose that $a^{\varphi(m)/d} \equiv 1 \bmod m$. Taking indices in base g we obtain an equivalent expression

$$(\varphi(m)/d) \cdot \mathrm{ind}_g(a) \equiv \mathrm{ind}_g(1) \equiv 0 \bmod \varphi(m),$$

which is equivalent to $\mathrm{ind}_g(a) \equiv 0 \bmod d$ by Proposition 4.3.1. This concludes the proof of the lemma. □

Back to the proof of Theorem 8.6.11, $a^{\varphi(m)/d} \equiv 1 \bmod m$ if and only if $\mathrm{ind}_g(a) \equiv 0 \bmod d$ if and only if $k \cdot \mathrm{ind}_g(x) \equiv \mathrm{ind}_g(a) \bmod \varphi(m)$ has d solutions for $\mathrm{ind}_g(x)$ and these correspond to d different solutions of $x^k \equiv a \bmod m$. □

Example 8.6.13. In Example 8.6.10 we saw that the congruence $3x^6 \equiv 4 \bmod 11$ has two solutions, namely $x \equiv 5, 6 \bmod 11$. Let us show that there are two solutions using Theorem 8.6.11. The congruence in question is equivalent to

$$x^6 \equiv 4 \cdot 3^{-1} \equiv 4 \cdot 4 \equiv 16 \equiv 5 \bmod 11.$$

Hence, Theorem 8.6.11 says that there are $\gcd(6, 10) = 2$ solutions if $5^{10/2} = 5^5 \equiv 1 \bmod 11$. So it only remains to calculate

$$5^5 \equiv 5 \cdot (5^2)^2 \equiv 5 \cdot (25)^2 \equiv 5 \cdot 3^2 \equiv 5 \cdot 9 \equiv 5 \cdot (-2) \equiv -10 \equiv 1 \bmod 11.$$

Next, we list a few corollaries of Theorem 8.6.11. If $m = p$ is prime, then we know the existence of a primitive root modulo p (by Theorem 8.4.1).

Corollary 8.6.14. *Let p be a prime and let $\gcd(a,p) = 1$. Then, a is congruent to a kth power in $\mathbb{Z}/p\mathbb{Z}$ if and only if*

$$a^{(p-1)/\gcd(k,p-1)} \equiv 1 \bmod p.$$

Corollary 8.6.15. *Suppose that there exists a primitive root modulo m. Then:*

(1) *The congruence $x^k \equiv 1 \bmod m$ has exactly $\gcd(k, \varphi(m))$ distinct solutions in $\mathbb{Z}/m\mathbb{Z}$. In particular, if k is a divisor of $\varphi(m)$, then $x^k \equiv 1 \bmod m$ has exactly k solutions.*

(2) *The number of distinct kth powers modulo m is $\varphi(m)/\gcd(k, \varphi(m))$.*

Proof. Part (1) follows directly from Theorem 8.6.11, with $a = 1$. For part (2), we note that b is a kth power if and only if $b^{\varphi(m)/\gcd(k,\varphi(m))} \equiv 1 \bmod m$ if and only if b is a solution of $x^{\varphi(m)/\gcd(k,\varphi(m))} \equiv 1 \bmod m$. By part (1), the latter congruence has exactly $\varphi(m)/\gcd(k, \varphi(m))$ solutions. □

Example 8.6.16. The congruences $x^6 \equiv 1$ and $x^7 \equiv 1 \bmod 43$ have, respectively, 6 solutions and 7 solutions, but $x^5 \equiv 1 \bmod 43$ only has one solution ($x \equiv 1 \bmod 43$), because $\gcd(6, \varphi(43)) = 6$, $\gcd(7, 42) = 7$, but $\gcd(5, 42) = 1$. Let us calculate the solutions to each of these congruences using indices. Recall that in Example 8.6.5 we have calculated a table of indices in base 3:

$a \bmod 43$	1	2	3	4	5	6	7	8	9	10	11	12	13	14
$\mathrm{ind}_3(a)$	42	27	1	12	25	28	35	39	2	10	30	13	32	20
$a \bmod 43$	15	16	17	18	19	20	21	22	23	24	25	26	27	28
$\mathrm{ind}_3(a)$	26	24	38	29	19	37	36	15	16	40	8	17	3	5
$a \bmod 43$	29	30	31	32	33	34	35	36	37	38	39	40	41	42
$\mathrm{ind}_3(a)$	41	11	34	9	31	23	18	14	7	4	33	22	6	21

Now, taking indices on the congruence $x^6 \equiv 1 \bmod 43$ we obtain

$$6\,\mathrm{ind}_3(x) \equiv \mathrm{ind}_3(1) \equiv 42 \equiv 0 \bmod 42,$$

and therefore $\mathrm{ind}_3(x) \equiv 0 \bmod 7$, so that $\mathrm{ind}_3(x) \equiv 7k \bmod 42$, for $0 \le k \le 5$. In other words, $\mathrm{ind}_3(x) \equiv 0, 7, 14, 21, 28, 35 \bmod 42$, and these correspond to

$$x \equiv 1, 37, 36, 42, 6, 7 \bmod 43,$$

respectively. Notice that to find x knowing $\mathrm{ind}_3(x)$, it is best to use the table of powers of 3 (as it appears in Example 8.6.5). Similarly, $x^7 \equiv 1 \bmod 43$ is equivalent to $7\,\mathrm{ind}_3(x) \equiv 0 \bmod 42$, which means that $\mathrm{ind}_3(x) \equiv 0 \bmod 6$, and the solutions satisfy $\mathrm{ind}_3(x) \equiv 6j \bmod 42$ for $0 \le j \le 6$. These correspond to

$$x \equiv 1, 41, 4, 35, 16, 11, 21 \bmod 43.$$

Last, $x^5 \equiv 1 \bmod 43$ translates to $5\,\mathrm{ind}_3(x) \equiv 0 \bmod 42$. Since $\gcd(5, 42) = 1$, this means that $\mathrm{ind}_3(x) \equiv 0 \bmod 42$, and there is a unique solution; namely, $x \equiv 1 \bmod 43$.

8.7. Existence of Primitive Roots Modulo m

In Section 8.5 we have shown that if $m = 2$, 4, p^k, or $2p^k$, for some odd prime p, then there exists a primitive root modulo m (see Theorem 8.5.7). In this section we will show that, in fact, no other integer $m > 2$ has a primitive root.

Example 8.7.1. The contrapositive of part (1) of Corollary 8.6.15 reads as follows: suppose k is a divisor of $\varphi(m)$ and the number of distinct solutions for the congruence $x^k \equiv 1 \bmod m$ is different from k. Then, there is no primitive root modulo m. In particular, note that if $m > 2$, then $\varphi(m)$ is always even (see Exercise 7.6.4). Thus, 2 is always a divisor of $\varphi(m)$, and the existence of a primitive root modulo m implies that $x^2 \equiv 1 \bmod m$ has exactly two solutions. This fact can be exploited to show that certain numbers do not have primitive roots.

For instance, let $m = 2^k$, for some $k \geq 3$. The units in $\mathbb{Z}/2^k\mathbb{Z}$ are the odd numbers between 1 and $2^k - 1$. Moreover,

$$(2n+1)^2 - 1 \equiv 4n^2 + 4n + 1 - 1 \equiv 4n(n+1) \bmod 2^k.$$

Therefore $x \equiv 2n+1 \bmod 2^k$ is a solution to $x^2 \equiv 1 \bmod 2^k$ if and only if $n(n+1) \equiv 0 \bmod 2^{k-2}$. Since $\gcd(n, n+1) = 1$, one of n and $n+1$ is even and the other one is odd.

- If n is even, then $n(n+1) \equiv 0 \bmod 2^{k-2}$ implies that $n \equiv 0 \bmod 2^{k-2}$ and thus $x \equiv 2n+1 \equiv 1 \bmod 2^{k-1}$. Hence, $x \equiv 1$ or $1 + 2^{k-1} \bmod 2^k$.
- If n is odd, then $n+1$ is even, and $n(n+1) \equiv 0 \bmod 2^{k-2}$ implies that $n+1 \equiv 0 \bmod 2^{k-2}$. Hence $x \equiv 2n+1 \equiv -1 \bmod 2^{k-1}$; i.e., $x \equiv -1$ or $2^{k-1} - 1 \bmod 2^k$.

Hence, we have found four solutions to $x^2 \equiv 1 \bmod 2^k$; namely,

$$x \equiv 1, 2^{k-1} - 1, 2^{k-1} + 1, 2^k - 1 \bmod 2^k.$$

Moreover, if $k \geq 3$, then

$$1 < 2^{k-1} - 1 < 2^{k-1} + 1 < 2^k - 1,$$

so they are distinct modulo 2^k (note that they are *not* distinct when $k = 2$). Therefore, we have shown the following result.

Lemma 8.7.2. *Let $k \geq 3$ and $m = 2^k$. Then, the congruence $x^2 \equiv 1 \bmod m$ has exactly four different solutions modulo m. In particular, there is no primitive root modulo m.*

We will use the same argument as in Example 8.7.1 to classify the integers $m \geq 2$ with a primitive root. First, we need to know the number of distinct solutions of $x^2 \equiv 1 \bmod m$ for each $m \geq 2$.

Proposition 8.7.3. *Let $m \geq 2$ be an integer, with prime factorization*

$$m = 2^{e_0} p_1^{e_1} p_2^{e_2} \cdots p_n^{e_n},$$

for some $e_0 \geq 0$ and $e_i \geq 1$ and distinct odd primes p_i, for $1 \leq i \leq n$. Then, the number of distinct solutions to the congruence $x^2 \equiv 1 \bmod m$ is

$$\begin{cases} 2^n & \text{if } e_0 = 0 \text{ or } 1, \\ 2^{n+1} & \text{if } e_0 = 2, \\ 2^{n+2} & \text{if } e_0 \geq 3. \end{cases}$$

In particular, the congruence $x^2 \equiv 1 \bmod m$ has exactly two solutions modulo m if and only if $m = 2$, 4, p^k, or $2p^k$.

Proof. By the Chinese remainder theorem (Theorem 4.5.9), a class $x \bmod m$ is a solution for the congruence $x^2 \equiv 1 \bmod m$ if and only if it also solves the system of equations

$$\begin{cases} x^2 \equiv 1 \bmod 2^{e_0}, \\ x^2 \equiv 1 \bmod p_1^{e_1}, \\ x^2 \equiv 1 \bmod p_2^{e_2}, \\ \vdots \\ x^2 \equiv 1 \bmod p_n^{e_n}. \end{cases}$$

Since each p_i is an odd prime, there is a primitive root modulo $p_i^{e_i}$ (by Theorem 8.5.7), and therefore $x^2 \equiv 1 \bmod p_i^{e_i}$ has exactly two solutions (by Corollary 8.6.15; notice again that $\varphi(m)$ is even for all $m > 2$, by Exercise 7.6.4).

If $e_0 = 1$, then $x^2 \equiv 1 \bmod 2$ has a unique solution; if $e_0 = 2$, then $x^2 \equiv 1 \bmod 4$ has two solutions; and if $e_0 \geq 3$, then $x^2 \equiv 1 \bmod 2^{e_0}$ has exactly four solutions, by Lemma 8.7.2.

Therefore, the number of solutions to the system of quadratic congruences is equal to the product of the number of solutions in each individual congruence, and this equals

$$\left(\begin{cases} 1 & \text{if } e_0 = 0 \text{ or } 1, \\ 2 & \text{if } e_0 = 2, \\ 2^2 & \text{if } e_0 \geq 3 \end{cases}\right) \cdot 2 \cdot 2 \cdots 2 = \left(\begin{cases} 1 & \text{if } e_0 = 0 \text{ or } 1, \\ 2 & \text{if } e_0 = 2, \\ 2^2 & \text{if } e_0 \geq 3 \end{cases}\right) \cdot 2^n,$$

as claimed. □

We are finally ready to classify all the numbers $m \geq 2$ with a primitive root.

Theorem 8.7.4. *Let $m \geq 2$ be an integer. Then, m has a primitive root if and only if $m = 2$, 4, p^k, or $2p^k$, for some odd prime p and some $k \geq 1$.*

Proof. By Theorem 8.5.7, a number of the form $m = 2$, 4, p^k, or $2p^k$, for some odd prime p and some $k \geq 1$, has a primitive root.

Now let $m > 2$ be a number with a primitive root. By Corollary 8.6.15, the congruence $x^2 \equiv 1 \bmod m$ has exactly two distinct solutions modulo m. By Proposition 8.7.3, the number m must be of the form $m = 2$, 4, p^k, or $2p^k$, for some odd prime p and some $k \geq 1$, as claimed. □

8.8. The Structure of $(\mathbb{Z}/p^k\mathbb{Z})^\times$

In this section we interpret the existence of primitive roots and the index function in terms of groups.

Theorem 8.8.1. *Let* $m = 2$, 4, p^k, *or* $2p^k$, *for some odd prime* p *and some* $k \geq 1$. *Then, there is a bijection*

$$\psi\colon (\mathbb{Z}/m\mathbb{Z})^\times \to \mathbb{Z}/\varphi(m)\mathbb{Z}$$

which respects group structures; i.e.,

$$\psi(a \cdot b \bmod m) = \psi(a \bmod m) + \psi(b \bmod m)$$

for any units a *and* $b \bmod m$. *In other words,* ψ *is an isomorphism of groups.*

Proof. Let $m = 2$, 4, p^k, or $2p^k$. Then, by Theorem 8.7.4, there is a primitive root g modulo m. For a unit $a \bmod m$, let us define $\psi(a \bmod m) \equiv \mathrm{ind}_g(a) \bmod \varphi(m)$. By Proposition 8.6.8, we know that $\mathrm{ind}_g\colon (\mathbb{Z}/m\mathbb{Z})^\times \to \mathbb{Z}/\varphi(m)\mathbb{Z}$ is a bijection. Moreover,

$$\psi(a \cdot b \bmod m) \equiv \mathrm{ind}_g(ab) \equiv \mathrm{ind}_g(a) \cdot \mathrm{ind}_g(b) \equiv \psi(a) \cdot \psi(b) \bmod \varphi(m),$$

by the properties of the index function (Proposition 8.6.8). □

So what about $m = 2^n$? We know that $\mathbb{Z}/2^n\mathbb{Z}$ does not have a primitive root. So, what is the structure of $(\mathbb{Z}/2^n\mathbb{Z})^\times$? What is the largest order of a unit modulo 2^n?

Example 8.8.2. What is the order of 3 modulo 2^n, for $n \geq 3$?

$$\begin{aligned}
3^2 &\equiv 9 \equiv 1 \bmod 8,\\
3^3 &\equiv 27 \equiv 11 \bmod 16,\\
3^4 &\equiv 33 \equiv 1 \bmod 16,\\
3^5 &\equiv 27 \cdot 9 \equiv -5 \cdot 9 \equiv -45 \equiv 19 \bmod 32,\\
3^6 &\equiv 19 \cdot 3 \equiv -39 \equiv 25 \bmod 32,\\
3^7 &\equiv 25 \cdot 3 \equiv 75 \equiv 11 \bmod 32,\\
3^8 &\equiv 11 \cdot 3 \equiv 33 \equiv 1 \bmod 32.
\end{aligned}$$

Hence, $\mathrm{ord}_8(3) = 2$, $\mathrm{ord}_{16}(3) = 4$, and $\mathrm{ord}_{32}(3) = 8$. Let us show that the order of $3 \bmod 2^n$ is 2^{n-2}. The statement is true for $n = 5$. Let us use induction. Assume that the statement is true for all $2, 3, 4, 5, \ldots, n$. By Lemma 8.5.1, the order of $3 \bmod 2^{n+1}$ is divisible by $\mathrm{ord}_{2^n}(3) = 2^{n-2}$. By Euler's theorem, the order of $3 \bmod 2^{n+1}$ is a divisor of $\varphi(2^{n+1}) = 2^n$, so there are three options: 2^{n-2}, 2^{n-1}, or 2^n. In addition, $3^{2^{n-3}} \equiv 1 \bmod 2^{n-1}$ and so $3^{2^{n-3}} = 1 + k2^{n-1}$. Moreover, k is odd, because the order of $3 \bmod 2^n$ is 2^{n-2} and not 2^{n-3}. Thus,

$$\begin{aligned}
3^{2^{n-2}} &= (3^{2^{n-3}})^2 = (1 + k2^{n-1})^2\\
&= 1 + k2^{n-2} + k^2 2^{2n-2}.
\end{aligned}$$

Since $n > 5$, we have $2n - 2 \geq n + 1$. Hence,

$$3^{2^{n-2}} \equiv 1 + k2^{n-2} \bmod 2^{n+1},$$

and it follows that $3^{2^{n-2}} \not\equiv 1 \bmod 2^{n+1}$, because k is odd. Hence, the order of $3 \bmod 2^{n+1}$ is not 2^{n-2}. It follows that the order is 2^{n-1} or 2^n. However, $\mathbb{Z}/2^{n+1}\mathbb{Z}$ does not have a primitive root (by Theorem 8.7.4), so the order cannot be $\varphi(2^{n+1}) = 2^n$. We conclude that the order of $3 \bmod 2^{n+1}$ is precisely $2^{(n+1)-2}$. Thus, by the principle of mathematical induction, $3 \bmod 2^n$ has order 2^{n-2} for all $n \geq 2$.

Since $\varphi(2^n) = 2^{n-1}$, the fact that $3 \bmod 2^n$ has order 2^{n-2} implies that the powers of 3 go over half of all the elements of $\mathbb{Z}/2^n\mathbb{Z}$. What units modulo 2^n are not powers of 3?

We claim that -1 is not a power of 3 modulo 2^n, for $n > 2$. Suppose for a contradiction that there is some t such that $1 < t < 2^{n-2}$ and $3^t \equiv -1 \bmod 2^n$. Then, $3^{2t} \equiv 1 \bmod 2^n$ and so the order of 3 is a divisor of $2t$; i.e., 2^{n-2} is a divisor of $2t$, so either $t = 2^{n-2}$ or $t = 2^{n-3}$. The former is impossible because $t < 2^{n-2}$ by assumption; hence $t = 2^{n-3}$. When $n = 3$, we have $3^1 \equiv 3 \not\equiv -1 \bmod 8$, so let us assume $n \geq 4$. Then, $t = 2^{n-3}$ is a positive power of 2. Since $3^t \equiv -1 \bmod 2^n$ and $n \geq 4$, then $3^t \equiv -1 \bmod 4$, but

$$3^t \equiv (3^2)^{2^{n-4}} \equiv 1^{2^{n-4}} \equiv 1 \bmod 4.$$

Hence, we have reached a contradiction, and -1 is not a power of 3 modulo 2^n, for any $n > 2$.

Now we claim that if $a \bmod 2^n$ is a power of 3, then $-a$ is not a power of 3. Indeed, if $a \equiv 3^t \bmod 2^n$ and $-a \equiv 3^s \bmod 2^n$, then $-1 \equiv 3^{s-t} \bmod 2^n$, which is a contradiction, since -1 is not a power of 3 modulo 2^n. We have shown that $\mathbb{Z}/2^n\mathbb{Z}$ is a disjoint union:

$$\mathbb{Z}/2^n\mathbb{Z} = \{3^t \bmod 2^n : 0 \leq t < 2^{n-2}\} \cup \{-3^t \bmod 2^n : 0 \leq t < 2^{n-2}\}.$$

We are ready to understand the structure of $(\mathbb{Z}/2^n\mathbb{Z})^\times$ as an abstract group.

Theorem 8.8.3. *Let $n \geq 3$. Then, there is a bijection*

$$\psi\colon (\mathbb{Z}/2^n\mathbb{Z})^\times \to (\mathbb{Z}/2\mathbb{Z}) \times (\mathbb{Z}/2^{n-2}\mathbb{Z})$$

which respects group structures; i.e.,

$$\psi(a \cdot b \bmod 2^n) = \psi(a \bmod 2^n) + \psi(b \bmod 2^n)$$

for any units a and $b \bmod 2^n$. (Note: addition on $(\mathbb{Z}/2\mathbb{Z}) \times (\mathbb{Z}/2^{n-2}\mathbb{Z})$ is performed coordinatewise.)

Proof. We define ψ as follows. Let $a \bmod 2^n$ be a unit. Then, by our results in Example 8.8.2, there is a unique t with $0 \leq t < 2^{n-2}$ such that either $a \equiv 3^t$ or $a \equiv -3^t \bmod 2^n$, so let us write $a \equiv (-1)^s \cdot 3^t \bmod 2^n$, where $s = 0$ or 1. Define

$$\psi(a \bmod m) \equiv (s \bmod 2, t \bmod 2^{n-2}).$$

By our previous remarks, $s \bmod 2$ and $t \bmod 2^{n-2}$ are uniquely determined for each unit $a \bmod 2^n$, and ψ is surjective because $\psi((-1)^s \cdot 3^t) = (s, t)$. Hence, ψ is a bijection.

Finally, suppose that $a \equiv (-1)^s \cdot 3^t$ and $b \equiv (-1)^u \cdot 3^v \bmod 2^n$. Then,

$$\begin{aligned}\psi(a \cdot b) &= \psi((-1)^{s+u} \cdot 3^{t+v})\\ &= (s + u \bmod 2, t + v \bmod 2^{n-2})\\ &= (s \bmod 2, t \bmod 2^{n-2}) + (u \bmod 2, v \bmod 2^{n-2})\\ &= \psi((-1)^s \cdot 3^t \bmod 2^n) + \psi((-1)^u \cdot 3^v \bmod 2^n)\\ &= \psi(a \bmod 2^n) + \psi(b \bmod 2^n).\end{aligned}$$

This concludes the proof of our theorem. □

Remark 8.8.4. In the previous result, for each $n \geq 3$, the number 3 can be substituted by any other congruence class $g \bmod 2^n$ of exact order 2^{n-2}. By Proposition 8.1.8, if d is odd, then $3^d \bmod 2^n$ has the same order as 3.

8.9. Applications

In this section we discuss applications of the concepts of multiplicative order and primitive root.

8.9.1. The Diffie–Hellman Key Exchange.

In Sections 4.6.4 and 7.5.3 we have seen our first applications of modular arithmetic to cryptography. In this section, we discuss a different application that solves a different problem: private communication through insecure public channels (such as the internet). Suppose Alice and Bob would like to communicate securely, but they live far apart. Before they can start communicating using some type of cipher (e.g., a Caesar cipher or a Vigenère cipher, as described in Section 4.6.4), they need to agree on a private key (or password) that they both know, but no one else knows. The problem is that Alice and Bob live far apart, so how can they agree on a key while communicating on an insecure channel?

The Diffie–Hellman key exchange protocol was first published by Whitfield Diffie and Martin Hellman in 1976 (although it had been discovered by the British intelligence agency GCHQ as early as in 1969). The goal of this scheme is for two parties (Alice and Bob) to be able to agree on a secret key, via public communication. It works as follows.

Diffie–Hellman key exchange:

(1) Alice and Bob agree on a (large) prime number p and a primitive root $g \bmod p$. The prime p and the primitive root g are usually public information.

(2) Alice chooses her secret key, an integer $1 < a < p - 1$, and Bob chooses his secret key, an integers $1 < b < p - 1$.

(3) Alice computes $A \equiv g^a \bmod p$, and Bob computes $B \equiv g^b \bmod p$.

(4) Alice sends A to Bob, and Bob sends B to Alice, through a public channel.

(5) Alice computes $K_A \equiv B^a \bmod p$, and Bob computes $K_B \equiv A^b \bmod p$.

(6) The secret key shared by Alice and Bob is $K \equiv K_A \equiv K_B \bmod p$.

Indeed, the keys K_A and K_B coincide modulo p:

$$K_A \equiv B^a \equiv (g^b)^a \equiv g^{ab} \equiv (g^a)^b \equiv A^b \equiv K_B \bmod p.$$

Remark 8.9.1. The security of the exchange relies on the fact that, given a large prime p, a primitive root $g \bmod p$, and a congruence class $A \bmod p$, it is computationally expensive (i.e., lengthy) to find $1 \leq a \leq p-1$ such that $g^a \equiv A \bmod p$. This is called the *discrete logarithm problem.* If p is small, then a spy can compute a table of indices for the primitive root g (as in Section 8.6) and use the properties of indices to find Alice's (resp. Bob's) secret key a (resp. b) from A (resp. B) and compute their secret key K. For example, suppose that a spy finds that $p = 11$ and intercepts the public messages that say $A = 9$ and $B = 7$. Then, the spy can build a table of indices as in Example 8.6.4 and find that Alice's secret key is $a = 6$. Hence, $K \equiv B^a \equiv 7^6 \equiv 4 \bmod 11$.

However, if p is large, then computing a full table of indices, or solving the discrete logarithm problem $g^x \equiv A \bmod p$, takes too long even in the most powerful computers. Currently, in practice, it is recommended that a Diffie–Hellman key exchange is done with a prime p of size $\approx 2^{1000}$.

Example 8.9.2. Suppose that Alice and Bob would like to start communicating with a Vigenère cipher with key $K = (k_1, k_2)$, but they first have to agree on a key K. They proceed with a Diffie–Hellman key exchange, as follows:

(1) Alice and Bob agree on $p = 101$ and a primitive root $g \equiv 2 \bmod 101$, through a public channel.

(2) Alice chooses her secret key, $a = 23$, and Bob chooses his secret key, $b = 17$.

(3) Alice computes $A \equiv g^a \equiv 2^{23} \equiv 53 \bmod 101$, and Bob computes $B \equiv g^b \equiv 2^{17} \equiv 75 \bmod 101$.

(4) Alice sends $A \equiv 53 \bmod 101$ to Bob, and Bob sends $B \equiv 75 \bmod 101$ to Alice, through a public channel.

(5) Alice computes $K_A \equiv B^a \equiv 75^{23} \equiv 29 \bmod 101$, and Bob computes $K_B \equiv A^b \equiv 53^{17} \equiv 29 \bmod 101$.

(6) The secret key shared by Alice and Bob is $K \equiv K_A \equiv K_B \equiv 29 \bmod 101$.

Now, they will use $K = 29$, interpreted as $K = (2, 9)$, as a key for a Vigenère cipher. For instance, Alice sends the message HELLO as JNNUQ, which Bob can decipher because he also knows K, the secret key.

8.9.2. Periods in Decimal Expansions. In this section we discuss an application of the concepts of multiplicative order and primitive roots to prove a formula for the length of the period in the decimal expansion of a rational number. Let us begin with an example.

Example 8.9.3. Let us calculate the decimal expansion of the rational number $3/7$. See Figure 8.2.

In the long division of 3 over 7, when we reach a remainder of 3, we can stop, because the whole process starts over again, and so

$$\frac{3}{7} = 0.428571428571428571\ldots.$$

Thus, the period of $3/7$ is 428571 and its length is 6. Let us see in more detail what calculations we are doing in the long division, in order to investigate why there

```
      0.4285714...
  7 ) 3.0000000...
      28
       20
       14
        60
        56
         40
         35
          50
          49
           10
            7
            30
            28
            .....
```

Figure 8.2. Long division to calculate the decimal expansion of 3/7.

should be a repeating period. The calculation starts with a long division of 30 by 7:

$$30 = 7 \cdot 4 + 2,$$

which means that the first digit of the decimal expansion, after "0." is a 4. In each of the following steps, we take the remainder of the previous long division, we multiply it by 10, and then divide by 7.

$$30 = 7 \cdot 4 + 2, \ \ 2 \cdot 10 = 7 \cdot 2 + 6, \ \ 6 \cdot 10 = 7 \cdot 8 + 4, \ldots$$

and so the expansion continues with 0.428.... In terms of congruences and forgetting about the digits of the expansion for a moment, the previous equations read

$$3 \cdot 10 \equiv 2 \bmod 7, \ \ 2 \cdot 10 \equiv 6 \bmod 7, \ \ 6 \cdot 10 \equiv 4 \bmod 7, \ldots,$$

or, in other words,

$$3 \cdot 10 \equiv 2, \ \ (3 \cdot 10) \cdot 10 \equiv 6, \ \ (3 \cdot 10^2) \cdot 10 \equiv 4 \bmod 7, \ldots.$$

Thus, each consecutive remainder is precisely the least non-negative residue of $3 \cdot 10^t \bmod 7$. The long division can stop as soon as we find a repetition in the sequence of remainders, i.e., whenever

$$3 \cdot 10^t \equiv 3 \cdot 10^s \bmod 7,$$

for some $t \geq s \geq 0$, and in this case the length of the period is $t - s$. Since $\gcd(3, 7) = 1$, we may divide both sides by 3 (Proposition 4.3.1 or multiply by $3^{-1} \equiv 5 \bmod 7$) and obtain

$$10^t \equiv 10^s \bmod 7.$$

Moreover, $\gcd(10, 7) = 1$ as well, so $10 \equiv 3$ is a unit modulo 7, and therefore it follows that $10^t \equiv 10^s \bmod 7$ if and only if

$$10^{t-s} \equiv 1 \bmod 7.$$

Since the multiplicative order of $10 \equiv 3 \bmod 7$ is 6 (Example 8.1.2), it follows that $10^{t-s} \equiv 1 \bmod 7$ if and only if $t - s$ is divisible by 6, and this first happens when $t = 6$ and $s = 0$. Hence, the length of the period in the decimal expansion of $3/7$ is 6, and the period begins right at the beginning of the decimal expansion (because $s = 0$ here).

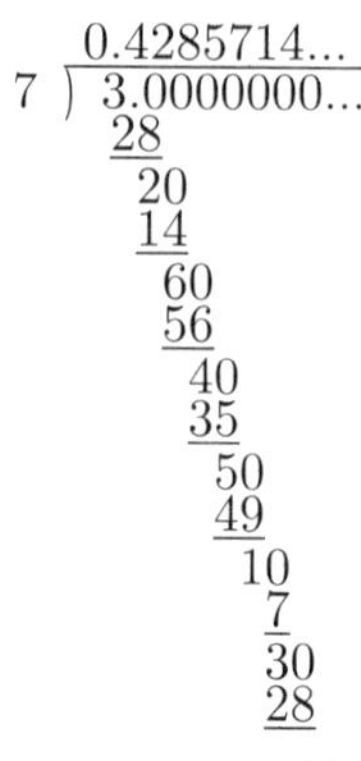

Let us begin with some basic results about decimal expansions.

Lemma 8.9.4. *Let a, b be natural numbers, with $\gcd(a, b) = 1$. Then, there are unique non-negative integers $q_0, q_1, \ldots$ such that*

$$\frac{a}{b} = q_0 + \frac{q_1}{10} + \frac{q_2}{10^2} + \cdots$$

where $0 \leq q_i \leq 9$ for all $i \geq 1$ and such that (q_i, r_i) satisfy the following recurrence relation: let $a = q_0 b + r_0$, with $0 \leq r_0 < b$, then q_{i+1} is the unique integer with $10r_i = q_{i+1}b + r_{i+1}$, for some $0 \leq r_{i+1} < b$, for every $i \geq 0$.

Proof. By the division theorem (Theorem 2.4.4), there exist unique integers q_0, r_0 such that $a = q_0 b + r_0$ and $0 \leq r_0 < b$, so that

$$\frac{a}{b} = q_0 + \frac{r_0}{b}.$$

We define a sequence of pairs $\{(q_i, r_i)\}_{i=1}^{\infty}$ recursively by

$$a = q_0 b + r_0 \quad \text{and} \quad 10r_i = q_{i+1}b + r_{i+1},$$

where q_{i+1} and r_{i+1} are as in the division theorem, with $0 \leq r_{i+1} < b$, for each $i \geq 0$. In particular, $10r_i < 10b$ and so $0 \leq q_i < 10$, for each $i \geq 1$. It follows that, for every $i \geq 1$, we have

$$\frac{a}{b} = q_0 + \frac{r_0}{b} = q_0 + \frac{q_1}{10} + \frac{r_1}{10b} = q_0 + \frac{q_1}{10} + \cdots + \frac{q_i}{10^i} + \frac{r_i}{10^i b}.$$

If we define $\{c_n\}_{n=0}^{\infty}$ by $c_n = \sum_{k=0}^{n} \frac{q_k}{10^k}$, then the equation above shows that

$$\left| \frac{a}{b} - c_n \right| = \frac{r_i}{10^n b} < \frac{1}{10^n}.$$

Therefore, $\lim_{n\to\infty} c_n = a/b$, as desired. □

Theorem 8.9.5. *Let a, b be natural numbers, with $\gcd(a, b) = 1$. Then, the decimal expansion of a/b is periodic, and the length of the period is precisely $t - s$, where s and t are non-negative integers $t > s \geq 0$ such that s and $t - s$ are minimal with the property that*

$$10^t \equiv 10^s \bmod b.$$

Moreover, the period begins at the $(s+1)$th digit of the fractional part of the decimal expansion; i.e., the decimal expansion of a/b is of the form

$$\frac{a}{b} = q_0.q_1 q_2 \cdots q_s \overline{q_{s+1} q_{s+2} \cdots q_t},$$

for some $q_0 \geq 0$ and $0 \leq q_i \leq 9$ for $i \geq 1$.

Proof. Let s and t be as in the statement of the proposition, i.e., non-negative integers $t > s \geq 0$ such that s and $t - s$ are minimal with the property that

$$10^t \equiv 10^s \bmod b.$$

Let $q_i \geq 0$ be the digits in a decimal expansion of a/b, as in Lemma 8.9.4; i.e., $a = q_0 b + r_0$ and $10r_i = q_{i+1}b + r_{i+1}$, with $0 \leq r_i < b$ and $0 \leq q_i < 10$ for each $i \geq 0$. It follows then from the definition of r_i that

$$r_i \equiv 10r_{i-1} \equiv 10^i a \bmod b,$$

for all $i \geq 0$. In particular,

$$r_t \equiv 10^t a \equiv 10^s a \equiv r_s \bmod b,$$

and since $0 \leq r_s, r_t < b$, they must be equal. Now, it follows from the definition of q_i and r_i that if $r_t = r_s$ for some $t \geq s$, then

$$q_{s+1}b + r_{s+1} = 10r_s = 10r_t = q_{t+1}b + r_{t+1},$$

and, therefore, by the uniqueness of the quotient and remainder in the division theorem (Theorem 2.4.4), we have $q_{s+1} = q_{t+1}$ and $r_{s+1} = r_{t+1}$. Hence, if $r_t = r_s$ for some $t \geq s$, then $(q_{t+k}, r_{t+k}) = (q_{s+k}, r_{s+k})$ for all $k \geq 1$, and therefore the decimal representation is periodic, with period $q_{s+1}q_{s+2}\cdots q_t$, and the length of the period is at most $t - s$.

It remains to show that the length of the period is indeed exactly equal to $t-s$. Since we have shown that a/b is periodic, there are numbers $u > w \geq 0$ and P, Q with $0 \leq P < 10^w$ and $0 \leq Q < 10^{u-w}$, such that the length of the period is $u - w$ (which is $\leq t - s$ by our remarks above) and

$$\frac{a}{b} = \frac{P}{10^w} + \frac{Q}{10^u} + \frac{Q}{10^{2u-w}} + \frac{Q}{10^{3(u-w)+w}} + \cdots,$$

or, equivalently,

$$10^w\left(\frac{a}{b} - \frac{P}{10^w}\right) = \frac{Q}{10^{u-w}} + \frac{Q}{10^{2(u-w)}} + \frac{Q}{10^{3(u-w)}} + \cdots.$$

This implies that

$$10^{u-w}\left(10^w\left(\frac{a}{b} - \frac{P}{10^w}\right) - \frac{Q}{10^{u-w}}\right) = \frac{Q}{10^{u-w}} + \cdots = 10^w\left(\frac{a}{b} - \frac{P}{10^w}\right).$$

In turn, this is equivalent to

$$\left(10^u\left(\frac{a}{b} - \frac{P}{10^w}\right) - Q\right) = 10^w\frac{a}{b} - P,$$

and therefore

$$(10^u - 10^w)\frac{a}{b} = 10^{u-w}P + 10^u Q,$$

or, equivalently, $(10^u - 10^w)a = b(10^{u-w}P + 10^u Q)$. In particular,

$$(10^u - 10^w)a \equiv 0 \bmod b.$$

Since $\gcd(a, b) = 1$ by assumption, a is a unit modulo b, and therefore $10^u \equiv 10^w \bmod b$. Since $t > s \geq 0$ are such that s and $t-s$ are smallest with the property $10^t \equiv 10^s \bmod b$, it follows that $w \geq s$ and $u - w \geq t - s$. This, together with the fact shown above that $u - w \leq t - s$ shows that $u - w = t - s$ (and so $w = s$ and $u = t$). Hence, the length of the period is $t - s$, as we wanted to prove.

Finally, note that by our remarks at the beginning of the proof, the period is formed by the digits $q_{s+1}q_{s+2}\cdots q_t$, as claimed. □

Example 8.9.6. Consider the rational number $3/14$. In order to calculate the length of the period in the decimal expansion, let us calculate (in Figure 8.3) first powers of 10 modulo 14 to find out the values of s and t as in Theorem 8.9.5.

Thus, we find that $10^7 \equiv 10^1 \bmod 14$, and this is the first time a repetition happens; i.e., $s = 1$, $t = 7$, and $t - s = 6$. Therefore, the period begins on the

n	0	1	2	3	4	5	6	7	8	9	10
$10^n \bmod 14$	1	10	2	6	4	12	8	10	2	6	4

Figure 8.3. Powers of 10 mod 14.

second digit of the decimal expansion of the fractional part of $3/14$, and the length of the period is 6 digits. Indeed,

$$0.214285714285714285714285714286\ldots = 0.2\overline{142857}.$$

Corollary 8.9.7. *Let a, b be natural numbers, with $\gcd(a, b) = 1$, and assume that $\gcd(b, 10) = 1$. Then, the length t of the period in the decimal expansion of a/b is exactly the multiplicative order of 10 modulo b, and t is a divisor of $\varphi(b)$. Moreover, the period begins with the first digit of the fractional part of the decimal expansion; i.e.,*

$$\frac{a}{b} = q_0.\overline{q_1 q_2 \cdots q_t},$$

for some $q_0 \geq 0$ and $0 \leq q_i \leq 9$ for $i \geq 1$.

Proof. Suppose that $\gcd(b, 10) = 1$ and $10^t \equiv 10^s \bmod b$, where $t > s \geq 0$ are such that s and $t - s$ are minimal with this property. Since b and 10 are relatively prime, the number 10 is a unit modulo b, and therefore we obtain

$$10^{t-s} \equiv 1 \bmod b.$$

By the minimality property of s and t, it follows that $s = 0$, and t is exactly the multiplicative order of 10 mod b. Finally, Corollary 8.1.6 implies that t is a divisor of $\varphi(b)$. □

Example 8.9.8. For instance, the multiplicative order of $10 \equiv 3 \bmod 7$ is 6, and therefore the length of the period of $3/7$ must be 6, and the period begins with the first digit of the fractional part. Indeed, as we saw in Example 8.9.3,

$$\frac{3}{7} = 0.428571428571428571\ldots = 0.\overline{428571}.$$

It remains to understand the decimal expansion of rational numbers a/b such that $\gcd(b, 10) \neq 1$. We need a lemma.

Lemma 8.9.9. *Let $h, k \geq 1$ and let $b = 2^h 5^k$. Let $M = \max\{h, k\}$; then $10^{M+1} \equiv 10^M \bmod b$, and $M + 1 \geq M \geq 0$ are the smallest values $t > s \geq 0$ with the property that $10^t \equiv 10^s \bmod b$.*

Proof. Let h, k, and M be as in the statement, and let $p = 2$ or 5 such that p^M is a divisor of b. Then, if $10^t \equiv 10^s \bmod b$, then $10^t \equiv 10^s \bmod p^M$, and therefore $p^t \equiv p^s \bmod p^M$. It follows that p^M is a divisor of $p^t - p^s = p^s(p^t - 1)$ and since p is relatively prime to $p^t - 1$, it follows that p^M divides p^s, and so $s \geq M$. Moreover, if $i \geq M$, then 10^i is divisible by $10^M = 2^M 5^M$ and also by $b = 2^h 5^k$, since $M = \max\{h, k\}$. Thus, $10^i \equiv 0 \bmod b$, and in particular $10^{M+1} \equiv 10^M \bmod b$. □

Theorem 8.9.10. *Let a, b be natural numbers, with $\gcd(a, b) = 1$, and suppose that $b = 2^h 5^k b'$ with $\gcd(b', 10) = 1$ and $h, k \geq 0$. Let $M = \max\{h, k\}$. Then, the length of the period in the decimal expansion of a/b is exactly t, where t is the* *multiplicative order of* 10 *modulo* b' *(as before,* t *is a divisor of* $\varphi(b')$*). Moreover, the period begins with the* $(M+1)$*th digit of the fractional part of the decimal expansion; i.e.,*

$$\frac{a}{b} = q_0.q_1q_2\cdots q_M\overline{q_{M+1}q_{M+2}\cdots q_{M+t}},$$

for some $q_0 \geq 0$ *and* $0 \leq q_i \leq 9$ *for* $i \geq 1$.

Proof. Let $t > s \geq 0$ be as in Theorem 8.9.5. In particular,

$$10^t \equiv 10^s \bmod b,$$

and, therefore, the same congruence is true modulo 2^h5^k and modulo b'. By Lemma 8.9.9, if $10^t \equiv 10^s \bmod 2^h5^k$, then $s \geq M = \max\{h,k\}$.

Now, consider $10^M a/b = a'/b'$, where $a' = 2^{M-h}5^{M-k}a$ and $b = 2^h5^kb'$ with $\gcd(b',10) = 1$. Thus, $\gcd(a',b') = 1$, and $\gcd(b',10) = 1$. Hence, by Corollary 8.9.7, the length of the period in the decimal expansion of a'/b' is t, the multiplicative order of 10 mod b', and the period starts with the first digit of the fractional part. In other words,

$$a'/b' = Q_0.\overline{Q_1Q_2\cdots Q_t},$$

for some $Q_0 \geq 0$ and $0 \leq Q_i \leq 9$ for $i \geq 1$. On the other hand, if

$$a/b = q_0.q_1q_2\cdots q_Mq_{M+1}q_{M+2}\cdots,$$

then

$$a'/b' = 10^M a/b = q_0 \cdot 10^M + q_1\cdots q_M.q_{M+1}q_{M+2}\cdots.$$

Hence, we conclude that the integer Q_0 equals $q_0 \cdot 10^M + (q_1\cdots q_M)_{10}$ and $Q_1 = q_{M+1}$, $Q_2 = q_{M+2}$, $Q_t = q_{M+t}$ form the period of the decimal expansion. In other words,

$$\frac{a}{b} = q_0.q_1q_2\cdots q_M\overline{q_{M+1}q_{M+2}\cdots q_{M+t}},$$

as desired. □

Example 8.9.11. Consider the rational number 4001/550. The numbers 4001 and 550 are relatively prime (in fact, 4001 is prime), and $550 = 2\cdot 5^2\cdot 11$. Moreover, the order of $10 \equiv -1 \bmod 11$ is $t = 2$. Hence, the period in the fractional part of the decimal expansion of 4001/550 starts with the 3rd digit ($M = \max\{1,2\}$ as in Theorem 8.9.10 equals 2), and the length of the period is $t = 2$. Indeed,

$$\frac{4001}{550} = 7.2745454545454545\ldots = 7.27\overline{45},$$

so the period is $\overline{45}$ of length 2, beginning with the third digit in the expansion, in agreement with the theory.

8.10. Exercises

Exercise 8.10.1. Find the (multiplicative) order of every non-zero element of $\mathbb{Z}/19\mathbb{Z}$.

Exercise 8.10.2. Show that the order of 10 mod 83 is at least 30, **without** calculating any power of 10 higher than $10^2 = 100$ (so you are not allowed to calculate 10^k or 10^k mod 83 for any $k \geq 3$).

Exercise 8.10.3. Let $m = 2^{15} - 1 = 32767$. Prove the following:

(1) The order of 2 mod m is 15; i.e., show that $2^n \not\equiv 1 \bmod m$ for any $1 \leq n < 15$ and $2^{15} \equiv 1 \bmod m$.

(2) The number 15 does not divide $m - 1 = 32766$.

(3) Use the previous parts to conclude that m is not prime (you are not allowed to find a factorization of m).

Exercise 8.10.4. The following is a table of powers of 2 modulo 13:

x	x^2	x^3	x^4	x^5	x^6	x^7	x^8	x^9	x^{10}	x^{11}	x^{12}
2	4	8	3	6	12	11	9	5	10	7	1

(a) From the table, we see that 2 mod 13 has order 12. Find all the units of order 12.

(b) Find the order of every unit in $\mathbb{Z}/13\mathbb{Z}$.

(c) Find all the squares modulo 13; i.e., what congruences are the square of another number? Which ones are not squares?

Exercise 8.10.5. Prove that 74 is a primitive root modulo 89.

Exercise 8.10.6. Find a primitive root modulo 61.

Exercise 8.10.7. Find a primitive root modulo 73.

Exercise 8.10.8. Show that 2 is a primitive root modulo 11^k, for every $k \geq 1$.

Exercise 8.10.9. Let p be an odd prime. Show that if g is a primitive root modulo p, then $g^{(p-1)/2} \equiv -1 \bmod p$.

Exercise 8.10.10. Let p be a prime, and let a be a non-zero integer.

(a) Show that if $a = -1$ is a primitive root modulo p, then $p = 2$ or $p = 3$.

(b) Show that if $a = b^2$ is a perfect square and a is a primitive root mod p, then $p = 2$ and $a = 1$. (Hint: use Exercise 8.10.9 and Fermat's little theorem.)

Exercise 8.10.11. Prove Wilson's theorem using primitive roots. (Hint: suppose that g is a primitive root mod p, and write every unit as a power of g.)

Exercise 8.10.12. Show that $x^3 + y^3 = n$ has no integer solutions if $n \equiv 3$ or $4 \bmod 7$. Conclude that 1001002001002 is not a sum of two cubes. (Hint: for the second part, use Proposition 4.6.4.)

Exercise 8.10.13. Show that $x^5 + y^5 = 3 + 11z^5$ has no solutions in integers x, y, z. (Hint: what congruence classes are fifth powers modulo 11?)

Exercise 8.10.14. The number $p = 4003$ is prime.

(1) The number 372 has exact order 2001 modulo 4003. Find a primitive root modulo p.

(2) The number 285 has exact order 87 modulo p, and the number 2163 has exact order 46 modulo p. Find another primitive root mod p.

(3) Given that $2^{87} \equiv 2163 \bmod 4003$ and $2^{46} \equiv 285 \bmod 4003$, prove that 2 is also a primitive root modulo 4003.

Exercise 8.10.15. A natural number of the form $M_n = 2^n - 1$, for some $n \geq 1$, is called a *Mersenne number*. If M_n is prime, then M_n is called a *Mersenne prime*. In this exercise we show a criterion to test the primality of M_n.

(1) Show that if M_n is prime, then n is prime. (Hint: Exercise 3.5.23.)

(2) For the rest of this exercise, assume that p is an odd prime, put $M_p = 2^p - 1$, and let q be a prime divisor of M_p. Show that the order of 2 modulo q is a divisor of p, and therefore the order is exactly p.

(3) Show that p is a divisor of $q - 1$. (Hint: use Corollary 8.1.6.)

(4) Conclude that $q = mp + 1$, for some even number $m \geq 2$.

(5) Thus, we have shown that if M_p is divisible by q, then $q = 2kp + 1$ for some $k \geq 1$.

Figure 8.4. Marin Mersenne (1588–1648) was a French theologian, natural philosopher, and mathematician. Image source: Wikimedia Commons.

Exercise 8.10.16. Using Exercise 8.10.15, determine if the following Mersenne numbers are primes: $2^7 - 1$, $2^{11} - 1$, and $2^{29} - 1$.

Exercise 8.10.17. Suppose that $\mathbb{Z}/m\mathbb{Z}$ has a primitive root, and let $\{a_1, \ldots, a_r\}$ be a complete residue system for the units modulo m (i.e., a complete residue system of $(\mathbb{Z}/m\mathbb{Z})^\times$). Show that $S = \{a_1^n, \ldots, a_r^n\}$ is also a complete residue system for the units modulo m if and only if $\gcd(n, \varphi(m)) = 1$.

Exercise 8.10.18. Show that 3^{k+1} is a divisor of $2^{3^k} + 1$ for all $k \geq 1$. (Hint: show that 2 is a primitive root modulo 3^{k+1}.)

Exercise 8.10.19. Find all the solutions of the congruence $11x^{14} \equiv 23 \bmod 43$. (Hint: use the table in Example 8.6.16.)

Exercise 8.10.20. Show that 2 is a primitive root modulo 19 and solve the following:

(a) Build a table of indices of modulo 19 in base 2.

(b) Use the table of indices to find all the possible values of x that satisfy the following congruences:
  (1) $9x \equiv 14 \bmod 19$.
  (2) $11x^7 \equiv 13 \bmod 19$.
  (3) $5x^6 \equiv 17 \bmod 19$.
  (4) $9^x \equiv 7 \bmod 19$.

Exercise 8.10.21. Find all the solutions for the following congruences:

(a) $x^2 \equiv 1 \bmod 35$.

(b) $x^2 \equiv 1 \bmod 140$.

(c) $x^2 \equiv 1 \bmod 105$.

Exercise 8.10.22. Decide whether there exists a primitive root modulo m for the following values of m:

$$m = 8,\ 14,\ 28,\ 35,\ 70,\ 162,\ 625,\ 1250.$$

Exercise 8.10.23. Let $m = 1250$.

(1) Show that 3 mod 1250 is a primitive root.

(2) How many primitive roots are there modulo 1250?

Exercise 8.10.24. Alice and Bob want to set up a private key using the Diffie–Hellman method (as in Section 8.9.1). They choose $p = 43$, $g = 3$, $a = 10$, and $b = 20$ for their prime, primitive root, and private keys, respectively. Compute the secret key K they will share. (Hint: the tables in Example 8.6.5 should help.)

Exercise 8.10.25. Show that 6 is a primitive root modulo 13. Then, find an integer $x \geq 1$ that solves the discrete logarithm problem $6^x \equiv 5 \bmod 13$.

Exercise 8.10.26. Han and Leia want to set up a secure communication channel using a Diffie–Hellman key exchange, with $p = 13$ and $g = 6$.

(1) Leia picks $a = 2$. Compute $A \equiv g^a \bmod p$.

(2) Leia receives $B = 2$ from Han. Compute the secret key produced by the Diffie–Hellman key exchange that Leia and Han will share.

Exercise 8.10.27. Governor Tarkin intercepts some messages from Han and Leia who are setting up a Diffie–Hellman key exchange. He finds that $p = 13$ and $g = 6$ and intercepts $A = 3$ from Leia and $B = 5$ from Han. Use this information to generate the secret key that Leia and Han are planning to share. (Note: it is not the same key that they were sharing in part (2) of Exercise 8.10.26.)

Exercise 8.10.28. Compute the length of the period for the following rational numbers and the position of the digit where the period starts (e.g., the period of $2/15 = 0.133333\ldots$ is of length 1 and the period starts with the second digit):

$$\frac{2}{7},\ \frac{15}{13},\ \frac{3}{19},\ \frac{7}{152},\ \frac{19}{65}.$$

Part 2

Quadratic Congruences and Quadratic Equations

CHAPTER 9

AN INTRODUCTION TO QUADRATIC EQUATIONS

Is euclidean geometry true? [...] We might as well ask if the metric system is true and if the old weights and measures are false; if cartesian coordinates are true and polar coordinates are false. One geometry cannot be more true than another; it can only be more convenient.

Henri Poincaré

In the first part of this book we have learned how to find all integral and rational solutions of polynomial equations in one variable $p(x) = 0$ (see Section 2.8 and Theorem 2.8.1) and linear equations in two variables (see Section 2.9 and Theorem 2.9.4). In the next few chapters we are interested in finding the integral and rational points on quadratic equations in two variables, i.e., equations of the form

$$ax^2 + bxy + cy^2 + dx + ey + f = 0,$$

where a, b, c, d, e, f are integers and a, b, or c is non-zero. As we will see, finding the rational solutions is easy (once we know *at least one* of them!), but finding all integral solutions can be rather complicated.

In this chapter, we will learn to distinguish "irreducible" quadratic equations from "reducible" ones. By this we mean that some quadratic equations can be factored into two linear equations that we already know how to solve (e.g., $x^2 - y^2 + x + y = 0$ is equivalent to $(x - y + 1)(x + y) = 0$, which in turn is equivalent to solving $x - y + 1 = 0$ and $x + y = 0$), while some equations cannot be reduced to the study of linear equations (e.g., $x^2 + 2y^2 = 1$). Thus, a first step in analysing a quadratic equation is to determine whether it can be factored.

Once we are certain that the quadratic equation is "irreducible" (it cannot be factored as a product of two lines), then a first step is to classify the equation as one of three possible *conic sections*: a parabola, an ellipse, or a hyperbola. We will close this chapter studying how to find all the rational and integral points on a parabola.

In Chapter 10, we continue our study of congruences, and we extend our theory to quadratic congruences. These congruences will be useful in studying quadratic equations in general and, in particular, they can be very useful to show that a certain equation does not have solutions (see Example 1.3.2).

In Chapter 11 we discuss the theorem of Hasse and Minkowski, which provides a criterion to determine when a quadratic equation has one rational solution. Once we know one rational solution, we can use rational parametrizations (as in Example 1.3.1) to find *all* the rational points on the equation. We will discuss rational parametrizations of quadratic equations in Section 9.3.

In Chapter 12 we discuss methods that are specific to finding points on circles and ellipses. In particular, we give a criterion to decide what circles have integral points, i.e., to decide whether $x^2 + y^2 = n$ has integral points. In other words, we classify what natural numbers are the sum of two squares.

Similarly, in Chapter 14 we discuss methods that are specific to hyperbolas. Special attention will be paid to what is known as a Pell equation, $x^2 - dy^2 = 1$ and $x^2 - dy^2 = n$ for an integer n. We will see that solving a Pell equation is closely related to finding rational approximations of the irrational number $\sqrt{d}$. Finally, Chapter 13 is an introduction to the theory of continued fractions, which will be applied to solving Pell's equation.

9.1. Product of Two Lines

Algebra is nothing more than geometry, in words;
geometry is nothing more than algebra, in pictures.

Sophie Germain

Let $L : ax + by = c$ and $L' : dx + ey = f$ be two lines in the plane. If we multiply both equations together, we produce the equation of a new geometric object C, whose *geometric locus* is the union of L and L'. In other words, if we consider the equation $C : (ax + by - c)(dx + ey - f) = 0$ and if $P = (x_0, y_0) \in C$, then either $ax_0 + by_0 - c = 0$ or $dx_0 + ey_0 - f = 0$; i.e., either $P \in L$ or $P \in L'$ so the points on C are indeed the union of the points on L and the points on L' (see Figure 9.1).

Definition 9.1.1. Let C be a quadratic equation given by a polynomial equation $f(x, y) = 0$, where $f(x, y) \in \mathbb{Z}[x, y]$. We say that C is a *product of two lines* L and L' if there are two non-constant polynomials $g(x, y)$ and $h(x, y)$ in $\mathbb{C}[x, y]$ such that $L : g(x, y) = 0$, the line $L' : h(x, y) = 0$, and

$$f(x, y) = g(x, y)h(x, y).$$

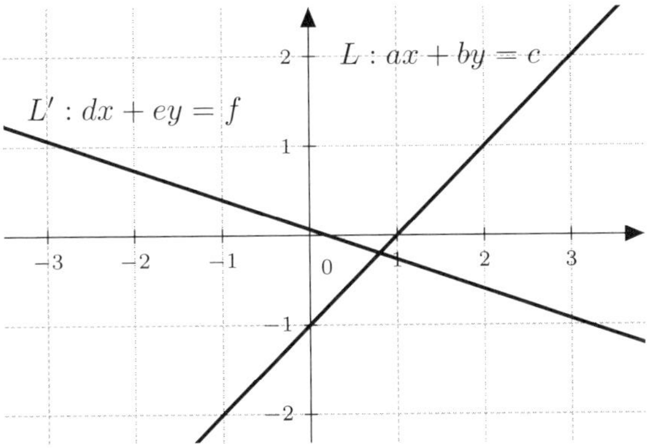

Figure 9.1. The geometric locus of $C : (ax + by - c)(dx + ey - f) = 0$.

Example 9.1.2. The equation $C : x^2 - y^2 + x + y = 0$ is a product of two lines. Indeed,

$$x^2 - y^2 + x + y = (x - y + 1)(x + y) = 0,$$

so C is the product of $L : x - y = -1$ and $L' : x + y = 0$. See Figure 9.2.

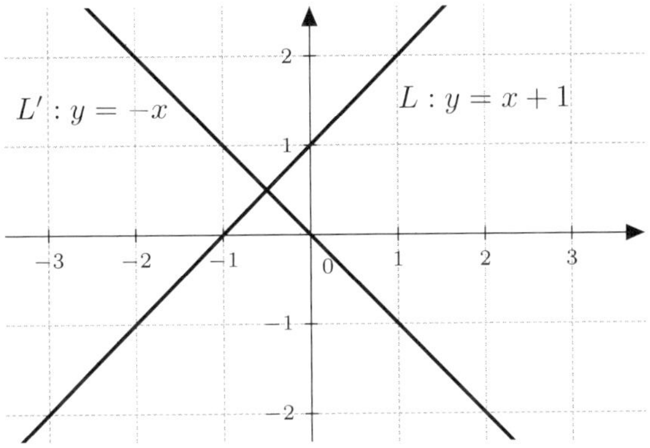

Figure 9.2. The geometric locus of $C : x^2 - y^2 + x + y = 0$.

Remark 9.1.3. The polynomial $f(x, y)$ may factor over $\mathbb{C}[x, y]$ but not over $\mathbb{Z}[x, y]$ or $\mathbb{Q}[x, y]$. For instance, consider

$$C : x^2 + xy + y^2 = 0.$$

We say that C is a product of two lines, because

$$f(x, y) = x^2 + xy + y^2 = \left(x - \frac{-1 + \sqrt{-3}}{2} \cdot y\right)\left(x - \frac{-1 - \sqrt{-3}}{2} \cdot y\right).$$

Thus, C is the product of two lines $L : y = (-1 + \sqrt{-3})/2 \cdot x$ and $L' : y = (-1 - \sqrt{-3})/2 \cdot x$ *in the complex plane* $\mathbb{C}$. In the real plane, however, there is only one point. Indeed, note that

$$x^2 + xy + y^2 = \left(x + \frac{y}{2}\right)^2 + \frac{3}{4}y^2 = 0,$$

so any real solution $(x_0, y_0) \in \mathbb{R}^2$ must satisfy $x + y/2 = 0$ and $y = 0$, so $(x_0, y_0) = (0, 0)$.

The easiest way to identify a quadratic equation as a product of two lines is via *normal and tangent vectors*: a quadratic equation C is a product of two lines if all the tangent vectors to the points on the curve are parallel to one or two fixed vectors v_1 and v_2. If all the tangent vectors are parallel to one single vector v_1, then C is the product of two parallel lines. If all the tangent vectors are parallel to either v_1 or v_2, then C is a product of two lines L and L', with direction vectors v_1 and v_2, respectively.

Next, we recall how to calculate tangent vectors, and we make the previous paragraph more concrete. We will, in fact, define both tangent and normal vectors to a curve, where the *normal* is a vector perpendicular to the tangent line.

Definition 9.1.4. Let $C : f(x, y) = 0$ be a curve on the plane given by a polynomial in two variables $f(x, y) \in \mathbb{Q}[x, y]$. Let $P = (x_0, y_0)$ be a point on C. The *normal vector* and the *tangent vector* of C at P, denoted, respectively, by $\vec{n}_C(P)$ and $\vec{t}_C(P)$, are given by

$$\vec{n}_C(P) = \left(\frac{\partial f}{\partial x}(P), \frac{\partial f}{\partial y}(P)\right) \quad \text{and} \quad \vec{t}_C(P) = \left(-\frac{\partial f}{\partial y}(P), \frac{\partial f}{\partial x}(P)\right).$$

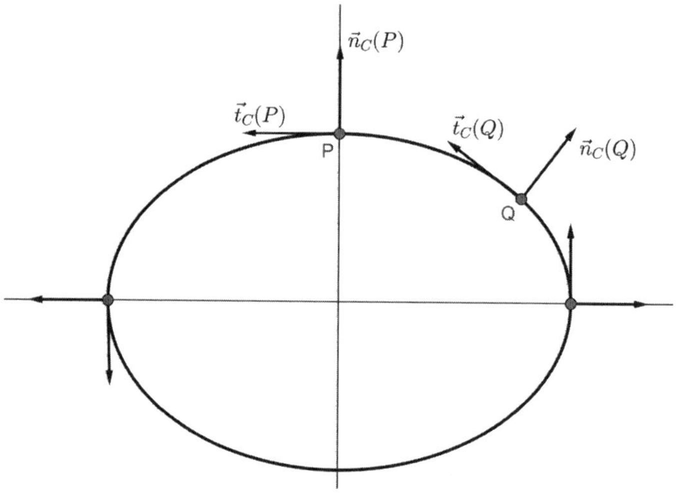

Figure 9.3. The normal and tangent vectors, $\vec{n}_C(P)$ and $\vec{t}_C(P)$, at the points $P = (0, 1)$ and $Q = (4/3, 1/3)$ on the ellipse $C : x^2 + 2y^2 = 2$.

Remark 9.1.5. Thus defined, the normal and the tangent vectors are perpendicular with respect to the usual dot product. Indeed,

$$\vec{n}_C(P) \cdot \vec{t}_C(P) = -\frac{\partial f}{\partial x}(P) \cdot \frac{\partial f}{\partial y}(P) + \frac{\partial f}{\partial x}(P) \cdot \frac{\partial f}{\partial y}(P) = 0.$$

Example 9.1.6. Let C be the ellipse given by $x^2 + 2y^2 - 2 = 0$ (see Figure 9.3). Then, the normal and tangent vectors are given by $\vec{n}_C = \left(\frac{\partial f}{\partial x}(P), \frac{\partial f}{\partial y}(P)\right) = (2x, 4y)$ and $\vec{t}_C = \left(-\frac{\partial f}{\partial y}(P), \frac{\partial f}{\partial x}(P)\right) = (-4y, 2x)$. For example, if we put $P = (0, 1)$ and $Q = (4/3, 1/3)$, then

$$\vec{n}_C(P) = (0, 4) \quad \text{and} \quad \vec{n}_C(Q) = (8/3, 4/3),$$

and

$$\vec{t}_C(P) = (-4, 0) \quad \text{and} \quad \vec{t}_C(Q) = (-4/3, 8/3).$$

Example 9.1.7. Let us calculate the normal vectors of $C : x^2 - y^2 + x + y = 0$. Here $f(x, y) = x^2 - y^2 + x + y$. Hence,

$$\vec{n}_C = \left(\frac{\partial f}{\partial x}, \frac{\partial f}{\partial y}\right) = (2x + 1, -2y + 1).$$

It is not immediately obvious from this expression for the normal vectors that, in fact, when evaluated at a point $P \in C$, all vectors are parallel to either $(1, -1)$ or $(1, 1)$. See Figure 9.4.

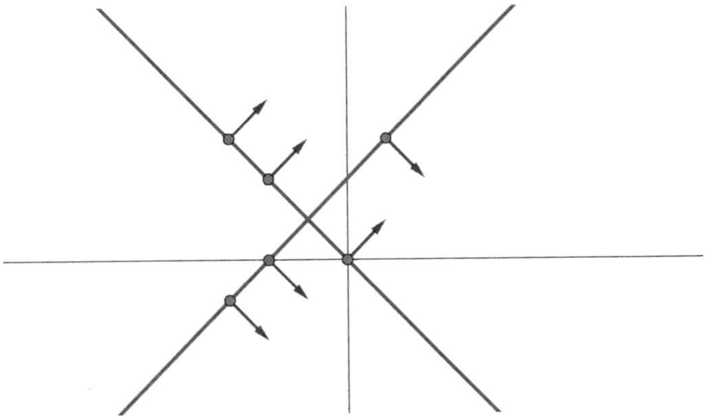

Figure 9.4. Several normal vectors at points on the curve $C : x^2 - y^2 + x + y = 0$. It turns out that C is a product of two lines, and all the normal vectors are parallel to either $(1, 1)$ or $(1, -1)$.

However, from Example 9.1.2 we know that we have a factorization $f(x, y) = x^2 - y^2 + x + y = (x - y + 1)(x + y)$. If we take partial derivatives using the product rule, we obtain

$$\begin{aligned}\vec{n}_C &= \left(\frac{\partial f}{\partial x}, \frac{\partial f}{\partial y}\right) = ((x + y) + (x - y + 1), -(x + y) + (x - y + 1)) \\ &= (x + y) \cdot (1, -1) + (x - y + 1) \cdot (1, 1),\end{aligned}$$

where $\cdot$ here denotes scalar multiplication; e.g., $(x + y) \cdot (1, -1) = (x + y, -(x + y))$.

Now, if $P \in C = L \cup L'$, either $P \in L : x - y = -1$ or $P \in L' : x + y = 0$. If $P = (x_0, y_0) \in L$, then $x_0 - y_0 + 1 = 0$ and

$$\vec{n}_C(P) = (x_0 + y_0) \cdot (1, -1),$$

which is a multiple of $(1, -1)$. If $P = (x_0, y_0) \in L'$, then $x_0 + y_0 = 0$ and

$$\vec{n}_C(P) = (x_0 - y_0 + 1) \cdot (1, 1),$$

which is a multiple of $(1, 1)$.

It is still unclear, however, how to decompose the normal vector appropriately, as in

$$\vec{n}_C = (2x + 1, -2y + 1) = (x + y) \cdot (1, -1) + (x - y + 1) \cdot (1, 1),$$

if we do not know how to factor the polynomial $f(x, y)$ in the first place. The first step is to find the point of intersection of the two lines, if there is one. The following proposition provides a criterion to find the intersection point.

Proposition 9.1.8. *Let $C : f(x, y) = 0$ be a quadratic equation (with $f(x, y) \in \mathbb{Z}[x, y]$) that is a product of two distinct lines L and L'. Suppose that L and L' intersect at a point $P = (x_0, y_0) \in C$. Then, the coordinates of P are rational numbers, and the normal vector of C at P vanishes; i.e., $\vec{n}_C(P) = (0, 0)$. Moreover:*

(1) *If m_1 and m_2 are, respectively, the slopes of L and L' and both are finite slopes (i.e., L, L' are not vertical), then*

$$C : f(x, y) = \lambda(y - y_0 - m_1(x - x_0))(y - y_0 - m_2(x - x_0)) = 0,$$

for some rational number $\lambda \in \mathbb{Q}$. In particular, after a change of variables $X = x - x_0$ and $Y = y - y_0$, we obtain an equation C' given by

$$C' : \lambda(Y^2 + \beta XY + \gamma X^2) = 0,$$

with $\beta = -(m_1 + m_2) \in \mathbb{Q}$ and $\gamma = m_1 m_2 \in \mathbb{Q}$. Hence, m_1 and m_2 can be retrieved from β and γ by

$$m_1 = \frac{-\beta + \sqrt{\beta^2 - 4\gamma}}{2} \quad \textit{and} \quad m_2 = \frac{-\beta - \sqrt{\beta^2 - 4\gamma}}{2}.$$

(2) *Otherwise, if one of the lines is vertical, say $L : x - x_0 = 0$, then*

$$C : f(x, y) = \lambda(x - x_0)(y - y_0 - m(x - x_0)) = 0.$$

In particular, after a change of variables $X = x - x_0$ and $Y = y - y_0$, we obtain an equation C' given by

$$C' : \lambda X(Y - mX) = 0,$$

where m is the slope of $L' : y - y_0 = m(x - x_0)$.

Proof. Let $C : f(x, y) = 0$ be a quadratic equation which is a product of two lines L and L' that intersect at a point $P = (x_0, y_0) \in C$. In particular, P belongs to both L and L'.

First assume that neither L nor L' is a vertical line. Then, there are finite slopes m_1 and m_2 such that L and L' are given, respectively, by $y - y_0 = m_i(x - x_0)$ for $i = 1, 2$. Therefore,

$$\begin{aligned} f(x, y) &= \lambda(y - y_0 - m_1(x - x_0))(y - y_0 - m_2(x - x_0)) \\ &= \lambda((y - y_0)^2 - (m_1 + m_2)(x - x_0)(y - y_0) + m_1 m_2 (x - x_0)^2) \end{aligned}$$

for some rational number $\lambda \in \mathbb{Q}$. Since $f(x,y)$ has integer coefficients, $\lambda \in \mathbb{Q}$, and $-\lambda(m_1+m_2)$ and $\lambda m_1 m_2$ are, respectively, the coefficients of xy and x^2, it follows that m_1+m_2 and m_1m_2 are rational numbers. We will use this below to show that x_0 and y_0 are also rational.

Now we may calculate the normal vector using the product rule of differentiation:

$$\begin{aligned}\vec{n}_C &= \lambda\cdot(-m_1(y-y_0-m_2(x-x_0))-m_2(y-y_0-m_1(x-x_0)),\\ &\qquad (y-y_0-m_2(x-x_0))+(y-y_0-m_1(x-x_0)))\\ &= \lambda\cdot(-(m_1+m_2)(y-y_0)+2m_1m_2(x-x_0),\\ &\qquad 2(y-y_0)-(m_1+m_2)(x-x_0)),\end{aligned}$$

and, therefore, $\vec{n}_C(P)=\vec{n}_C((x_0,y_0))=(0,0)$. Moreover, notice that if

$$C : ax^2+bxy+cy^2+dx+ey+f=0,$$

where a, b, c, d, e, f are integers, then $\vec{n}_C=(2ax+by+d, bx+2cy+e)$. Thus, (x_0,y_0), which is the unique intersection point of the lines L and L', is also the unique solution of the system

$$\begin{cases} 2ax+by=-d,\\ bx+2cy=-e,\end{cases}$$

which is linear and has integral coefficients, so the solution must have rational coefficients by Cramer's rule; i.e.,

$$(x_0,y_0)=\left(\frac{\det\begin{pmatrix}-d & b\\ -e & 2c\end{pmatrix}}{\det\begin{pmatrix}2a & b\\ b & 2c\end{pmatrix}}, \frac{\det\begin{pmatrix}2a & -d\\ b & -e\end{pmatrix}}{\det\begin{pmatrix}2a & b\\ b & 2c\end{pmatrix}}\right)=\left(\frac{-2cd+be}{4ac-b^2},\frac{-2ae+bd}{4ac-b^2}\right).$$

For part (2), notice that if the two distinct lines L and L' intersect at a point P, then they are not both vertical. Suppose one of them is vertical, say $L : x-x_0=0$. Then,

$$C : f(x,y)=\lambda(x-x_0)(y-y_0-m(x-x_0))=0.$$

Once again, we can calculate the normal vector of C:

$$\begin{aligned}\vec{n}_C &= \lambda\cdot((y-y_0-m(x-x_0))-m(x-x_0), x-x_0)\\ &= \lambda\cdot((y-y_0)-2m(x-x_0), x-x_0),\end{aligned}$$

and, therefore, $\vec{n}_C(P)=\vec{n}_C((x_0,y_0))=(0,0)$. The same argument as before shows that x_0 and y_0 are rational. Also, $-\lambda m$ is the coefficient of x^2 in $f(x,y)$, so m is also rational. The rest of the proof is straightforward. □

The previous proposition leaves only one case that remains to be treated: when C is a product of two parallel lines L and L'.

Proposition 9.1.9. *Let $C : f(x,y) = 0$ be a quadratic equation (with $f(x,y) \in \mathbb{Z}[x,y]$) that is a product of two parallel lines L and L' (i.e., L and L' have a common slope m).*

(1) *If both L and L' are vertical, then there are constants x_0, $x_1 \in \mathbb{Q}$ such that*
$$C : \lambda(x - x_0)(x - x_1) = 0,$$
for some $\lambda \in \mathbb{Q}$.

(2) *If L and L' have a common slope $m \in \mathbb{C}$, then*
$$-m \cdot \frac{\partial f}{\partial y} = \frac{\partial f}{\partial x},$$
the slope m is rational, and
$$C : \lambda(y - mx - b_1)(y - mx - b_2) = 0,$$
for some $\lambda \in \mathbb{Q}$ and $b_1, b_2 \in \mathbb{C}$. In particular, after a change of variables $X = y - mx$, we obtain an equation C' given by
$$C' : \lambda(X^2 + \beta X + \gamma) = 0,$$
with $\beta = -(b_1 + b_2) \in \mathbb{Q}$ and $\gamma = b_1 b_2 \in \mathbb{Q}$. Hence, b_1 and b_2 can be retrieved from β and γ by
$$b_1 = \frac{-\beta + \sqrt{\beta^2 - 4\gamma}}{2} \quad \text{and} \quad b_2 = \frac{-\beta - \sqrt{\beta^2 - 4\gamma}}{2}.$$

Proof. For part (1), if L and L' are vertical lines, then there exist $a, a', b, b' \in \mathbb{Q}$ such that $L : ax - b = 0$ and $L' : a'x - b' = 0$, and so C is given by $(ax - b)(a'x - b') = 0$. Notice that a and a' are non-zero. Thus, equivalently, C is given by
$$aa'\left(x - \frac{b}{a}\right)\left(x - \frac{b'}{a'}\right) = 0.$$
Thus, the result holds for $\lambda = aa'$, $x_0 = b/a$, and $x_1 = b'/a'$, which are rational numbers.

Let us assume that L and L' have a common slope $m \in \mathbb{C}$. Then, $L : y = mx + b_1$ and $L' : y = mx + b_2$, for some $b_1, b_2 \in \mathbb{C}$. Thus,
$$C : f(x,y) = \lambda(y - mx - b_1)(y - mx - b_2) = 0,$$
for some $\lambda \in \mathbb{C}$, which is the coefficient of y^2 in $f(x,y)$, so λ is in $\mathbb{Q}$. Also, the coefficient of xy is $-2\lambda m$; hence m is also rational. Similarly, $b_1 + b_2$ and $b_1 b_2$ are rational by considering, respectively, the coefficient of y and the constant term in $f(x,y)$.

In particular,
$$\frac{\partial f}{\partial x} = -m(2y - 2mx - b_1 - b_2),$$
$$\frac{\partial f}{\partial y} = 2y - 2mx - b_1 - b_2.$$
Hence, $-m \cdot \partial f/\partial y = \partial f/\partial x$, as claimed. The rest of the proof is clear. $\square$

Example 9.1.10. Let C be the quadratic equation given by

$$C : 14x^2 + 19xy - 33x - 3y^2 + 8y - 5 = 0.$$

Let us investigate whether C is a product of two lines. In order to use Propositions 9.1.8 or 9.1.9, we need to calculate the normal vector of C:

$$\vec{n}_C = (\partial f/\partial x, \partial f/\partial y) = (28x + 19y - 33, 19x - 6y + 8).$$

Since $\partial f/\partial x$ is not a constant multiple of $\partial f/\partial y$, if C is a product of two lines, then there must be a point of intersection P where $\vec{n}_C(P) = (0, 0)$. Thus, we need to solve the system

$$\begin{cases} 28x + 19y = 33, \\ 19x - 6y = -8. \end{cases}$$

This system has a unique solution; namely $P = (x_0, y_0) = (\frac{2}{23}, \frac{37}{23})$. Hence, we choose a change of variables $X = x - 2/23$ and $Y = y - 37/23$ and obtain

$$C' : 14X^2 + 19XY - 3Y^2 = 0,$$

or, equivalently, $C' : X^2 + 19/14XY - 3/14Y^2 = 0$. Hence, $\beta = -19/14$ and $\gamma = -3/14$ (as in Proposition 9.1.8) and the slopes m_1 and m_2 of L and L' are

$$m_1 = \frac{-\beta + \sqrt{\beta^2 - 4\gamma}}{2} = 7 \quad \text{and} \quad m_2 = \frac{-\beta - \sqrt{\beta^2 - 4\gamma}}{2} = -\frac{2}{3}.$$

It follows that $f(x, y) = 0$ is a multiple of

$$\left(y - \frac{37}{23} - 7\left(x - \frac{2}{23}\right)\right)\left(y - \frac{37}{23} + \frac{2}{3}\left(x - \frac{2}{23}\right)\right) = 0.$$

Indeed, if we multiply the latter equation by -3 and simplify inside the parentheses, we obtain

$$C : f(x, y) = (7x - y + 1)(2x + 3y - 5) = 0.$$

Hence, C is the product of the lines $L : 7x - y = -1$ and $L' : 2x + 3y = 5$. It follows that the rational points on C are the points in the set

$$C(\mathbb{Q}) = L(\mathbb{Q}) \cup L'(\mathbb{Q}) = \{(t, 7t + 1) : t \in \mathbb{Q}\} \cup \left\{\left(s, \frac{-2s + 5}{3}\right) : s \in \mathbb{Q}\right\},$$

while the integral points can be found using Theorem 2.9.4:

$$C(\mathbb{Z}) = L(\mathbb{Z}) \cup L'(\mathbb{Z}) = \{(1 + k, 8 + 7k) : k \in \mathbb{Z}\} \cup \{(1 + 3h, 1 - 2h) : h \in \mathbb{Z}\}.$$

Example 9.1.11. Let $C : f(x, y) = 0$ be the quadratic equation given by

$$C : f(x, y) = 12x^2 + 36xy - 32x + 27y^2 - 48y + 5 = 0.$$

Let us investigate whether C is a product of two lines. In order to use Propositions 9.1.8 or 9.1.9, we need to calculate the normal vector of C:

$$\vec{n}_C = (\partial f/\partial x, \partial f/\partial y) = (24x + 36y - 32, 36x + 54y - 48).$$

Notice that $\frac{2}{3} \cdot \partial f/\partial y = \partial f/\partial x$. By Proposition 9.1.9, this may be evidence that C is a product of two parallel lines L and L' of slope $m = -\frac{2}{3}$. Hence, we attempt a change of variables $X = y - mx$, i.e., $y = -\frac{2}{3}x + X$, and we obtain

$$C' : 27X^2 - 48X + 5 = 0,$$

or, equivalently, $C' : X^2 - \frac{48}{27}X + \frac{5}{27} = 0$. If we set $\beta = -48/27$ and $\gamma = 5/27$, then

$$b_1 = \frac{-\beta + \sqrt{\beta^2 - 4\gamma}}{2} = \frac{5}{3} \quad \text{and} \quad b_2 = \frac{-\beta - \sqrt{\beta^2 - 4\gamma}}{2} = \frac{1}{9}.$$

Hence C is a product of parallel lines $L : y = -\frac{2}{3}x + \frac{5}{3}$ and $L' : y = -\frac{2}{3}x - \frac{1}{9}$. Thus, $f(x, y) = 0$ is a constant multiple of

$$\left(y + \frac{2}{3}x - \frac{5}{3}\right)\left(y + \frac{2}{3}x - \frac{1}{9}\right) = 0.$$

And, indeed, if we multiply our previous equation by 27 and simplify, we obtain

$$C : f(x, y) = (2x + 3y - 5)(6x + 9y - 1) = 0.$$

Hence, C is the product of the lines $L : 2x + 3y = 5$ and $L' : 6x + 9y = 1$. It follows that the rational points on C are the set

$$C(\mathbb{Q}) = L(\mathbb{Q}) \cup L'(\mathbb{Q}) = \left\{\left(t, \frac{-2t+5}{3}\right) : t \in \mathbb{Q}\right\} \cup \left\{\left(s, \frac{-6s+1}{9}\right) : s \in \mathbb{Q}\right\},$$

while the integral points can be found using Theorem 2.9.4:

$$C(\mathbb{Z}) = L(\mathbb{Z}) \cup L'(\mathbb{Z}) = \{(1 + 3k, 1 - 2k) : k \in \mathbb{Z}\} \cup \emptyset,$$

where we have used Proposition 2.9.1 to show that $L'(\mathbb{Z}) = \emptyset$, because $\gcd(6, 9) = 3$ is not a divisor of 1.

Example 9.1.12. Let us determine all the rational and integral points that satisfy the quadratic equation given by

$$C : f(x, y) = 4x^2 - 12xy + 10x + 9y^2 - 15y + 5 = 0.$$

First, let us investigate if this equation is a product of two lines. We calculate

$$\vec{n}_C = (\partial f/\partial x, \partial f/\partial y) = (8x - 12y + 10, -12x + 18y - 15).$$

Notice that $-\frac{2}{3} \cdot \partial f/\partial y = \partial f/\partial x$. By Proposition 9.1.9, this may be evidence that C is a product of two parallel lines L and L' of slope $m = \frac{2}{3}$. Hence, we attempt a change of variables $X = y - mx$, i.e., $y = \frac{2}{3}x + X$, and we obtain

$$C' : 9X^2 - 15X + 5 = 0,$$

or, equivalently, $C' : X^2 - \frac{5}{3}X + \frac{5}{9} = 0$. If we set $\beta = -5/3$ and $\gamma = 5/9$, then

$$b_1 = \frac{-\beta + \sqrt{\beta^2 - 4\gamma}}{2} = \frac{5 + \sqrt{5}}{6} \quad \text{and} \quad b_2 = \frac{-\beta - \sqrt{\beta^2 - 4\gamma}}{2} = \frac{5 - \sqrt{5}}{6}.$$

Hence C is a product of parallel lines $L : y = \frac{2}{3}x + \frac{5+\sqrt{5}}{6}$ and $L' : y = \frac{2}{3}x + \frac{5-\sqrt{5}}{6}$. Thus, $f(x, y) = 0$ is a constant multiple of

$$\left(y - \frac{2}{3}x - \frac{5+\sqrt{5}}{6}\right)\left(y - \frac{2}{3}x - \frac{5-\sqrt{5}}{6}\right) = 0.$$

And, indeed, if we multiply our previous equation by 9 and simplify, we obtain

$$C : f(x, y) = \left(2x - 3y + \frac{5+\sqrt{5}}{2}\right)\left(2x - 3y + \frac{5-\sqrt{5}}{2}\right) = 0.$$

It follows that if $P = (x_0, y_0) \in C(\mathbb{Q}) = L(\mathbb{Q}) \cup L'(\mathbb{Q})$, then $P \in L(\mathbb{Q})$ or $L'(\mathbb{Q})$. However, if $x_0, y_0 \in \mathbb{Q}$, then $2x_0 - 3y_0 \in \mathbb{Q}$, and

$$2x_0 - 3y_0 + \frac{5+\sqrt{5}}{2} = 0 \quad \text{or} \quad 2x_0 - 3y_0 + \frac{5-\sqrt{5}}{2} = 0$$

would imply that either $\frac{5+\sqrt{5}}{2}$ or $\frac{5-\sqrt{5}}{2}$ is in $\mathbb{Q}$, which in turn implies that $\sqrt{5} \in \mathbb{Q}$, a contradiction (see Theorem 2.10.11 and Exercise 2.11.34). Hence, $C(\mathbb{Q}) = \emptyset$ in this case (notice, however, that there are infinitely many points over the reals).

We finish this section with a result that says that if a quadratic equation vanishes at all points on one line, then it is necessarily the product of two lines.

Proposition 9.1.13. *Let $C : f(x, y) = 0$ be a quadratic equation, and suppose that there is a line L such that $(\alpha, \beta) \in C(\mathbb{Q})$ for all $(\alpha, \beta) \in L$. Then, C is a product of two lines, L and another line L'.*

Proof. Let us first assume that L is not vertical, and let $L : l(x, y) = 0$ be a line given by $l(x, y) = y - y_0 - m(x - x_0)$ for some slope m and some point $P = (x_0, y_0) \in L$, and assume that $f(x, y)$ vanishes at every point on L. In other words, $f(x, m(x - x_0) + y_0) = 0$ for all x. Then, we claim that there is a linear polynomial $g(x, y)$ such that

$$f(x, y) = (y - y_0 - m(x - x_0)) \cdot g(x, y).$$

Indeed, let us write a Taylor expansion for $f(x, y)$, centered at $P = (x_0, y_0)$:

$$f(x, y) = A + B(x - x_0) + C(y - y_0) + D(x - x_0)^2 + E(x - x_0)(y - y_0) + F(y - y_0)^2.$$

Then,

$$0 = f(x, m(x - x_0) + y_0) = A + (B + Cm)(x - x_0) + (D + Em + Fm^2)(x - x_0)^2.$$

If this polynomial is identically zero, then each coefficient is zero; i.e.,

$$A = 0, \ B + Cm = 0, \ \text{and } D + Em + Fm^2 = 0.$$

Hence, if we let $l(x, y) = y - y_0 - m(x - x_0)$ as above, then we can use the fact that $A = 0$, $B = -Cm$, and $D = -Em - Fm^2$ to write $f(x, y)$ as follows:

$$\begin{aligned} f &= C \cdot l(x, y) + E(x - x_0)l(x, y) + F((y - y_0)^2 - m^2(x - x_0)^2) \\ &= (y - y_0 - m(x - x_0))(C + E(x - x_0) + F(y - y_0 + m(x - x_0))). \end{aligned}$$

Thus, if we write $g(x, y) = C + E(x - x_0) + F(y - y_0 + m(x - x_0))$, we have

$$f(x, y) = l(x, y)g(x, y),$$

as claimed. In particular, it follows from our claim that if $f(x, m(x - x_0) + y_0)$ is identically zero as a polynomial, then C is a product of two lines, $L : l(x, y) = 0$ and a second line $L' : g(x, y) = 0$, as desired.

If L is vertical, then $L : x = x_0$ for some $x_0 \in \mathbb{Q}$. If $f(x_0, y) = 0$ for all values of y, then we can proceed similarly to show that $f(x, y) = (x - x_0)g(x, y)$ for some linear polynomial $g(x, y)$, and so C is a product of L and $L' : g(x, y) = 0$. □

9.2. A Classification: Parabolas, Ellipses, and Hyperbolas

In this section we classify those quadratic equations that are *not* a product of two lines. Such a quadratic equation defines a *conic section*. As the term indicates, these curves are the result of intersecting a cone (a right conical surface $X^2 + Y^2 = Z^2$) with a plane at different angles (see Figure 9.5). Each quadratic equation can be classified as an ellipse, a hyperbola, or a parabola. In order to classify a quadratic equation, we simply "complete the squares" in the formula.

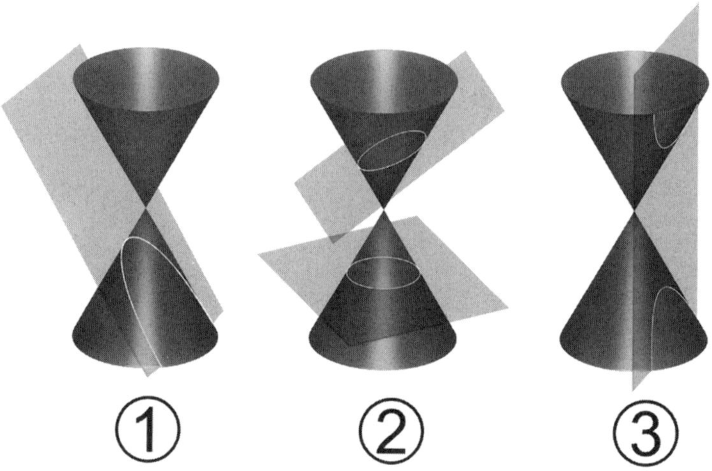

Figure 9.5. Conic sections arising from different intersections of a right conical surface and a plane: (1) parabolas, (2) circles and ellipses, and (3) hyperbolas. Image source: Pbroks13 (author), Wikimedia Commons, used under Creative Commons Attribution 3.0 Unported License.

Theorem 9.2.1. *Let $C : f(x, y) = 0$ be a quadratic equation, given by*

$$f(x, y) = ax^2 + bxy + cy^2 + dx + ey + f = 0,$$

where a, b, c, d, e, f are integers and a, b, or c is non-zero. Then, there is an invertible (linear) change of variables $X = \alpha x + \beta y + \kappa$ and $Y = \gamma x + \delta y + \rho$, with integer coefficients, such that f in terms of the new variables X and Y is of the reduced form

$$C' : \text{(i) } X^2 + BY^2 = D \text{ or (ii) } X^2 = Y \text{ or (iii) } X^2 = E,$$

for some integers B, D, and E, where B is non-zero.

Proof. Let C be a quadratic equation given by

$$f(x, y) = ax^2 + bxy + cy^2 + dx + ey + f,$$

such that a, b, and c are not all zero. Let us first assume that a or c are not zero. Without loss of generality, let us assume that a is non-zero, and consider $4af(x, y)$.

Then,

$$\begin{aligned}4af(x,y) &= 4a^2x^2 + 4abxy + 4acy^2 + 4adx + 4aey + 4af\\ &= (4a^2x^2 + 4abxy + 4adx) + 4acy^2 + 4aey + 4af\\ &= (4a^2x^2 + 4a(by+d)x) + 4acy^2 + 4aey + 4af\\ &= ((2ax + (by+d))^2 - (by+d)^2) + 4acy^2 + 4aey + 4af\\ &= (2ax + (by+d))^2 + c'y^2 + e'y + f',\end{aligned}$$

where

$$c' = 4ac - b^2,\ e' = 4ae - 2bd,\ f' = 4af - d^2.$$

Now we consider several different cases:

- Suppose $c' = e' = 0$. Then, a change of variables $X = 2ax + by + d$ and $Y = y$ leads to
$$C' : X^2 = E,$$
with $E = -f'$. The inverse change of variables is $y = Y$ and $x = \frac{1}{2a}(X - bY - d)$.
- If $c' = 0$ but $e' \neq 0$, then
$$f(x,y) = (2ax + (by+d))^2 + e'y + f' = 0.$$
Therefore a change of variables $X = 2ax + by + d$ and $Y = -(e'y + f')$ leads to
$$C' : X^2 = Y.$$
The inverse change of variables is given by $y = -\frac{1}{e'}(Y + f')$ and $x = \frac{1}{2a}(X - by - d) = \frac{1}{2a}(X + b(\frac{1}{e'}(Y + f')) - d)$.
- If $c' \neq 0$, then we proceed as before and we multiply through by $4c'$:
$$\begin{aligned}4c'(4af(x,y)) &= 4c'((2ax + (by+d))^2 + c'y^2 + e'y + f')\\ &= 4c'(2ax + (by+d))^2 + 4c'^2y^2 + 4c'e'y + 4c'f'\\ &= 4c'(2ax + (by+d))^2 + (2c'y + e')^2 - e'^2 + 4c'f'.\end{aligned}$$
Thus, a change of variables $Y = 2ax + (by + d)$ and $X = 2c'y + e'$ and putting $B = 4c'$ and $D = e'^2 - 4c'f'$ leads to
$$C' : X^2 + BY^2 = D.$$
The inverse change of variables is given by $y = \frac{1}{2c'}(X - e')$ and $x = \frac{1}{2a}(Y - b(\frac{1}{2c'}(X - e')) - d)$.

Finally, suppose that $a = c = 0$ and, therefore, $b \neq 0$. Since $xy = \frac{1}{4}((x+y)^2 - (x-y)^2)$, we have

$$\begin{aligned}f(x,y) &= bxy + dx + ey + f\\ &= \frac{b}{4}(x+y)^2 - \frac{b}{4}(x-y)^2 + dx + ey + f.\end{aligned}$$

Hence, a change of variables $V = x+y$ and $W = x-y$ (with inverse $x = (V+W)/2$ and $y = (V-W)/2$) leads to

$$\begin{aligned}C_0 : bV^2 - bW^2 + 2d(V+W) + 2e\,(V-W) + 4f\\ = bV^2 - bW^2 + 2(d+e)V + 2(d-e)W + 4f = 0.\end{aligned}$$

Since $b \neq 0$ and by the first part of this proof, there is a further change of variables that takes C_0 to a quadratic equation of the form $C' : X^2 + BY^2 = D$ or $X^2 = Y$ or $X^2 = E$. □

Example 9.2.2. Let $C : f(x,y) = 0$ be the quadratic equation given by

$$C : 3x^2 + 5xy + 7y^2 + x + y - 20 = 0.$$

Let us find a reduced form for C, following the method in the proof of Theorem 9.2.1. Let us first consider $4 \cdot 3 \cdot f(x,y)$:

$$\begin{aligned}12f(x,y) &= 36x^2 + 60xy + 12x + 84y^2 + 12y - 240\\ &= (36x^2 + (60y+12)x) + 84y^2 + 12y - 240\\ &= (6x+5y+1)^2 - (5y+1)^2 + 84y^2 + 12y - 240\\ &= (6x+5y+1)^2 + 59y^2 + 2y - 241.\end{aligned}$$

Next, we multiply through by $4 \cdot 59$:

$$\begin{aligned}2832f(x,y) &= 236(6x+5y+1)^2 + 4 \cdot 59^2y^2 + 472y - 241\\ &= 236(6x+5y+1)^2 + (118y+2)^2 - 56880.\end{aligned}$$

Therefore, the change of variables $X = 118y + 2$ and $Y = 6x + 5y + 1$ shows that C can be reduced to $C' : X^2 + 236Y^2 = 56880$.

Remark 9.2.3. The proof of Theorem 9.2.1 outlines a method to find a reduced form (and a change of variables of a special form that will be exploited later to find integral points; see Example 9.4.8) that works on all quadratic equations. However, one can simplify the equations in particular cases. For instance, if a is a square, say $a = n^2$, and n divides b and d, then there is no need to multiply $f(x,y)$ through by $4a$, and one can simply multiply by 4. If in addition, b and d are divisible by $2n$, then one does not need to multiply through by 4 either, and we can complete the square in x directly in the equation $f(x,y) = 0$. Similar considerations apply when completing the square in the y variable. We illustrate this in the following example (see also Example 9.2.8 below).

Example 9.2.4. Let us find a reduced form for the following equation:

$$C : f(x,y) = 4x^2 - 12xy + 10x + 9y^2 - 15y + 5 = 0.$$

We complete the squares in x and y, as outlined in the proof of Theorem 9.2.1. Since the coefficient of x^2 is $a = 4$, already a square, we do not need to multiply through by $4a$ and multiplying only by 4 is sufficient:

$$\begin{aligned}4f(x,y) &= 16x^2 - 48xy + 40x + 36y^2 - 60y + 20\\ &= (16x^2 + (40 - 48y)x) + 36y^2 - 60y + 20\\ &= (4x + 5 - 6y)^2 - (5 - 6y)^2 + 36y^2 - 60y + 20\\ &= (4x + 5 - 6y)^2 - 5.\end{aligned}$$

Hence, a linear change of variables $X = 4x+5-6y$, $Y = y$, reduces C to $C' : X^2 = 5$. Thus, C is a product of two lines; namely $(X - \sqrt{5})(X + \sqrt{5}) = 0$ which, in terms of the original variables x and y, reads

$$C : 4f(x, y) = \left(4x - 6y + 5 - \sqrt{5}\right)\left(4x - 6y + 5 + \sqrt{5}\right) = 0,$$

which is equivalent to the factorization we obtained in Example 9.1.12 using different methods.

Exercise 9.2.5. Let us find a reduced form for the equation $C : xy = 1$. In this case, since there is no x^2 or y^2 term, we use the equality $xy = \frac{1}{4}((x+y)^2 - (x-y)^2)$ to write

$$1 = xy = \frac{1}{4}((x+y)^2 - (x-y)^2) = \left(\frac{x+y}{2}\right)^2 - \left(\frac{x-y}{2}\right)^2.$$

Thus, if we change variables $X = (x + y)/2$ and $Y = (x - y)/2$, we find

$$1 = xy = X^2 - Y^2.$$

That is, a reduced form is given by $X^2 - Y^2 = 1$.

In Theorem 9.2.1 we have shown that a quadratic equation has a *reduced form*. In the next result we show that the reduced form is unique, which will allow us to classify quadratic equations as parabolas, ellipses, or hyperbolas.

Theorem 9.2.6. *With notation as in Theorem* 9.2.1*:*

(1) *A quadratic equation C is uniquely represented (via a linear change of variables) by a reduced form C' of type*

$$\text{(i) } X^2 + BY^2 = D \text{ or (ii) } X^2 = Y \text{ or (iii) } X^2 = E,$$

for some integers B, D, and E, where B is non-zero.

(2) *If C is represented by a reduced form of type* (i)*, i.e., $C' : X^2 + BY^2 = D$, then in every representation of C the sign of B is the same.*

Proof. For part (1), suppose that a quadratic equation C can be represented in two of the forms (i), (ii), and (iii). Then, there are invertible linear changes of variables (linear in X and Y) such that C changes to each of the forms. Thus, there is also an invertible linear change between two forms of the shape (i), (ii), and (iii).

First, suppose that there is a change of variables from an equation of the form $X'^2 = Y'$ to one of the form $X^2+BY^2 = D$, where $B \neq 0$. Then $X' = \alpha X+\beta Y+\kappa$ and $Y' = \gamma X + \delta Y + \rho$, and

$$\begin{aligned} X'^2 - Y' &= (\alpha X + \beta Y + \kappa)^2 - (\gamma X + \delta Y + \rho) \\ &= \alpha^2 X^2 + 2\alpha\beta XY + \beta^2 Y^2 + \cdots . \end{aligned}$$

But if this is of the form $X^2 + BY^2 = D$, then $2\alpha\beta = 0$, and therefore $\beta = 0$. Hence the change of variables would also eliminate the Y^2 coefficient and we reach a contradiction. Similarly, if this change of variables leads to an equation of the form $X^2 = E$, then we must have $\beta = 0$ and also $\gamma = 0$ to eliminate the Y term, but this leaves a change of variables $X' = \alpha X + \kappa$ and $Y' = \gamma X + \rho$ which is not invertible (the point (X_0, Y) maps to the same value (X_0', Y_0') *for any* value of Y).

It remains to see that there is no change of variables that takes an equation of the form $X'^2 = E$ to one of the form $X^2 + BY^2 = D$. As before, suppose the change of variables is given by $X' = \alpha X + \beta Y + \kappa$ and $Y' = \gamma X + \delta Y + \rho$. Then,

$$\begin{aligned} X'^2 - E &= (\alpha X + \beta Y + \kappa)^2 - E \\ &= \alpha^2 X^2 + 2\alpha\beta XY + \beta^2 Y^2 + \cdots \end{aligned}$$

and therefore $\alpha\beta = 0$, so $\alpha = 0$ or $\beta = 0$, which leads to either an equation in X or an equation in Y, but never to an equation of the form $X^2 + BY^2 = D$.

For part (2), suppose that C can be represented by $C' : X^2 + BY^2 = D$ and by $C'' : X'^2 + B'Y'^2 = D'$, where B and B' are non-zero and have different signs. Then, one of B or B' is positive, and we assume $B' > 0$ without loss of generality (and so $B < 0$). Moreover, there is a linear change of variables from C' to C'', given by $X' = \alpha X + \beta Y + \kappa$ and $Y' = \gamma X + \delta Y + \rho$. Thus,

$$(\alpha X + \beta Y + \kappa)^2 + B'(\gamma X + \delta Y + \rho)^2 - D' = X^2 + BY^2 - D.$$

Comparing the coefficients of X^2, we see that $\alpha^2 + B'\gamma^2 = 1$. Since $B' > 0$, we conclude that $\alpha = \pm 1$ and $\gamma = 0$. Comparing coefficients of XY, we see that $2\alpha\beta = 0$, and it follows that $\beta = 0$. Hence, $B = B'\delta^2 > 0$, which contradicts our assumption $B < 0$. Thus, C' and C'' cannot be equivalent via a linear change of variables. □

Definition 9.2.7. Let $C : f(x, y) = 0$ be a quadratic equation, and let C' be its reduced form (as in Theorem 9.2.1). Then:

- If $C' : X^2 + BY^2 = 0$ or if $C' : X^2 = E$, then C is a *product of two lines*.
- If $C' : X^2 + BY^2 = D$, with $B > 0$ and $D \neq 0$, then C is an *ellipse*.
- If $C' : X^2 + BY^2 = D$, with $B < 0$ and $D \neq 0$, then C is a *hyperbola*.
- If $C' : X^2 = Y$, then C is a *parabola*.

A quadratic equation that is not a product of two lines, i.e., a parabola, an ellipse, or a hyperbola, is called a *conic*, or *conic section*.

Example 9.2.8. Consider the quadratic equation $C : f(x, y) = 0$ given by

$$C : 9x^2 - 42xy + 30x + 49y^2 - 72y + 14 = 0.$$

Since the coefficient of x^2 is already a square $a = n^2 = 9$ and since the coefficients of xy and x are multiples of $2n = 6$, we can complete squares without multiplying through by any factor:

$$\begin{aligned} f(x, y) &= 9x^2 - 42xy + 30x + 49y^2 - 72y + 14 \\ &= (9x^2 + (30 - 42y)x) + 49y^2 - 72y + 14 \\ &= (3x + (5 - 7y))^2 - (5 - 7y)^2 + 49y^2 - 72y + 14 \\ &= (3x - 7y + 5)^2 - (2y + 11). \end{aligned}$$

Hence, $C : f(x, y) = 0$ can be reduced to $C' : X^2 = Y$ via a change of variables $X = 3x - 7y + 5$ and $Y = 2y + 11$. Therefore, C is a parabola.

Example 9.2.9. Let us consider the quadratic equation given by

$$C : f(x, y) = 7x^2 - 46xy + 42x + 22y^2 + 12y - 49 = 0.$$

Let us find a reduced form for C by completing the squares in x and y. First, we need to multiply through by $4 \cdot 7 = 28$:

$$\begin{aligned} 28f(x, y) &= 196x^2 - 1288xy + 1176x + 616y^2 + 336y - 1372 \\ &= (14^2x^2 + (1176 - 1288y)x) + 616y^2 + 336y - 1372 \\ &= (14x - 46y + 42)^2 - (42 - 46y)^2 + 616y^2 + 336y - 1372 \\ &= (14x - 46y + 42)^2 - 1500y^2 + 4200y - 3136. \end{aligned}$$

Next, we multiply through by 15 to make the coefficient of y^2 a perfect square:

$$\begin{aligned} 420f(x, y) &= 15(14x - 46y + 42)^2 - 22500y^2 + 63000y - 47040 \\ &= 15(14x - 46y + 42)^2 - (150y - 210)^2 + 210^2 - 47040 \\ &= 15(14x - 46y + 42)^2 - (150y - 210)^2 - 2940. \end{aligned}$$

Hence, $C : f(x, y) = 0$ can be reduced to $C' : X^2 - 15Y^2 = -2940$ via a change of variables $X = 150y - 210$ and $Y = 14x - 46y + 42$. Therefore, C is a hyperbola.

Note that we could have also simplified the equation, to obtain

$$21f(x, y) = 3(7x - 23y + 21)^2 - 5(15y - 21)^2 - 147.$$

In other words, C is also equivalent to $C'' : 3X'^2 - 5Y'^2 = 147$, via $X' = 7x - 23y + 21$ and $Y' = 15y - 21$.

Example 9.2.10. Let C be the quadratic equation given by $xy + x + y = 10$. Let us find a reduced form for C. Since the coefficients of x^2 and y^2 are zero, we first use a change of variables $\psi : C \to C_0$, from C to the curve C_0, given by

$$C_0 : V^2 - W^2 + 2V - 10 = 0$$

and $\psi(x, y) = (x + y, x - y)$. Now we can complete squares to find a reduced form for C_0:

$$V^2 - W^2 + 2V - 10 = (V + 1)^2 - Y^2 - 11.$$

Thus, if we put $C' : X^2 - Y^2 = 11$, then we have a map $\varphi : C_0 \to C'$ with $\varphi(V, W) = (V + 1, W)$. Hence, the reduced form of C is C', and so C is a hyperbola. The change of variables from C to C' is given by $\varphi \circ \psi : C \to C'$, which in coordinates is given by

$$\varphi \circ \psi(x, y) = \varphi(\psi(x, y)) = \varphi(x + y, x - y) = (x + y + 1, x - y).$$

The inverse map is given by

$$(\varphi \circ \psi)^{-1}(X, Y) = \left(\frac{X + Y}{2} - 1, \frac{X - Y}{2} - 1\right).$$

Remark 9.2.11. In Part 3 of this book we will introduce the projective plane (Section 15.1.2) and the projectivization of a curve (Section 15.1.4), and we will see that a fixed conic C can appear to be an ellipse, hyperbola, or parabola, depending on what affine chart we choose to represent C with. In other words, there *are* projective changes of variables that bring a parabola to look like a hyperbola, or a hyperbola to look like an ellipse. We refer the reader to Section 15.1.4 for more details, in particular Examples 15.1.4 and 15.1.5.

We finish this section with a summary of the bijections that we found in the proof of Theorem 9.2.1, which will be used in later proofs.

Corollary 9.2.12. *Let $C : f(x, y) = 0$ be a quadratic equation, given by*

$$f(x, y) = ax^2 + bxy + cy^2 + dx + ey + f = 0,$$

where a, b, c, d, e, f are integers and a, b, or c is non-zero.

(1) *Suppose that a is non-zero (the case of $c \neq 0$ is similar, by switching the roles of a and x, by c and y). Let $c' = 4ac - b^2$, $e' = 4ae - 2bd$, and $f' = 4af - d^2$.*
   (a) *Suppose $c' = e' = 0$. Then, C is a product of two lines, and there is a bijection $\varphi : C \to C'$, with $C' : X^2 = -f'$, of the form*

$$\varphi(x, y) = (2ax + by + d, y) \quad \text{and} \quad \varphi^{-1}(X, Y) = \left(\frac{1}{2a}(X - bY - d), Y\right).$$

   (b) *If $c' = 0$, but $e' = 4ae - 2bd \neq 0$, then C is a parabola, and there is a bijection $\varphi : C \to C'$, with $C' : X^2 = Y$, of the form*

$$\varphi(x, y) = (2ax + by + d, (4ae - 2bd)y + 4af - d^2),$$

   and the inverse change of variables is given by

$$\varphi^{-1}(X, Y) = \left(\frac{1}{2a}\left(X + \frac{b}{e'}(Y + f') - d\right), -\frac{1}{e'}(Y + f')\right).$$

   (c) *If $c' \neq 0$, then there is a bijection $\varphi : C \to C'$, with $C' : X^2 + BY^2 = D$ and $B = 4c'$ and $D = e'^2 - 4c'f'$, and a change of variables*

$$\varphi(x, y) = (2c'y + e', 2ax + (by + d)),$$

   and the inverse change of variables is given by

$$\varphi^{-1}(X, Y) = \left(\frac{1}{2a}\left(Y - \frac{b}{2c'}(X - e') - d\right), \frac{1}{2c'}(X - e')\right).$$

(2) *If $a = c = 0$, then there is a bijection $\psi : C \to C_0$ with $C_0 : a_0V^2 + b_0VW + c_0W^2 + \cdots = 0$ and $a_0 = b \neq 0$, defined by $\psi(x, y) = (x + y, x - y)$ and inverse $\psi^{-1}(V, W) = (\frac{V+W}{2}, \frac{V-W}{2})$, and a bijection $\varphi : C_0 \to C'$ of the form of part* (a), (b), *or* (c) *above. Thus, there is a bijection $\varphi \circ \psi : C \to C'$. Moreover,*

$$C_0 : bV^2 - bW^2 + 2(d + e)V + 2(d - e)W + 4f$$

so $c_0' = -4b^2$, $e_0' = 8b(d - e)$, and $f_0' = 16bf - 4(d + e)^2$.

Remark 9.2.13. As we pointed out in Remark 9.2.3, the equations we have produced in the proof of Theorem 9.2.1 and which are summarized in Corollary 9.2.12 are valid in the outmost general case, but in particular cases they can be simplified (see for instance Examples 9.2.4, and 9.2.8).

9.3. Rational Parametrizations of Conics

In this section, we determine **all** the rational points on a quadratic equation C, as long as we already know at least one rational point. As we will see in Chapter 11, there is a method (the Hasse–Minkowski theorem) to determine whether a quadratic equation has at least one rational point, so in this section we assume that we have found at least one such point on $C(\mathbb{Q})$.

In order to parametrize a conic we shall use the method known as "stereographic projection". We already saw an example of this method in the introduction, in Example 1.3.1, where we parametrized a hyperbola. In the following example, let us parametrize a circle to fix ideas, before we prove that the method works in more generality.

Example 9.3.1. Let N be a perfect square; i.e., $N = R^2$, for some $R \in \mathbb{Q}$. We would like to find all the rational points on the circle $C : x^2 + y^2 = N$, i.e., all the rational numbers $x_0, y_0 \in \mathbb{Q}$ such that $x_0^2 + y_0^2 = R^2$. The idea is to take the "north pole" of the circle, the point $P = (0, R)$, and trace a line L_m of slope m that passes through P. Since a line intersects a circle in either none or two points (perhaps equal, if the line is tangent) and since L_m already intersects the circle at P, it follows that there is a second point of intersection $Q_m \in C$. We shall see that Q_m has rational coordinates (this is a consequence of Proposition 5.5.22), and we will also show that every point Q in $C(\mathbb{Q})$, except for the south pole $(0, -R)$, is of the form $Q = Q_m$ for some $m \in \mathbb{Q}$. See Figure 9.6.

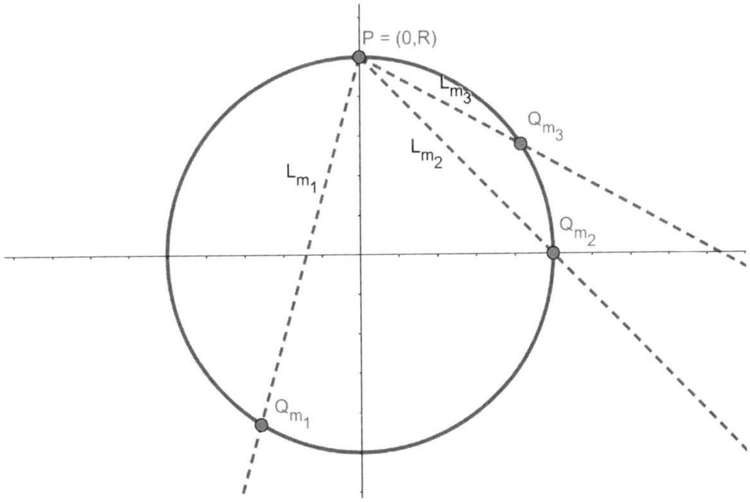

Figure 9.6. Stereographic projection of the circle of radius R. The graph shows the line L_m for three values of the slope $m = m_1$, m_2, and m_3 and their corresponding points on the circle Q_{m_1}, Q_{m_2}, and Q_{m_3}.

Let us begin with an equation for L_m. Since the slope is m and it passes through $P = (0, R)$, the equation of the line is

$$L_m : y - R = m(x - 0),$$

or $L_m : y = mx + R$. In order to find the intersection points of L_m and C, we need to solve the algebraic system:

$$\begin{cases} x^2 + y^2 = R^2, \\ y = mx + R. \end{cases}$$

Substituting the second equation of the system into the first one, we reduce the system to

$$R^2 = x^2 + (mx + R)^2 = x^2 + mx^2 + 2mRx + R^2,$$

or $x \cdot ((1 + m^2)x + 2Rm) = 0$. Therefore, either $x = 0$ or $x = -2Rm/(1 + m^2)$. Using the equation for L_m, we obtain that the points of intersection are

$$P = (0, R) \text{ and } Q_m = \left(-\frac{2Rm}{1+m^2}, \frac{R(1-m^2)}{1+m^2}\right).$$

Thus, Q_m is a new point on C and, since $R, m \in \mathbb{Q}$, the coordinates are rational numbers, so Q_m belongs to $C(\mathbb{Q})$.

Conversely, if $Q = (x_0, y_0)$ is another rational point on the circle, i.e., $Q \in C(\mathbb{Q})$, and Q is neither the north nor south pole, i.e., $Q \neq (0, \pm R)$, or, in other words, $x_0 \neq 0$, then the line L from $P = (0, R)$ to Q has a slope

$$m = \frac{y_0 - R}{x_0 - 0} = \frac{y_0 - R}{x_0},$$

which is a rational number. Hence, $L = L_m$ and $Q = Q_m$, because $L = L_m$ intersects C at exactly two points P and Q_m. Therefore, we have shown that

$$C(\mathbb{Q}) = \left\{\left(-\frac{2Rm}{1+m^2}, \frac{R(1-m^2)}{1+m^2}\right) : m \in \mathbb{Q}\right\} \cup \{(0, -R)\}.$$

We remark that $P = Q_0$, and L_0 is tangent to the circle at P; hence the intersection of L_0 and C consists of a single point, P, of multiplicity 2.

When a quadratic equation $C : f(x, y) = 0$ is a product of two lines L and L', the parametrization of rational points is simple, as $C(\mathbb{Q}) = L(\mathbb{Q}) \cup L'(\mathbb{Q})$ (see Examples 9.1.10, 9.1.11, and 9.1.12). We shall now prove that the method of projection demonstrated above (in Examples 1.3.1 and 9.3.1) works on any conic, i.e., on any quadratic equation that is not a product of two lines. First we treat the case of parabolas, which is much easier.

Proposition 9.3.2. *Let $C' : X^2 = Y$ be a parabola. Then, all the rational points on C are of the form $(X, Y) = (t, t^2)$, for some $t \in \mathbb{Q}$. If C is another quadratic equation which reduces to C' via a linear change of variables $\varphi : C \to C'$ with rational coefficients, then all the rational points on C are of the form $\varphi^{-1}((t, t^2))$ for some $t \in \mathbb{Q}$.*

Proof. Clearly, any point (t, t^2) belongs to $C'(\mathbb{Q})$, for any $t \in \mathbb{Q}$. Conversely, let $P = (x_0, y_0)$ be a rational point on the parabola $C'(\mathbb{Q})$. Then, $x_0^2 = y_0$, so y_0 is a perfect square, and $P = (x_0, x_0^2)$, so it is of the form (t, t^2) for $t = x_0 \in \mathbb{Q}$.

If C reduces to C' and $\varphi : C \to C'$ is a change of variables, then φ is a bijection, and since the coefficients are rational, then $\varphi(C(\mathbb{Q})) \subseteq C'(\mathbb{Q})$. Moreover, the coefficients of φ^{-1} are also rational, and so $\varphi^{-1}(C'(\mathbb{Q})) \subseteq C(\mathbb{Q})$. It follows that

$\varphi(C(\mathbb{Q})) = C'(\mathbb{Q})$, and therefore every rational point on C is of the form $\varphi^{-1}(P)$ for some rational point $P = (t, t^2)$ in $C'(\mathbb{Q})$. □

Example 9.3.3. Let C be the quadratic equation of Example 9.2.8, given by

$$C : 9x^2 - 42xy + 30x + 49y^2 - 72y + 14 = 0,$$

or, equivalently, $C : (3x - 7y + 5)^2 - (2y + 11) = 0$. The curve C is a parabola in reduced form, $C' : X^2 = Y$, via

$$\varphi(x, y) = (3x - 7y + 5, 2y + 11),$$

with inverse $\varphi^{-1}(X, Y) = \left(\frac{1}{3}(X + \frac{7}{2}(Y - 11) - 5), \frac{Y-11}{2}\right)$. Therefore, by Proposition 9.3.2, the rational points on $C(\mathbb{Q})$ are given by

$$C(\mathbb{Q}) = \left\{ \left(\frac{t}{3} + \frac{7t^2}{6} - \frac{29}{2}, \frac{t^2 - 11}{2} \right) : t \in \mathbb{Q} \right\}.$$

For instance, if we evaluate $t = 0$, we obtain $Q = (-29/2, -11/2)$ in $C(\mathbb{Q})$.

Theorem 9.3.4. *Let $C : f(x, y) = 0$ be a quadratic equation that is not the product of two lines, and let $P = (x_0, y_0)$ be a rational point on $C(\mathbb{Q})$. Let L_m be the line through P that has slope $m \in \mathbb{Q} \cup \{\infty\}$; i.e., $L_m : y - y_0 = m(x - x_0)$, for $m \in \mathbb{Q}$ and $L_\infty : x = x_0$. Then, L_m intersects C at exactly another point $Q_m \in C(\mathbb{Q})$ (we have $Q_m = P$ for some m), and*

$$C(\mathbb{Q}) = \{Q_m : m \in \mathbb{Q} \cup \{\infty\}\}.$$

Proof. Let $C : f(x, y) = 0$ be a quadratic equation that is not the product of two lines, and let $P = (x_0, y_0)$ be a rational point on $C(\mathbb{Q})$. Let L_m be the line through P that has slope $m \in \mathbb{Q} \cup \{\infty\}$; i.e.,

$$L_m : y - y_0 = m(x - x_0),$$

for $m \in \mathbb{Q}$ and $L_\infty : x = x_0$, when the slope is "infinite".

When $m = \infty$, the intersection of L_∞ and C is given by $p_\infty(y) = f(x_0, y)$, which is a polynomial of degree ≤ 2 in y. When $m \in \mathbb{Q}$, let $p_m(x) = f(x, m(x - x_0) + y_0)$, and note that $p_m(x) \in \mathbb{Q}[x]$. Then, $C \cap L_m = \{(x, y) \in L_m : p_m(x) = 0\}$. Suppose that $p_m(x)$, for $m \in \mathbb{Q} \cup \{\infty\}$, is identically zero as a polynomial, so that $p_m(x) = 0$ for all x. In other words, $f(x_0, x) = 0$ or $f(x, m(x - x_0) + y_0) = 0$ for all x. Then, it follows from Proposition 9.1.13 that C is a product of two lines, contradicting our initial hypothesis. Hence, $p_m(x)$ is not identically zero. Then, $p_m(x)$ is a non-zero polynomial of degree ≤ 2. Now Theorem 5.5.19 implies that $p_m(x)$ has at most two roots (counted with multiplicity). Since $P \in L_m \cap C$, the polynomial equation $p_m(x) = 0$ has either one or two distinct roots. If there is only one distinct root, we define $Q_m = P$; otherwise there is a second point $Q_m \in L_m \cap C$. Moreover, since $p_m(x) \in \mathbb{Q}[x]$ is of degree ≤ 2 and $x_0 \in \mathbb{Q}$ is a root of $p_m(x)$, it follows from Proposition 5.5.22 that a second root of $p_m(x)$ would also be defined over $\mathbb{Q}$ and this means that the x-coordinate of Q_m is rational. Since Q_m is also in L_m, and L_m is defined over $\mathbb{Q}$, it follows that the y-coordinate of Q_m is also rational. Hence, $Q_m \in L_m(\mathbb{Q}) \cap C(\mathbb{Q})$. We note here that the slope of the tangent line of C at P is rational; hence, P is of the form Q_m for some $m \in \mathbb{Q} \cup \{\infty\}$ (see Exercise 9.5.8).

Thus, for each slope $m \in \mathbb{Q} \cup \{\infty\}$, we have defined a rational point $Q_m \in C(\mathbb{Q})$. It remains to show that every rational point $Q \in C(\mathbb{Q})$ is of the form $Q = Q_m$ for some slope m. Indeed, let $Q = (x_1, y_1) \in C(\mathbb{Q})$ and let L be the line that passes through P and Q. Then, the slope of L is

$$m = \frac{y_1 - y_0}{x_1 - x_0},$$

if $x_1 \neq x_0$, or $m = \infty$ and $L : x = x_0$ if $x_1 = x_0$. In either case, the slope m is a rational number or ∞. Hence, $L = L_m$, and $Q = Q_m$ as defined above.

Therefore,

$$C(\mathbb{Q}) = \{Q_m : m \in \mathbb{Q} \cup \{\infty\}\},$$

as claimed. □

Example 9.3.5. Let $C : f(x, y) = 0$ be the quadratic equation given by

$$C : x^2 + xy + y^2 = 1,$$

and let us parametrize all the rational points on C. Luckily, it is easy to spot a point on P; namely $P = (1, 0)$. Let L_m be the line that passes through P with slope $m \in \mathbb{Q} \cup \{\infty\}$. When $m = \infty$, we have $L_\infty : x = 1$, so $L_m \cap C$ is given by those $(1, y)$ such that

$$f(1, y) = 1 + y + y^2 = 1,$$

so $y(y + 1) = 0$, and $y = 0$ or $y = -1$. Thus, the two points of intersection are $P = (1, 0)$ and $Q_\infty = (1, -1)$.

Now let $m \in \mathbb{Q}$, so that $L_m : y = m(x - 1)$. We find the intersection points of L_m and C'' by solving

$$\begin{cases} y = m(x-1), \\ x^2 + xy + y^2 = 1. \end{cases}$$

When we substitute the equation for L_m into the equation for C, we obtain

$$\begin{aligned} f(x, m(x-1)) &= x^2 + mx(x-1) + m^2(x-1)^2 - 1 \\ &= (1 + m + m^2)x^2 - (2m^2 + m)x + m^2 - 1. \end{aligned}$$

Thus, the solutions of $f(x, m(x - 1)) = 0$ are given by

$$\begin{aligned} x &= \frac{2m^2 + m \pm \sqrt{(2m^2+m)^2 - 4(1+m+m^2)(m^2-1)}}{2(1+m+m^2)} \\ &= \frac{2m^2 + m \pm \sqrt{m^2 + 4m + 4}}{2(1+m+m^2)} = \frac{2m^2 + m \pm \sqrt{(m+2)^2}}{2(1+m+m^2)} \\ &= \begin{cases} 1 \text{ or} \\ \frac{m^2-1}{m^2+m+1}. \end{cases} \end{aligned}$$

The solution $x = 1$ was to be expected as $P = (1, 0)$ is in the intersection of L_m and C, for all $m \in \mathbb{Q}$. The second solution is the x-coordinate of Q_m, the second point of intersection. Using the fact that Q_m is in L_m, we find

$$x(Q_m) = \frac{m^2 - 1}{m^2 + m + 1},$$

$$y(Q_m) = m(x(Q_m) - 1) = -\frac{m^2 + 2m}{m^2 + m + 1}.$$

Finally, by Theorem 9.3.4, we have

$$\begin{aligned} C(\mathbb{Q}) &= \{Q_m : m \in \mathbb{Q} \cup \{\infty\}\} \\ &= \left\{ \left(\frac{m^2-1}{m^2+m+1}, -\frac{m^2+2m}{m^2+m+1} \right) : m \in \mathbb{Q} \right\} \cup \{(1,-1)\}. \end{aligned}$$

For instance, when $m = 0$, 1, or 2, we have $Q_0 = (-1,0)$, $Q_1 = (0,-1)$, and $Q_2 = (3/7,-8/7)$, respectively. For what value of m do we get $Q_m = P$? Notice that the slope $\frac{dy}{dx}$ of the tangent line to C at a point (x,y) is given by implicit differentiation of $f(x,y)$ as follows:

$$2x + y + x\frac{dy}{dx} + 2y\frac{dy}{dx} = 0.$$

Thus, $\frac{dy}{dx} = -(2x+y)/(x+2y)$, which evaluated at $P = (1,0)$ gives $\frac{dy}{dx}(P) = -2$. Indeed, when $m = -2$ we obtain $Q_{-2} = P$.

Example 9.3.6. Let us find a parametrization of all the rational points on the quadratic equation given by

$$C : f(x,y) = 7x^2 - 46xy + 42x + 22y^2 + 12y - 49 = 0.$$

It is usually convenient to first find the reduced form of the curve C (as in Theorem 9.2.1), as it is easier to parametrize a reduced form. For instance, it is easier to find points on a reduced form.

In this case, we saw in Example 9.2.9 that C can be reduced to $C' : X^2 - 15Y^2 = -2940$ via a change of variables $X = 150y - 210$ and $Y = 14x - 46y + 42$. Therefore, C is a hyperbola. In fact the curve C is also equivalent to $C'' : 3X'^2 - 5Y'^2 = 147$, via $X' = 7x - 23y + 21$ and $Y' = 15y - 21$. The curve C'' has a rational (integral) point $P = (7,0)$, so let us parametrize C''.

Let L_m be the line through $P = (7,0)$ of slope $m \in \mathbb{Q} \cup \{\infty\}$. If $m = \infty$, then $L_\infty : X' = 7$, and there are no additional points in the intersection of L_∞ and C, because $f'(7,Y') = -5Y'^2 = 0$ is only possible when $Y' = 0$, which corresponds to P.

Thus, let $m \in \mathbb{Q}$ and let L_m be given by $Y' = m(X'-7)$. We find the intersection points of L_m and C'' by solving

$$\begin{cases} Y' = m(X'-7), \\ 3X'^2 - 5Y'^2 = 147. \end{cases}$$

Substitution yields

$$0 = 3X'^2 - 5(m(X'-7))^2 - 147 = (3-5m^2)X'^2 + 70m^2X' - 147 - 245m^2.$$

The solutions of the quadratic equation are

$$\begin{aligned} X' &= \frac{-70m^2 \pm \sqrt{(70m^2)^2 - 4(3-5m^2)(-147-245m^2)}}{6-10m^2} \\ &= \frac{-70m^2 \pm \sqrt{1764}}{6-10m^2} = \frac{-70m^2 \pm 42}{6-10m^2} = \frac{-35m^2 \pm 21}{3-5m^2} \\ &= \begin{cases} 7 \text{ or} \\ \frac{35m^2+21}{5m^2-3}. \end{cases} \end{aligned}$$

As is to be expected, $X' = 7$ is a solution, because $P = (7, 0)$ belongs to the intersection $L_m \cap C''$, for all $m \in \mathbb{Q}$. The second X' coordinate, however, corresponds to a second point of intersection Q'_m. Since Q'_m is also in L_m, we obtain

$$X'(Q'_m) = \frac{35m^2 + 21}{5m^2 - 3},$$
$$Y'(Q'_m) = m(X' - 7) = m\left(\frac{35m^2 + 21}{5m^2 - 3} - 7\right) = \frac{42m}{5m^2 - 3}.$$

Thus, $Q'_m = \left(\frac{35m^2+21}{5m^2-3}, \frac{42m}{5m^2-3}\right)$ and, by Theorem 9.3.4, we have

$$C''(\mathbb{Q}) = \left\{\left(\frac{35m^2 + 21}{5m^2 - 3}, \frac{42m}{5m^2 - 3}\right) : m \in \mathbb{Q}\right\}.$$

Since we have a bijection $\varphi : C \to C''$ given by $\varphi(x, y) = (7x - 23y + 21, 15y - 21)$ and inverse function

$$\varphi^{-1}(X', Y') = \left(\frac{1}{7}\left(X' + 23\left(\frac{Y' + 21}{15}\right) - 21\right), \frac{Y' + 21}{15}\right),$$

which send rational points to rational points, it follows that

$$C(\mathbb{Q}) = \varphi^{-1}(C''(\mathbb{Q})) = \varphi^{-1}(\{Q'_m : m \in \mathbb{Q}\}).$$

If we put $Q_m = \varphi^{-1}(Q'_m)$, then $C(\mathbb{Q}) = \{Q_m : m \in \mathbb{Q}\}$. Therefore, we obtain

$$C(\mathbb{Q}) = \{Q_m : m \in \mathbb{Q}\} = \left\{\left(\frac{65m^2 + 46m - 9}{25m^2 - 15}, \frac{35m^2 + 14m - 21}{25m^2 - 15}\right) : m \in \mathbb{Q}\right\}.$$

For instance, when $m = 0$ or $m = 1$, we have $Q_0 = (3/5, 7/5)$ or $Q_1 = (51/5, 14/5)$ $\in C(\mathbb{Q})$, respectively. When $m = 2$, we have $Q_2 = (343/85, 147/85)$ in $C(\mathbb{Q})$.

9.4. Integral Points on Quadratic Equations

In this section we use the reduced form of a quadratic equation C to find all of its integral points. We shall assume that we know all the integral points on the reduced form C' and deduce all the integral points on C from those in C'. In the later chapters of this second part of the book, we will explain how to find integral points on reduced forms. The key result that simplifies the search for integral points on a quadratic equation is next.

Theorem 9.4.1. *Let C be a quadratic equation and let C' be its reduced form. Then, there is a linear change of variables $\phi : C \to C'$ such that $\phi(C(\mathbb{Z})) \subseteq \phi(C'(\mathbb{Z}))$, and, in fact, $C(\mathbb{Z}) = \phi^{-1}(C'(\mathbb{Z})) \cap (\mathbb{Z} \times \mathbb{Z})$, or, equivalently,*

$$C(\mathbb{Z}) = \{(x_0, y_0) \in \mathbb{Z} \times \mathbb{Z} : \phi((x_0, y_0)) \in C'(\mathbb{Z})\}.$$

Moreover, the change of variables given in Corollary 9.2.12 satisfies this property.

Proof. Let $C : f(x, y) = 0$ be a quadratic equation given by a polynomial f with integer coefficients. Let C' be its reduced form, and let $\phi : C \to C'$ be the bijective change of variables given in Corollary 9.2.12. In every case, the map ϕ is given by $\phi(x, y) = (g(x, y), h(x, y))$ with $g = \alpha x + \beta y + \delta$, $h = \gamma x + \kappa y + \tau$ linear polynomials in x, y, with integer coefficients. Indeed, this is immediate when $a \neq 0$ or $c \neq 0$. When $a = c = 0$, then the map is given by the composition of

$\varphi(x, y) = (g_0(x, y), h_0(x, y))$, where g_0 and h_0 are linear with integer coefficients, with $\psi(x, y) = (x + y, x - y)$, and therefore

$$\phi(x, y) = \varphi(\psi(x, y)) = \varphi(x + y, x - y) = (g_0(x + y, x - y), h_0(x + y, x - y)),$$

and $g(x, y) = g_0(x + y, x - y)$, $h(x, y) = h_0(x + y, x - y)$ are linear in x and y, with integer coefficients, because g_0 and h_0 are linear and with integer coefficients:

$$\begin{aligned} g &= g_0(x + y, x - y) \\ &= \alpha(x + y) + \beta(x - y) + \delta \\ &= (\alpha + \beta)x + (\alpha - \beta)y + \delta, \end{aligned}$$

and similarly $h = h_0(x + y, x - y) = (\gamma + \kappa)x + (\gamma - \kappa)y + \tau$. Hence, it follows that if $(x_0, y_0) \in C(\mathbb{Z})$, then

$$\varphi(x_0, y_0) = (g(x_0, y_0), h(x_0, y_0)) \in C',$$

and $g(x_0, y_0) = (\alpha + \beta)x_0 + (\alpha - \beta)y_0 + \delta$ and $h(x_0, y_0) = (\gamma + \kappa)x_0 + (\gamma - \kappa)y_0 + \tau$ are integers, so $\varphi(x_0, y_0) \in C'(\mathbb{Z})$. This shows that $\varphi(C(\mathbb{Z})) \subseteq C'(\mathbb{Z})$.

It remains to show that $C(\mathbb{Z}) = \varphi^{-1}(C'(\mathbb{Z})) \cap (\mathbb{Z} \times \mathbb{Z})$. Since φ is a bijection and both φ and φ^{-1} send rational points to rational points, it follows that $C(\mathbb{Q}) = \varphi^{-1}(C'(\mathbb{Q}))$ and $C(\mathbb{Z}) = \varphi^{-1}(C'(\mathbb{Q})) \cap (\mathbb{Z} \times \mathbb{Z})$. Thus, it suffices to show that if $Q \in C'(\mathbb{Q})$ and $\varphi^{-1}(Q) \in C(\mathbb{Z})$, then $Q \in C'(\mathbb{Z})$.

Indeed, suppose that $Q \in C'(\mathbb{Q})$ and $P = \varphi^{-1}(Q) \in C(\mathbb{Z})$. By the first part of the proof, we have $\varphi(P) \in C'(\mathbb{Z})$, and, therefore, $\varphi(P) = \varphi(\varphi^{-1}(Q)) = Q \in C'(\mathbb{Z})$, as desired. In words, we have shown that every integral point on C comes from an integral point on C' via φ^{-1}. Equivalently, if $Q \in C'(\mathbb{Q})$ and $\varphi^{-1}(Q) \in C(\mathbb{Z})$, then $Q \in C'(\mathbb{Z})$. This shows that $C(\mathbb{Z}) = \varphi^{-1}(C'(\mathbb{Z})) \cap (\mathbb{Z} \times \mathbb{Z})$. □

Example 9.4.2. Let C be the quadratic equation of Examples 9.2.8 and 9.3.3, given by

$$C : 9x^2 - 42xy + 30x + 49y^2 - 72y + 14 = 0,$$

or, equivalently, $C : (3x - 7y + 5)^2 - (2y + 11) = 0$. The curve C is a parabola in reduced form, $C' : X^2 = Y$, via

$$\varphi(x, y) = (3x - 7y + 5, 2y + 11),$$

with inverse $\varphi^{-1}(X, Y) = \left(\frac{1}{3}(X + \frac{7}{2}(Y - 11) - 5), \frac{Y-11}{2}\right)$. In Example 9.3.3 we have found a parametrization of all the rational points on C. Here we intend to find all the *integral* points on C. First, we note that the rational points on C' are (t, t^2), for $t \in \mathbb{Q}$, so the integral points are (n, n^2) for $n \in \mathbb{Z}$. Thus, by Theorem 9.4.1, the integral points on C are those $(x_0, y_0) \in \mathbb{Z} \times \mathbb{Z}$ such that $\varphi((x_0, y_0)) = (n, n^2)$ for some $n \in \mathbb{Z}$; i.e.,

$$(1) \quad \begin{cases} 3x_0 - 7y_0 + 5 = n, \\ 2y_0 + 11 = n^2. \end{cases}$$

In order to verify the second equation of the system (1), the number n needs to satisfy $n^2 \equiv 11 \equiv 1 \bmod 2$; i.e., $n \equiv 1 \bmod 2$ so n is odd, so that $y_0 = (n^2 - 11)/2$ is an integer. Let us assume n is odd and write $n = 2k + 1$. In order for the first equation of (1) to have a solution, we need $n + 7y_0 - 5 \equiv 0 \bmod 3$, so that $x_0 = (n + 7y_0 - 5)/3$ is also an integer. In other words,

$$2k + 1 + 7((2k + 1)^2 - 11)/2 - 5 \equiv 0 \bmod 3,$$

or, equivalently,

$$0 \equiv 2k + 1 + 7(2k^2 + 2k - 5) - 5 \equiv 14k^2 + 16k - 39 \equiv 2k^2 + k \bmod 3,$$

where, in the last congruence, we have simply reduced the coefficients modulo 3. Hence, it suffices that $k(k + 2) \equiv 0 \bmod 3$, and therefore $k \equiv 0 \bmod 3$ or $k + 2 \equiv 0 \bmod 3$. Thus, $k = 3h$ or $k = 3h + 1$, for some $h \in \mathbb{Z}$. It follows that $n = 2k + 1 = 6h + 1$ or $6h + 3$ for some $h \in \mathbb{Z}$. For such n, we have

$$\begin{aligned}(x_0, y_0) &= \left(\frac{1}{3}\left(n + \frac{7}{2}(n^2 - 11) - 5\right), \frac{n^2 - 11}{2}\right)\\ &= \begin{cases}(42h^2 + 16h - 13, 18h^2 + 6h - 5) & \text{or}\\ (42h^2 + 44h - 3, 18h^2 + 18h - 1),\end{cases}\end{aligned}$$

for any $h \in \mathbb{Z}$. Therefore, we have shown

$$\begin{aligned}C(\mathbb{Z}) = &\{(42h^2 + 16h - 13, 18h^2 + 6h - 5) : h \in \mathbb{Z}\}\\ &\cup \{(42k^2 + 44k - 3, 18k^2 + 18k - 1) : k \in \mathbb{Z}\}.\end{aligned}$$

For instance, when $h = 0$ and $k = 0$ we obtain integral points $(-13, -5)$ and $(-3, -1) \in C(\mathbb{Z})$. When $h = 1$ and $k = 1$ we obtain points $(45, 19)$ and $(83, 35) \in C(\mathbb{Z})$, respectively.

In the previous example, in order to find all the integral points on a parabola, we had to solve two different *quadratic congruences* in one variable; namely $n^2 \equiv 1 \bmod 2$, and $2k^2 + k \equiv 0 \bmod 3$. In the following chapter, Chapter 10, we will study how to solve quadratic congruences in general and how to determine whether quadratic congruences in one variable have solutions. Here we show that, in fact, the problem of finding the integral points on a parabola can always be reduced to solving a finite number of quadratic congruences in one variable.

Theorem 9.4.3. *Let $C : f(x, y) = ax^2 + bxy + cy^2 + dx + ey + f = 0$ be a parabola, given by an equation with integer coefficients. Let $e' = 4ae - 2bd \neq 0$ and $f' = 4af - d^2$. Let $a \neq 0$ (or switch the roles of x and y). Then, C has an integral point if and only if*

(1) *the congruence $N^2 \equiv f' \bmod e'$ has a solution $N \equiv n_1 \bmod e'$ and*
(2) *the congruence*

$$be'K^2 + (2bn_1 - e')K + bt + d - n_1 \equiv 0 \bmod 2a$$

has a solution $K \equiv k_1 \bmod 2a$, where $t = (n_1^2 - f')/e'$.

Moreover, if $C(\mathbb{Z})$ is non-empty, then every integral point on C is of the form $\varphi^{-1}((n, n^2))$, for some $n \equiv n_1 + e'k_1 \bmod 2ae'$, for some n_1 and k_1 as in (1) and (2) above, where φ is the bijection given in Corollary 9.2.12, part (b).

Proof. Let $C : f(x, y) = ax^2 + bxy + cy^2 + \cdots = 0$ be a parabola, where the coefficients of f are integers. It follows from Corollary 9.2.12 that $c' = 4ac - b^2 = 0$

and either a or c is non-zero (if $a = c = 0$ and C is a parabola, then $c_0' = 8b^2 = 0$, so $b = 0$ as well and C would not be quadratic, a contradiction). Let us assume $a \neq 0$ (otherwise we switch the roles of a and x by c and y). Thus, there is a bijection from C to $C' : X^2 = Y$ of the form

$$\varphi(x, y) = (2ax + by + d, -(e'y + f')),$$

and the inverse change of variables is given by

$$\varphi^{-1}(X, Y) = \left(\frac{1}{2a}\left(X + \frac{b}{e'}(Y + f') - d\right), -\frac{1}{e'}(Y + f')\right),$$

where $e' = 4ae - 2bd \neq 0$ and $f' = 4af - d^2$. By Theorem 9.4.1, a point with coordinates $(x_0, y_0) \in \mathbb{Z} \times \mathbb{Z}$ is an integral point on $C(\mathbb{Z})$ if and only if there is some $n \in \mathbb{Z}$ such that

$$\text{(9.1)} \qquad \begin{cases} 2ax_0 + by_0 + d = n, \\ -(e'y_0 + f') = n^2. \end{cases}$$

The second equation of (9.1) has a solution if and only if $n^2 \equiv -f' \bmod e'$. Let us assume this is the case, and let $n \equiv n_1 \bmod e'$ be a solution. Then, the first equation of (9.1) also has a solution if and only if we have

$$2ax_0 + b(-(n^2 + f')/e') + d = n,$$

and if $n = n_1 + e'k$, for some $k \in \mathbb{Z}$, and $n_1^2 + f' = e't$, for some $t \in \mathbb{Z}$, we have

$$2ax_0 - b(t + 2n_1k + e'k^2) + d = n_1 + e'k.$$

Equivalently, we need the following quadratic equation in k to have a solution:

$$\text{(9.2)} \qquad be'k^2 + (e' - 2bn_1)k + bt - d + n_1 \equiv 0 \bmod 2a,$$

for some $k \in \mathbb{Z}$. Hence, if $k \equiv k_1 \bmod 2a$ is a solution to the quadratic congruence in (9.2), so that $k = k_1 + 2ah$ for some $h \in \mathbb{Z}$ is a solution, then $\varphi^{-1}((n, n^2))$ is an integral point on $C(\mathbb{Z})$ for any $n = n_1 + e'k = n_1 + e'(k_1 + 2ah) = n_1 + e'k_1 + 2ae'h$, for any $h \in \mathbb{Z}$, i.e., as long as $n \equiv n_1 + e'k_1 \bmod 2ae'$. And conversely, if $Q = (x_0, y_0) \in C(\mathbb{Z})$, then we have just shown that $Q = \varphi^{-1}((n, n^2))$ for some $n \equiv n_1 + e'k_1 \bmod 2ae'$, where $n_1^2 \equiv f' \bmod e'$ and k_1 satisfies the congruence in (9.2). □

Example 9.4.4. Let C be the quadratic equation given by

$$C : 4x^2 + 12xy + 20x + 9y^2 + 25y + 23 = 0.$$

The reader can verify that C is a parabola (Exercise 9.5.9). Let us find all the integral points on C. By Theorem 9.4.3, first we need to solve the congruence $N^2 \equiv f' \bmod e'$, where $e' = -80$ and $f' = -32$. In other words, we need to solve the congruence

$$N^2 \equiv -32 \bmod 80.$$

(Note that $x \equiv s \bmod t$ if and only if $x \equiv s \bmod (-t)$.) Since 80 is divisible by 5 if $N^2 \equiv -32 \bmod 80$, then we also have $N^2 \equiv -2 \equiv 3 \bmod 5$, where N is an integer. However, the squares modulo 5 are 0, 1, and 4 mod 5, but there is no integer N such that $N^2 \equiv 3 \bmod 5$. (We will revisit the question of what congruences are squares in Chapter 10, in much more detail.) Hence, there is no integer N such that $N^2 \equiv -32 \bmod 80$. Therefore, by Theorem 9.4.3, the parabola C has no integral points! Of course, it does have infinitely many rational points, and the reader can find a parametrization of all of them, using the methods of our previous chapter. For instance, $(-19/10, -2/5)$, $(-17/10, -1/5)$, and $(-107/30, 7/45)$ are points in $C(\mathbb{Q})$.

Example 9.4.5. Let C be the quadratic equation of Example 9.4.2, given by

$$C : 9x^2 - 42xy + 30x + 49y^2 - 72y + 14 = 0.$$

Let us once again find the integral points, now using Theorem 9.4.3. In this case, $c' = 4ac - b^2 = 0$, $e' = -72 \neq 0$, and $f' = -396$. First we need to solve the congruence

$$N^2 \equiv 396 \bmod (-72),$$

or, equivalently, $N^2 \equiv 36 \bmod 72$ (the sign in the modulus of a congruence is irrelevant), which has solutions, for instance $N \equiv \pm 6 \bmod 72$. In fact, a quick search yields all the solutions: $n_1 \equiv 6$, 18, 30, 42, 54, and 66 mod 72. For each one of these values of n_1, we obtain $t = (n_1^2 - 396)/(-72) = 5$, 1, -7, -19, -35, and -55, respectively. Now we need to solve the congruences:

(i) $3024K^2 + 432K - 234 \equiv 0 \bmod 18$, which holds for any K, because every coefficient is $\equiv 0 \bmod 18$;

(ii) $3024K^2 + 1440K - 54 \equiv 0 \bmod 18$, which holds for any K, because every coefficient is $\equiv 0 \bmod 18$;

(iii) $3024K^2 + 2448K + 294 \equiv 0 \bmod 18$, or $6 \equiv 0 \bmod 18$, which is impossible;

(iv) $3024K^2 + 3456K + 810 \equiv 0 \bmod 18$, which holds for any K;

(v) $3024K^2 + 4464K + 1494 \equiv 0 \bmod 18$, which holds for any K;

(vi) $3024K^2 + 5472K + 2346 \equiv 0 \bmod 18$, or $6 \equiv 0 \bmod 18$, which is impossible.

Hence, we need $n \equiv n_1 \equiv 6, 18, 42$, or $54 \bmod 72$, or, equivalently, $n_1 = 6k$ with $k \equiv 1, 3, 7$, or $9 \bmod 12$, or, equivalently, $k \equiv 1$ or $3 \bmod 6$. Thus, $n = 6(6h+1)$ or $6(6h+3)$ for some $h \in \mathbb{Z}$. Since the change of variables is given by

$$\varphi^{-1}(X, Y) = \left(\frac{1}{18}\left(X + \frac{7}{12}(Y - 396) - 30\right), \frac{1}{72}(Y - 396)\right),$$

we can calculate $\varphi^{-1}((n, n^2))$ when $n = 6(6h - 1)$:

$$(42h^2 + 16h - 13, 18h^2 + 6h - 5),$$

and when $n = 6(6h + 3)$:

$$(42h^2 + 44h - 3, 18h^2 + 18h - 1).$$

Therefore, we have shown

$$\begin{aligned} C(\mathbb{Z}) = {} & \{(42h^2 + 16h - 13, 18h^2 + 6h - 5) : h \in \mathbb{Z}\} \\ & \cup \{(42k^2 + 44k - 3, 18k^2 + 18k - 1) : k \in \mathbb{Z}\}, \end{aligned}$$

as we had already shown in Example 9.4.2.

Next, we treat the case of integral points on an ellipse or hyperbola.

Theorem 9.4.6. *Let $C : f(x,y) = ax^2 + bxy + cy^2 + dx + ey + f = 0$ be an ellipse or a hyperbola, given by an equation with integer coefficients. Let $c' = 4ac - b^2 \neq 0$, $e' = 4ae - 2bd$, and $f' = 4af - d^2$, and let $C' : X^2 + 4c'Y^2 = D$ be a reduced form of C, with $D = e'^2 - 4c'f'$.*

(1) *Let $a \neq 0$ (or switch the roles of x and y if $a = 0$ and $c \neq 0$). Then, C has an integral point if and only if there is an integral point $(X_0, Y_0) \in C'(\mathbb{Z})$ such that*

$$\begin{cases} X_0 \equiv e' \bmod 2c', \\ Y_0 \equiv b((X_0 - e')/2c') + d \bmod 2a. \end{cases}$$

(2) *Suppose $a = c = 0$. Then C has an integral point if and only if there is $(X_0, Y_0) \in C'(\mathbb{Z})$ such that*

$$\begin{cases} X_0 \equiv 8b(d-e) \bmod 8b^2, \\ Y_0 \equiv 2(d+e) \bmod 2b, \\ (Y_0 - 2(d+e))/2b \equiv (X_0 - 8b(d-e))/8b^2 \bmod 2. \end{cases}$$

Proof. Let $C : f(x,y) = ax^2 + bxy + cy^2 + dx + ey + f = 0$ be an ellipse or a hyperbola, with integer coefficients. Let us first assume that a or c is non-zero, and assume that $a \neq 0$ without loss of generality. Then, there is a non-zero B and D such that C is reduced to $C' : X^2 + BY^2 = D$, via a change of variables $\varphi : C \to C'$ of the form

$$\varphi(x,y) = (2c'y + e', 2ax + by + d),$$

and the inverse change of variables is given by

$$\varphi^{-1}(X,Y) = \left(\frac{1}{2a}\left(Y - \frac{b}{2c'}(X - e') - d\right), \frac{1}{2c'}(X - e')\right).$$

By Theorem 9.4.1, a point $(x_0, y_0) \in \mathbb{Z} \times \mathbb{Z}$ is an integral point on $C(\mathbb{Z})$ if and only if there is an integral point $(X_0, Y_0) \in C'(\mathbb{Z})$ such that $\varphi((x_0, y_0)) = (X_0, Y_0)$, or, equivalently, if and only if the system

$$\begin{cases} 2c'y_0 + e' = X_0, \\ 2ax_0 + by_0 + d = Y_0 \end{cases} \tag{9.3}$$

has a solution $(x_0, y_0) \in \mathbb{Z} \times \mathbb{Z}$. Given $(X_0, Y_0) \in C'(\mathbb{Z})$, the first equation of the system (9.3) has a solution if and only if $X_0 \equiv e' \bmod 2c'$. Let us assume this is the case, and put $t_0 = (X_0 - e')/2c' \in \mathbb{Z}$. Then, the second equation of (9.3) has a solution if and only if

$$Y_0 \equiv bt_0 + d \bmod 2a.$$

This proves part (1).

For part (2), let us assume that $a = c = 0$ and $b \neq 0$. Let C_0 be the conic defined in Corollary 9.2.12, given by

$$C_0 : bV^2 - bW^2 + 2(d+e)V + 2(d-e)W + 4f = 0$$

so $c'_0 = -4b^2 \neq 0$, $e'_0 = 8b(d-e)$, and $f'_0 = 16bf - (d+e)^2$. Let $C' : X^2 + BY^2 = D$ be the reduced form of C_0. Then, by part (1), a point $(V_0, W_0) \in \mathbb{Z} \times \mathbb{Z}$ is an integral point on $C_0(\mathbb{Z})$ if and only if there is an integral point $(X_0, Y_0) \in C'(\mathbb{Z})$ such that

$$(9.4) \qquad \begin{cases} X_0 \equiv e_0' \bmod 2c_0', \\ Y_0 \equiv b_0((X_0 - e_0')/2c_0') + d_0 \equiv d_0 \bmod 2a_0, \end{cases}$$

because $b_0 = 0$ in the equation for C_0. Since $\psi : C \to C_0$ is a bijection given by $\psi((x_0, y_0)) = (x_0 + y_0, x_0 - y_0)$ and it sends integer points to integer points, it remains to understand when $\psi^{-1}((V_0, W_0)) \in \mathbb{Z} \times \mathbb{Z}$, for some $(V_0, W_0) \in C_0(\mathbb{Z})$. Since $\psi^{-1}((V_0, W_0)) = ((V_0 + W_0)/2, (V_0 - W_0)/2)$, it follows that $\psi^{-1}((V_0, W_0)) \in \mathbb{Z} \times \mathbb{Z}$ if and only if $V_0 \equiv W_0 \bmod 2$.

Hence, $(x_0, y_0) \in \mathbb{Z} \cap \mathbb{Z}$ belongs to $C(\mathbb{Z})$ if and only if there is $(X_0, Y_0) \in C'(\mathbb{Z})$ such that $(\psi \circ \varphi)^{-1}(X_0, Y_0) = (x_0, y_0)$ if and only if $(X_0, Y_0) \in C'(\mathbb{Z})$ maps to $(V_0, W_0) \in C_0(\mathbb{Z})$ via φ^{-1} and (V_0, W_0) maps to (x_0, y_0) via ψ^{-1}, which in turn is equivalent to (9.4) and $V_0 \equiv W_0 \bmod 2$. Using the definition of φ^{-1}, we see that $(x_0, y_0) \in \mathbb{Z} \cap \mathbb{Z}$ belongs to $C(\mathbb{Z})$ if and only if there is $(X_0, Y_0) \in C'(\mathbb{Z})$ such that

$$(9.5) \qquad \begin{cases} X_0 \equiv e_0' \bmod 2c_0', \\ Y_0 \equiv d_0 \bmod 2a_0, \\ (Y_0 - d_0)/2a_0 \equiv (X_0 - e_0')/2c_0' \bmod 2, \end{cases}$$

as claimed. □

Example 9.4.7. Let $C : f(x, y) = 0$ be the quadratic equation given by

$$C : 3x^2 + 5xy + 7y^2 + x + y - 20 = 0.$$

In the notation of Corollary 9.2.12, we have $c' = 59 \neq 0$, $e' = 2$, and $f' = -241$, $B = 236$, and $D = 56880$. Therefore, C can be reduced to

$$C' : X^2 + 236Y^2 = 56880.$$

By Theorem 9.4.6, the ellipse C has an integral point if and only if C' has an integral point (X_0, Y_0) such that

$$\begin{cases} X_0 \equiv 2 \bmod 118, \\ Y_0 \equiv 5((X_0 - 2)/118) + 1 \bmod 6. \end{cases}$$

Let us first try to find such a point on C'. Since C' is an ellipse, we know from its equation for instance that $|X| \leq \sqrt{56880} \leq 239$. Since we need $X_0 \equiv 2 \bmod 118$, then there are only very limited options for X_0, namely $X_0 = 2$, 120, 238, -116, or -234. Out of these, we calculate for which values of X_0 we have that $Y_0 = \pm\sqrt{(56880 - X_0^2)/236}$ is an integer, and we find only four valid options $(X_0, Y_0) = (-234, \pm 3)$ and $(238, \pm 1)$. The point also needs to satisfy the congruence $Y_0 \equiv 5((X_0 - 2)/118) + 1 \bmod 6$, and only $(-234, 3)$, $(-234, -3)$, and $(238, -1)$ do. Now we may use our inverse map φ^{-1} to find the corresponding integral points on C, given by

$$\varphi^{-1}(X, Y) = \left(\frac{1}{6}\left(Y - \frac{5}{118}(X - 2) - 1\right), \frac{1}{118}(X - 2)\right).$$

Hence, the preimages of $(-234, -3)$, $(-234, 3)$, and $(238, -1)$ via φ are, respectively, $(1, -2)$, $(2, -2)$, and $(-2, 2)$, and these are **all** the integral points on C. Notice that the integral point $(238,1)$ on $C'(\mathbb{Z})$ is mapped back to $(-5/3,2)$ in $C(\mathbb{Q})$, which is not integral.

Example 9.4.8. Let $C : f(x,y) = 0$ be the quadratic equation given by

$$C : x^2 - 10xy + 14x - 128y^2 - 274y + 357 = 0.$$

In the notation of Corollary 9.2.12, we have $c' = -612 \neq 0$, $e' = -816$, and $f' = 1232$, $B = -2448$, and $D = 3681792$. Therefore, C can be reduced to

$$C' : X^2 - 2448Y^2 = 3681792.$$

By Theorem 9.4.6, the ellipse C has an integral point if and only if C' has an integral point (X_0, Y_0) such that

$$\begin{cases} X_0 \equiv -816 \bmod 1224, \\ Y_0 \equiv -10(-(X_0+816)/1224) + 14 \bmod 2. \end{cases}$$

The large coefficients in the equation of C' (and consequently the coefficients in the congruence for X_0) make the task of finding integral points on C' unnecessarily difficult in this case, because we can find a simpler reduced model (this comment goes back to our Remarks 9.2.3 and 9.2.13). Indeed,

$$\begin{aligned} f(x,y) &= x^2 - 10xy + 14x - 128y^2 - 274y + 357 \\ &= (x^2 + (-10y+14)x) - 128y^2 - 274y + 357 \\ &= (x-5y+7)^2 - (-5y+7)^2 - 128y^2 - 274y + 357 \\ &= (x-5y+7)^2 - 153y^2 - 204y + 308 \\ &= (x-5y+7)^2 - 17(9y^2+12y) + 308 \\ &= (x-5y+7)^2 - 17(3y+2)^2 + 376. \end{aligned}$$

Therefore, C can also be reduced to $C'' : X^2 - 17Y^2 = -376$, via

$$\varphi(x,y) = (x - 5y + 7, 3y + 2).$$

In particular, C has an integral point (x_0, y_0) if and only if C'' has an integral point (X_0, Y_0) such that

$$Y_0 \equiv 2 \bmod 3.$$

Notice that the congruence in X_0 would be modulo 1, so that is always verified! Moreover, C'' has an integral point $Q_1 = (7,5)$, and since $Y_0 = 5 \equiv 2 \bmod 3$, then Q_1 corresponds to an integral point in C; namely,

$$P_1 = (x,y) = \left(X_0 + 5\left(\frac{Y_0-2}{3}\right) - 7, \frac{Y_0-2}{3}\right) = (5,1).$$

Similarly, $Q_2 = (41,11)$ is a point on C'', and the Y-coordinate is $11 \equiv 2 \bmod 3$. Thus, Q_2 corresponds to a point on $C(\mathbb{Z})$; namely,

$$P_2 = (x,y) = \left(X_0 + 5\left(\frac{Y_0-2}{3}\right) - 7, \frac{Y_0-2}{3}\right) = (41+15-7,3) = (49,3).$$

In order to find *all* the integral points on $C(\mathbb{Z})$, however, we first need to determine all the integral points on $C''(\mathbb{Z})$ with Y-coordinate $\equiv 2 \bmod 3$. In Chapters 14 and 13 we will discuss methods to find all the integral points on hyperbolas, such as C''.

9.5. Exercises

Exercise 9.5.1. Find the normal and tangent vectors to each curve C below at the specified point P:

(a) $C : x^2 + y^2 = 1$ at $P = (0, 1)$.

(b) $C : x^2 + y^2 = 1$ at $P = (3/5, 4/5)$.

(c) $C : x^2 - y^2 = 0$ at $P = (2, 2)$.

(d) $C : x^2 + xy - 2y^2 + 4x - y + 3 = 0$ at $P = (3, 4)$.

(e) $C : x^2 + xy - 2y^2 + 4x - y + 3 = 0$ at $P = (4, 5)$.

Exercise 9.5.2. Each of the quadratic equations below is a product of two lines L and L'. Find the equations of the lines L and L'.

(a) $C_1 : 15x^2 - 8xy - 55y^2 - 29x + 87y - 14 = 0$.

(b) $C_2 : 18x^2 + 60xy + 50y^2 - 33x - 55y - 21 = 0$.

(c) $C_3 : 9x^2 - 42xy + 49y^2 - 6x + 14y - 2 = 0$.

(d) $C_4 : -14x^2 + 104xy + 152x + 71y^2 - 282y - 337 = 0$.

Exercise 9.5.3. For each of the quadratic equations C in Exercise 9.5.2, determine $C(\mathbb{Q})$ and $C(\mathbb{Z})$; i.e., find all the rational and integral points on C.

Exercise 9.5.4. Let $k \in \mathbb{Q}$, and let C be the curve given by the quadratic equation

$$C : 8x^2 + (2k + 12)xy + 3ky^2 + 2x + (k - 3)y - 1 = 0.$$

(a) Show that C is a product of two lines L and L', for all values of k, and find the equations for both lines.

(b) Find the values of k such that L and L' are parallel.

(c) Is there a value of k such that L and L' are perpendicular? (Two lines L and L' of slopes m and m' are perpendicular if $m' = -1/m$.)

Exercise 9.5.5. Classify the following quadratic equations C as products of two lines, parabolas, ellipses, or hyperbolas, and find a reduced form C' and a change of variables $\varphi : C \to C'$ such that $\varphi(C(\mathbb{Z})) \subseteq C'(\mathbb{Z})$.

(a) $x^2 + xy + y^2 = 0$.

(b) $x^2 + xy + y^2 = 1$.

(c) $x^2 - xy + y^2 = 1$.

(d) $xy + 1 = 0$.

(e) $xy - 1 = 0$.

(f) $x^2 + xy = 0$.

(g) $x^2 + xy + 1 = 0$.

(h) $x^2 + xy + y + 1 = 0$.

Exercise 9.5.6. For each of the quadratic equations C in Exercise 9.5.5, determine $C(\mathbb{Q})$; i.e., find all the rational points on C. In addition, if C is a product of two lines, a parabola, or an ellipse, then determine all the integral points on C as well.

Exercise 9.5.7. Let $k \in \mathbb{Q}$ and let C be the curve given by the equation

$$C : x^2 - 2xy + ky^2 - 4y - 1 = 0.$$

Find all the values of k, if any, such that C is

(a) a product of two lines,
(b) a parabola,
(c) an ellipse,
(d) a hyperbola.

Exercise 9.5.8. Let $C : f(x, y) = 0$ be a quadratic equation with $f(x, y) \in \mathbb{Z}[x, y]$. Let $P = (x_0, y_0)$ be a rational point on $C(\mathbb{Q})$. Show that the slope of the tangent line L of C at P is a rational number and the tangent line L itself is defined by a linear polynomial with rational coefficients.

Exercise 9.5.9. Let C be a quadratic equation given by

$$C : 4x^2 + 12xy + 20x + 9y^2 + 25y + 23 = 0.$$

(a) Show that C is a parabola.
(b) Find a change of variables from C to $C' : X^2 = Y$.
(c) Find a parametrization of $C(\mathbb{Q})$, the rational points on C.

Exercise 9.5.10. Find all integral solutions to the diophantine equation

$$48x^2 + 162xy + 105y^2 = 129.$$

(Hint: first, factor the left-hand side as $(ax + by)(cx + dy)$.)

Exercise 9.5.11. Find a rational point on the quadratic equation

$$C : 7x^2 - 10xy - 5y^2 - 32x - 20y = 39.$$

(Hint: first find a reduced form for C.)

Exercise 9.5.12. Let E be the ellipse given by $x^2 + 2y^2 = 738$.

(1) Find a parametrization for the rational points on the ellipse E.
(2) Find all the integral points on E.

Exercise 9.5.13. Let H be the hyperbola given by $x^2 - 5y^2 = 16$.

(1) Find a parametrization for the rational points on the hyperbola H.
(2) Can you find at least five distinct natural points on H?

Exercise 9.5.14. Let C be the parabola given by the equation

$$x^2 + 2xy + y^2 + x + 4y + 1 = 0.$$

(a) Parametrize all the rational points on C.
(b) Parametrize all the integral points on C.

Exercise 9.5.15. Find all the integral points on the conic given by the equation

$$x^2 + xy + y^2 - x - y = 1.$$

Exercise 9.5.16. Find all the integral points on the conic given by the equation

$$5x^2 + xy + y^2 - 3x - 7y + 6 = 0.$$

CHAPTER 10

QUADRATIC CONGRUENCES

It is not knowledge, but the act of learning, not possession but the act of getting there, which grants the greatest enjoyment.

In a letter from Carl Friedrich Gauss
to Farkas Bolyai (September 2, 1808)

In Chapter 4 we have learned how to solve linear congruences $ax \equiv b \bmod m$ (using Euclid's algorithm and Bezout's identity; see Section 4.4) and systems of linear congruences (using the Chinese remainder theorem; see Section 4.5). The goal of this chapter is to explain whether a quadratic congruence of the form

$$ax^2 + bx + c \equiv 0 \bmod m$$

has solutions and, if it has any, to explain how to find them. Before we begin, we remind the reader of a couple of examples we have already seen in previous chapters, where quadratic congruences appeared naturally and played an important role.

Example 10.0.1. In Example 4.2.8, we showed that the conic $C : x^2 - 5y^2 = 2$ has no integral points. The proof was based on the fact that the quadratic congruence $x^2 \equiv 2 \bmod 5$ has no solutions.

Example 10.0.2. Let C be a parabola, given by a quadratic equation of the form $f(x, y) = 0$. Are there integral points on C? In Chapter 9 (Section 9.4, Theorem 9.4.3) we saw that C has integral points if and only if certain quadratic congruences have solutions. For instance, let $C : 4x^2+12xy+20x+9y^2+25y+23 = 0$. One of the conditions needed for the existence of integral points on C is that $N^2 \equiv -32 \bmod 80$ has a solution, but this congruence has no solutions modulo 80. See Examples 9.4.4 and 9.4.5.

10.1. The Quadratic Formula

Let a, b, c and $m > 1$ be integers, with $a \neq 0$. If we want to find the roots of a quadratic polynomial $ax^2 + bx + c$ over the complex numbers, then we have the well-known quadratic formula

$$x = \frac{-b \pm \sqrt{b^2 - 4ac}}{2a}.$$

Does this formula work modulo m? Indeed, it does work, as long as $(2a, m) = 1$. Here is why. Suppose that x is an integer that satisfies $ax^2 + bx + c \equiv 0 \bmod m$. Then, we may complete the square to obtain an equivalent congruence:

$$\begin{aligned} 0 &\equiv x^2 + \frac{b}{a}x + \frac{c}{a} \\ &\equiv x^2 + 2 \cdot \left(\frac{b}{2a}\right) x + \left(\frac{b}{2a}\right)^2 - \left(\frac{b}{2a}\right)^2 + \frac{c}{a} \\ &\equiv \left(x + \frac{b}{2a}\right)^2 - \frac{b^2 - 4ac}{(2a)^2}. \end{aligned}$$

Notice that we were able to find an inverse for a and for $2a$ because $(2a, m) = 1$, by assumption. Thus,

$$\left(x + \frac{b}{2a}\right)^2 \equiv \frac{b^2 - 4ac}{(2a)^2} \bmod m. \tag{10.1}$$

Proposition 10.1.1. *Let $a, b, c \in \mathbb{Z}$, with $a \neq 0$, and let $m > 1$ be an integer relatively prime to $2a$.*

(a) *Suppose that α is an integer such that $\alpha^2 \equiv b^2 - 4ac \bmod m$. Then*

$$x \equiv \frac{-b \pm \alpha}{2a} \equiv (2a)^{-1} \cdot (-b \pm \alpha) \bmod m$$

are solutions of $ax^2 + bx + c \equiv 0 \bmod m$, where $(2a)^{-1}$ is the multiplicative inverse of $2a \bmod m$.

(b) *If $ax^2 + bx + c \equiv 0 \bmod m$ has a solution, then there exists $\alpha \in \mathbb{Z}$ such that $\alpha^2 \equiv b^2 - 4ac \bmod m$ and $x \equiv (-b + \alpha) \cdot (2a)^{-1} \bmod m$.*

Proof. Let a, b, c, m be as in the statement of the proposition. It follows from our previous discussion that $x \in \mathbb{Z}$ is a solution of $ax^2 + bx + c \equiv 0 \bmod m$ if and only if (10.1) holds.

(1) Suppose that α is some integer with $\alpha^2 \equiv b^2 - 4ac \bmod m$. Then, x is a solution of the quadratic congruence if and only if

$$\left(x + \frac{b}{2a}\right)^2 \equiv \frac{\alpha^2}{(2a)^2} \equiv \left(\frac{\alpha}{2a}\right)^2 \bmod m.$$

Therefore, if $x + b/(2a) \equiv \pm\alpha/(2a) \bmod m$, then x will also verify the original quadratic congruence. The latter congruence is equivalent to $x \equiv (-b \pm \alpha)/(2a) \bmod m$, as desired.

(2) Suppose $x \in \mathbb{Z}$ is a solution of $ax^2 + bx + c \equiv 0 \bmod m$, and therefore x is also a solution of (10.1). Then,

$$b^2 - 4ac \equiv (2a)^2 \cdot \left(x + \frac{b}{2a}\right)^2 \equiv \left((2a) \cdot \left(x + \frac{b}{2a}\right)\right)^2 \bmod m.$$

Therefore, $\alpha \equiv 2a(x + b/2a) \bmod m$ satisfies $\alpha^2 \equiv b^2 - 4ac \bmod m$, as claimed. In particular, $2ax + b \equiv \alpha \bmod m$, and therefore $x \equiv (-b + \alpha) \cdot (2a)^{-1} \bmod m$.

□

In particular, the previous proposition shows that m must be odd for the "quadratic formula" to work modulo m, because we need to invert 2 modulo m. The number α in the proposition is a square root of $b^2 - 4ac$ modulo m. If we write $\sqrt{b^2 - 4ac}$ instead of α in the formulas, then we obtain the usual quadratic formula. The quantity $b^2 - 4ac$ is called the *discriminant* of the quadratic equation.

Example 10.1.2. *Find all the solutions of the quadratic congruence*

$$3x^2 + 5x + 6 \equiv 0 \bmod 7.$$

By Proposition 10.1.1,

$$x \equiv \frac{-5 \pm \alpha}{6} \bmod 7$$

where $\alpha^2 \equiv 25 - 4 \cdot 3 \cdot 6 \equiv 4 + 5 \equiv 9 \bmod 7$. Thus, $\alpha \equiv 3 \bmod 7$ works. And $6^{-1} \equiv -1 \bmod 7$. Hence,

$$x \equiv -(-5 \pm 3) \equiv 2 \text{ or } 1 \bmod 7.$$

Let us check that, for example, $x \equiv 2 \bmod 7$ is a solution:

$$3(2)^2 + 5 \cdot 2 + 6 \equiv 12 + 10 + 6 \equiv 28 \equiv 0 \bmod 7.$$

Finally, $\mathbb{Z}/p\mathbb{Z}$ is a field and $f(x) = 3x^2 + 5x + 6$ is a polynomial of degree 2 over $\mathbb{Z}/p\mathbb{Z}[x]$. Hence, it has at most two distinct roots. Thus, $x \equiv 1$ and $2 \bmod 7$ are the only solutions modulo 7.

Example 10.1.3. *Show that the following quadratic congruence has no solutions:*

$$x^2 + 5x + 5 \equiv 0 \bmod 7.$$

By Proposition 10.1.1, if there was a solution, then the discriminant is a square modulo 7; i.e., $b^2 - 4ac \equiv 25 - 20 \equiv 5 \equiv \alpha^2 \bmod 7$, for some $\alpha \in \mathbb{Z}$. However, 5 is not a square modulo 7 because the only squares mod 7 are $0, 1, 2$, and 4.

If the modulus of a quadratic congruence is not prime, then we may use the Chinese remainder theorem to solve the congruence:

Example 10.1.4. *Find all the solutions of the following quadratic congruence:*

$$3x^2 + 3x + 17 \equiv 0 \bmod 35. \tag{10.2}$$

We will solve this problem in two different ways; see Example 10.1.5 for an alternative solution. By Proposition 10.1.1, every solution of (10.2) is of the form

$$x \equiv (-3 + \alpha) \cdot (2 \cdot 3)^{-1} \bmod 35,$$

where α satisfies $\alpha^2 \equiv 3^2 - 4 \cdot 3 \cdot 17 \equiv 15 \bmod 35$. Thus, in order to find all the roots of (10.2), we first need to find all the roots of $y^2 \equiv 15 \bmod 35$. We could do this by "brute force" (i.e., calculate y^2 mod 35 for every $0 \leq y < 35$), since 35 is small, but, instead, let us use the Chinese remainder theorem (Theorem 4.5.9).

The equation $y^2 \equiv 15 \bmod 35$ has solutions if and only if the system

$$\begin{cases} y^2 \equiv 0 \bmod 5, \\ y^2 \equiv 1 \bmod 7 \end{cases}$$

has a common solution, since $35 = 5 \cdot 7$, and $15 \equiv 0 \bmod 5$, and $15 \equiv 1 \bmod 7$. This system is equivalent to $y \equiv 0 \bmod 5$, and $y \equiv \pm 1 \bmod 7$. Hence, there are two possible solutions

$$\begin{cases} y \equiv 0 \bmod 5, \\ y \equiv 1 \bmod 7 \end{cases} \quad \text{and} \quad \begin{cases} y \equiv 0 \bmod 5, \\ y \equiv -1 \bmod 7. \end{cases}$$

Thus, $y \equiv 15 \bmod 35$, or $y \equiv 20 \bmod 15$. Therefore, there are two possibilities for $\alpha \equiv 15$ and $20 \bmod 15$, which correspond to the following solutions of (10.2):

$$x \equiv \frac{-3+15}{6} \equiv 2 \quad \text{and} \quad x \equiv \frac{-3+20}{6} \equiv 32 \bmod 35.$$

Hence, the unique solutions of $3x^2 + 3x + 17 \equiv 0 \bmod 35$ are $x \equiv 2$ and $32 \bmod 35$.

Example 10.1.5. *Find all the solutions of the following quadratic congruence:*

$$3x^2 + 3x + 17 \equiv 0 \bmod 35. \tag{10.3}$$

In Example 10.1.4 we saw one solution; here we provide an alternative approach. Let $f(x) \equiv 3x^2 + 3x + 17 \bmod 35$. Then, by the Chinese remainder theorem, $f(x) \equiv 0 \bmod 35$ if and only if $f(x) \equiv 0 \bmod 5$ and $f(x) \equiv 0 \bmod 7$, because $35 = 5 \cdot 7$ and $(5, 7) = 1$. Thus, we begin solving the congruence modulo 5 and 7, separately:

- Mod 5: (10.3) mod 5 is equivalent to $3x^2+3x+2 \equiv 0 \bmod 5$. The discriminant of this equation is $9-4\cdot3\cdot2 \equiv 0 \bmod 5$ and $6^{-1} \equiv 1 \bmod 5$. Thus, the solutions are $x \equiv -3 \pm 0 \equiv 2 \bmod 5$. This means that 2 mod 5 is a double root. Indeed,
$$3x^2 + 3x + 2 \equiv 3(x^2 + x + 4) \equiv 3(x-2)^2 \bmod 5.$$
- Mod 7: (10.3) mod 7 is equivalent to $3x^2 + 3x + 3 \equiv 3(x^2 + x + 1) \equiv 0 \bmod 7$ which, in turn, is equivalent to $x^2 + x + 1 \equiv 0 \bmod 7$, because $3 \not\equiv 0 \bmod 7$. The discriminant of the latter equation is $1 - 4 \cdot 1 \cdot 1 \equiv -3 \equiv 4 \equiv 2^2 \bmod 7$, and $2^{-1} \equiv 4 \bmod 7$. Thus, the solutions are $x \equiv 4(-1 \pm 2) \equiv 4$ or $2 \bmod 7$.

Now, we can reconstruct a solution modulo 35 using the Chinese remainder theorem. The possibilities are

$$\begin{cases} x \equiv 2 \bmod 5, \\ x \equiv 2 \bmod 7 \end{cases} \quad \text{or} \quad \begin{cases} x \equiv 2 \bmod 5, \\ x \equiv 4 \bmod 7. \end{cases}$$

The solutions of these systems are, respectively, $x \equiv 2 \bmod 35$ and $x \equiv 32 \bmod 35$. Therefore, $x \equiv 2$ and $x \equiv 32 \bmod 35$ are the unique solutions of (10.3).

Example 10.1.6. *Find all the solutions of the following quadratic congruence:*

$$x^2 + x + 8 \equiv 0 \bmod 35. \tag{10.4}$$

By Proposition 10.1.1, every solution of (10.4) is of the form

$$x \equiv (-1 + \alpha) \cdot 2^{-1} \bmod 35,$$

where α satisfies $\alpha^2 \equiv 1^2 - 4 \cdot 1 \cdot 8 \equiv -31 \equiv 4 \bmod 35$. Thus, in order to find all the roots of (10.4), we first need to find all the roots of $y^2 \equiv 4 \bmod 35$. Clearly, $y \equiv \pm 2 \bmod 35$ are solutions. Working modulo 5 and modulo 7 and using the Chinese remainder theorem, we can show that all the roots of $y^2 \equiv 4 \bmod 35$ are $y \equiv \pm 2$ and $\pm 12 \bmod 35$. Hence, the solutions of (10.4) are

$$x \equiv \frac{-1+2}{2} \equiv 18, \; x \equiv \frac{-1+33}{2} \equiv 16, \; x \equiv \frac{-1+12}{2} \equiv 23, \; x \equiv \frac{-1+23}{2} \equiv 11$$

modulo 35; i.e., the solutions are $x \equiv 11, 16, 18$, and $23 \bmod 35$.

Alternatively, let $f(x) \equiv x^2 + x + 8 \bmod 35$. We proceed as in the previous example. Thus, we begin solving the congruence modulo 5 and 7, separately:

- Mod 5: (10.4) mod 5 is equivalent to $x^2 + x + 3 \equiv 0 \bmod 5$. The discriminant of this equation is $1 - 4 \cdot 3 \equiv -11 \equiv 4 \bmod 5$ and $2^{-1} \equiv 3 \bmod 5$. Thus, the solutions are $x \equiv 3(-1 \pm 2) \equiv 3$ or $1 \bmod 5$.
- Mod 7: (10.4) mod 7 is equivalent to $x^2 + x + 1 \equiv 0 \bmod 7$. The discriminant is $1 - 4 \cdot 1 \cdot 1 \equiv -3 \equiv 4 \equiv 2^2 \bmod 7$ and $2^{-1} \equiv 4 \bmod 7$. Thus, the solutions are $x \equiv 4(-1 \pm 2) \equiv 4$ or $2 \bmod 7$.

Now, we can reconstruct a solution modulo 35 using the Chinese remainder theorem. The possibilities are

$$\begin{cases} x \equiv 1 \bmod 5, \\ x \equiv 2 \bmod 7 \end{cases} \quad \text{or} \quad \begin{cases} x \equiv 1 \bmod 5, \\ x \equiv 4 \bmod 7 \end{cases}$$

or

$$\begin{cases} x \equiv 3 \bmod 5, \\ x \equiv 2 \bmod 7 \end{cases} \quad \text{or} \quad \begin{cases} x \equiv 3 \bmod 5, \\ x \equiv 4 \bmod 7. \end{cases}$$

The solutions of these systems are, respectively, $x \equiv 16 \bmod 35$, $x \equiv 11 \bmod 35$, $x \equiv 23 \bmod 35$, and $x \equiv 18 \bmod 35$. Therefore, the (10.4) has exactly four distinct solutions: $x \equiv 11, 16, 18$, and $23 \bmod 35$.

Remark 10.1.7. Notice that $\mathbb{Z}/35\mathbb{Z}$ is not a field (3 and 5 are zero-divisors!) and, therefore, a polynomial of degree 2 in $\mathbb{Z}/35\mathbb{Z}[x]$ may have more than two distinct roots, as the previous example illustrates.

10.2. Quadratic Residues

In the previous section we have shown that, at least for an odd modulus $m > 1$, in order to solve any quadratic congruence $ax^2 + bx + c \equiv 0 \bmod m$, we need to solve equations of the form $x^2 \equiv d \bmod m$ (where d is the discriminant of the quadratic equation; i.e., we need to solve $x^2 \equiv b^2 - 4ac \bmod m$). Thus, we turn our attention to identifying the squares and non-squares modulo m.

Definition 10.2.1. Let $m > 1$ and let $a \in \mathbb{Z}$, relatively prime to m; i.e., a is a unit modulo m. We say that a is a *quadratic residue* modulo m (or a QR mod m) if there is an integer b such that $a \equiv b^2 \bmod m$. Otherwise, we say that a is a *quadratic non-residue* modulo m (or a QNR mod m).

Example 10.2.2. The quadratic residues modulo 7 are 1, 2, and 4, because $1 \equiv 1^2$, $2 \equiv 3^2$, and $4 \equiv 2^2 \bmod 7$. The quadratic non-residues are 3, 5, and 6 mod 7 because the equations $x^2 \equiv 3$, 5, or 6 mod 7 have no solutions with $x \in \mathbb{Z}$.

There is only one quadratic residue modulo 3, namely 1 mod 3. There is also only one quadratic non-residue modulo 3, namely 2 mod 3.

The quadratic residues modulo 5 are $\{1, 4 \bmod 5\}$ while the quadratic non-residues are $\{2, 3 \bmod 5\}$.

The quadratic residues modulo 11 are $\{1, 3, 4, 5, 9\}$ while the quadratic non-residues are $\{2, 6, 7, 8, 10\}$. This can be shown by squaring every congruence class modulo 11 and seeing what congruence classes appear upon squaring:

x mod 11	1	2	3	4	5	6	7	8	9	10
x^2 mod 11	1	4	9	5	3	3	5	9	4	1

We remark that in all cases above, $p > 2$ is prime and there are precisely $\frac{p-1}{2}$ quadratic residues and $\frac{p-1}{2}$ quadratic non-residues modulo p.

Proposition 10.2.3. *Let $p > 2$ be a prime. Then, there are precisely $\frac{p-1}{2}$ quadratic residues and $\frac{p-1}{2}$ quadratic non-residues modulo p.*

Proof. Let $p > 2$ be a prime. We claim that $S = \left\{1^2, 2^2, \ldots, \left(\frac{p-1}{2}\right)^2\right\}$ is a complete set of representative of all quadratic residues modulo p. Indeed:

- Suppose that a is a QR mod p. Then, $a \equiv b^2 \bmod p$, for some $1 \leq b \leq p-1$, and $b^2 \equiv (-b)^2 \bmod p$. Thus, either b or $-b$ is congruent to a number in the range $1, \ldots, \frac{p-1}{2}$. Hence we may assume $1 \leq b \leq \frac{p-1}{2}$ with $b^2 \equiv a \bmod p$ and, consequently, a is congruent to a number in S.
- All the elements of S are distinct modulo p. Suppose $1 \leq i, j \leq \frac{p-1}{2}$ and $i^2 \equiv j^2 \bmod p$. Then, $i \equiv \pm j \bmod p$ (because p is prime), but $1 \leq i, j \leq \frac{p-1}{2}$ forces $i \equiv j \bmod p$.

Hence, all the elements of S are quadratic residues (since they are squares of units, by definition), and we have shown that all quadratic residues have a representative in S and all elements of S are distinct modulo p. Thus, there are exactly $\frac{p-1}{2}$ quadratic residues in $\mathbb{Z}/p\mathbb{Z}$. Since there are $p-1$ units, there are $p-1-\frac{p-1}{2} = \frac{p-1}{2}$ quadratic non-residues. □

Remark 10.2.4. In Exercise 10.8.7 we outline an alternative proof of Proposition 10.2.3 that uses group theory.

Remark 10.2.5. Obviously, 1 mod p is always a quadratic residue modulo p, because $1^2 \equiv 1 \bmod p$. When is 2 mod p a quadratic residue modulo p? This is far from obvious and, more generally, deciding whether a mod p is a quadratic residue is difficult. However, Gauss's law of quadratic reciprocity will provide a very efficient method to settle this question (see Section 10.4).

Example 10.2.6. Suppose that $p > 2$ is an odd prime. When is $-1 \bmod p$ a quadratic residue?

- When $p = 3$, the congruence class of $-1 \equiv 2 \bmod 3$ is a quadratic non-residue, because -1 is not congruent to a square modulo 3.
- For $p = 5$, however, $-1 \equiv 4 \equiv 2^2 \bmod 5$ and so -1 is a QR mod 5.
- -1 is a QNR modulo 7, because, as we have seen above in Example 10.2.2, the list of QNR's modulo 7 is $\{3, 5, 6 \bmod 7\}$.
- -1 is also a QNR modulo 11, because the list of QNR's modulo 11 is 2, 6, 7, 8, and 10 mod 11.

Here is a small table of odd primes p and square roots of $-1 \bmod p$, if there is one:

p	3	5	7	11	13	17	23	29	31	37	41
$\sqrt{-1}$	QNR	± 2	QNR	QNR	± 5	± 4	QNR	± 12	QNR	± 6	± 9

Is there a pattern? It turns out that $-1 \bmod p$ is a quadratic residue if and only if $p \equiv 1 \bmod 4$. We shall prove this next in two different ways. In our first approach, we shall use primitive roots, and then we will offer a proof that does not rely on the existence of primitive roots.

Lemma 10.2.7. *Let $p > 2$ be a prime and let $a \in \mathbb{Z}$ be relatively prime to p. Let $g \in \mathbb{Z}$ be a primitive root modulo p and let n be the least positive integer such that $g^n \equiv a \bmod p$. Then, $a \bmod p$ is a quadratic residue if and only if n is even.*

Proof. Let p, a, g, and n be as in the statement of the lemma. Suppose first that n is even, with $n = 2m$. Then, $a \equiv g^n \equiv (g^m)^2 \bmod p$, and so $a \bmod p$ is a QR. Suppose now that a is a quadratic residue. Then, there is $d \in \mathbb{Z}$ such that $d^2 \equiv a \bmod p$. Let m be the least positive integer such that $g^m \equiv d \bmod p$. Then,

$$a \equiv d^2 \equiv g^{2m} \bmod p.$$

Thus, we have that $a \equiv g^{2m}$ and also $a \equiv g^n \bmod p$. It follows that $g^{2m-n} \equiv 1 \bmod p$. Since g is a primitive root, its order is $p-1$ and, by Proposition 8.1.5, we conclude that $p-1$ divides $2m-n$. In other words, there is some $k \in \mathbb{Z}$ such that $k(p-1) = 2m - n$, or $n = 2m - k(p-1)$. Since $p > 2$ is odd, $p - 1$ is even and so must be n, as claimed. □

Theorem 10.2.8. *Let $p > 2$ be an odd prime. Then, the congruence class of $-1 \bmod p$ is a quadratic residue modulo p if $p \equiv 1 \bmod 4$ and a quadratic non-residue if $p \equiv 3 \bmod 4$.*

Proof. Let $p > 2$ be a prime. By Theorem 8.4.1, there is a primitive root $g \bmod p$. Since g is a unit, by Fermat's little theorem we know that $g^{p-1} \equiv 1 \bmod p$. Since p is odd, the number $\frac{p-1}{2}$ is an integer, and $g^{(p-1)/2}$ is a root of $x^2 \equiv 1 \bmod p$, because

$$(g^{\frac{p-1}{2}})^2 \equiv g^{p-1} \equiv 1 \bmod p.$$

Since p is prime, by Theorem 5.5.19, the polynomial $x^2 - 1$ has at most two roots in $\mathbb{Z}/p\mathbb{Z}$, so the only roots are $\pm 1 \bmod p$. Hence, $g^{(p-1)/2} \equiv \pm 1 \bmod p$. However, $g^{(p-1)/2} \equiv 1 \bmod p$ is not possible because $\text{ord}_p(g) = p - 1$ and not $(p-1)/2$. It follows that $g^{(p-1)/2} \equiv -1 \bmod p$.

Thus, $n = \frac{p-1}{2}$ is the least positive integer such that $g^n \equiv -1 \bmod p$ and, therefore, by Lemma 10.2.7, the class of $-1 \bmod p$ is a quadratic residue if and only if n is even, i.e., if $\frac{p-1}{2} = 2k$, for some $k \in \mathbb{Z}$. It follows that $-1 \bmod p$ is a quadratic residue if and only if $p = 1 + 4k$, or $p \equiv 1 \bmod 4$. Similarly, $-1 \bmod p$ is a quadratic non-residue if and only if n is odd, i.e., if $\frac{p-1}{2} = 2k+1$, for some $k \in \mathbb{Z}$. It follows that $-1 \bmod p$ is a quadratic non-residue if and only if $p = 3 + 4k$, or $p \equiv 3 \bmod 4$, as claimed. □

Example 10.2.9. The proof of Theorem 10.2.8 tells us that $-1 \bmod p$ is a QR if and only if $p \equiv 1 \bmod 4$, but it also tells us how to find a square root in the case when $p \equiv 1 \bmod 4$. Indeed, a square root is given by $g^{(p-1)/4}$, where $g \bmod p$ is a primitive root.

For instance, let $p = 13$. The class of 2 mod 13 is a primitive root and, therefore, $2^{(13-1)/4}$ must be a square root of -1 mod 13. Indeed,

$$2^{\frac{13-1}{4}} \equiv 2^3 \equiv 8 \bmod 13,$$

and $8^2 \equiv (-5)^2 \equiv 25 \equiv 12 \equiv -1 \bmod 13$.

As promised, now we offer a second proof of Theorem 10.2.8 that does not use primitive roots. First, we shall use Wilson's theorem to compute the value of $((p-1)/2)! \bmod p$ for an odd prime p.

Lemma 10.2.10. *Let p be an odd prime, and let $r = (p-1)/2$. Then,*

$$(r!)^2 \equiv (-1)^{(p+1)/2} \bmod p.$$

Proof. By Wilson's theorem, Theorem 7.1.7, we have $(p-1)! \equiv -1 \bmod p$. Thus,

$$\begin{aligned} -1 &\equiv (p-1)! \\ &\equiv 1 \cdot 2 \cdot 3 \cdots \left(\frac{p-1}{2}\right) \cdot \left(\frac{p+1}{2}\right) \cdots (p-3)(p-2)(p-1) \\ &\equiv (-1)^{(p-1)/2} \cdot \left(1 \cdot 2 \cdot 3 \cdots \left(\frac{p-1}{2}\right)\right)^2, \end{aligned}$$

where we have used the fact that $p - k \equiv (-1) \cdot k \bmod p$, for all $1 \leq k \leq r$, where $r = (p-1)/2$. Hence, $-1 \equiv (-1)^{(p-1)/2}(r!)^2 \bmod p$, or, equivalently,

$$(r!)^2 \equiv (-1)^{(p+1)/2} \bmod p,$$

as desired. □

Now we are ready to write an alternative proof for Theorem 10.2.8.

Proof of Theorem 10.2.8. Suppose first that $a^2 \equiv -1 \bmod p$ for some integer $a \in \mathbb{Z}$ relatively prime to p. Raising both sides of the congruence to the power of $(p-1)/2$ we obtain on one hand

$$(a^2)^{(p-1)/2} \equiv a^{p-1} \equiv 1 \bmod p$$

by Fermat's little theorem and on the other hand

$$1 \equiv (a^2)^{(p-1)/2} \equiv (-1)^{(p-1)/2} \bmod p.$$

Since $p > 2$ is odd, $1 \not\equiv -1 \bmod p$, and therefore we must have that the exponent $(p-1)/2$ is even. This implies that $p \equiv 1 \bmod 4$.

For the converse, suppose that $p \equiv 1 \bmod 4$. Then, by Lemma 10.2.10 we have

$$(r!)^2 \equiv (-1)^{(p+1)/2} \bmod p,$$

where $r = (p-1)/2$. Since $p \equiv 1 \bmod 4$, it follows that $(p+1)/2$ is odd, and therefore $(r!)^2 \equiv -1 \bmod p$. In particular, -1 is a quadratic residue modulo p, as desired. □

Example 10.2.11. Lemma 10.2.10 gives a formula for the square root of $-1 \bmod p$, when $p \equiv 1 \bmod 4$. For instance, let $p = 13$. The lemma shows that $(r!)^2 \equiv -1 \bmod 13$, where $r = (13-1)/2 = 6$. We can compute

$$6! \equiv 6 \cdot 5 \cdot 4 \cdot 3 \cdot 2 \cdot 1 \equiv (6 \cdot 2) \cdot (4 \cdot 3) \cdot 5 \equiv (-1) \cdot (-1) \cdot 5 \equiv 5 \bmod 13,$$

and, indeed, $5^2 \equiv 25 \equiv -1 \bmod 13$.

10.3. The Legendre Symbol

How can we tell if a given integer a is a quadratic residue or a quadratic non-residue modulo a prime p? One could write down a complete list of all the quadratic residues following the ideas of the proof of Proposition 10.2.3. However, if p is large, this may be a daunting task! For instance, is 4699 a square modulo 4703? The following definition, lemmas, and theorems (including the law of quadratic reciprocity) will simplify this job enormously.

Definition 10.3.1. Let $p > 2$ be an odd prime and let a be an integer. The *Legendre symbol* (or quadratic residue symbol) is defined as follows:

$$\left(\frac{a}{p}\right) = \begin{cases} 0 & \text{if } p \mid a, \\ 1 & \text{if } a \text{ is a quadratic residue mod } p, \\ -1 & \text{if } a \text{ is a quadratic non-residue mod } p. \end{cases}$$

Remark 10.3.2. The Legendre symbol $\left(\frac{a}{p}\right)$ is only defined for odd primes p.

Example 10.3.3. Here are some values of the Legendre symbol:

$$\left(\frac{21}{7}\right) = 0, \ \left(\frac{23}{7}\right) = 1, \ \left(\frac{3}{7}\right) = -1, \ \left(\frac{73}{7}\right) = -1.$$

Reasons: 7 divides 21, $23 \equiv 2 \equiv 3^2 \bmod 7$, 3 is a QNR mod 7, and $73 \equiv 3 \bmod 7$, so 73 is also a QNR.

Here are some basic properties of the Legendre symbol:

Lemma 10.3.4. *Let $p > 2$ be a prime and let a, b be integers relatively prime to p. Then:*

(1) $\left(\frac{ab^2}{p}\right) = \left(\frac{a}{p}\right)$. *In particular,* $\left(\frac{b^2}{p}\right) = 1$.

(2) *If $a \equiv b \bmod p$, then* $\left(\frac{a}{p}\right) = \left(\frac{b}{p}\right)$.

(3) $\left(\frac{-1}{p}\right) = (-1)^{(p-1)/2}$.

Proof. Let $p > 2$ be a prime and let $\gcd(a,p) = \gcd(b,p) = 1$. Then:

(1) Since a, b are units modulo p, then ab^2 is also a unit. The number $ab^2 \equiv d^2 \bmod p$ if and only if $a \equiv d^2b^{-2} \equiv (db^{-1})^2 \bmod p$. Thus, ab^2 is a QR if and only if a is a QR mod p. Notice that b^2 is a square, and so it is a QR mod p, for any b relatively prime to p.

(2) Suppose $a \equiv b \bmod p$. Then $a \equiv d^2 \bmod p$ if and only if $b \equiv a \equiv d^2 \bmod p$. Hence, a is a QR if and only if b is a QR.

(3) In Theorem 10.2.8, we showed that -1 is a square modulo p if and only if $p \equiv 1 \bmod 4$. Since p is odd, $p \equiv 1$ or $3 \bmod 4$, and $(p-1)/2$ is even (and so $(-1)^{(p-1)/2} = 1$) if and only if $p \equiv 1 \bmod 4$. Thus, $\left(\frac{-1}{p}\right) = (-1)^{(p-1)/2}$, as claimed. □

Example 10.3.5. *Is 39 a quadratic residue modulo 43?* Equivalently, what is the value of $\left(\frac{39}{43}\right)$? Notice that $39 \equiv -4 \bmod 43$. Thus,

$$\left(\frac{39}{43}\right) = \left(\frac{-4}{43}\right) = \left(\frac{(-1)\cdot 4}{43}\right) = \left(\frac{-1}{43}\right) = (-1)^{\frac{42}{2}} = (-1)^{21} = -1,$$

where we have used properties (2), then (1), and then (3). Hence, 39 is a quadratic non-residue modulo 43.

Proposition 10.3.6 (Euler's criterion)**.** *Let $p > 2$ be a prime and let $a \in \mathbb{Z}$ be relatively prime to p. Then,*

$$\left(\frac{a}{p}\right) \equiv a^{\frac{p-1}{2}} \bmod p.$$

Proof. Let $p > 2$ be a prime, and let $\gcd(a,p) = 1$. Let $\beta \in \mathbb{Z}$ be a primitive root modulo p and let n be the least positive integer such that $\beta^n \equiv a \bmod p$.

- If $a \bmod p$ is a QR, then $n = 2m$ is even, by Lemma 10.2.7. Thus,

$$a^{\frac{p-1}{2}} \equiv (\beta^{2m})^{\frac{p-1}{2}} \equiv (\beta^{p-1})^m \equiv 1 \equiv \left(\frac{a}{p}\right) \bmod p.$$

- If $a \bmod p$ is a QNR, then $n = 2m+1$ is odd. Thus,

$$a^{\frac{p-1}{2}} \equiv (\beta^{2m+1})^{\frac{p-1}{2}} \equiv (\beta^{\frac{p-1}{2}})^{2m+1} \equiv (-1)^{2m+1} \equiv -1 \equiv \left(\frac{a}{p}\right) \bmod p$$

  where we have used the fact that $\beta^{(p-1)/2} \equiv -1 \bmod p$ because β is a primitive root.

Thus, in both cases $\left(\frac{a}{p}\right) \equiv a^{\frac{p-1}{2}} \bmod p$. □

Remark 10.3.7. Alternatively, since $p > 2$ is odd, Corollary 8.6.14 says that a is a quadratic residue mod p if and only if $a^{(p-1)/2} \equiv 1 \bmod p$. Since $a^{(p-1)/2} \equiv \pm 1 \bmod p$, for all units $a \bmod p$, Euler's criterion follows.

We offer yet another proof of Euler's criterion that does not rely on primitive roots.

Alternative proof of Euler's criterion. Suppose first that $a \in \mathbb{Z}$ is relatively prime to p and that it is a quadratic residue mod p, so that $a \equiv b^2 \bmod p$, for some integer b. Then,

$$a^{(p-1)/2} \equiv (b^2)^{(p-1)/2} \equiv b^{p-1} \equiv 1 \bmod p$$

by Fermat's little theorem.

Now suppose that a is not a quadratic residue. Since $(a^{(p-1)/2})^2 \equiv a^{p-1} \equiv 1 \bmod p$, it follows that $a^{(p-1)/2}$ is a root of $x^2 \equiv 1 \bmod p$, so $a^{(p-1)/2} \equiv \pm 1 \bmod p$ by Lemma 7.1.4. The polynomial $x^{(p-1)/2} \equiv 1 \bmod p$ has at most $(p-1)/2$ roots in $\mathbb{Z}/p\mathbb{Z}$ by Theorem 5.5.19, and the $(p-1)/2$ quadratic residues mod p are roots (we know there are $(p-1)/2$ quadratic residues by Proposition 10.2.3). Thus, a quadratic non-residue cannot be a root of $x^{(p-1)/2} \equiv 1 \bmod p$, and it follows that $a^{(p-1)/2} \equiv -1 \bmod p$, as claimed. □

Example 10.3.8. *Is* 2 *a quadratic residue modulo* 17*?* We shall use Euler's criterion which, in this case, says that

$$\left(\frac{2}{17}\right) \equiv 2^8 \bmod 17.$$

Clearly, $2^4 \equiv 16 \equiv -1 \bmod 17$ and so $2^8 \equiv (-1)^2 \equiv 1 \bmod 17$. Therefore, 2 is a quadratic residue.

The following result is a (surprising) corollary of Euler's criterion.

Corollary 10.3.9. *Let $p > 2$ be prime and let $a, b \in \mathbb{Z}$. Then,*

$$\left(\frac{a}{p}\right) \cdot \left(\frac{b}{p}\right) = \left(\frac{ab}{p}\right).$$

Proof. If a or $b \equiv 0 \bmod p$, then the equality is trivial $(0 = 0)$. Otherwise, we shall use Euler's criterion three times:

$$\begin{aligned}\left(\frac{a}{p}\right)\cdot\left(\frac{b}{p}\right) &\equiv a^{\frac{p-1}{2}}\cdot b^{\frac{p-1}{2}}\\ &\equiv (ab)^{\frac{p-1}{2}}\\ &\equiv \left(\frac{ab}{p}\right) \bmod p.\end{aligned}$$

Since the integers involved are all equal to ± 1, if they are congruent modulo $p > 2$, then they must be equal. □

Remark 10.3.10. If the reader is not surprised by the statement of Corollary 10.3.9, then try reading the statement once again. Part of the statement is not surprising at all: if a and b are quadratic residues, then $a \equiv d^2$ and $b \equiv e^2 \bmod p$ and, of course, $ab \equiv (de)^2 \bmod p$ and so ab is also a QR. The surprising part of the statement is that if a and b are both quadratic non-residues, then ab must be a quadratic residue! Thus, the previous corollary can be interpreted as follows:

$$\mathrm{QR}\times\mathrm{QR} = \mathrm{QR},\quad \mathrm{QR}\times\mathrm{QNR} = \mathrm{QNR},\quad \text{and}\quad \mathrm{QNR}\times\mathrm{QNR} = \mathrm{QR}.$$

Example 10.3.11. *Let $p = 17$. Find all the quadratic residues modulo p.*

We know that there are exactly $(17-1)/2 = 8$ quadratic residues, by Proposition 10.2.3. Clearly, $1, 4, 9$, and $16 \equiv -1 \bmod 17$ are quadratic residues. Thus, all products of two of them must also be a QR. So $4\cdot 9 \equiv 36 \equiv 2$, $-4 \equiv 13$, and $-9 \equiv 8 \bmod 17$ are also QRs. And $-2 \equiv 15$ must be a QR as well. Hence, the quadratic residues modulo 17 are $1, 2, 4, 8, 9, 13, 15$, and $16 \bmod 17$.

Notice that the set of congruences $\{1, 2, 4, 8, 9, 13, 15, 16\}$ modulo 17 is closed under multiplication. Now pick two quadratic non-residues, for example 3 and 5. Then, their product 15 must be a quadratic residue, and it is! Indeed $7^2 \equiv 15 \bmod 17$.

Lemma 10.3.4 determines when -1 is a square modulo p. When is 2 a quadratic residue mod p? Is 2 a quadratic residue modulo 4001? How about modulo 4003?

Theorem 10.3.12. *Let $p > 2$ be a prime. Then,*

$$\left(\frac{2}{p}\right) = (-1)^{\frac{p^2-1}{8}}.$$

In other words, 2 is a quadratic residue of p if $p \equiv \pm 1 \bmod 8$ and a quadratic non-residue if $p \equiv \pm 3 \bmod 8$.

Proof. First, we explain why the two formulations are equivalent. Let $p = a + 8k$. Then $(p^2-1)/8 = (a^2-1)/8 + 2ak + 8k^2$ is even for $a = 1, 7$ and odd for $a = 3, 5$. Hence, $(-1)^{(p^2-1)/8} = 1$ for $p \equiv 1, 7 \bmod 8$ and equals -1 for $p \equiv 3, 5 \bmod 8$.

In order to prove the theorem, we will use Euler's criterion, so we will calculate $2^{(p-1)/2} \bmod p$. We distinguish two cases:

- If $p \equiv 1$ or $5 \bmod 8$ (i.e., $p \equiv 1 \bmod 4$), we can write

$$\begin{aligned} 2^{\frac{p-1}{2}}\left(\frac{p-1}{2}\right)! &\equiv 2^{\frac{p-1}{2}} \cdot 1 \cdot 2 \cdot 3 \cdots \left(\frac{p-1}{2}\right) \\ &\equiv 2^{\frac{p-1}{2}} \cdot 1 \cdot 2 \cdot 3 \cdots \left(\frac{p-1}{4}\right) \cdot \left(\frac{p+3}{4}\right) \cdots \left(\frac{p-1}{2}\right) \\ &\equiv 2 \cdot 4 \cdot 6 \cdots \left(\frac{p-1}{2}\right) \cdot \left(\frac{p+3}{2}\right) \cdots (p-3) \cdot (p-1) \\ &\equiv 2 \cdot 4 \cdot 6 \cdots \left(\frac{p-1}{2}\right) \cdot \left(-\frac{p-3}{2}\right) \cdots (-3) \cdot (-1) \\ &\equiv 2 \cdot 4 \cdot 6 \cdots \left(\frac{p-1}{2}\right) \cdot \left(\frac{p-3}{2}\right) \cdots (3) \cdot (1) \cdot (-1)^{\frac{p-1}{4}} \\ &\equiv 1 \cdot 2 \cdot 3 \cdots \left(\frac{p-1}{2}\right) \cdot (-1)^{\frac{p-1}{4}} \\ &\equiv (-1)^{\frac{p-1}{4}} \cdot \left(\frac{p-1}{2}\right)! \bmod p \end{aligned}$$

  and since $((p-1)/2)! \bmod p$ is a unit, we obtain

$$\left(\frac{2}{p}\right) \equiv 2^{\frac{p-1}{2}} \equiv (-1)^{\frac{p-1}{4}} \bmod p$$

  which equals 1 if $p \equiv 1 \bmod 8$ and equals -1 if $p \equiv 5 \bmod 8$.

- If $p \equiv 3$ or $7 \bmod 8$ (i.e., $p \equiv 3 \bmod 4$), we can work similarly:

$$\begin{aligned} 2^{\frac{p-1}{2}}\left(\frac{p-1}{2}\right)! &\equiv 2 \cdot 4 \cdot 6 \cdots \left(\frac{p-3}{2}\right) \cdot \left(\frac{p+1}{2}\right) \cdots (p-3) \cdot (p-1) \\ &\equiv 2 \cdot 4 \cdot 6 \cdots \left(\frac{p-3}{2}\right) \cdot \left(-\frac{p-1}{2}\right) \cdots (-3) \cdot (-1) \\ &\equiv 1 \cdot 2 \cdot 3 \cdots \left(\frac{p-1}{2}\right) \cdot (-1)^{\frac{p+1}{4}} \\ &\equiv (-1)^{\frac{p+1}{4}} \cdot \left(\frac{p-1}{2}\right)! \bmod p. \end{aligned}$$

  Thus, $\left(\frac{2}{p}\right) \equiv 2^{\frac{p-1}{2}} \equiv (-1)^{\frac{p+1}{4}} \bmod p$ which equals 1 if $p \equiv 7 \bmod 8$ and equals -1 if $p \equiv 3 \bmod 8$.

□

Example 10.3.13. The numbers $p = 4001$ and $q = 4003$ are primes. The number 2 is a quadratic residue for 4001 (because $4001 \equiv 1 \bmod 8$) but it is not a quadratic residue modulo 4003. Notice that finding a square root of 2 modulo 4001 is not an easy task! But Theorem 10.3.12 says there is one. Here is one root: $1156^2 \equiv 2 \bmod 4001$.

Example 10.3.14. *Is 12 a quadratic residue modulo 43?*

One may be tempted to attack this problem as follows:

$$\left(\frac{12}{43}\right)=\left(\frac{3}{43}\right)\cdot\left(\frac{4}{43}\right)=\left(\frac{3}{43}\right).$$

However, it is not clear *a priori* whether 3 is a QR or a QNR. Instead, let us try another way:

$$\left(\frac{12}{43}\right)=\left(\frac{2}{43}\right)\cdot\left(\frac{6}{43}\right).$$

It is not hard to find a square root of 6 modulo 43 because $6 \equiv 49 \bmod 43$. Hence 6 is a QR. Moreover, by Theorem 10.3.12, the number 2 is a QNR modulo 43 (because $43 \equiv 3 \bmod 8$). Thus,

$$\left(\frac{12}{43}\right)=\left(\frac{2}{43}\right)\cdot\left(\frac{6}{43}\right)=(-1)\cdot 1=-1.$$

Therefore, 12 is not a square modulo 43. By the way, this implies that 3 is not a square either, because, as we saw above, $\left(\frac{12}{43}\right)=\left(\frac{3}{43}\right)$.

10.4. The Law of Quadratic Reciprocity

Thus far, we know how to calculate the following Legendre symbols:

$$\left(\frac{-1}{p}\right),\quad \left(\frac{2}{p}\right),\quad \text{and}\quad \left(\frac{a^2}{p}\right).$$

Suppose we want to calculate $\left(\frac{n}{p}\right)$. Since the Legendre symbol is multiplicative (by Corollary 10.3.9) and if n has a prime factorization $n=q_1^{e_1}q_2^{e_2}\cdots q_r^{e_r}$, for some primes q_i, then,

$$\left(\frac{n}{p}\right)=\left(\frac{q_1^{e_1}q_2^{e_2}\cdots q_r^{e_r}}{p}\right)=\left(\frac{q_1^{e_1}}{p}\right)\cdot\left(\frac{q_2^{e_2}}{p}\right)\cdots\left(\frac{q_r^{e_r}}{p}\right)=\left(\frac{q_1}{p}\right)^{e_1}\cdots\left(\frac{q_r}{p}\right)^{e_r}.$$

Therefore, if we knew how to calculate $\left(\frac{q}{p}\right)$, for any odd primes q and p, then we could calculate $\left(\frac{n}{p}\right)$ for any $n\in\mathbb{Z}$.

Example 10.4.1. Just to fix ideas, let us repeat the same discussion with some concrete numbers. Let $n=151875000$ and calculate

$$\left(\frac{151875000}{151875023}\right).$$

Yes, 151875023 is prime! First, factor n. It turns out that $n=151875000=2^3\cdot 3^5\cdot 5^7$. Thus,

$$\begin{aligned}\left(\frac{151875000}{151875023}\right) &= \left(\frac{2^3\cdot 3^5\cdot 5^7}{151875023}\right)\\ &= \left(\frac{2^3}{151875023}\right)\cdot\left(\frac{3^5}{151875023}\right)\cdot\left(\frac{5^7}{151875023}\right)\\ &= \left(\frac{2}{151875023}\right)^3\cdot\left(\frac{3}{151875023}\right)^5\cdot\left(\frac{5}{151875023}\right)^7.\end{aligned}$$

By Theorem 10.3.12, $\left(\frac{2}{1518775023}\right) = 1$ because the prime in the denominator is congruent to 7 mod 8. Now, if we were able to calculate $\left(\frac{3}{1518775023}\right)$ and $\left(\frac{5}{1518775023}\right)$, we would be done! This is precisely what the law of quadratic reciprocity will help us accomplish, in a simple way.

The law of quadratic reciprocity (see Theorem 10.4.2 below) was conjectured by Euler and Legendre and first proven by Gauss. He referred to it as the "fundamental theorem" in the *Disquisitiones Arithmeticae* and his papers, going as far as to write:

> *The fundamental theorem must certainly be regarded as one of the most elegant of its type.*

Gauss published six proofs in his lifetime, and two more were found in his posthumous papers. There are now over 200 published proofs.

Figure 10.1. Johann Carl Friedrich Gauss (1777–1855) was a German mathematician and physical scientist who contributed significantly to many fields, including number theory, algebra, statistics, analysis, differential geometry, geodesy, geophysics, electrostatics, astronomy, and optics. Image source: Wikimedia Commons.

Theorem 10.4.2 (Law of quadratic reciprocity). *Let p and q be two distinct odd primes. Then,*

$$\left(\frac{q}{p}\right) \cdot \left(\frac{p}{q}\right) = (-1)^{\frac{p-1}{2} \cdot \frac{q-1}{2}}.$$

In other words:

- *If $p \equiv 1 \bmod 4$ or $q \equiv 1 \bmod 4$, then*
$$\left(\frac{q}{p}\right) = \left(\frac{p}{q}\right).$$
- *If $p \equiv q \equiv 3 \bmod 4$, then*
$$\left(\frac{q}{p}\right) = -\left(\frac{p}{q}\right).$$

Example 10.4.3. Let $p = 37$ and $q = 41$. Since $41 \equiv 4 \equiv 2^2 \bmod 37$, it follows that

$$\left(\frac{q}{p}\right) = \left(\frac{41}{37}\right) = \left(\frac{4}{37}\right) = 1.$$

Using the fact that $41 \equiv 1 \bmod 4$, the law of quadratic reciprocity says that

$$1 = \left(\frac{41}{37}\right) = \left(\frac{37}{41}\right)$$

and, therefore, 37 is a quadratic residue modulo 41, which is not obvious. Indeed, $37 \equiv (\pm 18)^2 \bmod 41$.

Example 10.4.4. Let $q = 3$ and let p be a prime congruent to 3 mod 4 and 2 mod 3; i.e., $p \equiv 11 \bmod 12$. Dirichlet's theorem on primes in arithmetic progressions (Theorem 3.3.11) implies that there are infinitely many such primes p. For instance, $p = 11$ or $p = 563$ are primes congruent to 11 mod 12. Since $p \equiv 2 \bmod 3$, it follows that

$$\left(\frac{p}{q}\right) = \left(\frac{p}{3}\right) = \left(\frac{2}{3}\right) = -1.$$

And since $p \equiv q \equiv 3 \bmod 4$, the law of quadratic reciprocity shows that

$$\left(\frac{3}{p}\right) = \left(\frac{q}{p}\right) = -\left(\frac{p}{q}\right) = -\left(\frac{p}{3}\right) = -(-1) = 1.$$

Thus, 3 is a quadratic residue for all primes $p \equiv 11 \bmod 12$. For instance,

$$3 \equiv (\pm 5)^2 \bmod 11 \quad \text{and} \quad 3 \equiv (\pm 121)^2 \bmod 563.$$

See Proposition 10.4.8 for a more general result about the value of $\left(\frac{3}{p}\right)$.

Out of the many published proofs of the law of quadratic reciprocity, we chose to follow along the lines of a proof due to G. Rousseau (see [**Rou91**]) because it matches the techniques we have already employed to prove the supplementary laws for $\left(\frac{-1}{p}\right)$ and $\left(\frac{2}{p}\right)$ in Theorem 10.2.8 (see the alternative proof at the end of Section 10.2) and Theorem 10.3.12, respectively. For a different proof see [**Chi95**, p. 405], for example.

Proof. Consider $G = (\mathbb{Z}/p\mathbb{Z})^\times \times (\mathbb{Z}/q\mathbb{Z})^\times$. We want to describe a subset $H \subseteq G$ with the following property:

(♠) For every $g \in G$, either g or $-g$ belongs to H but not both.

Here $-(a \bmod p, b \bmod q) \equiv (-a \bmod p, -b \bmod q)$. We present two ways to define such a subset H:

(1) Let $H_1 = (\mathbb{Z}/p\mathbb{Z})^\times \times \{1, \ldots, (q-1)/2 \bmod q\}$; i.e., H_1 is the direct product of $(\mathbb{Z}/p\mathbb{Z})^\times$ times "the first half" of $(\mathbb{Z}/q\mathbb{Z})^\times$. In this case,

$$H_1 = \{(a, b) : 1 \leq a \leq p-1,\ 1 \leq b \leq (q-1)/2\} \subseteq G.$$

(2) Notice that $G = (\mathbb{Z}/p\mathbb{Z})^\times \times (\mathbb{Z}/q\mathbb{Z})^\times$ is in bijection with $(\mathbb{Z}/pq\mathbb{Z})^\times$, by Corollary 7.4.4. Let H_2 be "the first half" of $(\mathbb{Z}/pq\mathbb{Z})^\times$, i.e., those elements in G of

the form $(k \bmod p, k \bmod q)$, with $\gcd(k, pq) = 1$ and $1 \le k \le (pq-1)/2$. In other words,

$$H_2 = \{(k,k) : 1 \le k \le (pq-1)/2,\ \gcd(k,pq) = 1\} \subseteq G.$$

Now let us define π_1 as the product of all the elements in H_1 and similarly define π_2 as the product of all the elements in H_2. Before we go on, we prove a lemma about π_1 and π_2.

Lemma 10.4.5. *Let p and q be two distinct odd primes. Let $G = (\mathbb{Z}/p\mathbb{Z})^\times \times (\mathbb{Z}/q\mathbb{Z})^\times$, and let H_1 and H_2 be two subsets of G that satisfy property (♠). Let π_1 be the product of all elements in H_1, and define π_2 similarly. Then $\pi_2 \equiv u \cdot \pi_1$, where $u \equiv (1,1)$ or $(-1,-1) \bmod (p,q)$. In other words, either $\pi_2 \equiv \pi_1$ or $\pi_2 \equiv -\pi_1 \bmod (p,q)$.*

Proof. Let $H_1 = \{g_1, \ldots, g_n\}$ and $H_2 = \{h_1, \ldots, h_n\}$. Since H_2 satisfies (♠), it follows that for every $i = 1, \ldots, n$ there is a *unique* $h_j \in H_2$ such that $h_j \equiv \pm g_i$. Notice that h_j is unique, because if $h_k \not\equiv h_j$ also satisfies $h_k = \pm g_i$, then we must have $h_k \equiv -h_j$, which is a contradiction because H_2 satisfies (♠). After a reordering, we may assume that $h_i = (-1)^{e_i} g_i$, with $e_i = 0$ or 1. Hence,

$$\pi_2 \equiv \prod_{i=1}^{n} h_i \equiv \prod_{i=1}^{n} (-1)^{e_i} g_i \equiv (-1)^{(\sum_{i=1}^n e_i)} \cdot \prod_{i=1}^{n} g_i \equiv u \cdot \pi_1 \bmod (p,q),$$

where $u = (-1)^{(\sum_{i=1}^n e_i)}$. □

Let us resume the proof of the law of quadratic reciprocity. First, we calculate π_1:

$$\begin{aligned}\pi_1 &\equiv \left(\prod_{i=1}^{p-1} i^{(q-1)/2} \bmod p,\ \prod_{j=1}^{(q-1)/2} j^{p-1} \bmod q\right)\\ &\equiv \left(((p-1)!)^{\frac{q-1}{2}} \bmod p,\ \left(\left(\frac{q-1}{2}\right)!\right)^{p-1} \bmod q\right).\end{aligned}$$

Notice that $((q-1)/2)!^2 \equiv (-1)^{(q-1)/2}(q-1)! \bmod q$, so

$$\left(\left(\frac{q-1}{2}\right)!\right)^{p-1} \equiv ((-1)^{\frac{q-1}{2}}(q-1)!)^{\frac{p-1}{2}} \equiv (-1)^{\frac{p-1}{2}\cdot\frac{q-1}{2}} \cdot ((q-1)!)^{\frac{p-1}{2}} \bmod q.$$

Hence,

$$\pi_1 \equiv (((p-1)!)^{\frac{q-1}{2}} \bmod p, (-1)^{\frac{p-1}{2}\cdot\frac{q-1}{2}} \cdot ((q-1)!)^{\frac{p-1}{2}} \bmod q).$$

Let us now calculate the first coordinate of π_2 modulo p. We will multiply together all numbers that are not multiples of p and divide by the multiples of q, in the range $1 \le k \le (pq-1)/2$:

$$\prod_{\substack{1\le k\le (pq-1)/2\\ \gcd(k,pq)=1}} k \equiv \frac{(1\cdot 2\cdots(p-1))((p+1)(p+2)\cdots(p+(p-1)))\cdots}{q\cdot(2q)\cdot(3q)\cdots(\frac{p-1}{2}q)}$$

$$\equiv \frac{\left(\prod_{i=1}^{p-1} i\right)\left(\prod_{i=1}^{p-1} p+i\right)\cdots\left(\prod_{i=1}^{p-1}(\frac{q-1}{2}-1)p+i\right)\left(\prod_{i=1}^{(p-1)/2}\frac{q-1}{2}p+i\right)}{((p-1)/2)!\cdot q^{(p-1)/2}}$$

$$\equiv \frac{((p-1)!)^{\frac{q-1}{2}}\cdot((p-1)/2)!}{((p-1)/2)!\cdot q^{(p-1)/2}}$$

$$\equiv \frac{((p-1)!)^{\frac{q-1}{2}}}{q^{(p-1)/2}} \equiv ((p-1)!)^{\frac{q-1}{2}}\cdot\left(\frac{q}{p}\right) \bmod p,$$

where in the last congruence we have used Euler's criterion (Proposition 10.3.6), the fact that $((p-1)/2)!$ is a unit modulo p, and the trivial fact that $\frac{1}{\pm 1}=\pm 1$, so that $\left(\frac{q}{p}\right)^{-1}=\left(\frac{q}{p}\right)$. A similar calculation yields

$$\prod_{\substack{1\le k\le (pq-1)/2\\ \gcd(k,pq)=1}} k \equiv ((q-1)!)^{\frac{p-1}{2}}\cdot\left(\frac{p}{q}\right) \bmod q.$$

Therefore,

$$\pi_2 \equiv \left(((p-1)!)^{\frac{q-1}{2}}\cdot\left(\frac{q}{p}\right) \bmod p,\ ((q-1)!)^{\frac{p-1}{2}}\cdot\left(\frac{p}{q}\right) \bmod q\right).$$

By Lemma 10.4.5, there is a $u=\pm 1$ such that $\pi_2\equiv u\pi_1$. Thus,

$$((p-1)!)^{\frac{q-1}{2}}\cdot\left(\frac{q}{p}\right)\equiv u\cdot((p-1)!)^{\frac{q-1}{2}} \bmod p$$

and

$$((q-1)!)^{\frac{p-1}{2}}\cdot\left(\frac{p}{q}\right)\equiv u\cdot(-1)^{\frac{p-1}{2}\cdot\frac{q-1}{2}}\cdot((q-1)!)^{\frac{p-1}{2}} \bmod q.$$

If we cancel the appropriate terms in each of the two equations above, we obtain

$$\left(\frac{q}{p}\right)\equiv u \bmod p \quad\text{and}\quad \left(\frac{p}{q}\right)\equiv u\cdot(-1)^{\frac{p-1}{2}\cdot\frac{q-1}{2}} \bmod q.$$

Since all the numbers involved are by definition ± 1 and p and q are odd, these congruences are actually equalities. If we multiply both equalities together, we obtain

$$\left(\frac{q}{p}\right)\cdot\left(\frac{p}{q}\right)=u^2\cdot(-1)^{\frac{p-1}{2}\cdot\frac{q-1}{2}}=(-1)^{\frac{p-1}{2}\cdot\frac{q-1}{2}},$$

as desired (here we used the fact that $u=\pm 1$, so $u^2=1$). □

Example 10.4.6. Let us continue with the calculation of Example 10.4.1. We needed to calculate $\left(\frac{3}{151875023}\right)$ and $\left(\frac{5}{151875023}\right)$. Notice that $151875023\equiv 3 \bmod 4$.

Therefore, by the law of quadratic reciprocity,

$$\begin{aligned}\left(\frac{3}{151875023}\right) &= -\left(\frac{151875023}{3}\right) = -\left(\frac{2}{3}\right) = -(-1) = 1,\\ \left(\frac{5}{151875023}\right) &= \left(\frac{151875023}{5}\right) = \left(\frac{3}{5}\right) = -1.\end{aligned}$$

In the first line, besides the law of quadratic reciprocity, we used the facts that $151875023 \equiv 2 \bmod 3$ and $151875023 \equiv 3 \bmod 5$. Therefore, we now may conclude the exercise:

$$\begin{aligned}\left(\frac{151875000}{151875023}\right) &= \left(\frac{2}{151875023}\right)^3 \cdot \left(\frac{3}{151875023}\right)^5 \cdot \left(\frac{5}{151875023}\right)^7\\ &= 1^3 \cdot 1^5 \cdot (-1)^7 = -1.\end{aligned}$$

Hence, 151875000 is a quadratic non-residue modulo 151875023.

We can also use the law of quadratic reciprocity to find rules for particular primes. For example:

Example 10.4.7. *For what primes $p > 2$ is 5 a quadratic residue modulo p?* Let $p > 2$ and $p \neq 5$. By quadratic reciprocity,

$$\left(\frac{5}{p}\right) = \left(\frac{p}{5}\right) = \begin{cases} 1 & \text{if } p \equiv \pm 1 \bmod 5,\\ -1 & \text{if } p \equiv \pm 2 \bmod 5.\end{cases}$$

Here we have used the fact that $p \equiv 1 \bmod 5$ to be able to conclude $\left(\frac{5}{p}\right) = \left(\frac{p}{5}\right)$.

Proposition 10.4.8. *Let $p > 2$, with $p \neq 3$, be a prime. Then,*

$$\left(\frac{3}{p}\right) = \begin{cases} 1 & \text{if } p \equiv \pm 1 \bmod 12,\\ -1 & \text{if } p \equiv \pm 5 \bmod 12.\end{cases}$$

Proof. Let $p > 2$ be prime.

- If $p \equiv 1 \bmod 4$, then

$$\left(\frac{3}{p}\right) = \left(\frac{p}{3}\right) = \begin{cases} 1 & \text{if } p \equiv 1 \bmod 3,\\ -1 & \text{if } p \equiv 2 \bmod 3.\end{cases}$$

  Thus, if $p \equiv 1 \bmod 4$ and $p \equiv 1 \bmod 3$ (so $p \equiv 1 \bmod 12$, by the Chinese remainder theorem), then $\left(\frac{3}{p}\right) = 1$. If $p \equiv 5 \bmod 12$, then $\left(\frac{3}{p}\right) = -1$.

- If $p \equiv 3 \bmod 4$, then

$$\left(\frac{3}{p}\right) = -\left(\frac{p}{3}\right) = \begin{cases} -1 & \text{if } p \equiv 1 \bmod 3,\\ 1 & \text{if } p \equiv 2 \bmod 3.\end{cases}$$

  Thus, if $p \equiv 3 \bmod 4$ and $p \equiv 1 \bmod 3$ (so $p \equiv 7 \bmod 12$, by the Chinese remainder theorem), then $\left(\frac{3}{p}\right) = -1$. If $p \equiv 11 \bmod 12$, then $\left(\frac{3}{p}\right) = 1$. $\square$

Example 10.4.9. *Is 40 a square modulo 43?*

$$\left(\frac{40}{43}\right) = \left(\frac{-3}{43}\right) = \left(\frac{-1}{43}\right)\left(\frac{3}{43}\right) = (-1)\cdot(-1) = 1$$

where we have used the fact that $43 \equiv 3 \bmod 4$ and the fact that $41 \equiv 7 \bmod 12$. Thus, 40 is a square modulo 43. Indeed, $13^2 \equiv 40 \bmod 43$.

10.5. The Jacobi Symbol

The Jacobi symbol is an extension of the Legendre symbol that allows (odd) composite numbers in the lower part of the notation. For instance, $\left(\frac{-1}{21}\right)$ is not a well-defined Legendre symbol (as 21 is not prime), but it will be a perfectly fine Jacobi symbol. The Jacobi symbol is very useful in practice for fast computation of Legendre symbols because its properties help us avoid factoring large numbers as we carry out a calculation of a symbol (see Example 10.5.10).

Definition 10.5.1. Let a be an integer, and let n be an odd natural number. The *Jacobi symbol* $\left(\frac{a}{n}\right)$ is defined by

$$\left(\frac{a}{n}\right) = \left(\frac{a}{p_1}\right)^{e_1}\cdot\left(\frac{a}{p_2}\right)^{e_2}\cdots\left(\frac{a}{p_t}\right)^{e_t},$$

where $n = p_1^{e_1}p_2^{e_2}\cdots p_t^{e_t}$ is a factorization of n as a product of prime numbers $p_1,\ldots,p_t \geq 1$, with $t \geq 1$, and $\left(\dfrac{a}{p_i}\right)$ is the Legendre symbol, for each $1 \leq i \leq t$. If $n = 1$, we define $\left(\frac{a}{1}\right) = 1$ for $a \neq 0$ and $\left(\frac{0}{1}\right) = 0$.

Figure 10.2. Carl Gustav Jacob Jacobi (1804–1851) was a German mathematician who made fundamental contributions to elliptic functions, dynamics, differential equations, and number theory. He introduced the Jacobi symbol in 1837. Image source: Wikimedia Commons.

Example 10.5.2. We can calculate the Jacobi symbol

$$\left(\frac{-1}{21}\right) = \left(\frac{-1}{3}\right) \cdot \left(\frac{-1}{7}\right) = (-1) \cdot (-1) = 1,$$

where we have used Lemma 10.3.4 to calculate the Legendre symbols $\left(\frac{-1}{3}\right) = -1$ and $\left(\frac{-1}{7}\right) = -1$.

Remark 10.5.3. It is important to notice that a positive value $\left(\frac{a}{n}\right) = 1$ of the Jacobi symbol *does not imply* that a is a square modulo n (i.e., that $x^2 \equiv a \bmod n$ has a solution), as it would be the case for the Legendre symbol. For instance, consider Example 10.5.2 where we calculated $\left(\frac{-1}{21}\right) = 1$. However, -1 is not a square modulo 21, because if $x^2 \equiv -1 \bmod 21$ had a solution, then, by the Chinese remainder theorem (Theorem 4.5.9), the congruences $x^2 \equiv -1 \bmod 3$ and $x^2 \equiv -1 \bmod 7$ would have a solution. But -1 is not a quadratic residue modulo 3 or 7.

However, if the Jacobi symbol $\left(\frac{a}{n}\right) = -1$, then it does follow that a is necessarily a quadratic non-residue modulo n. Indeed, if $n = p_1^{e_1} p_2^{e_2} \cdots p_t^{e_t}$ is the unique factorization of n and

$$\left(\frac{a}{n}\right) = \left(\frac{a}{p_1}\right)^{e_1} \cdot \left(\frac{a}{p_2}\right)^{e_2} \cdots \left(\frac{a}{p_t}\right)^{e_t} = -1,$$

then there must be a prime p in the factorization of n such that $\left(\frac{a}{p}\right) = -1$, so that a is a quadratic non-residue mod p and p divide n. Suppose now for a contradiction that a is a quadratic residue mod n. Then, there is another integer b such that $a \equiv b^2 \bmod n$. Since $p \mid n$, it follows that $a \equiv b^2 \bmod p$, which contradicts $\left(\frac{a}{p}\right) = -1$. Thus, a must be a quadratic non-residue mod n.

Remark 10.5.4. There is a further generalization of the Legendre and Jacobi symbols, called the Kronecker symbol, that gives the expression $\left(\frac{a}{n}\right)$ a value, for any integers a and n (including negative or even values of n). However, we will not need the Kronecker symbol and we will not define it here.

Let us show the basic properties of the Jacobi symbol, which follow directly from the definition and the analogous properties of the Legendre symbol.

Proposition 10.5.5. *Let a, b be integers, and let m, n be odd natural numbers. Then, the following properties of the Jacobi symbol hold:*

(1) *If* $a \equiv b \bmod n$, *then* $\left(\frac{a}{n}\right) = \left(\frac{b}{n}\right)$.

(2) $\left(\frac{a}{mn}\right) = \left(\frac{a}{m}\right) \cdot \left(\frac{a}{n}\right)$.

(3) $\left(\frac{a}{n}\right) = 0$ *if and only if* $a = 0$ *or* $\gcd(a, n) > 1$.

(4) $\left(\frac{ab}{n}\right) = \left(\frac{a}{n}\right) \cdot \left(\frac{b}{n}\right)$.

Proof. Let $m = q_1^{f_1} \cdots q_s^{f_s}$ and $n = p_1^{e_1} \cdots p_t^{e_t}$ be the unique prime factorizations of m and n, respectively. Then:

(1) Notice that $a \equiv b \bmod n$ and $p \mid n$ imply that $a \equiv b \bmod p$. Thus,
$$\left(\frac{a}{n}\right) = \left(\frac{a}{p_1}\right)^{e_1} \cdots \left(\frac{a}{p_t}\right)^{e_t} = \left(\frac{b}{p_1}\right)^{e_1} \cdots \left(\frac{b}{p_t}\right)^{e_t} = \left(\frac{b}{n}\right),$$
where we have used the fact that $\left(\frac{a}{p}\right) = \left(\frac{b}{p}\right)$ for the Legendre symbol, whenever $a \equiv b \bmod p$, which was shown in Lemma 10.3.4.

(2) Note that $mn = q_1^{f_1} \cdots q_s^{f_s} p_1^{e_1} \cdots p_t^{e_t}$, where some primes may be repeated (i.e., $p_i = q_j$ is possible for some $i, j \geq 1$). Thus,
$$\left(\frac{a}{mn}\right) = \left(\frac{a}{q_1}\right)^{f_1} \cdots \left(\frac{a}{q_s}\right)^{f_s} \cdot \left(\frac{a}{p_1}\right)^{e_1} \cdots \left(\frac{a}{p_t}\right)^{e_t} = \left(\frac{a}{m}\right) \cdot \left(\frac{a}{n}\right).$$

(3) If $\gcd(a, n) = 1$, then $\left(\dfrac{a}{p}\right) = \pm 1$ for each prime divisor p of n, and therefore $\left(\dfrac{a}{n}\right) = \pm 1$ as well. Conversely, if the Jacobi symbol of a on n vanishes, then there must be a prime $p \mid n$ such that the Jacobi symbol of a on p vanishes and, therefore, $\gcd(a, n) \geq p$.

(4) If either $\gcd(a, n)$ or $\gcd(b, n) \neq 1$, then both sides of the equation are zero by part (3). Thus, we may assume $\gcd(a, n) = \gcd(b, n) = 1$, and then
$$\begin{aligned}\left(\frac{ab}{n}\right) &= \left(\frac{ab}{p_1}\right)^{e_1} \cdots \left(\frac{ab}{p_t}\right)^{e_t} \\ &= \left(\frac{a}{p_1}\right)^{e_1} \cdot \left(\frac{b}{p_1}\right)^{e_1} \cdots \left(\frac{a}{p_t}\right)^{e_t} \cdot \left(\frac{b}{p_t}\right)^{e_t} = \left(\frac{a}{n}\right) \cdot \left(\frac{b}{n}\right),\end{aligned}$$
where we have used the multiplicativity of the Legendre symbol, as in Corollary 10.3.9.

This conclude the proof of the proposition. □

Our next goal is to show that the Jacobi symbol also satisfies a version of the law of quadratic reciprocity. Before we can prove this, we need a technical lemma.

Lemma 10.5.6. *Let a, b, and c be odd positive integers. Then:*

(1) $(ab - 1)/2 \equiv (a - 1)/2 + (b - 1)/2 \bmod 2$.

(2) $(a^2b^2 - 1)/8 \equiv (a^2 - 1)/8 + (b^2 - 1)/8 \bmod 2$.

(3) *If* $\left(\frac{a}{c}\right)\left(\frac{c}{a}\right) = (-1)^{\frac{a-1}{2} \cdot \frac{c-1}{2}}$ *and* $\left(\frac{b}{c}\right)\left(\frac{c}{b}\right) = (-1)^{\frac{b-1}{2} \cdot \frac{c-1}{2}}$, *then*
$$\left(\frac{ab}{c}\right)\left(\frac{c}{ab}\right) = (-1)^{\frac{ab-1}{2} \cdot \frac{c-1}{2}}.$$

Proof. Let a, b, c be odd positive integers.

(1) Since $a - 1$ and $b - 1$ are even, it follows that $(a - 1)(b - 1) \equiv 0 \bmod 4$, and so $ab - a - b + 1 \equiv 0 \bmod 4$. Rearranging the terms in the congruence we obtain
$$ab - 1 \equiv (a - 1) + (b - 1) \bmod 4.$$

Since $ab-1$, $a-1$, and $b-1$ are even, we may divide through by 2 (see Proposition 4.3.1) and we obtain $(ab-1)/2 \equiv (a-1)/2 + (b-1)/2 \bmod 2$, as desired.

(2) Since a and b are odd, it follows that a^2-1 and $b^2-1 \equiv 0 \bmod 8$ and $(a^2-1)(b^2-1) \equiv 0 \bmod 16$. Hence $a^2b^2 - a^2 - b^2 + 1 \equiv 0 \bmod 16$, and it follows that

$$a^2b^2 - 1 \equiv (a^2-1) + (b^2-1) \bmod 16.$$

Since $(ab)^2 - 1$ is also divisible by 8, we may divide the congruence through by 8 and obtain

$$\frac{a^2b^2-1}{8} \equiv \frac{a^2-1}{8} + \frac{b^2-1}{8} \bmod 2,$$

as claimed.

(3) Suppose that $\left(\frac{a}{c}\right)\left(\frac{c}{a}\right) = (-1)^{\frac{a-1}{2}\frac{c-1}{2}}$ and $\left(\frac{b}{c}\right)\left(\frac{c}{b}\right) = (-1)^{\frac{b-1}{2}\frac{c-1}{2}}$. Then

$$\begin{aligned}\left(\frac{ab}{c}\right)\left(\frac{c}{ab}\right) &= \left(\frac{a}{c}\right)\left(\frac{b}{c}\right)\left(\frac{c}{a}\right)\left(\frac{c}{b}\right) = \left(\frac{a}{c}\right)\left(\frac{c}{a}\right)\left(\frac{b}{c}\right)\left(\frac{c}{b}\right)\\ &= (-1)^{\frac{a-1}{2}\frac{c-1}{2}+\frac{b-1}{2}\frac{c-1}{2}} = (-1)^{\left(\frac{a-1}{2}+\frac{b-1}{2}\right)\frac{c-1}{2}}\\ &= (-1)^{\frac{ab-1}{2}\frac{c-1}{2}},\end{aligned}$$

where we have first used parts (2) and (3) of Proposition 10.5.5 and then part (1) of this lemma. □

Lemma 10.5.7. *Let m be an odd natural number, and let $p \geq 3$ be a prime number relatively prime to m. Then,*

$$\left(\frac{m}{p}\right)\left(\frac{p}{m}\right) = (-1)^{\frac{m-1}{2}\frac{p-1}{2}}.$$

Proof. We shall prove the statement using complete induction on all odd numbers m. The base case $m = 1$ holds, since every term in the equation equals 1. Now suppose that the statement is true for all odd numbers t in the range $1 \leq t \leq m$, and consider the next odd number $m+2$. If $m+2 = q$ is prime and $q \neq p$, then the statement holds by the law of quadratic reciprocity for p and q using Legendre symbols (Theorem 10.4.2). If $m+2$ is composite, then there are odd numbers a and b, such that $m+2 = ab$ and $1 < a, b < m+2$. Hence, the statement holds for a and b; i.e.,

$$\left(\frac{a}{p}\right)\left(\frac{p}{a}\right) = (-1)^{\frac{a-1}{2}\frac{p-1}{2}} \quad \text{and} \quad \left(\frac{b}{p}\right)\left(\frac{p}{b}\right) = (-1)^{\frac{b-1}{2}\frac{p-1}{2}},$$

and, therefore, by Lemma 10.5.6, we have

$$\left(\frac{m+2}{p}\right)\left(\frac{p}{m+2}\right) = \left(\frac{ab}{p}\right)\left(\frac{p}{ab}\right) = (-1)^{\frac{ab-1}{2}\frac{p-1}{2}} = (-1)^{\frac{(m+2)-1}{2}\frac{c-1}{2}},$$

as we needed to prove. Hence, by the principle of (complete) mathematical induction, the result is true for all odd $m \geq 1$ and any $p \geq 3$. □

We are ready to prove the most important properties of the Jacobi symbol, including the law of quadratic reciprocity.

Theorem 10.5.8. *Let m and n be odd natural numbers that are relatively prime. Then:*

(1) $\left(\dfrac{-1}{n}\right) = (-1)^{(n-1)/2}$.

(2) $\left(\dfrac{2}{n}\right) = (-1)^{(n^2-1)/8}$.

(3) $\left(\dfrac{m}{n}\right)\left(\dfrac{n}{m}\right) = (-1)^{\frac{m-1}{2}\frac{n-1}{2}}$.

Proof. We shall prove all three properties using complete induction on the odd numbers n. The base case $n = 1$ holds because by definition $\left(\dfrac{a}{1}\right) = 1$ for all non-zero integer a, so we will concentrate in proving the induction step in each case.

(1) Suppose part (1) of the theorem is true for all odd numbers $1 \leq t \leq n$. If the next odd number, $n + 2 = p \geq 3$, is prime, then $\left(\frac{-1}{p}\right) = (-1)^{(p-1)/2}$ is true by Lemma 10.3.4. If the odd number $n + 2$ is composite, then there are odd numbers a and b, such that $n + 2 = ab$ and $1 < a, b < n + 2$. Thus, by the induction hypothesis,

$$\left(\frac{-1}{a}\right) = (-1)^{(a-1)/2} \quad \text{and} \quad \left(\frac{-1}{b}\right) = (-1)^{(b-1)/2}.$$

Hence,

$$\begin{aligned}\left(\frac{-1}{n+2}\right) = \left(\frac{-1}{ab}\right) = \left(\frac{-1}{a}\right)\left(\frac{-1}{b}\right) &= (-1)^{(a-1)/2+(b-1)/2}\\ &= (-1)^{(ab-1)/2} = (-1)^{((n+2)-1)/2},\end{aligned}$$

as desired, where first we have used part (2) of Proposition 10.5.5 and then part (1) of Lemma 10.5.6. Hence, by the principle of (complete) mathematical induction, the result is true for all odd $n \geq 1$.

(2) The proof of part (2) of the theorem is very similar to that of part (1) of the theorem, so we have left it as an exercise for the reader (Exercise 10.8.33).

(3) Suppose part (3) of the theorem, the law of quadratic reciprocity, is true for all odd numbers $1 \leq t \leq n$ and any odd $m \geq 1$ relatively prime to t. Let us consider the law for the next odd number $n+2$. If $n+2 = p$ is prime, then the law holds by Lemma 10.5.7. If $n+2$ is composite, then there are odd numbers a and b, such that $n + 2 = ab$ and $1 < a, b < n + 2$. Hence, the statement holds for a and b; i.e.,

$$\left(\frac{m}{a}\right)\left(\frac{a}{m}\right) = (-1)^{\frac{m-1}{2}\frac{a-1}{2}} \quad \text{and} \quad \left(\frac{m}{b}\right)\left(\frac{b}{m}\right) = (-1)^{\frac{m-1}{2}\frac{b-1}{2}},$$

and, therefore, by Lemma 10.5.6, we have

$$\left(\frac{m}{n+2}\right)\left(\frac{n+2}{m}\right) = \left(\frac{m}{ab}\right)\left(\frac{ab}{m}\right) = (-1)^{\frac{ab-1}{2}\frac{m-1}{2}} = (-1)^{\frac{(n+2)-1}{2}\frac{m-1}{2}}.$$

Hence, the law also holds for $n+2$ and any odd $m \geq 1$ relatively prime to $n+2$. Thus, by the principle of (complete) mathematical induction, the statement holds for any odd $n \geq 1$ and any odd $m \geq 1$ relatively prime to n.

This concludes the proof of the theorem. □

The Jacobi symbol can be used to simplify calculations of the Legendre symbol, using the properties of Theorem 10.5.8.

Example 10.5.9. Let us calculate $\left(\frac{539}{541}\right)$ in a few different ways.

(1) The number 541 is prime, so $\left(\frac{539}{541}\right)$ is a Legendre symbol. Thus,

$$\left(\frac{539}{541}\right) = \left(\frac{-2}{541}\right) = \left(\frac{-1}{541}\right)\left(\frac{2}{541}\right) = 1 \cdot (-1) = -1,$$

where we have used the fact that $541 \equiv 1 \bmod 4$ and $\equiv 5 \bmod 8$, together with Lemma 10.3.4 and Theorem 10.3.12.

(2) We can also calculate $539 = 7^2 \cdot 11$ and calculate

$$\left(\frac{539}{541}\right) = \left(\frac{7^2}{541}\right)\left(\frac{11}{541}\right) = \left(\frac{11}{541}\right) = \left(\frac{541}{11}\right) = \left(\frac{2}{11}\right) = -1,$$

where we have used the law of quadratic reciprocity (for Legendre symbols) and the fact that $11 \equiv 3 \bmod 4$.

(3) In the first two methods, we need to know first that 541 is prime, and for the second method, a factorization of 539. Instead, we can use Jacobi symbols without the need to check that 541 is prime or factoring 539:

$$\left(\frac{539}{541}\right) = \left(\frac{541}{539}\right) = \left(\frac{2}{539}\right) = -1,$$

where we have used the law of quadratic reciprocity for Jacobi symbols (note that $541 \equiv 1 \bmod 4$), and part (2) of Theorem 10.5.8 together with $539 \equiv 3 \bmod 8$.

Example 10.5.10. Let us calculate the value of the Legendre symbol $\left(\frac{12345}{104729}\right)$. While it is given that 104729 is a prime number, a factorization of 12345 is not given. Instead of factoring the number 12345, we interpret the Legendre symbol as a Jacobi symbol, and we proceed using Theorem 10.5.8 to compute its value. Notice that $12345 \equiv 1 \bmod 4$. Then,

$$\left(\frac{12345}{104729}\right) = \left(\frac{104729}{12345}\right) = \left(\frac{5969}{12345}\right) = \left(\frac{12345}{5969}\right) = \left(\frac{407}{5969}\right),$$

where we have used the fact that $104729 \equiv 5969 \bmod 12345$ and $12345 \equiv 407 \bmod 5969$. Since $5969 \equiv 1 \bmod 4$, we continue using the law of quadratic reciprocity for Jacobi symbols:

$$\left(\frac{12345}{104729}\right) = \cdots = \left(\frac{407}{5969}\right) = \left(\frac{5969}{407}\right) = \left(\frac{207}{407}\right) = -\left(\frac{407}{207}\right) = -\left(\frac{200}{207}\right).$$

Now, since $200 = 2 \cdot 10^2$, we obtain

$$\left(\frac{12345}{104729}\right) = \cdots = -\left(\frac{200}{207}\right) = -\left(\frac{2}{207}\right) \cdot \left(\frac{10^2}{207}\right) = -1 \cdot 1 \cdot 1 = -1,$$

by Theorem 10.5.8, because $207 \equiv 7 \bmod 8$. Thus, 12345 is a quadratic non-residue modulo $p = 104729$.

Remark 10.5.11. Euler's criterion does not hold for Jacobi symbols. For instance $\left(\frac{2}{15}\right) = 1$ because $15 \equiv -1 \bmod 8$, but

$$2^{(15-1)/2} \equiv 2^7 \equiv 2^5 \cdot 2^2 \equiv 32 \cdot 4 \equiv 2 \cdot 4 \equiv 8 \bmod 15.$$

The failure of Euler's criterion for Jacobi symbols will lead in Section 10.7.1 to the Solovay–Strassen primality test.

10.6. Cipolla's Algorithm

Let p be an odd prime, and let a be an integer relatively prime to p. As shown in the examples above, the Legendre symbol (extended via the Jacobi symbol) is a very efficient tool in determining whether a is a quadratic residue modulo p. However, in the applications (such as finding the roots of a quadratic polynomial, as in Proposition 10.1.1, or finding the integral points on a conic, as in Example 9.4.4), if a is a quadratic residue, then we would like to find a "square root" of a modulo p, i.e., an integer b such that $b^2 \equiv a \bmod p$. Unfortunately, the Legendre and Jacobi symbols do not provide a clue of what the square root of $a \bmod p$ may be when $\left(\frac{a}{p}\right) = 1$. In this section we present an application of finite fields (see Chapter 6) to the problem of finding a square root modulo p, usually known as Cipolla's algorithm (named after Michele Cipolla, an Italian mathematician who described the method in 1907).

Figure 10.3. Michele Cipolla (1880–1947) was an Italian mathematician. Image source: Wikimedia Commons.

The algorithm is based on the following theorem.

Theorem 10.6.1 (Cipolla's algorithm). *Let $p > 2$ be a prime, let s be a quadratic residue modulo p, and let $t \in \mathbb{Z}$ such that $t^2 - s$ is a quadratic non-residue modulo p.*

Let

$$\alpha = \left(t + \sqrt{t^2 - s}\right)^{(p+1)/2} \in \mathbb{F}_p\left[\sqrt{t^2 - s}\right].$$

Then, $\alpha \in \mathbb{F}_p$, and $\alpha^2 \equiv s \bmod p$.

Proof. Let p, s, and t be as in the statement, so that $t^2 - s$ is a quadratic non-residue modulo p. Then, by Theorem 6.4.1, there is a field of p^2 elements given by $\mathbb{F}_p[x]/(x^2 - (t^2 - s))$, which we identify with $\mathbb{F}_p[\sqrt{t^2 - s}]$ as in Section 6.5. Let us write $\omega = \sqrt{t^2 - s}$.

Our first claim is that $\omega^p \equiv -\omega \bmod p$. This equality follows from Exercise 7.6.28 (as a property of the Frobenius automorphism), but we will show it here for the sake of completeness. Indeed,

$$\omega^{p-1} \equiv (\omega^2)^{(p-1/2)} \equiv (t^2 - s)^{(p-1)/2} \equiv -1 \bmod p,$$

by Euler's criterion (Proposition 10.3.6), since $t^2 - s$ is a quadratic non-residue. And $\omega^{p-1} \equiv -1$ implies that $\omega^p \equiv -\omega \bmod p$. In particular, it follows from Exercise 4.7.27 that if $a, b \in \mathbb{F}_p$, then

$$(a + b\omega)^p \equiv a^p + b^p\omega^p \equiv a^p - b^p\omega \bmod p.$$

In addition, if $a, b \in \mathbb{F}_p$, then $a^p \equiv a$ and $b^p \equiv b \bmod p$, by Fermat's (little) theorem (Theorem 7.2.1) and, therefore,

$$(a + b\omega)^p \equiv a^p - b^p\omega \equiv a - b\omega \bmod p.$$

In particular, if we put $a = t \in \mathbb{F}_p$ and $b = 1$, then we obtain $(t+\omega)^p \equiv (t-\omega) \bmod p$. Finally, let $\alpha = (t + \omega)^{(p+1)/2} \in \mathbb{F}_p[\omega]$. Then,

$$\begin{aligned}\alpha^2 &\equiv (t + \omega)^{(p+1)} \equiv (t + \omega)(t + \omega)^p\\ &\equiv (t + \omega)(t - \omega) \equiv t^2 - \omega^2\\ &\equiv t^2 - (t^2 - s) \equiv s \bmod p.\end{aligned}$$

Hence, $\alpha^2 \equiv s \bmod p$, as claimed. This concludes the proof of the theorem. □

Example 10.6.2. Let us illustrate Cipolla's algorithm with one example. Let $p = 13$. The integer $a = 3$ is a quadratic residue modulo 13 because

$$\left(\frac{3}{13}\right) = \left(\frac{13}{3}\right) = \left(\frac{1}{3}\right) = 1.$$

Let us find a square root of 3 modulo 13. The reader probably realizes that $(\pm 4)^2 \equiv 3 \bmod 13$, but here we are simply illustrating how Cipolla's algorithm will arrive at the same conclusion.

The first step of the algorithm is to find a number t such that $t^2 - a$ is a quadratic non-residue modulo 13. In this case $t = 1$ works, because

$$\left(\frac{1^2 - 3}{13}\right) = \left(\frac{-2}{13}\right) = \left(\frac{-1}{13}\right)\left(\frac{2}{13}\right) = 1 \cdot (-1) = -1,$$

because $13 \equiv 5 \bmod 8$. Since -2 is not a square modulo 13, we may form the field $\mathbb{F}_{13}[\sqrt{-2}]$, as in Section 6.5. Cipolla's algorithm says that

$$\alpha \equiv \left(t + \sqrt{t^2 - a}\right)^{(p+1)/2} \equiv \left(1 + \sqrt{-2}\right)^7 \in \mathbb{F}_{13}\left[\sqrt{-2}\right]$$

lives in fact in $\mathbb{F}_{13}$ and it is a square root of $a = 3$. Let us calculate α and verify that $\alpha^2 \equiv 3 \bmod 13$:

$$\begin{aligned}(1+\sqrt{-2})^2 &\equiv 1-2+2\sqrt{-2} \equiv -1+2\sqrt{-2} \bmod 13,\\ (1+\sqrt{-2})^3 &\equiv (1+\sqrt{-2})(1+\sqrt{-2})^2 \equiv (1+\sqrt{-2})(-1+2\sqrt{-2})\\ &\equiv -5+\sqrt{-2} \bmod 13,\\ (1+\sqrt{-2})^4 &\equiv ((1+\sqrt{-2})^2)^2 \equiv (-1+2\sqrt{-2})^2 \equiv -7-4\sqrt{-2},\\ (1+\sqrt{-2})^7 &\equiv (1+\sqrt{-2})^3(1+\sqrt{-2})^4 \equiv (-5+\sqrt{-2})(-7-4\sqrt{-2})\\ &\equiv 35+8+(20-7)\sqrt{-2} \equiv 9+0\sqrt{-2}\\ &\equiv 9 \bmod 13.\end{aligned}$$

Thus, $\alpha \equiv 9 \bmod 13$, and $\alpha^2 \equiv 81 \equiv 3 \bmod 13$. Therefore, the square roots of 3 modulo 13 are ± 9, or, equivalently, $\pm 4 \bmod 13$.

10.7. Applications

In this section we discuss applications of the Legendre and Jacobi symbols and quadratic reciprocity to primality testing and graph theory.

10.7.1. The Solovay–Strassen Primality Test. In Section 7.5.1 we saw a primality test based on Fermat's little theorem. The Solovay–Strassen test for the primality of an integer n is based on Euler's criterion (Proposition 10.3.6), which says that

$$a^{(p-1)/2} \equiv \left(\frac{a}{p}\right) \bmod p,$$

for any prime $p > 2$ and any integer a relatively prime to p. We state the contrapositive of Euler's theorem which will be used as the primality test.

Theorem 10.7.1 (Solovay–Strassen primality test). *Let $n > 1$ be an odd integer, and suppose that a is an integer $1 \leq a \leq n-1$, relatively prime to n, such that*

$$a^{(n-1)/2} \not\equiv \left(\frac{a}{n}\right) \bmod n,$$

where $\left(\frac{a}{n}\right)$ is the Jacobi symbol. Then, n is not prime.

Example 10.7.2. Is $n = 561$ a prime number? Fermat's primality test is inconclusive for $a = 5$, because

$$5^{560} \equiv 1 \bmod 561.$$

In fact, $n = 561$ is a Carmichael number (see Example 7.5.4 and Exercises 7.6.19 and 7.6.20), so the Fermat's primality test fails for many values of a. However, Theorem 10.7.1 correctly identifies 561 as a composite number, for $a = 5$. Indeed, $5^{280} \equiv 67 \bmod 561$, and

$$\left(\frac{5}{561}\right) = \left(\frac{561}{5}\right) = \left(\frac{1}{5}\right) = 1,$$

where we have used the law of quadratic reciprocity for the Jacobi symbol (Theorem 10.5.8) and part (1) of Proposition 10.5.5.

The strength of the Solovay–Strassen primality test is that if n is not prime, then at least 50% of all integers a in the interval $[1, n-1]$ either satisfy that $\gcd(a,n) > 1$ or $a^{(n-1)/2} \not\equiv \left(\frac{a}{n}\right) \bmod n$. It follows that it is relatively easy to find a suitable value of a to use Theorem 10.7.1.

Proposition 10.7.3. *Let $n > 1$ be an odd natural number. Then, the set*

$$H = \left\{ a \bmod n : \gcd(a,n) = 1 \textit{ and } a^{(n-1)/2} \equiv \left(\frac{a}{n}\right) \bmod n \right\}$$

forms a subgroup of $(\mathbb{Z}/n\mathbb{Z})^\times$.

Proof. Let a be an integer relatively prime to n, and let a' be a multiplicative inverse for $a \bmod n$. Then,

$$\left(\frac{a}{n}\right)\left(\frac{a'}{n}\right) = \left(\frac{aa'}{n}\right) = \left(\frac{1}{n}\right) = 1,$$

and so $\left(\frac{a}{n}\right) = \left(\frac{a'}{n}\right)$ and $\left(\frac{a}{n}\right)^{-1} = \left(\frac{a'}{n}\right)$. Thus, if $a \in H$, it follows that

$$a'^{(n-1)/2} \equiv (a^{(n-1)/2})^{-1} \equiv \left(\frac{a}{n}\right)^{-1} \equiv \left(\frac{a'}{n}\right) \bmod n.$$

Hence, $a' \in H$. Now it suffices to show that if a and b are in H, then $ab \in H$. Indeed,

$$(ab)^{(n-1)/2} \equiv a^{(n-1)/2} b^{(n-1)/2} \equiv \left(\frac{a}{n}\right)\left(\frac{b}{n}\right) \equiv \left(\frac{ab}{n}\right) \bmod n,$$

and so $ab \in H$ if a and b are in H, where we have used Proposition 10.5.5. □

We are now ready to prove the assertion that either $\gcd(a,n) > 1$ or Theorem 10.7.1 works for at least 50% of all values of a in $[1, n-1]$. Note that we can determine whether $\gcd(a,n) = 1$ efficiently using Euclid's algorithm (Section 2.6), which does not require factoring a or n. If $\gcd(a,n) > 1$, then we have found a non-trivial factor of n and, therefore, n would not be prime.

We shall prove the assertion for square-free n, and we leave the general case (when n has a square factor) for the exercises (see Exercise 10.8.34).

Corollary 10.7.4. *Let $n > 1$ be an odd natural number, and assume that n is composite and square-free. Then, the set*

$$C = \left\{ a : 1 \le a \le n-1, \textit{ such that } a^{(n-1)/2} \not\equiv \left(\frac{a}{n}\right) \bmod n \right\}$$
$$\cup \left\{ a : 1 \le a \le n-1, \textit{ such that } \left(\frac{a}{n}\right) \equiv 0 \bmod n \right\}$$

has size at least $n - 1 - \frac{\varphi(n)}{2} \ge \frac{n-1}{2}$.

Proof. Let $n > 1$ be odd and composite. Notice that C contains all zero-divisors $Z_n = \{a : 1 \le a \le n-1 \text{ and } \gcd(a,n) \ne 1\}$, because $\left(\frac{a}{n}\right) \equiv 0 \bmod n$ if and only if $\left(\frac{a}{n}\right) = 0$ if and only if $\gcd(a,n) \ne 1$. Moreover, if we define H by

$$H = \left\{ a \bmod n : \gcd(a,n) = 1 \text{ and } a^{(n-1)/2} \equiv \left(\frac{a}{n}\right) \bmod n \right\},$$

then,

$$C = Z_n \cup ((\mathbb{Z}/n\mathbb{Z})^\times - H),$$

and, therefore,

$$|C| = (n - 1 - \varphi(n)) + (\varphi(n) - |H|) = n - 1 - |H|.$$

By Proposition 10.7.3, the set H is a subgroup of $(\mathbb{Z}/n\mathbb{Z})^\times$. If we assume that H is a *proper subgroup*, i.e., $H \subsetneq G$, then $|H|$ is a proper divisor of $|G|$ (by Lagrange's theorem, Theorem 5.2.19), and therefore $|H| \leq |G|/2 = \varphi(n)/2$. Hence,

$$|C| = n - 1 - |H| \geq n - 1 - \varphi(n)/2.$$

Moreover, $\varphi(n) \geq n-1$, and we conclude that $|C| \geq n-1-(n-1)/2 = (n-1)/2$, as claimed. Thus, in order to prove the corollary it suffices to show that H is a proper subgroup or, in other words, it suffices to show that there is some a mod n in $(\mathbb{Z}/n\mathbb{Z})^\times$ that is not in H.

From now on, we assume that n is square-free (the non-square-free case is dealt with in Exercise 10.8.34). Let p be a prime divisor of n. Then $n = pn'$ for some $n' > 1$ relatively prime to p. Let b mod p be a quadratic non-residue modulo p, and let a mod n be a solution of the system of congruence

$$\begin{cases} x \equiv b \bmod p, \\ x \equiv 1 \bmod n'. \end{cases}$$

Notice a unique solution a mod n of the system exists by the Chinese remainder theorem (Theorem 4.5.9), because $\gcd(p, n') = 1$. First, we calculate the Jacobi symbol $\left(\frac{a}{n}\right)$:

$$\left(\frac{a}{n}\right) = \left(\frac{a}{pn'}\right) = \left(\frac{a}{p}\right)\left(\frac{a}{n'}\right) = \left(\frac{b}{p}\right)\left(\frac{1}{n'}\right) = (-1)\cdot 1 = -1,$$

where we have used the properties of the Jacobi symbol, together with the facts that $a \equiv b \bmod p$ and b is a quadratic non-residue, and $a \equiv 1 \bmod n'$.

Suppose for a contradiction that a mod n as constructed in the previous paragraph is in H, and so $a^{(n-1)/2} \equiv -1 \bmod n$. Since n' is a divisor of n, it follows that $a^{(n-1)/2} \equiv -1 \bmod n'$ as well, but $a \equiv 1 \bmod n'$, and therefore $a^{(n-1)/2} \equiv 1 \bmod n'$, so we have reached a contradiction. Hence, a mod n is not in H, and we have shown that $H \subsetneq (\mathbb{Z}/n\mathbb{Z})^\times$. This concludes the proof. □

Example 10.7.5. We exemplify the sets H and C as in the proof of Corollary 10.7.4, for $n = 15$. Here

$$Z_{15} = \{a : 1 \leq a \leq 14 \text{ and } \gcd(a, 15) \neq 1\} = \{3, 5, 6, 9, 10, 12 \bmod 15\},$$

and comparing a^7 mod 15 and $\left(\frac{a}{15}\right)$ mod 15 for each $a \in (\mathbb{Z}/15\mathbb{Z})^\times$ yields

$$H = \{1, 14 \bmod 15\}.$$

Consequently,

$$C = \{2, 3, 4, 5, 6, 7, 8, 9, 10, 11, 12, 13 \bmod 15\},$$

and so $|C| = 12$. Hence, the value of $|C|$ satisfies the bounds of Corollary 10.7.4. Indeed,

$$|C| = 12 \geq 14 - \varphi(15)/2 = 14 - 4 = 10 \geq (15-1)/2 = 7.$$

Thus, if we tried to verify $a^7 \equiv \left(\frac{a}{15}\right) \bmod 15$ for any of the twelve values of a in C (out of the fourteen possible values of a), we would conclude that 15 is not prime.

Example 10.7.6. Let $n = 1729$, which is sometimes called the Hardy–Ramanujan number (see Example 15.0.1). Suppose we want to prove that 1729 is not prime using Fermat's primality test. It turns out that

$$a^{1728} \not\equiv 1 \bmod 1729$$

for 432 values of a in the range $1 \leq a \leq 1728$. That is, the probability that we would find a value of a that shows that n is composite is $432/1728 = 1/4 = 0.25$, or 25%.

If instead we use the Solovay–Strassen primality test, we find that

$$a^{(1729-1)/2} = a^{864} \not\equiv \left(\frac{a}{1729}\right) \bmod 1729$$

for 1080 values of a in $[1, 1728]$. Thus, the probability of finding a value of a that shows that 1729 is not prime is $1080/1728 = 5/8 = 0.625$, or 62.5%.

It is worth pointing out that Fermat's primality test has a particularly poor performance in this example because $n = 1729$ is a Carmichael number (see Example 7.5.4).

10.7.2. The Goldwasser–Micali Cryptosystem. In this section we discuss an application of the quadratic residue symbol to cryptography. We have seen cryptographic applications of congruences in Sections 7.5.3 and 8.9.1, which describe RSA cryptography and the Diffie–Hellman key exchange, respectively. The Goldwasser–Micali (GM) cryptosystem solves a common problem: encryption of messages that can only be one of a small list of possibilities (e.g., yes or no; 0 or 1; plus or minus one; a color; a US state capital, etc.), through public channels. The GM system was proposed by Shafrira Goldwasser and Silvio Micali in 1982 (see Figure 10.4).

Suppose that Alice is sending 1-digit binary messages, either 1 or -1, to Bob, through an insecure public channel. Since the cryptosystem being used is public information, spies would know how to encrypt messages (but not necessarily how to decrypt messages), so they only need to encrypt 1 and -1 to know their encrypted versions and be able to decipher each of Alice's messages. In order to prevent this from happening, the Goldwasser–Micali (GM) cryptosystem embeds the binary message into a large ciphertext that makes the entire message look random once encrypted. The setup for GM is as follows.

Goldwasser–Micali cryptosystem:

(1) Bob sets up the system to receive transmissions from Alice. Bob chooses distinct large primes p and q, computes $N = pq$, and chooses a number b relatively prime to N, such that b is a quadratic non-residue modulo p and also modulo q; i.e.,

$$\left(\frac{b}{p}\right) = \left(\frac{b}{q}\right) = -1.$$

He publishes b and N through a public channel. (See Exercise 10.8.35 for the existence of such a b.)

Figure 10.4. Shafrira Goldwasser (left) is an American-Israeli computer scientist, and Silvio Micali (right) is an Italian computer scientist. They were awarded the 2012 Turing Award for their work in cryptography. Image credit: Jason Dorfman, CSAIL/MIT.

(2) Alice chooses a message $m = 1$ or -1, an integer $1 < a < N$, and computes the encrypted message

$$e \equiv \begin{cases} a^2 \bmod N & \text{if } m = 1 \text{ or} \\ ba^2 \bmod N & \text{if } m = -1. \end{cases}$$

Alice sends her encrypted message e to Bob. (Note: Alice will select a new value of a for each consecutive message.)

(3) Bob decrypts e by computing $m = \left(\frac{e}{p}\right)$.

Indeed, Bob retrieves Alice's message in this way:

$$\left(\frac{e}{p}\right) = \begin{cases} \left(\frac{a^2}{p}\right) = 1 & \text{if } m = 1 \text{ or} \\ \left(\frac{ba^2}{p}\right) = \left(\frac{b}{p}\right)\left(\frac{a^2}{p}\right) = -1 & \text{if } m = -1, \end{cases}$$

where we have used the fact that p is a divisor of N, and therefore $e \equiv a^2 \bmod N$ (resp. $e \equiv ba^2 \bmod N$) implies that $e \equiv a^2 \bmod p$ (resp. $e \equiv ba^2 \bmod p$).

Example 10.7.7. Bob sets up a Goldwasser–Micali cryptosystem with $p = 37$ and $q = 41$, so that $N = 1517$. He chooses an integer $b = 13$, such that b is a quadratic non-residue modulo 37 and modulo 41 (the reader should verify Bob is right, using quadratic reciprocity). He publishes the pair $(N, b) = (1517, 13)$.

Now, Alice would like to send three messages to Bob, namely $m_1 = 1$, $m_2 = 1$, and $m_3 = -1$. For this, she picks three integers $a_1 = 59$, $a_2 = 61$, and $a_3 = 65$ and computes their GM-encrypted versions:

$$\begin{cases} e_1 & \equiv a_1^2 \equiv 59^2 \equiv 447 \bmod 1157, \\ e_2 & \equiv a_2^2 \equiv 61^2 \equiv 687 \bmod 1157, \text{ and} \\ e_3 & \equiv b \cdot a_3^2 \equiv 13 \cdot 65^2 \equiv 313 \bmod 1157. \end{cases}$$

She sends 447, 687, and 313 to Bob.

Now Bob can decrypt these messages computing Legendre symbols:

$$\begin{cases} m_1 & = \left(\frac{e_1}{37}\right) = \left(\frac{447}{37}\right) = 1, \\ m_2 & = \left(\frac{e_2}{37}\right) = \left(\frac{687}{37}\right) = 1, \text{ and} \\ m_3 & = \left(\frac{e_3}{37}\right) = \left(\frac{313}{37}\right) = -1. \end{cases}$$

We remark that $m_1 = m_2 = 1$, but their encrypted versions are different, $e_1 = 447$ and $e_2 = 687$, respectively, because Alice picked different values of a for each message.

Remark 10.7.8. As in the case of the RSA cryptosystem (see Remark 7.5.12), the security of the Goldwasser–Micali system depends on the difficulty of factoring $N = pq$. Indeed, if spies do not know how to factor N, then they cannot retrieve the message. Notice that a spy would know N, b, and an encrypted message e, but

$$\left(\frac{e}{N}\right) = \begin{cases} \left(\frac{a^2}{N}\right) = 1 \text{ and} \\ \left(\frac{ba^2}{N}\right) = \left(\frac{b}{N}\right)\left(\frac{a^2}{N}\right) = \left(\frac{b}{pq}\right) = \left(\frac{b}{p}\right)\left(\frac{b}{q}\right) = (-1)(-1) = 1, \end{cases}$$

so computing the Legendre symbol $\left(\frac{e}{N}\right)$ yields no information to our spy, where we have used the properties of the Jacobi symbol (see Section 10.5).

10.7.3. The Rado Graph. In graph theory, the *Rado graph* (also known as the *random graph*, or the *Erdös–Rényi graph*) is an infinite graph that contains all finite and countably infinite graphs as subgraphs. Formally, the Rado graph is defined as follows: it is the unique (up to isomorphism) countable graph R such that for every finite graph G and every vertex v of G, every embedding of $G - \{v\}$ as a subgraph of R can be extended to an embedding of G into R.

The Rado graph was first constructed by Ackermann in 1937 and, in 1964, Richard Rado rediscovered the graph. The graph can be described using binary expansions of non-negative integers. The vertices of the graph are labelled by the numbers 0, 1, 2, ... and an edge connects the vertices x and y in the graph (with $x < y$) whenever the xth bit of the binary representation of y is non-zero.

For example, 0 and 3 are connected because $3 = 1 + 2$ is $(11)_2$ in base 2; i.e., the 0th binary digit of 3 is non-zero. Similarly, every odd number is connected to 0. The number 1 is also connected to 3, because the 1st digit of $3 = (11)_2$ is also non-zero.

The Rado graph satisfies the following *extension property*: for any finite disjoint sets of vertices U and V, there exists a vertex x in R connected to everything in U and to nothing in V. For instance, let R be the Rado graph as defined by Rado above, in terms of binary digits. Let U and V be two finite disjoint sets of vertices

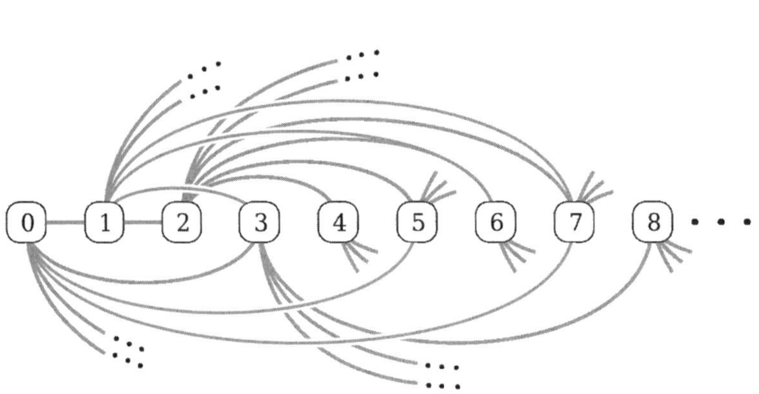

Figure 10.5. The Rado graph, as constructed by Richard Rado in 1964. Image source: Wikimedia Commons.

of R. We can define a number x by

$$x = \sum_{u \in U} 2^u + 2^{(\max\{U \cup V\}+1)}.$$

Then, every u in U is connected to x, because $x > u$ for every $u \in U \cup V$, and the uth digit of x is 1. However, every v in V is disconnected from x, because the vth digit of x is zero. As it turns out, it can be shown that the Rado graph is (up to isomorphism) the unique countable graph with the extension property.

In 2001, Peter Cameron **[Cam01]** showed that the Rado graph can also be constructed via quadratic reciprocity. Let R be a countable graph, such that the vertices are labelled by the prime numbers that are congruent to 1 modulo 4 (e.g., 5, 13, 17, 29, 37, 41, 53, 61, 73, 89, 97, etc.), and connect two vertices p and q by an edge whenever one of the two numbers is a quadratic residue modulo the other, i.e., when $\left(\frac{p}{q}\right) = \left(\frac{q}{p}\right) = 1$ (by quadratic reciprocity and the restriction of the vertices to primes congruent to 1 mod 4, this is a symmetric relation). For instance, 5 and 29 are connected, but 5 is not connected to 13 or 17.

Let us show that the graph R defined via quadratic reciprocity is the Rado graph by showing that it satisfies the extension property. Let $U = \{p_1, \ldots, p_n\}$ and $V = \{q_1, \ldots, q_m\}$ be any two finite disjoint sets of primes that are 1 mod 4. For each $1 \leq j \leq m$, let a_j be a (non-zero) quadratic non-residue modulo q_j, and consider the following system of congruences:

$$\begin{cases} x \equiv 1 \bmod 4, & \\ x \equiv 1 \bmod p_i & \text{for } 1 \leq i \leq n, \\ x \equiv a_j \bmod q_j & \text{for } 1 \leq j \leq m. \end{cases}$$

By the Chinese remainder theorem (Theorem 4.5.9), the system has a unique solution $x \equiv a \bmod 4N$, where $N = p_1 \cdots p_n q_1 \cdots q_m$, for some integer a (notice that $\gcd(a, N) = 1$). Hence, by Dirichlet's theorem on primes in arithmetic progressions (Theorem 3.3.11), there is a prime number p such that $p \equiv a \bmod 4N$. Hence, this prime p is 1 mod 4 (hence a vertex in R), and it satisfies that $p \equiv 1 \bmod p_i$, so $\left(\frac{p}{p_i}\right) = 1$ and p is connected to everything in U, but $p \equiv a_j \bmod q_j$ and

$\left(\frac{p}{q_j}\right) = \left(\frac{a_j}{q_j}\right) = -1$, so p is not connected to anything in V. Hence, this graph R has the extension property and it is therefore isomorphic to the Rado graph.

10.8. Exercises

Exercise 10.8.1. Find the solutions of $x^2 + x + 1 \equiv 0 \bmod 13$.

Exercise 10.8.2. Find the solutions of $3x^2 + 3x - 1 \equiv 0 \bmod 17$.

Exercise 10.8.3. Find the solutions of $2x^2 + 9x + 10 \equiv 0 \bmod 11$.

Exercise 10.8.4. Find all solutions (if any) of the following equations:

(1) $x^2 + 21x + 82 \equiv 0 \bmod 137$,

(2) $x^2 + 5x + 3 \equiv 0 \bmod 37$,

(3) $x^2 + 5x + 7 \equiv 0 \bmod 37$.

Exercise 10.8.5. Use the Chinese remainder theorem to find the solutions of $2x^2 + 9x + 10 \equiv 0 \bmod 77$.

Exercise 10.8.6. Write down lists of all the quadratic residues modulo 2, 3, 5, 7, 11, and 13.

Exercise 10.8.7. In this exercise we give an alternative proof for Proposition 10.2.3, using group theory. We shall show that if $p > 2$ is prime, then there are $(p-1)/2$ quadratic residues and $(p-1)/2$ quadratic non-residues modulo p.

(a) Let $\psi\colon (\mathbb{Z}/p\mathbb{Z})^\times \to (\mathbb{Z}/p\mathbb{Z})^\times$ be the map given by $\psi(x \bmod p) \equiv x^2 \bmod p$. Show that ψ is a well-defined group homomorphism (see Section 5.2.1).

(b) Show that the image of ψ is the set of quadratic residues modulo p. Conclude that the set of quadratic residues is a subgroup of $(\mathbb{Z}/p\mathbb{Z})^\times$, which we will denote by $((\mathbb{Z}/p\mathbb{Z})^\times)^2$.

(c) Show that the kernel of ψ, $\operatorname{Ker}(\psi) = \{x \bmod p : \psi(x) \equiv 1 \bmod p\}$, has exactly two elements.

(d) Let $(\mathbb{Z}/p\mathbb{Z})^\times/\{\pm 1 \bmod p\}$ be the quotient group of $(\mathbb{Z}/p\mathbb{Z})^\times$ by its subgroup $\{\pm 1\}$, as defined in Exercise 5.6.7. Show that ψ induces an isomorphism of groups
$$(\mathbb{Z}/p\mathbb{Z})^\times/\{\pm 1\} \cong ((\mathbb{Z}/p\mathbb{Z})^\times)^2.$$

(e) Conclude that the size of $((\mathbb{Z}/p\mathbb{Z})^\times)^2$ is $(p-1)/2$.

Exercise 10.8.8. Let p be an odd prime, and let $a \in \mathbb{Z}$ be an integer relatively prime to p. Prove that the congruence $x^2 \equiv a \bmod p$ has either two distinct solutions or none.

Exercise 10.8.9. Prove that of any 23 integers, two can always be found such that the difference of their squares is divisible by 100. (Hint: list all the squares modulo 100. See also Exercise 2.11.15.)

Exercise 10.8.10. Is 45 a quadratic residue modulo 47?

Exercise 10.8.11. Is -13 a square modulo 37?

Exercise 10.8.12. Is 14 a quadratic residue modulo 65?

Exercise 10.8.13. The following is a table of powers of 2 modulo 13:

x	x^2	x^3	x^4	x^5	x^6	x^7	x^8	x^9	x^{10}	x^{11}	x^{12}
2	4	8	3	6	12	11	9	5	10	7	1

(1) Without finding them, how many primitive roots are there in $\mathbb{Z}/13\mathbb{Z}$? Find **all** primitive roots of 13.

(2) Use the table to find **all** quadratic residues modulo 13.

Exercise 10.8.14. Continue the table that appears in Example 10.2.6 up to all the primes ≤ 100. In other words, for every prime $p \leq 100$, determine whether -1 is a quadratic residude, and if it is, determine the square roots of -1 modulo p.

Exercise 10.8.15. Calculate the following values of the Legendre symbol:

$$\left(\frac{2}{3}\right), \left(\frac{2}{5}\right), \left(\frac{3}{5}\right), \left(\frac{-2}{7}\right), \left(\frac{-21}{7}\right), \left(\frac{7}{11}\right), \left(\frac{11}{7}\right), \text{ and } \left(\frac{-2}{7}\right).$$

Exercise 10.8.16. Find the following values of the Legendre symbol:

$$\left(\frac{113}{127}\right), \quad \left(\frac{113}{131}\right), \quad \left(\frac{113}{137}\right), \quad \left(\frac{210}{229}\right).$$

The numbers $113, 127, 131, 137$ are primes.

Exercise 10.8.17. Prove that the equation $x^2 - 137y^2 = 113$ has no integer solutions.

Exercise 10.8.18. Let $p \geq 2$ be a prime, such that $p \equiv 1 \bmod 8$. Calculate the following values of the Legendre symbol:

$$\left(\frac{p+1}{p}\right), \left(\frac{p-1}{p}\right), \left(\frac{-2p}{p}\right), \left(\frac{(p-1)^2}{p}\right), \left(\frac{(p-1)^3}{p}\right), \left(\frac{(p-1)(p+2)}{p}\right).$$

Exercise 10.8.19. Calculate the following values of the Legendre symbol:

$$\left(\frac{15}{23}\right), \left(\frac{30}{37}\right), \left(\frac{-30}{41}\right), \left(\frac{60}{97}\right), \left(\frac{53}{97}\right), \text{ and } \left(\frac{59}{61}\right).$$

Exercise 10.8.20. For what odd primes is 7 a quadratic residue? Proceed as in Example 10.4.7.

Exercise 10.8.21. Let p be an odd prime. Show that

$$\left(\frac{-2}{p}\right) = \begin{cases} 1 & \text{if } p \equiv 1, 3 \bmod 8, \\ -1 & \text{if } p \equiv 5, 7 \bmod 8. \end{cases}$$

Also, find examples of primes p in each unit congruence class modulo 8, and find an integer that squares to $-2 \bmod p$ or prove there are none.

Exercise 10.8.22. Let $p > 3$ be a prime. Show that

$$\left(\frac{-3}{p}\right) = \begin{cases} 1 & \text{if } p \equiv 1 \bmod 6, \\ -1 & \text{if } p \equiv 5 \bmod 6. \end{cases}$$

Also, find examples of primes p in each unit congruence class modulo 6, and find an integer that squares to $-3 \bmod p$ or prove there are none.

Exercise 10.8.23. Let $p \neq 5$ be an odd prime. Show that

$$\left(\frac{-5}{p}\right) = \begin{cases} 1 & \text{if } p \equiv 1, 3, 7, 9 \bmod 20, \\ -1 & \text{if } p \equiv 11, 13, 17, 19 \bmod 20. \end{cases}$$

Exercise 10.8.24. For what primes p is -10 a quadratic residue? Are there infinitely many primes p such that -10 is a quadratic residue modulo p? (Hint: use quadratic reciprocity!)

Exercise 10.8.25. For what odd primes is -7 a quadratic residue?

Exercise 10.8.26. For what odd primes is 11 a quadratic residue?

Exercise 10.8.27. Find a prime p such that

$$\left(\frac{2}{p}\right) = \left(\frac{3}{p}\right) = \left(\frac{4}{p}\right) = \left(\frac{5}{p}\right) = \left(\frac{6}{p}\right) = \left(\frac{7}{p}\right) = \left(\frac{8}{p}\right) = \left(\frac{9}{p}\right) = \left(\frac{10}{p}\right) = 1.$$

Exercise 10.8.28. Find the value of the following Legendre symbol: $\left(\frac{4699}{4703}\right)$. (Note: 4703 is prime but 4699 is not!)

Exercise 10.8.29. Are there two odd primes p, q such that $p \neq q$, $p \equiv q \equiv 3 \bmod 4$ and such that p is a quadratic residue modulo q and q is a quadratic residue modulo p? What is the smallest odd prime q such that 3 is a quadratic residue modulo q and q is a quadratic residue modulo 3?

Exercise 10.8.30. Use induction to show that, for all n, there exists a set of n distinct odd primes $\{p_1, \ldots, p_n\}$ such that

$$\left(\frac{p_i}{p_j}\right) = 1$$

for all $1 \leq i, j \leq n$ with $i \neq j$; i.e., every prime in the list is a quadratic residue modulo any other prime in the list.

Exercise 10.8.31. Suppose that p and q are twin primes. Is it possible that 2 is a quadratic residue for both p and q? Is 2 necessarily a quadratic residue of p or q? Find twin primes p and q such that 2 is a quadratic residue modulo p but not modulo q.

Exercise 10.8.32. Recall that if b is a primitive root for p, then $b^{(p-1)/2} \equiv -1 \bmod p$.

(1) Use the fact above to show that all primitive roots must be quadratic non-residues modulo p.

(2) Let p be a prime such that $p = 2q+1$ (a.k.a. a Sophie Germain prime), where q is another odd prime. (For example, $23 = 2 \cdot 11 + 1$.) Justify the following assertion: p has exactly q quadratic non-residues and $q-1$ primitive roots. (Hint: use your answer to part (1).)

(3) Prove that the primitive roots of $p = 2q+1$ are just the quadratic non-residues of p, with one exception. (Hint: use parts (1) and (2).)

(4) What quadratic non-residue of p is not a primitive root? (Hint: prove first that $p \equiv 3 \bmod 4$, because q is an odd prime. Use this to find a quadratic non-residue. Then use (3).)

(5) Calculate the Legendre symbol $\left(\frac{17}{23}\right)$ and use this value to find a primitive root of $p = 23$. Explain. (Hint: use (3) and (4).)

Exercise 10.8.33. Show that for any odd natural number n, the Jacobi symbol satisfies $\left(\dfrac{2}{n}\right) = (-1)^{(n^2-1)/8}$. (Hint: use Theorem 10.3.12, Proposition 10.5.5, and Lemma 10.5.6.)

Exercise 10.8.34. Let $n > 1$ be an odd natural number. Then, the set

$$H = \left\{a \bmod n : \gcd(a,n) = 1 \text{ and } a^{(n-1)/2} \equiv \left(\frac{a}{n}\right) \bmod n\right\}$$

forms a subgroup of $(\mathbb{Z}/n\mathbb{Z})^\times$, by Proposition 10.7.3. In the proof of Corollary 10.7.4 we showed that $H \subsetneq G$ when n is square-free. The goal of this exercise is to prove that $H \subsetneq G$ even if n has a square factor.

(1) Let n be an odd natural number that is not square-free, and let p be a prime such that p^2 is a divisor of n. Let $n = pn'$. Use the binomial theorem (Exercise 2.11.14) to show that

$$(1 + n')^p \equiv 1 \bmod n.$$

(2) Show that the multiplicative order of $a \equiv 1 + n' \bmod n$ is exactly p. Conclude that $a^{n-1} \not\equiv 1 \bmod n$. (Hint: show that p does not divide $n - 1$; then use Proposition 8.1.5.)

(3) Show that $\left(\dfrac{a}{n}\right) = 1$ using the properties of the Jacobi symbol. Conclude that $a \bmod n$ is not in H.

Exercise 10.8.35. Let p and q be distinct odd primes, and let $N = pq$.

(1) How many numbers $1 \leq b \leq N$, relatively prime to N, are quadratic non-residues modulo p and also modulo q?

(2) Let $N = 143$. Find all the numbers $1 \leq b \leq N$, relatively prime to N, that are quadratic residues mod 11 and also mod 13.

Exercise 10.8.36. Let $p = 13$ and $q = 47$.

(1) Show that $b = 11$ is a quadratic non-residue for both $p = 13$ and $q = 47$.

(2) C3PO sets up a Goldwasser–Micali cryptosystem with $N = 13 \cdot 47$ and $b = 11$. R2D2 sends the encrypted messages $m_1 = 126$ and $m_2 = 164$ back to C3PO. Do these messages correspond to 1 or -1?

(3) K2SO intercepts an Imperial Goldwasser–Micali transmission with $b = 17$ and $N = 62773913$ and an encrypted message $e = 34567$. Can you break the code and decrypt the message?

Exercise 10.8.37. Let R be the Rado graph defined using quadratic reciprocity (as in Section 10.7.3). Draw the subgraph of R formed by the vertices that correspond to the prime numbers 5, 13, 17, 29, 37, 41, 53, 61, 73, 89, and 97.

CHAPTER 11

THE HASSE–MINKOWSKI THEOREM

This lecture is particularly interesting, for it contains the first example of the method which Minkowski would develop some years later in his famous "geometry of numbers".

Jean Alexandre Eugène Dieudonné, referring to Minkowski's *Habilitationsschrift*

In Section 5.1.1 (Theorem 5.1.11) we saw that if a diophantine equation has an integral solution, then it also has solutions in $\mathbb{Z}/m\mathbb{Z}$, for every $m > 1$, and we asked ourselves whether the converse is true. In this chapter, we state the Hasse–Minkowski theorem, which shows that a converse is true for quadratic equations and rational points, as long as there exists a solution over $\mathbb{R}$ and a *compatible* family of congruence classes modulo m, for every $m > 1$, that are solutions of the given diophantine equation. However, as we shall see, this is not true in general for curves of higher degree (more precisely, this is not true in higher *genus*, as in Section 1.5).

11.1. Quadratic Forms

In order to state the Hasse–Minkowski theorem, we first need to define the concept of quadratic form. The impatient reader can skip to Remark 11.1.12 for the relationship between quadratic forms and quadratic equations.

Definition 11.1.1. A *quadratic form* $q(X)$ over $\mathbb{Z}$ in $n \geq 2$ variables is a function $q(X_1, \ldots, X_n)$ given by a homogeneous polynomial of degree 2 in the variables $X_1, \ldots, X_n$; i.e.,

$$q(X_1, \ldots, X_n) = a_{1,1}X_1^2 + a_{1,2}X_1X_2 + \cdots + a_{n,n}X_n^2 = \sum_{1 \leq i \leq j \leq n} a_{i,j}X_iX_j,$$

with $a_{i,j} \in \mathbb{Z}$ for every $1 \leq i \leq j \leq n$.

Example 11.1.2. Let q be the quadratic form in three variables given by $q(X,Y,Z) = X^2+Y^2-Z^2$. Then, $(a,b,c) \in \mathbb{Z}^3$ is a non-trivial integral solution of $q(X,Y,Z) = 0$ if and only if $(a,b,c) \in \mathbb{Z}^3$ is a pythagorean triple; i.e., $a^2+b^2=c^2$.

Next, we state a basic property of quadratic forms, which explains the word "quadratic" in the terminology.

Lemma 11.1.3. *Let $q(X_1,\ldots,X_n)$ be a quadratic form. Then,*

$$q(\alpha X_1,\ldots,\alpha X_n) = \alpha^2 q(X_1,\ldots,X_n),$$

for every $\alpha \in \mathbb{Q}$.

Proof. Let $\alpha \in \mathbb{Q}$, and let $q = \sum_{i\leq j} a_{i,j}X_iX_j$. Then,

$$q(\alpha X_1,\ldots,\alpha X_n) = \sum_{i\leq j} a_{i,j}(\alpha X_i)(\alpha X_j) = \alpha^2 \sum_{i\leq j} a_{i,j}X_iX_j = \alpha^2 q(X_1,\ldots,X_n),$$

as claimed. □

Example 11.1.4. Let q be the quadratic form in three variables given by $q(X,Y,Z) = X^2+Y^2-Z^2$, and let $\alpha = 3$. Then,

$$q(3X,3Y,3Z) = (3X)^2+(3Y)^2-(3Z)^2 = 9X^2+9Y^2-9Z^2 = 9q(X,Y,Z).$$

Remark 11.1.5. Let $q(X_1,\ldots,X_n) = \sum_{i,j} a_{i,j}X_iX_j$ be a quadratic form, and let A_q be the matrix

$$\begin{pmatrix} a_{1,1} & \frac{1}{2}a_{1,2} & \cdots & \frac{1}{2}a_{1,n} \\ \frac{1}{2}a_{1,2} & a_{2,2} & \cdots & \frac{1}{2}a_{2,n} \\ \vdots & \vdots & \ddots & \vdots \\ \frac{1}{2}a_{n,1} & \frac{1}{2}a_{n,2} & \cdots & a_{n,n} \end{pmatrix}.$$

Let $\vec{X} = (X_1,\ldots,X_n)$ and let $\vec{X}^t$ be the same vector regarded as a column vector. Then, the reader can verify that

$$\vec{X}\cdot A_q \cdot \vec{X}^t = q(X),$$

where the operator $\cdot$ represents matrix multiplication.

Definition 11.1.6. Let $q(X_1,\ldots,X_n)$ be a quadratic form. The matrix A_q defined in Remark 11.1.5 is called the *Gram matrix* of the quadratic form q. The determinant of A_q is called the *discriminant (or determinant)* of the quadratic form q, and it will be denoted by $\operatorname{disc}(q)$. Finally, we say that a quadratic form is *regular* if its discriminant is non-zero.

Example 11.1.7. Let $q(X,Y,Z) = X^2+Y^2-Z^2$. The Gram matrix of q is given by

$$A_q = \begin{pmatrix} 1 & 0 & 0 \\ 0 & 1 & 0 \\ 0 & 0 & -1 \end{pmatrix},$$

and $\operatorname{disc}(q) = \det(A_q) = -1$. Hence, q is a regular quadratic form.

Example 11.1.8. Let $q(X,Y,Z) = X^2 - Y^2 + XZ + YZ$. Its Gram matrix is given by

$$A_q = \begin{pmatrix} 1 & 0 & \frac{1}{2} \\ 0 & -1 & \frac{1}{2} \\ \frac{1}{2} & \frac{1}{2} & 0 \end{pmatrix},$$

and $\operatorname{disc}(q) = \det(A_q) = -1/4 + 1/4 = 0$. Hence, q is not a regular quadratic form. (Note that $X^2 - Y^2 + XZ + YZ = (X - Y + Z)(X + Y)$.)

Proposition 11.1.9. *Let $q(X_1, X_2, X_3) = (aX + bY + cZ)(dX + eY + fZ)$ be a quadratic form. Then, q is not regular.*

Proof. This is left as an exercise for the reader; see Exercise 11.7.2. □

Let us turn to the question of whether the equation $q(X_1, \ldots, X_n) = 0$ has solutions.

Lemma 11.1.10. *Let $q(X_1, \ldots, X_n)$ be a quadratic form. Then, the equation $q = 0$ has a rational solution $(t_1, \ldots, t_n) \in \mathbb{Q}^n$ if and only if $q = 0$ has an integral solution $(a_1, \ldots, a_n) \in \mathbb{Z}^n$. Moreover, if $q = 0$ has a non-trivial integral solution (i.e., one coordinate is non-zero), then there exists a solution $(b_1, \ldots, b_n) \in \mathbb{Z}^n$ such that* $\gcd(b_1, \ldots, b_n) = 1$.

Proof. Since $\mathbb{Z}^n \subset \mathbb{Q}^n$, one direction is clear (an integral solution is also rational). Now, suppose that $(t_1, \ldots, t_n) \in \mathbb{Q}^n$ is a solution of $q(X_1, \ldots, X_n) = 0$, where each $t_i = c_i/d_i \in \mathbb{Q}$, for $1 \le i \le n$, with $\gcd(c_i, d_i) = 1$. Let m be the least common multiple of the denominators d_i, so that, for each $1 \le i \le n$, there is an integer f_i such that $m = d_i f_i$. We claim that $(c_1 f_1, \ldots, c_n f_n) \in \mathbb{Z}^n$ is a solution of $q(X_1, \ldots, X_n) = 0$. Indeed,

$$\begin{aligned} q(c_1 f_1, \ldots, c_n f_n) &= q\left(\frac{c_1}{d_1} m, \ldots, \frac{c_n}{d_n} m\right) = m^2 q\left(\frac{c_1}{d_1}, \ldots, \frac{c_n}{d_n}\right) \\ &= m^2 q(t_1, \ldots, t_n) = 0, \end{aligned}$$

where we have used Lemma 11.1.3 with $\alpha = m$.

Finally, let $(a_1, \ldots, a_n) \in \mathbb{Z}^n$ be a solution of $q = 0$, with some $a_i \neq 0$, and let $d = \gcd(a_1, \ldots, a_n)$. Then, for each $1 \le i \le n$, there is a b_i such that $a_i = db_i$. Then,

$$\gcd(b_1, \ldots, b_n) = \gcd\left(\frac{a_1}{d}, \ldots, \frac{a_n}{d}\right) = 1,$$

by Exercise 2.11.25, and $(b_1, \ldots, b_n)$ is a solution of $q = 0$, because

$$q(b_1, \ldots, b_n) = q\left(\frac{a_1}{d}, \ldots, \frac{a_n}{d}\right) = \frac{1}{d^2} q(a_1, \ldots, a_n) = 0,$$

by Lemma 11.1.3. □

Definition 11.1.11. Let $q(X_1, \ldots, X_n)$ be a quadratic form. We say that $(b_1, \ldots, b_n) \in \mathbb{Z}^n$ is a *primitive solution* of $q = 0$ if $q(b_1, \ldots, b_n) = 0$ and $\gcd(b_1, \ldots, b_n) = 1$.

Remark 11.1.12. The relationship between quadratic forms and quadratic equations is as follows. Suppose $f(x, y) = ax^2 + bxy + cy^2 + dx + ey + f = 0$, with coefficients in $\mathbb{Z}$, and define

$$q(X, Y, Z) = Z^2 \cdot f\left(\frac{X}{Z}, \frac{Y}{Z}\right) = aX^2 + bXY + cY^2 + dXZ + eYZ + fZ^2.$$

Then, q is a quadratic form with the following property:

- There is a bijection between primitive (integral) solutions $(u, v, w) \in \mathbb{Z}^3$ of $q = 0$ with $w \neq 0$ and rational solutions $(a, b) \in \mathbb{Q}^2$ of $f(x, y) = 0$.

The bijection ψ is given by

$$(u, v, w) \mapsto \left(\frac{u}{w}, \frac{v}{w}\right) \quad \text{and} \quad \left(\frac{a}{b}, \frac{c}{d}\right) \mapsto \left(\frac{an}{b}, \frac{cn}{d}, n\right),$$

where a/b and c/d are rational numbers written in reduced form and $n = \text{lcm}(b, d)$. We leave it as Exercise 11.7.3 to prove that ψ is indeed a bijection.

Example 11.1.13. Let $f(x, y) = x^2 + y^2 - 1$. The solutions of $f(x, y) = 0$ correspond to the points on the curve $C : x^2 + y^2 = 1$. The associated quadratic form is $q(X, Y, Z) = X^2 + Y^2 - Z^2$. A rational point $(3/5, 4/5)$ on C corresponds to a primitive solution $(3, 4, 5)$ of the equation $q(X, Y, Z) = X^2 + Y^2 - Z^2 = 0$. Conversely, a solution of $q = 0$, such as $(5, 12, 13)$, corresponds to a rational point on C, namely $(5/13, 12/13)$.

Proposition 11.1.14. *Let $f(x, y) = ax^2 + bxy + cy^2 + dx + ey + f = 0$, with coefficients in $\mathbb{Z}$, and suppose that $b^2 - 4ac$ is not a square. Let $q(X, Y, Z) = Z^2 \cdot f(X/Z, Y/Z)$ be the quadratic form associated to f. Then, the following are equivalent:*

(1) *The quadratic form q has a non-trivial integral point; i.e., there are integers a_1, a_2, a_3, not all zero, such that $q(a_1, a_2, a_3) = 0$.*

(2) *The curve $C : f(x, y) = 0$ has a rational point; i.e., there are rational numbers t_1, t_2 such that $f(t_1, t_2) = 0$.*

Proof. By Remark 11.1.12 there is a bijection between the rational solutions of $f(x, y) = 0$ and the primitive integral solutions $(u, v, w) \in \mathbb{Z}^3$ of $q = 0$ with $w \neq 0$. By Lemma 11.1.3, if there is an integral solution of $q = 0$, then there is a primitive solution. Hence, in order to conclude the proof, we need to show that if $b^2 - 4ac$ is not a square, then every integral solution (u, v, w) of $q = 0$ has $w \neq 0$. Suppose for a contradiction that (u, v, w) is an integral solution with $w = 0$. Then,

$$0 = q(u, v, 0) = au^2 + buv + cv^2.$$

If $ac = 0$, then $b^2 - 4ac = b^2$ would be a square. Thus, a and c are non-zero. Then, $au^2 + buv + cv^2 = 0$ implies that

$$u = \frac{-bv \pm \sqrt{b^2v^2 - 4acv^2}}{2a} = \left(\frac{-b \pm \sqrt{b^2 - 4ac}}{2a}\right) \cdot v.$$

It follows that $\pm\sqrt{b^2 - 4ac} = 2au/v + b$, and so $b^2 - 4ac = (2au/v + b)^2$, a rational square. But we have assumed that $b^2 - 4ac$ is not a square, and we have reached a contradiction. Therefore, if (u, v, w) is an integral solution of $q = 0$, then $w \neq 0$. □

11.2. The Hasse–Minkowski Theorem

Before we state the Hasse–Minkowski theorem, we need to introduce the notion of a compatible family of congruence classes.

Example 11.2.1. Let us consider the congruence class of 8 mod 15:

$$8 \bmod 15 = \{\ldots,\ -22,\ -7,\ 8,\ 23,\ 38,\ 53, \ldots\}.$$

Since 3 is a divisor of 15, a number in the congruence class of 8 mod 15 is also in the class of $8 \equiv 2 \bmod 3$. Thus, we say that 8 mod 15 and 2 mod 3 are compatible (the concept of compatible and incompatible congruences already appeared when we studied systems of congruences; see, for instance, Example 4.5.8). Similarly, since 5 divides 15, a number in the congruence class of 8 mod 15 is also in the class of $8 \equiv 3 \bmod 5$, and we say that the congruences 8 mod 15 and 3 mod 5 are compatible.

Now consider the set of congruence classes

$$S = \{2 \bmod 3,\ 3 \bmod 5,\ 4 \bmod 7,\ 8 \bmod 15,\ 18 \bmod 35\}.$$

We say that the set S is a system of compatible congruence classes because the following condition is satisfied:

- If $\gcd(m, n) = d$ and $a \bmod n$ and $b \bmod m$ are in S, then $a \equiv b \bmod d$.

For instance, $n = 7$ is a divisor of $m = 35$, and $18 \equiv 4 \bmod 7$. For instance, 8 mod 15 and 18 mod 35 are in S, and $18 \equiv 8 \bmod 5$, where $5 = \gcd(15, 35)$. Equivalently, the system S is compatible because the system of linear congruences

$$\begin{cases} x \equiv 2 \bmod 3, \\ x \equiv 3 \bmod 5, \\ x \equiv 4 \bmod 7, \\ x \equiv 8 \bmod 15, \\ x \equiv 18 \bmod 35 \end{cases}$$

has solutions in the integers. For example, $x = 53$ is a solution (and the reader can verify that the solutions are $x \equiv 53 \bmod 105$).

Example 11.2.2. The curve $C : x^2 - 61y^2 = 1$ is an example of a *Pell's equation* which we will study in Chapter 14, when we discuss hyperbolas. The curve C has infinitely many integral points but their coefficients are quite large, so they are not easy to find by brute force. For instance, the integral point with the smallest positive coefficients is $P = (1766319049, 226153980)$. Indeed,

$$1766319049^2 - 61 \cdot 226153980^2 = 1.$$

The existence of P implies that there exist solutions to $x^2 - 61y^2 \equiv 1 \bmod m$, for all integers $m \geq 2$. Indeed, $S = \{(1766319049, 226153980) \bmod m\}_{m \geq 2}$ is such a collection of solutions. For example,

$$(1, 0) \bmod 4,\ (4, 0) \bmod 5,\ (5, 3) \bmod 7,\ (1, 4) \bmod 8,\ (19, 3) \bmod 21,$$

and $(19, 10) \bmod 35$ are all examples of solutions $(a, b) \bmod m$ of C over $\mathbb{Z}/m\mathbb{Z}$, such that $P \equiv (a, b) \bmod m$. Moreover, the family S has an important property: it is compatible, in the sense that if $\gcd(m, n) = d$, then the solution mod m and the solution mod n reduce to the solution mod d. For example $(1,4)$ mod 8 reduces to $(1,0)$ modulo 4. The solution $(19,10)$ mod 35 reduces to $(5,3)$ mod 7 and to $(4,0)$ mod 5. The solutions $(19,3)$ mod 21 and $(19,10)$ mod 35 both reduce to the solution $(5,3)$ modulo $\gcd(21,35)=7$.

The Hasse–Minkowski theorem will answer the following question: if C is a curve defined over $\mathbb{Q}$, with points over $\mathbb{R}$, and S is a compatible family of solutions for C modulo m, for every $m \geq 2$, is there a rational solution for C? Notice that, as we explained in Remark 5.1.14, we cannot expect the existence of an *integral* solution in general, so the Hasse–Minkowski theorem will predict the existence of *rational* solutions.

First, we write down a formal definition of compatible family of solutions.

Definition 11.2.3. Let $\{m_k\}_{k\geq 1}$ be a sequence of positive integers. We say that $S = \{a_k \bmod m_k\}_{k\geq 1}$ forms a *compatible (or coherent) family* of congruence classes if the following condition is satisfied: if $\gcd(m_j, m_k) = d$, then $a_k \equiv a_j \bmod d$. Equivalently, S is compatible if every system of finitely many linear congruences in S has an integer solution.

In particular, if p is a prime and $m_k = p^k$ for all k, then we say $\{a_k \bmod p^k\}_{k\geq 1}$ forms a compatible family of congruences if $a_k \equiv a_h \bmod p^h$ for every $1 \leq h < k$.

Example 11.2.4. Let $\{m_k\} = \{m \geq 2\}$ be the sequence of all positive integers ≥ 2. The sequence

$$\{(m-1) \bmod m\}_{m=2}^{\infty} = \{1 \bmod 2,\ 2 \bmod 3,\ 3 \bmod 4, \ldots\}$$

forms a compatible family of congruences, because if $\gcd(n,m) = d$ and $m = dk$ and $n = dj$, then

$$m - 1 \equiv dk - 1 \equiv 0 - 1 \equiv dj - 1 \equiv n - 1 \bmod d,$$

as desired. Equivalently, if $\{m_1, \ldots, m_k\}$ is a finite set of integers ≥ 2, then the system

$$\begin{cases} x \equiv (m_1 - 1) \bmod m_1, \\ x \equiv (m_2 - 1) \bmod m_2, \\ \vdots \\ x \equiv (m_k - 1) \bmod m_k \end{cases}$$

has an integer solution; namely $x = -1$.

Example 11.2.5. Let $p = 5$. Then, the sequence $\{a_k \bmod 5^k\}$, where $a_k = 3 + 5 + 5^2 + \cdots + 5^{k-1}$, i.e.,

$$\{3 \bmod 5,\ 8 \bmod 25,\ 33 \bmod 125,\ 158 \bmod 625, \ldots\},$$

is a compatible family of congruence classes, because if $h < k$, then 5^h is a divisor of 5^k, and

$$a_k \equiv 3+5+5^2+\cdots+5^{h-1}+5^h+\cdots+5^{k-1} \equiv 3+5+5^2+\cdots+5^{h-1} \equiv a_h \bmod 5^h.$$

Definition 11.2.6. Let q be a quadratic form in n variables. A sequence of solutions $\{(a_{1,m}, \ldots, a_{n,m}) \bmod m\}_{m\geq 1}$ of $q \equiv 0 \bmod m$, for each $m \geq 1$, forms a *compatible family of solutions* if each coordinate forms a compatible family of congruences; i.e., $\{a_{i,m} \bmod m\}_{m\geq 1}$ forms a compatible family of congruences, for each $i = 1, \ldots, n$.

We say that a compatible family of solutions $\{(a_{1,m}, \ldots, a_{n,m}) \bmod m\}_{m\geq 1}$ is *non-trivial* if there is some $m \geq 2$ such that the solution $(a_{1,m}, \ldots, a_{n,m}) \bmod m$ is not congruent to $(0, \ldots, 0) \bmod m$.

We are now ready to state the Hasse–Minkowski theorem, which was originally proved by Hermann Minkowski and generalized by Helmut Hasse.

Figure 11.1. Hermann Minkowski (1864–1909) and Helmut Hasse (1898–1979). Images source (left): Wikimedia Commons. Image author (right): Konrad Jacobs (Erlangen). Image source: Archives of the Mathematisches Forschungsinstitut Oberwolfach.

Theorem 11.2.7 (Hasse–Minkowski theorem). *Let $q(X_1, \ldots, X_n)$ be a regular quadratic form defined over $\mathbb{Q}$. Then, $q = 0$ has a non-trivial integral solution (i.e., not all coordinates are zero) if and only if there is a non-trivial solution over $\mathbb{R}$ and the congruences $q \equiv 0 \bmod m$, for all $m > 1$, have a non-trivial compatible system of solutions.*

Example 11.2.8. Let n be a natural number. In Example 11.1.7 we have shown that $q = X^2+Y^2-nZ^2$ is a regular quadratic form. Now, Theorem 11.2.7 says that $X^2 + Y^2 = nZ^2$ has a non-trivial integral solution if and only if $X^2 + Y^2 = nZ^2$ has a non-trivial solution over $\mathbb{R}$ and the congruences $X^2 + Y^2 \equiv nZ^2 \bmod m$, for all $m > 1$, have a non-trivial compatible system of solutions.

The proof of the Hasse–Minkowski theorem is, unfortunately, beyond the scope of this book (see **[Ger08]** or **[Ser73]** for a proof). In the rest of this section, we will rephrase the theorem in different ways, and we will state consequences of the theorem in certain particular cases that we are interested in.

Our first simplification of Theorem 11.2.7 makes use of the Chinese remainder theorem to reduce the Hasse–Minkowski theorem to a statement about congruences modulo prime powers, instead of natural numbers $m \geq 2$.

Proposition 11.2.9. *The congruences* $q \equiv 0 \bmod m$, *for every* $m > 1$, *have a non-trivial compatible system of solutions if and only if for each prime* p, *the congruences* $q \equiv 0 \bmod p^k$, *for each* $k \geq 1$, *have a non-trivial compatible system of solutions.*

Proof. Suppose first that $S = \{(a_{1,m}, \ldots, a_{n,m}) \bmod m\}_{m \geq 1}$ is a compatible system of solutions of $q \equiv 0 \bmod m$, and let p be a fixed prime. Then, for each $k \geq 1$ and $m = p^k$, the sequence $S_p = \{(a_{1,p^k}, \ldots, a_{n,p^k}) \bmod p^k\}_{k \geq 1}$ is a system of solutions of $q \equiv 0 \bmod p^k$ and, moreover, it is compatible. Indeed, if $1 \leq h < k$, then p^h is a divisor of p^k, and by the compatibility of S it follows that $a_{i,p^k} \equiv a_{i,p^h} \bmod p^h$ for all $i = 1, \ldots, n$. Hence, for every p there is a compatible system of solutions S_p.

For the converse, suppose that for every p there is a compatible system of solutions $S_p = \{(b_{1,p^k}, \ldots, b_{n,p^k}) \bmod p^k\}_{k \geq 1}$ for the congruences $q \equiv 0 \bmod p^k$. Let $m > 1$ be arbitrary. We want to construct a compatible system of solutions for $q \equiv 0 \bmod m$. For this, write the unique prime factorization of m, namely,

$$m = p_1^{e_1} \cdots p_r^{e_r},$$

for primes $p_1 < \cdots < p_r$ and $e_i \geq 1$ for $i = 1, \ldots, r$. Let S_{p_i} be the compatible systems of solutions that exist by hypothesis. By the Chinese remainder theorem (Theorem 4.5.9) and for each $i = 1, \ldots, n$, the system

$$\begin{cases} x \equiv b_{i,p_1^{e_1}} \bmod p_1^{e_1}, \\ x \equiv b_{i,p_2^{e_2}} \bmod p_2^{e_2}, \\ \vdots \\ x \equiv b_{i,p_r^{e_r}} \bmod p_r^{e_r} \end{cases}$$

has a unique solution $x \equiv a_{i,m} \bmod m$. We claim that thus constructed, $S = \{(a_{1,m}, \ldots, a_{n,m}) \bmod m\}_{m \geq 1}$ is a compatible system of solutions for $q \equiv 0 \bmod m$, for all $m \geq 2$. Indeed, for a fixed m and for each $j = 1, \ldots, r$, we have

$$q(a_{1,m}, \ldots, a_{n,m}) \equiv q(b_{1,p_i^{e_i}}, \ldots, b_{n,p_i^{e_i}}) \equiv 0 \bmod p_i^{e_i},$$

because each S_{p_i} is a system of solutions. Consequently, $q(a_{1,m}, \ldots, a_{n,m}) \equiv 0 \bmod p_1^{e_1} \cdots p_r^{e_r} = m$, and S is a system of solutions of $q \equiv 0 \bmod m$.

It remains to show that S is a compatible system. For this, suppose that ℓ and m are integers with GCD equal to d. Then $d = p_1^{f_1} \cdots p_r^{f_r}$ for some $0 \leq f_j \leq e_j$, for each $j = 1, \ldots, r$. Let $i \in \{1, \ldots, n\}$ be fixed. Then, for each $j = 1, \ldots, r$, we have

$$a_{i,m} \equiv b_{i,p_j^{e_j}} \equiv b_{i,p_j^{f_j}} \equiv a_{i,\ell} \bmod p_j^{f_j},$$

where we have used the fact that S_{p_j} is compatible. In particular, $a_{i,m} \equiv a_{i,\ell} \bmod p_j^{f_j}$ for each $j = 1, \ldots, r$ and, again by the Chinese remainder theorem, we conclude that $a_{i,m} \equiv a_{i,\ell} \bmod d$, as desired. Hence S is a compatible system of solutions of $q \equiv 0 \bmod m$, for all $m \geq 2$, and the proof is done. □

Therefore, the theorem of Hasse–Minkowski has the following equivalent formulation.

Theorem 11.2.10 (Hasse–Minkowski, second version). *Let* $q(X_1, \ldots, X_n)$ *be a regular quadratic form defined over* $\mathbb{Q}$. *Then,* $q = 0$ *has a non-trivial integral* *solution if and only if there is a non-trivial solution over $\mathbb{R}$ and, for each prime p, there is a non-trivial compatible system of solutions of $q \equiv 0 \bmod p^k$, for all $k \geq 1$.*

Example 11.2.11. Let n be a natural number. The quadratic form $q = X^2+Y^2-nZ^2$ is regular. Now, Theorem 11.2.10 says that $X^2+Y^2 = nZ^2$ has a non-trivial integral solution if and only if it has a non-trivial solution over $\mathbb{R}$ and for each prime p the congruences $X^2+Y^2 \equiv nZ^2 \bmod p^k$, for all $k \geq 1$, have a non-trivial compatible system of solutions.

In Section 11.4 we will show that, in fact, for most odd primes, it suffices to check for solutions in $\mathbb{Z}/p\mathbb{Z}$, and one does not need to verify the existence of solutions in $\mathbb{Z}/p^k\mathbb{Z}$ for $k > 1$. In certain cases, it also suffices to find solutions in $\mathbb{Z}/8\mathbb{Z}$, and there is no need to worry about $\mathbb{Z}/2^k\mathbb{Z}$ for $k \geq 4$. However, this is still a criterion that would be difficult to use, as we would need to check infinitely many conditions, at least one for each prime p. As it turns out, Theorem 11.2.10 is equivalent to the following formulation, in which the criterion is reduced to a condition that only involves finitely many primes (see [**Ger08**, Theorems 4.3 and 5.7 and Remark 5.11].)

Theorem 11.2.12 (Hasse–Minkowski, third version). *Let $q = \sum_{i\leq j} a_{i,j}X_iX_j$ be a regular quadratic form defined over $\mathbb{Z}$. Then, $q = 0$ has a non-trivial integral solution if and only if there is a non-trivial solution over $\mathbb{R}$ and for each prime p equal to 2 or dividing $Q = \prod_{i,j} a_{i,j}$, there is a non-trivial compatible system of solutions of $q \equiv 0 \bmod p^k$, for all $k \geq 1$.*

Example 11.2.13. Let n be a natural number. The quadratic form $q = X^2+Y^2-nZ^2$ is regular. Now, Theorem 11.2.12 says that $X^2+Y^2 = nZ^2$ has a non-trivial integral solution if and only if it has a non-trivial solution over $\mathbb{R}$ and, for $p = 2$ and all primes p dividing n, the congruence $X^2+Y^2 \equiv nZ^2 \bmod p^k$, for each $k \geq 1$, have a non-trivial compatible system of solutions.

Finally, the Hasse–Minkowski theorem is often stated in terms of *p-adic numbers*. In Section 11.5 we will define the p-adic integers $\mathbb{Z}_p$ and the p-adic numbers $\mathbb{Q}_p$ as compatible sequences of integers and rational numbers, respectively. Here we simply state the p-adic version of the theorem, for completeness.

Theorem 11.2.14 (Hasse–Minkowski, p-adic version). *Let $q(X_1, \ldots, X_n)$ be a regular quadratic form defined over $\mathbb{Q}$. Then, $q = 0$ has a non-trivial integral solution if and only if there is a non-trivial solution over $\mathbb{R}$ and over $\mathbb{Q}_p$ for each prime p.*

Similarly to Theorem 11.2.12, one can improve the p-adic version of the Hasse–Minkowski theorem so that we only need to check for solvability in $\mathbb{Q}_p$ for $p = 2$ and those primes dividing a coefficient of the quadratic form q.

Example 11.2.15. Let n be a natural number. The quadratic form $q = X^2+Y^2-nZ^2$ is regular. Now, Theorem 11.2.14 says that $X^2+Y^2 = nZ^2$ has a non-trivial integral solution if and only if it has a non-trivial solution over $\mathbb{R}$ and over $\mathbb{Q}_p$ for $p = 2$ and each prime p dividing n.

11.3. An Example of Hasse–Minkowski

In this section we shall explore the usage and consequences of the theorem of Hasse–Minkowski in a particular example. We will come back and answer the same question in Section 12.1 (see Theorem 12.1.10) without using the Hasse–Minkowski theorem.

Question 11.3.1. *Let $n > 1$ be a fixed natural number. Are there rational points on the circle $x^2 + y^2 = n$? Alternatively, is n a sum of two (rational) squares?*

For instance,

$$\begin{aligned} 1 &= 1^2 + 0^2 = (3/5)^2 + (4/5)^2, \\ 2 &= 1^2 + 1^2 = (7/13)^2 + (17/13)^2, \\ 5 &= 1^2 + 2^2 = (22/17)^2 + (31/17)^2, \\ 8 &= 2^2 + 2^2 = (2/29)^2 + (82/29)^2, \\ 9 &= 3^2 + 0^2 = (9/5)^2 + (12/5)^2. \end{aligned}$$

Is 3 a sum of two squares? How about 6 or 7?

Example 11.3.2. Let us show that $x^2 + y^2 = 3$ has no rational solutions. Let $f(x, y) = x^2 + y^2 - 3$. Since $b^2 - 4ac = -4$ is not a square, Remark 11.1.12 and Proposition 11.1.14 imply that $f(x, y) = 0$ has a rational solution if and only if $q(X, Y, Z) = X^2 + Y^2 - 3Z^2 = 0$ has a non-trivial primitive integral solution. By the Hasse–Minkowski theorem (Theorem 11.2.7), this is equivalent to $q = 0$ having solutions over $\mathbb{R}$ and the existence of a non-trivial compatible system of solutions for each $X^2 + Y^2 \equiv 3Z^2 \bmod m$, for all $m > 1$. Clearly $q = 0$ has real solutions, namely $(\pm\sqrt{3}, 0, \pm 1)$.

However, $X^2 + Y^2 \equiv 3Z^2 \bmod m$ has no non-trivial compatible system of solutions. Indeed, when $m = 2$, the only non-trivial solutions of $X^2 + Y^2 \equiv 3Z^2 \bmod 2$ are $(1, 1, 0)$, $(1, 0, 1)$, and $(0, 1, 1)$. When $m = 4$, the squares modulo 4 are $0, 1 \bmod 4$, and the only possible non-trivial solutions of $X^2 + Y^2 \equiv 3Z^2 \bmod 4$ are $(2, 0, 0)$, $(0, 2, 0)$, $(2, 0, 2)$, and $(0, 2, 2)$. Since the solutions modulo 4 reduce to the trivial solution $(0, 0, 0) \bmod 2$, there is no solution that is compatible modulo 2 and modulo 4.

First, let us simplify Question 11.3.1 by noticing that we can reduce it to the case of a square-free odd number n.

Lemma 11.3.3. *Let $n > 1$. The following statements are equivalent:*

(1) *The number n is a sum of two (rational) squares.*

(2) *The number $2^k n$ is a sum of two (rational) squares, for every $k \geq 1$.*

(3) *The number nt^2 is a sum of two (rational) squares, for any $t \in \mathbb{Z}$.*

Proof. Let us first show that (1) and (2) are equivalent. If n is a sum of two squares, say $n = x^2 + y^2$, then

$$2n = 2(x^2 + y^2) = (x - y)^2 + (x + y)^2$$

is also a sum of two squares. Now, by induction, we can show that $2^k n$ is a sum of two squares as well. Conversely, if $m = 2n$ is a sum of two squares, say $2n = a^2 + b^2$, then

$$n = \left(\frac{a+b}{2}\right)^2 + \left(\frac{a-b}{2}\right)^2$$

is also a sum of two squares (see Exercise 11.7.8). Hence, if $2^k n$ is a sum of squares, using induction we can show that n is also a sum of two squares.

It remains to show that (1) and (3) are equivalent. Clearly, if $n = x^2 + y^2$, then $nt^2 = (xt)^2 + (yt)^2$. Conversely, if $nt^2 = a^2 + b^2$, then $n = (a/t)^2 + (b/t)^2$. Hence (1), (2), and (3) are all equivalent conditions. □

Example 11.3.4. Let $n = 13 = 9 + 4 = 3^2 + 2^2$. Thus, $2n = 26$ is also a sum of two squares, and Lemma 11.3.3 tells us how to find such squares:

$$26 = 2 \cdot 13 = 2(3^2 + 2^2) = (3-2)^2 + (3+2)^2 = 1^2 + 5^2.$$

Conversely, $m = 74 = 2 \cdot 37$ is a sum of two squares, namely $a^2 + b^2 = 49 + 25$, so Lemma 11.3.3 says that 37 is also a sum of two squares (see Exercise 11.7.8); namely

$$37 = \left(\frac{7+5}{2}\right)^2 + \left(\frac{7-5}{2}\right)^2 = 6^2 + 1^2.$$

Finally, $n = 425 = 17 \cdot 5^2$ is a sum of squares; for instance, $a^2 + b^2 = 19^2 + 8^2$. This implies that 17 is also a sum of two (rational) squares,

$$17 = \left(\frac{19}{5}\right)^2 + \left(\frac{8}{5}\right)^2,$$

and, in fact, 17 is also a sum of two integral squares, as $17 = 4^2 + 1^2$.

In Section 11.4 we will show the following lemma (Theorems 11.4.9 and 11.4.7), which we will use here to answer our Question 11.3.1 on rational points on the circle.

Lemma 11.3.5. *Let c be an integer not divisible by a prime p. Then:*

(1) *Let $p = 2$. For every $k \geq 1$, the congruence $x^2 \equiv c \bmod 2^k$ has a non-trivial compatible system of solutions if and only if $c \equiv 1 \bmod 8$.*
(2) *Let $p > 2$. For every $k \geq 1$, the congruence $x^2 \equiv c \bmod p^k$ has a non-trivial compatible system of solutions for every $k \geq 1$ if and only if $x^2 \equiv c \bmod p$ has a solution.*

Example 11.3.6. Let $p = 7$. The congruence $x^2 \equiv 2 \bmod 7$ has solutions; namely $x \equiv \pm 3 \bmod 7$. Let $s \equiv 3 \bmod 7$. Is there a solution of $x^2 \equiv 2 \bmod 49$ such that $x \equiv 3 \bmod 7$? Lemma 11.3.5 says that the answer must be yes and, indeed,

$$10^2 \equiv 100 \equiv 2 \bmod 49,$$

so $s_2 \equiv 10 \bmod 49$ is a solution of $x^2 \equiv 2 \bmod 49$ such that $s_2 \equiv s \bmod 7$. Moreover, Lemma 11.3.5 says that, for each $k \geq 1$, there is a solution $s_k \bmod p^k$, with $s_k \equiv s_{k-1} \bmod p^{k-1}$. One such compatible system of solutions is given by

$$\{s_k\}_{k\geq 1} = \{3 \bmod 7,\ 10 \bmod 49,\ 108 \bmod 7^3,\ 2166 \bmod 7^4,\ 4567 \bmod 7^5, \ldots\}.$$

In Section 11.4 we will prove Lemma 11.3.5 and show how to find compatible systems of solutions for polynomial congruences in general.

Let us assume Lemma 11.3.5 for now and deduce some consequences.

Corollary 11.3.7. *Let n be an odd integer. Then, the congruences $X^2 + Y^2 \equiv nZ^2 \bmod 2^k$, for each $k \geq 1$, have a non-trivial compatible system of solutions for all $k \geq 1$ if and only if $n \equiv 1 \bmod 4$.*

Proof. Let n be odd. The pair of congruences $X^2 + y^2 \equiv nZ^2 \bmod 2$ and mod4 have non-trivial compatible solutions if and only if $n \equiv 1 \bmod 4$ (see Exercise 11.7.12). Thus, if $n \equiv 3 \bmod 4$, there is no non-trivial compatible system of solutions for the congruences $X^2 + y^2 \equiv nZ^2 \bmod 2^k$, for $k \geq 1$.

Now, let us assume that $n \equiv 1 \bmod 4$. Let $k \geq 3$ be fixed and let us consider whether $X^2+Y^2 \equiv nZ^2 \bmod 2^k$ has a compatible system of solutions. If $n \equiv 1 \bmod 8$, then there is a compatible system, namely $(x_{0,k}, 0, 1) \bmod 2^k$ where $\{x_{0,k} \bmod 2^k\}_{k\geq 1}$ is a non-trivial compatible system of solutions of $x^2 \equiv n \bmod 2^k$ (whose existence is guaranteed by Lemma 11.3.5). If $n \equiv 5 \bmod 8$, then $n - 4 \equiv 1 \bmod 8$, and therefore $x^2 \equiv n - 4 \bmod 2^k$ has a non-trivial compatible system of solutions $\{x_{0,k} \bmod 2^k\}$. It follows that $X^2+Y^2 \equiv nZ^2 \bmod 2^k$ has a non-trivial compatible system of solutions $\{(x_{0,k}, 2, 1) \bmod 2^k\}$. Since we have shown that there is a non-trivial compatible system whenever $n \equiv 1$ or $5 \bmod 8$, it follows that it is also true when $n \equiv 1 \bmod 4$.

Hence, we have shown that $X^2 + Y^2 \equiv nZ^2 \bmod 2^k$ for all $k \geq 1$ has a non-trivial compatible system of solutions if and only if $n \equiv 1 \bmod 4$. □

Corollary 11.3.8. *Let n be an integer, and let p be a prime divisor of n, such that p^2 is not a divisor of n (or assume that n is square-free). Then, the congruences $X^2 + Y^2 \equiv nZ^2 \bmod p^k$, for each $k \geq 1$, have a non-trivial compatible system of solutions if and only if $p \equiv 1 \bmod 4$.*

Proof. Suppose first that $p \equiv 1 \bmod 4$. Then, -1 is a square modulo p (by Lemma 10.3.4); i.e., $x^2 \equiv -1 \bmod p$ has a solution. It follows that -1 is also a square modulo p^k, for every $k \geq 1$ and, in fact, by Lemma 11.3.5, there is a compatible system of solutions $\{b_k \bmod p^k\}$ for $x^2 \equiv -1 \bmod p^k$, for each $k \geq 1$. Hence, $\{(1, b_k, 0) \bmod p^k\}$ is a non-trivial compatible system of solutions of $X^2 + Y^2 \equiv nZ^2 \bmod p^k$, for each $k \geq 1$, as desired.

Conversely, if $p \equiv 3 \bmod 4$ and $V = \{v_k \equiv (a_k, b_k, c_k) \bmod p^k\}$ was a non-trivial compatible system of solutions of $X^2 + Y^2 \equiv nZ^2 \bmod p^k$, for all $k \geq 1$, then v_1 would be a solution modulo p, and therefore $a_1^2 + b_1^2 \equiv 0 \bmod p$. If $a_1 b_1 \not\equiv 0 \bmod p$, then $(a_1 b_1^{-1})^2 \equiv -1 \bmod p$, but this is impossible because $p \equiv 3 \bmod 4$ and so -1 is not a quadratic residue modulo p (again by Lemma 10.3.4). Thus, we must have that one of a_1 or $b_1 \equiv 0 \bmod p$, but then $a_1^2 + b_1^2 \equiv 0 \bmod p$ implies that both are zero modulo p. Since v_1 was non-trivial, we conclude that $v_1 \equiv (0, 0, c_1) \bmod p$, for some $c_1 \not\equiv 0 \bmod p$. Now consider $v_2 \equiv (a_2, b_2, c_2) \bmod p^2$. Since V is compatible, we know that $v_2 \equiv v_1 \bmod p$, and so $a_2 \equiv a_1 \equiv 0 \equiv b_1 \equiv b_2 \bmod p$, and $c_2 \equiv c_1 \not\equiv 0 \bmod p$. In particular $a_2^2 \equiv b_2^2 \equiv 0 \bmod p^2$, and so

$$a_2^2 + b_2^2 \equiv nc_2^2 \bmod p^2$$

implies that $nc_2^2 \equiv 0 \bmod p^2$. Since n is divisible by p but not p^2, then $n = pn'$ with $n' \not\equiv 0 \bmod p$, and so $n'c_2^2 \equiv 0 \bmod p$, and since n' is invertible modulo p, we conclude that $c_2^2 \equiv 0 \bmod p$. But this is impossible because $c_2 \not\equiv 0 \bmod p$. Hence, we conclude that the system V cannot exist. □

Finally, we shall use the Hasse–Minkowski theorem (specifically Theorem 11.2.12) in order to answer Question 11.3.1. In Section 12.1 we will see an alternative self-contained proof, which does not use the Hasse–Minkowski theorem (Theorem 12.1.10).

Theorem 11.3.9. *Let n be a natural number. The circle $C_n : x^2 + y^2 = n$ has a rational point if and only if every prime divisor p of n with $p \equiv 3 \bmod 4$ appears to an even power in the prime factorization of n. Equivalently, the circle C_n has a rational point if and only if the square-free part of n is not divisible by any prime p of the form $p \equiv 3 \bmod 4$.*

Proof. It follows from Lemma 11.3.3 that we may assume n is odd and square-free, and the theorem is reduced to proving that C_n has a rational point if and only if n is not divisible by primes of the form $p \equiv 3 \bmod 4$.

Let $f(x,y) = x^2 + y^2 - n$. Since $b^2 - 4ac = -4$ is not a square, Proposition 11.1.14 implies that $f(x,y) = 0$ has a rational solution if and only if $q(X,Y,Z) = X^2 + Y^2 - nZ^2 = 0$ has a non-trivial integral solution. By Theorem 11.2.7, this is equivalent to $X^2 + Y^2 = nZ^2$ having solutions over $\mathbb{R}$ and the congruences $X^2 + Y^2 \equiv nZ^2 \bmod m$ for all $m > 1$ having a compatible system of solutions. Clearly, there are solutions over $\mathbb{R}$, for example $(\pm\sqrt{n}, 0, \pm 1)$ or $(0, \pm\sqrt{n}, \pm 1)$. Now, Theorem 11.2.12 says that it suffices to check whether there are compatible systems of solutions over $\mathbb{Z}/2^k\mathbb{Z}$ and $\mathbb{Z}/p^k\mathbb{Z}$, for each $k \geq 1$ and for each p dividing n.

Finally, by Corollary 11.3.7 and 11.3.8, the equations $X^2 + Y^2 \equiv nZ^2 \bmod p^k$, for all $k \geq 1$, have a non-trivial compatible system of solutions, for $p = 2$ and every $p|n$, if and only if $n \equiv 1 \bmod 4$, and every prime divisor of n is of the form $p \equiv 1 \bmod 4$ (notice that the condition on the prime divisors implies the first condition that $n \equiv 1 \bmod 4$). Hence, the proof of the theorem is finished. □

Now that we have answered Question 11.3.1 about *rational* points on the circle of radius n, we might ask ourselves about *integral* points as well.

Question 11.3.10. *Let $n > 1$ be a fixed natural number. Are there integral points on the circle $x^2 + y^2 = n$? Alternatively, is n a sum of two (integral) squares?*

The following proposition shows that Questions 11.3.1 and 11.3.10 are, in fact, equivalent.

Proposition 11.3.11. *The circle $C_n : x^2 + y^2 = n$ has an integral solution if and only if it has a rational solution.*

Proof. An integral solution is rational, so the forward direction is clear. Let us assume that C_n has a rational solution, and let us show that there is also an integral solution. It suffices to show that the square-free part n' of n is a sum of two integral squares (for if $n = n's^2$ and $n' = x^2 + y^2$, then $n = (xs)^2 + (ys)^2$). By Lemma 11.3.3, the circle C_n has a rational point if and only if $C_{n'}$ has a rational point. So suppose $n' = x^2 + y^2$ has a solution $(a/c, b/d)$ with $\gcd(a,c) = \gcd(b,d) = 1$, so that

$$n'(cd)^2 = (ad)^2 + (bc)^2.$$

Thus, there are integers μ, α, β with $n'\mu^2 = \alpha^2 + \beta^2$, with $\gcd(\alpha, \beta) = 1$. Indeed, if $\gcd(ad, bc) = \delta$, then δ^2 must divide $(cd)^2$ (because n' is square-free), so we may pick $\mu = cd/\delta$, $\alpha = ad/\delta$, and $\beta = bc/\delta$.

If $n' = 1$, then $n = s^2$, and C_n has an integral point $(s, 0)$. So let us assume n' is square-free and greater than 1. Let t be the unique positive integer such that $t^2 < n' < (t+1)^2$. Since there are $(t+1)^2 > n'$ integers of the form $\alpha u + \beta v$ with $0 \le u, v \le t$, it follows that n' divides the difference $\alpha(u - u') + \beta(v - v')$ of two of them (see Exercise 4.7.12). Setting $x = u - u'$ and $y = v - v'$, we have the inequalities $|x|, |y| \le t$. The integer n' then divides $(\alpha x + \beta y)(\alpha x - \beta y) = \alpha^2 x^2 - \beta^2 y^2$; since it also divides $n'\mu^2 = (\alpha^2 + \beta^2)y^2 = \alpha^2 y^2 + \beta^2 y^2$, it divides their sum, which is equal to $\alpha^2(x^2 + y^2)$. Now, the integers α and n' being coprime (because $\gcd(\alpha, \beta) = 1$), it follows that n' divides $x^2 + y^2$ (by Corollary 2.7.6). The inequalities $0 < x^2 + y^2 \le 2t^2 < 2n'$ finally imply that $n' = x^2 + y^2$. Hence

$$n = n's^2 = (x^2 + y^2)s^2 = (xs)^2 + (ys)^2$$

is also a sum of two integral squares, and the proof is done. □

Let us illustrate the method above to find an integral point on $C : x^2 + y^2 = 13$ starting from a given rational point on C.

Example 11.3.12. Let $n = 13$. The circle of radius 13, given by $C : x^2 + y^2 = 13$, has a rational point $\left(\frac{6}{5}, \frac{17}{5}\right)$. Since 13 is square-free, here $n = n' = 13$, and $13 \cdot 5^2 = 6^2 + 17^2$, so that $\alpha = 6$, $\beta = 17$, and $\mu = 5$. Also, $3^2 < 13 < 4^2$, so $t = 3$. Consider

$$\begin{aligned} S &= \{6u + 17v : 0 \le u, v \le 3\} \\ &= \{0, 6, 12, 17, 18, 23, 29, 34, 35, 40, 46, 51, 52, 57, 63, 69\}. \end{aligned}$$

In particular, $12 - 51 = -39$ is divisible by 13, so let $x = 2 - 0 = 2$ and $y = 0 - 3 = -3$. Then, 13 divides $(6 \cdot 2 - 17 \cdot 3)(6 \cdot 2 + 17 \cdot 3) = 6^2 2^2 - 17^2 3^2$ and it also divides $(6^2 + 17^2) \cdot 3^2$, so it divides their sum $6^2 2^2 + 6^2 3^2 = 6^2(2^2 + 3^2)$. Since $\gcd(13, 6) = 1$, it follows that 13 divides $2^2 + 3^2$, and since $0 < 2^2 + 3^2 < 2 \cdot 13$, we conclude that $13 = 2^2 + 3^2$.

In the following example we give a geometric construction that produces an integral point on a circle starting from a rational point. This method can be made into an alternative proof of Proposition 11.3.11 (see Theorem 2.1 in [**Con3**]).

Example 11.3.13. The circle $C : x^2 + y^2 = 193$ has a rational point P_1 with coordinates $\left(\dfrac{933}{101}, \dfrac{1048}{101}\right)$. Let us find an integral point on C. First, we find the point Q_1 on the plane with integral coordinates closest to P_1:

$$P_1 = \left(\frac{933}{101}, \frac{1048}{101}\right) \approx (9.237\ldots, 10.376\ldots) \approx (9, 10) = Q_1.$$

The point Q_1 is not on C. We define L_1 as the line that goes through P_1 and Q_1. Thus, L_1 is given by the equation

$$L_1 : y = \frac{19}{12}x - \frac{17}{4},$$

and the points of intersection of C and L_1 are P_1 and a second point P_2 given by

$$P_2 = \left(-\frac{27}{5}, -\frac{64}{5}\right).$$

The point P_2 is on C, and the denominator is smaller than that of P_1, but P_2 is not integral. Hence, we repeat the process, and we find a point Q_2 on the plane, with integral coefficients, and closest to P_2:

$$P_1 = \left(-\frac{27}{5}, -\frac{64}{5}\right) = (-5.4, -12.8) \approx (-5, -13) = Q_2.$$

We define L_2 as the line that goes through P_2 and Q_2, which is thus given by $L_2 : y = -\frac{1}{2}x - \frac{31}{2}$. The points of intersection of C and L_2 are P_2 and a point $P_3 = (-7, -12)$ which is on C and has integral points, as desired.

11.3.1. Another Example of Hasse–Minkowski. In this section we will use Hasse–Minkowski to investigate the rational points on the curve $C : x^2 - 29y^2 = -1$. By Theorem 11.2.12, we only need to concern ourselves with $\mathbb{R}$, $p = 2$, and $p = 29$.

($\mathbb{R}$) Clearly, C has points over the real numbers; for instance, $(0, 1/\sqrt{29})$ is a real point on C.

(2) We shall use Lemma 11.3.5 to show that C has a system of solutions that form a compatible system modulo 2^k for all $k \geq 1$. Indeed, let $P_k = (2x_k, 3)$, with x_k an integer. Then, P_k is a point on C modulo 2^k if and only if $4x_k^2 \equiv 29 \cdot 9 - 1 \equiv 4 \cdot 65 \bmod 2^k$. Thus, if $x_k^2 \equiv 65 \bmod 2^k$, then P_k is a point on C modulo 2^k. Since $65 \equiv 1 \bmod 8$, Lemma 11.3.5 implies that there is a compatible system $x_k \bmod 2^k$, for each $k \geq 1$, such that $x_k^2 \equiv 65 \bmod 2^k$. Hence, $P_k = (2x_k, 3)$ is a compatible system of points on C modulo 2^k. For example, we can pick

$$P_1 = (0, 1),\ P_2 = (2, 3) = P_3 = P_4 = P_5 = P_6,$$
$$P_7 = (66, 3) = P_8 = P_9 = P_{10},\ P_{11} = (3138, 3), \ldots.$$

(29) Finally, Lemma 11.3.5 shows that $x^2 \equiv -1 \bmod 29^k$ has a compatible system of solutions x_k modulo 29^k, because $x^2 \equiv -1 \bmod 29$ has a solution ($29 \equiv 1 \bmod 4$, so -1 is a quadratic non-residue). Hence, the points $P_k = (x_k, 0)$ form a compatible system of points on C modulo 29^k for all $k \geq 1$. For instance, one such system is given by

$$P_1 = (12, 0),\ P_2 = (41, 0),\ P_3 = (10133, 0),\ P_4 = (34522, 0), \ldots.$$

Since there are solutions of C in $\mathbb{R}$ and compatible systems of solutions modulo 2^k and 29^k, Theorem 11.2.12 shows that there are rational points on C. For instance, the points

$$\left(\frac{5}{2}, \frac{1}{2}\right),\ \left(\frac{2}{5}, \frac{1}{5}\right),\ \left(\frac{26}{7}, \frac{5}{7}\right),\ \left(\frac{23}{14}, \frac{5}{14}\right), \ldots$$

are rational points on C. Are there any *integral* points on C? The theorem of Hasse–Minkowski does not say either way. However, we will show later that there are also integral points on the hyperbola C; for instance, $(70, 13)$ or $(1372210, 254813)$ is on C. We will use the theory of continued fractions (Chapter 13) to find and determine all of the integral points (see Example 14.3.1).

11.4. Polynomial Congruences for Prime Powers

In Section 11.3 we discussed the need for Lemma 11.3.5, where we deduce the existence of certain solutions to quadratic congruences $x^2 \equiv c$ modulo p^k from the existence of a solution modulo p (see also Example 11.3.6). We will prove Lemma 11.3.5 below, in Theorems 11.4.7 and 11.4.9, but we will begin with a discussion of the same problem for polynomial congruences in general; i.e., if $p(x)$ is a polynomial in $\mathbb{Z}[x]$ such that $s \in \mathbb{Z}$ is a root modulo p, is there a solution $s_k \in \mathbb{Z}$ of $p(x) \equiv 0 \bmod p^k$, for each $k \geq 1$? If so, how do we find such solutions?

Example 11.4.1. Let $f(x) = x^2 - 2$. Can we find a compatible system of solutions for the congruences $f(x) \equiv 0 \bmod 7^k$, for every $k \geq 1$? In other words, can we find a sequence $\{s_k \bmod 7^k\}$ such that $s_k^2 \equiv 2 \bmod 7^k$ and $s_{k+1} \equiv s_k \bmod 7^k$, for all $k \geq 1$?

In order to find such a sequence, we begin with one root of $f(x) \equiv 0 \bmod 7$, namely $s_1 \equiv 3 \bmod 7$, and then we shall use the algorithm usually known as Newton's method (or Newton–Raphson's method) to construct consecutive terms of the sequence. The successive terms in the sequence are given by the formula

$$s_{k+1} \equiv s_k - f(s_k) \cdot (f'(s_k))^{-1} \bmod 7^{k+1},$$

where $f'(x)$ is the derivative of the polynomial $f(x)$ and $(f'(s_k))^{-1}$ is the multiplicative inverse of $f'(s_k) \bmod 7^{k+1}$, if such an inverse exists (we will give conditions below that guarantee this method to work). For instance,

$$s_2 \equiv 3 - (3^2 - 2) \cdot (2 \cdot 3)^{-1} \equiv 3 - 7 \cdot 41 \equiv 10 \bmod 49.$$

Now we can check that $10^2 \equiv 100 \equiv 2 \bmod 49$ and $s_2 \equiv 10 \equiv 3 \equiv s_1 \bmod 7$, so s_2 and s_1 are compatible solutions. Similarly, we can construct a few more terms in the sequence:

$$s_3 \equiv 10 - 98 \cdot (20)^{-1} \equiv 10 - 98 \cdot 223 \equiv 108 \bmod 7^3,$$
$$s_4 \equiv 108 - 11662 \cdot (216)^{-1} \equiv 108 - 11662 \cdot 1056 \equiv 2166 \bmod 7^4,$$
$$s_5 \equiv 2166 - 4691554 \cdot (4332)^{-1} \equiv 2166 - 4691554 \cdot 1742 \equiv 4567 \bmod 7^5.$$

The reader can verify that $\{s_1, s_2, s_3, s_4, s_5\}$ form a compatible system of solutions for $f(x) \equiv 0 \bmod 7^k$, for $k = 1, 2, 3, 4, 5$.

When $f(x)$ is a quadratic equation, finding compatible systems of solutions modulo 2^k for all $k \geq 1$ is particularly tricky and needs special care. Similarly, if $f(x)$ is of degree p, then finding solutions modulo p^k is a special case. Let us see a couple of examples.

Example 11.4.2. Let $f(x) = x^2 - 41$. Can we find a compatible system of solutions for the congruences $f(x) \equiv 0 \bmod 2^k$, for every $k \geq 1$? In other words, can we find a sequence $\{s_k \bmod 2^k\}$ such that $s_k^2 \equiv 41 \bmod 2^k$ and $s_{k+1} \equiv s_k \bmod 2^k$, for all $k \geq 1$?

Let us try to apply Newton's method as in Example 11.4.1. We begin by finding a root modulo 2, and this is simple enough as $s_1 \equiv 1 \bmod 2$ works. However, if we try to find the successive terms in the sequence s_k using the formula

$$s_{k+1} \equiv s_k - f(s_k) \cdot (f'(s_k))^{-1} \bmod 2^{k+1},$$

we will run into trouble when calculating $(f'(s_k))^{-1} \bmod 2^{k+1}$, because $f'(x) = 2x$, and therefore $f'(s_k)$ is not invertible for any $s_k \in \mathbb{Z}$. However, if we consider $f(s_k) \cdot (f'(s_k))^{-1}$ as a rational number and the denominator turns out to be odd, then we can proceed ahead with the formula. In our case,

$$s_2 \equiv s_1 - f(s_1) \cdot (f'(s_1))^{-1} \equiv 1 - (-40) \cdot \frac{1}{2} \equiv 21 \equiv 1 \bmod 4.$$

Similarly,

$$\begin{aligned} s_3 &\equiv 1 - (-40) \cdot 2^{-1} \equiv 21 \equiv 5 \bmod 8, \\ s_4 &\equiv 5 - (-16) \cdot (10)^{-1} \equiv 5 + 8 \cdot 13 \equiv 109 \equiv 13 \bmod 16, \\ s_5 &\equiv 13 - 128 \cdot (26)^{-1} \equiv 13 - 64 \cdot 5 \equiv 13 \bmod 32, \end{aligned}$$

which produces the first few terms of a compatible system $\{1, 1, 5, 13, 13, \ldots\}$ of solutions of $x^2 \equiv 41 \bmod 2^k$.

Example 11.4.3. Let $f(x) = x^2 - 21$. Can we find a compatible system of solutions for the congruences $f(x) \equiv 0 \bmod 2^k$, for every $k \geq 1$? Let us try Newton's method once again, with $s_1 \equiv 1 \bmod 2$. Now,

$$s_2 \equiv s_1 - f(s_1) \cdot (f'(s_1))^{-1} \equiv 1 - (-20) \cdot \frac{1}{2} \equiv 11 \equiv 3 \bmod 4.$$

So far, so good. Let us compute s_3:

$$s_3 \equiv s_2 - f(s_2) \cdot (f'(s_2))^{-1} \equiv 3 - (-12) \cdot \frac{1}{6} \equiv 5 \bmod 8.$$

Something went wrong, because $5^2 - 21 \equiv 4 \bmod 8$, so $5 \bmod 8$ is not a solution of $x^2 \equiv 21 \bmod 8$. In fact, the reader can check that there are no solutions of $x^2 \equiv 21 \bmod 8$! So the method was doomed to fail.

In the next few results we will explain when Newton's method is guaranteed to work and why.

Lemma 11.4.4. *Let p be a prime, let k be a positive integer, and let $f(x) \in \mathbb{Z}[x]$ be a polynomial. Then, for any integer t we have a congruence of polynomials*

$$f(x + p^k t) \equiv f(x) + f'(x) p^k t \bmod p^{k+1}\mathbb{Z}[x].$$

In other words, there is a polynomial $h(x) \in \mathbb{Z}[x]$ such that $f(x + p^k t) - (f(x) + f'(x)p^k t)$ is divisible by $p^{k+1}h(x)$, for any $t \in \mathbb{Z}$.

Proof. We will use induction on the degree of the polynomial $f(x)$. If the degree is 0, then $f(x) = f_0$ is constant, thus $f'(x) = 0$, and $f(x+p^k t) = f_0 = f(x) + f'(x)p^k t$, so the result holds. Let us assume, then, the result for polynomials of degree n and suppose $f(x)$ has degree $n + 1$. Then, we may write $f(x) = a + xg(x)$, for some $a \in \mathbb{Z}$ and a polynomial $g(x)$ of degree n. It follows that $f'(x) = g(x) + xg'(x)$. Moreover, by the induction hypothesis, the lemma is true for $g(x)$, and therefore

$$\begin{aligned} f(x + p^k t) &\equiv a + (x + p^k t) g(x + p^k t) \equiv a + (x + p^k t)(g(x) + g'(x) p^k t) \\ &\equiv a + xg(x) + (xg'(x) + g(x)) p^k t + g'(x) t^2 p^{2k} \\ &\equiv a + xg(x) + (xg'(x) + g(x)) p^k t \\ &\equiv f(x) + f'(x) p^k t \bmod p^{k+1}\mathbb{Z}[x], \end{aligned}$$

as desired. This proves the induction step and, therefore, by the principle of mathematical induction, the result is true for polynomials of any degree. □

Theorem 11.4.5. *Let p be a prime, let k be a positive integer, and let $f(x) \in \mathbb{Z}[x]$ be a polynomial. Suppose that $s_k \in \mathbb{Z}$ is a solution of $f(x) \equiv 0 \bmod p^k$.*

(1) *If $f'(s_k)$ is not divisible by p, then there is precisely one solution s_{k+1} of $f(x) \equiv 0 \bmod p^{k+1}$ such that $s_{k+1} \equiv s_k \bmod p^k$. Moreover, s_{k+1} is given by*

$$s_{k+1} \equiv s_k - f(s_k) \cdot (f'(s_k))^{-1} \bmod p^{k+1}.$$

(2) *If $p \mid f'(s_k)$ and $p^{k+1} \mid f(s_k)$, then there are p solutions of $f(x) \equiv 0 \bmod p^{k+1}$ that are congruent to s_k modulo p, given by $s_k + p^k \cdot j$ for $j = 0, 1, \ldots, p-1$.*

(3) *If $p \mid f'(s_k)$ but $f(s_k)$ is not divisible by p^{k+1}, then there are no solutions of $f(x) \equiv 0 \bmod p^{k+1}$ that are congruent to s_k modulo p^k.*

In particular, if $s_1 \in \mathbb{Z}$ is a solution of $f(x) \equiv 0 \bmod p$ and $f'(s_1) \not\equiv 0 \bmod p$, then there is an integer $s_k \in \mathbb{Z}$, unique modulo p^k, such that $f(s_k) \equiv 0 \bmod p^k$ for all $k \geq 1$ and such that $s_{k+1} \equiv s_k \bmod p^k$.

Proof. Let s_{k+1} be a solution of $f(x) \equiv 0 \bmod p^{k+1}$ such that $s_{k+1} \equiv s_k \bmod p^k$. It follows that $s_{k+1} \equiv s_k + p^k t$ for some integer t. Lemma 11.4.4 implies that

$$0 \equiv f(s_{k+1}) \equiv f(s_k + p^k t) \equiv f(s_k) + f'(s_k) p^k t \bmod p^{k+1}.$$

Since $f(s_k) \equiv 0 \bmod p^k$, it follows that $f(s_k)/p^k$ is an integer and $f'(s_k)t \equiv -f(s_k)/p^k \bmod p$. We distinguish three cases:

(1) If $f'(s_k)$ and p are relatively prime, then Corollary 4.4.4 shows that $f'(s_k)t \equiv -f(s_k)/p^k \bmod p$ has a unique solution t modulo p, namely

$$t \equiv -f(s_k)/(f'(s_k)p^k) \bmod p,$$

and therefore $s_{k+1} \equiv s_k - f(s_k) \cdot (f'(s_k))^{-1} \bmod p^{k+1}$, as claimed.

(2) If $f'(s_k)$ is divisible by p, then $f(s_k)/p^k \equiv 0 \bmod p$. Thus, we must have $f(s_k) \equiv 0 \bmod p^{k+1}$, and any value of $t \bmod p$ works. Thus $s_{k+1} \equiv s_k + p^k t \bmod p^{k+1}$, with $t = 0, 1, \ldots, p-1$.

(3) Finally, if $p \mid f'(s_k)$ but $f(s_k)$ is not divisible by p^{k+1}, then $f'(s_k)t \equiv -f(s_k)/p^k \bmod p$ is impossible and there is no solution s_{k+1} of the required form.

Finally, if $s_1 \in \mathbb{Z}$ is a solution of $f(x) \equiv 0 \bmod p$ and $f'(s_1) \not\equiv 0 \bmod p$, then part (1) says that there is a unique solution $s_2 \bmod p^2$. Moreover, since $s_2 \equiv s_1 \bmod p$, it follows that $f'(s_2) \equiv f'(s_1) \not\equiv 0 \bmod p$. Hence, we may apply part (1) recursively to show the existence of unique values $s_k \bmod p^k$ of the required form. □

Example 11.4.6. Let us review Example 11.4.3 using the results in Theorem 11.4.5. There we had $f(x) = x^2 - 21$ and $p = 2$. We begin with $s_1 \equiv 1 \bmod 2$. Since 2 divides $f'(1) = 2$ and $f(1) = -20 \equiv 0 \bmod 4$, there are two possibilities for s_2; namely $s_2 = 1 + 0 = 1$ and $s_2' = 1 + 2 = 3$. However, $f'(s_2) = 2$ and $f'(s_2') = 6$ are divisible by 2, but neither $f(s_2) = -20$ nor $f(s_2') = -12$ is divisible by 8. Thus, Theorem 11.4.5 shows that there is no compatible system of solutions of $x^2 - 21 \bmod 2^k$ for all $k \geq 1$.

In the rest of this section we will apply Newton's method (i.e., Theorem 11.4.5) to the particular case of polynomials of the form $f(x) = x^2 - a$, for $a \in \mathbb{Z}$, and determine conditions on a such that there exists a compatible system of solutions modulo p^k. As illustrated in Example 11.4.1, the case of $p \neq 2$ is easier to handle. See Exercise 11.7.18 for an example of Newton's method applied to finding roots of a cubic congruence.

Theorem 11.4.7. *Let p be an odd prime, and let $a \in \mathbb{Z}$ be an integer relatively prime to p. If $x^2 \equiv a \bmod p$ is solvable, then $x^2 \equiv a \bmod p^k$ has exactly two compatible systems of solutions for all $k \geq 1$. Otherwise, if $x^2 \equiv a \bmod p$ has no solutions, then there are no solutions for any $x^2 \equiv a \bmod p^k$.*

Proof. By Exercise 10.8.8 we know that $x^2 \equiv a \bmod p$ has either two solutions or none. Since any solution of $x^2 \equiv a \bmod p^k$ would reduce to a solution of $x^2 \equiv a \bmod p$, we deduce that if there are none modulo p, then there are none modulo p^k for all $k \geq 1$. Thus, let us assume that there are two solutions modulo p; namely $s_{1,1}$ and $s_{1,2} \equiv -s_{1,1} \bmod p$. Notice that $s_{1,i}^2 \equiv a \bmod p$ and $a \not\equiv 0 \bmod p$, so $s_{1,i} \not\equiv 0 \bmod p$ either for $i = 1, 2$. Since $f(s_{1,i}) \equiv 0 \bmod p$ and $f'(s_{1,i}) = 2s_{1,i} \not\equiv 0 \bmod p$ because p is odd and $s_{1,i} \not\equiv 0 \bmod p$, our Theorem 11.4.5 implies that there exist, for each $i = 1, 2$, a unique compatible system of solutions $\{s_{1,i}, s_{2,i}, \ldots\}$ of $x^2 \equiv a \bmod p^k$ such that $s_{k,i} \equiv s_{1,i}$ for each $k \geq 1$. Hence, the equation $x^2 \equiv a \bmod p^k$ has exactly two compatible systems of solutions, as claimed. □

Example 11.4.8. For instance, $x^2 \equiv 2 \bmod 7$ has two solutions; namely $x \equiv 3$ and $-3 \equiv 4 \bmod 7$. Thus, the congruences $x^2 \equiv 2 \bmod 7^k$ have two compatible systems of solutions x_1 and $x_2 = -x_1$, which are

$$x_1 = (3 \bmod 7, 10 \bmod 7^2, 108 \bmod 7^3, 2166 \bmod 7^4, 4567 \bmod 7^5, \ldots),$$
$$x_2 = (4 \bmod 7, 39 \bmod 7^2, 235 \bmod 7^3, 235 \bmod 7^4, 12240 \bmod 7^5, \ldots).$$

Finally, we need to prove a version of Theorem 11.4.7 for powers of $p = 2$.

Theorem 11.4.9. *Let $a \in \mathbb{Z}$ be odd. Then $x^2 \equiv a \bmod 2$ has exactly one solution, and*

(1) *the equation $x^2 \equiv a \bmod 4$ is solvable if and only if $a \equiv 1 \bmod 4$, in which case there are two solutions modulo 4, and*
(2) *the equation $x^2 \equiv a \bmod 2^k$, with $k \geq 3$, is solvable if and only if $a \equiv 1 \bmod 8$, in which case there are exactly four solutions modulo 2^k. In particular, if s is any solution, then all of the solutions are given by $\pm s$ and $\pm s + 2^{k-1}$.*

Proof. Part (1) is a simple exercise in $\mathbb{Z}/4\mathbb{Z}$ arithmetic, so we will concentrate on proving part (2) . The square of any odd number $2n+1$ is $(2n+1)^2 \equiv 4n(n+1)+1 \equiv 1 \bmod 8$. Thus, if $a \not\equiv 1 \bmod 8$, the equation $x^2 \equiv a \bmod 2^k$ cannot be solvable for $k \geq 3$.

Let us now assume that $a \equiv 1 \bmod 8$, and let $k \geq 3$ be fixed. First, notice that there are 2^{k-3} numbers in the interval $[1, 2^k]$ that are congruent to $1 \bmod 8$ (see Exercise 4.7.28). Since every square of an odd number is $\equiv 1 \bmod 8$, it suffices to show that there are 2^{k-3} squares of odd numbers that are distinct modulo 2^k. Let $S = \{a_1, \ldots, a_{2^{k-3}}\}$ be the set of the 2^{k-3} odd numbers in the interval $[1, 2^{k-2}]$.

We claim that the squares of elements in S are distinct modulo 2^k. Indeed, if $a^2 \equiv b^2 \bmod 2^k$ with $a, b \in S$ and $a > b$, then $2^k | (a-b)(a+b)$. Since a and b are odd, either $(a-b)$ or $(a+b)$ is $\equiv 2 \bmod 4$ and the other one is $\equiv 0 \bmod 4$ (see Exercise 4.7.9). Hence, either $(a-b)$ or $(a+b)$ is divisible by 2 but not by 4, and the other one is divisible by 2^{k-1}. However, $1 \leq b < a \leq 2^{k-3} < 2^{k-1}$, so neither $a-b$ nor $a+b$ can be divisible by 2^{k-1}, and we have reached a contradiction. Hence, the squares of elements in S are 2^{k-3} distinct values modulo 2^k, and therefore every odd number $a \equiv 1 \bmod 8$ is congruent to a square of a number in S modulo 2^k. Therefore, $x^2 \equiv a \bmod 2^k$ has solutions.

It remains to count the number of distinct solutions to $x^2 \equiv a \bmod 2^k$. Clearly, if $s \bmod 2^k$ is a solution, then $\pm s + \lambda \cdot 2^{k-1} \bmod 2^k$, with $\lambda \in \mathbb{Z}$ (whose value only matters modulo 2), are also solutions, because

$$(\pm s + \lambda \cdot 2^{k-1})^2 \equiv s^2 + \lambda \cdot 2^k + \lambda^2 \cdot 2^{2(k-1)} \equiv a \bmod 2^k.$$

If a is odd, then the numbers $s, -s, s + 2^{k-1}, -s + 2^{k-1}$ are all distinct modulo 2^k, and so, the equation $x^2 \equiv a \bmod 2^k$ has at least four distinct solutions. Notice that if $a \not\equiv a' \bmod 2^k$, then the solutions of $x^2 \equiv a$ and $x^2 \equiv a' \bmod 2^k$ are necessarily distinct. Since there are 2^{k-3} numbers $a \equiv 1 \bmod 8$ in $[1, 2^k]$, then putting together all the solutions of systems $x^2 \equiv a \bmod 8$, they account for at least $4 \cdot 2^{k-3} = 2^{k-1}$ distinct odd numbers in $[1, 2^k]$, which is in fact the total number of odd numbers in $[1, 2^k]$. Hence, there cannot be more than four distinct solutions for each equation $x^2 \equiv a \bmod 2^k$, and therefore we have shown that there are exactly four distinct solutions modulo 2^k. □

11.5. The p-Adic Numbers

In this section we briefly introduce the p-adic integers $\mathbb{Z}_p$ and the p-adic numbers $\mathbb{Q}_p$. We strongly recommend **[Gou97]** to learn more about the p-adics.

Let $p \geq 2$ be a prime. The p-adic numbers may be thought of as a generalization of $\mathbb{Z}/p\mathbb{Z}$. The main difference is that the p-adic numbers form a ring of characteristic zero (see Definition 6.6.2), while $\mathbb{Z}/p\mathbb{Z}$ has characteristic p. In $\mathbb{Z}/p\mathbb{Z}$ we only consider congruences modulo p, while in $\mathbb{Z}_p$ we consider congruences modulo p^n for all $n > 0$, simultaneously.

Definition 11.5.1. The *p-adic integers*, denoted by $\mathbb{Z}_p$, are defined as follows:

$$\mathbb{Z}_p = \{(a_1, a_2, \ldots) : a_n \in \mathbb{Z}/p^n\mathbb{Z} \text{ such that } a_{n+1} \equiv a_n \bmod p^n\}.$$

In other words, a p-adic integer is an infinite vector $(a_n)_{n=1}^{\infty} = (a_n \bmod p^n)_{n=1}^{\infty}$ such that the nth coordinate is a congruence class in $\mathbb{Z}/p^n\mathbb{Z}$ and the sequence is compatible under congruences; i.e., $a_{n+1} \in \mathbb{Z}/p^{n+1}\mathbb{Z}$ reduces to the previous term a_n modulo p^n; that is, $a_{n+1} \equiv a_n \bmod p^n$. For instance,

$$(2, 2, 29, 29, 272, 758, \ldots)$$

are the first few terms of a 3-adic integer. Notice that all the coordinates are compatible with the previous terms under congruences modulo powers of 3. The vector $(2, 2, 2, 2, \ldots)$ is another element of $\mathbb{Z}_3$ (which we will denote simply by 2).

The p-adic integers have addition and multiplication operations, defined coordinate by coordinate:

$$(a_n)_{n=1}^{\infty} + (b_n)_{n=1}^{\infty} = (a_n + b_n \bmod p^n)_{n=1}^{\infty},$$

and

$$(a_n)_{n=1}^{\infty} \cdot (b_n)_{n=1}^{\infty} = (a_n \cdot b_n \bmod p^n)_{n=1}^{\infty}.$$

The reader should check that the sum and product of two compatible vectors of congruences are also compatible under congruences and, therefore, a new element of $\mathbb{Z}_p$. These operations make $\mathbb{Z}_p$ a commutative ring with identity element $1 = (1, 1, 1, 1, \ldots)$ and zero element $0 = (0, 0, 0, 0, \ldots)$.

Remark 11.5.2. For any prime $p \geq 2$, the p-adic integers contain a copy of $\mathbb{Z}$, where the integer m is represented by the element

$$m = (m \bmod p,\ m \bmod p^2,\ m \bmod p^3, \ldots).$$

For example, the number 200 in $\mathbb{Z}_3$ is given by

$$200 = (2, 2, 11, 38, 200, 200, 200, 200, 200, 200, \ldots).$$

Thus, we may write $\mathbb{Z} \subseteq \mathbb{Z}_p$ (see Exercise 11.7.19). However, there are elements in $\mathbb{Z}_p$ that are not in $\mathbb{Z}$, so $\mathbb{Z} \subsetneq \mathbb{Z}_p$, as the following example shows.

Example 11.5.3. Let $p = 7$. We are going to show that $\mathbb{Z}_7$, unlike $\mathbb{Z}$, contains an element whose square is 2 (which we will denote by "$\sqrt{2}$"). Indeed, 2 is a quadratic residue in $\mathbb{Z}/7\mathbb{Z}$, and 2 has two square roots, namely 3 and 4 modulo 7. By Example 11.4.8, we know that 2 is, in fact, a quadratic residue modulo 7^n for all $n \geq 1$. Thus, there exist integers a_n such that $a_n^2 \equiv 2 \bmod 7^n$ for all $n \geq 1$. Moreover, the compatible systems $\{a_n \bmod 7^n\}$ constructed in Example 11.4.8 are such that $a_n^2 \equiv 2 \bmod 7^n$ and $a_{n+1} \equiv a_n \bmod 7^n$ (we say that a_n can be *lifted* to $\mathbb{Z}/7^{n+1}\mathbb{Z}$; see Exercise 11.7.20). Indeed, here are the first few coordinates of an element α of $\mathbb{Z}_7$ such that $\alpha^2 = (2, 2, 2, \ldots)$:

$$\alpha = (3, 10, 108, 2166, 4567, \ldots).$$

Thus, α should be regarded as "$\sqrt{2}$" inside $\mathbb{Z}_7$, and $-\alpha$ is another square root of 2.

Remark 11.5.4. The usual integers, $\mathbb{Z}$, are not a field because not every non-zero element has a multiplicative inverse (only ± 1 have inverses!). Similarly, the p-adic integers $\mathbb{Z}_p$ do not form a field either. For example, $p = (p, p, p, \ldots)$ is not invertible in $\mathbb{Z}_p$ and non-zero ($p \not\equiv 0 \bmod p^2$), but many elements of $\mathbb{Z}_p$ *are* invertible. For instance, if $p > 2$, then 2 is invertible in $\mathbb{Z}_p$ (in other words, there is a number $\frac{1}{2} \in \mathbb{Z}_p$). Indeed, the inverse of 2 is given by

$$\frac{1}{2} = \left(\frac{1+p}{2} \bmod p,\ \frac{1+p^2}{2} \bmod p^2, \ldots,\ \frac{1+p^n}{2} \bmod p^n, \ldots \right).$$

For example, in $\mathbb{Z}_5$, the inverse of 2 is given by $(3, 13, 63, 313, \ldots)$. One can show that if $\alpha = (a_n)_{n=1}^{\infty}$ with $a_1 \not\equiv 0 \bmod p$, then α is invertible in $\mathbb{Z}_p$, while if $a_1 \equiv 0 \bmod p$, then α is not invertible. Moreover, for any $\alpha \in \mathbb{Z}_p$ there is an $r \geq 0$ such that $\alpha = p^r \beta$, where $\beta \in \mathbb{Z}_p$ is invertible. We summarize these facts in the following statement.

Lemma 11.5.5. *Let $\mathbb{Z}_p^\times$ be the group of all invertible p-adic integers.*

(a) $\mathbb{Z}_p^\times = \{z \in \mathbb{Z}_p : z \not\equiv 0 \bmod p\}$.

(b) *Every element of $\alpha \in \mathbb{Z}_p$ can be written as $\alpha = p^r\beta$ with $r \geq 0$ and an invertible $\beta \in \mathbb{Z}_p^\times$. Moreover, this decomposition is unique; i.e., there is a unique $r \in \mathbb{Z}$ and $\beta \in \mathbb{Z}_p^\times$ such that $\alpha = p^r\beta$.*

We leave the proof of the lemma as Exercise 11.7.24 for the reader. Here we will provide an example to illustrate the lemma.

Example 11.5.6. Let $p = 3$, and consider the p-adic integer α that has the following expansion:

$$\alpha = (0 \bmod 3,\ 0 \bmod 9,\ 18 \bmod 27,\ 72 \bmod 3^4,\ 153 \bmod 3^5,\ 639 \bmod 3^6, \ldots).$$

Let us write α as $p^r\beta$, with β invertible and $r \geq 0$. Since $\alpha \equiv 0 \bmod 9$ but $\alpha \not\equiv 0 \bmod 27$, we set $r = 2$. Let us write $\beta = (b_n \bmod 3^n)_{n\geq 1}$. Since $18 \equiv \alpha \equiv 9\beta \bmod 27$, it follows that $18 \equiv 9\beta \bmod 27$ and, therefore, $2 \equiv \beta \bmod 3$ (by Proposition 4.3.1). Thus, $\beta \equiv b_1 \equiv 2 \bmod 3$, and this is the unique value of $b_1 \bmod 3$ that is valid. Similarly, $72 \equiv 9\beta \bmod 3^4$ implies that $\beta \equiv b_2 \equiv 8 \bmod 9$. If we continue in this manner, we obtain the first few coordinates of β:

$$\beta = (2 \bmod 3,\ 8 \bmod 9,\ 17 \bmod 27,\ 71 \bmod 81, \ldots).$$

It remains to show that $\beta = (b_n \bmod 3^n)_{n\geq 1}$ is invertible. Since $b_1 \equiv 2 \bmod 3$ and $b_n \equiv b_1 \equiv 2 \not\equiv 0 \bmod 3$, it follows that each $b_n \bmod 3^n$ is invertible, so $\beta^{-1} = (b_n^{-1} \bmod 3^n)_{n\geq 1}$. Thus,

$$\beta^{-1} = (2 \bmod 3,\ 8 \bmod 9,\ 8 \bmod 27,\ 8 \bmod 81, \ldots).$$

Even though $\mathbb{Z}_p$ is not a field, we can embed $\mathbb{Z}_p$ in a field in the same way that $\mathbb{Z}$ sits inside $\mathbb{Q}$. We define the field of p-adic numbers as follows.

Definition 11.5.7. We define the *field of p-adic numbers* by

$$\mathbb{Q}_p = \left\{p^r \cdot \beta : r \in \mathbb{Z} \text{ and } \beta \in \mathbb{Z}_p^\times\right\}.$$

Thus, $\mathbb{Z}_p = \{p^r \cdot \beta : r \geq 0 \text{ and } \beta \in \mathbb{Z}_p^\times\}$ is a subring of $\mathbb{Q}_p$. We now show that the second version and the p-adic version of the Hasse–Minkowski theorem (Theorems 11.2.10 and 11.2.14) are equivalent. Indeed, it suffices to show that a quadratic form equation has a compatible system of solutions modulo p^k, for every $k \geq 1$, if and only if it has a p-adic solution.

Proposition 11.5.8. *Let $q(x_1, \ldots, x_n) = 0$ be a diophantine equation (i.e., $q \in \mathbb{Z}[x_1, \ldots, x_n]$), and let p be a prime number. Then, $q = 0$ has a compatible system of solutions modulo p^k, for every $k \geq 1$, if and only if there is a p-adic solution $(z_1, \ldots, z_n) \in (\mathbb{Q}_p)^n$ of $q = 0$.*

Proof. Suppose first that $(s_{1,k} \bmod p^k, \ldots, s_{n,k} \bmod p^k)_{k\geq 1}$ is a compatible system of solutions modulo p^k of $q = 0$. In other words, there are integers $s_{i,k}$, for $k \geq 1$ and $1 \leq i \leq n$, such that

$$q(s_{1,k}, \ldots, s_{n,k}) \equiv 0 \bmod p^k$$

and $s_{i,k+1} \equiv s_{i,k} \bmod p^k$. Therefore, the sequences

$$z_i = (s_{i,1} \bmod p,\ s_{i,2} \bmod p^2, \ldots,\ s_{i,k} \bmod p^k, \ldots)$$

define p-adic integers $z_1, \ldots, z_n$ because the compatibility condition $s_{i,k+1} \equiv s_{i,k} \bmod p^k$ is satisfied. Moreover, since each n-tuple $(s_{1,k}, \ldots, s_{n,k})$ satisfy the equation $q \equiv 0 \bmod p^k$, it follows that $q(z_1, \ldots, z_n) = 0$.

Conversely, suppose $z_1, \ldots, z_n \in \mathbb{Q}_p$ are p-adic numbers such that

$$q(z_1, \ldots, z_n) = 0,$$

and let $m = \min\{\nu_p(z_i) : 1 \leq i \leq n\}$. Then, $z_i' = p^{-m} z_i$ are p-adic integers (see Exercise 11.7.26), and

$$q(z_1', \ldots, z_n') = q(p^{-m} z_1, \ldots, p^{-m} z_n) = p^{-2m} q(z_1, \ldots, z_n) = 0,$$

where we have used Lemma 11.1.3. Thus, $q = 0$ also has a solution with coordinates in $\mathbb{Z}_p$, so we may assume that $z_1, \ldots, z_n$ are in $\mathbb{Z}_p$ in the first place. Now, the p-adic coordinates mod p^k of each $z_i = (s_{i,k} \bmod p^k)_{k \geq 1}$ form a compatible system of solutions of $q \equiv 0 \bmod p^k$, for all $k \geq 1$, as desired. □

11.6. Hensel's Lemma

The following results are used to show the existence of a solution to polynomial equations over $\mathbb{Q}_p$. Here we will only discuss the application to the p-adics, $\mathbb{Q}_p$. Before we state Hensel's lemma, we define the p-adic valuation of a p-adic number.

Definition 11.6.1. Let $p \geq 2$, let $\mathbb{Q}_p$ be the field of p-adic numbers, and let $\mathbb{Z}_p$ be the p-adic integers. The *p-adic valuation* is a function $\nu_p \colon \mathbb{Q}_p^* = \mathbb{Q}_p \setminus \{0\} \to \mathbb{Z}$ defined as follows: if $t \in \mathbb{Q}_p^*$ is written as $t = p^r \cdot \beta$, with $\beta \in \mathbb{Z}_p^\times$, then $\nu_p(t) = r$.

Example 11.6.2. Here are some values of p-adic valuations:

$$\nu_3(27) = 3,\ \nu_2(384) = \nu_2(2^7 \cdot 3) = 7,\ \nu_7(27/49) = -2.$$

Theorem 11.6.3 (Hensel's lemma)**.** *Let $p \geq 2$, let $\mathbb{Q}_p$ be the field of p-adic numbers, and let $\mathbb{Z}_p$ be the p-adic integers. Let ν_p be the p-adic valuation (as in Definition* 11.6.1*). Let $f(x)$ be a polynomial with coefficients in $\mathbb{Z}_p$ and suppose there exists $\alpha_0 \in \mathbb{Z}_p$ such that*

$$\nu_p(f(\alpha_0)) > \nu_p(f'(\alpha_0)^2).$$

Then, there exists a root $\alpha \in \mathbb{Q}_p$ of $f(x)$. Moreover, the sequence

$$\alpha_{i+1} = \alpha_i - \frac{f(\alpha_i)}{f'(\alpha_i)}$$

converges to α. Furthermore,

$$\nu_p(\alpha - \alpha_0) \geq \nu_p\left(\frac{f(\alpha_i)}{f'(\alpha_i)}\right) > 0.$$

Notice the similarities with Newton's method (in particular, with Theorem 11.4.5). We will not prove the most general case of Hensel's lemma here. For a proof, see [**Con2**]. The so-called "trivial case" of Hensel's lemma, stated below, is just a rephrasing of the first part of Theorem 11.4.5.

Corollary 11.6.4 (Trivial case of Hensel's lemma). *Let $p \geq 2$, and let $\mathbb{Z}_p$ and $\mathbb{Q}_p$ be as before. Let $f(x)$ be a polynomial with coefficients in $\mathbb{Z}_p$ and suppose there exists $\alpha_0 \in \mathbb{Z}_p$ such that*

$$f(\alpha_0) \equiv 0 \bmod p, \quad f'(\alpha_0) \not\equiv 0 \bmod p.$$

Then, there exists a root $\alpha \in \mathbb{Q}_p$ of $f(x)$; i.e., $f(\alpha) = 0$.

Example 11.6.5. Let p be a prime number greater than 2 and let $p \neq 7$. Are there solutions to $x^2 + 7 = 0$ in $\mathbb{Z}_p$? Suppose $\alpha \in \mathbb{Z}_p$ is a solution. If we write $\alpha = (a_n \bmod p^n)_{n\geq 1}$, then we would have $a_1^2 + 7 \equiv 0 \bmod p$. Thus, -7 must be a quadratic residue modulo p. Thus, let $p \neq 7$ be an odd prime such that

$$\left(\frac{-7}{p}\right) = 1,$$

where $(\frac{\cdot}{p})$ is Legendre's quadratic residue symbol. Hence, there exist $\alpha_0 \in \mathbb{Z}$ such that $\alpha_0^2 \equiv -7 \bmod p$. We claim that $x^2 + 7 = 0$ has a solution in $\mathbb{Z}_p$ if and only if -7 is a quadratic residue modulo p. Indeed, if we let $f(x) = x^2 + 7$ (so $f'(x) = 2x$), the element $\alpha_0 \in \mathbb{Z}_p$ satisfies the conditions of the (trivial case of) Hensel's lemma. Therefore, there exists a root $\alpha \in \mathbb{Z}_p$ of $x^2 + 7 = 0$. We leave it to the reader to show that $x^2 + 7 = 0$ has no solutions in $\mathbb{Z}_7$ (nor in $\mathbb{Q}_7$; see Exercise 11.7.27).

For example, -7 is a square modulo 11, since $-7 \equiv 4 \bmod 11$. There are two roots α_1 and α_2 of $x^2 + 7 = 0$ in $\mathbb{Z}_{11}$; namely

$$\alpha_1 = (2, 90, 1058, 10375, 156785, 317836, \ldots),$$
$$\alpha_2 = -\alpha_1 = (9, 31, 273, 4266, 4266, 1453725, \ldots).$$

Example 11.6.6. Let $p = 2$. Are there any solutions to $x^2 + 7 = 0$ in $\mathbb{Q}_2$? Notice that if we let $f(x) = x^2 + 7$, then $f'(x) = 2x$ and, for any $\alpha_0 \in \mathbb{Z}_2$, the number $f'(\alpha_0) = 2\alpha_0$ is congruent to 0 modulo 2. Thus, we cannot use the trivial case of Hensel's lemma (i.e., Corollary 11.6.4).

Let $\alpha_0 = 1 \in \mathbb{Z}_2$. Notice that $f(1) = 8$ and $f'(1) = 2$. Thus,

$$3 = \nu_2(8) > \nu_2(2^2) = 2$$

and the general case of Hensel's lemma applies. Hence, there exists a 2-adic solution to $x^2 + 7 = 0$.

In the following result, we use Hensel's lemma to classify all the square numbers in $\mathbb{Q}_2$. Below, we will use congruences between p-adic integers, which we define as follows: if $\alpha = (a_n \bmod p^n)_{n\geq 1}$ and $\beta = (b_n \bmod p^n)_{n\geq 1}$ and $k \geq 1$ is fixed, then we say that $\alpha \equiv \beta \bmod p^k$ if $a_k \equiv b_k \bmod p^k$.

Theorem 11.6.7. *Let $\alpha \in \mathbb{Q}_2$. Then, α is a perfect square (i.e., $\alpha \in (\mathbb{Q}_2)^2$) if and only if $\alpha = 2^k u$ where $k \in 2\mathbb{Z}$ and $u \in \mathbb{Z}_2$ with $u \equiv 1 \bmod 8$.*

Proof. Let us suppose first that $\alpha \in \mathbb{Q}_2$ is a perfect square; i.e., $\alpha = \beta^2$ for some $\beta \in \mathbb{Q}_2$. Then, by the definition of $\mathbb{Q}_2$, we have $\beta = 2^h v$, for some $h \in \mathbb{Z}$ and some $v \in \mathbb{Z}_2$ with $v \equiv 1 \bmod 2$. In particular,

$$\alpha = \beta^2 = (2^h v)^2 = 2^{2h} v^2.$$

Moreover, since $v \equiv 1 \bmod 2$, we have $v = 1 + 2w$, for some $w \in \mathbb{Z}_2$, and so

$$u = v^2 = (1 + 2w)^2 = 1 + 4w + 4w^2 = 1 + 4w(w + 1).$$

But either w or $w + 1 \equiv 0 \bmod 2$, and so $w(w + 1) \equiv 0 \bmod 2$. It follows that $4w(w + 1) \equiv 0 \bmod 8$, and so

$$v^2 = 1 + 4(w(w + 1)) \equiv 1 \bmod 8.$$

For the converse, assume that $\alpha = 2^k u$ where $k = 2h \in 2\mathbb{Z}$ and $u \in \mathbb{Z}_2$ with $u \equiv 1 \bmod 8$. It suffices to show that $u = v^2$ for some $v \in (\mathbb{Z}_2)^2$, because then $\alpha = (2^h v)^2$. In order to show that u is a square, we use Hensel's lemma. Consider $f(x) = x^2 - u$, and let $\alpha_0 = 1 \in \mathbb{Z}_2$. Then, $f(1) = 1 - u \equiv 0 \bmod 8$, and $f'(1) = 2$. Thus,

$$3 = \nu_2(8) > \nu_2(2^2) = 2,$$

and the general case of Hensel's lemma applies. Hence, there exists a 2-adic solution v to $x^2 - u = 0$, or, in other words, $v^2 = u$. Hence, u is a square, as claimed. This concludes the proof of the theorem. □

11.7. Exercises

Exercise 11.7.1. Compute the Gram matrix for the following quadratic forms and decide whether they are regular:

(a) $q(X, Y, Z) = X^2 + Y^2 + Z^2 + XY + YZ + XZ$.

(b) $q(X, Y, Z, T) = X^2 + Y^2 + Z^2 - T^2$. (Note: this quadratic form is related to the theory of special relativity.)

(c) $q(X, Y, Z, T) = XY - ZT$.

(d) $q(X, Y, Z, T) = X^2 - XZ + XT - Y^2 + YZ + YT - ZT$.

(e) $q(X, Y) = X^2 - XY - Y^2$.

Exercise 11.7.2. Let $q(X_1, X_2, X_3) = (aX + bY + cZ)(dX + eY + fZ)$ be a quadratic form. Show that q is not regular.

Exercise 11.7.3. Show that the map ψ defined in Remark 11.1.12 is a bijection.

Exercise 11.7.4. Determine whether the following sequences of congruences form a compatible system in the sense of Definition 11.2.3:

(a) $\{2 \bmod 3^k\}_{k\geq 1} = \{2 \bmod 3,\ 2 \bmod 9,\ 2 \bmod 27, \ldots\}$.

(b) $\{p^{k-1} \bmod p^k\}_{k\geq 1}$ where p is a prime number.

(c) $\{p \cdot (p^{k-1} - 1) \bmod p^k\}_{k\geq 1}$ where p is a prime number.

(d) $\{1 + (m - 1)! \bmod (m!)\}_{m\geq 2} = \{0 \bmod 2,\ 3 \bmod 6,\ 7 \bmod 24, \ldots\}$.

(e) $\{1! + 2! + 3! + \cdots + (m - 1)! \bmod (m!)\}_{m\geq 2} = \{1 \bmod 2,\ 3 \bmod 6, \ldots\}$.

Exercise 11.7.5. Show that the equation $5X^2 + 13Y^2 - 7Z^2 = 0$ has no non-trivial integral solutions. (Hint: work mod 2 and mod 4.)

Exercise 11.7.6. Show that the equation $X^2 + Y^2 + Z^2 + XY + XZ + YZ = 0$ has no non-trivial integral solutions. (Hint: complete squares, and work over $\mathbb{R}$.)

Exercise 11.7.7. Show that the equation $Y^2 - X^2 - Z^2 + XY - 3XZ + YZ = 0$ has no non-trivial integral solutions.

Exercise 11.7.8. Let n be an odd number such that $2n$ is a sum of two integral squares; i.e., there are $a, b \in \mathbb{N}$ such that $2n = a^2 + b^2$. Show that

$$n = \left(\frac{a+b}{2}\right)^2 + \left(\frac{a-b}{2}\right)^2$$

is a representation of n as a sum of integral squares.

Exercise 11.7.9. Find the first five terms of a compatible system of congruences $\{s_k \bmod 5^k\}_{k\geq 1}$ that are a solution of $x^2 \equiv 11 \bmod 5^k$. In other words, find integers $s_1, \ldots, s_5$ such that $s_k^2 \equiv 11 \bmod 5^k$ and $s_{k+1} \equiv s_k \bmod 5^k$.

Exercise 11.7.10. Find the first six terms of a compatible system of congruences $\{s_k \bmod 2^k\}_{k\geq 1}$ that are a solution of $x^2 \equiv 17 \bmod 2^k$. In other words, find integers $s_1, \ldots, s_6$ such that $s_k^2 \equiv 17 \bmod 2^k$ and $s_{k+1} \equiv s_k \bmod 2^k$.

Exercise 11.7.11. Decide whether the following systems of congruences have a non-trivial compatible system of solutions, for all $k \geq 1$:

(a) $x^2 \equiv 13 \bmod 2^k$.
(b) $x^2 \equiv 68 \bmod 2^k$.
(c) $x^2 \equiv 7 \bmod 13^k$.
(d) $5x^2 \equiv 11 \bmod 13^k$.
(e) $x^2 \equiv 1289 \bmod 4001^k$.

Exercise 11.7.12. Let n be an odd integer. Show that the congruences $X^2 + Y^2 \equiv nZ^2$ modulo 2 and modulo 4 have non-trivial compatible solutions (i.e., $u \equiv (a_2, b_2, c_2) \bmod 2$ with $a_2^2 + b_2^2 \equiv nc_2^2 \bmod 2$ and a_2, b_2, or $c_2 \not\equiv 0 \bmod 2$, and $v \equiv (a_4, b_4, c_4) \bmod 4$ with $a_4^2 + b_4^2 \equiv nc_4^2 \bmod 4$, such that $v \equiv u \bmod 2$, coordinate-wise) if and only if $n \equiv 1 \bmod 4$.

Exercise 11.7.13. Use the Hasse–Minkowski theorem to show that the hyperbola $C : x^2 - 4001y^2 = 1289$ has a rational point. (Hint: proceed as in Section 11.3.1.)

Exercise 11.7.14. Show that the ellipse $C : x^2 + 1289y^2 = 4001$ has a rational point.

Exercise 11.7.15. Show that the hyperbola $C : x^2 - 4001y^2 = 1249$ does not have a rational point.

Exercise 11.7.16. The circle $C : x^2 + y^2 = 298$ has a rational point $(\frac{121}{13}, \frac{189}{13})$. Use the geometric method of Example 11.3.13 to find an integral point on C.

Exercise 11.7.17. Use Newton's method (Theorem 11.4.5) to find the first four terms of a compatible system of congruences $\{s_k \bmod 13^k\}$ that are solutions of $x^2 \equiv 10 \bmod 13^k$.

Exercise 11.7.18. Use Newton's method to find the first three terms of a compatible system of congruences $\{s_k \bmod 17^k\}$ that are solutions of $x^3 + 2x + 1 \equiv 0 \bmod 17^k$.

Exercise 11.7.19. Show that if q and t are distinct integers (in $\mathbb{Z}$), then their representatives in $\mathbb{Z}_p$ for any prime $p \geq 2$, given by $q = (q \bmod p^n)_{n=1}^{\infty}$ and $t = (t \bmod p^n)_{n=1}^{\infty}$, are also distinct in $\mathbb{Z}_p$.

Exercise 11.7.20. Let $p > 2$ be a prime number.

(1) Let $b \in \mathbb{Z}$ with $\gcd(b, p) = 1$, and let $n \geq 1$. Suppose $a_n \in \mathbb{Z}$ such that $a_n^2 \equiv b \bmod p^n$. Show that there exists $a_{n+1} \in \mathbb{Z}$ such that $a_{n+1}^2 \equiv b \bmod p^{n+1}$ and $a_{n+1} \equiv a_n \bmod p^n$. (Hint: write $a_n^2 = b + kp^n$ and consider $f(x) = a_n + xp^n$. Find x such that $f(x)^2 \equiv b \bmod p^{n+1}$.)

(2) Suppose $a_1^2 \equiv b \bmod p$, where $\gcd(b,p) = 1$. Show that the vector $\alpha = (a_n)_{n=1}^{\infty}$, defined recursively by

$$a_{n+1} \equiv a_n - \frac{a_n^2 - b}{2a_n} \bmod p^{n+1},$$

is a well-defined element of $\mathbb{Z}_p$ and, moreover, $\alpha^2 = b$; i.e.,

$$\alpha^2 = (b \bmod p,\ b \bmod p^2,\ b \bmod p^3, \ldots),$$

so α is a square root of b.

Exercise 11.7.21. Find the first four coordinates of the 5-adic expansion of $\frac{1}{3}$ in $\mathbb{Z}_5$.

Exercise 11.7.22. Let $p \neq 3$ be a prime. Show that $\frac{1}{3} \in \mathbb{Z}_p$, and find a formula (as in Remark 11.5.4) for $1/3$ in terms of p. (Hint: write a formula for primes $p \equiv 1 \bmod 3$ and another formula for primes $p \equiv 2 \bmod 3$.)

Exercise 11.7.23. Find the first four coordinates of the 5-adic expansions of $\pm\sqrt{6}$ in $\mathbb{Z}_5$; i.e., find the first four coordinates of α and $-\alpha$ such that $\alpha^2 = 6$ in $\mathbb{Z}_5$.

Exercise 11.7.24. In this exercise we prove Lemma 11.5.5.

(a) Show that $\mathbb{Z}_p^{\times}$, i.e., the group of invertible p-adic integers, is given by $\{z \in \mathbb{Z}_p : z \not\equiv 0 \bmod p\}$.

(b) Let $z \in \mathbb{Z}_p$, and let $n \geq 1$ be an integer such that $z \equiv 0 \bmod p^n$ but $z \not\equiv 0 \bmod p^{n+1}$ (i.e., $\nu_p(z) = n$). Show that there is another element $y \in \mathbb{Z}_p$ such that $p^n y = z$. Moreover, show that y is unique and invertible in $\mathbb{Z}_p$. (Hint: describe each coordinate $y = (y_m \bmod p^m)_{m \geq 1}$ in terms of the coordinates of z.)

Exercise 11.7.25. Let ν_p be the p-adic valuation, as in Definition 11.6.1. Show the following properties, for any non-zero $v, w \in \mathbb{Q}_p$:

(a) $\nu_p(v \cdot w) = \nu_p(v) + \nu_p(w)$.

(b) $\nu_p(v + w) \geq \min\{\nu_p(v), \nu_p(w)\}$.

(c) If $\nu_p(v) \neq \nu_p(w)$, then $\nu_p(v + w) = \min\{\nu_p(v), \nu_p(w)\}$.

Exercise 11.7.26. Let $z \in \mathbb{Q}_p$, and let $m = \nu_p(z)$. Show that $p^{-m} z \in \mathbb{Z}_p$ (i.e., $\nu_p(p^{-m} z) = 0$).

Exercise 11.7.27. Show that the equation $x^2 + 7 = 0$ has no solutions over $\mathbb{Q}_7$. (Hint: use the 7-adic valuation.)

Exercise 11.7.28. Use Hensel's lemma to show that the following equations have a p-adic solution:

(a) $x^2 = 41$ over $\mathbb{Q}_5$.

(b) $x^2 = 41$ over $\mathbb{Q}_2$.

(c) $x^3 + x + 2 = 0$ over $\mathbb{Q}_7$.

(d) $x^2 - 4001y^2 = 1289$ over $\mathbb{Q}_{4001}$.

(e) $x^2 - 4001y^2 = 1289$ over $\mathbb{Q}_{1289}$.

CHAPTER 12

CIRCLES, ELLIPSES, AND THE SUM OF TWO SQUARES PROBLEM

The description of right lines and circles, upon which geometry is founded, belongs to mechanics. Geometry does not teach us to draw these lines, but requires them to be drawn.

Sir Isaac Newton, from *Principia Mathematica*

In Chapter 9 we saw that for any quadratic equation $C : f(x, y) = 0$ with integer coefficients that is classified as an ellipse (as in Definition 9.2.7), there is a change of variables $\phi : C \to C'$, where $C' : X^2 + BY^2 = D$ for some integers $B > 0$ and $D \neq 0$, that sends $C(\mathbb{Z})$ to $C'(\mathbb{Z})$, integral points to integral points (see Theorem 9.4.1; the map φ is explicitly described in Corollary 9.2.12). Therefore, our task in this chapter is to describe the rational and integral points on ellipses of the form $C' : X^2 + BY^2 = D$. In Section 12.1 we analyze the special case of circles (i.e., $B = 1$).

12.1. Rational and Integral Points on a Circle

In this section we study the rational points and integral points on circles, i.e., on quadratic curves of the form $x^2 + y^2 = r$, for some rational number $r \in \mathbb{Q}$.

Question 12.1.1. *Let $r \in \mathbb{Q}$ be a fixed non-zero rational number. Are there rational points on the circle $x^2+y^2 = r$? Alternatively, is r a sum of two (rational) squares? Are there integral points on the circle $x^2 + y^2 = r$?*

The first thing to notice is that we may assume r is an integer.

Lemma 12.1.2. *Let $r \in \mathbb{Q}$ be a non-zero rational number, such that $r = n/m$ where $n \in \mathbb{Z}$ and $m > 1$ are relatively prime. Then, the circle $C_{n/m} : x^2 + y^2 = n/m$ has rational points if and only if the circle $C_{nm} : x^2 + y^2 = nm$ has rational points.*

Proof. The proof follows from the fact that the map $\psi : C_{n/m} \to C_{nm}$ given by $\psi((a,b)) = (ma, mb)$ is a bijection between $C_{n/m}(\mathbb{Q})$ and $C_{nm}(\mathbb{Q})$. We leave it to the reader to check the details in Exercise 12.8.1. □

For example, if we want to find a rational point on $x^2 + y^2 = 13/5$, it suffices to find a point on $x^2 + y^2 = 65$. The latter has an integral point $(7, 4)$ which corresponds to a rational point $(7/5, 4/5)$ in the circle $x^2 + y^2 = 13/5$. Next, we show that it suffices to look for integral points.

Lemma 12.1.3. *Let r be a non-zero rational number, and let C_r be the circle with equation $x^2 + y^2 = r$.*

(1) *If C_r has an integral point, then r is an integer.*

(2) *Let r be an integer. Then, C_r has an integral point if and only if C_r has a rational point.*

Proof. If (a, b) is an integral point, then $r = a^2 + b^2 \in \mathbb{Z}$, which shows the first part. The second part was shown in Proposition 11.3.11. □

Therefore, we have reduced Question 12.1.1 to the following equivalent question:

Question 12.1.4. *Let $n > 1$ be a fixed natural number. Are there integral points on the circle $x^2 + y^2 = n$? In other words, can n be written as the sum of two (integral) squares?*

We have already answered this question, in Chapter 11 (in particular, see Section 11.3 and Theorem 11.3.9), but we used the deep theorem of Hasse and Minkowski to do so (Theorem 11.2.12), which we will not prove in this book as it is beyond our scope. In this section, however, we provide a self-contained answer to Question 12.1.4, i.e., a proof of Theorem 11.3.9 that does not make use of the Hasse–Minkowski theorem.

Let us begin by considering the case when $n = p$ is a prime number. What primes p can be written as the sum of two (integral) squares? Clearly $p = 2 = 1^2 + 1^2$, so we may concentrate on odd primes $p \geq 3$. Since for a fixed prime p there are only finitely many possibilities for $a, b \in \mathbb{Z}$ such that $a^2 + b^2 = p$ (because a and b must be $\leq \sqrt{p}/2$), we can find out what odd primes ≤ 100 are a sum of two squares:

$$5 = 2^2 + 1^2,\ 13 = 3^2 + 2^2,\ 17 = 4^2 + 1^2,\ 29 = 5^2 + 2^2,$$
$$37 = 6^2 + 1^2,\ 41 = 5^2 + 4^2,\ 53 = 7^2 + 2^2,\ 61 = 6^2 + 5^2,$$
$$73 = 8^2 + 3^2,\ 89 = 8^2 + 5^2,\ \text{and } 97 = 9^2 + 4^2,$$

while the primes $3, 7, 11, 19, 23, 31, 43, 47, 59, 67, 71, 79$, and 83 are not the sum of two squares. The reader may have realized that the primes (at least those ≤ 100) that are a sum of two squares satisfy $p \equiv 1 \bmod 4$, while every prime $p \equiv 3 \bmod 4$ cannot be expressed as a sum of two squares. Thus, our data suggests that an odd prime p is a sum of two squares if and only if $p \equiv 1 \bmod 4$. The following theorem was claimed and stated by Pierre de Fermat in 1640, but the first published proof is due to Euler, in 1749. Nonetheless, it is usually known as Fermat's theorem on sums of two squares.

Theorem 12.1.5. *An odd prime number p is a sum of two squares if and only if $p \equiv 1 \bmod 4$.*

Proof. Let p be an odd prime and suppose that $p = a^2 + b^2$ is a sum of two squares, for some $a, b \in \mathbb{Z}$. Notice that $\gcd(ab, p) = 1$, because if $p|a$, say, then $p|(p-a^2) = b^2$, and so $p|b$ which would imply that $p^2|(a^2+b^2) = p$, a contradiction. Thus, a and b are units modulo p and therefore invertible. Since $a^2+b^2 \equiv 0 \bmod p$, it follows that $a^2 \equiv -b^2 \bmod p$, and therefore $(a/b)^2 \equiv -1 \bmod p$. We conclude that -1 is a square modulo p and, by Lemma 10.3.4, we conclude that $(p-1)/2$ is even; i.e., $p \equiv 1 \bmod 4$.

For the converse, let us assume that $p \equiv 1 \bmod 4$. Then, Lemma 10.3.4 shows that -1 is a square modulo p. Thus, there is some integer s such that $s^2 \equiv -1 \bmod p$. Let $\lfloor\sqrt{p}\rceil$ be the floor of $\sqrt{p}$, i.e., the largest integer $\leq \sqrt{p}$, and consider the set of integers

$$S = \{(x, y) : 0 \leq x, y < \lfloor\sqrt{p}\rfloor\}.$$

We claim that there are two different pairs (x_1, y_1) and $(x_2, y_2) \in S$ such that $sx_1 - y_1 \equiv sx_2 - y_2 \bmod p$. Indeed, if all the possible values $sx - y$ for $(x, y) \in S$ were different modulo p, then there would be $(\lfloor\sqrt{p}\rfloor + 1)^2$ distinct values modulo p in S, but

$$(\lfloor\sqrt{p}\rfloor + 1)^2 > (\sqrt{p})^2 = p.$$

Since there are exactly p distinct values in a complete set of representatives modulo p, this is a contradiction.

Hence, there are two distinct pairs (x_1, y_1) and $(x_2, y_2) \in S$ such that $sx_1 - y_1 \equiv sx_2 - y_2 \bmod p$, or, equivalently, $sx_0 \equiv y_0 \bmod p$ where $x_0 = x_1 - x_2$ and $y_0 = y_1 - y_2$. Since $(x_1, y_1) \neq (x_2, y_2)$, either $x_0 \neq 0$ or $y_0 \neq 0$. Moreover, since $sx_0 \equiv y_0 \bmod p$, it follows that $s^2x_0^2 \equiv y_0^2 \bmod p$, and therefore $-x_0^2 \equiv y_0^2 \bmod p$, or $x_0^2 + y_0^2 \equiv 0 \bmod p$. Thus, $x_0^2 + y_0^2$ is a non-zero multiple of p and

$$0 < x_0^2 + y_0^2 \leq (\lfloor\sqrt{p}\rfloor)^2 + (\lfloor\sqrt{p}\rfloor)^2 = 2(\lfloor\sqrt{p}\rfloor)^2 < 2(\sqrt{p})^2 = 2p.$$

The only non-zero multiple of p strictly between 0 and $2p$ is precisely p, and therefore $x_0^2 + y_0^2 = p$, as desired. □

Example 12.1.6. The proof of Theorem 12.1.5 is constructive. Let $p = 41 \equiv 1 \bmod 4$ and let us find $x, y \in \mathbb{Z}$ such that $x^2 + y^2 = 41$, using the method outlined in the proof. We first find s such that $s^2 \equiv -1 \bmod 41$. Since $81 \equiv 40 \equiv -1 \bmod 41$, we see that $s = 9$ works. Now we calculate $\lfloor\sqrt{41}\rfloor = 6$ and we construct

$$S = (x, y) : 0 \leq x, y \leq 6.$$

The set S has 49 elements and we need to find two different pairs (x_1, y_1) and (x_2, y_2) such that $9x_1 - y_1 \equiv 9x_2 - y_2 \bmod 41$. For instance, we find

$$9 \cdot 4 - 0 \equiv 36 \equiv 9 \cdot 0 - 5 \bmod 41,$$

and we can take $x_0 = |4 - 0| = 4$ and $y_0 = |0 - 5| = 5$. Indeed,

$$4^2 + 5^2 = 16 + 25 = 41,$$

as desired.

Theorem 12.1.5 confirms our suspicion that only the primes that are 1 mod 4 can be written as the sum of two squares. Now we turn to the case of composite numbers. The following identity shows that if m and n are sums of two squares, then their product mn is also the sum of two squares.

Lemma 12.1.7. *Let $m, n \in \mathbb{Z}$ such that $m = a^2 + b^2$ and $n = c^2 + d^2$, for some $a, b, c, d \in \mathbb{Z}$. Then,*

$$mn = (ac + bd)^2 + (ad - bc)^2 = (ac - bd)^2 + (ad + bc)^2.$$

Proof. The proof amounts to verifying the validity of the algebraic equality, so we leave it as an exercise for the reader. See also Exercise 12.8.3 for an alternative proof that explains a reason for the equality to hold. □

Example 12.1.8. Let $m = 5$ and $n = 41$. Then, our previous lemma says that $205 = 5 \cdot 41$ is also a sum of two squares. Indeed, $m = 5 = 2^2 + 1^2$ and $n = 41 = 5^2 + 4^2$. Thus, using the formula given by the lemma, we obtain

$$\begin{aligned} 205 &= (2 \cdot 5 + 1 \cdot 4)^2 + (2 \cdot 4 - 1 \cdot 5)^2 = (2 \cdot 5 - 1 \cdot 4)^2 + (2 \cdot 4 + 1 \cdot 5)^2 \\ &= 14^2 + 3^2 = 6^2 + 13^2. \end{aligned}$$

So, in fact, we have obtained two distinct representations of 205 as a sum of two squares.

As a consequence of Lemma 12.1.7 and Theorem 12.1.5, every integer whose prime factorization contains only primes that are $\equiv 1 \bmod 4$ is a sum of two squares. However, $45 = 6^2 + 3^2$ is also a sum of two squares, and $45 = 3^2 \cdot 5$, so there is a prime $\equiv 3 \bmod 4$ in the prime factorization. Next, we show that if n is a sum of two squares and $q \equiv 3 \bmod 4$ is a prime divisor of n, then q^2 is a divisor of n.

Lemma 12.1.9. *Let n be an integer such that $n = a^2 + b^2$ for some $a, b \in \mathbb{Z}$, and suppose q is a prime such that $q \equiv 3 \bmod 4$.*

(1) *If $q|n$, then $q|a$ and $q|b$. In particular, $q^2|n$.*

(2) *If $q|n$, then q appears to an even power in the prime factorization of n.*

Proof. For (1), let $q \equiv 3 \bmod 4$ be a prime divisor of $n = a^2 + b^2$. Suppose for a contradiction that a is not divisible by q. Then, a is a unit modulo q, and there is $a' \in \mathbb{Z}$ such that $aa' \equiv 1 \bmod q$. Since $q|n$, it follows that $a^2 + b^2 \equiv 0 \bmod q$, and therefore $(aa')^2 + (a'b)^2 \equiv 0 \bmod q$, or, equivalently, $(a'b)^2 \equiv -1 \bmod q$, which is impossible because -1 is not a square modulo q, by Lemma 10.3.4. Hence a is divisible by q, and the same argument shows that b is also divisible by q.

For (2), suppose for a contradiction that $n = a^2 + b^2$ and $n = n'q^{2e+1}$, with $\gcd(n', q) = 1$. By part (1), the prime q divides a and b, so that $a = qa_1$ and $b = qb_1$, for some integers a_1, b_1. Therefore

$$n_1 = n/q^2 = n'q^{2(e-1)+1} = a_1^2 + b_1^2.$$

Applying part (1) repeatedly, we obtain a sequence

$$n_i = n/(q^{2i}) = n'q^{2(e-i)+1} = a_i^2 + b_i^2,$$

for some integers a_i, b_i and for every $i = 1, \ldots, e$. In particular,

$$n_e = n/(q^{2e}) = n'q = a_n^2 + b_n^2,$$

and therefore, once again by part (1), the prime q would divide a_n and b_n, and this would imply that q^2 is a divisor of $n'q$, and so $q|n'$. But this is impossible because $\gcd(n', q) = 1$. Hence, the power of a prime $q \equiv 3 \bmod 4$ dividing a sum of squares n must be even. □

We are finally ready to state and prove a complete classification of what natural numbers can be expressed as the sum of two squares, which does not use the Hasse–Minkoswki theorem (as in Theorem 11.3.9).

Theorem 12.1.10. *Let $n > 1$ be a natural number. The circle $C_n : x^2+y^2 = n$ has an integral point if and only if every prime divisor p of n with $p \equiv 3 \bmod 4$ appears to an even power in the prime factorization of n. Equivalently, n can be written as a sum of two squares if and only if the square-free part of n is not divisible by any prime p of the form $p \equiv 3 \bmod 4$.*

Proof. Suppose first that $n > 1$ is a sum of two squares, i.e., $n = a^2 + b^2$, for some $a, b \in \mathbb{Z}$, and suppose that n has a prime divisor $q \equiv 3 \bmod 4$. Then, Lemma 12.1.9, the prime q appears to an even power in the prime factorization of n.

For the converse, suppose that $n = n'm^2$, where n' is square-free, and assume that n' is not divisible by any prime p of the form $p \equiv 3 \bmod 4$. Then, $n' = 2^e p_1 p_2 \cdots p_t$, for $e = 0$ or 1, and some primes $p_i \equiv 1 \bmod 4$, for $i = 1, \ldots, t$. Clearly $2 = 1^2+1^2$, and by Theorem 12.1.5, each p_i is a sum of two squares, say $p_i = a_i^2+b_i^2$, for some $a_i, b_i \in \mathbb{Z}$. Hence, applying Lemma 12.1.7 repeatedly, we see that n' can also be written as the sum of two squares, say $n' = a'^2 + b'^2$. Consequently,

$$n = n'm^2 = (a'^2 + b'^2)m^2 = (a'm)^2 + (b'm)^2,$$

and therefore n is also a sum of two squares, as desired. □

Example 12.1.11. The number 3978 factors as $2 \cdot 3^2 \cdot 13 \cdot 17$. Since 13 and 17 are congruent to 1 mod 4, it follows from Theorem 12.1.10 that 3978 must be the sum of two squares. Let us find $a, b \in \mathbb{Z}$ such that $3978 = a^2 + b^2$ following the proof of the theorem. First, we easily find

$$2 = 1^2 + 1^2, \quad 13 = 3^2 + 2^2, \quad \text{and} \quad 17 = 4^2 + 1^2,$$

so, using the formula in Lemma 12.1.7, we obtain

$$26 = 2 \cdot 13 = (1^2 + 1^2)(3^2 + 2^2) = (3 + 2)^2 + (2 - 3)^2 = 5^2 + 1^2$$

and

$$442 = 26 \cdot 17 = (5^2 + 1^2)(4^2 + 1^2) = (20 + 1)^2 + (5 - 4)^2 = 21^2 + 1^2.$$

Finally,

$$3978 = 3^2 \cdot 442 = 3^2 \cdot (21^2 + 1^2) = 63^2 + 3^2.$$

In fact, 3978 can also be represented as $729 + 3249 = 27^2 + 57^2$.

We finish this section with the statement of a theorem on the number of representations of an integer as the sum of two squares. Here, we will count

$$\begin{aligned} 5 &= 2^2 + 1^2 = (-2)^2 + 1^2 = 2^2 + (-1)^2 = (-2)^2 + (-1)^2 \\ &= 1^2 + 2^2 = 1^2 + (-2)^2 = (-1)^2 + 2^2 = (-1)^2 + (-2)^2 \end{aligned}$$

as eight distinct representations of 5 as a sum of two squares, but we will say they are all *essentially* the same representation.

Theorem 12.1.12. *Suppose n can be represented as a sum of two squares and write*

$$n = 2^a \cdot p_1^{a_1} \cdots p_s^{a_s} \cdot q_1^{b_1} \cdots q_t^{b_t},$$

where each p_i (resp. q_j) is a prime congruent to $1 \bmod 4$ (resp. $3 \bmod 4$) and $a_i, b_j \geq 1$, with b_j even. Then, n can be written as a sum of two squares in

$$N(n) = 4\prod_{i=1}^{s}(a_i + 1)$$

ways. Moreover, if $N(n)$ is divisible by 8, then there are exactly $N(n)/8$ essentially distinct representations of n as a sum of two squares. If $N(n)$ is not divisible by 8, then $N(n) = 8k + 4$ for some $k \geq 0$, and there are exactly $k+1$ essentially distinct representations.

A proof of Theorem 12.1.12 can be found, for example, in [**AC95**, Theorem 8.11].

Example 12.1.13. A prime $p \equiv 1 \bmod 4$ can be represented as a sum of two squares in eight different ways (which are essentially the same representation). For instance,

$$\begin{aligned} 13 &= 3^2 + 2^2 = (-3)^2 + 2^2 = 3^2 + (-2)^2 = (-3)^2 + (-2)^2 \\ &= 3^2 + 2^2 = 3^2 + (-2)^2 = (-3)^2 + 2^2 = (-3)^2 + (-2)^2. \end{aligned}$$

In Example 12.1.11 we saw that 3978 is the sum of two squares. Since $3978 = 2 \cdot 13 \cdot 17 \cdot 3^2$, Theorem 12.1.12 says that 3978 can be represented as a sum of two squares in

$$4 \cdot (1+1) \cdot (1+1) = 16$$

different ways. Indeed,

$$\begin{aligned} 3978 &= 63^2 + 3^2 = (-63)^2 + 3^2 = 63^2 + (-3)^2 = (-63)^2 + (-3)^2 \\ &= 3^2 + 63^2 = 3^2 + (-63)^2 = (-3)^2 + 63^2 = (-3)^2 + (-63)^2 \\ &= 57^2 + 27^2 = (-57)^2 + 27^2 = 57^2 + (-27)^2 = (-57)^2 + (-27)^2 \\ &= 27^2 + 57^2 = 27^2 + (-57)^2 = (-27)^2 + 57^2 = (-27)^2 + (-57)^2. \end{aligned}$$

Theorem 12.1.12 can be reinterpreted in terms of integral points as follows.

Corollary 12.1.14. *Let n be a natural number such that the circle $C_n : x^2+y^2 = n$ has at least one integral point, and let $n' = 2^a p_1^{a_1} \cdots p_s^{a_s}$ be the square-free part of n. Then, C_n has exactly*

$$4\prod_{i=1}^{s}(a_i + 1)$$

integral points.

Example 12.1.15. Let $n = 15925$ and consider the circle $C : x^2 + y^2 = 15925$. Since $n = 5^2 \cdot 7^2 \cdot 13$, it follows that n can be written as the sum of two squares, by Theorem 12.1.10. Hence, C has integral points, and by Corollary 12.1.14, there must be

$$4 \cdot (2+1) \cdot (1+1) = 24$$

integral points on C. Indeed, the set $C(\mathbb{Z})$ of integral points on C consists of

$$\begin{gathered}(-126,7),(-126,-7),(-119,42),(-119,-42),(-105,70),(-105,-70),\\(-70,105),(-70,-105),(-42,119),(-42,-119),(-7,126),(-7,-126),\\(7,126),(7,-126),(42,119),(42,-119),(70,105),(70,-105),\\(105,70),(105,-70),(119,42),(119,-42),(126,7),\quad \text{and} \quad (16,-7)\end{gathered}$$

and these are all the points in the set $C(\mathbb{Z})$ of integral points on the circle.

12.2. Pythagorean Triples

In this section we use a parametrization of a circle (as seen in Example 9.3.1) to determine all the pythagorean triples, i.e., the integer solutions of the diophantine equation $x^2 + y^2 = z^2$. Let us begin with a proof of Pythagoras's theorem.

Theorem 12.2.1 (Pythagoras of Samos (c. 570 – c. 495 BC))**.** *Let T be a right triangle whose hypotenuse's length is c, and let the other two sides have lengths a and b. Then,*

$$a^2 + b^2 = c^2.$$

Conversely, every triple of positive numbers (a, b, c) such that $a^2 + b^2 = c^2$ corresponds to the lengths of the sides of a right triangle.

Proof. We present here the proof that is commonly attributed to Pythagoras. The idea is to express the area of a square of side length $a + b$ (i.e., $(a + b)^2$) in two different ways, as depicted in Figure 12.1.

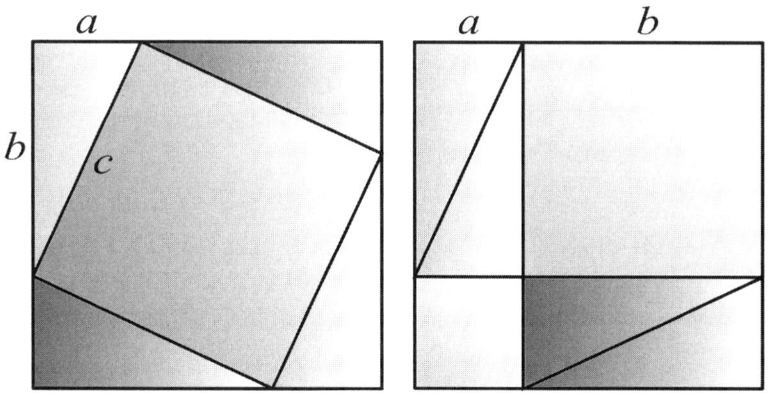

Figure 12.1. Pythagoras's proof by "rearrangement" of his famous theorem.

Let A be the area of the triangle T described by the statement of the theorem. The picture on the right-hand side of Figure 12.1 has area

$$(a+b)^2 = a^2 + 2ab + b^2 = 4 \cdot \frac{ab}{2} + a^2 + b^2 = 4A + a^2 + b^2.$$

In order to calculate the area of the square in the left-hand side of the picture, it needs to be shown that the tilted region in the center is a square. It suffices to verify that the angles are $\pi/2$ each. However, this follows from the fact that the angle of T where the sides of length a and c meet and the angle where b and c meet add up to $\pi/2$ (since the third angle in T is $\pi/2$, and all three angles should add up to π). Therefore, the center is a tilted square with side length c, and the area of the larger square is given by

$$(a+b)^2 = 4A + c^2.$$

Hence, we must have

$$4A + a^2 + b^2 = (a+b)^2 = 4A + c^2$$

and the equation $a^2 + b^2 = c^2$ follows.

For the converse, let T be a right triangle with sides of length a and b next to the $\pi/2$ angle. Since we just proved that the length h of the hypotenuse of a right triangle is given by $\sqrt{a^2+b^2}$, it follows that

$$h = \sqrt{a^2+b^2} = \sqrt{c^2} = |c| = c.$$

Thus, the lengths of the sides of T are a, b, and c, as desired. □

Certainly, integral solutions of the diophantine equation $x^2 + y^2 = z^2$ exist, and some are well known; e.g., $3^2 + 4^2 = 5^2$. The question, however, is whether we can describe *all* the integral solutions to Pythagoras's equation, or, equivalently, whether we can find all the right triangles with sides of integer length. Such solutions are called pythagorean triples, which we define formally next.

Definition 12.2.2. A triple of natural numbers (a, b, c) is called a *pythagorean triple* if $a^2 + b^2 = c^2$. The triples (a, b, c) and (b, a, c) are considered as the same pythagorean triple.

Example 12.2.3. The triple $(3, 4, 5)$ is a pythagorean triple, because $3^2 + 4^2 = 9 + 16 = 25 = 5^2$. If we multiply both sides of the equation by a square, then we obtain a new pythagorean triple. For instance,

$$4 \cdot (3^2 + 4^2) = 4 \cdot 5^2$$

yields $6^2+8^2 = 10^2$. Thus, $(6, 8, 10)$ is also a pythagorean triple and so is $(3\lambda, 4\lambda, 5\lambda)$ for any $\lambda \geq 1$. The triple, $(3, 4, 5)$, that generates this family of pythagorean triples is called *primitive*. Note that the coordinates of the pythagorean triple $(3\lambda, 4\lambda, 5\lambda)$ share a greatest common divisor of λ, and the primitive one corresponds to a $\gcd = 1$.

Definition 12.2.4. A triple of natural numbers (a, b, c) is called a *primitive pythagorean triple* if $a^2 + b^2 = c^2$ and $\gcd(a, b, c) = 1$.

Remark 12.2.5. If (a,b,c) is a primitive pythagorean triple, then a,b,c are also pairwise coprime. Indeed, if any two coordinates share a common factor d, then d also divides the third coordinate in the triple, because

$$a^2 = c^2 - b^2, \quad b^2 = c^2 - a^2, \quad \text{and } c^2 = a^2 + b^2.$$

See also Exercise 2.11.37. This implies that

$$\gcd(a,b) = \gcd(b,c) = \gcd(a,c) = \gcd(a,b,c)$$

for any pythagorean triple (a,b,c).

Remark 12.2.6. If (a,b,c) is a pythagorean triple, then it is a multiple of a primitive pythagorean triple. Indeed, if $d = \gcd(a,b,c)$, then $(a',b',c') = (a/d, b/d, c/d)$ is a triple of integers that satisfy

$$a'^2 + b'^2 = \left(\frac{a}{d}\right)^2 + \left(\frac{b}{d}\right)^2 = \frac{1}{d^2}\left(a^2+b^2\right) = \frac{1}{d^2}\cdot c^2 = \left(\frac{c}{d}\right)^2 = c'^2,$$

and $\gcd(a',b',c') = \gcd(a/d, b/d, c/d) = 1$ by Exercise 2.11.25. Hence, (a',b',c') is a primitive pythagorean triple, and $(a,b,c) = (da', db', cd')$.

For instance, the pythagorean triple $(15,36,39)$ shares a $\gcd(15,36,39) = 3$ and is a multiple of the primitive triple $(5,12,13)$.

Example 12.2.7. The smallest primitive pythagorean triples (such that $c \leq 100$) are the following:

$$(3,4,5), (5,12,13), (8,15,17), (7,24,25), (20,21,29), (12,35,37),$$
$$(9,40,41), (28,45,53), (11,60,61), (16,63,65), (33,56,65), (48,55,73),$$
$$(13,84,85), (36,77,85), (39,80,89), (65,72,97).$$

Let us now tackle the problem of describing all pythagorean triples. First, notice that if (a,b,c) is a pythagorean triple, then $a^2+b^2=c^2$ by definition, and it follows that $\left(\frac{a}{c}\right)^2 + \left(\frac{b}{c}\right)^2 = 1$. Thus, $\left(\frac{a}{c}, \frac{b}{c}\right)$ is a point on the circle $x^2+y^2=1$ of radius 1. Conversely, we can start with a rational point on a circle of radius 1 and obtain a pythagorean triple. If (r,s) is a rational point in the circle, then we may write $r = \frac{e}{f}$ and $s = \frac{g}{h}$ as rational numbers for some integers e,f,g,h (since $(\pm r, \pm s)$ are also points in the circle, we may assume e,f,g,h are non-negative integers). Then, $(e/f)^2 + (g/h)^2 = 1$ implies

$$(eh)^2 + (gf)^2 = (fh)^2$$

and therefore (eh, gf, fh) is a pythagorean triple. In the next theorem, we shall use our parametrization of the points on a circle of radius 1 given by Example 9.3.1 to give a description of all pythagorean triples.

Remark 12.2.8. Before we state the theorem, we note that if (a,b,c) is a primitive pythagorean triple, then one of a or b is even and the other one is odd (see Exercise 4.7.10).

Theorem 12.2.9. *The set of all primitive pythagorean triples (a,b,c), with a odd and b even, is the set of all triples*

$$(n^2 - m^2, 2nm, n^2+m^2)$$

where n, m are positive integers of opposite parity, with $n > m > 0$, and $\gcd(n, m) = 1$. *Thus, every pythagorean triple is of the form*

$$(\lambda(n^2 - m^2), \lambda(2nm), \lambda(n^2 + m^2))$$

for some $\lambda \geq 1$ and some $n > m > 0$.

Proof. Let us first verify that $\tau = (n^2 - m^2, 2nm, n^2 + m^2)$ is a pythagorean triple: $(n^2 - m^2)^2 + (2nm)^2 = n^4 - 2n^2m^2 + m^2 + 4n^2m^2 = n^4 + 2n^2m^2 + m^4 = (n^2 + m^2)^2$. Moreover, τ is primitive when n, m are positive integers of opposite parity, with $n > m > 0$, and $\gcd(n, m) = 1$ (this was left as an exercise for the reader in Exercise 1.8.16, but we will provide a proof here for completeness). Suppose p is a common prime divisor of a, b, c. Then, p is a divisor of $b = 2nm$ and therefore $p = 2$, or p divides n or m. Since n or m is even, it suffices to show the case when p divides n or m. Without loss of generality, let us assume that p divides n. Then, p is a divisor of n and a divisor of $a = n^2 - m^2$. It follows that p divides m as well, which contradicts $\gcd(n, m) = 1$.

Let us show now that every primitive pythagorean triple can be written as stated in the theorem. Let (a, b, c) be a primitive pythagorean triple. Then, $(a/c, b/c)$ is a point on the circle of radius $R = 1$. By our work in Example 9.3.1 (changing the slope value m of the line for $-t$ in the notation of the example), there is a value of $t \in \mathbb{Q}$ such that

$$\left(\frac{a}{c}, \frac{b}{c}\right) = \left(\frac{2t}{1+t^2}, \frac{1-t^2}{1+t^2}\right).$$

Since $a/c > 0$, the number t must be positive. And since $t \in \mathbb{Q}$, there are integers $m, n \geq 1$ such that $t = m/n$. Moreover, we may require m/n to be written in lowest terms, so that $\gcd(m, n) = 1$. Thus,

$$\begin{aligned}\left(\frac{a}{c}, \frac{b}{c}\right) &= \left(\frac{2t}{1+t^2}, \frac{1-t^2}{1+t^2}\right)\\ &= \left(\frac{2 \cdot \frac{m}{n}}{1+\left(\frac{m}{n}\right)^2}, \frac{1-\left(\frac{m}{n}\right)^2}{1+\left(\frac{m}{n}\right)^2}\right) = \left(\frac{2nm}{n^2+m^2}, \frac{n^2-m^2}{n^2+m^2}\right).\end{aligned}$$

Since $\gcd(a, c) = \gcd(b, c) = 1$ because (a, b, c) is primitive and since we have shown above that $\gcd(2nm, n^2 + m^2) = \gcd(n^2 - m^2, n^2 + m^2) = 1$, it follows that

$$a = 2nm, \quad b = n^2 - m^2, \quad \text{and} \quad c = n^2 + m^2.$$

Notice that $n > m$ because $b > 0$ and that n and m must have different parities; otherwise we would have $\gcd(b, c) = \gcd(n^2 - m^2, n^2 + m^2) \geq 2$.

Finally, it follows from Remark 12.2.6 that every pythagorean triple is a multiple of a primitive one. Hence, every pythagorean triple is of the form $(\lambda(n^2 - m^2), \lambda(2nm), \lambda(n^2 + m^2))$. □

Example 12.2.10. Let $n = 2$ and $m = 1$, which are positive, of opposite parity, and relatively prime. Then, the pair $(n, m) = (2, 1)$ produces a primitive pythagorean triple

$$(n^2 - m^2, 2nm, n^2 + m^2) = (2^2 - 1, 2 \cdot 2 \cdot 1, 2^2 + 1^2) = (3, 4, 5).$$

Table 12.1. Primitive pythagorean triples corresponding to $0 < m < n \leq 6$.

		n				
		2	3	4	5	6
	1	$(3, 4, 5)$		$(15, 8, 17)$		$(35, 12, 37)$
	2		$(5, 12, 13)$		$(21, 20, 29)$	
m	3			$(7, 24, 25)$		$(27, 36, 45)$
	4				$(9, 40, 41)$	
	5					$(11, 60, 61)$

In Table 12.1 we have used the formula given by Theorem 12.2.9 to calculate all the primitive pythagorean triples of the form $(n^2 - m^2, 2nm, n^2 + m^2)$ with $0 < m < n \leq 6$.

Remark 12.2.11. If n, m are any positive integers with $n > m > 0$, then $(n^2 - m^2, 2nm, n^2 + m^2)$ is a pythagorean triple, but perhaps not a primitive one. However, not every non-primitive pythagorean triple can be constructed in this way. For instance, consider the pythagorean triple $(9, 12, 15)$, which is a multiple of the primitive triple $(3, 4, 5)$. There are no n, m such that

$$(9, 12, 15) = (n^2 - m^2, 2nm, n^2 + m^2).$$

Indeed, if there were such n, m, then 15 would be a sum of two squares, which is impossible (this can be seen directly or by using Theorem 12.1.10).

12.3. Fermat's Last Theorem for $n = 4$

As an application of our classification of pythagorean triples in Theorem 12.2.9, we will prove Fermat's last theorem for $n = 4$; i.e., we will show that $X^4 + Y^4 = Z^4$ has no integer solutions with $XYZ \neq 0$. We will, in fact, prove something stronger: the diophantine equation $x^4 + y^4 = z^2$ has no integral solutions with $xyz \neq 0$.

Theorem 12.3.1. *The diophantine equation $x^4 + y^4 = z^2$ has no integral solutions with $xyz \neq 0$.*

Proof. Suppose for a contradiction that there is a solution (x_0, y_0, z_0), for some $x_0, y_0, z_0 \in \mathbb{Z}$, with $x_0 y_0 z_0 \neq 0$. Then, $(\pm x_0, \pm y_0, \pm z_0)$ are also solutions, so we may assume that $x_0, y_0, z_0 > 0$. By the well-ordering principle (see Section 2.1), we may also assume that (x_0, y_0, z_0) is the solution with positive coordinates and smallest value of z_0.

We claim that (x_0^2, y_0^2, z_0) is a primitive pythagorean triple. Indeed, we have $((x_0)^2)^2 + ((y_0)^2)^2 = z_0^2$, and if p is a common prime divisor of x_0^2, y_0^2, and z_0, then p divides x_0, y_0, and z_0 as well, and so $((x_0/p)^2, (y_0/p)^2, z_0/p)$ is another integral solution, with positive coordinates, but $0 < z_0/p < z_0$, which contradicts the minimality of z_0. Thus, (x_0^2, y_0^2, z_0) is a primitive pythagorean triple, and by

Theorem 12.2.9, there are $m, n > 0$ of opposite parity and relatively prime such that

$$x_0^2 = n^2 - m^2, \quad y_0^2 = 2nm, \quad z_0 = n^2 + m^2.$$

In particular, $x_0^2 + m^2 = n^2$, and since $\gcd(m, n) = 1$, it follows that (x_0, m, n) is another primitive pythagorean triple. Thus, there are $u, v > 0$ of opposite parity and relatively prime such that

$$x_0 = v^2 - u^2, \quad m = 2uv, \quad n = v^2 + u^2.$$

Notice that $y_0^2 = 2nm = 4nuv$ and that $\gcd(n, m) = \gcd(u, v) = 1$, together with Exercise 2.11.36, implies that u, v, and n must each be a perfect square, say $u = a^2$, $v = b^2$, and $n = c^2$. Hence, the equation $v^2 + u^2 = n$ becomes $a^4 + b^4 = c^2$. Thus, we have found a new solution (a, b, c) of $x^4 + y^4 = z^4$ with positive coordinates $a, b, c > 0$. Moreover, the z-coordinate of this solution is c, and

$$c \le c^2 = n < n^2 + m^2 = z_0,$$

but $c < z_0$ contradicts the minimality of the z-coordinate of the solution (x_0, y_0, z_0) we started from. This contradiction means that (x_0, y_0, z_0) cannot exist in the first place and concludes the proof of the theorem. □

Corollary 12.3.2. *Fermat's last theorem is true for $n = 4$; i.e., the equation $X^4 + Y^4 = Z^4$ does not have integral solutions with $XYZ \neq 0$.*

Proof. Clearly, a solution (a, b, c) of $X^4 + Y^4 = Z^4$, with $abc \neq 0$, would yield a solution (a, b, c^2) of $x^4 + y^4 = z^2$, contradicting Theorem 12.3.1. Hence, no such integral solution can exist. □

12.4. Ellipses

In the previous section we analyzed the particular case of circles. In this section, we move on to the general question of whether an ellipse $C : X^2 + BY^2 = D$, with $B, D > 0$, has rational or integral solutions. In the case of a circle $X^2 + Y^2 = D$, we showed that the existence of rational solutions is equivalent to the existence of integral solutions (Proposition 11.3.11). However, the situation is more subtle when dealing with ellipses, as the following example shows.

Example 12.4.1. Let C be the ellipse given by $X^2 + 7Y^2 = 2$. Clearly, C does not have any integral solutions, for if (x_0, y_0) was an integral point, then $|x_0| \le 1$ and $y_0 = 0$, but $(\pm 1, 0)$ are not on C. However,

$$\left(\frac{1}{2}\right)^2 + 7 \cdot \left(\frac{1}{2}\right)^2 = 2,$$

and so $(1/2, 1/2) \in C(\mathbb{Q})$. Using the methods of Section 9.3, we can find a parametrization of the rational points on C:

$$C(\mathbb{Q}) = \left\{ \left(\frac{7m^2 - 14m - 1}{14m^2 + 2}, \frac{-7m^2 - 2m + 1}{14m^2 + 2} \right) : m \in \mathbb{Q} \right\} \cup \left\{ \left(\frac{1}{2}, -\frac{1}{2} \right) \right\}.$$

Thus, C has infinitely many rational points but no integral points.

Given some fixed $B, D > 0$, we can decide whether $C : X^2 + BY^2 = D$ has a rational solution using the Hasse–Minkowski theorem (Theorem 11.2.12 or its p-adic formulation, Theorem 11.2.14; see also Section 11.3 for an example of an application). Thus, for the remainder of the section, we will concentrate on the question of whether an ellipse has an integral point and how to find one (or all). As the following lemma points out, determining if an ellipse $X^2 + BY^2 = D$ has integral points is a finite computation that could be tackled by a "brute force" search (testing whether $\lfloor\sqrt{D/B}\rfloor + 1$ numbers are perfect squares).

Lemma 12.4.2. *Let $B, D > 0$ be integers. The ellipse $C : X^2 + BY^2 = D$ has an integral point if and only if the finite subset of integers*

$$\mathcal{S} = \{D - B \cdot n^2 : n \in \mathbb{Z} \text{ with } 0 \leq n \leq \sqrt{D/B}\}$$

contains a perfect square. The set S has $\lfloor\sqrt{D/B}\rfloor + 1$ elements, and if $D - Bn^2 = m^2$ is a perfect square, for $n, m \in \mathbb{Z}$, then $(m, n) \in C(\mathbb{Z})$.

Proof. Suppose first that $\mathcal{S}$ contains a square m^2. Then, there is n such that $D - Bn^2 = m^2$, and therefore $m^2 + Bn^2 = D$. It follows that $(m, n) \in C(\mathbb{Z})$.

Conversely, if $x_0^2 + By_0^2 = D$, for some $x_0, y_0 \in \mathbb{Z}$, then we may assume $x_0, y_0 \geq 0$, and then

$$B \cdot y_0^2 \leq D - |x_0|^2 \leq D.$$

Thus, $0 \leq y_0 \leq \lfloor\sqrt{D/B}\rfloor$, and $D - By_0^2 = x_0^2$ is a perfect square. Thus, $x_0 \in \mathcal{S}$. □

A brute force search can be tedious (when $\sqrt{D/B}$ is large), so we are interested in results to search for integral points in a smarter way and results to prove the existence (or non-existence) of such points. In the following example we show how the theory of quadratic residues (Chapter 10) and quadratic reciprocity (Section 10.4) can be used to prove the non-existence of integral points in certain cases (see also previous related Examples 4.2.8, 9.4.4, and 9.4.5).

Example 12.4.3. Let $C : X^2 + 15Y^2 = 4001$. By Lemma 12.4.2 one would only have to check $\lfloor\sqrt{4001/15}\rfloor + 1 = 17$ possible values of Y in order to determine whether C has an integral point. Instead of doing this, we notice that if $a^2 + 15b^2 = 4001$ is a solution, for some $a, b \in \mathbb{Z}$, then $a^2 \equiv 4001 \bmod 3$ and $a^2 \equiv 4001 \bmod 5$. Although $4001 \equiv 1 \bmod 5$ is a quadratic residue modulo 5, we also have $4001 \equiv 2 \bmod 3$, which is a quadratic non-residue modulo 3. Thus, the congruence $a^2 \equiv 4001 \equiv 2 \bmod 3$ has no solutions, and it follows that (a, b) with $a^2 + 15b^2 = 4001$ cannot exist. In other words, $C(\mathbb{Z})$ is empty (and, in fact, $C(\mathbb{Q})$ is empty as well; see Exercise 12.8.15).

Example 12.4.4. The theory of quadratic residues can also be used to find an integral solution (when there is one). Let $C : X^2 + 23Y^2 = 223$. If (a, b) is an integral point, for some $a, b \geq 0$, then $a^2 \equiv 223 \equiv 16 \bmod 23$. Since 16 is a square modulo 23, there is no contradiction as in the previous example. However, this tells us that $a \equiv \pm 4 \bmod 23$. Also, we know that $0 \leq a \leq \lfloor\sqrt{223}\rfloor = 14$. Thus, if we put these two constraints together, we reach the conclusion that the only possibility is $a = 4$. Since $(223 - 16)/23 = 9$, it follows that $(4, 3)$ is an integral point on C and, in fact, $C(\mathbb{Z}) = \{(\pm 4, \pm 3)\}$.

In Section 12.6 we will reduce the question of whether an ellipse has an integral point to the case of $X^2 + BY^2 = p$, where p is prime, but first, we need to take a necessary detour and introduce quadratic fields and their norms.

12.5. Quadratic Fields and Norms

Let $B > 0$ be fixed, and suppose that $X^2 + BY^2 = n$ and $X^2 + BY^2 = m$ have integral solutions, for some $n, m \geq 1$. Is there an integral solution of $X^2 + BY^2 = nm$ as well? For instance $X^2 + 3Y^2 = 13$ and $X^2 + 3Y^2 = 7$ have integral solutions $P = (1, 2)$ and $Q = (2, 1)$, respectively, and $X^2 + 3Y^2 = 91$ has an integral solution $R = (4, 5)$. Is there a systematic way to find R in terms of P and Q?

In this subsection we show that the answers to these questions are always positive. Moreover, we introduce the theory of quadratic fields, which gives a nice theoretical framework for our answer. First, however, we observe that there is a purely algebraic formula that explains this "multiplicative" behavior of integral solutions of ellipses (and hyperbolas), which generalizes the formula shown in Lemma 12.1.7 for the sum of two squares.

Lemma 12.5.1. *Let $B \neq 0$ be fixed, and let a, b, c, d be arbitrary integers. Then,*

$$(a^2 + Bb^2)(c^2 + Bd^2) = (ac - Bbd)^2 + B(ad + bc)^2.$$

In particular, if (a, b) and (c, d) are solutions of $X^2+BY^2 = n$ and $X^2+BY^2 = m$, respectively, then $(ac - Bbd, ad + bc)$ is a solution of $X^2 + BY^2 = nm$.

The proof of the lemma is a simple algebraic calculation and has been left to the reader as an exercise (Exercise 12.8.16).

Example 12.5.2. The points $P = (1, 2)$ and $Q = (2, 1)$ lie on the conics $X^2+3Y^2 = 13$ and $X^2 + 3Y^2 = 7$, respectively. Thus, Lemma 12.5.1 shows that

$$(1 \cdot 2 - 3 \cdot 2 \cdot 1,\ 1 \cdot 1 + 2 \cdot 2) = (-4, 5)$$

belongs to $X^2 + 3Y^2 = 91$, and so do $(\pm 4, \pm 5)$. Moreover, since $P' = (1, -2)$ is also a solution of $X^2 + BY^2 = 13$, together with $Q = (2, 1)$, we find

$$(1 \cdot 2 - 3 \cdot (-2) \cdot 1,\ 1 \cdot 1 - 2 \cdot 2) = (8, -3)$$

also satisfies $X^2 + 3Y^2 = 91$, and so do $(\pm 8, \pm 3)$.

The reader is probably wondering about the origin of the formula in Lemma 12.5.1. As we shall see next, the formula is a consequence of the multiplicative properties of norms in quadratic fields.

Proposition 12.5.3. *Let d be a non-zero square-free integer. Let $\mathbb{Q}(\sqrt{d})$ be the subring of complex numbers $\mathbb{C}$ generated by 1 and $\sqrt{d}$, with the addition and multiplication operations inherited from $\mathbb{C}$. Then:*

(1) *The ring elements of $\mathbb{Q}(\sqrt{d})$ are precisely*

$$\mathbb{Q}(\sqrt{d}) = \left\{a + b\sqrt{d} : a, b \in \mathbb{Q}\right\}.$$

(2) *The ring $\mathbb{Q}(\sqrt{d})$ is a field.*

Proof. By definition, $\mathbb{Q}(\sqrt{d})$ is the subring of $\mathbb{C}$ generated by 1 and $\sqrt{d}$. Thus,

$$\mathcal{R} = \left\{a + b\sqrt{d} : a, b \in \mathbb{Q}\right\} \subseteq \mathbb{Q}(\sqrt{d}).$$

In order to show equality, it suffices to show that the set $\mathcal{R}$ on the left is closed under addition and multiplication (and therefore the smallest ring containing 1 and $\sqrt{d}$). Indeed,

$$a + b\sqrt{d} + e + f\sqrt{d} = (a+e) + (b+f)\sqrt{d} \in \mathcal{R},$$

and

$$(a + b\sqrt{d})(e + f\sqrt{d}) = (ae + dbf) + (af + be)\sqrt{d} \in \mathcal{R},$$

for any $a, b, e, f \in \mathbb{Q}$. Hence, $\mathcal{R} = \mathbb{Q}(\sqrt{d})$.

Since $\mathbb{Q}(\sqrt{d})$ is a ring, in order to show that it is a field, it suffices to show that every non-zero element of $\mathbb{Q}(\sqrt{d})$ is invertible. Indeed,

$$\frac{1}{a + b\sqrt{d}} = \frac{(a - b\sqrt{d})}{(a + b\sqrt{d})(a - b\sqrt{d})} = \frac{a - b\sqrt{d}}{a^2 - db^2} = \frac{a}{a^2 - db^2} - \frac{b}{a^2 - db^2} \cdot \sqrt{d},$$

for any $a, b \in \mathbb{Q}$. Moreover, $a^2 - db^2 = 0$ if and only if $a = b = 0$ because d is assumed to be square-free. Hence, if $a + b\sqrt{d} \neq 0$, then $(a, b) \neq (0, 0)$, and it follows that $(a + b\sqrt{d})^{-1} \in \mathbb{Q}(\sqrt{d})$, as desired. □

Definition 12.5.4. Let d be a non-zero square-free integer. The field $\mathbb{Q}(\sqrt{d})$ is called a *quadratic field*. If $d > 0$, we say $\mathbb{Q}(\sqrt{d})$ is a *real quadratic field*, and if $d < 0$, then we say it is an *imaginary quadratic field*.

Example 12.5.5. Let $d = -3$ and let $\alpha = \frac{-1+\sqrt{-3}}{2}$. Then,

$$\frac{1}{\alpha} = \frac{1}{\frac{-1+\sqrt{-3}}{2}} = \frac{2 \cdot (-1 - \sqrt{-3})}{(-1 + \sqrt{-3})(-1 - \sqrt{-3})} = \frac{2 \cdot (-1 - \sqrt{-3})}{4} = -\frac{1 + \sqrt{-3}}{2}.$$

Thus, α^{-1} also belongs to the field $\mathbb{Q}(\sqrt{-3})$. It is interesting to note that

$$\alpha^2 = \left(\frac{-1 + \sqrt{-3}}{2}\right)^2 = -\frac{1 + \sqrt{-3}}{2} = \frac{1}{\alpha}.$$

Therefore, $\alpha^2 = 1/\alpha$ and so $\alpha^3 = 1$. In other words, α is a third root of unity.

The quantity $a^2 - db^2$ that appears in the denominator of the inverse of $\alpha = a + b\sqrt{d}$ is called the norm of α in $\mathbb{Q}(\sqrt{d})$.

Definition 12.5.6. Let d be a non-zero square-free integer, let $\mathbb{Q}(\sqrt{d})$ be the associated quadratic field, and let $\alpha = a + b\sqrt{d} \in \mathbb{Q}(\sqrt{d})$, for some $a, b \in \mathbb{Q}$. Then, the *(absolute) norm* of α, denoted by $N(\alpha)$, is defined by

$$N(\alpha) = N(a + b\sqrt{d}) = a^2 - db^2.$$

Example 12.5.7. The norm of $\alpha = \frac{-1+\sqrt{-3}}{2}$ in $\mathbb{Q}(\sqrt{-3})$ is

$$N(\alpha) = \left(\frac{-1}{2}\right)^2 + 3 \cdot \left(\frac{1}{2}\right)^2 = \frac{1}{4} + \frac{3}{4} = 1.$$

The norm of $\beta = 1 + i$, where $i = \sqrt{-1}$, in the quadratic field $\mathbb{Q}(i)$ is

$$N(\beta) = 1^2 + 1^2 = 2.$$

Remark 12.5.8. When $d < 0$, the norm $N(a + b\sqrt{d}) = a^2 - db^2$ is precisely the square of the usual complex norm (also called the complex modulus) of the complex number $a + b\sqrt{d}$. Indeed, if $\alpha = s + t \cdot i$ is a complex number, for some $s, t \in \mathbb{R}$ and $i = \sqrt{-1}$, then the usual complex norm is

$$|\alpha| = |s + t \cdot i| = \sqrt{s^2 + t^2}.$$

Hence, if we put $d = -n$, for some $n > 0$, then

$$\begin{aligned} |a + b\sqrt{d}|^2 &= |a + b\sqrt{-n}|^2 = |a + b\sqrt{n} \cdot i|^2 = a^2 + (b\sqrt{n})^2 \\ &= a^2 + nb^2 = a^2 - db^2 = N(a + b\sqrt{d}). \end{aligned}$$

Since the complex norm $|\cdot|$ is known to be multiplicative (see Exercise 12.8.3), i.e., $|\alpha \cdot \beta| = |\alpha| \cdot |\beta|$ for any $\alpha, \beta \in \mathbb{C}$, we deduce that the norm $N(\cdot)$ in the quadratic field $\mathbb{Q}(\sqrt{d})$ is also multiplicative, at least when $d < 0$. Next, we show that the norm in quadratic fields is always multiplicative, which will explain the origin of the algebraic identity in Lemma 12.5.1.

Lemma 12.5.9. *Let d be a non-zero square-free integer, let $a, b, e, f \in \mathbb{Q}$, and define elements $\alpha = a + b\sqrt{d}$ and $\beta = e + f\sqrt{d}$ in $\mathbb{Q}(\sqrt{d})$. Then,*

$$N(\alpha\beta) = N(\alpha)N(\beta).$$

Proof. Let $\alpha, \beta \in \mathbb{Q}(\sqrt{d})$ be as in the statement of the lemma. We calculate

$$\alpha \cdot \beta = (a + b\sqrt{d})(e + f\sqrt{d}) = (ae + dbf) + (af + be)\sqrt{d}.$$

Thus,

$$\begin{aligned} N(\alpha\beta) &= N((ae + dbf) + (af + be)\sqrt{d}) \\ &= (ae + dbf)^2 - d(af + be)^2 \\ &= (a^2 - db^2)(e^2 - df^2) = N(\alpha)N(\beta), \end{aligned}$$

where we have used Lemma 12.5.1 with $B = -d$. □

Example 12.5.10. Consider the quadratic field $\mathbb{Q}(\sqrt{2})$, and let $\alpha = 5 + \sqrt{2}$ and $\beta = 3 + 2\sqrt{2}$. Then,

$$N(\alpha) = 5^2 - 2 = 23, \quad N(\beta) = 3^2 - 2 \cdot 2^2 = 1.$$

Therefore, $N(\alpha\beta) = N(\alpha)N(\beta) = 23$ as well. Let us verify this directly:

$$\alpha\beta = (5 + \sqrt{2})(3 + 2\sqrt{2}) = 19 + 13\sqrt{2},$$

and

$$N(\alpha\beta) = N(19 + 13\sqrt{2}) = 19^2 - 2 \cdot 13^2 = 361 - 2 \cdot 169 = 23,$$

as claimed. This means that both $(3, 2)$ and $(19, 13)$ are points on the hyperbola $X^2 - 2Y^2 = 23$.

We will come back to the topic of quadratic fields in Section 14.3.1, where we will discuss quadratic rings of algebraic integers and their unit subgroups.

12.6. Integral Points on Ellipses

The multiplicative property of the norm for a quadratic field implies that, in order to find an integral point on $C : X^2 + BY^2 = D$ and if D factors as D_1D_2 for some $D_1, D_2 \geq 1$, then it suffices to find integral points on the auxiliary ellipses $C_1 : X^2 + BY^2 = D_1$ and $C_2 : X^2 + BY^2 = D_2$. In fact, we show next that it suffices to find an integral point on $X^2 + BY^2 = D'$, where D' is the square-free part of D.

Proposition 12.6.1. *Let $B, D > 0$ be fixed integers, and suppose that there is a factorization $D = D_1D_2(D')^2$, for some $D_1, D_2, D' \geq 1$, such that the conics $C_1 : X^2 + BY^2 = D_1$ and $C_2 : X^2 + BY^2 = D_2$ have integral points. Then, $X^2 + BY^2 = D$ also has an integral point.*

Proof. Let B and $D = D_1D_2(D')^2$ be as in the statement, and suppose that (a, b) and (c, d) are integral points on the curves C_1 and C_2, respectively. Then,

$$D_1D_2 = (a^2 + Bb^2)(c^2 + Bd^2) = (ac - Bbd)^2 + B(ad + bc)^2,$$

or, in other words, $(ac - Bbd, ad + bc)$ is an integral point on $X^2 + BY^2 = D_1D_2$. Thus,

$$((ac - Bbd)D', (ad + bc)D')$$

is an integral point on $X^2 + BY^2 = D_1D_2(D')^2 = D$, as desired. □

Example 12.6.2. Let us find an integral point on the ellipse $C : X^2 + 6Y^2 = 2625$. Note that $2625 = 3 \cdot 5^3 \cdot 7$. The equation $X^2 + 6Y^2 = 3$ has no integral solutions; however

$$1^2 + 6 \cdot 1^2 = 7 \quad \text{and} \quad 3^2 + 6 \cdot 1^2 = 15.$$

Since $2625 = 7 \cdot 15 \cdot 5^2$, we conclude that C does have an integral point, by Proposition 12.6.1, and the formula in the proof gives

$$(3 - 6 \cdot 1, 1 + 3) = (-3, 4)$$

for an integral point on $X^2 + 6Y^2 = 105$ and

$$(-3 \cdot 5, 4 \cdot 5) = (-15, 20)$$

as a point on $X^2 + 6Y^2 = 2625$. Thus, $(\pm 15, \pm 20)$ are integral points on C. Note that if instead we use the points $(1, -1)$ and $(3, 1)$ in the formula, we obtain

$$((3 + 6) \cdot 5, (1 - 3) \cdot 5) = (45, -10),$$

and so we find four additional integral points, $(\pm 45, \pm 10)$, on C.

12.7. Primes of the Form $X^2 + BY^2$

As a consequence of Proposition 12.6.1 and for a fixed $B > 0$, we are interested to know for which square-free numbers D the ellipse $X^2 + BY^2 = D$ has integral solutions. We are particularly interested in the case when $D = p$ is a prime. In other words, we are interested to determine what primes can be expressed in the form $x^2 + By^2$, for some $x, y \in \mathbb{Z}$. When $B = 1$, we have shown:

(i) A prime p is of the form $x^2 + y^2$ if and only if $p = 2$, or $p \equiv 1 \bmod 4$.

This was shown in Theorem 11.3.9, using the Hasse–Minkowski theorem, and directly in Theorem 12.1.5. Using similar techniques, one can show analogous theorems for expressions of the form $x^2 + By^2$. For instance:

(ii) A prime p is of the form $x^2 + 2y^2$ if and only if $p = 2$, or $p \equiv 1$ or $3 \bmod 8$.

(iii) A prime p is of the form $x^2 + 3y^2$ if and only if $p = 3$, or $p \equiv 1 \bmod 6$.

These two statements were claimed by Fermat (in a letter to Pascal in 1654), correctly, but without proof. Euler proved statements (i), (ii), and (iii) above, i.e., the characterizations of primes of the form $p = X^2 + BY^2$, for $B = 1, 2, 3$, and conjectured the case of $B = 5$:

(v) A prime p is of the form $x^2 + 5y^2$ if and only if $p = 5$, or $p \equiv 1$ or $9 \bmod 20$.

The reader may have noticed that we skipped over (iv):

(iv) A prime p is of the form $x^2 + 4y^2$ if and only if $p \equiv 1 \bmod 4$,

which of course is equivalent to (i), i.e., to Theorem 12.1.5. Indeed, if p is odd and of the form $x^2 + y^2$, then one of x or y must be even, say $y = 2y'$, and therefore $p = x^2 + 4y'^2$.

The most general result of the type (i)–(v) is given by the following theorem. We remind the reader that we defined the discriminant of a polynomial in Section 5.5.1.

Theorem 12.7.1. *Let $n > 0$ be an integer. Then, there is a monic irreducible polynomial $f_n(x) \in \mathbb{Z}[x]$ such that if an odd prime p is not a divisor of n or of the discriminant of $f_n(x)$, then p is of the form $x^2 + ny^2$ if and only if $-n$ is a quadratic residue mod p and the congruence equation $f_n(x) \equiv 0 \bmod p$ has a solution.*

Unfortunately, the proof of Theorem 12.7.1 is far beyond the scope of this book (see [**Cox13**, Theorem 9.2] for a proof—at the graduate level).

Remark 12.7.2. The polynomial $f_n(x)$ of Theorem 12.7.1 may be computed via an alternative description (it may be taken to be the minimal polynomial of a real algebraic integer α for which $L = K(\alpha)$ is the ring class field of the order $\mathbb{Z}[\sqrt{-n}]$ in the imaginary quadratic field $K = \mathbb{Q}(\sqrt{-n})$). Here is a table for $1 \leq n \leq 20$ that provides a polynomial $f_n(x)$ that works for n and the discriminant Δ of $f_n(x)$:

n	$f_n(x)$	Δ
$1, 2, 3, 4, 7$	x	1
$5, 10, 15$	$x^2 + x - 1$	5
$6, 8, 16$	$x^2 - 2$	8
$9, 12$	$x^2 - 3$	12
11	$x^3 + x^2 + x - 1$	-44
13	$x^2 + x - 3$	13
14	$x^4 + 2x^3 + x^2 + 2x + 1$	-448
17	$x^4 + x^3 - 2x^2 + x + 1$	-1156
18	$x^2 - 6$	24
19	$x^3 + x^2 + 3x + 1$	-76
20	$x^4 - x^2 - 1$	-400

Example 12.7.3. Let $n = 1$, 2, or 3. According to Theorem 12.7.1 and Remark 12.7.2, a prime p is of the form $x^2 + ny^2$ if and only if $-n$ is a quadratic residue modulo p and the congruence equation $f_n(x) = x \equiv 0 \bmod p$ has a solution. Clearly, $x \equiv 0 \bmod p$ has a solution for every prime p (namely, $x \equiv 0 \bmod p$). Moreover:

(i) Let $n = 1$. The number -1 is a quadratic residue modulo p if and only if $p = 2$, or $p \equiv 1 \bmod 4$. Hence, $p = x^2 + y^2$ if and only if $p = 2$, or $p \equiv 1 \bmod 4$. Thus, we have recovered the conclusion of Theorem 12.1.5.

(ii) Let $n = 2$. The number -2 is a quadratic residue modulo $p > 2$ if and only if $p \equiv 1$ or $3 \bmod 8$ (see Exercise 10.8.21). Thus, a prime p is of the form $x^2 + 2y^2$ if and only if $p = 2$, or $p \equiv 1, 3 \bmod 8$, which recovers the statement (ii) we mentioned earlier in this section.

(iii) Let $n = 3$. The number -3 is a quadratic residue modulo $p > 3$ if and only if $p \equiv 1 \bmod 6$ (see Exercise 10.8.22). Thus, a prime p is of the form $x^2 + 3y^2$ if and only if $p = 3$, or $p \equiv 1 \bmod 6$, which recovers the statement (iii) we mentioned earlier in this section.

Example 12.7.4. Let $n = 5$. According to Theorem 12.7.1 and Remark 12.7.2, a prime p is of the form $x^2 + 5y^2$ if and only if -5 is a quadratic residue modulo p and the congruence equation $f_5(x) = x^2 + x - 1 \equiv 0 \bmod p$ has a solution. Using the law of quadratic reciprocity (see Exercise 10.8.23) one can show that if $p \neq 2, 5$, then

$$\left(\frac{-5}{p}\right) = \begin{cases} 1 & \text{if } p \equiv 1, 3, 7, 9 \bmod 20, \\ -1 & \text{if } p \equiv 11, 13, 17, 19 \bmod 20. \end{cases}$$

In addition, the quadratic congruence equation $x^2 + x - 1 \equiv 0 \bmod p$ will have a solution for a prime $p > 2$ if and only if its discriminant $\Delta \equiv 5 \bmod p$ is a quadratic residue (by Proposition 10.1.1). Notice that -5 and $5 \bmod p$ are quadratic residues if and only if -5 and $-1 \bmod p$ are quadratic residues. Thus, we need $p \equiv 1 \bmod 4$

and $p \equiv 1, 3, 7, 9 \bmod 20$, which leaves $p \equiv 1, 9 \bmod 20$ as the only possibilities. It follows that p is of the form $x^2 + 5y^2$ if and only if $p = 5$, or $p \equiv 1, 9 \bmod 20$, as claimed in the statement (v) mentioned earlier in this section.

Example 12.7.5. Let C be the ellipse given by $X^2+6Y^2 = 144175$. Let us attempt to find an integral point on C using Theorem 12.7.1 and Proposition 12.6.1. Since

$$144175 = 5^2 \cdot 73 \cdot 79,$$

it would suffice to find points on $C_1 : X^2 + 6Y^2 = 73$ and $C_2 : X^2 + 6Y^2 = 79$. The coefficients here are small enough that we could determine whether there are points by "brute force", but where is the fun in that? Let us instead determine what primes are of the form $x^2 + 6y^2$. By Theorem 12.7.1, a prime $p \neq 2, 3$ is of the form $x^2 + 6y^2$ if and only if -6 is a quadratic residue modulo p and $x^2 - 2 \equiv 0 \bmod p$ has a solution (i.e., 2 is a quadratic residue mod p). These conditions are equivalent to 2 and -3 being quadratic residues simultaneously mod p, which in turn are equivalent to $p \equiv \pm 1 \bmod 8$ and $p \equiv 1 \bmod 6$. Using the Chinese remainder theorem, we find that $p \neq 2, 3$ is of the form $x^2 + 6y^2$ if and only if $p \equiv 1, 7 \bmod 24$.

Since 73 and 79 are, respectively, congruent to 1 and 7 mod 24, it follows that $X^2 + 6Y^2 = 73$ and $X^2 + 6Y^2 = 79$ must have integral points. Indeed,

$$7^2 + 6 \cdot 2^2 = 73 \quad \text{and} \quad 5^2 + 6 \cdot 3^2 = 79.$$

Hence, by the method of Proposition 12.6.1 we have

$$\begin{aligned} 73 \cdot 79 &= (7^2 + 6 \cdot 2^2) \cdot (5^2 + 6 \cdot 3^2) \\ &= (7 \cdot 5 - 6 \cdot 2 \cdot 3)^2 + 6 \cdot (7 \cdot 3 + 2 \cdot 5)^2 \\ &= (-1)^2 + 6 \cdot 31^2 = 1^2 + 6 \cdot 31^2 = 5767. \end{aligned}$$

Finally, we can multiply through by 5^2 to find an integral point on C:

$$144175 = 5^2 \cdot 73 \cdot 79 = 5^2 \cdot (1^2 + 6 \cdot 31^2) = 5^2 + 6 \cdot 155^2,$$

and so $(5, 155) \in C(\mathbb{Z})$, as desired.

Remark 12.7.6. Note, however, that not every integral point on an ellipse $C : X^2 + BY^2 = D$ can be found using the method outlined in Example 12.7.5. For instance, take $C : X^2 + 6Y^2 = 2695$. The ellipse C has an integral point $(31, 17) \in C(\mathbb{Z})$. However,

$$2695 = 5 \cdot 7^2 \cdot 11,$$

and neither 5 nor 11 is of the form $x^2 + 6y^2$. But $5 \cdot 11$ is of this form; for instance, $55 = 7^2 + 6 \cdot 1^2$. Notice also that although 7^2 is a divisor of 2695, neither coefficient of $(31, 17)$ is divisible by 7.

12.8. Exercises

Exercise 12.8.1. For $r \in \mathbb{Q}$, let $C_r : x^2 + y^2 = r$. Let $\psi : C_{n/m} \to C_{nm}$ be given by $\psi((a, b)) = (ma, mb)$.

(1) Prove that ψ is well-defined; i.e., $\psi((a, b)) \in C_{nm}$, for any $(a, b) \in C_{n/m}$.

(2) Prove that ψ has a well-defined inverse function and therefore it is a bijection.

(3) Prove that ψ provides a bijection between $C_{n/m}(\mathbb{Q})$ and $C_{nm}(\mathbb{Q})$.

Exercise 12.8.2. Find $a, b \in \mathbb{Z}$ such that $p = a^2 + b^2$ following the method described in the proof of Theorem 12.1.5 (as in Example 12.1.6), for

(1) $p = 29$,
(2) $p = 37$,
(3) $p = 53$.

Exercise 12.8.3. The norm of a complex number $\alpha = a + bi$ is defined by $N(\alpha) = a^2 + b^2$.

(1) Let $\alpha = a + bi$ and $\beta = c + di$ be two complex numbers. Show that $N(\alpha\beta) = N(\alpha)N(\beta)$.
(2) Use the first part to show that if n and m are natural numbers that can be written as the sum of two (integral) squares, then nm can also be written as the sum of two (integral) squares.

Exercise 12.8.4. Use the fact that $26 = 25 + 1$ and $50 = 49 + 1$ to write 1300 as a sum of two squares. (Hint: use Lemma 12.1.7.)

Exercise 12.8.5. Determine which of the following numbers can be written as a sum of two squares and, if so, in how many (essentially distinct) ways:

(a) 1450.
(b) 1451.
(c) 1452.
(d) 1453.
(e) 1215445.

Exercise 12.8.6. Use the following illustration to come up with a proof of Pythagoras's theorem. (Hint: the area of a large square equals the sum of the areas of a small square and four rectangles.)

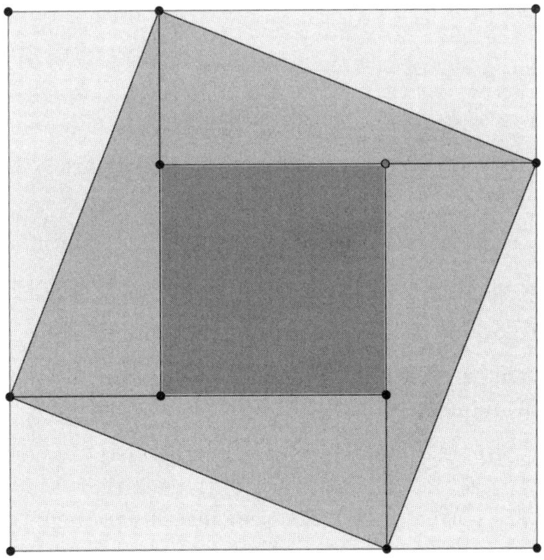

Exercise 12.8.7. Use the following illustration to come up with a proof of Pythagoras's theorem. (Hint: the area of a large square equals the sum of the areas of a small square and two rectangles.)

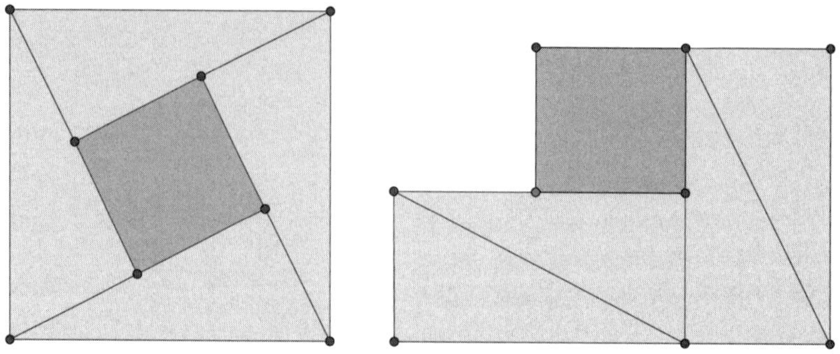

Exercise 12.8.8. Use the following illustration to come up with a proof of Pythagoras's theorem. (Hint: use the formula for the area of a trapezoid.)

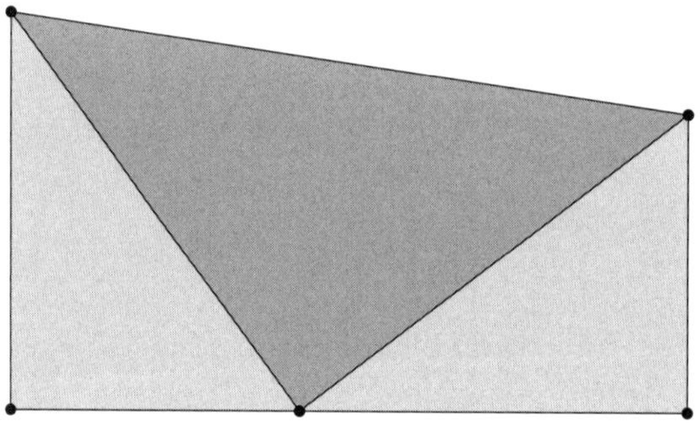

Note: this proof is due to the United States' President James A. Garfield (1831–1881).

Exercise 12.8.9. Are there infinitely many pythagorean triples of the form $(a, b, b+1)$, such as $(3, 4, 5)$ or $(5, 12, 13)$? If so, give a parametrization of all such triples.

Exercise 12.8.10. (a) Let (a, b, c) be a primitive pythagorean triple. Show that c is a sum of two squares.

(b) Suppose $c \geq 5$ is a number that is a sum of two squares. Is c the hypotenuse of a right triangle?

Exercise 12.8.11. Can two (distinct) perfect squares average to be a perfect square? In other words, are there integers $0 < a < b < c$ such that $\dfrac{a^2 + b^2}{2} = c^2$? If so, find an infinite family of such squares. (Hint: Lemma 11.3.3.)

Exercise 12.8.12. Show that the diophantine equation $X^2 + Y^2 = 3Z^2$ has only one integral solution, namely $(0, 0, 0)$.

Exercise 12.8.13. Suppose that Fermat's last theorem is known for prime exponents (i.e., $X^p + Y^p = Z^p$ has no integral solutions with $XYZ \neq 0$, when $p > 2$ is prime). Then, deduce Fermat's last theorem for all exponents $n \geq 3$. (Hint: use Corollary 12.3.2.)

Exercise 12.8.14. Show that the diophantine equation $x^4 - y^4 = z^2$ has no integral solutions with $xyz \neq 0$. (Hint: use an argument similar to the proof of Theorem 12.3.1.)

Exercise 12.8.15. Show that $C : X^2 + 15Y^2 = 4001$ does not have any rational solutions. (Hint: show that there are no integral solutions of $X^2 + 15Y^2 = 4001Z^2$, other than $X = Y = Z = 0$.)

Exercise 12.8.16. Verify directly the algebraic identity

$$(a^2 + Bb^2)(c^2 + Bd^2) = (ac - Bbd)^2 + B(ad + bc)^2$$

stated in Lemma 12.5.1.

Exercise 12.8.17. Find five different elements of $\mathbb{Q}(\sqrt{2})$ with norm 1 and five elements with norm -1.

Exercise 12.8.18. Find an integral point on each of the following ellipses or show that no such point exists:

(a) $x^2 + 6y^2 = 7$.
(b) $x^2 + 6y^2 = -7$.
(c) $x^2 + 6y^2 = 83$.
(d) $x^2 + 6y^2 = 97$.
(e) $x^2 + 6y^2 = 103$.
(f) $x^2 + 6y^2 = 629433$.

Exercise 12.8.19. Find an integral point on each of the following ellipses or show that no such point exists:

(a) $x^2 + 5y^2 = 29$.
(b) $x^2 + 5y^2 = 6$.
(c) $x^2 + 5y^2 = 4001$. (Hint: 4001, 4003, and 4013 are primes.)
(d) $x^2 + 5y^2 = 4003$.
(e) $x^2 + 5y^2 = 4013$.
(f) $x^2 + 5y^2 = 34112526$.

Exercise 12.8.20. Use Theorem 12.7.1 (and Remark 12.7.2) to characterize those primes $p \neq 2, 3$ such that p is of the form:

(a) $x^2 + 8y^2$.
(b) $x^2 + 16y^2$.
(c) $x^2 + 9y^2$.
(d) $x^2 + 12y^2$.

CHAPTER 13

CONTINUED FRACTIONS

"Can you do Addition?" the White Queen asked.
"What's one and one and one and one and one and one and one and one and one and one?"
"I don't know," said Alice. "I lost count."
"She can't do Addition," the Red Queen interrupted. "Can you do Subtraction? Take nine from eight."
"Nine from eight I can't, you know," Alice replied very readily: "but—"
"She can't do Subtraction," said the White Queen.

Lewis Carroll, from *Through the Looking Glass*

In this section we take a detour to study *continued fractions* which we will use to find rational and integral points on hyperbolas in the next chapter. A finite continued fraction is a rational number $\frac{p}{q}$ written as a sequence of iterated fractions of the form $\frac{p}{q} = a + \frac{1}{b}$, where b is another rational number also of the form $a' + \frac{1}{b'}$. For instance,

$$\frac{83}{13} = 6 + \cfrac{1}{2 + \cfrac{1}{1 + \cfrac{1}{1 + \cfrac{1}{2}}}}.$$

The first thing to notice is that if $b > 1$, then $a + 1/b$ is approximately a. Thus, a continued fraction expression provides several approximations of $\frac{p}{q}$. For instance, $\frac{83}{13} = 6.\overline{384615}$ is approximately 6. It is also approximately

$$6 + \frac{1}{2} = \frac{13}{2} = 6.5.$$

The continued fraction expansion provides the following approximations of $\frac{83}{13}$:

$$6,\ 6+\frac{1}{2}=\frac{13}{2}=6.5,\ 6+\cfrac{1}{2+\cfrac{1}{1}}=\frac{19}{3}=6.\overline{3},\quad \text{and } 6+\cfrac{1}{2+\cfrac{1}{1+\cfrac{1}{1}}}=\frac{32}{5}=6.4.$$

If we write the approximations in increasing order, we find them ordered as follows:

$$6<6.\overline{3}<\frac{83}{13}<6.4<6.5$$

and if we write $c_0=6$, $c_1=13/2$, $c_2=19/3$, and $c_3=32/5$, then

$$c_0<c_2<\frac{83}{13}<c_3<c_1,$$

and $|\frac{83}{13}-c_k|$ decreases as k increases. We shall see that every rational number has a continued fraction expansion and that the associated approximations satisfy similar inequalities.

More importantly, we will show that an arbitrary real number can be written as an infinite continued fraction expansion. For example, in Example 13.2.7 we shall see that

$$\pi=3+\cfrac{1}{7+\cfrac{1}{15+\cfrac{1}{1+\cfrac{1}{292+\dots}}}}.$$

In particular, π is approximately 3 or $3+\frac{1}{7}=\frac{22}{7}$ or $3+\frac{1}{7+1/15}=\frac{333}{106}$ or $3+\frac{1}{7+1/16}=\frac{355}{113}$. The approximation $\pi\cong 22/7$ is well known. Indeed, as we shall see, these rational approximations of π are *best possible*, in the sense that each is closer to π than any other fraction with the same or a smaller denominator.

Why are continued fractions important in arithmetic geometry? Let us see an example. Consider the hyperbola $C: x^2-23y^2=1$ and suppose $(p,q)\in C(\mathbb{Z})$ is an integral solution, so that $p^2-23q^2=1$. This implies that

$$\frac{p^2-1}{q^2}=23,$$

and if we take square roots of both sides, we obtain

$$\sqrt{23}=\sqrt{\frac{p^2}{q^2}-\frac{1}{q^2}}=\frac{\sqrt{p^2-1}}{q}.$$

Since $\sqrt{p^2-1}$ is approximately p, we obtain a rational approximation $\sqrt{23}\approx\frac{p}{q}$. This suggests that a place to look for integral points (p,q) in $C(\mathbb{Z})$ is among the rational approximations $\frac{p}{q}$ of $\sqrt{23}$. As we will see later in the chapter, the irrational number $\sqrt{23}$ has an infinite continued fraction expansion that begins as follows:

$$\sqrt{23} = 4 + \cfrac{1}{1 + \cfrac{1}{3 + \cfrac{1}{1 + \cfrac{1}{8 + \dots}}}}.$$

In particular, $\sqrt{23}$ is approximately

$$4 < \frac{19}{4} < \cdots < \sqrt{23} < \cdots < \frac{24}{5} < 5,$$

where each approximation is computed by truncating the infinite expansion into a finite one. For instance,

$$\sqrt{23} < \frac{24}{5} = 4 + \cfrac{1}{1 + \cfrac{1}{3 + \cfrac{1}{1}}}.$$

Do any of these approximations give us an integral point on C? The candidates so far are $(4,1)$, $(5,1)$, $(19,4)$, and $(24,5)$. Let us check:

$$4^2 - 23 = -7,\ 5^2 - 23 = 2,\ 19^2 - 23 \cdot 4^2 = -7, \text{ and } 24^2 - 23 \cdot 5^2 = 1.$$

Hence, we have found one integral point, namely $(24,5)$. In fact, there are other approximations coming from the continued fraction expansion that also provide rational points on C; for instance, $\sqrt{23} \approx \frac{1151}{240}$ and $1151^2 - 23 \cdot 240^2 = 1$. Can we find all the points on C in this way?

13.1. Finite Continued Fractions

Let us begin with the definition of a finite continued fraction.

Definition 13.1.1. Let $a_0, a_1, \dots, a_n$ be real numbers, all positive except perhaps a_0. We define the *continued fraction* $[a_0, a_1, \dots, a_n]$ by

$$[a_0, a_1, \dots, a_n] = a_0 + \cfrac{1}{a_1 + \cfrac{1}{\ddots + \cfrac{1}{a_{n-1} + \cfrac{1}{a_n}}}}.$$

We say that a continued fraction is *simple* if $a_0 \in \mathbb{Z}$ and $a_i \in \mathbb{N}$ for all $i > 0$.

Example 13.1.2. Let c be the continued fraction $[1,2,3,4]$. Then,

$$c = 1 + \cfrac{1}{2 + \cfrac{1}{3 + \cfrac{1}{4}}} = 1 + \cfrac{1}{2 + \cfrac{4}{13}} = 1 + \frac{13}{30} = \frac{43}{30}.$$

Let us see how one can use Euclid's algorithm to find a continued fraction expression for c:

$$\begin{aligned} 43 &= 30 \cdot 1 + 13, \\ 30 &= 13 \cdot 2 + 4, \\ 13 &= 4 \cdot 3 + 1. \end{aligned}$$

Now we can find a continued fraction expansion, starting from the first line of the work of Euclid's algorithm:

$$\frac{43}{30} = 1 + \frac{13}{30} = 1 + \frac{1}{\frac{30}{13}} = 1 + \frac{1}{2 + \frac{4}{13}} = 1 + \frac{1}{2 + \frac{1}{\frac{13}{4}}} = 1 + \frac{1}{2 + \frac{1}{3 + \frac{1}{4}}}.$$

In the following result we show that every rational number a/b has a continued fraction expansion, which can be found from the work of Euclid's algorithm as in the previous example.

Theorem 13.1.3. *Every rational number has a simple continued fraction expansion.*

Proof. Let $c \in \mathbb{Q}$. Suppose first that $c < 0$. Then, $c = \lfloor c \rfloor + \{c\}$ where $\lfloor c \rfloor < 0$ is the greatest integer function (and so that $c - 1 < \lfloor c \rfloor \leq c$) and $\{c\}$ is the fractional part of c (and so $0 \leq \{c\} < 1$). Thus, either $c \in \mathbb{Z}$ or $\{c\} \neq 0$ and we may write

$$c = \lfloor c \rfloor + \frac{1}{\frac{1}{\{c\}}}.$$

Since $\frac{1}{\{c\}}$ is a positive rational number, it follows that we have reduced the proof to the case of positive rational numbers.

Let us assume that $c = a/b$ is a positive rational number, with relatively prime positive integers a and b. We carry out Euclid's algorithm for the pair (a, b) and obtain positive integers q_i, r_i for $i = 1, \ldots, n$ (except for $r_n = 0$) such that

$$\begin{aligned} a &= b \cdot q_1 + r_1, \\ b &= r_1 q_2 + r_2, \\ r_1 &= r_2 q_3 + r_3, \\ &\vdots \\ r_{n-2} &= r_{n-1} q_n + r_n = r_{n-1} q_n. \end{aligned}$$

Then,

$$c = \frac{a}{b} = q_1 + \frac{r_1}{b} = q_1 + \cfrac{1}{\cfrac{b}{r_1}} = q_1 + \cfrac{1}{q_2 + \cfrac{r_2}{r_1}} = q_1 + \cfrac{1}{q_2 + \cfrac{1}{q_3 + \cfrac{r_3}{r_2}}}$$

$$= \cdots = q_1 + \cfrac{1}{q_2 + \cfrac{1}{\ddots + \cfrac{1}{q_{n-1} + \cfrac{1}{q_n}}}}.$$

Therefore, we have shown that $c = [q_1, q_2, \ldots, q_n]$, where the numbers q_i are the quotients that appear in Euclid's algorithm for a and b. $\square$

Example 13.1.4. Let us find a simple continued fraction for $c = -57/20$. We begin with Euclid's algorithm for -57 and 20:

$$\begin{aligned} -57 &= 20 \cdot (-3) + 3, \\ 20 &= 3 \cdot 6 + 2, \\ 3 &= 2 \cdot 1 + 1, \\ 2 &= 1 \cdot 2 + 0. \end{aligned}$$

This means that

$$c = -\frac{57}{20} = -3 + \cfrac{1}{\cfrac{20}{3}} = -3 + \cfrac{1}{6 + \cfrac{1}{\cfrac{3}{2}}} = -3 + \cfrac{1}{6 + \cfrac{1}{1 + \cfrac{1}{2}}}.$$

Thus, we find that $-57/20 = [-3, 6, 1, 2]$. Notice, however, that we could have done one additional step:

$$c = -\frac{57}{20} = -3 + \cfrac{1}{\cfrac{20}{3}} = -3 + \cfrac{1}{6 + \cfrac{1}{\cfrac{3}{2}}} = -3 + \cfrac{1}{6 + \cfrac{1}{1 + \cfrac{1}{2}}} = -3 + \cfrac{1}{6 + \cfrac{1}{1 + \cfrac{1}{1 + \cfrac{1}{1}}}}.$$

This shows that $-57/20 = [-3, 6, 1, 2] = [-3, 6, 1, 1, 1]$ has two simple continued fractions. In the next result, we will show that there are no other simple continued fraction expansions.

Proposition 13.1.5. *Let c be a real number.*

(1) *If $c \in \mathbb{Z}$, then there are exactly two simple continued fraction expansions for c, namely $[c]$ and $[c-1, 1]$.*

(2) *If c is not an integer, then there is a unique integer $n = \lfloor c \rfloor$ and a unique positive real number t, with $t > 1$, such that $c = n + \frac{1}{t}$.*

(3) *If c is a rational number, then there are exactly two simple continued fractions for c.*

Proof. For (1), we always have $c = [c]$. Now, suppose we have another simple continued fraction expansion for c of the form $[a_0, a_1, \dots, a_n]$ with $a_i \in \mathbb{Z}$ and with $a_i > 0$ for $i \geq 1$. Then, we have $c = a_0 + 1/t$, where $t = [a_1, \dots, a_n] > 0$. Since c is an integer, we have $c - a_0 = 1/t$ is an integer, which can only happen if $t = 1$. In this case, $c = [c-1, 1]$, and the result follows.

For (2), if c is not an integer, the existence of a pair $(n, t) \in \mathbb{Z} \times \mathbb{R}^{>1}$ such that $c = n + 1/t$ follows simply from the fact that $n = \lfloor c \rfloor$ and $t = 1/(c - \lfloor c \rfloor)$ work (note that $0 < c - \lfloor c \rfloor < 1$ and therefore $t > 1$).

For the uniqueness, suppose that $c = n + 1/t = m + 1/s$, with $m, n \in \mathbb{Z}$ and $s, t \in \mathbb{R}^{>1}$. Then, $n - m = 1/t - 1/s$ is an integer. But $s, t > 1$ implies that $|1/t - 1/s| < 1$ and so $1/t - 1/s = 0$ and $t = s$. Thus, $n = m$ also.

Let us show (3). We suppose that c is rational, but not an integer. Hence, if c is given by the continued fraction $[a_0, \dots, a_{n-1}, a_n]$, then a_0 is uniquely determined. Similarly, $a_1, \dots, a_{n-2}$ are uniquely determined. There are two remaining cases to consider:

- If $a_n > 1$, then the expression $a_{n-1} + \frac{1}{a_n}$ uniquely determines a_{n-1}, and there are two continued fractions for a_n, namely $[a_n]$ and $[a_n - 1, 1]$. Thus,
$$c = [a_0, \dots, a_{n-1}, a_n] = [a_0, \dots, a_{n-1}, a_n - 1, 1]$$
are the two unique continued fraction expansions for c.
- If $a_n = 1$, then $a_{n-2} + \frac{1}{a_{n-1} + \frac{1}{a_n}} = a_{n-2} + \frac{1}{a_{n-1}+1}$ uniquely determines a_{n-2} and $(a_{n-1} + 1)$. Thus,
$$c = [a_0, \dots, a_{n-2}, a_{n-1}, a_n] = [a_0, \dots, a_{n-2}, a_{n-1} + 1]$$
are the two unique continued fraction expansions for c.

In all cases we have described exactly two continued fractions for c, and this concludes the proof of the proposition. $\square$

Remark 13.1.6. A rational number can have more than two continued fractions, if we do not require our expansions to be simple. For instance,
$$4 = [4] = [3, 1] = [\pi, (4 - \pi)^{-1}].$$
That is,
$$4 = 3 + \frac{1}{1} = \pi + \cfrac{1}{\cfrac{1}{4 - \pi}}.$$
In fact, if α is any real number with $3 < \alpha < 4$, then $4 = [\alpha, (4 - \alpha)^{-1}]$ is a continued fraction. For the most part, in this book we will be only concerned with simple continued fractions.

Given a continued fraction $[a_0, a_1, \dots, a_n]$, we are interested in the values of the truncated continued fractions $[a_0, a_1, \dots, a_k]$ for $0 \leq k \leq n$.

Definition 13.1.7. Let $c = [a_0, a_1, \dots, a_n]$ be a continued fraction. For each $0 \leq k \leq n$, the number $c_k = [a_0, a_1, \dots, a_k]$ is called the kth *convergent* of c.

Example 13.1.8. Let $c = -57/20 = [-3, 6, 1, 2]$. Then, the convergents of c are

$$c_0 = -3,\ c_1 = -3 + \frac{1}{6} = -\frac{17}{6},\ c_2 = -3 + \frac{1}{6 + \frac{1}{1}} = -\frac{20}{7},\ \text{and}\ c_3 = c = -\frac{57}{20}.$$

The following result provides an algorithm to find the values (as a reduced fraction) of the convergents of a continued fraction.

Theorem 13.1.9. *Let $c = [a_0, a_1, \dots, a_n]$ be a simple continued fraction, and define sequences p_k and q_k, for $k \geq -1$, by*

$$\begin{cases} p_{-1} = 1, \\ p_0 = a_0, \\ p_k = a_k p_{k-1} + p_{k-2} & \text{for } k \geq 1, \end{cases} \quad \text{and} \quad \begin{cases} q_{-1} = 0, \\ q_0 = 1, \\ q_k = a_k q_{k-1} + q_{k-2} & \text{for } k \geq 1. \end{cases}$$

Then:

(1) *The number $q_k \geq k$ for all $k \geq 0$.*

(2) *For any real number x and any $k \geq 0$,*

$$[a_0, a_1, \dots, a_k, x] = \frac{x \cdot p_k + p_{k-1}}{x \cdot q_k + q_{k-1}}.$$

(3) *The kth convergent of c is given by $[a_0, a_1, \dots, a_k] = \dfrac{p_k}{q_k}$.*

Proof. Part (1) follows directly from the definition of q_k. Indeed, $q_0 = 1$, $a_k \geq 1$ for all $k \geq 1$ because c is simple, $q_1 = a_1 \geq 1$, and $q_k = a_k q_{k-1} + q_{k-2} > q_{k-1}$ for all $k \geq 2$. Thus, by induction, it follows that $q_k \geq k$ for all $k \geq 0$.

We shall prove (2) by induction on k. When $k = 0$, we have

$$\frac{x \cdot p_0 + p_{-1}}{x \cdot q_0 + q_{-1}} = \frac{xa_0 + 1}{x} = a_0 + \frac{1}{x} = [a_0, x].$$

Now assume the results holds for k, for any real x. Then,

$$\begin{aligned} [a_0, a_1, \dots, a_k, a_{k+1}, x] &= \left[a_0, a_1, \dots, a_k, a_{k+1} + \frac{1}{x}\right] \\ &= \frac{(a_{k+1} + \frac{1}{x}) \cdot p_k + p_{k-1}}{(a_{k+1} + \frac{1}{x}) \cdot q_k + q_{k-1}} \\ &= \frac{x(a_{k+1}p_k + p_{k-1}) + p_k}{x(a_{k+1}q_k + q_{k-1}) + q_k} = \frac{xp_{k+1} + p_k}{xq_{k+1} + q_k}, \end{aligned}$$

where we have used the induction hypothesis for k and $x = a_{k+1} + 1/x$ and the definition of the sequences p_k and q_k. This completes the proof of the induction step and the proof of (2).

For (3), we simply evaluate the expression in (2) for $k - 1$ at $x = a_k$:

$$[a_0, a_1, \dots, a_{k-1}, a_k] = \frac{a_k \cdot p_{k-1} + p_{k-2}}{a_k \cdot q_{k-1} + q_{k-2}} = \frac{p_k}{q_k},$$

as desired. □

Example 13.1.10. Consider the continued fraction $c = [1, 2, 1, 2, 1, 2]$. Then, $p_{-1} = 1$, $p_0 = 1$, $p_1 = 2 \cdot 1 + 1 = 3$, and $q_{-1} = 0$, $q_0 = 1$, and $q_1 = 2 \cdot 1 + 0 = 2$. For efficiency and ease, we usually compute the terms in the sequences p_k and q_k using a table as the one below:

k	-1	0	1	2	3	4	5
a_k		1	2	1	2	1	2
p_k	1	1	3	4	11	15	41
q_k	0	1	2	3	8	11	30

Thus, $c = c_5 = 41/30$, and the rest of the convergents are $c_0 = 1$, $c_1 = 3/2$, $c_2 = 4/3$, $c_3 = 11/8$, and $c_4 = 15/11$.

The reader may have noticed a curious arithmetic pattern in the convergents: $1 \cdot 2 - 3 \cdot 1 = -1$ and $3 \cdot 3 - 4 \cdot 2 = 1$ and $4 \cdot 8 - 11 \cdot 3 = -1$, etc. In other words, $p_k \cdot q_{k+1} - p_{k+1}q_k$ alternates between 1 and -1. Let us prove this fact.

Proposition 13.1.11. *Let $c = [a_0, \ldots, a_n]$ be a continued fraction, and let $\{p_k\}$ and $\{q_k\}$ be the sequences defined in Theorem* 13.1.9. *Then,*

$$p_k q_{k+1} - p_{k+1} q_k = (-1)^{k+1}$$

holds for all $k \geq -1$.

Proof. We proceed by induction on k. For $k = -1$, we have

$$1 \cdot 1 - a_0 \cdot 0 = 1 = (-1)^{-1+1}.$$

Now suppose that the result holds for k. Then,

$$\begin{aligned} p_{k+1}q_{k+2} - p_{k+2}q_{k+1} &= p_{k+1}(a_{k+2}q_{k+1} + q_k) - (a_{k+2}p_{k+1} + p_k)q_{k+1} \\ &= p_{k+1}q_k - p_k q_{k+1} \\ &= -(p_k q_{k+1} - p_{k+1}q_k) = -(-1)^{k+1} = (-1)^{k+2}, \end{aligned}$$

as desired. This completes the induction step and the proof. □

As a corollary we show that the expression $c_k = p_k/q_k$ is actually given in reduced terms; that is, the numerator and denominator are relatively prime.

Corollary 13.1.12. *Let $c = [a_0, \ldots, a_n]$ be a continued fraction, and let $\{p_k\}$ and $\{q_k\}$ be the sequences defined in Theorem* 13.1.9. *Then,* $\gcd(p_k, q_k) = 1$ *for all $k \geq 0$.*

Proof. By Proposition 13.1.11, we have $p_k q_{k+1} - p_{k+1} q_k = (-1)^{k+1}$. In particular, $\gcd(p_k, q_k)$ is a positive divisor of 1 (by Proposition 2.9.1), and therefore it is equal to 1. □

It turns out that consecutive even (resp. odd) convergents also satisfy a similar equation to the one in Proposition 13.1.11.

Corollary 13.1.13. *Let $c = [a_0, \ldots, a_n]$ be a continued fraction, and let $\{p_k\}$ and $\{q_k\}$ be the sequences defined in Theorem* 13.1.9. *Then,*

$$p_k q_{k+2} - p_{k+2} q_k = (-1)^{k+1} a_{k+2}$$

holds for all $k \geq -1$.

Proof. The result follows from Proposition 13.1.11:

$$\begin{aligned} p_k q_{k+2} - p_{k+2} q_k &= p_k(a_{k+2}q_{k+1} + q_k) - (a_{k+2}p_{k+1} + p_k)q_k \\ &= a_{k+2}(p_k q_{k+1} - p_{k+1}q_k) \\ &= (-1)^{k+1} a_{k+2}, \end{aligned}$$

as desired. □

Example 13.1.14. For example, if $c = [1,2,1,2,1,2]$, we have $c_1 = 3/2$, $c_3 = 11/8$, and we have

$$3 \cdot 8 - 11 \cdot 2 = 2 = (-1)^2 \cdot 2.$$

In Example 13.1.10 we calculated c_0 through $c_5 = c$. Notice that, as we noticed in the introduction, here the convergents are also ordered in the following manner:

$$1 < \frac{4}{3} < \frac{15}{11} < \frac{41}{30} < \frac{11}{8} < \frac{3}{2};$$

i.e., $c_0 < c_2 < c_4 < c = c_5 < c_3 < c_1$. Next, we prove that the convergents are always ordered in a similar manner.

Theorem 13.1.15. *Let $c = [a_0, \ldots, a_n]$ be a continued fraction. Then, the convergents c_k for $k = 0, \ldots, n$ satisfy*

$$c_0 < c_2 < c_4 < \cdots < c < \cdots < c_5 < c_3 < c_1.$$

Proof. By Corollary 13.1.13, we have that

$$p_k q_{k+2} - p_{k+2} q_k = (-1)^{k+1} a_{k+2}$$

holds for all $k \geq -1$. If we divide the expression above by $q_k q_{k+2}$ throughout, we obtain, for $k \geq 0$,

$$c_k - c_{k+2} = (-1)^{k+1} \frac{a_{k+2}}{q_k q_{k+2}}$$

where we have used Theorem 13.1.9 to write $c_k = p_k/q_k$. Since a_{k+2} and q_k are positive for $k \geq 0$ (see part (1) of Theorem 13.1.9), it follows that $c_k - c_{k+2}$ is positive for odd k and negative for even k. In other words, $c_{k+2} < c_k$ when $k \geq 1$ is odd, and $c_k < c_{k+2}$ when $k \geq 0$ is even.

Next, we show that $c_k < c_{k+1}$ if k is even. For this, we divide the expression in Proposition 13.1.11 by $q_k q_{k+1}$ to obtain

$$c_k - c_{k+1} = (-1)^{k+1} \cdot \frac{1}{q_k q_{k+1}} \tag{13.1}$$

for $k \geq 0$. In particular, if k is even, then $c_k - c_{k+1} < 0$, as desired.

It remains to show that $c_k < c_j$ if k is even and j is odd. Let us write $k = 2n$ and $j = 2m+1$, and let us assume $n \leq m$ (the case $n > m$ is very similar). The facts we have shown imply that

$$c_k = c_{2n} < c_{2m} < c_{2m+1} = c_j$$

where we have used that $c_{2n} < c_{2m}$ because the indices are even and $2n \leq 2m$ and that $c_{2m} < c_{2m+1}$ because the indices are consecutive with $2m$ even. Therefore, we have shown $c_k < c_j$, as desired.

Finally, since $c = c_n$, then $c_k < c < c_j$ for any even $k \leq n$ and any odd $j \geq n$. This completes the proof of the theorem. □

13.2. Infinite Continued Fractions

In this section we begin exploring *infinite* continued fractions. In other words, we would like to build a theory for expressions of the form

$$[a_0, a_1, \dots, a_n, \dots] = a_0 + \cfrac{1}{a_1 + \cfrac{1}{\ddots + \cfrac{1}{a_n + \cfrac{1}{\ddots}}}}.$$

We will define such a continued fraction as a limit of its convergents, but first we need to show that the convergents... converge!

Theorem 13.2.1. *Let a_0 be an integer, and let $\{a_i\}_{i\geq 1}$ be a sequence of positive integers. Let c_k be the (finite, simple) continued fraction $[a_0, a_1, \dots, a_k]$. Then:*

(1) *The limit $\alpha = \lim_{k\to\infty} c_k$ exists and it is finite (i.e., $\alpha \in \mathbb{R}$).*

(2) *For any $k, j \geq 0$, we have $c_{2k} < \alpha < c_{2j+1}$.*

Proof. Let us first consider two sequences $\{c_{2k}\}_{k\geq 0}$ and $\{c_{2k+1}\}_{k\geq 0}$ and prove that both sequences converge. Indeed, the sequence $\{c_{2k}\}$ is increasing because if $k < j$, then $c_{2j} = [a_0, a_1, \dots, a_{2j}]$ is a simple continued fraction and c_{2k} is a convergent of c_{2j}. By Theorem 13.1.15, we have $c_{2k} < c_{2j}$. Moreover, the same theorem implies that $c_{2k} < c_1$ for any $k \geq 0$. Hence, the sequence $\{c_{2k}\}$ is strictly increasing and bounded above and therefore convergent by the monotone convergence theorem. Similarly, $\{c_{2k+1}\}$ converges because it is strictly decreasing and bounded below.

Let $\alpha_1 = \lim_{k\to\infty} c_{2k}$ and $\alpha_2 = \lim_{j\to\infty} c_{2j+1}$. Let j be fixed, let $k \geq 0$ be arbitrary, and put $m = \max\{k, j\}$. Consider the finite simple continued fraction $c_{2m+1} = [a_0, \dots, a_{2m+1}]$. By Theorem 13.1.15, c_{2k} and c_{2j+1} are convergents for c_{2m+1} and $c_{2k} < c_{2j+1}$. Since this holds for any k, it follows that $\alpha_1 = \lim_{k\to\infty} c_{2k} \leq c_{2j+1}$. Moreover, since this holds for any $j \geq 0$, it follows that $\alpha_1 \leq \lim_{j\to\infty} c_{2j+1} = \alpha_2$. Hence, $\alpha_1 \leq \alpha_2$ and, further,

$$c_{2k} < \alpha_1 \leq \alpha_2 < c_{2j+1}$$

for any $k, j \geq 0$. In particular, $c_{2k} < \alpha_1 \leq \alpha_2 < c_{2k+1}$, and for $k \geq 1$ we have

$$|c_k - c_{k+1}| = \frac{1}{q_k q_{k+1}} \leq \frac{1}{k(k+1)}$$

where we have used (13.1) from the proof of Theorem 13.1.15 and Theorem 13.1.9 to show that $q_k \geq k$ for $k \geq 1$. It follows that $|c_k - c_{k+1}|$ goes to 0 as $k \to \infty$, and therefore we must have $\alpha_1 = \alpha_2$.

We have shown that all three sequence $\{c_k\}$, $\{c_{2k}\}$, and $\{c_{2j+1}\}$ converge to $\alpha = \alpha_1 = \alpha_2 \in \mathbb{R}$, and $c_{2k} < \alpha < c_{2j+1}$ for any $k, j \geq 0$, as desired. □

We are now ready to give a formal definition of infinite continued fraction.

Definition 13.2.2. Let $a_0 \in \mathbb{Z}$ and let $\{a_k\}_{k\geq 1}$ be a sequence of positive integers. We define the *infinite simple continued fraction* $[a_0, a_1, \dots, a_k, \dots]$ to be the value of the limit $\lim_{k\to\infty} c_k$, where c_k is the kth convergent $[a_0, a_1, \dots, a_k]$.

Example 13.2.3. Let us attempt to calculate the value of an infinite continued fraction. Let

$$\alpha = 1 + \cfrac{1}{1 + \cfrac{1}{\ddots + \cfrac{1}{1 + \cfrac{1}{\ddots}}}};$$

i.e., $\alpha = [1, 1, \ldots, 1, \ldots]$. In particular, notice that

$$\alpha = 1 + \frac{1}{\alpha}.$$

It follows that $\alpha^2 - \alpha - 1 = 0$, and there are two possible values for α, namely $\alpha = \frac{1+\sqrt{5}}{2}$ or $\frac{1-\sqrt{5}}{2}$. However, α is positive (in fact, we know that $\alpha > c_0 = 1$) and $(1-\sqrt{5})/2 = -0.61\ldots$ is negative. We conclude that α must be $\varphi = \frac{1+\sqrt{5}}{2}$, the *golden ratio*.

We may now use the convergents of the continued fraction expansion to find rational approximations of the golden ratio. We shall extend the technique we used in Example 13.1.10 to compute the sequences of p_k and q_k to infinite continued fractions as follows:

k	-1	0	1	2	3	4	5	6	7	$\cdots$
a_k		1	1	1	1	1	1	1	1	$\cdots$
p_k	1	1	2	3	5	8	13	21	34	$\cdots$
q_k	0	1	1	2	3	5	8	13	21	$\cdots$

Thus,

$$1 < \frac{3}{2} < \frac{8}{5} < \frac{21}{13} < \frac{1+\sqrt{5}}{2} < \frac{34}{21} < \frac{13}{8} < \frac{5}{3} < 2,$$

where the closest convergent is $34/21 = 1.619047\ldots$ while $\frac{1+\sqrt{5}}{2} = 1.618033\ldots$.

The reader may have recognized the numbers in the sequences p_k and q_k as Fibonacci numbers F_n. Recall that the Fibonacci numbers are defined by $F_0 = 1$, $F_1 = 1$, and $F_{n+1} = F_n + F_{n-1}$. Since $a_k = 1$ for all $k \geq 0$, the recursive equations that define p_k and q_k are

$$p_{k+1} = p_k + p_{k-1} \quad \text{and} \quad q_{k+1} = q_k + q_{k-1}.$$

Moreover, $p_{-1} = 1 = p_0$ and $q_0 = 1 = q_1$, and it follows that $q_k = F_k$ and $p_k = F_{k+1}$ are Fibonacci numbers. In particular, Theorem 13.2.1 implies that the limit of the sequence of convergents $p_k/q_k = F_{k+1}/F_k$ is the golden ratio; i.e., the sequence of ratios of consecutive Fibonacci numbers converges to the golden ratio. This provides an alternative proof of Exercise 2.11.39.

We continue investigating the basic properties of infinite simple continued fractions. First, we shall establish that the limit of an infinite simple continued fraction is an irrational number (e.g., $(1+\sqrt{5})/2$ as in Example 13.2.3).

Theorem 13.2.4. *The value of any infinite simple continued fraction is irrational.*

Proof. Let $\alpha = [a_0, \ldots, a_k, \ldots]$ be a continued fraction with $a_0 \in \mathbb{Z}$ and $a_i \in \mathrm{N}$ for all $i \geq 1$, and suppose for a contradiction that α is a rational number m/n, for some relatively prime integers m, n with $n > 0$.

Fix an even number k for the moment. Then, by Theorem 13.2.1, we have $c_k < \alpha < c_{k+1}$ where c_k, c_{k+1} are consecutive convergents. In particular,

$$0 < \frac{m}{n} - c_k = \frac{m}{n} - \frac{p_k}{q_k} < \frac{p_{k+1}}{q_{k+1}} - \frac{p_k}{q_k} = \frac{1}{q_k q_{k+1}}$$

where the last equality comes from (13.1). Next we multiply through by nq_k to obtain

$$0 < mq_k - np_k < \frac{n}{q_{k+1}} \leq \frac{n}{k+1}$$

where we have used the fact that $q_k \geq k$ for all $k \geq 1$ (see Theorem 13.1.9). Hence, for a sufficiently large value of k (that is, when $k \geq n$) we will have inequalities $0 < mq_k - np_k < 1$, but $mq_k - np_k$ is an integer, so we have reached a contradiction. Thus, α cannot be rational. □

Remark 13.2.5. Theorem 13.2.4 provides an infinite supply of irrational numbers that are easy to describe, one for each sequence of integers $a_0 \in \mathbb{Z}$ and $a_k \geq 1$. In fact, this is an uncountable set of irrational numbers (see Exercise 13.4.10).

Next, we will show that if a real number α has an infinite simple continued fraction, then this expression is unique.

Proposition 13.2.6. *Let* $\alpha = [a_0, a_1, \ldots, a_k, \ldots] = [b_0, b_1, \ldots, b_k, \ldots]$ *be infinite simple continued fractions for the same real number* α. *Then,* $a_k = b_k$ *for all* $k \geq 0$.

Proof. Suppose that α has two infinite continued fraction expansions as in the statement. Then,

$$\alpha = a_0 + \frac{1}{\beta_1} = b_0 + \frac{1}{\beta_2}$$

where $\beta_1 = [a_1, \ldots, a_k, \ldots]$ and $\beta_2 = [b_1, \ldots, b_k, \ldots]$. Since a_i, b_i are integers ≥ 1 for all $i \geq 1$, it follows that β_1, β_2 are positive real numbers $> a_1, b_1 \geq 1$, by Theorem 13.2.1, and $a_0, b_0 \in \mathbb{Z}$ by definition. By Theorem 13.2.4, the number α is irrational (thus, definitely not an integer). Hence, Proposition 13.1.5 implies that $a_0 = b_0 = \lfloor \alpha \rfloor$ and $\beta_1 = \beta_2$. The same argument shows that $a_1 = b_1 = \lfloor \beta_1 \rfloor$ and $[a_2, \ldots] = [b_2, \ldots]$. If we proceed by induction, then we can easily show that $a_i = b_i$ for all $i \geq 0$, as claimed. □

Now it remains to show that *every* real number α has at least one infinite simple continued fraction expansion. Let us first see an example of how to find such a continued fraction.

Example 13.2.7. Let us find a continued fraction expansion for π; i.e., let us find $[a_0, a_1, \ldots, a_k, \ldots]$ such that the limit of convergents is π. As in the proof of Proposition 13.2.6, or by Proposition 13.1.5, we must have $a_0 = \lfloor \pi \rfloor = 3$. We shall write $\alpha_0 = \alpha$ and then define α_1 such that

$$\pi = \alpha_0 = a_0 + \frac{1}{\alpha_1} = 3 + \frac{1}{\alpha_1}.$$

In other words, $\alpha_1 = 1/(\alpha_0 - a_0) = 1/(\pi - 3) = 7.0625133059310457700\ldots$. Now, we need to find a continued fraction expansion for α_1, and by the same argument the expansion should start with $\lfloor \alpha_1 \rfloor = 7$; i.e.,

$$\alpha_1 = 7 + \frac{1}{\alpha_2}$$

where $\alpha_2 = 1/(\alpha_1 - 7) = 15.9965944066857198 32\ldots$. We continue by defining $\alpha_3 = 1.0034\ldots$, $\alpha_4 = 292.6345\ldots$, etc., in this manner, such that

$$\alpha_{k+1} = 1/(\alpha_k - \lfloor \alpha_k \rfloor)$$

and $a_k = \lfloor \alpha_k \rfloor$. Thus, the first few terms of the infinite simple continued fraction for $\alpha = \pi$ are

$$\pi = 3 + \cfrac{1}{7 + \cfrac{1}{15 + \cfrac{1}{1 + \cfrac{1}{292 + \cfrac{1}{\alpha_5}}}}};$$

i.e., $\pi = [3, 7, 15, 1, 292, \alpha_5]$. A few more terms of the continued fraction would be given by

$$\pi = [3, 7, 15, 1, 292, 1, 1, 1, 2, 1, 3, 1, 14, 2, 1, 1, 2, 2, 2, 2, 1, 84, 2, 1, 1, 15, 3, 13, \ldots].$$

It is important to realize that here we have only shown that *if* π has a continued fraction, then it must start as given above. However, we must show that the limit of convergents actually equals π. We will show this next and extend the same technique to any real number.

Theorem 13.2.8. *Let $\alpha \in \mathbb{R}$ be an irrational number and define sequences $\{\alpha_k\}$ and $\{a_k\}$ for $k \geq 0$ by $\alpha_0 = \alpha$, $a_0 = \lfloor \alpha_0 \rfloor$, and*

$$\alpha_{k+1} = \frac{1}{\alpha_k - a_k}, \quad \textit{where } a_k = \lfloor \alpha_k \rfloor,$$

for every $k \geq 0$. Then:

(1) *For any $k \geq 0$, we have $\alpha = [a_0, a_1, \ldots, a_k, \alpha_{k+1}]$ and $a_k \geq 1$ when $k \geq 1$.*

(2) *Let $c_k = [a_0, \ldots, a_k] = \frac{p_k}{q_k}$. Then,*

$$\alpha - \frac{p_k}{q_k} = \frac{(-1)^k}{q_k(\alpha_{k+1} q_k + q_{k-1})}.$$

(3) *We have $\alpha = [a_0, \ldots, a_k, \ldots]$.*

(4) *More generally, $\alpha_k = [a_k, a_{k+1}, \ldots]$.*

Proof. Let us start with a proof of (1) using induction on k. By Proposition 13.1.5, if α is any irrational number, then there is a unique integer $a = \lfloor \alpha \rfloor$ and a unique irrational number $\alpha' > 1$ such that $\alpha = a + 1/\alpha'$, so that $\alpha' = 1/(\alpha - a)$. Using this fact for $\alpha = \alpha_0$, we obtain $\alpha = \alpha_0 = [a_0, \alpha_1]$. This proves the base case of $k = 0$. Now assume that $\alpha = [a_0, \ldots, a_k, \alpha_{k+1}]$. Then, we can apply Proposition 13.1.5 to α_{k+1} to obtain $\alpha_{k+1} = a_{k+1} + 1/\alpha_{k+2}$, and therefore $\alpha = [a_0, \ldots, a_k, a_{k+1}, \alpha_{k+2}]$, as needed. This completes the proof by induction of (1). Notice that $0 < \alpha_k - \lfloor \alpha_k \rfloor < 1$ for any $k \geq 0$, and so $\alpha_{k+1} > 1$ and $a_{k+1} = \lfloor \alpha_{k+1} \rfloor \geq 1$, as claimed.

For (2) and (3), let us define c as the value of the infinite simple continued fraction defined by $[a_0, a_1, \ldots, a_k, \ldots]$ where $a_k = \lfloor \alpha_k \rfloor$ for all $k \geq 0$. We want to prove that $c = \alpha$. Let $c_k = p_k/q_k$ be the kth convergent of c. By part (1), we have $\alpha = [a_0, \ldots, a_k, \alpha_{k+1}]$, and part (2) of Theorem 13.1.9 with $x = \alpha_{k+1}$ implies that

$$\alpha = \frac{\alpha_{k+1} p_k + p_{k-1}}{\alpha_{k+1} q_k + q_{k-1}}.$$

Hence,

$$\begin{aligned} \alpha - \frac{p_k}{q_k} &= \frac{\alpha_{k+1} p_k + p_{k-1}}{\alpha_{k+1} q_k + q_{k-1}} - \frac{p_k}{q_k} = \frac{q_k(\alpha_{k+1} p_k + p_{k-1}) - p_k(\alpha_{k+1} q_k + q_{k-1})}{q_k(\alpha_{k+1} q_k + q_{k-1})} \\ &= \frac{p_{k-1} q_k - p_k q_{k-1}}{q_k(\alpha_{k+1} q_k + q_{k-1})} = \frac{(-1)^k}{q_k(\alpha_{k+1} q_k + q_{k-1})}, \end{aligned}$$

where the last equation follows from Proposition 13.1.11. Since we know that $\alpha_{k+1} > 1$ and $q_k \geq k$ for $k \geq 1$, then

$$|\alpha - c_k| = \frac{1}{q_k(\alpha_{k+1} q_k + q_{k-1})} < \frac{1}{k^2}$$

and so, $\lim_{k\to\infty} c_k = \alpha$. It follows that $c = \alpha$, as desired.

Finally, to prove (4), if we set $\alpha' = \alpha_k$, then we can apply (3) to α' to show that $\alpha' = [a'_0, a'_1, \ldots] = [a_k, a_{k+1}, \ldots]$, because $a'_0 = \lfloor \alpha' \rfloor = \lfloor \alpha_k \rfloor = a_k$, and $\alpha'_1 = 1/(\alpha' - a_k) = \alpha_{k+1}$, and so $\alpha'_j = \alpha_{k+j}$, and $a'_j = a_{k+j}$ for all $j \geq 0$. $\square$

Example 13.2.9. Let us find the infinite continued fraction of $\alpha = e$ (the base of the natural logarithm) using the sequences α_k and a_k as described in Theorem 13.2.8:

$$\begin{aligned} \alpha_0 &= e = 2.7182818284\ldots, \\ a_0 &= \lfloor \alpha_0 \rfloor = \lfloor e \rfloor = 2, \\ \alpha_1 &= \frac{1}{\alpha_0 - a_0} = \frac{1}{e-2} = 1.3922111911\ldots, \\ a_1 &= \lfloor \alpha_1 \rfloor = 1, \\ \alpha_2 &= \frac{1}{\alpha_1 - a_1} = \frac{1}{0.3922111911\ldots} = 2.5496467783\ldots, \\ a_2 &= \lfloor \alpha_2 \rfloor = 2, \\ \alpha_3 &= \frac{1}{\alpha_2 - a_2} = \frac{1}{0.5496467783\ldots} = 1.8193502435\ldots, \\ a_3 &= \lfloor \alpha_3 \rfloor = 1, \\ \alpha_4 &= \frac{1}{\alpha_3 - a_3} = \frac{1}{0.8193502435\ldots} = 1.2204792857\ldots, \\ a_4 &= \lfloor \alpha_4 \rfloor = 1, \\ \alpha_5 &= \frac{1}{\alpha_4 - a_4} = \frac{1}{0.2204792857\ldots} = 4.5355734730\ldots, \\ a_5 &= \lfloor \alpha_5 \rfloor = 4. \end{aligned}$$

Thus, the first few coefficients of the continued fraction of e are

$$e = [2, 1, 2, 1, 1, 4, \ldots] = 2 + \cfrac{1}{1 + \cfrac{1}{2 + \cfrac{1}{1 + \cfrac{1}{1 + \cfrac{1}{4 + \cfrac{1}{\ddots}}}}}}.$$

In fact,

$$e = [2, 1, 2, 1, 1, 4, 1, 1, 6, 1, 1, 8, 1, 1, 10, 1, 1, 12, 1, 1, 14, \ldots],$$

a fact due to Euler (see [**Coh06**] for a proof).

Example 13.2.10. Let us find the infinite continued fraction of $\alpha = \sqrt{2}$:

$$\begin{aligned}
\alpha_0 &= \sqrt{2}, \\
a_0 &= \lfloor \alpha_0 \rfloor = \lfloor \sqrt{2} \rfloor = 1, \\
\alpha_1 &= \frac{1}{\alpha_0 - a_0} = \frac{1}{\sqrt{2} - 1} = 2.4142135623\ldots, \\
a_1 &= \lfloor \alpha_1 \rfloor = 2, \\
\alpha_2 &= \frac{1}{\alpha_1 - a_1} = \frac{1}{0.4142135623\ldots} = 2.4142135623\ldots, \\
a_2 &= \lfloor \alpha_2 \rfloor = 2.
\end{aligned}$$

We observe a numerical coincidence, as it seems that $\alpha_1 = \alpha_2$. Is that so? Let us check:

$$\alpha_2 = \frac{1}{\alpha_1 - a_1} = \frac{1}{\frac{1}{\sqrt{2}-1} - 2} = \frac{1}{\frac{1-2(\sqrt{2}-1)}{\sqrt{2}-1}} = \frac{1}{\frac{3-2\sqrt{2}}{\sqrt{2}-1}} = \frac{1}{\sqrt{2} - 1} = \alpha_1,$$

where the last equality follows from $(\sqrt{2}-1)^2 = 3-2\sqrt{2}$. Thus, we have shown that $\alpha_1 = \alpha_2$ and $a_1 = a_2$. It also follows that $\alpha_3 = 1/(\alpha_2 - a_2) = 1/(\alpha_1 - a_1) = \alpha_2$, and therefore $\alpha_3 = \alpha_2 = \alpha_1$ and $a_3 = a_2 = a_1 = 2$. Hence $\alpha_k = \alpha_1$ and $a_k = a_1 = 2$ for all $k \geq 1$. We have shown that

$$\sqrt{2} = [1, 2, 2, 2, 2, 2, \ldots, 2, \ldots] = 1 + \cfrac{1}{2 + \cfrac{1}{2 + \cfrac{1}{\ddots}}}.$$

In the next section, we will explore periodic continued fractions in depth.

13.2.1. Periodic Continued Fractions.

In Example 13.2.10 we computed the continued fraction expansion of $\sqrt{2}$ and noted that it is periodic; namely, $\sqrt{2} = [1, 2, 2, 2, 2, \ldots] = [1, \overline{2}]$. The number $\sqrt{2}$ is an example of a quadratic irrational number.

Definition 13.2.11. A number $\alpha \in \mathbb{C}$ is said to be a *quadratic irrational number* if it is of the form $a + b\sqrt{d}$, for some $u, v \in \mathbb{Q}$ and a non-zero integer d that is not a perfect square. If $\alpha = u + v\sqrt{d}$, then the *conjugate* of α is $\overline{\alpha} = u - v\sqrt{d}$.

In Exercise 13.4.13 we show that a number α is a quadratic irrational if and only if there is a quadratic equation with integer coefficients whose roots are precisely α and $\overline{\alpha}$. In this section we will show that the continued fraction of any quadratic irrational number is periodic and, conversely, any periodic continued fraction corresponds to a quadratic irrational number.

Definition 13.2.12. We say that an infinite continued fraction $c = [a_0, a_1, \ldots]$ is *periodic* if there are numbers $n \geq 0$ and $m \geq 1$ such that $a_k = a_{k+m}$ for all $k \geq n$. In other words, c is of the form

$$c = [a_0, a_1, \ldots, a_{n-1}, b_1, \ldots, b_m, b_1, \ldots, b_m, \ldots]$$

where the sequence $b_1, \ldots, b_m$ repeats indefinitely. In this case, we write $c = [a_0, a_1, \ldots, a_{n-1}, \overline{b_1, \ldots, b_m}]$. If n and m are the smallest numbers with this property, then we say that $b_1, \ldots, b_m$ is the *period* of c and m is the *length of the period.* If $n = 0$, then we say that c is *purely periodic.*

Let us work out another example of computing a continued fraction of a quadratic irrational.

Example 13.2.13. Let us compute the continued fraction of $\alpha = 1 + \sqrt{6}$. Once again, we follow the notation of Theorem 13.2.8. This time, however, we will not write decimal expansions. Instead, we will simplify every quadratic number in the form $(a+b\sqrt{c})/d$. In order to compute $\lfloor \alpha_k \rfloor$ at each stage, we will only use the fact that $2 < \sqrt{6} < 3$:

$$\begin{aligned}
\alpha_0 &= 1 + \sqrt{6},\\
a_0 &= \lfloor \alpha_0 \rfloor = \lfloor 1 + \sqrt{6} \rfloor = 3,\\
\alpha_1 &= \frac{1}{\alpha_0 - a_0} = \frac{1}{1 + \sqrt{6} - 3} = \frac{1}{\sqrt{6} - 2} = \frac{2 + \sqrt{6}}{2},\\
a_1 &= \lfloor \alpha_1 \rfloor = 2,\\
\alpha_2 &= \frac{1}{\alpha_1 - a_1} = \frac{1}{\frac{2+\sqrt{6}}{2} - 2} = \frac{2}{\sqrt{6} - 2} = 2 + \sqrt{6},\\
a_2 &= \lfloor \alpha_2 \rfloor = 4,\\
\alpha_3 &= \frac{1}{\alpha_2 - a_2} = \frac{1}{2 + \sqrt{6} - 4} = \frac{1}{\sqrt{6} - 2} = \frac{2 + \sqrt{6}}{2} = \alpha_1,\\
a_3 &= \lfloor \alpha_3 \rfloor = \lfloor \alpha_1 \rfloor = a_1 = 2.
\end{aligned}$$

Thus, we have reached a repetition in the sequence; namely $\alpha_1 = \alpha_3$. It follows that

$$\alpha_4 = \frac{1}{\alpha_3 - a_3} = \frac{1}{\alpha_1 - a_1} = \alpha_2$$

and also $a_4 = a_2$. Therefore, $\alpha_{2j+1} = \alpha_1$ and $\alpha_{2j+2} = \alpha_2$ for all $j \geq 0$, and so

$$a_{2j+1} = a_1 = 2 \text{ and } a_{2j+2} = a_2 = 4$$

for all $j \geq 0$. Hence,

$$1+\sqrt{6} = [3,2,4,2,4,2,4,\ldots] = [3,\overline{2,4}].$$

Example 13.2.14. Conversely, let us calculate the value of a periodic continued fraction. Suppose that $c = [\overline{1,2,3}]$. This, in particular, means that $c = [1,2,3,c]$. Computing convergents for $[1,2,3,c]$ we find that

k	-1	0	1	2	3
a_k		1	2	3	c
p_k	1	1	3	10	$10c+3$
q_k	0	1	2	7	$7c+2$

and therefore $c = \frac{10c+3}{7c+2}$, or, equivalently, $7c^2 - 8c - 3 = 0$. Thus,

$$c = \frac{8 \pm \sqrt{64+84}}{14} = \frac{8 \pm \sqrt{148}}{14} = \frac{4 \pm \sqrt{37}}{7}.$$

Since $c > 0$, we conclude that $c = (4+\sqrt{37})/7$.

Example 13.2.15. As another example, let us calculate the value of $\alpha = [1,2,\overline{3}]$. First, we compute the value of the period, i.e., the value of $\beta = [\overline{3}]$. Note that

$$\beta = 3 + \frac{1}{\beta}$$

and therefore $\beta^2 - 3\beta - 1 = 0$. Since $\beta > 3$ we must have $\beta = \frac{3+\sqrt{13}}{2}$. Now we are ready to compute the value of α using convergents of $\alpha = [1,2,\beta]$, i.e.,

k	-1	0	1	2
a_k		1	2	β
p_k	1	1	3	$3\beta+1$
q_k	0	1	2	$2\beta+1$

Thus, we conclude

$$\begin{aligned} \alpha &= \frac{3\beta+1}{2\beta+1} = \frac{3\cdot\frac{3+\sqrt{13}}{2}+1}{2\cdot\frac{3+\sqrt{13}}{2}+1} = \frac{3\cdot(3+\sqrt{13})+2}{2\cdot(4+\sqrt{13})} \\ &= \frac{(11+3\sqrt{13})(4-\sqrt{13})}{6} = \frac{5+\sqrt{13}}{6}. \end{aligned}$$

Let us generalize the strategy of Examples 13.2.14 and 13.2.15 to show that every periodic continued fraction corresponds to a quadratic irrational number.

Theorem 13.2.16. *Let $\alpha = [a_0, a_1, \ldots, a_n, \overline{b_1, b_2, \ldots, b_m}]$ be a simple, periodic continued fraction. Then, the number α is a quadratic irrational number; i.e., $\alpha = a + b\sqrt{d}$, for some rationals a, b and a square-free integer d.*

Proof. Let us first show that $\beta = [\overline{b_1, b_2, \ldots, b_m}]$ is a quadratic irrational. Indeed, the definition of a periodic continued fraction implies an equality of the form $\beta = [b_1, b_2, \ldots, b_m, \beta]$. Let $c_k = [b_1, \ldots, b_k] = p_k/q_k$ and note that c_k is a finite simple continued fraction. Then, Theorem 13.1.9 implies that

$$\beta = [b_1, b_2, \ldots, b_m, \beta] = \frac{\beta \cdot p_k + p_{k-1}}{\beta \cdot q_k + q_{k-1}}.$$

In particular, $q_k\beta^2 + (q_{k-1} - p_k)\beta - p_{k-1} = 0$. Thus, β satisfies a quadratic equation and it is irrational (by Theorem 13.2.4), and so β is a quadratic irrational number, say of the form $\beta = a + b\sqrt{d}$ and, in particular, β belongs to the field $\mathbb{Q}(\sqrt{d})$ (see Section 12.5 to read more about these number fields, and in particular see Proposition 12.5.3 to see that $\mathbb{Q}(\sqrt{d})$ is a field).

Now we write $\alpha = [a_0, a_1, \ldots, a_n, \beta]$. Then, if $d_k = p'_k/q'_k = [a_0, a_1, \ldots, a_k]$, then Theorem 13.1.9 again implies that

$$\alpha = \frac{\beta \cdot p'_n + p'_{n-1}}{\beta \cdot q'_n + q'_{n-1}}.$$

Since $\beta \in \mathbb{Q}(\sqrt{d})$ and p'_k, q'_k are integers, it follows that α belongs to the field $\mathbb{Q}(\sqrt{d})$. Hence, α is again of the form $a' + b'\sqrt{d}$ for some rational numbers a', b'. This means that α is also a quadratic irrational number. □

The goal for the rest of this section is two-fold: first, we want to show that the converse of Theorem 13.2.16 also holds; i.e., we want to show that the continued fraction of a quadratic irrational number is periodic. Second, we want to describe a method to simplify the calculations of the continued fraction of a quadratic irrational (such as in Example 13.2.13). We begin with a preparatory lemma.

Lemma 13.2.17. *Let α be a quadratic irrational number. Then, there exist integers r and s and a positive integer d that is not a perfect square, such that s is a divisor of $d - r^2$ and $\alpha = (r + \sqrt{d})/s$.*

Proof. Suppose that α is a quadratic irrational number. Then, by Exercise 13.4.13, the number α satisfies a quadratic equation $a\alpha^2 + b\alpha + c = 0$, for some $a, b, c \in \mathbb{Z}$ with $b^2 - 4ac$ not zero and not a perfect square. Then, $\alpha = \frac{-b\pm\sqrt{b^2-4ac}}{2a}$. Let $d = b^2 - 4ac$ (in particular, d is not a perfect square and it is non-zero) and let r, s be defined by

$$(r, s) = \begin{cases} (-b, 2a) & \text{if } \alpha = (-b + \sqrt{d})/2a, \\ (b, -2a) & \text{if } \alpha = (-b - \sqrt{d})/2a. \end{cases}$$

The reader can check that the choices of d, r, and s are such that $s|(d - r^2)$ and $\alpha = (r + \sqrt{d})/s$. □

Example 13.2.18. Let $\alpha = \frac{4+\sqrt{37}}{7}$. Then, α is already in the form $(r + \sqrt{d})/s$ of Lemma 13.2.17, because $7|(37 - 16) = 21$.

Let $\beta = \frac{1+2\sqrt{5}}{3}$. Then, β is a quadratic irrational number but not in the form of Lemma 13.2.17, so let us find r, s, and d as in the lemma. First, note that

$3\beta - 1 = 2\sqrt{5}$ and therefore

$$9\beta^2 - 6\beta - 19 = 0.$$

Thus, $\beta = \frac{6+\sqrt{720}}{18}$ and we can pick $d = 720$, $r = 6$, and $s = 18$.

Remark 13.2.19. If $\alpha = \frac{n+m\sqrt{t}}{q}$ for some $n, m, t, q \in \mathbb{Z}$, then $\alpha = \frac{n+\sqrt{tm^2}}{q}$. If q is not a divisor of $tm^2 - n^2$, then we may pick

$$d = tm^2q^2, \quad r = n \cdot |q|, \quad \text{and} \quad s = q \cdot |q|,$$

and now $\alpha = (r + \sqrt{d})/s$ and s is a divisor of $(d - r^2)$, by Exercise 13.4.15. For instance, let $\alpha = (1 + 2\sqrt{5})/3$ as in Example 13.2.18. If we pick

$$d = 5 \cdot 2^2 \cdot 3^2 = 180, \quad r = 1 \cdot |3| = 3, \quad \text{and} \quad s = 3 \cdot |3| = 9,$$

then $\alpha = \frac{3+\sqrt{180}}{9}$ and 9 is a divisor of $180 - 9 = 171$.

We can now state a theorem that describes an algorithm to find expressions for $\alpha = \alpha_0$, α_1, α_2, etc., so that we can find a continued fraction for a quadratic irrational α.

Theorem 13.2.20. *Let α be a quadratic irrational number, and let α_k and a_k be as in Theorem* 13.2.8. *Let $\alpha = \alpha_0 = (r_0 + \sqrt{d})/s_0$, where d is not a perfect square and s_0 is a divisor of $d - r_0^2$. We define sequences $\{r_k\}_{k\geq 0}$ and $\{s_k\}_{k\geq 0}$ by*

$$r_{k+1} = a_k s_k - r_k \quad \text{and} \quad s_{k+1} = (d - r_{k+1}^2)/s_k.$$

Then, for all $k \geq 0$, we have $\alpha_k = \frac{r_k+\sqrt{d}}{s_k}$, with $r_k, s_k \in \mathbb{Z}$ where s_k is a non-zero-divisor of $d - r_k^2$.

Proof. As usual, we will proceed by induction on k. By Lemma 13.2.17, we can write $\alpha = \alpha_0$ as $(r_0 + \sqrt{d})/s_0$, where d is not a perfect square and r_0, s_0 are integers with $s_0|(d - r_0^2)$. This shows the case of $k = 0$.

Let us now assume that $\alpha_k = \frac{r_k+\sqrt{d}}{s_k}$, with $r_k, s_k \in \mathbb{Z}$ where s_k is a non-zero-divisor of $d - r_k^2$. Clearly, the number $r_{k+1} = a_k s_k - r_k$ is an integer, since $a_k, r_k, s_k \in \mathbb{Z}$ by the induction hypothesis and the fact that $\alpha = [a_0, a_1, \ldots]$ is simple. Moreover, $r_{k+1} \equiv -r_k \bmod s_k$. Hence,

$$d - r_{k+1}^2 \equiv d - r_k^2 \equiv 0 \bmod s_k$$

because $s_k|(d - r_k^2)$ by the induction hypothesis for k. Hence, $s_{k+1} = (d - r_{k+1}^2)/s_k$ is an integer as claimed and non-zero because d is not a perfect square. Further, the definition of s_{k+1} implies that $s_{k+1}s_k = d - r_{k+1}^2$, and therefore s_{k+1} is also a divisor of $d - r_{k+1}^2$. It remains to show that $\alpha_{k+1} = (r_{k+1} + \sqrt{d})/s_{k+1}$. Indeed,

$$\begin{aligned}
\alpha_{k+1} &= \frac{1}{\alpha_k - a_k} = \frac{1}{\frac{r_k+\sqrt{d}}{s_k} - a_k} = \frac{1}{\frac{r_k - a_k s_k + \sqrt{d}}{s_k}} \\
&= \frac{1}{\frac{-r_{k+1}+\sqrt{d}}{s_k}} = \frac{(r_{k+1} + \sqrt{d})}{\frac{(-r_{k+1}+\sqrt{d})(r_{k+1}+\sqrt{d})}{s_k}} \\
&= \frac{r_{k+1} + \sqrt{d}}{\frac{d-r_{k+1}^2}{s_k}} = \frac{r_{k+1} + \sqrt{d}}{s_{k+1}}.
\end{aligned}$$

Hence, we have proved the induction step, and this concludes the proof by induction. □

Example 13.2.21. Let us compute the continued fraction expansion of $1+\sqrt{6}$ as in Example 13.2.13, but this time using the algorithm outlined by Theorem 13.2.20. In order to do so, we set up a table similar to the one in Example 13.1.10, but this time we calculate values of r_k, s_k, and a_k, instead of p_k and q_k. We proceed as follows. We begin with $\alpha = \alpha_0 = 1+\sqrt{6}$ and we bring α_0 to the form $(r_0+\sqrt{d})/s_0$, with $s_0|(d-r_0^2)$. In our case, $d=6$, $r_0=1$, and $s_0=1$, so we are done. In each consecutive step, we will determine r_{k+1}, s_{k+1}, and $a_k = \lfloor (r_k+\sqrt{d})/s_k \rfloor$, in that order. For example, we begin with $a_0 = \lfloor 1+\sqrt{6} \rfloor = 3$ and then compute

$$r_1 = 3\cdot 1 - 1 = 2,\ s_1 = (6-4)/1 = 2, \text{ and } a_1 = \left\lfloor \frac{2+\sqrt{6}}{2} \right\rfloor = 2.$$

We continue computing terms in the sequence in the table below. In our computations of floor functions, we use that $2 < \sqrt{6} < 3$:

k	0	1	2	3	$\cdots$
r_k	1	2	2	2	$\cdots$
s_k	1	2	1	2	$\cdots$
a_k	3	2	4	2	$\cdots$

As soon as there is a repetition in a triple (r_k, s_k, a_k) we can stop. We notice that $(r_1, s_1, a_1) = (2,2,2) = (r_3, s_3, a_3)$. It follows that

$$r_4 = a_3 s_3 - r_3 = a_1 s_1 - r_1 = r_2 \text{ and } s_4 = \frac{d-r_4^2}{s_3} = \frac{d-r_2^2}{s_1} = s_2.$$

It follows also that $a_4 = \lfloor (r_4+\sqrt{d})/s_4 \rfloor = \lfloor (r_2+\sqrt{d})/s_2 \rfloor = a_2$, and we begin a periodic cycle of coefficients in the continued fraction. Hence,

$$1+\sqrt{6} = [3,2,4,2,4,2,4,\ldots] = [3,\overline{2,4}].$$

Example 13.2.22. Let us find the continued fraction of $\alpha = \frac{1+2\sqrt{5}}{3}$. Since α is not in the form $(r_0+\sqrt{d})/s_0$, we first need to bring it to that form: $\alpha = \frac{3+\sqrt{180}}{9}$ by Remark 13.2.19. Also note that $13 < \sqrt{180} < 14$; therefore $a_0 = 1$. We begin our table with the parameters $d=180$, $r_0=3$, $s_0=9$, and $a_0=1$. We obtain

k	0	1	2	3	4	5	6	7	$\cdots$
r_k	3	6	10	10	6	12	12	6	$\cdots$
s_k	9	16	5	16	9	4	9	16	$\cdots$
a_k	1	1	4	1	2	6	2	1	$\cdots$

Since $(r_1, s_1, a_1) = (6, 16, 1) = (r_7, s_7, a_7)$, we can stop here. Thus,

$$\alpha = \frac{1+2\sqrt{5}}{3} = [1,1,4,1,2,6,2,1,4,1,2,6,2,\ldots] = [1,\overline{1,4,1,2,6,2}].$$

In particular, the continued fraction is periodic (and the period is of length six). Notice that it would suffice to look for a repetition of a pair (r_k, s_k) because a_k is determined by r_k and s_k. In the following theorem we will show that every quadratic irrational number has a periodic continued fraction by proving that the integers r_k, s_k are bounded. Therefore, there are a finite number of possible combinations for (r_k, s_k) and a repetition must occur.

Before we prove our next theorem, we shall need a technical lemma.

Lemma 13.2.23. *Let α be a quadratic irrational number, and let α_k and a_k be the sequences defined in Theorem 13.2.8.*

(1) *There is some $m > 0$ such that $\overline{\alpha_m} < 0$, where $\overline{\alpha_m}$ is the conjugate of α_m (see Definition 13.2.11 and Exercise 13.4.14).*
(2) *If $m > 0$ is such that $\overline{\alpha_m} < 0$, then $-1 < \overline{\alpha_n} < 0$ for every $n > m$.*

Proof. Let $c_k = [a_0, \ldots, a_k] = \frac{p_k}{q_k}$. Then, by Theorem 13.2.8, we have

$$\alpha - \frac{p_k}{q_k} = \frac{(-1)^k}{q_k(\alpha_{k+1}q_k + q_{k-1})} \quad \text{and} \quad \overline{\alpha} - \frac{p_k}{q_k} = \frac{(-1)^k}{q_k(\overline{\alpha_{k+1}}q_k + q_{k-1})},$$

where we have used the properties of conjugates to derive the second equality from the first (see Exercise 13.4.14). Thus,

$$\left(\alpha - \frac{p_k}{q_k}\right)\left(\overline{\alpha} - \frac{p_k}{q_k}\right) = \frac{1}{q_k^2(\alpha_{k+1}q_k + q_{k-1})(\overline{\alpha_{k+1}}q_k + q_{k-1})}. \tag{13.2}$$

Since α is a quadratic irrational (in particular, not rational), we have $\alpha \neq \overline{\alpha}$. Moreover, $c_k = p_k/q_k$ converges to α by Theorem 13.2.1, and $c_{2k} < \alpha < c_{2j+1}$ for any $k, j \geq 0$. Therefore, there is some $k \geq 0$ such that c_k is strictly between α and $\overline{\alpha}$. For such k, the quantity $(\alpha - c_k)(\overline{\alpha} - c_k)$ is negative. Since $q_k \geq 1$ and $\alpha_{k+1} > 0$ for any $k \geq 0$ (recall $a_k \geq 1$ for all $k \geq 1$), it follows from (13.2) that $\overline{\alpha_{k+1}}q_k + q_{k-1} < 0$, and therefore $\alpha_{k+1} < 0$. Thus, $m = k+1$ works and proves (1).

For (2), it suffices to show that if $\overline{\alpha_m} < 0$, then $-1 < \overline{\alpha_{m+1}} < 0$. By definition $\alpha_{m+1} = 1/(\alpha_m - a_m)$, and therefore $\overline{\alpha_{m+1}} = 1/(\overline{\alpha_m} - a_m) < 0$. Since $a_m \geq 1$ for $m \geq 1$, it follows that $\overline{\alpha_m} - a_m < -1$ and it follows that $-1 < \overline{\alpha_{m+1}} < 0$, as desired. □

The following theorem is originally due to Lagrange (see Figure 5.1).

Theorem 13.2.24. *The continued fraction expansion of a quadratic irrational number is periodic.*

Proof. Let α be a quadratic irrational number and let $\{\alpha_k\}$ be the sequence defined in Theorem 13.2.8. For each $k \geq 0$, we shall write $\alpha_k = (r_k + \sqrt{d})/s_k$ as in Theorem 13.2.20. In order to prove the theorem, we will first show that there is an integer $m > 0$ such that r_k and s_k are uniformly bounded for all $k > m$ (i.e., independently of k).

By Lemma 13.2.23, there is an integer $m > 0$ such that $-1 < \overline{\alpha_k} < 0$ for all $k > m$. Since $\alpha_k = [a_k, a_{k+1}, \ldots] > 1$ for all $k \geq 1$ (because $a_k \geq 1$ for $k \geq 1$) and if we write $\alpha_k = (r_k + \sqrt{d})/s_k$, then

$$1 < \alpha_k - \overline{\alpha_k} = \frac{2\sqrt{d}}{s_k}$$

for all $k > m$. In particular, it follows that $0 < s_k < 2\sqrt{d}$, and so s_k is bounded. Moreover,

$$\alpha_k \cdot \overline{\alpha_k} = \frac{r_k + \sqrt{d}}{s_k} \cdot \frac{r_k - \sqrt{d}}{s_k} = \frac{r_k^2 - d}{s_k^2}.$$

Hence, if $k > m$, then $\alpha_k \cdot \overline{\alpha_k} < 0$, and so $r_k^2 - d < 0$ as well. Hence, $-\sqrt{d} < r_k < \sqrt{d}$. Moreover, $\alpha_k > 1$ and $\overline{\alpha_k} > -1$ and so

$$0 < \alpha_k + \overline{\alpha_k} = \frac{2r_k}{s_k}.$$

This means that $r_k > 0$, and therefore $0 < r_k < \sqrt{d}$.

Hence, we have shown that if $k > m$, then there are at most $2d$ possibilities for the pair (r_k, s_k). Since the sequence $\{(r_k, s_k)\}_{k>m}$ is infinite, there must be a repetition, i.e., $(r_j, s_j) = (r_{j+n}, s_{j+n})$ for some $n \geq 1$, and therefore $(r_j, s_j, a_j) = (r_{j+n}, s_{j+n}, a_{j+n})$ since a_j is determined by r_j and s_j. Now, by the recurrence relations that define r_k and s_k, we see that

$$(r_{j+t}, s_{j+t}, a_{j+t}) = (r_{j+n+t}, s_{j+n+t}, a_{j+n+t})$$

for all $t \geq 0$, and therefore the continued fraction is periodic, as desired. □

In the proof of Theorem 13.2.24 we first had to find a value of m such that $-1 < \overline{\alpha_k} < 0$ for all $k > m$. It turns out that if $\alpha > 1$ is a quadratic irrational and $-1 < \overline{\alpha} < 0$, then the continued fraction of α is *purely* periodic and, in fact, the converse is also true.

Example 13.2.25. In order to illustrate the phenomenon to be proved in the next proposition, let us consider $\alpha = (4 + \sqrt{37})/7$. In Example 13.2.14 we showed that $\alpha = [\overline{1, 2, 3}]$. Let us compute the continued fraction of $\beta = -1/\overline{\alpha}$. First, note that

$$\beta = -\frac{1}{\overline{\alpha}} = -\frac{1}{\frac{4-\sqrt{37}}{7}} = -\frac{7 \cdot (4 + \sqrt{37})}{(4 - \sqrt{37})(4 + \sqrt{37})} = \frac{4 + \sqrt{37}}{3}.$$

Since 3 is a divisor of $37 - 4^2 = 21$, we can use the algorithm of Theorem 13.2.20, where $r_0 = 4$, $s_0 = 3$, and $d = 37$ (and $6 < \sqrt{37} < 7$):

k	0	1	2	3	$\cdots$
r_k	4	5	3	4	$\cdots$
s_k	3	4	7	3	$\cdots$
a_k	3	2	1	3	$\cdots$

We see that $(r_0, s_0) = (r_3, s_3)$, and therefore the continued fraction is periodic with length of period equal to 3. Thus, $\beta = -1/\overline{\alpha} = [\overline{3, 2, 1}]$ while $\alpha = [\overline{1, 2, 3}]$.

Definition 13.2.26. A quadratic irrational number α is said to be *reduced* if $\alpha > 1$ and $-1 < \overline{\alpha} < 0$.

Proposition 13.2.27. *Let α be a quadratic irrational number, and let α_k be the sequence defined in Theorem 13.2.8. Then:*

(1) *If α is reduced, then α_k is reduced for all $k \geq 0$.*

(2) *If $\alpha = [\overline{a_0, a_1, \ldots, a_{m-1}}]$ is purely periodic, then $-1/\overline{\alpha} = [\overline{a_{m-1}, \ldots, a_1, a_0}]$.*

(3) *The number* α *has a purely periodic continued fraction (i.e., a continued fraction of the form* $[\overline{a_0, \ldots, a_{n-1}}]$*) if and only if* α *is reduced.*

Proof. If α is reduced, then $\alpha_0 = \alpha$ is reduced and, in particular, $\overline{\alpha_k} < 0$. Then, by Lemma 13.2.23, the numbers α_k are reduced for all $k \geq 0$ (note that α_k is always greater than 1 for all $k \geq 1$). This proves (1).

For (2), the equation $\alpha_{k+1} = 1/(\alpha_k - a_k)$ implies that $\overline{\alpha_{k+1}} = 1/(\overline{\alpha_k} - a_k)$ and, in turn, this is equivalent to

$$\overline{\alpha_k} = a_k + \frac{1}{\overline{\alpha_{k+1}}}.$$

If we put $\beta_k = -1/\overline{\alpha_k}$, then $-1/\beta_k = a_k - \beta_{k+1}$, or, equivalently, $\beta_{k+1} = a_k + 1/\beta_k$. Now suppose that $\alpha = \alpha_0 = [\overline{a_0, \ldots, a_{m-1}}]$. We will show that $\beta_0 = -1/\overline{\alpha_0} = [\overline{a_{m-1}, \ldots, a_0}]$. If $m = 1$, then $\alpha = [\overline{a_0}]$, and $\alpha = a_0 + \frac{1}{\alpha}$. Thus,

$$-\frac{1}{\alpha} = a_0 - \alpha = a_0 + \frac{1}{-\frac{1}{\alpha}}.$$

Taking conjugates, we obtain $\beta_0 = a_0 + 1/\beta_0$, and therefore $\beta_0 = [\overline{a_0}]$, as desired. Now let us assume that $m > 1$. Since the expansion is periodic of length m, we have $\alpha_0 = \alpha_m$ and in particular $a_0 \geq 1$, and so $a_k \geq 1$ for all $k \geq 0$. Moreover, if we use the equation $\beta_{k+1} = a_k + 1/\beta_k$ for $k = m - 1$ and $k = m - 2$, we obtain

$$\beta_0 = \beta_m = a_{m-1} + \frac{1}{\beta_{m-1}} = a_{m-1} + \cfrac{1}{a_{m-2} + \cfrac{1}{\beta_{m-2}}} = [a_{m-1}, a_{m-2}, \beta_{m-2}].$$

If we repeat this process, we obtain

$$\beta_0 = [a_{m-1}, a_{m-2}, \ldots, a_0, \beta_0] = [\overline{a_{m-1}, a_{m-2}, \ldots, a_0}],$$

as claimed.

For (3), let us first assume that α is reduced. Since α_k is reduced by part (1), we know that $-1 < \overline{\alpha_k} < 0$, and therefore $0 < 1/\beta_k < 1$. Hence, using the equation $\beta_{k+1} = a_k + 1/\beta_k$ from part (2), we see that $\lfloor \beta_{k+1} \rfloor = a_k$. Moreover, by Theorem 13.2.24, the continued fraction for α is periodic with period of, say, length $m \geq 1$. This means that $\alpha_{m+j} = \alpha_j$ for some $j \geq 0$. Let j be minimal with this property. If $j = 0$, then we are done because this means the continued fraction is purely periodic. Otherwise, suppose $j > 0$ and j is minimal with the property $\alpha_{m+j} = \alpha_j$. Taking conjugates, we also obtain $\beta_{m+j} = \beta_j$. In particular, if $j > 0$, we have

$$a_{j-1} + \frac{1}{\beta_{j-1}} = \beta_j = \beta_{m+j} = a_{m+j-1} + \frac{1}{\beta_{m+j-1}}.$$

It follows that $a_{j-1} = \lfloor \beta_j \rfloor = \lfloor \beta_{m+j} \rfloor = a_{m+j-1}$ and, therefore, $1/\beta_{j-1} = 1/\beta_{m+j-1}$. But this means that $\alpha_{j-1} = \alpha_{m+j-1}$, which contradicts the minimality of j. Hence, we have reached a contradiction with $j > 0$, and we must have $j = 0$; i.e., the continued fraction is purely periodic.

For the converse, assume that $\alpha = [\overline{a_0, \ldots, a_{m-1}}]$ is purely periodic. First note that $a_0 = a_m \geq 1$, and therefore $\alpha > 1$. Moreover, part (2) implies that the continued fraction of $-1/\overline{\alpha}$ is also purely periodic, and in particular $-1/\overline{\alpha} > 1$. This implies that $-1 < \overline{\alpha} < 0$, as needed. □

Example 13.2.28. Let $\alpha = \sqrt{7}$. Then, $\overline{\alpha} = -\sqrt{7}$ is between -3 and -2. Thus, $-1 < 2 - \sqrt{7} < 0$, and $\alpha' = 2 + \sqrt{7}$ is reduced. Hence, Proposition 13.2.27 says that the continued fraction of α' must be purely periodic. Indeed,

$$2 + \sqrt{7} = [\overline{4, 1, 1, 1}].$$

It follows that $\sqrt{7} = [2, \overline{1, 1, 1, 4}]$.

In the next section we will investigate the continued fraction of $\sqrt{d}$ for any positive integer d that is not a perfect square.

13.2.2. The Continued Fraction of $\sqrt{d}$.

Let us begin with an example.

Example 13.2.29. Let $\alpha = \sqrt{19}$. We may apply our algorithm in Theorem 13.2.20 starting with $r_0 = 0$, $s_0 = 1$, and $d = 19$:

k	0	1	2	3	4	5	6	7	$\cdots$
r_k	0	4	2	3	3	2	4	4	$\cdots$
s_k	1	3	5	2	5	3	1	3	$\cdots$
a_k	4	2	1	3	1	2	8	2	$\cdots$

Since $(r_1, s_1) = (4, 3) = (r_7, s_7)$, we may stop here and conclude that

$$\sqrt{19} = [4, \overline{2, 1, 3, 1, 2, 8}].$$

The number $\sqrt{19}$ is not reduced (as in Definition 13.2.26), but $4 + \sqrt{19}$ is reduced and, indeed, the continued fraction of $4 + \sqrt{19}$ is purely periodic:

$$4 + \sqrt{19} = [\overline{8, 2, 1, 3, 1, 2}] = [8, 2, 1, 3, 1, 2, 8, 2, 1, 3, 1, 2, 8, \ldots].$$

The reader has probably noticed that there is a certain symmetry to the coefficients of the continued fraction (the expansion is a palindrome centered at 3 if we ignore the number 8). The same symmetry occurs for the continued fraction of a quadratic irrational of the form $\sqrt{d}$; for instance,

$$\sqrt{6} = [2, \overline{2, 4}], \ \sqrt{15} = [3, \overline{1, 6}], \ \sqrt{23} = [4, \overline{1, 3, 1, 8}], \ \sqrt{29} = [5, \overline{2, 1, 1, 2, 10}].$$

Let us see why this is so.

Proposition 13.2.30. *Let d be a positive integer that is not a perfect square, and let $a_0 = \lfloor \sqrt{d} \rfloor$. Then, the simple continued fraction of $\sqrt{d}$ is given by*

$$\sqrt{d} = [a_0, \overline{a_1, a_2, \ldots, a_2, a_1, 2a_0}].$$

In other words, $\sqrt{d} = [a_0, \overline{a_1, \ldots, a_m}]$, with $a_m = 2a_0$ and $a_i = a_{m-i}$ for all $1 \leq i \leq m-1$.

Proof. Let $d \geq 2$ be an integer that is not a perfect square, let $\alpha = \sqrt{d}$, and put $a_0 = \lfloor \sqrt{d} \rfloor$. Then, $\beta = a_0 + \alpha$ is reduced, because $\beta > 1$ and $\overline{\beta} = a_0 - \sqrt{d} = \lfloor \sqrt{d} \rfloor - \sqrt{d}$ is between -1 and 0 by the definition of the floor function. It follows from Proposition 13.2.27 that the (simple) continued fraction of β is purely periodic, say $\beta = [\overline{b_0, \ldots, b_{m-1}}]$. It follows that

$$[\overline{b_0, \ldots, b_{m-1}}] = \beta = a_0 + \alpha = [2a_0, a_1, \ldots, a_{m-1}, \ldots]$$

and so $\sqrt{d} = [a_0, \overline{a_1, \ldots, a_m}]$ with $b_0 = a_m = 2a_0$ and $a_i = b_i$ for $1 \leq i \leq m$. It remains to show the symmetry property $a_i = a_{m-i}$ for all $1 \leq i \leq m-1$.

By Proposition 13.2.27, we know that

$$-1/\overline{\beta} = [a_{m-1}, a_{m-2}, \ldots, a_1, 2a_0].$$

Moreover, $-1/\overline{\beta} = -1/(a_0 - \alpha) = 1/(\alpha - a_0)$. Also, from the expansion for $\alpha = \sqrt{d}$ we deduce that $\alpha - a_0 = [0, \overline{a_1, \ldots, a_{m-1}, 2a_0}]$. In particular

$$[a_{m-1}, a_{m-2}, \ldots, a_1, 2a_0] = -1/\overline{\beta} = 1/(\alpha - a_0) = [\overline{a_1, \ldots, a_{m-1}, 2a_0}].$$

Hence, $a_1 = a_{m-1}$, $a_2 = a_{m-2}$, and $a_i = a_{m-i}$ for all $1 \leq i \leq m-1$, as desired. □

The proposition we just proved says that the period of $\sqrt{d}$ begins with a_1 and ends with some $a_m = 2a_0$. In the next proposition we shall prove that the very first time we have $a_k = 2a_0$ is actually the end of the period. First, we need a lemma.

Lemma 13.2.31. *Let d be a positive integer that is not a perfect square, let $\alpha = \sqrt{d}$, and let a_k, α_k, r_k, and s_k be defined as in Theorem 13.2.20. Then:*

(1) *For all $k \geq 1$, we have $0 < r_k < \sqrt{d}$ and $0 < s_k < 2\sqrt{d}$.*

(2) *$s_k = 1$ if and only if k is a positive multiple of m, where m is the length of the period of the continued fraction of $\sqrt{d}$.*

Proof. Part (1) was proved already in the proof of Theorem 13.2.24. For (2), let m be the length of the period of the continued fraction of $\sqrt{d}$, and let us assume first that k is a positive multiple of m. By Proposition 13.2.30, the continued fraction of $\sqrt{d}$ is of the form

$$\sqrt{d} = [a_0, \overline{a_1, a_2, \ldots, a_{m-1}, 2a_0}].$$

By the definition of α_m, we have $\sqrt{d} = [a_0, \ldots, a_{m-1}, \alpha_m]$, and from the shape of the period of $\sqrt{d}$ we see that $\alpha_m = [\overline{2a_0, a_1, \ldots, a_{m-1}}]$, which equals $\sqrt{d} + a_0$. Similarly, $\sqrt{d} + a_0 = \alpha_m = \alpha_{2m} = \cdots = \alpha_{jm}$ for all $j \geq 1$. This shows that if k is a positive multiple of m, then $\alpha_k = a_0 + \sqrt{d}$ and therefore $r_k = a_0$ and $s_k = 1$.

For the converse, suppose that $s_k = 1$ for some $k \geq 1$. The shape of the continued fraction of $\sqrt{d}$ implies that every α_k, for $k \geq 1$, has a purely periodic continued fraction. Thus, Proposition 13.2.27 says that α_k must be reduced, and therefore $-1 < \overline{\alpha_k} < 0$. If $\alpha_k = (r_k + \sqrt{d})/s_k$ with $s_k = 1$, then $\overline{\alpha_k} = (r_k - \sqrt{d})/s_k = r_k - \sqrt{d}$. It follows that $-1 < r_k - \sqrt{d} < 0$, or, equivalently, that $0 < \sqrt{d} - r_k < 1$, and so r_k must be $\lfloor\sqrt{d}\rfloor$. In other words,

$$\alpha_k = \lfloor\sqrt{d}\rfloor + \sqrt{d} = a_0 + \sqrt{d} = [\overline{2a_0, a_1, \ldots, a_{m-1}}].$$

On the other hand, from the continued fraction of $\sqrt{d}$, we know that the continued fraction of α_k is of the form

$$\alpha_k = [\overline{a_k, a_{k+1}, \ldots, a_{m-1}, 2a_0, a_1, \ldots, a_{k-1}}].$$

But then we would have $a_1 = a_{k+1}$, $a_2 = a_{k+2}$, and $a_j = a_{j+k}$ for all $j \geq 1$. Since the length of the period of $\sqrt{d}$ is m, the minimality of m (see Definition 13.2.12) implies that k must be a multiple of m, and the proof is complete. □

Proposition 13.2.32. *Let d be a positive integer that is not a perfect square, and suppose that the continued fraction of $\sqrt{d}$ is of the form*

$$\sqrt{d} = [a_0, \overline{a_1, a_2, \ldots, a_2, a_1, 2a_0}].$$

Then, $a_k \leq a_0$ for all $k \leq m - 1$. In particular, $a_k = 2a_0$ if and only if k is a positive multiple of m.

Proof. Since $a_0 \leq a_0$, we can assume $1 \leq k \leq m - 1$. Let $\alpha = \sqrt{d}$, and let a_k, α_k, r_k, and s_k be defined as in Theorem 13.2.20. By Lemma 13.2.31 we have $2 \leq s_k < 2\sqrt{d}$ and $0 < r_k < \sqrt{d}$. Thus,

$$\alpha_k = \frac{r_k + \sqrt{d}}{s_k} < \frac{\sqrt{d} + \sqrt{d}}{2} = \sqrt{d}.$$

It follows that $a_k = \lfloor \alpha_k \rfloor \leq \lfloor \sqrt{d} \rfloor = a_0$, as desired. □

Example 13.2.33. The statement of Proposition 13.2.32 is exemplified in the continued fractions we already found in Example 13.2.29; namely

$$\sqrt{6} = [2, \overline{2, 4}], \quad \sqrt{15} = [3, \overline{1, 6}], \quad \sqrt{23} = [4, \overline{1, 3, 1, 8}], \quad \sqrt{29} = [5, \overline{2, 1, 1, 2, 10}].$$

13.3. Approximations of Irrational Numbers

In the introduction to this chapter we mentioned the well-known approximation $\pi \approx 22/7$, and in Example 13.2.7 we calculate the continued fraction expansion for π, from which one can deduce other approximations using convergents, such as

$$\frac{22}{7}, \ \frac{333}{106}, \ \frac{355}{113}, \quad \text{or} \quad \frac{103993}{33102}.$$

In this section we want to address the following question: what can we say about the approximations that appear as convergents of π? There are, of course, other approximations of π, such as $\frac{31}{10}$, or $\frac{314}{100} = \frac{157}{50}$. Why is $\frac{157}{50}$ not a convergent of π? More generally, given a rational approximation m/n of an irrational number α, can we decide whether m/n is a convergent of α? In our first result in this section, we show that convergents are some of the most "economical" approximations, in terms of the size of the denominator.

Proposition 13.3.1. *Let $k \geq 0$ and let $c_k = p_k/q_k$ and $c_{k+1} = p_{k+1}/q_{k+1}$ be, respectively, the kth and $(k+1)$th convergents of an irrational number α. Then, for every $k \geq 0$ we have*

(1) $|c_k - c_{k+1}| = \dfrac{1}{q_k q_{k+1}}$,

(2) $|\alpha - c_k| < \dfrac{1}{q_k q_{k+1}}$,

(3) $|\alpha - c_k| < 1/q_k^2$, *and*

(4) $|q_{k+1}\alpha - p_{k+1}| < |q_k \alpha - p_k|$.

Proof. Let $k \geq 0$ be fixed. Part (1) follows directly from (13.1) in the proof of Theorem 13.1.15. Moreover, Theorem 13.2.1 says that either $c_k < \alpha < c_{k+1}$ or $c_{k+1} < \alpha < c_k$ according to the parity of k. Thus, $|\alpha - c_k| < |c_k - c_{k+1}|$ and (2) follows from (1). Finally, since the sequence $\{q_k\}_{k\geq -1}$ is increasing, it follows that $|\alpha - c_k| < 1/q_k^2$, as claimed by (3).

For (4), we use Theorem 13.2.8 to write

$$\left|\alpha - \frac{p_{k+1}}{q_{k+1}}\right| = \frac{1}{q_{k+1}(\alpha_{k+2}q_{k+1} + q_k)}.$$

Multiplying throughout by q_{k+1} yields $|q_{k+1}\alpha - p_{k+1}| = 1/(\alpha_{k+2}q_{k+1} + q_k)$. Since $a_j = \lfloor \alpha_j \rfloor \geq 0$ for $j \geq 1$, we have

$$\begin{aligned} \alpha_{k+2}q_{k+1} + q_k &> a_{k+2}q_{k+1} + q_k \geq q_{k+1} + q_k = (a_{k+1}q_k + q_{k-1}) + q_k \\ &= (a_{k+1} + 1)q_k + q_{k-1} > \alpha_{k+1}q_k + q_{k-1}. \end{aligned}$$

It follows that

$$|q_{k+1}\alpha - p_{k+1}| = \frac{1}{\alpha_{k+2}q_{k+1} + q_k} < \frac{1}{\alpha_{k+1}q_k + q_{k-1}} = |q_k\alpha - p_k|.$$

This shows (4) and completes the proof of the proposition. □

Example 13.3.2. Let $\alpha = \pi$. Then, the first convergents are

$$c_0 = 3,\ c_1 = \frac{22}{7},\ c_2 = \frac{333}{106},\ c_3 = \frac{355}{113}, \ldots.$$

The distance from π to $c_1 = 22/7$ is

$$|\pi - c_1| = \left|\pi - \frac{22}{7}\right| = 0.00126448\ldots$$

and $1/7^2 = 1/49 = 0.02040816\ldots$, so it follows that $|\pi - 22/7| < 1/7^2$ as we expected from Proposition 13.3.1. Similarly, $|\pi - 333/106| = 0.00008321\ldots$ is smaller than the theoretical bound of $1/106^2 = 0.00008899\ldots$.

Is $31/10$ a convergent of π? Suppose it was. Then, Proposition 13.3.1 would imply that $|\pi - 31/10| < 1/10^2$. However, $|\pi - 31/10| = 0.04159265\ldots$ and $1/10^2 = 0.01$. Thus, the bound of $|\pi - 31/10| < 1/10^2$ is not satisfied. It follows that $31/10$ cannot be a convergent. Similarly, $|\pi - 314/100| = |\pi - 157/50| = 0.00159265\ldots$ which is larger than $1/50^2 = 0.0004$. Hence, $314/100 = 157/50$ cannot be a convergent.

In the following result, we show that if an approximation is "economical enough", then it is in fact a convergent. The key point is to spell out what "economical" means in this context.

Theorem 13.3.3. *Let α be an irrational number, and suppose that s/t is a rational number with $t > 1$ and with the property that, for all integers u, v,*

$$|t\alpha - s| < |v\alpha - u| \quad \textit{whenever } 1 \leq v < t. \tag{13.3}$$

Then, s/t is a convergent of α.

Proof. Let α be irrational, and let $c_k = p_k/q_k$, for each $k \geq 0$, be the convergents associated to the infinite continued fraction of α. Recall that the convergents are ordered as

$$c_0 < c_2 < \cdots < c_{2j} < \cdots < \alpha < \cdots < c_{2j+1} < \cdots < c_3 < c_1$$

by Theorems 13.1.15 and 13.2.1. Let $s/t \in \mathbb{Q}$ be as in the statement of the theorem. Let us assume for a contradiction that $s/t \neq c_k$ for all $k \geq 0$.

Suppose first that $s/t < c_0$. Then,

$$|t\alpha - s| \geq \frac{1}{t} \cdot |t\alpha - s| = \left|\alpha - \frac{s}{t}\right| > |\alpha - c_0| = |q_0\alpha - p_0|$$

where we have used the fact that $c_0 = p_0/q_0$ and $q_0 = 1 < t$. This contradicts the property in equation (13.3) and so we must have $s/t > c_0$ instead.

If $s/t > c_1$, then

$$|t\alpha - s| > |tc_1 - s| = \frac{1}{q_1} \cdot |tp_1 - sq_1| \geq \frac{1}{q_1} > |q_0\alpha - p_0|$$

where we have used $s/t \neq c_1$ to bound $|tp_1 - sq_1| \geq 1$ and Proposition 13.3.1, part (2), for the inequality $|q_0\alpha - p_0| < 1/q_1$. The inequality $|t\alpha - s| > |q_0\alpha - p_0|$ contradicts (13.3) and so we must have $c_0 < s/t < c_1$.

It follows that s/t is located between two convergents c_{n-1} and c_{n+1} for some $n \geq 1$. Thus, $|s/t - c_{n-1}| < |c_n - c_{n-1}|$. If we multiply this inequality through by tq_nq_{n-1}, then we obtain

$$q_n|sq_{n-1} - tp_{n-1}| < t|p_nq_{n-1} - p_{n-1}q_n| = t,$$

where we have used Proposition 13.1.11 to show that $|p_nq_{n-1} - p_{n-1}q_n| = 1$. Since $s/t \neq c_{n-1}$, it follows that $sq_{n-1} - tp_{n-1} \neq 0$ and therefore $q_n < t$. Moreover, $|\alpha - s/t| > |c_{n+1} - s/t|$ and so

$$|t\alpha - s| > \frac{1}{q_{n+1}} \cdot |tp_{n+1} - sq_{n+1}| \geq \frac{1}{q_{n+1}} > |q_n\alpha - p_n|$$

where the last inequality comes from Proposition 13.3.1, part (2). Since we have shown that $q_n < t$, the inequality $|t\alpha - s| > |q_n\alpha - p_n|$ contradicts (13.3). Thus, it is impossible for s/t to be between c_{n-1} and c_{n+1} and we have reached a contradiction. □

Theorem 13.3.3 has the following important corollaries, which imply, in particular, that the convergents are the "most economical" rational approximations in terms of the size of their numerators.

Corollary 13.3.4. *Let α be an irrational number, let $c_k = p_k/q_k$ be the convergents of α, and let s/t be a rational approximation of α, with $t > 1$.*

(1) *If $|t\alpha - s| < |q_k\alpha - p_k|$, then $d \geq q_{k+1}$.*
(2) *If $|\alpha - s/t| < |\alpha - p_k/q_k|$ for some $k \geq 1$, then $d > q_k$.*

Proof. Let $d \geq 2$ be the smallest positive integer for which there is an integer c such that $|d\alpha - c| < |q_k\alpha - p_k|$. Theorem 13.3.3 implies that $c/d = c_n$ for some $n \geq 0$. Since the convergent c/d is closer to α than $c_k = p_k/q_k$, Proposition 13.3.1 implies that $n > k$. Thus, $d \geq q_{k+1}$. Since d is the smallest such value, it follows that $t \geq d \geq q_{k+1}$, as claimed in (1).

For (2), suppose for a contradiction that $|\alpha - s/t| < |\alpha - p_k/q_k|$ but we have $t \leq q_k$. Multiplying by t on the left and by q_k on the right leads to

$$|t\alpha - s| \leq |q_k\alpha - p_k|.$$

Then, by part (1), we have $t \geq q_{k+1}$. Hence, we would have $q_k \geq t \geq q_{k+1}$, but this is impossible by Theorem 13.1.9 because $q_k < q_{k+1}$ for all $k \geq 1$. This shows (2). □

We end this section with a bound that assures that a rational approximation is a convergent, which does not rely on a comparison with any other convergent.

Theorem 13.3.5. *Let α be an irrational number, and let s/t be a rational approximation of α, with $t \geq 1$, and such that*

$$\left|\alpha - \frac{s}{t}\right| < \frac{1}{2t^2}.$$

Then, s/t is one of the convergents of the infinite continued fraction of α.

Proof. If $t = 1$ and $|\alpha - s| < 1/2$, then $s = \lfloor\alpha\rfloor$ or $\lfloor\alpha\rfloor + 1$, and the result follows from Exercise 13.4.11, so we shall assume $t > 1$. We shall apply Theorem 13.3.3. To do so, we need to show that $|t\alpha - s| < |v\alpha - u|$ whenever $1 \leq v < t$, or, equivalently, that if $|v\alpha - u| \leq |t\alpha - s|$, then $t \leq v$. Suppose the latter occurs for some $u/v \neq s/t$. Then,

$$\left|\frac{u}{v} - \frac{s}{t}\right| \leq \left|\alpha - \frac{u}{v}\right| + \left|\alpha - \frac{s}{t}\right|$$

by the triangle inequality. By assumption, $|t\alpha - s| < 1/(2t)$ and this, together with $|v\alpha - u| \leq |t\alpha - s|$, implies that $|\alpha - u/v| < 1/(2tv)$. Thus,

$$\left|\frac{u}{v} - \frac{s}{t}\right| \leq \left|\alpha - \frac{u}{v}\right| + \left|\alpha - \frac{s}{t}\right| < \frac{1}{2tv} + \frac{1}{2t^2}.$$

If we multiply throughout by tv, we obtain $|tu - sv| < 1/2 + v/(2t)$. But $u/v \neq s/t$ implies that $|tu - sv| \geq 1$ and therefore $v/(2t) \geq 1/2$, and so $t \leq v$, as desired. □

Example 13.3.6. Let $\alpha = \pi$ and consider the rational approximation $22/7$. Then, $|\pi - 22/7| = 0.0016244\ldots$ and $1/(2 \cdot 49) = 0.010204081\ldots$. Thus,

$$\left|\pi - \frac{22}{7}\right| < \frac{1}{2 \cdot 7^2}$$

and $22/7$ must be a convergent of π, by Theorem 13.3.5.

However, the bound in Theorem 13.3.5 is not a necessary condition; i.e., the theorem cannot be extended to an if and only if statement. For instance, take $333/106$ which is another convergent of π. Then $|\pi - 333/106| = 0.0000832\ldots$ but $1/(2 \cdot 106^2) = 0.0000444\ldots$ is smaller. So not every convergent of α satisfies the bound of Theorem 13.3.5.

13.4. Exercises

Exercise 13.4.1. Find the continued fraction expansion for each of the following rational numbers:

$$\text{(a)}\ \frac{35}{31}, \quad \text{(b)}\ \frac{17}{31}, \quad \text{(c)}\ -\frac{17}{31}, \quad \text{(d)}\ \frac{101}{37}.$$

Exercise 13.4.2. Compute the rational numbers with the following continued fraction expansions:

$$\text{(a)}\ [1,1,1], \quad \text{(b)}\ [1,1,1,1], \quad \text{(c)}\ [1,2,1], \quad \text{(d)}\ [-2,1,2,1].$$

Exercise 13.4.3. Find the convergents $c_0, \ldots, c_5$ for the rational number with continued fraction expansion $[1, 2, 3, 4, 5, 6]$.

Exercise 13.4.4. Find the continued fraction expansion of $q = 225/157$ and then use a table (as in Example 13.1.10) to compute the convergents of q.

Exercise 13.4.5. Let $c = [a_0, a_1, \ldots, a_n]$ be a simple continued fraction, and define sequences p_k and q_k as in Theorem 13.1.9. Show that $q_k \geq 2^{k/2}$ for all $k \geq 2$, as follows:

(1) Show the result for $k = 2$ and $k = 3$.
(2) Show that $q_{k+2} > 2q_k$ for all $k \geq 1$.
(3) Show that if $q_k \geq 2^{k/2}$, then $q_{k+2} \geq 2^{(k+2)/2}$.
(4) Use induction to prove $q_k \geq 2^{k/2}$ for all even integers $k \geq 2$ (respectively, all odd integers $k \geq 3$).

Exercise 13.4.6. Let a and b be relatively prime natural numbers, let $[a_0, \ldots, a_n]$ be the continued fraction expansion of a/b, and let $c_k = p_k/q_k$ be the kth convergent of a/b. Show that

$$aq_{n-1} - bp_{n-1} = (-1)^{n-1}.$$

In other words, $(q_{n-1}, -p_{n-1})$ is an integer solution of $ax + by = (-1)^{n-1}$, and $(-q_{n-1}, p_{n-1})$ is a solution of $ax + by = (-1)^n$. (Hint: use Proposition 13.1.11.)

Exercise 13.4.7. Use Exercises 13.4.4 and 13.4.6 to find an integer point on the line L: $225x + 157y = 1$, and then determine/parametrize all the integral points on L.

Exercise 13.4.8. Find an infinite simple continued fraction $\alpha = [a_0, a_1, \ldots]$ such that $0.001 < \alpha < 0.002$.

Exercise 13.4.9. Find the first six coefficients of the (infinite) continued fraction expansion of the following numbers, using the methods of Section 13.2:

$$\text{(a) } \pi^2, \quad \text{(b) } e + \pi, \quad \text{(c) } \sqrt{2} + \sqrt{3}, \quad \text{(d) } \sqrt{2 + \sqrt{2}}, \quad \text{(e) } \log(2).$$

Exercise 13.4.10. Use Cantor's diagonalization argument (and Proposition 13.2.6) to show that the set of values of all infinite simple continued fractions is uncountable.

Exercise 13.4.11. Let α be an irrational number such that $\alpha - \lfloor \alpha \rfloor > 1/2$.

(1) Show that the integers $\lfloor \alpha \rfloor$ and $\lfloor \alpha \rfloor + 1$ are the first two convergents of α.
(2) Find a square-free number $d > 0$ such that $\sqrt{d} - \lfloor \sqrt{d} \rfloor > 1/2$, and determine the continued fraction expansion of $\sqrt{d}$.

Exercise 13.4.12. Let α be an irrational number with $\alpha > 1$ and an infinite continued fraction expansion $\alpha = [a_0, a_1, a_2, \ldots]$.

(1) Show that $\frac{1}{\alpha} = [0, a_0, a_1, a_2, \ldots]$.
(2) Let $c_k = p_k/q_k$ and $C_k = P_k/Q_k$ be, respectively, the kth convergent of α and $1/\alpha$. Show that $C_k = P_k/Q_k = q_{k-1}/p_{k-1} = 1/c_{k-1}$.

Exercise 13.4.13. By definition, a number $\alpha \in \mathbb{C}$ is a quadratic irrational number if it is of the form $u + v\sqrt{d}$, for some $u, v \in \mathbb{Q}$ and a non-zero integer d that is not a perfect square. Show that α is a quadratic irrational if and only if there are integers $a, b, c \in \mathbb{Z}$ such that $a\alpha^2 + b\alpha + c = 0$ where $b^2 - 4ac \neq 0$ is not a perfect square. Moreover, show that α and $\overline{\alpha}$ (the conjugate of α, as in Definition 13.2.11) satisfy the same quadratic equation.

Exercise 13.4.14. Let u, v be rational numbers and let d be a non-zero integer that is not a perfect square. The conjugate of $\alpha = u + v\sqrt{d}$ is defined to be $\overline{\alpha} = u - v\sqrt{d}$. Prove the following properties of conjugates for quadratic irrational numbers $\alpha = u + v\sqrt{d}$ and $\beta = x + y\sqrt{e}$, where $u, v, x, y \in \mathbb{Q}$ and d, e are non-zero integers that are not perfect squares:

(1) $\overline{\alpha} \neq \alpha$.
(2) $\overline{\alpha + \beta} = \overline{\alpha} + \overline{\beta}$.
(3) $\overline{\alpha \cdot \beta} = \overline{\alpha} \cdot \overline{\beta}$
(4) $\overline{\alpha/\beta} = \overline{\alpha}/\overline{\beta}$ if $\beta \neq 0$.

Exercise 13.4.15. Let $\alpha = \frac{n+m\sqrt{t}}{q}$ for some $n, m, t, q \in \mathbb{Z}$, and suppose that q is not a divisor of $tm^2 - n^2$. Show that if we define

$$d = tm^2q^2, \quad r = n \cdot |q|, \quad \text{and} \quad s = q \cdot |q|,$$

then $\alpha = (r + \sqrt{d})/s$ and s is a divisor of $(d - r^2)$.

Exercise 13.4.16. For each quadratic irrational number α below, use Exercise 13.4.15 to find r, d, and s such that $\alpha = (r + \sqrt{d})/s$ and s is a divisor of $(d - r^2)$:

$$\text{(a)}\ 2 + \sqrt{5}, \quad \text{(b)}\ \frac{2 + \sqrt{5}}{3}, \quad \text{(c)}\ \frac{2 + \sqrt{5}}{5}, \quad \text{(d)}\ \frac{1 + 3\sqrt{5}}{6}.$$

Exercise 13.4.17. Find the continued fraction expansion of the following irrational numbers (as in Examples 13.2.21 and 13.2.22):

$$\text{(a)}\ 2 + \sqrt{5}, \quad \text{(b)}\ \frac{2 + \sqrt{5}}{3}, \quad \text{(c)}\ \frac{2 + \sqrt{5}}{5}, \quad \text{(d)}\ \frac{1 + 3\sqrt{5}}{6}.$$

Exercise 13.4.18. Find the continued fraction expansion of the following irrational numbers:

$$\text{(a)}\ \sqrt{7}, \quad \text{(b)}\ \sqrt{10}, \quad \text{(c)}\ \sqrt{11}, \quad \text{(d)}\ \sqrt{13}, \quad \text{(e)}\ \sqrt{14}.$$

Exercise 13.4.19. Determine the values of the following continued fractions:

$$\text{(a)}\ [\overline{4,5}], \quad \text{(b)}\ [1, \overline{4,5}], \quad \text{(c)}\ [1, 1, \overline{4,5}], \quad \text{(d)}\ [1, 2, 4, \overline{1,2}], \quad \text{(e)}\ [\overline{1,3,5}].$$

Exercise 13.4.20. Let d be a positive integer such that $d = n^2 + 1$, for some integer $n > 1$. Show that the integers $\lfloor\sqrt{d}\rfloor$ and $\lfloor\sqrt{d}\rfloor + 1$ are the first two convergents of the continued fraction of $\sqrt{d}$.

Exercise 13.4.21. Prove that $\sqrt{9n^2 + 6} = [3n, \overline{n, 6n}]$, for all $n \geq 1$. Use this to compute the continued fraction of $\sqrt{15}$, $\sqrt{42}$, and $\sqrt{231}$.

Exercise 13.4.22. Find the best rational approximation a/b of $\sqrt{2}$ with denominator not exceeding 100. Without calculating $\sqrt{2} - a/b$, explain why $|\sqrt{2} - a/b| < 0.000085$. (Hint: use Corollary 13.3.4, part (2), to find a/b, and use Proposition 13.3.1, part (2), for the error estimate.)

Exercise 13.4.23. Find the best rational approximation a/b of e with denominator not exceeding 100. Without calculating $e - a/b$, estimate $|e - a/b|$. (The continued fraction of e was given in Example 13.2.9.)

Exercise 13.4.24. Find the best rational approximation a/b of $\log(2)$ with denominator not exceeding 100. Without calculating $\log(2) - a/b$, estimate $|\log(2) - a/b|$.

Exercise 13.4.25. Without computing any convergents, decide whether $22/9$ is a convergent of $\sqrt{6}$. (Hint: Theorem 13.3.5.)

CHAPTER 14

HYPERBOLAS AND PELL'S EQUATION

Out of nothing I have created a strange new universe.

János Bolyai, in reference to the creation of a non-euclidean geometry

In Chapter 9 we saw that for any quadratic equation $C : f(x, y) = 0$ with integer coefficients that is classified as a hyperbola (as in Definition 9.2.7), there is a change of variables $\phi : C \to C'$, where $C' : X^2 - BY^2 = D$ for some integers $B > 0$ and $D \neq 0$, that sends $C(\mathbb{Z})$ to $C'(\mathbb{Z})$, i.e., integral points to integral points (see Theorem 9.4.1; the map φ is explicitly described in Corollary 9.2.12). Therefore, our task in this chapter is to describe the rational and integral points on ellipses of the form $C' : X^2 - BY^2 = D$. Our analysis begins in Section 14.1 with the special case of square hyperbolas (i.e., the case of $B = 1$).

14.1. Square Hyperbolas

In this section we treat the special case of hyperbolas of the form $x^2 - By^2 = D$, when $B = E^2$ is a perfect square. In this case,

$$(x - Ey)(x + Ey) = D,$$

so if we put $X = x - Ey$ and $Y = x + Ey$, we obtain $XY = D$. This shows that the integral points on $x^2 - By^2 = D$ are intimately related to the divisors of D.

Proposition 14.1.1. *Let D, E be non-zero integers, and let C and C' be the hyperbolas given by $x^2 - E^2y^2 = D$ and $XY = D$, respectively. Then, there is a bijective map $\phi : C \to C'$ defined by*

$$\phi(x, y) = (x - Ey, x + Ey)$$

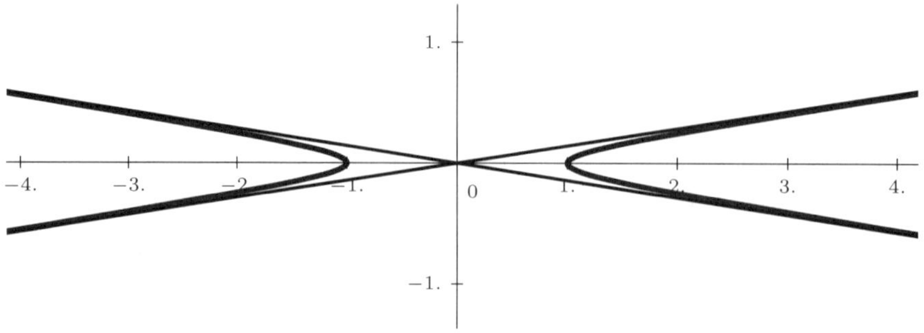

Figure 14.1. The square hyperbola $x^2 - 49y^2 = 1$ and its two main tangent lines $y = x/7$ and $y = -x/7$.

with inverse

$$\phi^{-1}(X,Y) = \left(\frac{X+Y}{2}, \frac{X-Y}{2E}\right)$$

such that

(1) *$\phi(C(\mathbb{Z})) \subseteq C'(\mathbb{Z})$ and $C(\mathbb{Z}) \subseteq \phi^{-1}(C'(\mathbb{Z}))$ and*

(2) *if $(a,b) \in C'(\mathbb{Z})$, then $\phi^{-1}((a,b)) \in C(\mathbb{Z})$ if and only if $a \equiv b \bmod 2E$.*

Proof. The map ϕ as defined in the statement is well-defined, as it sends points on C to points on C'. The map $(X,Y) \mapsto ((X+Y)/2, (X-Y)/(2E))$ is easily checked to be the inverse map of ϕ, and therefore ϕ is a bijection. Since E is an integer, the fact that $\phi(C(\mathbb{Z})) \subseteq C'(\mathbb{Z})$ is immediate from the definition of ϕ. Since ϕ is a bijection, it follows that $C(\mathbb{Z}) \subseteq \phi^{-1}(C'(\mathbb{Z}))$ as claimed in (1).

Finally, let $(a,b) \in C'(\mathbb{Z})$. Then, $\phi^{-1}((a,b)) \in C(\mathbb{Z})$ if and only if $(a+b)/2$ and $(a-b)/(2E)$ are integers. The quantity $(a-b)/(2E)$ is an integer if and only if $a \equiv b \bmod 2E$. Moreover, if $a \equiv b \bmod 2E$, then, in particular, $a \equiv b \bmod 2$ and so $a \equiv -b \bmod 2$ as well, so $(a+b)/2$ is an integer. This concludes the proof. □

The following corollary gives us a criterion for the existence of integral points on square hyperbolas.

Corollary 14.1.2. *Let $C : x^2 - By^2 = D$ be a hyperbola, such that B, D are non-zero integers and $B = E^2$ is a perfect square. Then:*

(1) *The curve C always has rational points $(\pm(1+D)/2, \pm(1-D)/(2E))$ and therefore infinitely many rational points.*

(2) *The curve C has an integral point if and only if there is a factorization $a \cdot b = D$ with integers $a, b \in \mathbb{Z}$ such that $a \equiv b \bmod 2E$. In such case, the point on C is given by $((a+b)/2, (a-b)/(2E))$.*

(3) *The square hyperbola C has only finitely many integral points.*

Proof. Let $C : x^2 - E^2y^2 = D$ and $C' : XY = D$, and let $\phi : C \to C'$ be the map defined in Proposition 14.1.1. Since $(1, D)$ is on C', we obtain a rational point $\phi^{-1}((1,D))$ on C, which coincides with the point given in (1). Thus, there are infinitely many points by Theorem 9.3.4, since $x^2 - By^2 = D$ is not the product of two lines (because $B, D \neq 0$; see Definition 9.2.7).

For part (2), note that Proposition 14.1.1 says that $C(\mathbb{Z}) \subseteq \phi^{-1}(C'(\mathbb{Z}))$, so every integral point on C comes from an integral point on C', which in turn corresponds to factorizations $a \cdot b = D$ with integers a, b. Further, $\phi^{-1}((a,b)) \in C(\mathbb{Z})$ if and only if $a \equiv b \bmod 2E$, as claimed.

Finally, the finiteness of the possible factorizations of D in the integers implies that there are only finitely many integral points on C. □

Example 14.1.3. Let $C : x^2 - 49y^2 = 51$. Then, Corollary 14.1.2 gives the rational points $(\pm 26, \pm 25/7)$ on C, and the reader can use the methods of Section 9.3 to find a parametrization of all rational points on C (see Exercise 14.4.1). Moreover, our result also provides a method to find the integral points. The possible factorizations of 51 are

$$1 \cdot 51, \ 3 \cdot 17, \ (-1)(-51), \ (-3)(-17).$$

Since $51 \not\equiv 1 \bmod 14$, the pairs $(1, 51)$ and $(-1, -51)$ do not provide integral points. However, $17 \equiv 3 \bmod 14$, so the factorizations $3 \cdot 17 = 17 \cdot 3 = (-3)(-17) = (-17)(-3)$ produce the four integral points $(\pm 10, \pm 1)$ on C. By Corollary 14.1.2, these are the only integral points on C.

Example 14.1.4. Let $C : x^2 - 49y^2 = 52$. Then, Corollary 14.1.2 gives the rational points $(\pm 53/2, \pm 51/14)$ on C. Also, we can show that there are no integral points. Indeed, the possible factorizations of $D = 52$ are

$$1 \cdot 52, \ 2 \cdot 26, \ 4 \cdot 13, \ (-1) \cdot (-52), \ (-2) \cdot (-26), \ (-4) \cdot (-13).$$

However, 51, 24, and 9 are not 0 mod 14. Thus, Corollary 14.1.2 implies that there are no integral points on C.

14.2. Pell's Equation $x^2 - By^2 = 1$

For the rest of the chapter, the number B will be a positive integer that is not a perfect square.

Our second step in understanding the points on hyperbolas is analyzing the particular case of curves of the form $x^2 - By^2 = 1$. Such curves are called Pell's equations, named after the English mathematician John Pell (see Figure 14.2). However, Pell's equation predates Pell, as it appeared in texts of Bhahmagupta as early as the 7th century. In correspondence with Fermat, the mathematician Lord William Brouncker rediscovered an efficient method to solve Pell's equation that had essentially appeared in the work of Bhaskara in the 12th century. The method of Bhaskara and Brouncker is essentially the same as the method used to compute the infinite continued fraction of $\sqrt{B}$. We will describe this method in detail in this section.

The connection between Pell's equation and the continued fraction of $\sqrt{B}$ is as follows. Suppose that (a, b) is an integral point on $x^2 - By^2 = 1$, with $a, b > 0$. Then, $a^2 - Bb^2 = 1$ and if follows that

$$B = \frac{a^2 - 1}{b^2} = \left(\frac{a}{b}\right)^2 - \frac{1}{b^2}.$$

Figure 14.2. John Pell (1611–1685) was an English diplomat and mathematician. Image source: Wikimedia Commons.

In particular,

$$\sqrt{B} = \sqrt{\left(\frac{a}{b}\right)^2 - \frac{1}{b^2}}$$

and it follows that a/b is a rational approximation of $\sqrt{B}$. In Chapter 13 we studied continued fractions and we have seen that they can be used to find good rational approximations of irrational numbers, called convergents. In particular, in Section 13.3, we saw that the convergents of an irrational number are the "most economical" rational approximations in terms of the size of the denominator (see Corollary 13.3.4 for the precise meaning of this). Our goal here is to prove that the approximation of $\sqrt{B}$ given by an integral solution of Pell's equation $x^2 - By^2 = 1$ is indeed a convergent.

Definition 14.2.1. Let $B \in \mathbb{N}$ but not a perfect square. A pair of integers $(a, b) \in \mathbb{Z} \times \mathbb{Z}$ is a *positive integral solution (or positive solution)* of the Pell equation $x^2 - By^2 = 1$ if $a^2 - Bb^2 = 1$ and $a, b > 0$.

Remark 14.2.2. Suppose that (a, b) and (c, d) are positive integral solutions of $x^2 - By^2 = 1$. Then, $a < c$ if and only if $a^2 < c^2$ if and only if $b^2 < d^2$, which is in turn equivalent to $b < d$. Thus, if there are integral solutions, then there is a *least positive solution*, a second positive solution, a third, etc.

Theorem 14.2.3. *Suppose (a, b) is a positive integral solution of $x^2 - By^2 = 1$. Then, a/b is one of the convergents of the infinite continued fraction of $\sqrt{B}$.*

Proof. Let $a, b > 0$ such that $a^2 - Bb^2 = 1$. We shall use Theorem 13.3.5. In order to apply the theorem we must prove that

$$\left|\sqrt{B} - \frac{a}{b}\right| < \frac{1}{2b^2}.$$

Since $a^2 - Bb^2 = 1$, it follows that $(a - b\sqrt{B}) = 1/(a + b\sqrt{B})$, and so $a/b - \sqrt{B} = 1/b(a + b\sqrt{B})$. Moreover, $a^2 = 1 + Bb^2$, so $a > b\sqrt{B}$. Thus,

$$\frac{a}{b} - \sqrt{B} = \frac{1}{b(a + b\sqrt{B})} < \frac{1}{2b^2\sqrt{B}} < \frac{1}{2b^2}.$$

Hence, Theorem 13.3.5 applies and a/b must be a convergent of $\sqrt{B}$. □

Example 14.2.4. Consider $C : x^2 - 5y^2 = 1$. Then, the pairs $(9, 4)$ and $(161, 72)$ are positive (integral) solutions of C. The first few convergents of $\sqrt{5}$ are

$$2, \frac{9}{4}, \frac{38}{17}, \frac{161}{72}, \frac{682}{305}, \ldots,$$

so, in agreement with Theorem 14.2.3, the quotient of the coordinates of positive solutions, namely $9/4$ and $161/72$, are convergents. However, it is important to notice that not all convergents are solutions. For instance, $2^2 - 5 \cdot 1^2 = -1 = 38^2 - 5 \cdot 17^2$.

Example 14.2.5. Let $C : x^2 - 7y^2 = 1$. We shall use Theorem 14.2.3 to try to find positive solutions. First, we find a few convergents $c_k = p_k/q_k$ of $\sqrt{7}$:

$$2, 3, \frac{5}{2}, \frac{8}{3}, \frac{37}{14}, \frac{45}{17}, \frac{82}{31}, \frac{127}{48}, \frac{590}{223}, \ldots.$$

Next, we check if (p_k, q_k) is an integral point on C. In Table 14.1 we calculate c_k and $p_k^2 - 7 \cdot q_k^2$ for $0 \leq k \leq 8$. Thus, out of the convergents c_k with $0 \leq k \leq 8$, only the convergents $c_3 = 8/3$ and $c_7 = 127/48$ provide positive solutions $(8, 3)$, $(127, 48)$. However, there seems to be a pattern in the values of $p_k^2 - 7 \cdot q_k^2$. If it holds, then c_{11} and c_{15}, for instance, should also provide positive solutions. Indeed, $c_{11} = 2024/765$, $c_{15} = 32257/12192$, and

$$2024^2 - 7 \cdot 765^2 = 1 = 32257^2 - 7 \cdot 12192^2.$$

Thus, $(2024, 765)$ and $(32257, 12192)$ are also positive solutions of C.

Table 14.1. Convergents p_k/q_k for $\sqrt{7}$ and the values of $p_k^2 - 7 \cdot q_k^2$.

k	0	1	2	3	4	5	6	7	8
p_k/q_k	2	3	$\frac{5}{2}$	$\frac{8}{3}$	$\frac{37}{14}$	$\frac{45}{17}$	$\frac{82}{31}$	$\frac{127}{48}$	$\frac{590}{223}$
$p_k^2 - 7 \cdot q_k^2$	-3	2	-3	1	-3	2	-3	1	-3

Theorem 14.2.3 and Example 14.2.5 raise two questions. First, all positive solutions are convergents, but which convergents are solutions? And, second, is there always at least one positive solution of Pell's equation $x^2 - By^2 = 1$ among the convergents of $\sqrt{B}$? Before we answer these questions, let us dig deeper into Example 14.2.5.

Example 14.2.6. Let C be the conic $x^2 - 7y^2 = 1$ as in Example 14.2.5. Where are the values of $p_k^2 - 7q_k^2$ (see Table 14.1) coming from? Let us step back and compute the continued fraction of $\sqrt{7}$ using the method of Theorem 13.2.20 in Section 13.2.1.

In order to begin, we write $\sqrt{7} = (0+\sqrt{7})/1$ so that $r_0 = 1$ and $s_0 = 1$:

k	0	1	2	3	4	$\cdots$
r_k	0	2	1	1	2	$\cdots$
s_k	1	3	2	3	1	$\cdots$
a_k	2	1	1	1	4	$\cdots$

By Proposition 13.2.32, we are done when we reach $a_m = 2a_0$, and so we have found that $\sqrt{7} = [2, \overline{1,1,1,4}]$. Now we see that the values of $p_k^2 - 7q_k^2$ seem to be given by $(-1)^{k+1}s_{k+1}$. If that is the case and if (p_k, q_k) is an integral solution, then we must have $(-1)^{k+1}s_{k+1} = 1$ and, in particular, $s_{k+1} = 1$. By Lemma 13.2.31, the value $s_{k+1} = 1$ only happens when $k+1$ is a multiple of m. Thus, we should have (p_k, q_k) is a solution whenever $k+1 \equiv 0 \bmod m$ and $k+1$ is even. In our particular case of $\sqrt{7}$ we have $m = 4$, so every multiple of m is even, and therefore (p_k, q_k) should be an integral point on $x^2 - 7y^2 = 1$ if and only if $k \equiv 3 \bmod 4$. Indeed, we saw in Example 14.2.5 that c_3, c_7, c_{11}, and c_{15}, for instance, provide points. We shall prove these facts next, in a more general context.

Lemma 14.2.7. *Let B be a positive integer that is not a perfect square, let $\alpha = \sqrt{B}$, and let a_k, α_k, r_k, and s_k be defined as in Theorem 13.2.20. Let $c_k = p_k/q_k$ be the convergents of the continued fraction of $\sqrt{d}$. Then, for every $k \geq 0$ we have*

(1) $\dfrac{p_k + q_k\sqrt{B}}{p_{k-1} + q_{k-1}\sqrt{B}} = \dfrac{r_{k+1} + \sqrt{B}}{s_k}$,

(2) $p_{k-1}^2 - Bq_{k-1}^2 = (-1)^k s_k$.

Proof. We prove (1) by induction on k. For $k = 0$, we have $\sqrt{B} = (0+\sqrt{B})/1$ and therefore $r_0 = 0$ and $s_0 = 1$. It follows from the definition of the sequence r_k in Theorem 13.2.20 that $r_1 = a_0 s_0 - r_0 = a_0$. Moreover, from the definitions of p_k and q_k (see Theorem 13.1.9) we know that $p_{-1} = 1$, $q_{-1} = 0$, $p_0 = a_0$, and $q_0 = 1$. Hence,

$$\frac{p_0 + q_0\sqrt{B}}{p_{-1} + q_{-1}\sqrt{B}} = a_0 + \sqrt{B} = \frac{r_1 + \sqrt{B}}{s_0}.$$

This proves the base case $k = 0$. Next, suppose (1) holds for $k \geq 0$. By the recurrence relations defining p_k and q_k we have

$$p_{k+1} + q_{k+1}\sqrt{B} = a_{k+1}(p_k + q_k\sqrt{B}) + p_{k-1} + q_{k-1}\sqrt{B},$$

and if we divide both sides of the previous equation by $p_k + q_k\sqrt{B}$, then we obtain

$$\frac{p_{k+1} + q_{k+1}\sqrt{B}}{p_k + q_k\sqrt{B}} = a_{k+1} + \frac{p_{k-1} + q_{k-1}\sqrt{B}}{p_k + q_k\sqrt{B}}.$$

If we use our induction hypothesis (i.e., (1) in the case of k), then we obtain

$$\frac{p_{k+1} + q_{k+1}\sqrt{B}}{p_k + q_k\sqrt{B}} = a_{k+1} + \frac{s_k}{r_{k+1} + \sqrt{B}}.$$

Next we rationalize the denominator in the right-hand side of the previous equation to obtain that

$$\frac{s_k}{r_{k+1}+\sqrt{B}} = s_k\frac{\sqrt{B}-r_{k+1}}{B-r_{k+1}^2} = \frac{\sqrt{B}-r_{k+1}}{s_{k+1}},$$

where we have used the definition of s_{k+1} in terms of s_k and r_{k+1}. Hence,

$$\frac{p_{k+1}+q_{k+1}\sqrt{B}}{p_k+q_k\sqrt{B}} = a_{k+1} + \frac{\sqrt{B}-r_{k+1}}{s_{k+1}} = \frac{a_{k+1}s_{k+1}-r_{k+1}+\sqrt{B}}{s_{k+1}} = \frac{r_{k+2}+\sqrt{B}}{s_{k+1}},$$

which is equation (1) in the case of $k+1$. This completes the proof of the induction step, and therefore our proof by induction of (1) is complete.

In order to prove part (2), let us multiply the equation in (1) by its conjugate to obtain

$$\frac{p_k^2 - Bq_k^2}{p_{k-1}^2 - Bq_{k-1}^2} = \frac{r_{k+1}^2 - B}{s_k^2} = -\frac{s_k s_{k+1}}{s_k^2} = -\frac{s_{k+1}}{s_k}, \tag{14.1}$$

where we have used the identity $s_k s_{k+1} = B - r_{k+1}^2$. If we write $t_k = p_k^2 - Bq_k^2$, then (14.1) reads $t_k/t_{k-1} = -s_{k+1}/s_k$. It follows that

$$\frac{t_{k-1}}{t_{-1}} = \prod_{j=0}^{k-1}\frac{t_j}{t_{j-1}} = \prod_{j=0}^{k-1}\left(-\frac{s_{j+1}}{s_j}\right) = (-1)^k\frac{s_k}{s_0}.$$

Finally, using the values $p_{-1} = s_0 = 1$ and $q_{-1} = 0$, we obtain that $t_{k-1} = p_{k-1}^2 - Bq_{k-1}^2 = (-1)^k s_k$, which proves (2). □

We are now ready to prove the patterns we observed in Examples 14.2.4, 14.2.5, and 14.2.6 by putting together our results of Lemma 13.2.31, Theorem 14.2.3, and Lemma 14.2.7.

Theorem 14.2.8. *Let B be a positive integer that is not a perfect square, assume that the period of the continued fraction of $\sqrt{B}$ has length m, and let p_k/q_k be the associated convergents.*

(1) *If m is even, then all the positive integral solutions of $x^2 - By^2 = 1$ are given by $(x, y) = (p_{jm-1}, q_{jm-1})$ for all $j \geq 1$.*
(2) *If m is odd, then all the positive integral solutions of $x^2 - By^2 = 1$ are given by $(x, y) = (p_{2jm-1}, q_{2jm-1})$ for all $j \geq 1$.*

In particular, Pell's equation $x^2 - By^2 = 1$ always has integral solutions.

Proof. As we have seen in Theorem 14.2.3, every positive integral solution (a, b) of $x^2 - By^2 = 1$ arises from a convergent a/b of $\sqrt{B}$. That is, $(a, b) = (p_k, q_k)$ for some $k \geq 0$, where p_k/q_k is the kth convergent of $\sqrt{B}$. By Lemma 14.2.7, part (2), we have that

$$p_k^2 - Bq_k^2 = (-1)^{k+1}s_{k+1}.$$

Moreover, recall that $s_k \geq 1$ for all k (by part (1) of Lemma 13.2.31). Thus, (p_k, q_k) is a solution of Pell's equation if and only if k is odd and $s_{k+1} = 1$. Further, Lemma 13.2.31 shows that $s_{k+1} = 1$ if and only if $k + 1$ is a positive multiple of m.

Hence, if m is even, (a, b) is a solution of Pell's equation if and only if $(a, b) = (p_k, q_k)$ with $k = jm - 1$ for some $j \geq 1$, and if m is odd, then $k = 2jm - 1$ for some $j \geq 1$. This completes the proof of the theorem. □

Remark 14.2.9. Recall from Theorem 13.1.9 that $p_k = a_k p_{k-1} + p_{k-2}$ and $q_k = a_k q_{k-1} + q_{k-2}$ for $k \geq 1$. Thus, the sequences $\{p_k\}$ and $\{q_k\}$ are increasing. It follows from Theorem 14.2.8 that the *smallest* positive integral solution of a Pell's equation $x^2 - By^2 = 1$ is given by (p_{m-1}, q_{m-1}) if m is even and by (p_{2jm-1}, q_{2jm-1}) if m is odd. The least positive solution is called the fundamental solution of the equation.

Example 14.2.10. Let us find the two smallest positive integral solutions of the equation $x^2 - 19y^2 = 1$. The first step is to find the continued fraction of $\sqrt{19}$, which we already computed in Example 13.2.29:

$$\sqrt{19} = [4, \overline{2, 1, 3, 1, 2, 8}].$$

In particular, we note that the length of the period is $m = 6$, even. Hence, Theorem 14.2.8 tells us that the two smallest positive solutions of the equation are given by

$$(p_5, q_5) \quad \text{and} \quad (p_{11}, q_{11}).$$

Thus, it remains to compute the convergents c_5 and c_{11}:

k	-1	0	1	2	3	4	5	...
a_k		4	2	1	3	1	2	
p_k	1	4	9	13	48	61	170	
q_k	0	1	2	3	11	14	39	

k	...	6	7	8	9	10	11
a_k		8	2	1	3	1	2
p_k		1421	3012	4433	16311	20744	57799
q_k		326	691	1017	3742	4759	13260

We deduce that $c_5 = 170/39$ and $c_{11} = 57799/13260$, and therefore the first two positive integral solutions are $(170, 39)$ and $(57799, 13260)$.

Example 14.2.11. Let us now find the first two positive solutions of $x^2 - 13y^2 = 1$. First, we compute the continued fraction of $\sqrt{13}$, which turns out to be $\sqrt{13} = [3, \overline{1, 1, 1, 1, 6}]$. Thus, the length of the period is $m = 5$, odd, and the first two solutions will correspond to the convergents c_k with $k = 2 \cdot 5 - 1 = 9$ and $k = 2 \cdot 2 \cdot 5 - 1 = 19$, i.e., (p_9, q_9) and (p_{19}, q_{19}). We will not compute a table of convergents here, and we leave it up to the reader to verify that the solutions are $(649, 180)$ and $(842401, 233640)$.

We remark here that the $c_{m-1} = c_4$ convergent does not provide a solution of $x^2 - 13y^2 = 1$. If m is odd, then Lemma 14.2.7 says that (p_{m-1}, q_{m-1}) is a solution of $x^2 - 13y^2 = -1$ instead. Indeed, $c_4 = 18/5$ and $18^2 - 13 \cdot 5^2 = -1$. In the next section we shall explore the solutions of more general Pell equations of the form $x^2 - By^2 = N$.

14.3. Generalized Pell's Equations $x^2 - By^2 = N$

In the previous section we found all solutions to Pell's equation $x^2 - By^2 = 1$. We will begin this section by showing that our previous work also determines the solutions of $x^2 - By^2 = -1$ and then study even more general Pell's equations of the form $x^2 - By^2 = N$, for some non-zero integer N.

Example 14.3.1. After we discussed the Hasse–Minkowski theorem, in Section 11.3.1, we showed that the curve $C : x^2 - 29y^2 = -1$ has rational points (e.g., $(5/2, 1/2)$). Are there any integral solutions of $x^2 - 29y^2 = -1$? Here we shall use continued fractions to determine all the integral points on C. Suppose that (a, b) is an integral solution. Then, Exercise 14.4.4 shows that a/b must be a convergent of $\sqrt{29}$; i.e., $a/b = p_k/q_k$ for some $k \geq 0$. By Lemma 14.2.7 we know that $p_k^2 - Bq_k^2 = (-1)^{k+1}s_{k+1}$ so we must have $s_{k+1} = 1$ and $k+1$ odd. However, by Lemma 13.2.31 we have $s_{k+1} = 1$ if and only if $k+1$ is a multiple of m, the length of the period of the continued fraction of $\sqrt{29}$. Said continued fraction is $[5, \overline{2, 1, 1, 2, 10}]$ and so $m = 5$. Hence, all the (posivite) integral points on C are of the form $(p_{5(2j+1)-1}, q_{5(2j+1)-1}) = (p_{10j+4}, q_{10j+4})$ for any $j \geq 0$. For instance,

$$(p_4, q_4) = (70, 13) \quad \text{and} \quad (p_{14}, q_{14}) = (1372210, 254813)$$

are the two smallest positive solutions of $x^2 - 29y^2 = -1$. The next solution is given by the 24th convergent, with coordinates $(26898060350, 4994844413)$.

Example 14.3.2. Of course, some Pell's equations of the form $C : x^2 - By^2 = -1$ lack integral points altogether due to simple congruence considerations. For instance, if p is a prime and $p \equiv 3 \bmod 4$ and B is an integer divisible by p, then there cannot be any integral points on C (see Exercise 14.4.10). For instance, let $B = 3$ and consider $x^2 - 3y^2 = -1$. Since the continued fraction of $\sqrt{3}$ is $[1, \overline{1, 2}]$, the length of the period is $m = 2$, even, and it follows from our considerations in Example 14.3.1 that there cannot be integral points. In the next result we formalize our observations about the equation $x^2 - By^2 = -1$.

Theorem 14.3.3. *Let B be a positive integer that is not a perfect square, assume that the period of the continued fraction of $\sqrt{B}$ has length m, and let p_k/q_k be the associated convergents.*

(1) *If m is even, then the equation $x^2 - By^2 = -1$ has no integral solutions.*
(2) *If m is odd, then all the positive integral solutions of $x^2 - By^2 = 1$ are given by $(x, y) = (p_{2jm+(m-1)}, q_{2jm+(m-1)})$ for all $j \geq 0$.*

Since the proof is very similar to that of Theorem 14.2.8, we leave it as an exercise for the reader (see Exercise 14.4.11).

Definition 14.3.4. Let B be a positive integer that is not a perfect square, and let N be a non-zero integer. The equation $x^2 - By^2 = N$ is known as a *generalized Pell's equation, or of Pell type.* The equation $x^2 - By^2 = -1$ is sometimes called a *negative Pell's equation.*

Our next example shows the range of possibilities in the integral solvability of generalized Pell's equations.

Example 14.3.5. Consider the generalized Pell's equations $x^2 - 3y^2 = N$, for $N = 1, -1, 7$, and 13.

- When $N = 1$, Theorem 14.2.8 says that the equation has solutions and, in fact, determines all of them. Since $\sqrt{3} = [1, \overline{1, 2}]$, the length of the period is $m = 2$, and the solutions are given by (p_{2j-1}, q_{2j-1}) for all $j \geq 1$, where p_k/c_k is the kth convergent. For instance, the first few solutions are $(2, 1)$, $(7, 4)$, $(26, 15)$, etc.
- When $N = -1$, Example 14.3.2 shows that there are no integral solutions.
- When $N = 7$, the equation $x^2 - 3y^2 = 7$ has no solutions. To see this, we reduce modulo 7 to reach $x^2 \equiv 3y^2 \bmod 7$. Since 3 is a quadratic non-residue modulo 7, it turns out the only solution would be $x \equiv y \equiv 0 \bmod 7$. However, if both x and y are divisible by 7, then 7^2 divides $x^2 - 3y^2 = 7$ and we have a contradiction.
- When $N = 13$, we can find at least one point by inspection, namely $(4, 1)$. Are there others? Recall that in Lemma 12.5.1 we have shown that out of a solution of equations $x^2 - By^2 = n$ and $x^2 - By^2 = m$, we can construct a solution of $x^2 - By^2 = nm$. More precisely, if $(4, 1)$ solves $x^2 - 3y^2 = 13$ and (a, b) solves $x^2 - 3y^2 = 1$, then
$$(4 \cdot a + 3 \cdot 1 \cdot b,\ 4 \cdot b + 1 \cdot a) = (4a + 3b, 4b + a)$$
is a solution of $x^2 - 3y^2 = 1 \cdot 13 = 13$. Since we know that (p_{2j-1}, q_{2j-1}) is a point on the Pell's equation with $N = 1$, we conclude that
$$(4p_{2j-1} + 3q_{2j-1}, 4q_{2j-1} + p_{2j-1})$$
is a point on $x^2 - 3y^2 = 13$ for all $j \geq 1$. For instance, $(11, 6)$, $(40, 23)$, and $(149, 86)$ are the points we obtain for $j = 1$, 2, and 3, respectively.

More generally, we can show the following result about generalized Pell's equations, as consequences of Lemma 12.5.1 and Theorem 14.2.8.

Proposition 14.3.6. *Let N and M be non-zero integers, and let B be a non-zero integer that is not a square. Then:*

(1) *If $x^2 - By^2 = N$ and $x^2 - By^2 = M$ have integral solutions, so does $x^2 - By^2 = NM$.*
(2) *If $x^2 - By^2 = N$ has at least one integral solution, then it has infinitely many.*

The proof is simple, so we leave it as Exercise 14.4.12. We note, however, that Example 14.3.5 and Proposition 14.3.6 prompt the following natural question: can we construct all the positive integral solutions of a generalized Pell's equation $C : x^2 - By^2 = N$ using the smallest positive solution of C and all the solutions of $x^2 - By^2 = 1$? For instance, are the solutions we found in Example 14.3.5 for $N = 13$ all the positive integral solutions of $x^2 - 3y^2 = 13$? The answer to this question is, in fact, yes, and we will discuss the reason in the next section by taking a different point of view (i.e., quadratic fields).

We conclude this section by showing that, just like in the cases of $N = 1$ and $N = -1$, the positive solutions of certain generalized Pell's equations $x^2 - By^2 = N$ come from convergents of $\sqrt{B}$.

Theorem 14.3.7. *If $|N| \leq \sqrt{B}$ and (a,b) is a positive integer solution of $x^2 - By^2 = N$, then a/b is a convergent associated to the continued fraction expansion of $\sqrt{B}$.*

Proof. Let us first assume that N is positive. Then,

$$\frac{a}{b} - \sqrt{B} = \frac{a - b\sqrt{B}}{b} = \frac{(a-b\sqrt{B})(a+b\sqrt{B})}{b(a+b\sqrt{B})} = \frac{N}{b(a+b\sqrt{B})},$$

where we have used the fact that $a^2 - Bb^2 = N$. Since $(a+b\sqrt{B})(a-b\sqrt{B}) = N > 0$, we must have $a - b\sqrt{B} > 0$ as well, and so $a > b\sqrt{B}$. Thus, we conclude

$$0 < \frac{a-b\sqrt{B}}{b} = \frac{a}{b} - \sqrt{B} = \frac{N}{b(a+b\sqrt{B})} < \frac{\sqrt{B}}{b(2b\sqrt{B})} = \frac{1}{2b^2}.$$

Thus, Theorem 13.3.5 applies here and a/b must be a convergent of $\sqrt{B}$.

Now suppose that N is negative. Then,

$$-\frac{N}{B} = b^2 - \frac{1}{B}a^2.$$

A similar argument to the one carried out above for positive N yields

$$0 < \frac{b}{a} - \sqrt{\frac{1}{B}} < \frac{1}{2a^2}$$

and Theorem 13.3.5 now implies that b/a is a convergent of $1/\sqrt{B}$. Finally, Exercise 13.4.12 shows that if b/a is a convergent of $1/\sqrt{B}$, then a/b must be a convergent of $\sqrt{B}$. This completes the proof of the theorem. □

Example 14.3.8. Consider the generalized Pell's equation $C : x^2 - 7y^2 = 2$ and the point $(45, 17)$ on C. Since $N = 2 < \sqrt{7}$, it follows from Theorem 14.3.7 that $45/17$ must be a convergent of $\sqrt{7}$. Indeed, in Example 14.2.5, we listed a few convergents, and $45/17$ is one of them (c_5 to be precise). Notice that we also computed s_{k+1}, and Lemma 14.2.7 says that a convergent $c_k = p_k/q_k$ of $\sqrt{7}$ will be a solution of $x^2 - 7y^2 = 2$ precisely when $s_{k+1} = 2$ and $k+1$ is even (k is odd). Thus, Table 14.1 shows that $c_1 = 3/1 = 3$ also provides a point $(3, 1)$ on $x^2 - 7y^2 = 2$.

Remark 14.3.9. As a word of caution, we note here that Theorem 14.3.7 does not say that if $|N| \leq \sqrt{B}$, then $x^2 - By^2 = N$ must have a solution. For example, if we consider $x^2 - 7y^2 = -2$, then there cannot be any integral solutions, as -2 is a quadratic non-residue modulo 7.

14.3.1. Quadratic Rings and Units. In this section we will reinterpret the solutions of Pell's equation and generalized Pell's equations as elements in quadratic fields with a certain norm. We encountered quadratic fields for the first time in Section 12.5, when discussing ellipses. Recall that the set

$$\mathbb{Q}(\sqrt{B}) = \{a + b\sqrt{B} : a, b \in \mathbb{Q}\}$$

is, in fact, a field (Proposition 12.5.3). In this section we will work with a subring of $\mathbb{Q}(\sqrt{B})$, namely $\mathbb{Z}[\sqrt{B}]$, which we define as follows:

$$\mathbb{Z}[\sqrt{B}] = \{a + b\sqrt{B} : a, b \in \mathbb{Z}\}.$$

We leave it to the reader to verify that $\mathbb{Z}[\sqrt{B}]$ is a ring (Exercise 14.4.14). We say that $\mathbb{Z}[\sqrt{B}]$ is a *quadratic ring*. Also recall (Definition 12.5.6) that we have a norm map $\mathrm{N} : \mathbb{Q}(\sqrt{B}) \to \mathbb{Q}$ given by

$$\mathrm{N}(a + b\sqrt{B}) = a^2 - Bb^2.$$

Recall that the norm is multiplicative; that is, $\mathrm{N}(\alpha\beta) = \mathrm{N}(\alpha) \cdot \mathrm{N}(\beta)$ for all α and $\beta \in \mathbb{Q}(\sqrt{B})$ (see Lemma 12.5.9). Also, note that it follows from the definition of the norm that if we restrict the norm map to $\mathbb{Z}[\sqrt{B}]$, then the image of the norm is contained in $\mathbb{Z}$; that is, $\mathrm{N}(a + b\sqrt{B}) = a^2 - Bb^2$ is an integer, for all $a, b \in \mathbb{Z}$.

Now, let M be a non-zero integer. From the definitions above we see that an element $\alpha = a + b\sqrt{B} \in \mathbb{Z}[\sqrt{B}]$ has norm M (i.e., $\mathrm{N}(\alpha) = M$) if and only if $a^2 - Bb^2 = M$ if and only if (a, b) is an integral point in the generalized Pell's equation $x^2 - By^2 = M$. Let us first identify the elements of norm 1 as the units of the ring $\mathbb{Z}[\sqrt{B}]$.

Proposition 14.3.10. *Let B be a non-zero integer that is not a square, and let $\mathbb{Z}[\sqrt{B}]$ be the quadratic ring defined by*

$$\mathbb{Z}[\sqrt{B}] = \{a + b\sqrt{B} : a, b \in \mathbb{Z}\}.$$

Then:

(1) *An element $\alpha \in \mathbb{Z}[\sqrt{B}]$ is a unit if and only if $\mathrm{N}(\alpha) = \pm 1$.*

(2) *If $B < 0$, then the only units in $\mathbb{Z}[\sqrt{B}]$ are ± 1, unless $B = -1$, in which case the units are ± 1 and $\pm i$.*

(3) *If $B > 0$, then the ring $\mathbb{Z}[\sqrt{B}]$ has infinitely many units.*

Proof. Let us suppose first that $u = a + b\sqrt{B} \in \mathbb{Z}[\sqrt{B}]$ is a unit. Thus, by Definition 5.3.7, there is $v = c + d\sqrt{B} \in \mathbb{Z}[\sqrt{B}]$ such that $u \cdot v = 1$. Since the norm is multiplicative, we obtain

$$1 = \mathrm{N}(1) = \mathrm{N}(u \cdot v) = \mathrm{N}(u) \cdot \mathrm{N}(v),$$

and since $u, v \in \mathbb{Z}[\sqrt{B}]$, their norms $N(u), N(v)$ are integers that multiply to 1. Hence, $N(u)$ and $N(v)$ are ± 1.

Conversely, suppose that $u = a + b\sqrt{B}$ has norm $\varepsilon = \pm 1$. Then,

$$\varepsilon = \mathrm{N}(u) = a^2 - Bb^2 = (a + b\sqrt{B})(a - b\sqrt{B}).$$

Hence, if we let $v = \varepsilon \cdot (a - b\sqrt{B})$, then $u \cdot v = \varepsilon^2 = 1$, and therefore u is a unit.

Now, let us assume that $B < 0$. Then, $x^2 - By^2 = -1$ is impossible for $x, y \in \mathbb{Z}$, and therefore by part (1) there are no units of norm -1. If $x^2 - By^2 = 1$ and since $|B| > 1$, it follows that $y = 0$ and $x = \pm 1$. Thus, the only units are ± 1, as claimed.

If $B > 0$, then Theorem 14.2.8 says that $x^2 - By^2 = 1$ has infinitely many integral solutions (and Theorem 14.3.3 says that $x^2 - By^2 = -1$ has infinitely many integral solutions as well, under certain restrictions). Thus, part (1) says that there exist infinitely many units in $\mathbb{Z}[\sqrt{B}]$ of norm 1. □

Example 14.3.11. In this example we shall use Proposition 14.3.10 as an alternative method to find solutions to Pell's equation. Let us consider first $x^2 - 19y^2 = 1$. In Example 14.2.10 we produced two positive solutions, namely $P = (170, 39)$

and $Q = (57799, 13260)$, that correspond to the 5th and 11th convergent, respectively. In light of Proposition 14.3.10, the point P also corresponds to a unit $u = 170 + 39\sqrt{19}$ with norm 1. Since units form a multiplicative subgroup of $\mathbb{Z}[\sqrt{B}]$, it follows that u^2 is also a unit (and $\mathrm{N}(u^2) = \mathrm{N}(u)^2 = 1^2 = 1$ because of the multiplicativity of the norm). Thus,

$$u^2 = (170 + 39\sqrt{19})^2 = 57799 + 13260\sqrt{19}$$

is also a unit of norm 1, and therefore $(57799, 13260)$ is another point on $x^2 - 19y^2 = 1$ (the point Q that corresponds to the 11th convergent). We can now generate a third point by considering u^3:

$$u^3 = (57799 + 13260\sqrt{19}) \cdot (170 + 39\sqrt{19}) = 19651490 + 4508361\sqrt{19}.$$

Since $\mathrm{N}(u^3) = \mathrm{N}(u)^3 = 1^3 = 1$, it follows that $R = (19651490, 4508361)$ is another point on $x^2 - 19y^2 = 1$. The reader can verify that $19651490/4508361$ is, in fact, the 17th convergent of $\sqrt{19}$.

Example 14.3.12. The method used in Example 14.3.11 can also be applied to finding points on the negative Pell's equation. For instance, let us consider $x^2 - 13y^2 = -1$. In Example 14.2.11 we found two solutions of $x^2 - 13y^2 = 1$ and also noticed that $(18, 5)$ is a point on $x^2 - 13y^2 = -1$. This means that $18 + 5\sqrt{13}$ is a unit of norm -1 in the ring $\mathbb{Z}[\sqrt{13}]$. Therefore u^2 is another unit of norm 1, and u^3 is a unit of norm -1. Indeed,

$$\mathrm{N}(u^2) = \mathrm{N}(u)^2 = (-1)^2 = 1 \quad \text{and} \quad \mathrm{N}(u^3) = \mathrm{N}(u)^3 = (-1)^3 = -1.$$

Thus, $u^3 = 23382 + 6485\sqrt{13}$ is associated to a point $(23382, 6485)$ on $x^2 - 13y^2 = -1$. The reader can check that $23382/6485$ is the 14th convergent of $\sqrt{13}$.

Suppose (a, b) is a positive integer solution of $x^2 - By^2 = 1$. Then, by Theorem 14.2.8, the rational number a/b is a convergent of $\sqrt{B}$. On the other hand, our Proposition 14.3.10 says that $u = a + b\sqrt{B}$ is a unit in $\mathbb{Z}[\sqrt{B}]$. In Examples 14.3.11 and 14.3.12 we have seen how to find other solutions as powers of a given unit, and they raise the following natural question: is there a unit u such that every solution of $x^2 - By^2 = 1$ is associated to a power of u? The answer is yes, and the following theorem demonstrates this fact.

Theorem 14.3.13. *Let $P_1 = (a, b)$ be the smallest positive integer solution of $x^2 - By^2 = 1$, and let $u = a + b\sqrt{B}$ be the associated unit in $\mathbb{Z}[\sqrt{B}]$.*

(1) *If $n \geq 1$ and $u^n = a_n + b_n\sqrt{B}$ with a_n, $b_n \in \mathbb{Z}$, then (a_n, b_n) is an integral point on $x^2 - By^2 = 1$.*
(2) *Conversely, if (c, d) is a positive integer solution of $x^2 - By^2 = 1$, then there is some $n \geq 1$ such that $u^n = c + d\sqrt{B}$.*

Proof. Part (1) follows from the fact that $\mathrm{N}(u^n) = \mathrm{N}(u)^n = 1^n = 1$ by Lemma 12.5.9 and, on the other hand, we have $\mathrm{N}(a_n + b_n\sqrt{B}) = a_n^2 - Bb_n^2$. For part (2), let (c, d) be an arbitrary positive integer solution of Pell's equation, and let $\beta = c + d\sqrt{B}$. Note that $\mathrm{N}(\beta) = c^2 - Bd^2 = 1$. Let $\alpha_n = u^n = a_n + b_n\sqrt{B}$ and denote its conjugate by $\alpha_n' = (u^n)' = a_n - b_n\sqrt{B}$. Since α_n has norm 1, it follows that $\alpha_n\alpha_n' = 1$ and therefore $\alpha_n' = 1/\alpha_n > 0$. Also, $\alpha_1 = u = a + b\sqrt{B}$ with

$a, b > 0$, and so $\alpha_n < \alpha_{n+1}$ for all $n \geq 1$. Since β is also a positive integer solution, it follows that $\beta > 1$ and there is some $n \geq 1$ such that

$$\alpha_n < \beta \leq \alpha_{n+1}.$$

If we multiply through by α_n', then we obtain

$$1 = \alpha_n \alpha_n' < \beta \alpha_n' \leq \alpha_{n+1} \alpha_n' = \alpha_1 \alpha_n \alpha_n' = \alpha_1 = u.$$

If we write $\gamma = \beta \alpha_n' = e + f\sqrt{B}$ for some integers $e, f \in \mathbb{Z}$, then the preceding equation reads $1 < \gamma \leq u$. Since

$$\mathrm{N}(\gamma) = \mathrm{N}(\beta \alpha_n') = \mathrm{N}(\beta)\,\mathrm{N}(\alpha_n') = 1 \cdot 1 = 1,$$

it follows that (e, f) is also an integral point on Pell's equation, where $\gamma = e + f\sqrt{B}$, but perhaps not a *positive* integral solution. Let us show that indeed $e, f > 0$. From $\gamma = \beta \alpha_n' > 1$ it follows that $0 < \gamma' = 1/\gamma < 1$, and so $1 < e + f\sqrt{B}$ and $0 < e - f\sqrt{B} < 1$. In particular,

$$1 < (e + f\sqrt{B}) + (e - f\sqrt{B}) = 2e \text{ and } 1 < (e + f\sqrt{B}) - (e - f\sqrt{B}) = 2f\sqrt{B}$$

imply that e and f are both positive. Therefore, (e, f) is a positive integral solution of $x^2 - By^2 = 1$, and

$$1 < e + f\sqrt{B} = \gamma \leq u = a + b\sqrt{B}$$

where (a, b) was supposed to be, by assumption, the smallest positive integer solution of Pell's equation. Thus, we must have $\gamma = u$ and $a = e$ and $b = f$. It follows that $\gamma = \beta \alpha_n' = \alpha_1 = u$, and if we multiply through by α_n, we obtain

$$\beta = \beta \alpha_n' \alpha_n = \alpha_1 \alpha_n = u^{n+1}.$$

Hence, we have shown that $c + d\sqrt{B} = \beta = u^{n+1}$; i.e., β is a power of u, as desired. □

In fact, every integer solution of Pell's equation (not only the positive ones) is given by some unit of the form $\pm u^n$ in $\mathbb{Z}[\sqrt{B}]$, with $n \in \mathbb{Z}$ (see Exercise 14.4.15). More generally, one can show the following theorem (see for instance [**AC95**, Theorem 10.14]).

Theorem 14.3.14. *Let $C : x^2 - By^2 = -1$, and suppose there are integral points on C. Further, suppose that (a, b) is the smallest positive solution of C, let $u = a + b\sqrt{B}$, and write $u^n = a_n + b_n\sqrt{B}$ for each $n \in \mathbb{Z}$. Then:*

(1) *If n is odd, then $P_n = (a_n, b_n)$ is an integral point on C, and if n is even, then P_n is an integral point on Pell's equation $x^2 - By^2 = 1$.*

(2) *Conversely, if $Q = (c, d)$ is an integral point on C, then there is an odd integer n such that $c + d\sqrt{B} = u^n$ or $-u^n$. If Q is an integral point on $x^2 - By^2 = 1$, then $c + d\sqrt{B} = u^n$ or $-u^n$, for some even integer n.*

In light of Theorems 14.3.13 and 14.3.14, we define the concept of fundamental unit as follows.

Definition 14.3.15. Let B be a natural number that is not a perfect square, and let $\mathbb{Z}[\sqrt{B}]$ be the associated quadratic ring. A *fundamental unit* $u \in \mathbb{Z}[\sqrt{B}]$ is a unit defined as follows:

(1) If $x^2 - By^2 = -1$ has no integral solutions, then a fundamental unit $u = a + b\sqrt{B}$ is the unit in $\mathbb{Z}[\sqrt{B}]^\times$ such that (a, b) is the smallest positive integer solution of Pell's equation $x^2 - By^2 = 1$.

(2) If $x^2 - By^2 = -1$ has integral solutions, then a fundamental unit $u = a + b\sqrt{B}$ is the unit in $\mathbb{Z}[\sqrt{B}]^\times$ such that (a, b) is the smallest positive integer solution of the negative Pell's equation $x^2 - By^2 = -1$.

Example 14.3.16. By Example 14.3.1, the fundamental unit of $\mathbb{Z}[\sqrt{29}]$ is $u = 170 + 13\sqrt{29}$, with $\mathrm{N}(u) = -1$. By Example 14.3.2, there are no integral points on $x^2 - 3y^2 = -1$, and therefore the fundamental unit of $\mathbb{Z}[\sqrt{3}]$ is $u = 2 + \sqrt{3}$, with $\mathrm{N}(u) = 1$.

We may reinterpret our work up to this point as a description of the unit group of $\mathbb{Z}[\sqrt{B}]$ that goes further than Proposition 14.3.10 in the case of $B > 0$. Before we state the theorem, we recall that we defined the concept of group isomorphism in Section 5.2.1.

Theorem 14.3.17. *Let $B > 0$ be an integer that is not a perfect square, let $\mathbb{Z}[\sqrt{B}]$ be a quadratic ring, let U be the group of units in $\mathbb{Z}[\sqrt{B}]$, and let u be a fundamental unit for $\mathbb{Z}[\sqrt{B}]$. Then:*

(1) $U = \{\pm u^n : n \in \mathbb{Z}\}$.

(2) *There is an isomorphism of groups $\psi : \mathbb{Z}/2\mathbb{Z} \times \mathbb{Z} \to U$ defined by*

$$\psi((a \bmod 2, n)) = (-1)^a \cdot u^n.$$

Proof. Part (1) is covered by Theorem 14.3.14, so it remains to show that the map ψ in part (2) is an isomorphism of groups (Definition 5.2.31). Note that $\mathbb{Z}/2\mathbb{Z} \times \mathbb{Z}$ is a group under componentwise addition of coordinates. First, we check that ψ is well-defined. If $a \equiv b \bmod 2$, then

$$\psi((a \bmod 2, n)) = (-1)^a \cdot u^n = (-1)^b \cdot u^n = \psi((b \bmod 2, n)).$$

Also, we check that ψ is a group homomorphism:

$$\begin{aligned}\psi((a \bmod 2, n) + (b \bmod 2, m)) &= \psi((a + b \bmod 2, n + m)) \\ &= (-1)^{a+b} \cdot u^{n+m} \\ &= (-1)^a \cdot u^n \cdot (-1)^b \cdot u^m \\ &= \psi((a \bmod 2, n)) + \psi((b \bmod 2, m)).\end{aligned}$$

Theorem 14.3.14 shows that ψ is surjective, i.e., every unit $v \in U$ can be written as $v = \pm u^n$, and therefore it is in the image of ψ. It remains to show that ψ is injective. This is equivalent to showing that all the elements of $\{\pm u^n : n \in \mathbb{Z}\}$ are distinct in $\mathbb{Z}[\sqrt{B}]$. For this, note that the fundamental unit $u = a + b\sqrt{B} > 1$, and therefore $u^n < u^{n+1}$ for all $n \geq 1$. Thus, we have

$$\cdots < -u^{n+1} < -u^n < \cdots < -u < -1 < 1 < u < \cdots < u^n < u^{n+1} < \cdots$$

which shows that all the units in the set $\{\pm u^n : n \in \mathbb{Z}\}$ are distinct. Thus, ψ is injective and therefore an isomorphism of groups. □

Now that we have related the solutions of Pell's equation to the units in $\mathbb{Z}[\sqrt{B}]$, we can also prove again Proposition 14.3.6 in terms of quadratic rings.

Proposition 14.3.18. *Let B be a positive integer that is not a square, and let N and M be non-zero integers.*

(1) *If $x^2 - By^2 = N$ and $x^2 - By^2 = M$ have positive integer solutions, then $x^2 - By^2 = NM$ also has positive solutions.*
(2) *If $x^2 - By^2 = N$ has one positive integral solution, then it has infinitely many.*

Proof. Part (1) follows directly from Lemma 12.5.1, but we shall use Lemma 12.5.9 instead. Indeed, suppose $a^2 - Bb^2 = N$ and $c^2 - Bd^2 = M$, for some positive integers a, b, c, d. Then, $\mathrm{N}(a + b\sqrt{B}) = N$ and $\mathrm{N}(c + d\sqrt{B}) = M$. Thus,

$$\mathrm{N}((a + b\sqrt{B})(c + d\sqrt{B})) = \mathrm{N}(ac + Bbd + (ad + bc)\sqrt{B}) = NM.$$

Hence, $(ac + Bbd, ad + bc)$ is a positive integral point on $x^2 - By^2 = NM$.

For (2), if $\alpha = (a + b\sqrt{B})$ with $\mathrm{N}(\alpha) = N$ and if v is any unit of norm 1, then $\mathrm{N}(\alpha \cdot v) = \mathrm{N}(\alpha)\,\mathrm{N}(v) = N \cdot 1 = N$. By Theorem 14.3.17, if u is a fundamental unit for $\mathbb{Z}[\sqrt{B}]$, then the units of norm 1 with positive coordinates are given by either $\{u^n\}$ or $\{u^{2n}\}$ according to whether $\mathrm{N}(u) = 1$ or -1, respectively. Further, Theorem 14.3.17 shows that $u^n \neq u^m$ for any $n \neq m$. Thus, either $\alpha \cdot u^n$ or $\alpha \cdot u^{2n}$, for all $n \geq 1$, is an infinite collection of elements of norm N, which in turn provide infinitely many positive integer solutions of $x^2 - By^2 = N$. □

Example 14.3.19. Consider the equation $x^2 - 3y^2 = 6$. Clearly, there is a positive solution $(3, 1)$, since $3^2 - 3 = 6$. Moreover, the fundamental unit of $\mathbb{Z}[\sqrt{3}]$ is $u = 2 + \sqrt{3}$. Thus, if we write $\alpha = 3 + \sqrt{3}$, then $\alpha \cdot u^n$ will produce infinite positive solutions of $x^2 - 3y^2 = 6$. For instance,

$$\alpha \cdot u = 9 + 5\sqrt{3},$$
$$\alpha \cdot u^2 = 33 + 19\sqrt{3},$$
$$\alpha \cdot u^3 = 123 + 71\sqrt{3},$$
$$\alpha \cdot u^4 = 459 + 265\sqrt{3}.$$

Therefore, $(3, 1)$, $(9, 5)$, $(33, 19)$, $(123, 71)$, and $(459, 265)$ are positive solutions of $x^2 - 3y^2 = 6$.

Example 14.3.20. Consider the equation $x^2 - 3y^2 = 13$ and its least positive solution $(4, 1)$. As in the previous example, we can find more positive solutions by multiplying $\alpha = 4 + \sqrt{3}$ by the fundamental unit $u = 2 + \sqrt{3}$:

$$\alpha \cdot u = 11 + 6\sqrt{3},$$
$$\alpha \cdot u^2 = 40 + 23\sqrt{3},$$
$$\alpha \cdot u^3 = 149 + 86\sqrt{3},$$
$$\alpha \cdot u^4 = 556 + 321\sqrt{3}.$$

Therefore, $(4,1)$, $(11,6)$, $(40,23)$, $(149,86)$, and $(556,321)$ are positive solutions of $x^2 - 3y^2 = 13$. However, unlike in Theorem 14.3.13 and the case of Pell's equation, this method does not yield all positive solutions. For instance, consider $\alpha' = 4 - \sqrt{3}$ and

$$\alpha' \cdot u = 5 + 2\sqrt{3}$$

yields a point $(5,2)$ that would not appear as coordinates of an element $\alpha \cdot u^n$ for $n \geq 1$. Similarly, $\alpha' \cdot u^2 = 16 + 9\sqrt{3}$ which yields a point $(16,9)$ that we had skipped in our previous list.

14.4. Exercises

Exercise 14.4.1. Find parametrizations of all the rational points on the square hyperbolas $C : x^2 - 49y^2 = 51$ and $C' : x^2 - 49y^2 = 52$.

Exercise 14.4.2. Determine whether the following square hyperbolas have any integral points, and if so, find all of them:

(a) $xy = 17$.
(b) $x^2 - 9y^2 = 13$.
(c) $x^2 - 9y^2 = 14$.
(d) $x^2 - 25y^2 = 119$.
(e) $x^2 - 169y^2 = 560$.

Exercise 14.4.3. Let p and q be primes, and let $C : x^2 - p^2y^2 = q$.

(a) Show that C has infinitely many rational points.
(b) Show that C has integral points if and only if $q \equiv 1 \bmod 2p$.

Exercise 14.4.4. Suppose (a,b) is a positive solution of $x^2 - By^2 = -1$. Show that a/b is one of the convergents of the infinite continued fraction of $\sqrt{B}$. (Hint: adapt the proof of Theorem 14.2.3.)

Exercise 14.4.5. Find a positive solution for each of the following Pell's equations:

(a) $x^2 - 7y^2 = 1$.
(b) $x^2 - 10y^2 = 1$.
(c) $x^2 - 11y^2 = 1$.
(d) $x^2 - 13y^2 = 1$.
(e) $x^2 - 14y^2 = 1$.

(Hint: Exercise 13.4.18.)

Exercise 14.4.6. For each of the Pell's equations $x^2 - By^2 = 1$ in Exercise 14.4.5, find the first three positive solutions.

Exercise 14.4.7. Find the first five positive solutions for $x^2 - 10y^2 = 1$.

Exercise 14.4.8. Let $n \geq 1$ and let $B = 9n^2 + 6$. Find an expression (in terms of n) for the first and second positive solutions for the Pell's equation $x^2 - By^2 = 1$. (Hint: Exercise 13.4.21.)

Exercise 14.4.9. For each of the Pell's equations $x^2 - By^2 = 1$ in Exercise 14.4.5, determine whether the negative Pell's equation $x^2 - By^2 = -1$ has a positive solution, and if so, find one.

Exercise 14.4.10. Let p be a prime that is congruent to 3 mod 4, and let B be an integer divisible by p.

(1) Prove that $C : x^2 - By^2 = -1$ has no integral points.

(2) Conclude that if B is a positive integer, with a prime divisor p that is congruent to 3 mod 4, then the length of the period of the continued fraction of $\sqrt{B}$ must be even.

(3) Verify that the length of the period of the continued fraction of $\sqrt{3}$, $\sqrt{6}$, $\sqrt{7}$, $\sqrt{11}$, $\sqrt{15}$, and $\sqrt{21}$ is even.

Exercise 14.4.11. Prove Theorem 14.3.3. (Hint: see Examples 14.3.1 and 14.3.2, and follow the proof of Theorem 14.2.8.)

Exercise 14.4.12. Prove Proposition 14.3.6. (Hint: use Lemma 12.5.1 and Theorem 14.2.8. See also Example 14.3.5.)

Exercise 14.4.13. Find the first three positive solutions of $x^2 - 10y^2 = 9$ and $x^2 - 10y^2 = 15$.

Exercise 14.4.14. Show that if B is a non-zero integer that is not a perfect square, then the set $\mathbb{Z}[\sqrt{B}]$ is a (commutative) ring.

Exercise 14.4.15. Show that if $u = a + b\sqrt{B}$ is the smallest positive integer solution of Pell's equation $x^2 - By^2 = 1$ and (c, d) is an integer solution of $x^2 - By^2 = 1$ (where c, d are not necessarily positive), then there is some integer $n \in \mathbb{Z}$ such that $c + d\sqrt{B}$ equals u^n or $-u^n$. (Hint: use Theorem 14.3.13 and the fact that $(a + b\sqrt{B})^{-1} = a - b\sqrt{B}$.)

Exercise 14.4.16. The first positive solution (or fundamental solution) of the Pell's equation $x^2 - 103y^2 = 1$ is $(227528, 22419)$. Find the second positive solution. (Hint: use norms.)

Exercise 14.4.17. Find three positive solutions for the negative Pell's equation $x^2 - 101y^2 = -1$.

Exercise 14.4.18. Find the fundamental unit for each of the following quadratic rings: $\mathbb{Z}[\sqrt{7}]$, $\mathbb{Z}[\sqrt{10}]$, $\mathbb{Z}[\sqrt{11}]$, $\mathbb{Z}[\sqrt{13}]$, and $\mathbb{Z}[\sqrt{14}]$.

Exercise 14.4.19. Describe all the integral points on the hyperbola

$$C : 31x^2 + 96xy + 74y^2 = -1.$$

Exercise 14.4.20. Decide whether the following hyperbola has any integral points:

$$C : 27x^2 - 30xy + 6y^2 + 82x - 44y + 61 = 0.$$

Part 3

Cubic Equations and Elliptic Curves

CHAPTER 15

AN INTRODUCTION TO CUBIC EQUATIONS

> *"No," he replied, "1729 is a very interesting number; it is the smallest number expressible as the sum of two cubes in two different ways."*
>
> Srinivasa Ramanujan, 1920

Previously in this book, we have learned how to find all rational and integral points on curves in the plane given by linear and quadratic equations. In this third part, we give a brief introduction to the theory of cubic equations. We begin with an example that comes from a famous anecdote in the history of mathematics.

Example 15.0.1. The Indian mathematician Srinivasa Ramanujan died at an early age, when he was only 32 years old. G. H. Hardy, a British mathematician who arranged for Ramanujan to move to Cambridge (England), was his mentor and a close collaborator. See Figure 15.1. As Hardy recounted:

> *Every positive integer was one of [Ramanujan's] personal friends. [...] I remember once going to see him when he was ill at Putney. I had ridden in taxi cab number 1729 and remarked that the number seemed to me rather a dull one, and that I hoped it was not an unfavorable omen. "No," he replied, "it is a very interesting number; it is the smallest number expressible as the sum of two cubes in two different ways."*

Indeed,

$$1729 = 1^3 + 12^3 = 9^3 + 10^3.$$

The number 1729 is referred to as the Hardy–Ramanujan number and has become famous after the anecdote related above (e.g., the number 1729 and other taxicab numbers appear repeatedly in the TV show *Futurama*; for instance, Bender is his mother's 1729th descendant in the episode *Xmas Story*). It is worth pointing out

Figure 15.1. S. Ramanujan (1887–1920) in the center and G. H. Hardy (1877–1947) in the far right, at Cambridge University. Image source: Wikimedia Commons.

that Ramanujan was referring to sums of *positive* cubes, for if we allow positive and negative solutions, then 1729 is *not* the smallest integer that can be expressed as the sum of two cubes in two different ways:

$$91 = (-5)^3 + 6^3 = 3^3 + 4^3.$$

More generally, the smallest integer that can be written as the sum of two (positive) cubes in n different ways is called the nth taxicab number. For instance, the 3rd taxicab number is

$$87539319 = 167^3 + 436^3 = 228^3 + 423^3 = 255^3 + 414^3.$$

In terms of diophantine equations, a taxicab number yields a positive integer d such that the cubic equation $x^3 + y^3 = d$ has several integral positive solutions. We recommend [**Kan91**] for more about the life and work of Ramanujan.

Before we begin our introduction to cubic curves, we shall introduce projective geometry. The reason for this digression is that curves in the plane are sometimes better understood once they are "projectivized". For instance, if we projectivize $C : x^3 + y^3 = d$, we obtain a projective equation $X^3 + Y^3 = dZ^3$ which will allow us to find a change of variables from C to a curve $E : y^2 = x^3 - 432d^2$. The curve E is given by an equation (a so-called Weierstrass form) that is easier to work with in order to determine its rational and integral points (see Chapter 16 and Example 16.1.3 in particular).

15.1. The Projective Line and Projective Space

> *Projective geometry has opened up for us with the greatest facility new territories in our science, and has rightly been called the royal road to our particular field of knowledge.*
>
> Felix Klein

In this section we introduce projective geometry by constructing the projective line and plane. For a thorough introduction to projective geometry, see [**SK52**].

15.1.1. The Projective Line. Let us begin with an example. Consider the function $f(x) = \frac{1}{x}$. We know from calculus that f is continuous (and differentiable) on all of the real numbers except at $x = 0$. Would it be possible to extend the real line so that $f(x)$ is continuous everywhere? The answer is yes, it is possible, and the solution is to *glue* the "end" of the real line at ∞ with the other "end" at $-\infty$. We will describe the solution in detail below. Formally, we need the *projective line*, which is a line with points $\mathbb{R} \cup \{\infty\}$, i.e., a real line plus a single point at infinity that ties the line together (to form a shape resembling a circle).

The formal definition of the projective line is as follows (it may seem a little confusing at first, but it is a fairly efficient definition to work and compute with). First, we need to define a relation between vectors of real numbers in the plane. Let a, b, x, y be real numbers such that neither (x, y) nor (a, b) is the zero vector. We say that $(x, y) \sim (a, b)$ if the vector (x, y) is a non-zero multiple of the vector (a, b). In other words, if we consider (a, b) and (x, y) as points in the plane, we say that $(a, b) \sim (x, y)$ if they both lie in one line on the plane that passes through the origin. Let us write $\mathbb{R}^* = \mathbb{R} - \{0\}$. Then,

$$(x, y) \sim (a, b) \text{ if and only if there is } \lambda \in \mathbb{R}^* \text{ such that } x = \lambda a,\ y = \lambda b.$$

For instance, $(\sqrt{2}, \sqrt{2}) \sim (1, 1)$. We denote by $[x, y]$ the set of all vectors (a, b) such that $(x, y) \sim (a, b)$:

$$[x, y] = \{(a, b) : a, b \in \mathbb{R} \text{ such that } (a, b) \neq (0, 0) \text{ and } (x, y) \sim (a, b)\}.$$

Finally, we define the *real projective line* by

$$\mathbb{P}^1(\mathbb{R}) = \{[x, y] : x, y \in \mathbb{R} \text{ with } (x, y) \neq (0, 0)\}.$$

The reader can verify that $\sim$ is an example of an equivalence relation (see Remark 4.2.5) and each set $[x, y]$ is an equivalence class for the relation $\sim$.

The projective line $\mathbb{P}^1(\mathbb{R})$ can be regarded as the set of all lines through the origin (each class $[x, y]$ consists of all points except the origin on the line that goes through (x, y) and $(0, 0)$). It is important to notice that if $[x, y] \in \mathbb{P}^1(\mathbb{R})$ and $y \neq 0$, then $(x, y) \sim (\frac{x}{y}, 1)$, so the class of $[x, y]$ contains a unique representative of the form $(a, 1)$ for some $a = \frac{x}{y} \in \mathbb{R}$. This allows the following decomposition of $\mathbb{P}^1(\mathbb{R})$:

$$\mathbb{P}^1(\mathbb{R}) = \{[x, 1] : x \in \mathbb{R}\} \cup \{[1, 0]\}.$$

The set of points $\{[x, 1]\}$ is in bijection with $\mathbb{R}$ and, therefore, forms a real line (or *affine* line). The point $[1, 0]$, which is the only point in $\mathbb{P}^1(\mathbb{R})$ that does not belong to the real line $\{[x, 1]\}$, is called the *point at infinity* (see Figure 15.2).

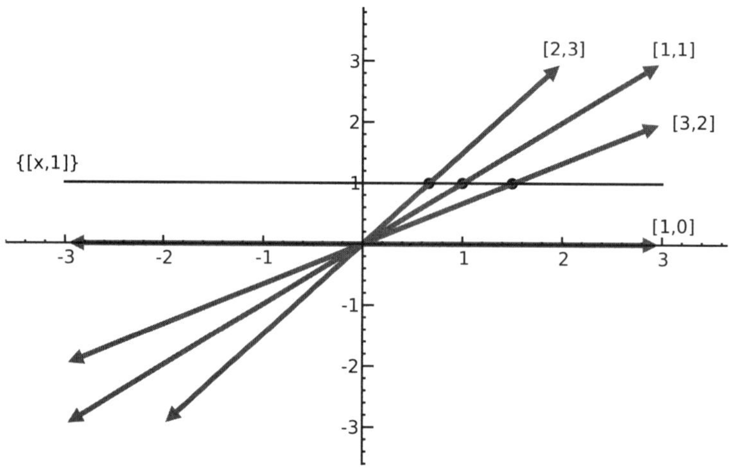

Figure 15.2. Some points in the projective line, e.g., $[2,3] \in \mathbb{P}^1(\mathbb{R})$, and their representatives of the form $[x,1]$, e.g., $[\frac{2}{3},1]$, except for $[1,0]$.

Notice that when $x \in \mathbb{R}$ gets "large" (i.e., far away from 0, that is, when $x \to \infty$ or $x \to -\infty$), the point $[x,1] \in \mathbb{P}^1(\mathbb{R})$ corresponds to a line in the real plane that is closer and closer to the horizontal line. Since the horizontal line corresponds to the point $[1,0] \in \mathbb{P}^1(\mathbb{R})$, we see that as x gets large (in either the positive or negative direction!), the points $[x,1]$ get closer and closer to $[1,0]$, the point at infinity. This is what we meant at the beginning of this section by "glueing" both ends of the real line, ∞ and $-\infty$, at one point.

Let us see that, with this definition, the function $f\colon \mathbb{R} \to \mathbb{R}$, $f(x) = 1/x$ is continuous everywhere when extended to $\mathbb{P}^1(\mathbb{R})$. We define instead an extended function $F\colon \mathbb{P}^1(\mathbb{R}) \to \mathbb{P}^1(\mathbb{R})$ by

$$F([x,y]) = [y,x].$$

Notice that a point on the real line of $\mathbb{P}^1$, i.e., a point of the form $[x,1]$, is sent to the point $[1,x]$ of $\mathbb{P}^1$, and $(1,x) \sim (\frac{1}{x},1)$ as long as $x \neq 0$. So $[x,1]$ with $x \neq 0$ is sent to $[\frac{1}{x},1]$ via F (i.e., the real point x is sent to $\frac{1}{x}$). Hence, F coincides with f on $\mathbb{R} - \{0\}$. But F is perfectly well-defined on $x = 0$, i.e., on the point $[0,1]$, and $F([0,1]) = [1,0]$ so that $[0,1]$ is sent to the point at infinity. Moreover, limits from both sides coincide:

$$\lim_{x\to 0^+} F([x,1]) = \lim_{x\to 0^-} F([x,1]) = F([0,1]) = [1,0].$$

15.1.2. The Projective Plane. We may generalize the construction above of the projective line in order to construct a projective plane that will consist of a real plane plus a number of points at infinity, one for each direction in the plane; i.e., the projective plane will be a real plane plus a projective line of points at infinity.

Let $a,b,c,x,y,z \in \mathbb{R}$ such that neither (a,b,c) nor (x,y,z) is the zero vector, and write $\mathbb{R}^* = \mathbb{R} - \{0\}$ as before. We define a relation $\sim$ as follows:

$(x,y,z) \sim (a,b,c)$ if and only if there is $\lambda \in \mathbb{R}^*$ such that $x = \lambda a$, $y = \lambda b$, $z = \lambda c$.

We also define classes of similar vectors by

$$[x, y, z] = \{(a, b, c) : a, b, c \in \mathbb{R} \text{ such that } (a, b, c) \neq \vec{0} \text{ and } (x, y, z) \sim (a, b, c)\}.$$

Notice that, as in the case of the projective line, the class $[x, y, z]$ contains all the points in the line in $\mathbb{R}^3$ that goes through (x, y, z) and $(0, 0, 0)$ except the origin. We define the *projective plane* to be the collection of all such lines:

$$\mathbb{P}^2(\mathbb{R}) = \{[x, y, z] : x, y, z \in \mathbb{R} \text{ such that } (x, y, z) \neq (0, 0, 0)\}.$$

If $z \neq 0$, then $(x, y, z) \sim (\frac{x}{z}, \frac{y}{z}, 1)$. Thus,

$$\mathbb{P}^2(\mathbb{R}) = \{[x, y, 1] : x, y \in \mathbb{R}\} \cup \{[a, b, 0] : a, b \in \mathbb{R}\}.$$

The points of the set $\{[x, y, 1] : x, y \in \mathbb{R}\}$ are in 1-to-1 correspondence with the real plane $\mathbb{R}^2$ (or *affine plane*), and the points in $\{[a, b, 0] : a, b \in \mathbb{R}\}$ are called the points at infinity and form a $\mathbb{P}^1(\mathbb{R})$, a projective line.

Remark 15.1.1. One interesting consequence of the definitions is that any two parallel lines in the real plane $\{[x, y, 1]\}$ intersect at a point at infinity $[a, b, 0]$. Indeed, let $L : y = mx + b$ and $L' : y = mx + b'$ be distinct parallel lines in the real plane. If points in the real plane $\{[x, y, 1]\}$ correspond to lines in $\mathbb{R}^3$, then lines in the real plane correspond to *planes* in $\mathbb{R}^3$:

$$L = \{[x, y, z] : mx - y + bz = 0\}, \quad L' = \{[x, y, z] : mx - y + b'z = 0\}.$$

What is $L \cap L'$? The intersection points are those $[x, y, z]$ such that $mx - y + bz = mx - y + b'z = 0$, which implies that $(b - b')z = 0$. Since $L \neq L'$, we have $b \neq b'$ and, therefore, we must have $z = 0$. Hence

$$L \cap L' = \{[x, mx, 0] : x \in \mathbb{R}\} = \{[1, m, 0]\},$$

and so the intersection consists of a single point at infinity: $[1, m, 0]$.

A subset of $\mathbb{P}^2(\mathbb{R})$ that is in bijection with a real plane $\mathbb{R}^2$ is called an *affine chart* of the projective plane. For instance, the subset $\{[x, y, 1] : x, y \in \mathbb{R}\}$ is an affine chart. The three principal affine charts are those given by

$$\{[x, y, 1] : x, y \in \mathbb{R}\}, \ \{[x, 1, z] : x, z \in \mathbb{R}\}, \ \text{ and } \ \{[1, y, z] : y, z \in \mathbb{R}\},$$

but of course there are many other affine charts; for instance, $\{[x, y, 1 - x - y] : x, y \in \mathbb{R}\}$ defines a different affine chart.

Example 15.1.2. Consider the subset of the projective plane given by

$$C = \{[x, y, z] \in \mathbb{P}^2(\mathbb{R}) : x^2 + y^2 = z^2\}.$$

First, let us verify that C is well-defined in the projective plane. For this, we need to show that if $[x, y, z] \in C$, then $[\lambda x, \lambda y, \lambda z]$ is also in C, for any $\lambda \in \mathbb{R}^*$, since $[x, y, z] = [\lambda x, \lambda y, \lambda z]$. Indeed, if $[x, y, z] \in C$, then $x^2 + y^2 = z^2$, and therefore,

$$(\lambda x)^2 + (\lambda y)^2 = \lambda^2 x^2 + \lambda^2 y^2 = \lambda^2 \cdot (x^2 + y^2) = \lambda^2 z^2 = (\lambda z)^2.$$

Thus, $[\lambda x, \lambda y, \lambda z]$ is also in C.

Let us understand the geometry of C by using an affine chart. In particular, we shall use the decomposition

$$\mathbb{P}^2(\mathbb{R}) = \{[x, y, 1] : x, y \in \mathbb{R}\} \cup \{[a, b, 0] : a, b \in \mathbb{R}\}.$$

Let us see what $C \cap \{[x, y, 1]\}$ is:

$$C \cap \{[x, y, 1]\} = \{[x, y, 1] \in \mathbb{P}^2(\mathbb{R}) : x^2 + y^2 = 1\}.$$

Thus, $C \cap \{[x, y, 1]\}$ is a circle of radius 1 in the affine chart. It remains to see what $C \cap \{[a, b, 0] : a, b \in \mathbb{R}\}$ is:

$$C \cap \{[a, b, 0] : a, b \in \mathbb{R}\} = \{[a, b, 0] : a^2 + b^2 = 0\},$$

but $a^2 + b^2 = 0$ for $a, b \in \mathbb{R}$ implies $a = b = 0$. Thus, $C \cap \{[a, b, 0] : a, b \in \mathbb{R}\} = \emptyset$.

We conclude that $C \subseteq \mathbb{P}^2(\mathbb{R})$ is in bijection with a circle in the affine plane. Notice, however, that if we choose a different chart, we might see a different shape. For instance, if we instead decompose

$$\mathbb{P}^2(\mathbb{R}) = \{[x, 1, z] : x, z \in \mathbb{R}\} \cup \{[a, 0, c] : a, c \in \mathbb{R}\},$$

then $C \cap \{[x, 1, z]\} = \{[x, 1, z] \in \mathbb{P}^2(\mathbb{R}) : z^2 - x^2 = 1\}$ is a hyperbola and not a circle as we saw when we used a different affine chart.

15.1.3. Working over an Arbitrary Field. The projective line and plane can be defined over any field. Let K be a field (e.g., $K = \mathbb{Q}, \mathbb{R}, \mathbb{C}$, or $\mathbb{F}_p$). The usual *affine plane* (or euclidean plane) is defined by

$$\mathbb{A}^2(K) = \{(x, y) : x, y \in K\}.$$

The projective plane over K is defined by

$$\mathbb{P}^2(K) = \{[x, y, z] : x, y, z \in K \text{ such that } (x, y, z) \neq (0, 0, 0)\}.$$

As before, $(x, y, z) \sim (a, b, c)$ if and only if there is $\lambda \in K$ such that $(x, y, z) = \lambda \cdot (a, b, c)$.

We will work with curves defined in projective space over $\mathbb{F}_p$ in Section 16.5.

15.1.4. Curves in the Projective Plane. Let K be a field and let C be a curve in affine space, given by a polynomial in two variables:

$$C : f(x, y) = 0$$

for some $f(x, y) \in K[x, y]$; e.g., $C : y^2 - x^3 - 1 = 0$. We want to extend C to a curve in the projective plane $\mathbb{P}^2(K)$. In order to do this, we consider the points on the curve (x, y) to be points in the plane $[\frac{x}{z}, \frac{y}{z}, 1]$ of $\mathbb{P}^2(K)$. Thus, we have

$$C : \left(\frac{y}{z}\right)^2 - \left(\frac{x}{z}\right)^3 - 1 = 0,$$

or, equivalently, $zy^2 - x^3 - z^3 = 0$. Notice that the polynomial $F(x, y, z) = zy^2 - x^3 - z^3$ is homogeneous in its variables (each monomial has the same degree, in this case 3) and $F(x, y, 1) = f(x, y)$. The curve in $\mathbb{P}^2(K)$, given by

$$\widehat{C} : F(x, y, z) = zy^2 - x^3 - z^3 = 0,$$

is the curve we were looking for, which extends our original curve C in the affine plane, and that we will call the *projectivization* of C. Notice that if the point $(x, y) \in C$, then $[x, y, 1] \in \widehat{C}$. However, there may be some extra points in $\widehat{C}$ which were not present in C, namely those points of $\widehat{C}$ at infinity. Recall that the points at infinity are those with $z = 0$, so $F(x, y, 0) = -x^3 = 0$ implies that $x = 0$ also, and the only point at infinity in $\widehat{C}$ is $[0, 1, 0]$.

In general, if $C \subseteq \mathbb{A}^2(K)$ is given by $f(x, y) = 0$ and d is the highest degree of a monomial in f, then $\widehat{C} \in \mathbb{P}^2(K)$ is given by

$$\widehat{C} : F(x, y, z) = 0,$$

where $F(x, y, z) = z^d \cdot f\left(\frac{x}{z}, \frac{y}{z}\right)$. Conversely, if $\widehat{C} : F(x, y, z) = 0$ is a curve in the projective plane, then $C : F(x, y, 1) = 0$ is a curve in the affine plane. In this case, C is the projection of $\widehat{C}$ onto the affine chart $\{[x, y, 1]\}$; we may also look at other charts, e.g., $\{[1, y, z]\}$, which would yield an affine curve $C' : F(1, y, z) = 0$.

Example 15.1.3. Let C be given by

$$C : y - x^2 = 0$$

so that C is a parabola in the affine plane $\mathbb{A}^2(\mathbb{R})$. Then, the projectivization $\widehat{C}$ of C is given by

$$\widehat{C} : F(x, y, z) = z^2 f\left(\frac{x}{z}, \frac{y}{z}\right) = zy - x^2 = 0.$$

The curve $\widehat{C}$ has a unique point at infinity, namely $[0, 1, 0]$. This means that the two "arms" of the parabola meet at a single point at infinity. Thus, a parabola has the shape of an ellipse in $\mathbb{P}^2(K)$. How about hyperbolas? Let

$$C : x^2 - y^2 = 1.$$

Then $\widehat{C} : x^2 - y^2 = z^2$ and there are two points at infinity, namely $[1, 1, 0]$ and $[1, -1, 0]$. Thus, the four arms of the hyperbola in the affine plane meet in two points, and the hyperbola also has the shape of an ellipse in the projective plane $\mathbb{P}^2(K)$.

Example 15.1.4. Let C be the parabola $y = x^2$. Then, the projectivization of C is given by $\widehat{C} : YZ - X^2 = 0$. Our original curve C corresponds to the points of $\widehat{C}$ in the affine chart $\{[X, Y, Z] = [x, y, 1]\}$ of $\mathbb{P}^2$. If instead we look at the points on $\widehat{C}$ that belong to the affine chart $\{[X, Y, Z] = [1, y, z]\}$, then we obtain a curve $yz = 1$ in the affine plane, which is a square hyperbola.

Example 15.1.5. Let C be the hyperbola $x^2 - 2y^2 = 1$. Then, the projectivization of C is given by $\widehat{C} : X^2 - 2Y^2 = Z^2$. Our original curve C corresponds to the points of $\widehat{C}$ in the affine chart $\{[X, Y, Z] = [x, y, 1]\}$ of $\mathbb{P}^2$. If instead we look at the points on $\widehat{C}$ that belong to the affine chart $\{[1, Y, Z] = [1, y, z]\}$, then we obtain a curve $1 = z^2 + 2y^2$ in the affine plane, which is the equation of an ellipse.

Example 15.1.6. Let $d \geq 1$, and let C be the curve $x^3 + y^3 = d$ that already appeared in Example 15.0.1. Then, the projectivization of C is $\widehat{C} : X^3 + Y^3 = dZ^3$. We note that if we decompose the projective plane as

$$\mathbb{P}^2(\mathbb{R}) = \{[x, y, 1] : x, y \in \mathbb{R}\} \cup \{[a, b, 0] : a, b \in \mathbb{R}\},$$

then the intersection of $\widehat{C}$ with the affine chart $\{[x, y, 1]\}$ is in bijection with C, while the intersection of $\widehat{C}$ with $\{[a, b, 0] : a, b \in \mathbb{R}\}$ gives

$$\widehat{C} \cap \{[a, b, 0] : a, b \in \mathbb{R}\} = \{[a, b, 0] : a^3 + b^3 = 0\} = \{[1, -1, 0]\}$$

which consists of one projective point at infinity. It is worth pointing out that if we were working over $\mathbb{C}$, then

$$\widehat{C} \cap \{[a, b, 0] : a, b \in \mathbb{C}\} = \{[a, b, 0] : a^3 + b^3 = 0\} = \{[1, \theta, 0] : \theta^3 = -1\}$$

and there are three choices for $\theta \in \mathbb{C}$ such that $\theta^3 = -1$, so there are three complex points at infinity on $\widehat{C}$, and only one of them is real.

15.1.5. Singular and Smooth Curves.

In so far as the statements of geometry speak about reality, they are not certain, and in so far as they are certain, they do not speak about reality.

Albert Einstein

Definition 15.1.7. We say that a projective curve $C : F(x, y, z) = 0$ is *singular* at a point $P \in C$ if and only if $\frac{\partial F}{\partial x}(P) = \frac{\partial F}{\partial y}(P) = \frac{\partial F}{\partial z}(P) = 0$. In other words, C is singular at P if the normal vector at P vanishes. Otherwise, we say that C is *non-singular (or smooth)* at P. If C is non-singular at every point, we say that C is a *smooth (or non-singular)* curve.

Example 15.1.8. Let d be a fixed positive integer, and let $\widehat{C} : X^3 + Y^3 = dZ^3$ be the projective curve that appeared in Example 15.1.6 associated to the taxicab numbers. Let $F(X, Y, Z) = X^3 + Y^3 - dZ^3$. Since

$$\frac{\partial F}{\partial X} = 3X^2, \quad \frac{\partial F}{\partial Y} = 3Y^2, \quad \frac{\partial F}{\partial Z} = -3dZ^2,$$

it follows that the only point P with $\frac{\partial F}{\partial X}(P) = \frac{\partial F}{\partial Y}(P) = \frac{\partial F}{\partial Z}(P) = 0$ would have $P = [0, 0, 0]$ which is not an allowed projective point. Hence, $\widehat{C}$ is non-singular at all points of $\mathbb{P}^2(\mathbb{C})$.

Example 15.1.9. For example, $C : zy^2 = x^3$ is singular at $P = [0, 0, 1]$ because $F(x, y, z) = zy^2 - x^3$ and

$$\frac{\partial F}{\partial x} = -x^2, \quad \frac{\partial F}{\partial y} = 2yz, \quad \frac{\partial F}{\partial z} = y^2.$$

Thus, $\frac{\partial F}{\partial x}(P) = \frac{\partial F}{\partial y}(P) = \frac{\partial F}{\partial z}(P) = 0$ for $P = [0, 0, 1]$. See Figure 15.3.

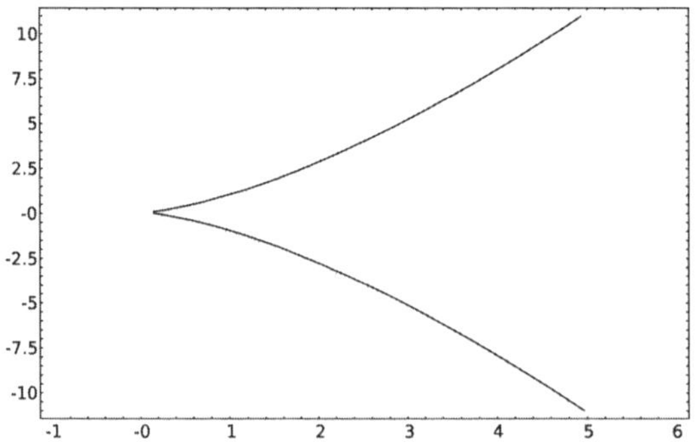

Figure 15.3. The chart $\{[x, y, 1]\}$ of the curve $zy^2 = x^3$.

Example 15.1.10. The curve $D : z^2y^2 = x^4 + z^4$ has partial derivatives

$$\frac{\partial F}{\partial x} = -4x^3, \quad \frac{\partial F}{\partial y} = 2yz^2, \quad \frac{\partial F}{\partial z} = 2y^2z - 4z^3.$$

Thus, if $P = [x, y, z] \in D(\mathbb{Q})$ is singular, then

$$-4x^3 = 0, \quad 2yz^2 = 0, \quad \text{and} \quad 2y^2z - 4z^3 = 0.$$

The first two equalities imply that $x = 0$ and $yz = 0$ (what would happen if we were working over a field of characteristic 2, such as $\mathbb{F}_2$?). If $y = 0$, then $z = 0$ by the third equation, but $[0, 0, 0]$ is not a well-defined point in $\mathbb{P}^2(\mathbb{Q})$, so this is impossible. However, if $x = z = 0$, then y may take any value. Hence, $P = [0, 1, 0]$ is a singular point. Notice that the affine curve that corresponds to the chart $z = 1$ of D, given by $y^2 = x^4 + 1$, is non-singular at all points in the affine plane but is singular at a point at infinity; namely $P = [0, 1, 0]$. See Figure 15.4.

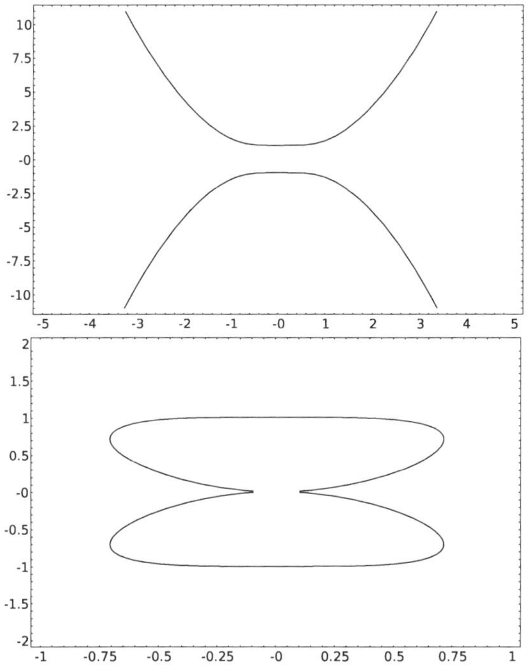

Figure 15.4. The chart $\{[x, y, 1]\}$ of the curve $z^2y^2 = x^4 + z^4$ (top, non-singular) and the chart $\{[x, 1, z]\}$ (bottom, singular).

Example 15.1.11. As we shall see in Chapter 16, a curve of the form $E : y^2 = x^3 + Ax + B$, or in projective coordinates given by $zy^2 = x^3 + Axz^2 + Bz^3$, is an example of an *elliptic curve* as long as it is non-singular. The reader can show that E is non-singular if and only if $4A^3 + 27B^2 \neq 0$ (use Exercise 5.6.34). The quantity $\Delta = -16 \cdot (4A^3 + 27B^2)$ is called the discriminant of E.

15.2. Singular Cubic Curves

In the third part of this book, we are interested in describing the rational points on cubic curves

$$C : f(x, y) = ax^3 + bx^2y + cxy^2 + dy^3 + ex^2 + fxy + gy^2 + hx + jy + k = 0.$$

We distinguish two cases according to whether C has a singularity (as a projective curve; see Section 15.1.5 for the definition of singular and non-singular curves). In this section we suppose that the curve $C : f(x, y) = 0$ is singular and will show that, in this case, we can parametrize all the rational points on C.

In this section we will assume that the singularity occurs at the given point $P = (x_0, y_0) \in C(\mathbb{Q})$. If none of the singular points have $\mathbb{Q}$-coordinates, then we will treat the curve in the next subsection together with the non-singular curves. By taking an appropriate affine patch (see Example 15.2.4 below), we may assume that the singularity occurs at an affine point $P = (x_0, y_0)$, and in fact, by a linear change of variables $(x, y) \mapsto (x + x_0, y + y_0)$ we can assume $P = (0, 0)$. Since P is singular, it follows that $\partial f/\partial x(P) = \partial f/\partial y(P) = 0$. In particular, the Taylor expansion of $f(x, y)$ around $(x, y) = (0, 0)$ is given by

$$f(x, y) = f(0, 0) + \frac{\partial f}{\partial y}(0, 0) \cdot x + \frac{\partial f}{\partial y}(0, 0) \cdot y + (\text{higher-degree terms}).$$

It follows that the constant and linear terms of $f(x, y)$ are zero. Thus,

$$f(x, y) = ax^3 + bx^2y + cxy^2 + dy^3 + ex^2 + fxy + gy^2 = 0,$$

or, equivalently,

$$\frac{1}{y^3} \cdot f(x, y) = a \cdot \frac{x^3}{y^3} + b \cdot \frac{x^2}{y^2} + c \cdot \frac{x}{y} + d + e \cdot \frac{x^2}{y^2} \cdot \frac{1}{y} + f \cdot \frac{x}{y} \cdot \frac{1}{y} + g \cdot \frac{1}{y} = 0.$$

Now, we apply a change of variables $X = x/y$ and $Y = 1/y$ to obtain

$$C' : g(X, Y) = aX^3 + bX^2 + cX + d + eX^2Y + fXY + gY = 0$$

and a map $C \to C'$ defined by

$$\phi(x_1, y_1) = \left(\frac{x_1}{y_1}, \frac{1}{y_1}\right),$$

as long as $y_1 \neq 0$. However, note that not much is lost by this restriction: if $y_1 = 0$ and $f(x_1, y_1) = 0$, then $ax_1^3 + ex_1^2 = x_1^2(ax_1 + e) = 0$, and so $x_1 = 0$ or $x_1 = -e/a$. Thus, we obtain a well-defined map ϕ: $C \setminus \{(0, 0), (-e/a, 0)\} \to C'$ defined as above such that

$$\phi^{-1}(X_1, Y_1) = \left(\frac{X_1}{Y_1}, \frac{1}{Y_1}\right) = \phi(X_1, Y_1).$$

Notice, however, that we can parametrize C', since, for any value of X_1 we have

$$Y_1 = -\frac{aX_1^3 + bX_1^2 + cX_1 + d}{eX_1^2 + fX_1 + g}$$

as long as $eX_1^2 + fX_1 + g \neq 0$. Otherwise, we must have $aX_1^3 + bX_1^2 + cX_1 + d = 0 = eX_1^2 + fX_1 + g$. Thus, we have shown the following theorem.

Theorem 15.2.1. *Let $C : f(x,y) = 0$ be a cubic curve with a singularity at $(0,0)$, given by*

$$C : ax^3 + bx^2y + cxy^2 + dy^3 + ex^2 + fxy + gy^2 = 0.$$

Then, $C(\mathbb{Q})$ is formed by the points $\{(0,0), (-e/a, 0)\}$ and

$$\left\{\left(-\frac{t(et^2+ft+g)}{at^3+bt^2+ct+d}, -\frac{et^2+ft+g}{at^3+bt^2+ct+d}\right) : t \in \mathbb{Q},\ at^3+bt^2+ct+d \neq 0\right\}$$

together with all points of the form $\left\{\left(\frac{\alpha}{s}, \frac{1}{s}\right) : s \in \mathbb{Q}\right\}$, for each $\alpha \in \mathbb{Q}$ such that $e\alpha^2 + f\alpha + g = 0 = a\alpha^3 + b\alpha^2 + c\alpha + d$.

Let us see a few examples.

Example 15.2.2. Let $C : y^2 = x^3$, which is singular at $(0,0)$. Instead of applying Theorem 15.2.1 directly, let us follow the ideas of the proof to find a parametrization. We begin with a curve $C : -x^3 + y^2 = 0$ that is singular at $(0,0)$. If we divide through by y^3, we obtain

$$-\left(\frac{x}{y}\right)^3 + \frac{1}{y} = 0,$$

so a change of variables $X = x/y$ and $Y = 1/y$ (with inverse $x = X/Y$, $y = 1/Y$) leads to

$$C' : -X^3 + Y = 0,$$

or, equivalently, $Y = X^3$. Clearly, all the rational points on C' are of the form $\{(t, t^3) : t \in \mathbb{Q}\}$, and reversing the change of variables, we obtain points

$$\left\{\left(\frac{1}{t^2}, \frac{1}{t^3}\right) : t \in \mathbb{Q},\ t \neq 0\right\} \subseteq C(\mathbb{Q}).$$

Moreover, any rational point (x_0, y_0) in $C(\mathbb{Q})$ gives a point in $(x_0/y_0, 1/y_0) \in C'(\mathbb{Q})$, unless $y_0 = 0$, in which case $x_0^3 = 0$ and so $x_0 = 0$. Hence, we have shown that all the rational points on C are given by

$$\begin{aligned} C(\mathbb{Q}) &= \{(0,0)\} \cup \left\{\left(\frac{1}{t^2}, \frac{1}{t^3}\right) : t \in \mathbb{Q},\ t \neq 0\right\} \\ &= \{(0,0)\} \cup \{(s^2, s^3) : s \in \mathbb{Q},\ s \neq 0\} = \{(r^2, r^3) : r \in \mathbb{Q}\}, \end{aligned}$$

which coincides with the conclusion of Theorem 15.2.1.

Example 15.2.3. Let $C : x^3 - y^3 + x^2 - 2xy + y^2 = 0$. Then, C is singular at $(0,0)$. Dividing through by y^3, we obtain

$$C' : X^3 - 1 + X^2Y - 2XY + Y - 0$$

with a map $\phi\colon C(\mathbb{Q}) \setminus \{(0,0), (-1,0)\} \to C'(\mathbb{Q})$ given by $\phi(x,y) = (x/y, 1/y)$ with inverse $\phi^{-1}(X,Y) = (X/Y, 1/Y)$, as long as $Y \neq 0$. Since $C' : X^3 - 1 + (X^2 - 2X + 1)Y = 0$, the points on C' are either of the form

$$\left\{\left(t, -\frac{t^3-1}{t^2-2t+1}\right) : t^2 - 2t + 1 \neq 0\right\} = \left\{\left(t, -\frac{t^2+t+1}{t-1}\right) : t \neq 1\right\}$$

or $(t-1)^2 = 0$ and $t^3 - 1 = 0$. In this case, $t = 1$ is a common solution, so there is an additional set of solutions $\{(1, s) : s \in \mathbb{Q}\} \subset C'(\mathbb{Q})$. Hence,

$$\begin{aligned} C(\mathbb{Q}) &= \{(0,0), (-1,0)\} \cup \phi^{-1}(C'(\mathbb{Q})) \\ &= \{(0,0), (-1,0)\} \cup \left\{\left(-\frac{t(t-1)}{t^2+t+1}, -\frac{t-1}{t^2+t+1}\right)\right\} \cup \left\{\left(\frac{1}{s}, \frac{1}{s}\right) : s \in \mathbb{Q}^*\right\}. \end{aligned}$$

We note now that $(r, r) \in C(\mathbb{Q})$ for all $r \in \mathbb{Q}$. This is a consequence of the fact that

$$C : (x-y)(x^2 + xy + y^2 + x - y) = 0,$$

or, in other words, C is the union of the line $L : x = y$ and the quadratic curve $Q : x^2 + xy + y^2 + x - y = 0$. Using our methods from Part 2, we find out that Q is in fact a non-singular conic, and from Corollary 9.2.12, we see that Q is an ellipse (with a bijection to $X^2 + 3Y^2 = 12$), and our previous work in this example implies a parametrization

$$Q(\mathbb{Q}) = \left\{\left(-\frac{t(t-1)}{t^2+t+1}, -\frac{t-1}{t^2+t+1}\right) : t \in \mathbb{Q}\right\}.$$

Example 15.2.4. Let $C : yx + 1 = x^3$. Let us first see if C is singular. For this, we calculate the normal vector

$$\vec{n}_C = (y - 3x^2, x).$$

Thus, $\vec{n}_C = (0,0)$ if and only if $(x, y) = (0,0)$, but this point does not belong to C. It follows that C is non-singular in the affine plane. However, we need to be concerned about the possibility of singularities at infinity. Let us find a projective equation for C (see Section 15.1.4 for more details):

$$C : XYZ + Z^3 - X^3 = 0.$$

The normal vector is now given by

$$\vec{n}_C = (YZ - 3X^2, XZ, XY + 3Z^2),$$

and $\vec{n}_C = (0,0,0)$ if and only if $X = Z = 0$. Since the point $[0,1,0]$ belongs to C, it follows that C is indeed singular at $P = [0,1,0]$. Let us choose a different affine chart so that P belongs to it. Since the Y-coordinate of P is non-zero, let us choose the chart $\{[x, 1, z]\}$, so that C is now given by

$$C : xz + z^3 - x^3 = 0.$$

We may now parametrize $C : -x^3 + z^3 + xz = 0$, using Theorem 15.2.1 (first divide through by z^3, etc.). The result is

$$\left\{\left(\frac{t^2}{t^3-1}, \frac{t}{t^3-1}\right) : t \in \mathbb{Q},\ t \neq 1\right\},$$

or, in projective coordinates,

$$\{[t^2, t^3 - 1, t] : t \in \mathbb{Q},\ t \neq 1\}.$$

Finally, if we return to the affine chart $\{[x, y, 1]\}$, then we obtain a parametrization

$$\left\{\left(t, \frac{t^3-1}{t}\right) : t \in \mathbb{Q},\ t \neq 0, 1\right\}.$$

Notice, however, that this is a parametrization of those points in C with $y \neq 0$. Since $(x, y) = (1, 0) \in C$ is the only point in C with $y = 0$, we finally obtain

$$C(\mathbb{Q}) = \left\{ \left(t, \frac{t^3 - 1}{t} \right) : t \in \mathbb{Q},\ t \neq 0 \right\}.$$

15.3. Weierstrass Equations

In this section we will show that a non-singular cubic curve

$$C : f(x, y) = ax^3 + bx^2y + cxy^2 + dy^3 + ex^2 + fxy + gy^2 + hx + jy + k = 0$$

with at least one rational solution $P = (x_0, y_0) \in C(\mathbb{Q})$ can be simplified to a much simpler equation via a certain change of variables. We will find a transformation to a model of the form

$$C' : y^2 = x^3 + Ax + B,$$

which is usually called a (short) *Weierstrass model* for the curve.

Definition 15.3.1. An equation of the form

$$y^2 + a_1xy + a_3y = x^3 + a_2x^2 + a_4x + a_6$$

for some constants $a_i \in \mathbb{Q}$ is called a *Weierstrass equation*, or *Weierstrass form*. When $a_1 = a_2 = a_3 = 0$, the equation reduces to the form

$$y^2 = x^3 + a_4x + a_6$$

and it is called a *short Weierstrass equation* (or form).

15.3.1. From Long to Short Weierstrass Forms.

Let us first show that any cubic equation given by a (long) Weierstrass form can be reduced to a short Weierstrass form. Let C be given by

$$C : y^2 + a_1xy + a_3y = x^3 + a_2x^2 + a_4x + a_6$$

for some constants $a_i \in \mathbb{Q}$. Then, we can complete the square on the y variable

$$\left(y + \frac{a_1x + a_3}{2} \right)^2 = x^3 + a_2x^2 + a_4x + a_6 + \left(\frac{a_1x + a_3}{2} \right)^2.$$

Now, a change of variables $u = x$ and $v = y + (a_1x + a_3)/2$ puts the equation in the form

$$v^2 = u^3 + au^2 + bu + c$$

for some $a, b, c \in \mathbb{Q}$. Finally, $s = u + a/3$, $t = v$ brings the equation to the form

$$t^2 = s^3 + As + B$$

for some $A, B \in \mathbb{Q}$.

Example 15.3.2. Let $C : y^2 + xy + y = x^3 - 2x^2$. Then,

$$(y + (x + 1)/2)^2 = x^3 - 2x^2 + ((x + 1)/2)^2$$

and so $(y + (x+1)/2)^2 = x^3 - 2x^2 + x^2/4 + x + 1/4 = x^3 - 7/4x^2 + x + 1/4$. Thus, a change of variables

$$t = y + (x + 1)/2 \quad \text{and} \quad s = x - 7/12$$

brings the equation to

$$t^2 = (s+7/12)^3 - 7/4(s+7/12)^2 + (s+7/12) + 1/4 = s^3 - 1/48s + 377/864.$$

Hence, we obtain $C' : t^2 = s^3 - 1/48s + 377/864$ and a map $\phi\colon C \to C'$ given by

$$\phi(x,y) = \left(x - \frac{7}{12},\ y + \frac{x+1}{2}\right),$$

with inverse $\phi^{-1}\colon C' \to C$ given by

$$\phi^{-1}(s,t) = \left(s + \frac{7}{12},\ t - \frac{s+\frac{7}{12}+1}{2}\right) = \left(s + \frac{7}{12},\ t - \frac{12s+19}{24}\right).$$

In Example 15.3.3 we will see that the equation can be further simplified to $t^2 = s^3 - 27s + 20358$.

15.3.1.1. *From Rational to Integral Coefficients.* Another way one can simplify Weierstrass equations is by changing variables to obtain another Weierstrass form with integer coefficients (instead of just rational). In particular, the change of variables $(s,t) = (\lambda^2 x, \lambda^3 y)$ brings a usual Weierstrass equation to the curve

$$C' : t^2 + a_1\lambda st + a_3\lambda^3 t = s^3 + a_2\lambda^2 s^2 + a_4\lambda^4 s + a_6\lambda^6.$$

If λ is the least common multiple of all denominators of the coefficients a_i, then the coefficients in the new curve C' are integers. We note here that the fact that a_i changes to $a_i\lambda^i$ is one of the reasons for the surprising numbering of the coefficients a_i, which skips a_5.

Example 15.3.3. Let $C : y^2 = x^3 - 1/48x + 377/864$ (this curve appeared in Example 15.3.2). Since $48 = 2^4 \cdot 3$ and $864 = 2^5 \cdot 3^3$, let us choose $\lambda = 6$ and a change of variables $(s,t) = (6^2x, 6^3y)$. Then, the curve C becomes

$$\frac{t^2}{6^6} = \frac{s^3}{6^6} - \frac{1}{2^4 \cdot 3} \cdot \frac{s}{6^2} + \frac{377}{2^5 \cdot 3^3},$$

or, equivalently,

$$t^2 = s^3 - 3^3 s + 377 \cdot 2 \cdot 3^3 = s^3 - 27s + 20358.$$

Hence, the change of variables brings C to the form $t^2 = s^3 - 27s + 20358$, which has integer coefficients, as desired.

The same type of change of variables $(s,t) = (\lambda^2 x, \lambda^3 y)$ can be used to simplify integral coefficients in a Weierstrass form.

Example 15.3.4. Let $C : y^2 = x^3 + 1458$. Since $1458 = 2 \cdot 3^6$, the change of variables $(s,t) = (x/9, y/27)$ brings the equation to

$$3^6 t^2 = 3^6 s^3 + 2 \cdot 3^6$$

which simplifies to $t^2 = s^3 + 2$.

15.3.2. Non-Singular Cubics. In this section we discuss a method to simplify cubic curves that applies to non-singular cubics. The same method also applies to those cubics that are singular but have other non-singular points defined over $\mathbb{Q}$. We begin as before with a curve

$$C : f(x, y) = ax^3 + bx^2y + cxy^2 + dy^3 + ex^2 + fxy + gy^2 + hx + jy + k = 0$$

with at least one rational solution $P \in C(\mathbb{Q})$, and we assume that C is non-singular at P (the curve C or its projectivization may be singular at other points $Q \neq P$). We will work in projective coordinates. Hence, C is given by $F(X, Y, Z) = 0$, where F is the polynomial

$$aX^3 + bX^2Y + cXY^2 + dY^3 + eX^2Z + fXYZ + gY^2Z + hXZ^2 + jYZ^2 + kZ^3.$$

Let $P = [x_0, y_0, z_0] \in \mathbb{P}^2(\mathbb{Q})$ in projective coordinates, and let

$$L : \frac{\partial F}{\partial X}(P) \cdot X + \frac{\partial F}{\partial Y}(P) \cdot Y + \frac{\partial F}{\partial Z}(P) \cdot Z = rX + sY + tZ = 0$$

be the tangent line to C at P. Since L is tangent to C at P, it follows that P is in fact a double point in the intersection $L \cap C$. We distinguish two cases, according to whether the point P is of double or triple intersection at $L \cap P$.

15.3.2.1. *Points of Triple Intersection.* If P is a triple point of intersection, then $L \cap C = \{P\}$, since a line and a cubic can only intersect at most at three points, even when counted with multiplicity. If $P = [x_0, y_0, z_0]$ and $L : rX + sY + tZ = 0$ as before, then we make a change of variables

$$U = r_1X + s_1Y + t_1Z, \quad V = r_2X + s_2Y + t_2Z, \quad W = rX + sY + tZ,$$

for some constants $r_i, s_i, t_i \in \mathbb{Q}$ for $i = 1, 2$, chosen so that the matrix

$$\begin{pmatrix} r_1 & s_1 & t_1 \\ r_2 & s_2 & t_2 \\ r & s & t \end{pmatrix}$$

is invertible (so that the change of variables is invertible as well) and so that $P = [u_0, v_0, 0]$ in the new U, V, W-coordinates. Notice, also, that if $Q \in C$ and $Q = [u, v, 0]$ in the new coordinates, then $Q = [x, y, z]$ with $W = rx + sy + tz = 0$. Hence, $Q \in C$ and $Q \in L$, and therefore we must have $Q = P$. In other words, the only point in C with $W = 0$ is P.

Now, if the model for C in the new U, V, W-coordinates is given by $G = 0$, where $G = G(U, V, W)$ is the polynomial

$$aU^3 + bU^2V + cUV^2 + dV^3 + eU^2W + fUVW + gV^2W + hUW^2 + jVW^2 + kW^3$$

(the constants $a, \ldots, k$ here are not necessarily the same as for F), then the points on C with $W = 0$ are those points $[U, V, 0]$ such that

$$aU^3 + bU^2V + cUV^2 + dV^3 = 0.$$

If $V \neq 0$ (one proceeds similarly if $U \neq 0$), then $[U/V, 1, 0] \in C$ and thus

$$a(U/V)^3 + b(U/V)^2 + c(U/V) + d = 0.$$

In particular, if the polynomial $f(x) = ax^3 + bx^2 + cx + d$ has more than one root, we would have more than one point on C with $W = 0$. Hence, $f(x) = a(x - \omega)^3$, for some constant a and $\omega = u_0/v_0$. In particular, a further change of variables

$$R = U - \omega V, \quad S = V, \quad T = W$$

brings C to an equation of the form

$$aR^3 + eR^2T + fRST + gS^2T + hRT^2 + jST^2 + kT^3 = 0.$$

Taking now the affine chart $\{[x, y, 1]\}$, we obtain an equation for C of the form

$$ax^3 + ex^2 + fxy + gy^2 + hx + jy + k = 0,$$

or, equivalently,

$$C : gy^2 + fxy + jy = -ax^3 - ex^2 - hx - k.$$

A further change $(x, y) \mapsto (-agx, (-a)^2gy)$ brings C to the form

$$C : y^2 + a_1xy + a_3y = x^3 + a_2x^2 + a_4x + a_6$$

for some constants $a_i \in \mathbb{Q}$, for $i = 1, 2, 3, 4, 6$. Let us illustrate this with an example.

Example 15.3.5. Let $C : x^3 + y^3 - y = 0$ with $P = (0, 0)$ or, in projective coordinates, $\widehat{C} : X^3 + Y^3 - YZ^2 = 0$. The curve is non-singular, since

$$\vec{n}_{\widehat{C}} = (3X^2, 3Y^2 - Z^2, -2YZ)$$

vanishes if and only if $X = Y = Z = 0$ (but $[0, 0, 0]$ is not a point in projective space). At $P = [0, 0, 1]$, the normal vector is given by

$$\vec{n}_{\widehat{C}}(P) = (0, -1, 0)$$

and so the tangent line L at P is given by $L : -Y = 0$, or, equivalently, $Y = 0$. Next, we find the intersection of L and $\widehat{C}$:

$$L \cap \widehat{C} = \begin{cases} X^3 + Y^3 - YZ^2 = 0, \\ Y = 0, \end{cases}$$

and so $X = Y = 0$. Hence, $L \cap \widehat{C} = \{P\}$ and P must be a triple point of intersection. The next step is to build an invertible change of coordinates with $W = Y$. We choose $U = X$, $V = Z$, $W = Y$, so that the matrix

$$\begin{pmatrix} r_1 & s_1 & t_1 \\ r_2 & s_2 & t_2 \\ r & s & t \end{pmatrix} = \begin{pmatrix} 1 & 0 & 0 \\ 0 & 0 & 1 \\ 0 & 1 & 0 \end{pmatrix}$$

is invertible. With respect to the U, V, W-coordinates, the equation for $\widehat{C}$ becomes

$$\widehat{C} : U^3 + W^3 - WV^2 = 0.$$

Now we take the affine chart $\{[x, y, 1]\}$ and the affine equation is

$$C' : x^3 + 1 - y^2 = 0,$$

or, equivalently, $C' : y^2 = x^3 + 1$, which is a (short) Weierstrass equation.

Example 15.3.6. Let us use the method explained in this section to find a Weierstrass form for $C : x^3+y^3 = d$, the curve that we have discussed in Examples 15.0.1, 15.1.6, and 15.1.8. We already know that $\widehat{C} : X^3 + Y^3 - dZ^3 = 0$ is non-singular for all $d \neq 0$. Moreover, there is a point $P = [\sqrt[3]{d}, 0, 1]$ on $\widehat{C}$, with tangent line

$$L : \sqrt[3]{d^2}X - dZ = 0,$$

or, equivalently, $L : X - \sqrt[3]{d}Z = 0$. It follows that P is a point of triple intersection between L and $\widehat{C}$ because the system

$$L \cap \widehat{C} = \begin{cases} X^3 + Y^3 - dZ^3 = 0, \\ X - \sqrt[3]{d}Z = 0 \end{cases}$$

implies that $Y^3 = 0$, and therefore $P = [\sqrt[3]{d}, 0, 1]$ is the only solution. The next step is to build an invertible change of coordinates with $W = X - \sqrt[3]{d}Z$. We choose $U = X$, $V = Y$, $W = X - \sqrt[3]{d}Z$, so that the matrix

$$\begin{pmatrix} r_1 & s_1 & t_1 \\ r_2 & s_2 & t_2 \\ r & s & t \end{pmatrix} = \begin{pmatrix} 1 & 0 & 0 \\ 0 & 1 & 0 \\ 1 & 0 & -\sqrt[3]{d} \end{pmatrix}$$

is invertible. With respect to the U, V, W-coordinates, the equation for $\widehat{C}$ becomes

$$\widehat{C} : U^3 + V^3 - d\left(\frac{1}{\sqrt[3]{d}}(U - W)\right)^3 = 0,$$

or, equivalently,

$$\widehat{C} : U^3 + V^3 - (U - W)^3 = V^3 + 3U^2W - 3UW^2 + W^3 = 0.$$

Now we take the affine chart $\{[y, -x, 1] : x, y \in \mathbb{R}\}$ and the affine equation is now given by

$$3y^2 - 3y - x^3 + 1 = 0,$$

or $3y^2 - 3y = x^3 - 1$. If we do a further change of variables $(s, t) = (x/3, y/3)$, we obtain

$$3(3t)^2 - 3(3t) = (3s)^3 - 1,$$

or $27t^2 - 9t = 27s^3 - 1$, which simplifies to $t^2 - t/3 = s^3 - 1/27$, a Weierstrass form. We leave it to the reader to verify that this model can be simplified further to $y^2 = x^3 - 432$.

Thus, we have found a sequence of changes of variables from $C : x^3 + y^3 = d$ to $C' : y^2 = x^3 - 432$. This change of variables is, though, undesirable for our purposes, because it is defined over $\mathbb{R}$, so it does not map rational points to rational points. In Exercise 15.4.13 we suggest a different point P' to produce a change of variables over $\mathbb{Q}$.

15.3.2.2. *Points of (Exactly) Double Intersection.* As before, we assume C is given by $F(X, Y, Z) = 0$, where F is the polynomial

$$aX^3 + bX^2Y + cXY^2 + dY^3 + eX^2Z + fXYZ + gY^2Z + hXZ^2 + jYZ^2 + kZ^3.$$

Let $P = [x_0, y_0, z_0] \in \mathbb{P}^2(\mathbb{Q})$ in projective coordinates, and let

$$L : \frac{\partial F}{\partial X}(P) \cdot X + \frac{\partial F}{\partial Y}(P) \cdot Y + \frac{\partial F}{\partial Z}(P) \cdot Z = rX + sY + tZ = 0$$

be the tangent line to C at P. Since L is tangent to C at P, it follows that P is in fact a double point in the intersection $L \cap C$, and here we assume that P is just a double point of intersection. This means that $L \cap C = \{P, Q\}$ where $Q \neq P$, and the intersection of L and C at Q is simple. Since F, L, and P are defined over $\mathbb{Q}$, it follows that Q is also a point of C defined over $\mathbb{Q}$ (see Exercise 15.4.14).

Let $M : r_1X + s_1Y + t_1Z = 0$ be a line passing through Q different from L. In particular, (r_1, s_1, t_1) is not a multiple of (r, s, t). Finally, let $N : r_2X + s_2Y + t_2Z = 0$ be another line passing through P, chosen so that the matrix

$$\begin{pmatrix} r_1 & s_1 & t_1 \\ r_2 & s_2 & t_2 \\ r & s & t \end{pmatrix}$$

is invertible. Thus, we may define an invertible change of variables by

$$U = r_1X + s_1Y + t_1Z, \qquad V = r_2X + s_2Y + t_2Z, \qquad W = rX + sY + tZ.$$

It is important to note that in the U, V, W-coordinates, we have $P = [1, 0, 0]$ because $P \in L$, so $W(P) = 0$, and $P \in N$, so $V(P) = 0$, and similarly we have $Q = [0, 1, 0]$ because $Q \in L \cap M$.

With respect to the new variables U, V, W, the curve C is now given by $G = 0$, where $G = G(U, V, W) = F(X, Y, Z)$, where X, Y, Z are given in terms of U, V, W by the inverse of the change of variables, i.e., by

$$\begin{pmatrix} X \\ Y \\ Z \end{pmatrix} = \begin{pmatrix} r_1 & s_1 & t_1 \\ r_2 & s_2 & t_2 \\ r & s & t \end{pmatrix}^{-1} \begin{pmatrix} U \\ V \\ W \end{pmatrix}.$$

Let us write $G(U, V, W)$ as the polynomial

$$aU^3 + bU^2V + cUV^2 + dV^3 + eU^2W + fUVW + gV^2W + hUW^2 + jVW^2 + kW^3$$

(the constants $a, \ldots, k$ here are not necessarily the same as for F). We claim that $a = b = d = 0$ in these coordinates:

- Since $P \in C$, then $G(P) = G([1, 0, 0]) = a \cdot 1^3 = 0$, and it follows that $a = 0$.
- Since $Q \in C$, then $G(Q) = G([0, 1, 0]) = d \cdot 1^3 = 0$, and it follows that $d = 0$.
- Consider $C \cap \{W = 0\} : bU^2V + cUV^2 = 0$. Notice that $W = 0$ corresponds to the tangent line L in the old coordinates, so P must be a doble point of intersection in $C \cap \{W = 0\}$, and Q must be a simple point of intersection. Since
$$bU^2V + cUV^2 = UV(bU + cV)$$
and P is a simple root of UV, we must have that $bU + cV$ vanishes at P, and it follows that $b \cdot 1 + c \cdot 0 = b$ must be zero.

Hence, we have shown that in the U, V, W-coordinates, the equation for C is of the form $G(U, V, W) = 0$ with G given by

$$cUV^2 + eU^2W + fUVW + gV^2W + hUW^2 + jVW^2 + kW^3 = 0.$$

We now take the affine chart $\{[s, t, 1]\}$ and obtain an affine curve

$$C : cst^2 + es^2 + fst + gt^2 + hs + jt + k = 0.$$

After a change of variables $(s,t) \mapsto (cs+g,t)$ if necessary, we may assume that C is given by

$$C: st^2 + es^2 + fst + hs + jt + k = 0.$$

Now, we may multiply the equation of C through by s and obtain

$$(st)^2 + es^3 + f(st)s + hs^2 + j(st) + ks = 0$$

and a change of variables $x = s$ and $y = st$ (with inverse $s = x$, $t = y/x$ defined except when $x = 0$) brings the equation to the form

$$C' : y^2 + ex^3 + fxy + hx^2 + jy + kx = 0,$$

or, equivalently, $y^2 + fxy + jy = ex^3 + hx^2 + kx$. A further change $(x,y) \mapsto (ex, e^2y)$ brings the equation to a Weierstrass form (as in Definition 15.3.1):

$$C' : y^2 + a_1xy + a_3y = x^3 + a_2x^2 + a_4x + a_6,$$

for some rational coefficients $a_1, \ldots, a_4, a_6$. Using the techniques from Section 15.3.1 we can further simplify the equation to a short Weierstrass form.

Example 15.3.7. Let C be the curve given by $x^3 + xy^2 - 1 = 0$. Before we launch into using our fancy machinery, let us see if there is a direct way to change variables to reach a Weiestrass form. First, we find the projectivization $\widehat{C} : X^3 + XY^2 - Z^3 = 0$, and now we look for an affine chart where the curve may be given in Weierstrass form. For instance, let us look at the chart $\{[1, y, z]\}$, where the equation becomes

$$C' : 1 + y^2 - z^3 = 0,$$

or $y^2 = z^3 - 1$, which is a short Weierstrass form. Thus, we have found a map $C \to C'$ given by

$$(x,y) \mapsto \left(\frac{y}{x}, \frac{1}{x}\right),$$

which is defined everywhere on C, because there are no points on C with $x = 0$.

Example 15.3.8. Let C be the curve given by

$$C : 9x^2y + 28xy^2 - 6y^3 - 6x^2 - 15xy + 31y^2 - 3x - 24y + 3 = 0,$$

with a given rational point $P = (-1, 0)$. We homogenize the equation to

$$9X^2Y + 28XY^2 - 6Y^3 - 6X^2Z - 15XYZ + 31Y^2Z - 3XZ^2 - 24YZ^2 + 3Z^3 = 0,$$

with $P = [-1, 0, 1]$ in projective coordinates. The tangent line to P is given by

$$L : \frac{\partial F}{\partial X}(P) \cdot X + \frac{\partial F}{\partial Y}(P) \cdot Y + \frac{\partial F}{\partial Z}(P) \cdot Z = 9X + 9Z = 0,$$

or, equivalently, $L : X + Z = 0$. Next, we find $C \cap L$. We substitute $Z = -X$ in the equation for C and obtain $-3XY^2 - 6Y^3 = 0$; i.e., $3Y^2(X + 2Y) = 0$. Thus, the points in the intersection $C \cap L$ are $P = [-1, 0, 1]$ and $Q = [-2, 1, 2]$.

In order to complete the change of variables to U, V, W-coordinates, we need two more lines M and N, passing through Q and P, respectively. We can pick $M : X + 2Y = 0$, and then we can choose $N : Y = 0$ because the matrix

$$A = \begin{pmatrix} 1 & 2 & 0 \\ 0 & 1 & 0 \\ 1 & 0 & 1 \end{pmatrix}, \quad A^{-1} = \begin{pmatrix} 1 & -2 & 0 \\ 0 & 1 & 0 \\ -1 & 2 & 1 \end{pmatrix}$$

is invertible (in fact, its determinant is 1, which is even more desirable as the inverse matrix also has integer coefficients). With these choices of lines M and N, we set up a change of variables

$$\begin{pmatrix} U \\ V \\ W \end{pmatrix} = \begin{pmatrix} 1 & 2 & 0 \\ 0 & 1 & 0 \\ 1 & 0 & 1 \end{pmatrix} \begin{pmatrix} X \\ Y \\ Z \end{pmatrix}, \quad \text{or} \quad \begin{pmatrix} X \\ Y \\ Z \end{pmatrix} = \begin{pmatrix} 1 & -2 & 0 \\ 0 & 1 & 0 \\ -1 & 2 & 1 \end{pmatrix} \begin{pmatrix} U \\ V \\ W \end{pmatrix}.$$

In the U, V, W-coordinates, the equation for C is given by

$$\begin{aligned} F(X,Y,Z) &= F(U-2V, V, -U+2V+W) \\ &= 9U^2W - 3UV^2 - 3UVW - 12UW^2 + V^2W + 3W^3. \end{aligned}$$

We now de-homogenize by picking an affine chart $\{[s,t,1]\}$ and obtain an affine curve with model

$$9s^2 - 3st^2 - 3st - 12s + t^2 + 3 = 0,$$

and with a change of variables $(s', t') = (-3s+1, t)$ this brings the curve to

$$C' : s't'^2 + s'^2 + s't' + 2s' - t' = 0,$$

or, equivalently, $(s't')^2 + s'^3 + (s't')s' + 2s'^2 - (s't') = 0$. Now with a change of variables $(u, v) = (s', s't')$ we have a curve

$$C'' : v^2 + u^3 + uv + 2u^2 - v = 0,$$

or $C'' : v^2 + uv - v = -u^3 - 2u^2$. Finally, if we change $(u', v') = (-u, -v)$, we obtain a (long) Weierstrass equation

$$C'' : v'^2 + u'v' + v' = u'^3 - 2u'^2.$$

Note that in Examples 15.3.2 and 15.3.3 we have seen how to reduce this equation to a short Weierstrass form.

It remains to find a map $C \to C''$. The composition of all the changes of variables yields

$$\begin{aligned} (u', v') &= (-u, -v) = (-s', -s't') = (3s-1, (3s-1)t) \\ &= \left(3\frac{U}{W} - 1, \left(3\frac{U}{W} - 1\right)\frac{V}{W}\right) = \left(\frac{3U - W}{W}, \left(\frac{3U-W}{W}\right)\frac{V}{W}\right) \\ &= \left(\frac{3(X+2Y) - (X+Z)}{X+Z}, \left(\frac{3(X+2Y)-(X+Z)}{X+Z}\right)\frac{Y}{X+Z}\right) \\ &= \left(\frac{2X + 6Y - Z}{X+Z}, \left(\frac{2X+6Y-Z}{X+Z}\right)\frac{Y}{X+Z}\right) \\ &= \left(\frac{2\frac{X}{Z} + 6\frac{Y}{Z} - 1}{\frac{X}{Z}+1}, \left(\frac{2\frac{X}{Z} + 6\frac{Y}{Z} - 1}{\frac{X}{Z}+1}\right)\frac{\frac{Y}{Z}}{\frac{X}{Z}+1}\right) \\ &= \left(\frac{2x+6y-1}{x+1}, \left(\frac{2x+6y-1}{x+1}\right)\frac{y}{x+1}\right) \\ &= \left(\frac{2x+6y-1}{x+1}, \frac{(2x+6y-1)y}{(x+1)^2}\right). \end{aligned}$$

Hence, we have found a change of variables $\phi\colon C \to C''$ defined by

$$\phi(x,y) = \left(\frac{2x+6y-1}{x+1}, \frac{(2x+6y-1)y}{(x+1)^2}\right).$$

The inverse is given by

$$\begin{aligned}
(x,y) = \left(\frac{X}{Z}, \frac{Y}{Z}\right) &= \left(\frac{U-2V}{-U+2V+W}, \frac{V}{-U+2V+W}\right)\\
&= \left(\frac{s-2t}{-s+2t+1}, \frac{t}{-s+2t+1}\right)\\
&= \left(\frac{\frac{1-s'}{3}-2t'}{-\frac{1-s'}{3}+2t'+1}, \frac{t'}{-\frac{1-s'}{3}+2t'+1}\right)\\
&= \left(\frac{\frac{1-u}{3}-2\frac{v}{u}}{-\frac{1-u}{3}+2\frac{v}{u}+1}, \frac{\frac{v}{u}}{-\frac{1-u}{3}+2\frac{v}{u}+1}\right)\\
&= \left(\frac{u-u^2-6v}{-(u-u^2)+6v+3u}, \frac{3v}{-(u-u^2)+6v+3u}\right)\\
&= \left(\frac{u-u^2-6v}{u^2+2u+6v}, \frac{3v}{u^2+2u+6v}\right)\\
&= \left(\frac{-u'-u'^2+6v'}{u'^2-2u'-6v'}, \frac{-3v'}{u^2-2u'-6v'}\right),
\end{aligned}$$

or, in other words, $\phi^{-1}\colon C'' \to C$ is given by

$$\phi^{-1}(u',v') = \left(\frac{-u'-u'^2+6v'}{u'^2-2u'-6v'}, \frac{-3v'}{u^2-2u'-6v'}\right).$$

15.4. Exercises

Exercise 15.4.1. Show that the relation $\sim$ defined in Section 15.1.1 is an equivalence relation (see Remark 4.2.5) on the set of non-zero vectors $\mathbb{R}^2 \setminus \{(0,0)\}$.

Exercise 15.4.2. Find the projectivization of the following affine curves:

(a) $x^2 + y^2 = 2$.

(b) $xy = 1$.

(c) $y^2 + y = x^3 + x$.

(d) $x^3 + xy^2 + 2x^2 + y + 7 = 0$.

(e) $x^n + y^n = 1$, for any $n \geq 2$.

Exercise 15.4.3. Let $C : y = x^3$ be an affine curve in the plane $\mathbb{A}^2(\mathbb{R})$.

(a) Is C smooth in $\mathbb{A}^2(\mathbb{R})$?

(b) Find the projectivization $\widehat{C} \subseteq \mathbb{P}^2(\mathbb{R})$ of the curve $C : y = x^3$.

(c) Is $\widehat{C}$ smooth in $\mathbb{P}^2(\mathbb{R})$? If not, find all the singular points on $\widehat{C}$.

Exercise 15.4.4. Determine all the singular points (if any) on the following projective curves. If the curve is given in affine form, first find a projectivization.

(a) $XY + YZ + XZ = 0$.

(b) $X^2 + 2XY + Y^2 - Z^2 = 0$.

(c) $y^2 + y = x^3$.

(d) $y^2 = x(x-1)^2$.

(e) $y^2 = x^3 - 6x^2 + 12x - 8$.

(f) $y^2 + xy = x^3$.

Exercise 15.4.5. Parametrize all the rational points on the curves that appear in parts (d) and (e) of Exercise 15.4.4.

Exercise 15.4.6. The cubic curve $C : x^3 + x^2y + 2xy^2 + 2y^3 - 2x^2 - 2xy = 0$ is singular. Use the method of Example 15.2.3 to parametrize all the rational points on C.

Exercise 15.4.7. The cubic curve $C : x^3 + 3x^2y + 3xy^2 + y^3 - x + y = 0$ is non-singular in the affine plane, but singular at a point at infinity. Use the method of Example 15.2.4 to parametrize all the rational points on C.

Exercise 15.4.8. Find a change of variables that brings the following curves from long to short Weierstrass form:

(a) $y^2 = x^3 + x^2 + 1$.

(b) $y^2 + y = x^3$.

(c) $y^2 + xy = x^3 + 1$.

(d) $y^2 + xy + 3y = x^3 + 2x^2 + 4x + 6$.

(e) $t^2 - t/3 = s^3 - 1/27$.

Exercise 15.4.9. Find a change of variables that brings the following curves to a model with integral coefficients:

(a) $y^2 = x^3 + \frac{5}{2}x + \frac{3}{4}$.

(b) $y^2 + xy + \frac{1}{7}y = x^3 + 1$.

(c) $y^2 + \frac{1}{2}y = x^3 + \frac{1}{25}$.

Exercise 15.4.10. Find a change of variables that brings the following curves to a model with smaller integral coefficients:

(a) $y^2 = x^3 + 4x^2 + 320$.

(b) $y^2 + 7xy + 1029y = x^3$.

(c) $y^2 + 10xy + 7000y = x^3 + 200x^2 + 370000x$.

Exercise 15.4.11. The curve $C : y = x^3$ is singular in projective coordinates, but non-singular at $(0,0)$.

(a) Find the projectivization $\widehat{C}$ of C.

(b) Find the tangent line to $\widehat{C}$ at $P = [0,0,1]$, and show that P is a triple point of intersection with its tangent line.

(c) Use the procedure of Section 15.3.2.1 to find an affine chart where $\widehat{C}$ is given by a (singular) short Weierstrass model.

Exercise 15.4.12. The curve $C : x^3 + 2x^2y + xy^2 - y = 0$ is singular in projective coordinates, but non-singular at $(0,0)$.

(a) Find the projectivization $\widehat{C}$ of C.

(b) Find the tangent line to $\widehat{C}$ at $P = [0,0,1]$, and show that P is a triple point of intersection with its tangent line.

(c) Use the procedure of Section 15.3.2.1 to find an affine chart where $\widehat{C}$ is given by a (singular) short Weierstrass model.

Exercise 15.4.13. Find a change of variables from $C : x^3 + y^3 = d$ to $C' : y^2 = x^3 - 432d^2$. (Hint: follow the method of Example 15.3.6, but use the point $P' = [1,-1,0]$ instead of P.)

Exercise 15.4.14. Let $C : f(x,y) = 0$ be a curve such that $f(x,y)$ is a polynomial with rational coefficients of degree $n \geq 1$ (i.e., the largest degree of a monomial in f is n). Let $L : g(x,y) = 0$ be a line defined over $\mathbb{Q}$ (i.e., $g(x,y) = ax + by - c$ with $a, b, c \in \mathbb{Q}$) such that $L \cap C = \{P_1, \ldots, P_{n-1}, P_n\}$, and suppose that $P_1, \ldots, P_{n-1}$ are points defined over $\mathbb{Q}$ (not necessarily distinct, since some of them may have multiplicity of intersection ≥ 1). Show that P_n is also defined over $\mathbb{Q}$.

Exercise 15.4.15. Let $C/\mathbb{Q}$ be an affine curve.

(a) Suppose that $C/\mathbb{Q}$ is given by an equation of the form

$$C : xy^2 + ax^2 + bxy + cy^2 + dx + ey + f = 0. \tag{15.1}$$

Find an invertible change of variables that takes the equation of C onto one of the form $xy^2 + gx^2 + hxy + jx + ky + l = 0$. (Hint: consider a change of variables $X = x + \lambda$, $Y = y$.)

(b) Suppose that $C'/\mathbb{Q}$ is given by an equation of the form

$$C' : xy^2 + ax^2 + bxy + cx + dy + e = 0. \tag{15.2}$$

Find an invertible change of variables that takes the equation of C' onto one of the form $y^2 + \alpha xy + \beta y = x^3 + \gamma x^2 + \delta x + \eta$. (Hint: multiply (15.2) by x and consider the change of variables $X = x$ and $Y = xy$. Make sure that, at the end, the coefficients of y^2 and x^3 equal 1.)

(c) Suppose that $C''/\mathbb{Q}$ is a curve given by an equation of the form

$$C'' : y^2 + axy + by = x^3 + cx^2 + dx + e. \tag{15.3}$$

Find an invertible change of variables that takes the equation of C'' onto one of the form $y^2 = x^3 + Ax + B$. (Hint: do it in two steps. First eliminate the xy and y terms. Then eliminate the x^2 term.)

(d) Let $E/\mathbb{Q} : y^2 + 43xy - 210y = x^3 - 210x^2$. Find an invertible change of variables that takes the equation of E to one of the form $y^2 = x^3 + Ax + B$.

Exercise 15.4.16. Find a Weierstrass form for the following curves by finding their projectivization and then choosing a different affine chart:

(a) $xy^2 + x^2y = 1$.
(b) $x^3 - 2x^2y - xy - y^2 - y = 0$.
(c) $xy^2 + x^2y = 1 + x^2$.

Exercise 15.4.17. Use the methods of Section 15.3.2.2 to find a Weierstrass form for each of the following non-singular curves:

(a) $2X^3 + 3X^2Z - XY^2 + 3XZ^2 - Y^2Z + Z^3 = 0$, with $P = [0, 1, 1]$.
(b) $X^3 + 3X^2Y + 3XY^2 + Y^3 - Y^2Z - YZ^2 = 0$, with $P = [0, 0, 1]$.
(c) $X^3 + XY^2 + XZ^2 + Z^3 + YZ^2 = 0$, with $P = [0, 1, -1]$.
(d) $X^3 + Y^3 + Z^3 + 3XYZ = 0$.

Exercise 15.4.18. Let C and E be curves defined, respectively, by $C : V^2 = U^4 + 1$ and $E : y^2 = x^3 - 4x$. Let ψ be the map defined by

$$\psi(U, V) = \left(\frac{2(V+1)}{U^2}, \frac{4(V+1)}{U^3}\right).$$

(a) Show that if $U \neq 0$ and $(U, V) \in C(\mathbb{Q})$, then $\psi(U, V) \in E(\mathbb{Q})$.
(b) Find an inverse function for ψ; i.e., find $\varphi : E \to C$ such that $\varphi(\psi(U, V)) = (U, V)$.

Next, we work in projective coordinates. Let $C : W^2V^2 = U^4 + W^4$ and $E : zy^2 = x^3 + z^3$.

(c) Write down the definition of ψ in projective coordinates, or, in other words, what is $\psi([U, V, W])$?
(d) Show that $\psi([0, 1, 1]) = [0, 1, 0] = \mathcal{O}$.
(e) Show that $\psi([0, -1, 1]) = [0, 0, 1]$.
(Hint: show that $\psi([U, V, W]) = [2U^2, 4UW, W(V - W)]$.)

CHAPTER 16

ELLIPTIC CURVES

It is possible to write endlessly on elliptic curves. (This is not a threat.)

Serge Lang,
from *Elliptic Curves: Diophantine Analysis*

Elliptic curves are ubiquitous in number theory, algebraic geometry, complex analysis, cryptography, physics, and beyond. While the basic theory of elliptic curves stands on a solid foundation, there are many aspects of elliptic curves that are not known (yet) and are nowadays the central subject of much research. Entire conferences are dedicated to the study of elliptic curves (see Figure 16.1).

Figure 16.1. A group photo of an instructional conference on elliptic curves that was held at UConn in 2014.

In this chapter we summarize the main aspects of the theory of elliptic curves. Unfortunately, we will not be able to provide many of the proofs because they are beyond the scope of this book. The contents of this chapter are largely based on the book [**Loz11**], and we refer the reader to that reference to learn more about elliptic curves and their connection to modular forms and Fermat's last theorem. Silverman and Tate's book **[ST92]** is an excellent introduction to elliptic curves for undergraduates. Washington's book **[Was08]** is also accessible for undergraduates and emphasizes the cryptographic applications of elliptic curves. Stein's book **[Ste08]** also has an interesting chapter on elliptic curves. There are several graduate-level texts on elliptic curves. Silverman's book **[Sil86]** is the standard reference, but Milne's **[Mil06]** is also an excellent introduction to the theory of elliptic curves (and also includes a chapter on modular forms).

16.1. Definition

Definition 16.1.1. An *elliptic curve* over $\mathbb{Q}$ is a smooth cubic projective curve E defined over $\mathbb{Q}$ with at least one rational point $\mathcal{O} \in E(\mathbb{Q})$ that we call the *origin.*

In other words, an elliptic curve is a curve E in the projective plane (see Section 15.1.2) given by a cubic polynomial $F(X, Y, Z) = 0$ with rational coefficients; i.e.,

$$\begin{aligned} F(X,Y,Z) &= aX^3 + bX^2Y + cXY^2 + dY^3 \\ &\quad +eX^2Z + fXYZ + gY^2Z \\ &\quad +hXZ^2 + jYZ^2 + kZ^3 = 0, \end{aligned} \tag{16.1}$$

with coefficients $a, b, c, \ldots \in \mathbb{Q}$ and such that E is smooth; i.e., the normal vector $\left(\frac{\partial F}{\partial X}(P), \frac{\partial F}{\partial Y}(P), \frac{\partial F}{\partial Z}(P)\right)$ does not vanish at any $P \in E$ (see Section 15.1.5 for a brief introduction to singularities and non-singular or smooth curves). If the coefficients $a, b, c, \ldots$ are in a field K, then we say that E is defined over K (and write E/K).

Even though the fact that E is a projective curve is crucial, we usually consider just affine charts of E, e.g., those points of the form $\{[X, Y, 1]\}$, and study instead the affine curve given by

$$\begin{aligned} &aX^3 + bX^2Y + cXY^2 + dY^3 \\ &+eX^2 + fXY + gY^2 + hX + jY + k = 0 \end{aligned} \tag{16.2}$$

but with the understanding that in this new model we may have left out some points of E *at infinity* (i.e., those points $[X, Y, 0]$ satisfying (16.1)).

In general, one can find a change of coordinates that simplifies (16.2) enormously:

Proposition 16.1.2. *Let E be an elliptic curve, given by (16.1), defined over a field K of characteristic different from 2 or 3. Then, there exist a curve $\hat{E}$ given by*

$$zy^2 = x^3 + Axz^2 + Bz^3, \quad A, B \in K \quad \textit{with} \quad 4A^3 + 27B^2 \neq 0,$$

and an invertible change of variables $\psi : E \to \hat{E}$ of the form

$$\psi([X, Y, Z]) = \left[\frac{f_1(X,Y,Z)}{g_1(X,Y,Z)}, \frac{f_2(X,Y,Z)}{g_2(X,Y,Z)}, \frac{f_3(X,Y,Z)}{g_3(X,Y,Z)}\right]$$

where f_i and g_i are polynomials with coefficients in K for $i = 1, 2, 3$ and the origin $\mathcal{O}$ is sent to the point $[0, 1, 0]$ of $\hat{E}$; i.e., $\psi(\mathcal{O}) = [0, 1, 0]$.

The existence of such a change of variables is a consequence of the Riemann–Roch theorem of algebraic geometry (for a proof of the proposition see [**Sil86**, Chapter III.3]). In practice, we follow the procedures outlined in Section 15.3.

A projective equation of the form $zy^2 = x^3+Axz^2+Bz^3$, or $y^2 = x^3+Ax+B$ in affine coordinates, is called a *Weierstrass equation*, or Weierstrass form (see Section 15.3). From now on, we will often work with an elliptic curve in this form. Notice that a curve E given by a Weierstrass equation $y^2 = x^3 + Ax + B$ is non-singular if and only if $4A^3 + 27B^2 \neq 0$, and it has a unique point at infinity, namely $[0, 1, 0]$, which we shall call the origin $\mathcal{O}$ or the point at infinity of E.

Sometimes we shall use a more general Weierstrass equation

$$y^2 + a_1xy + a_3y = x^3 + a_2x^2 + a_4x + a_6$$

with $a_i \in \mathbb{Q}$ (the funky choice of notation for the coefficients was explained in Section 15.3.1.1), but most of the time we will work with equations of the form $y^2 = x^3 + Ax + B$. The reader can learn how to find a change of variables from one form to the other in Exercise 15.4.15.

Example 16.1.3. Let $d \in \mathbb{Z}$, $d \neq 0$, and let E be the elliptic curve given by the cubic equation

$$X^3 + Y^3 = dZ^3$$

with $\mathcal{O} = [1, -1, 0]$. This curve already appeared in Examples 15.0.1, 15.1.6, 15.1.8, and 15.3.6. In particular, we verified that E is a smooth curve in Example 15.1.8. We wish to find a Weierstrass equation for E. Note that if we change $X = U + V$, $Y = -V$, $Z = W$, then we obtain a new equation

$$U^3 + 3U^2V + 3UV^2 = dW^3. \tag{16.3}$$

Since this equation is quadratic in V and cubic in W, with no other cubic monomials that involve W, the variable W will end up playing the role of x, and the variable V will play the role of y in our Weierstrass model. Next, we change variables to obtain a coefficient of 1 in front of V^2 and W^3. If we multiply (16.3) through by d^2, we obtain

$$d^2U^3 + 3d^2U^2V + 3d^2UV^2 = d^3W^3, \tag{16.4}$$

and now we change variables $x = 3dW$, $y = 9dV$, and $z = U$. Then, (16.4) becomes

$$d^2z + \frac{dyz}{3} + \frac{y^2z}{27} = \frac{x^3}{27}, \tag{16.5}$$

or, equivalently, $y^2z + 9dyz = x^3 - 27d^2z$, which is a Weierstrass equation. Thus, $[x, y, z] = [3dW, 9dV, U] = [3dZ, -9dY, X + Y]$ and we have found a change of variables $\psi : E \to \widehat{E}$ given by

$$\psi([X, Y, Z]) = [3dZ, -9dY, X + Y]$$

such that the image lands on the curve in Weierstrass equation $\widehat{E} : y^2z + 9dyz = x^3 - 27d^2z$. The map ψ is invertible; the inverse map $\psi^{-1} : \widehat{E} \to E$ is

$$\psi^{-1}([x, y, z]) = \left[\frac{9dz + y}{9d}, -\frac{y}{9d}, \frac{x}{3d}\right].$$

In affine coordinates, the change of variables is going from $X^3 + Y^3 = d$ to the curve $y^2 + 9dy = x^3 - 27d^2$ via the maps

$$\psi(X,Y) = \left(\frac{3d}{X+Y}, -\frac{9dY}{X+Y}\right),$$
$$\psi^{-1}(x,y) = \left(\frac{9d+y}{3x}, -\frac{y}{3x}\right).$$

We leave it as an exercise for the reader to verify that the model can be further simplified to the form $y^2 = x^3 - 432d^2$.

Definition 16.1.4. Let $E : f(x,y) = 0$ be an elliptic curve with origin $\mathcal{O}$, and let $E' : g(X,Y) = 0$ be an elliptic curve with origin $\mathcal{O}'$. We say that E and E' are *isomorphic over* $\mathbb{Q}$ if there is an invertible change of variables $\psi : E \to E'$, defined by rational functions with coefficients in $\mathbb{Q}$, such that $\psi(\mathcal{O}) = \mathcal{O}'$.

Example 16.1.5. Sometimes, a curve given by a quartic polynomial can be isomorphic over $\mathbb{Q}$ to another curve given by a cubic polynomial. For instance, consider the curves

$$C/\mathbb{Q} : V^2 = U^4 + 1 \quad \text{and} \quad E/\mathbb{Q} : y^2 = x^3 - 4x.$$

The map $\psi : C \to E$ given by

$$\psi(U,V) = \left(\frac{2(V+1)}{U^2}, \frac{4(V+1)}{U^3}\right)$$

is an invertible rational map, defined over $\mathbb{Q}$, that sends $(0,1)$ to $\mathcal{O}$, and $\psi(0,-1) = (0,0)$. See Exercise 15.4.18. More generally, any quartic

$$C : V^2 = aU^4 + bU^3 + cU^2 + dU + q^2$$

for some $a,b,c,d,q \in \mathbb{Z}$ is isomorphic over $\mathbb{Q}$ to a curve of the form $E : y^2 + a_1xy + a_3y = x^3 + a_2x^2 + a_4x + a_6$, also defined over $\mathbb{Q}$. The isomorphism is given in [**Was08**, Theorem 2.17, p. 37].

Let E be an elliptic curve over $\mathbb{Q}$ given by a Weierstrass equation

$$E\colon y^2 + a_1xy + a_3y = x^3 + a_2x^2 + a_4x + a_6, \quad a_i \in \mathbb{Q}.$$

With a change of variables $(x,y) \mapsto (u^{-2}x,\ u^{-3}y)$, we can find the equation of an elliptic curve isomorphic to E given by

$$y^2 + (a_1u)xy + (a_3u^3)y = x^3 + (a_2u^2)x^2 + (a_4u^4)x + (a_6u^6)$$

with coefficients $a_iu^i \in \mathbb{Z}$ for $i = 1,2,3,4,6$. By the way, *this* is one of the reasons for the peculiar numbering of the coefficients a_i.

Example 16.1.6. Let E be given by $y^2 = x^3 + \frac{x}{2} + \frac{5}{3}$. We may change variables by $x = \frac{X}{6^2}$ and $y = \frac{Y}{6^3}$ to obtain a new equation $Y^2 = X^3 + 648X + 77760$ with integral coefficients.

16.2. Integral Points

In 1929, Siegel proved the following result about integral points $E(\mathbb{Z})$, i.e., about those points on E with integer coordinates:

Theorem 16.2.1 (Siegel's theorem; [**Sil86**, Ch. IX, Thm. 3.1]). *Let $E/\mathbb{Q}$ be an elliptic curve given by $y^2 = x^3 + Ax + B$, with $A, B \in \mathbb{Z}$. Then E has only a finite number of integral points.*

Siegel's theorem is a consequence of a well-known theorem of Roth on diophantine approximation. Unfortunately, Siegel's theorem is not effective and provides neither a method to find the integral points on E nor a bound on the number of integral points. However, in [**Bak90**], Alan Baker found an alternative proof that provides an explicit upper bound on the size of the coefficients of an integral solution. More concretely, if $x, y \in \mathbb{Z}$ satisfy $y^2 = x^3 + Ax + B$, then

$$\max(|x|, |y|) < \exp((10^6 \cdot \max(|A|, |B|))^{10^6}).$$

Obviously, Baker's bound is not a very sharp bound, but it is theoretically interesting nonetheless.

16.3. The Group Structure on $E(\mathbb{Q})$

From now on, we will concentrate on trying to find all rational points on a curve $E : y^2 = x^3 + Ax + B$. We will use the following notation for the rational points on E:

$$E(\mathbb{Q}) = \{(x, y) \in E \mid x, y \in \mathbb{Q}\} \cup \{\mathcal{O}\}$$

where $\mathcal{O} = [0, 1, 0]$ is the point at infinity.

One of the aspects that makes the theory of elliptic curves so rich is that the set $E(\mathbb{Q})$ can be equipped with a group structure, geometric in nature. The (addition) operation on $E(\mathbb{Q})$ can be defined as follows (see Figure 16.2). Let E be given by a Weierstrass equation $y^2 = x^3 + Ax + B$ with $A, B \in \mathbb{Q}$. Let P and Q be two rational points in $E(\mathbb{Q})$ and let $\mathfrak{L} = \overline{PQ}$ be the line that goes through P and Q (if $P = Q$, then we define $\mathfrak{L}$ to be the tangent line to E at P). Since the curve E is defined by a cubic equation and since we have defined $\mathfrak{L}$ so it already intersects E at two rational points, there must be a third point of intersection R in $\mathfrak{L} \cap E$, which is also defined over $\mathbb{Q}$ (this is due to Proposition 5.5.22), and

$$\mathfrak{L} \cap E(\mathbb{Q}) = \{P, Q, R\}.$$

The sum of P and Q, denoted by $P + Q$, is by definition the second point of intersection with E of the vertical line that goes through R, or, in other words, the reflection of R across the x-axis.

The addition operation that we have defined on points of $E(\mathbb{Q})$ is commutative because $\mathfrak{L} = \overline{PQ} = \overline{QP}$; i.e., the line through P and Q is the line through Q and P. The origin $\mathcal{O}$ is the zero element, and for every $P \in E(\mathbb{Q})$ there exists a point $-P$ such that $P + (-P) = \mathcal{O}$. If E is given by $y^2 = x^3 + Ax + B$ and $P = (x_0, y_0)$, then $-P = (x_0, -y_0)$. The addition is also associative (but this is not obvious, and it is tedious to prove) and, therefore, $(E, +)$ is an abelian group.

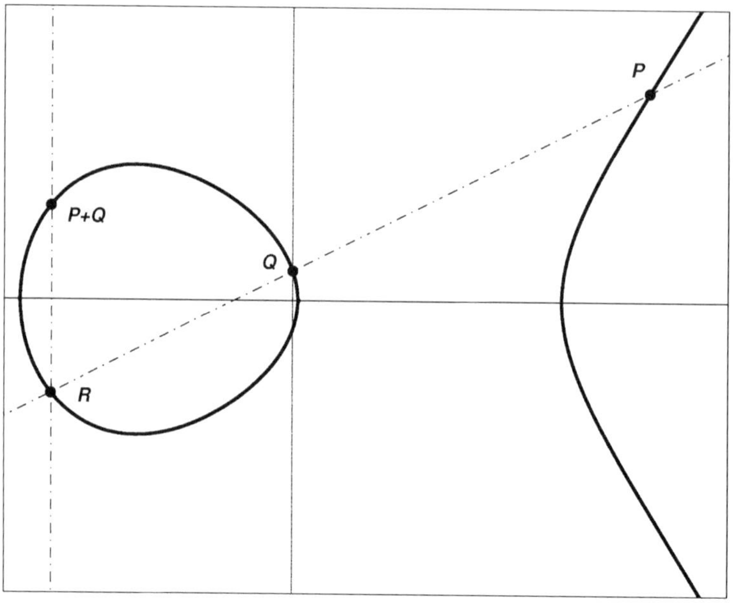

Figure 16.2. Addition of points on an elliptic curve.

Example 16.3.1. Let E be the elliptic curve $y^2 = x^3 - 25x$. The points $P = (5, 0)$ and $Q = (-4, 6)$ belong to $E(\mathbb{Q})$. Let us find $P + Q$. First, we find the equation of the line $\mathfrak{L} = \overline{PQ}$. The slope must be

$$m = \frac{0 - 6}{5 - (-4)} = -\frac{6}{9} = -\frac{2}{3}$$

and the line is $\mathfrak{L} : y = -\frac{2}{3}(x - 5)$. Now we find the third point of intersection of $\mathfrak{L}$ and E by solving

$$\begin{cases} y = -\frac{2}{3}(x - 5), \\ y^2 = x^3 - 25x. \end{cases}$$

Plugging the first equation into the second one, we obtain an equation

$$x^3 - \frac{4}{9}x^2 - \frac{185}{9}x - \frac{100}{9} = 0,$$

which factors as $(x-5)(x+4)(9x+5) = 0$. The first two factors are expected, since we already knew that $P = (5, 0)$ and $Q = (-4, 6)$ are in $\mathfrak{L} \cap E$. The third point of intersection must have $x = -\frac{5}{9}$, $y = -\frac{2}{3}(x - 5) = \frac{100}{27}$ and, indeed, $R = (-\frac{5}{9}, \frac{100}{27})$ is a point in $\mathfrak{L} \cap E(\mathbb{Q})$. Thus, $P + Q$ is the reflection of R across the x-axis; i.e., $P + Q = (-\frac{5}{9}, -\frac{100}{27})$.

Let us find $Q + Q = 2Q$. The line $\mathfrak{L}$ in this case is the tangent line to E at Q. The slope of $\mathfrak{L}$ can be found using implicit differentiation on $y^2 = x^3 - 25x$:

$$2y\frac{dy}{dx} = 3x^2 - 25, \quad \text{so} \quad \frac{dy}{dx} = \frac{3x^2 - 25}{2y}.$$

Hence, the slope of $\mathfrak{L}$ is $m = \frac{23}{12}$ and $\mathfrak{L} : y = \frac{23}{12}(x+4)+6$. In order to find R we need to solve

$$\begin{cases} y = \frac{23}{12}(x+4)+6, \\ y^2 = x^3 - 25x. \end{cases}$$

Simplifying yields $x^3 - \frac{529}{144}x^2 - \frac{1393}{18}x - \frac{1681}{9} = 0$, which factors as

$$(x+4)^2(144x - 1681) = 0.$$

Once again, two factors were expected: $x = -4$ *needs* to be a double root because $\mathfrak{L}$ is *tangent* to E at $Q = (-4, 6)$. The third factor tells us that the x-coordinate of R is $x = \frac{1681}{144}$, and $y = \frac{23}{12}(x+4)+6 = \frac{62279}{1728}$. Thus, $Q + Q = 2Q = (\frac{1681}{144}, -\frac{62279}{1728})$.

Example 16.3.2. Let $E : y^2 = x^3 + 1$ and put $P = (2, 3)$. Let us find P, $2P$, $3P$, etc.

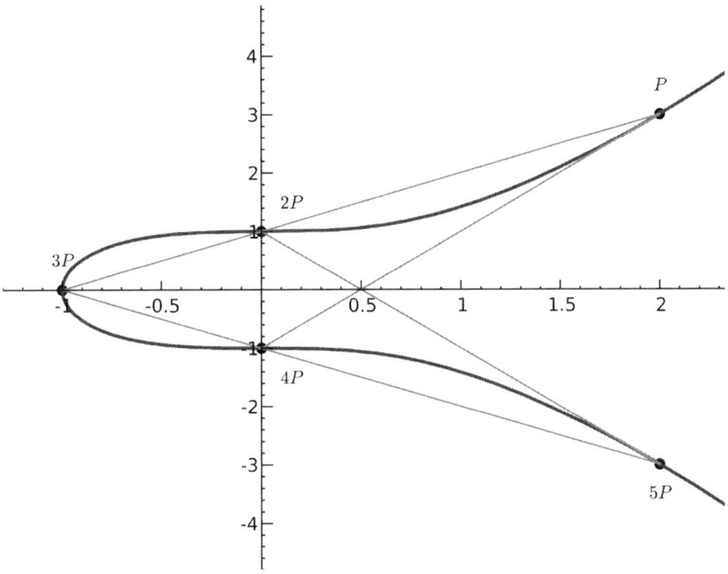

Figure 16.3. The rational points on $y^2 = x^3 + 1$, except the point at ∞.

- In order to find $2P$, first we need to find the tangent line to E at P, which is $y - 3 = 2(x-2)$ or $y = 2x - 1$. The third point of intersection is $R = (0, -1)$, so $2P = (0, 1)$.
- To find $3P$, we add P and $2P$. The third point of intersection of E with the line that goes through P and $2P$ is $R' = (-1, 0)$; hence, $3P = (-1, 0)$.
- The point $4P$ can be found by adding $3P$ and P. The third point of intersection of E and the line through P and $3P$ is $R'' = 2P = (0, 1)$, and so $4P = P + 3P = (0, -1)$.
- We find $5P$ by adding $4P$ and P. Notice that the line that goes through $4P = (0, -1)$ and $P = (2, 3)$ is tangent at $(2, 3)$, so the third point of intersection is P. Thus, $5P = 4P + P = (2, -3)$.

- Finally, $6P = P + 5P$ but $5P = (2, -3) = -P$. Hence, $6P = P + (-P) = \mathcal{O}$, the point at infinity.

This means that P is a point of finite order, and its order equals 6. See Figure 16.3.

The addition law can be defined more generally on any smooth projective cubic curve $E : f(X, Y, Z) = 0$, with a given rational point $\mathcal{O}$. Let $P, Q \in E(\mathbb{Q})$ and let $\mathfrak{L}$ be the line that goes through P and Q. Let R be the third point of intersection of $\mathfrak{L}$ and E. Then R is also a rational point in $E(\mathbb{Q})$. Let $\mathfrak{L}'$ be the line through R and $\mathcal{O}$. We define $P + Q$ to be the third point of intersection of $\mathfrak{L}'$ and E. Notice that any vertical line $x = a$ in the affine plane passes through $[0, 1, 0]$, because the same line in projective coordinates is given by $x = az$ and $[0, 1, 0]$ belongs to such line. Thus, if E is given by a model $y^2 = x^3 + Ax + B$ and $\mathcal{O}$ is chosen to be the point $[0, 1, 0]$, then $\mathfrak{L}'$ is always a vertical line, so $P + Q$ is always the reflection of R with respect to the x-axis.

The next step in the study of the structure of $E(\mathbb{Q})$ was conjectured by Henri Poincaré in 1908, proved by Louis Mordell in 1922, and generalized by André Weil in his thesis in 1928 (see Figure 1.7 in Section 1.4).

Theorem 16.3.3 (Mordell–Weil theorem). *Let $E/\mathbb{Q}$ be an elliptic curve. Then, $E(\mathbb{Q})$ is a finitely generated abelian group. In other words, there are points $P_1, \ldots,$ P_n such that any other point Q in $E(\mathbb{Q})$ can be expressed as a linear combination*

$$Q = a_1 P_1 + a_2 P_2 + \cdots + a_n P_n$$

for some $a_i \in \mathbb{Z}$.

The group $E(\mathbb{Q})$ is usually called the Mordell–Weil group of E, in honor of the two mathematicians who proved the theorem.

Example 16.3.4. Consider the elliptic curve $E/\mathbb{Q}$ given by the Weierstrass equation

$$y^2 + y = x^3 - 7x + 6.$$

The set of rational points $E(\mathbb{Q})$ for this elliptic curve is infinite. For instance, the following points are on the curve:

$$(1, 0),\ (2, 0),\ (0, -3),\ (-3, -1),\ (8, -22),\ (-2, -4),\ (3, -4),$$
$$(3, 3),\ (-1, -4),\ (1, -1),\ (0, 2),\ (2, -1),\ (-2, 3),\ (-1, 3),$$
$$\left(\frac{1}{4}, \frac{13}{8}\right),\ \left(\frac{25}{9}, -\frac{91}{27}\right),\ \left(-\frac{26}{9}, \frac{28}{27}\right),\ \left(\frac{7}{9}, \frac{17}{27}\right), \ldots .$$

At first glance, it may seem very difficult to describe all the points on $E(\mathbb{Q})$, including those listed above, in a succinct manner. However, the Mordell–Weil theorem tells us that there must be a finite set of points that generate the whole group. Indeed, it can be proved that the three points

$$P = (1, 0), \quad Q = (2, 0), \quad \text{and} \quad R = (0, -3)$$

are generators of $E(\mathbb{Q})$. This means that *any other point* on $E(\mathbb{Q})$ can be expressed as a $\mathbb{Z}$-linear combination of P, Q, and R. In other words,

$$E(\mathbb{Q}) = \{a \cdot P + b \cdot Q + c \cdot R : a, b, c \in \mathbb{Z}\}.$$

For instance,

$$(-3,-1) = P+Q, \quad (8,-22) = P+R, \quad (-2,-4) = P-Q,$$

$$(-1,-4) = Q-R, \quad \text{and} \quad (3,3) = P-R.$$

The proof of the Mordell–Weil theorem has three fundamental ingredients:

(a) the so-called *weak* Mordell–Weil theorem, which says that $E(\mathbb{Q})/mE(\mathbb{Q})$ is finite for any $m \geq 2$ (see below),

(b) the concept of height functions on abelian groups,

(c) the *descent theorem*, which establishes that an abelian group A with a height function h, such that A/mA is finite (for some $m \geq 2$), is finitely generated.

Here is a precise statement for the weak Mordell–Weil theorem.

Theorem 16.3.5 (Weak Mordell–Weil). *$E(\mathbb{Q})/mE(\mathbb{Q})$ is a finite group for all $m \geq 2$.*

We will discuss the proof of a special case of the weak Mordell–Weil theorem in Section 16.7 (see Corollary 16.7.7).

It follows from the Mordell–Weil theorem and the general structure theory of finitely generated abelian groups that

$$E(\mathbb{Q}) \cong E(\mathbb{Q})_{\text{torsion}} \oplus \mathbb{Z}^{R_E}. \tag{16.6}$$

In other words, $E(\mathbb{Q})$ is isomorphic to the direct sum of two abelian groups (notice however that this decomposition *is not* canonical! See (4) in Example 16.3.6 below.). The first summand is a finite group formed by all *torsion* elements, i.e., those points on E of finite order:

$$E(\mathbb{Q})_{\text{torsion}} = \{P \in E(\mathbb{Q}) : \text{ there is } n \in \mathrm{N} \text{ such that } nP = \mathcal{O}\}.$$

The second summand of (16.6), sometimes called the *free part*, is $\mathbb{Z}^{R_E}$, i.e., R_E copies of $\mathbb{Z}$ for some integer $R_E \geq 0$. It is generated by R_E points of $E(\mathbb{Q})$ of infinite order (i.e., $P \in E(\mathbb{Q})$ such that $nP \neq \mathcal{O}$ for all non-zero $n \in \mathbb{Z}$). The number R_E is called the *rank* of the elliptic curve $E/\mathbb{Q}$. Notice, however, that the set

$$F = \{P \in E(\mathbb{Q}) : P \text{ is of infinite order}\} \cup \{\mathcal{O}\}$$

is not a subgroup of $E(\mathbb{Q})$ if the torsion subgroup is non-trivial. For instance, if T is a torsion point and P is of infinite order, then P and $P+T$ belong to F but $T = (P+T) - P$ does not belong to F. This fact makes the isomorphism of (16.6) not canonical because the subgroup of $E(\mathbb{Q})$ isomorphic to $\mathbb{Z}^{R_E}$ cannot be chosen, in general, in a unique way.

Example 16.3.6. The following are some examples of elliptic curves and their Mordell–Weil groups:

(1) The curve $E_1/\mathbb{Q} : y^2 = x^3 + 6$ has no rational points, other than the point at infinity $\mathcal{O}$. Therefore, there are no torsion points (other than $\mathcal{O}$) and no points of infinite order. In particular, the rank is 0, and $E_1(\mathbb{Q}) = \{\mathcal{O}\}$.

(2) The curve $E_2/\mathbb{Q} : y^2 = x^3 + 1$ has only six rational points. As we saw in Example 16.3.2, the point $P = (2, 3)$ has exact order 6. Therefore $E_2(\mathbb{Q}) \cong \mathbb{Z}/6\mathbb{Z}$ is an isomorphism of groups. Since there are no points of infinite order, the rank of $E_2/\mathbb{Q}$ is 0, and

$$E_2(\mathbb{Q}) = \{\mathcal{O},\ P,\ 2P,\ 3P,\ 4P,\ 5P\} = \{\mathcal{O}, (2, \pm 3), (0, \pm 1), (-1, 0)\}.$$

(3) The curve $E_3/\mathbb{Q} : y^2 = x^3 - 2$ does not have any rational torsion points other than $\mathcal{O}$ (as we shall see in the next section). However, the point $P = (3, 5)$ is a rational point. Thus, P must be a point of infinite order and $E_3(\mathbb{Q})$ contains infinitely many distinct rational points. In fact, the rank of E_3 is equal to 1 and P is a generator of all of $E_3(\mathbb{Q})$; i.e.,

$$E_3(\mathbb{Q}) = \{nP : n \in \mathbb{Z}\} \quad \text{and} \quad E_3(\mathbb{Q}) \cong \mathbb{Z}.$$

(4) The elliptic curve $E_4/\mathbb{Q} : y^2 = x^3 + 7105x^2 + 1327104x$ features both torsion and infinite order points. In fact, $E_4(\mathbb{Q}) \cong \mathbb{Z}/4\mathbb{Z} \oplus \mathbb{Z}^3$. The torsion subgroup is generated by the point $T = (1152, 111744)$ of order 4. The free part is generated by three points of infinite order:

$$P_1 = (-6912, 6912), \quad P_2 = (-5832, 188568), \quad P_3 = (-5400, 206280).$$

Hence

$$E_4(\mathbb{Q}) = \{aT + bP_1 + cP_2 + dP_3 : a = 0, 1, 2 \text{ or } 3 \text{ and } b, c, d \in \mathbb{Z}\}.$$

As we mentioned above, the isomorphism $E_4(\mathbb{Q}) \cong \mathbb{Z}/4\mathbb{Z} \oplus \mathbb{Z}^3$ is not canonical. For instance, $E_4(\mathbb{Q}) \cong \langle T \rangle \oplus \langle P_1, P_2, P_3 \rangle$ but also $E_4(\mathbb{Q}) \cong \langle T \rangle \oplus \langle P_1', P_2, P_3 \rangle$ with $P_1' = P_1 + T$.

Example 16.3.7. Let C be the curve $x^3 + y^3 = 1729$, which is related to the famous Hardy–Ramanujan number 1729 (see Example 15.0.1). In Example 16.1.3 we saw that C is isomorphic to $E : y^2 + 9 \cdot 1729y = x^3 - 27 \cdot (1729)^2$. One can show that the Mordell–Weil group of E is isomorphic to $\mathbb{Z}^2$, generated by the points

$$P = (273, -7371) \quad \text{and} \quad Q = (399, -1197).$$

The map $\psi^{-1} \colon E \to C$ of Example 16.1.3, given by

$$\psi^{-1}(x, y) = \left(\frac{9d + y}{3x}, -\frac{y}{3x} \right),$$

maps P and Q, respectively, to the points $P' = (10, 9)$ and $Q' = (12, 1)$, which correspond to the well-known representations of 1729 as the sum of two cubes:

$$1729 = 10^3 + 9^3 = 12^3 + 1^3,$$

that Ramanujan referred to during Hardy's visit. The addition on E gives new rational solutions. For instance, $P + Q = (1729, -79534)$ and $\psi^{-1}(P + Q)$ yields

$(-37/3, 46/3)$ on C. Indeed,

$$1729 = \left(-\frac{37}{3}\right)^3 + \left(\frac{46}{3}\right)^3$$

is an expression of 1729 as the sum of two (rational) cubes.

The rank of $E/\mathbb{Q}$ is, in a sense, a measurement of the arithmetic complexity of the elliptic curve. It is not known if there is an upper bound for the possible values of R_E (the largest rank known, to date, is 28, discovered by Noam Elkies; see Andrej Dujella's website **[Duj09]** for up-to-date records and examples of curves with "high" ranks). It has been conjectured (with some controversy) that ranks can be arbitrarily large; i.e., for all $n \in \mathbb{N}$ there exists an elliptic curve E over $\mathbb{Q}$ with $R_E \geq n$. We state this as a conjecture for future reference:

Conjecture 16.3.8 (Conjecture of the rank). *Let $N \geq 0$ be a natural number. Then there exists an elliptic curve E defined over $\mathbb{Q}$ with rank $R_E \geq N$.*

One of the key pieces of evidence in favor of such a conjecture was offered by Shafarevich and Tate, who proved that there exist elliptic curves defined over function fields $\mathbb{F}_p(T)$ and with arbitrarily large ranks ($\mathbb{F}_p(T)$ is a field that shares many similar properties with $\mathbb{Q}$; see **[ShT67]**). In any case, the problem of finding elliptic curves of high rank is particularly interesting because of its arithmetic and computational complexity.

16.4. The Torsion Subgroup

In this section we concentrate on the torsion points of an elliptic curve:

$$E(\mathbb{Q})_{\text{torsion}} = \{P \in E(\mathbb{Q}) : \text{ there is } n \in \mathbb{N} \text{ such that } nP = \mathcal{O}\}.$$

Example 16.4.1. The curve $E_n : y^2 = x^3 - n^2x = x(x-n)(x+n)$ has three rational points that are easy to find, namely $P = (0,0), Q = (-n,0), T = (n,0)$, and one can check (see Exercise 16.10.5) that each one of these points is torsion of order 2; i.e., $2P = 2Q = 2T = \mathcal{O}$, and $P + Q = T$. In fact $E_n(\mathbb{Q})_{\text{torsion}} = \{\mathcal{O}, P, Q, T\} \cong \mathbb{Z}/2\mathbb{Z} \oplus \mathbb{Z}/2\mathbb{Z}$.

Note that the Mordell–Weil theorem implies that $E(\mathbb{Q})_{\text{torsion}}$ is always finite. This fact prompts a natural question: *what abelian groups can appear in this context?* The answer was conjectured by Levi and by Ogg and was proven by Mazur:

Theorem 16.4.2 (Ogg's conjecture; Mazur, **[Maz78]**). *Let $E/\mathbb{Q}$ be an elliptic curve. Then, $E(\mathbb{Q})_{torsion}$ is isomorphic to one of the following groups:*

$$\begin{aligned} \mathbb{Z}/N\mathbb{Z} \quad &\textit{with} \quad 1 \leq N \leq 10 \textit{ or } N = 12 \textit{ or} \\ \mathbb{Z}/2\mathbb{Z} \oplus \mathbb{Z}/2M\mathbb{Z} \quad &\textit{with} \quad 1 \leq M \leq 4. \end{aligned} \tag{16.7}$$

Moreover, each group in the lists above occurs for infinitely many non-isomorphic elliptic curves over $\mathbb{Q}$.

Curve	Torsion	Generators
$y^2 = x^3 - 2$	trivial	$\mathcal{O}$
$y^2 = x^3 + 8$	$\mathbb{Z}/2\mathbb{Z}$	$(-2, 0)$
$y^2 = x^3 + 4$	$\mathbb{Z}/3\mathbb{Z}$	$(0, 2)$
$y^2 = x^3 + 4x$	$\mathbb{Z}/4\mathbb{Z}$	$(2, 4)$
$y^2 - y = x^3 - x^2$	$\mathbb{Z}/5\mathbb{Z}$	$(0, 1)$
$y^2 = x^3 + 1$	$\mathbb{Z}/6\mathbb{Z}$	$(2, 3)$
$y^2 = x^3 - 43x + 166$	$\mathbb{Z}/7\mathbb{Z}$	$(3, 8)$
$y^2 + 7xy = x^3 + 16x$	$\mathbb{Z}/8\mathbb{Z}$	$(-2, 10)$
$y^2 + xy + y = x^3 - x^2 - 14x + 29$	$\mathbb{Z}/9\mathbb{Z}$	$(3, 1)$
$y^2 + xy = x^3 - 45x + 81$	$\mathbb{Z}/10\mathbb{Z}$	$(0, 9)$
$y^2 + 43xy - 210y = x^3 - 210x^2$	$\mathbb{Z}/12\mathbb{Z}$	$(0, 210)$
$y^2 = x^3 - 4x$	$\mathbb{Z}/2\mathbb{Z} \oplus \mathbb{Z}/2\mathbb{Z}$	$\binom{(2,0)}{(0,0)}$
$y^2 = x^3 + 2x^2 - 3x$	$\mathbb{Z}/4\mathbb{Z} \oplus \mathbb{Z}/2\mathbb{Z}$	$\binom{(3,6)}{(0,0)}$
$y^2 + 5xy - 6y = x^3 - 3x^2$	$\mathbb{Z}/6\mathbb{Z} \oplus \mathbb{Z}/2\mathbb{Z}$	$\binom{(-3,18)}{(2,-2)}$
$y^2 + 17xy - 120y = x^3 - 60x^2$	$\mathbb{Z}/8\mathbb{Z} \oplus \mathbb{Z}/2\mathbb{Z}$	$\binom{(30,-90)}{(-40,400)}$

Figure 16.4. Examples of each of the possible torsion subgroups over $\mathbb{Q}$.

Example 16.4.3. For instance, the torsion subgroup of the elliptic curve with Weierstrass equation $y^2 + 43xy - 210y = x^3 - 210x^2$ is isomorphic to $\mathbb{Z}/12\mathbb{Z}$ and it is generated by the point $(0, 210)$. The elliptic curve $y^2 + 17xy - 120y = x^3 - 60x^2$ has a torsion subgroup isomorphic to $\mathbb{Z}/2\mathbb{Z} \oplus \mathbb{Z}/8\mathbb{Z}$, generated by the rational points $(30, -90)$ and $(-40, 400)$. See Figure 16.4 for a complete list of examples with each possible torsion subgroup.

Furthermore, it is known that if G is any of the groups in 16.7, there are infinitely many elliptic curves whose torsion subgroup is isomorphic to G. See, for example, [**Kub76**, Table 3, p. 217] or Appendix E of [**Loz11**].

Example 16.4.4. Let $E_t : y^2 + (1 - t)xy - ty = x^3 - tx^2$ with non-zero $t \in \mathbb{Q}$. Then, the torsion subgroup of $E_t(\mathbb{Q})$ contains a subgroup isomorphic to $\mathbb{Z}/5\mathbb{Z}$, and $(0, 0)$ is a point of exact order 5. Conversely, if $E : y^2 = x^3 + Ax + B$ is an elliptic curve with torsion subgroup equal to $\mathbb{Z}/5\mathbb{Z}$, then there is an invertible change of variables that takes E to an equation of the form E_t for some $t \in \mathbb{Q}$.

A useful and simple consequence of Mazur's theorem is that if the order of a rational point $P \in E(\mathbb{Q})$ is larger than 12, then P must be a point of infinite order and, therefore, $E(\mathbb{Q})$ contains an infinite number of distinct rational points. Except for this criterion, Mazur's theorem is not very helpful in effectively computing the torsion subgroup of a given elliptic curve. However, the following result, proven independently by Trygve Nagell (in 1935) and by Élisabeth Lutz (in 1937), provides a simple algorithm to determine $E(\mathbb{Q})_{\text{torsion}}$:

Theorem 16.4.5 (Nagell–Lutz, [**Nag35**], [**Lut37**]). *Let $E/\mathbb{Q}$ be an elliptic curve with Weierstrass equation*

$$y^2 = x^3 + Ax + B, \quad A, B \in \mathbb{Z}.$$

Then, every torsion point $P \neq \mathcal{O}$ of E satisfies:

(1) *The coordinates of P are integers; i.e., $x(P), y(P) \in \mathbb{Z}$.*
(2) *If P is a point of order $n \geq 3$, then $4A^3 + 27B^2$ is divisible by $y(P)^2$.*
(3) *If P is of order 2, then $y(P) = 0$ and $x(P)^3 + Ax(P) + B = 0$.*

For a proof, see [**Sil86**, Ch. VIII, Corollary 7.2] or [**Mil06**, Ch. II, Theorem 5.1].

Example 16.4.6. Let $E/\mathbb{Q} : y^2 = x^3 - 2$, so that $A = 0$ and $B = -2$. The polynomial $x^3 - 2$ does not have any rational roots, so $E(\mathbb{Q})$ does not contain any points of order 2. Also, $4A^3 + 27B^2 = 27 \cdot 4$. Thus, if $(x(P), y(P))$ are the coordinates of a torsion point in $E(\mathbb{Q})$, then $y(P)$ is an integer and $y(P)^2$ divides $27 \cdot 4$. This implies that $y(P) = \pm 1, \pm 2, \pm 3$, or ± 6. In turn, this implies that $x(P)^3 = 3, 6, 11$, or 38, respectively. However, $x(P)$ is an integer, and none of 3, 6, 11, or 38 is a perfect cube. Thus, $E(\mathbb{Q})_{\text{torsion}}$ is trivial (i.e., the only torsion point is $\mathcal{O}$).

Example 16.4.7. Let $p \geq 2$ be a prime number and let us define a curve $E_p \colon y^2 = x^3 + p^2$. Since $x^3 + p^2 = 0$ does not have any rational roots, $E_p(\mathbb{Q})$ does not contain points of order 2. Let P be a torsion point on $E_p(\mathbb{Q})$. The list of all squares dividing $4A^3 + 27B^2 = 27p^4$ is short, and by the Nagell–Lutz theorem the possible values for $y(P)$ are

$$y = \pm 1, \ \pm p, \ \pm p^2, \ \pm 3p, \ \pm 3p^2, \text{ and } \pm 3.$$

Clearly, $(0, \pm p) \in E_p(\mathbb{Q})$ and one can show that those two points and $\mathcal{O}$ are the only torsion points; see Exercise 16.10.7. Thus, the torsion subgroup of $E_p(\mathbb{Q})$ is isomorphic to $\mathbb{Z}/3\mathbb{Z}$ for any prime $p \geq 2$.

Remark 16.4.8. It is important to note that the conclusions of the Nagell–Lutz theorem depend on the Weierstrass equation being of the form $y^2 = x^3 + Ax + B$. For instance, consider the elliptic curve

$$E : y^2 + xy = x^3 + 4x + 1.$$

Then, E has a torsion point $P = (-1/4, 1/8)$ of order 2 and, clearly, the coordinates of P are not integral. However, we can change variables on E to bring it to a short Weierstrass model $y^2 = x^3 + 5157x + 31158$ and in this equation the 2-torsion point P has coordinates $(-6, 0)$, which are integral as predicted by Nagell–Lutz.

16.5. Elliptic Curves over Finite Fields

Let $p \geq 2$ be a prime and let $\mathbb{F}_p$ be the finite field with p elements (see Section 5.1 and Chapter 6); i.e.,

$$\mathbb{F}_p = \mathbb{Z}/p\mathbb{Z} = \{a \bmod p : a = 0, 1, 2, \ldots, p-1\}.$$

$\mathbb{F}_p$ is a field and we may consider elliptic curves defined over $\mathbb{F}_p$. As for elliptic curves over $\mathbb{Q}$, there are three conditions that need to be satisfied: the curve needs to be given by a cubic equation, the curve needs to be smooth, and there must be a point on the curve defined over $\mathbb{F}_p$ (the existence of an $\mathbb{F}_p$-point, however, will be automatic by Hasse's theorem, Theorem 16.5.11).

Example 16.5.1. For instance, $E : y^2 \equiv x^3 + 1 \bmod 5$ is an elliptic curve defined over $\mathbb{F}_5$. It is given by a cubic equation ($zy^2 \equiv x^3 + z^3 \bmod 5$ in the projective plane $\mathbb{P}^2(\mathbb{F}_5)$) and it is smooth, because for $F \equiv zy^2 - x^3 - z^3 \bmod 5$, the partial derivatives are

$$\frac{\partial F}{\partial x} \equiv -3x^2, \quad \frac{\partial F}{\partial y} \equiv 2yz, \quad \frac{\partial F}{\partial z} \equiv y^2 - 3z^2 \bmod 5.$$

Thus, if the partial derivatives are congruent to 0 modulo 5, then $x \equiv 0 \bmod 5$ and $yz \equiv 0 \bmod 5$. The latter congruence implies that y or $z \equiv 0 \bmod 5$, and $\partial F/\partial z \equiv 0$ implies that $y \equiv z \equiv 0 \bmod 5$. Since $[0, 0, 0]$ is not a point in the projective plane, we conclude that there are no singular points on $E/\mathbb{F}_5$.

However, $C/\mathbb{F}_3 : y^2 \equiv x^3 + 1 \bmod 3$ is not an elliptic curve because it is not smooth. Indeed, the point $P = (2 \bmod 3, 0 \bmod 3) \in C(\mathbb{F}_3)$ is a singular point:

$$\begin{aligned} \frac{\partial F}{\partial x}(P) &\equiv -3 \cdot 2^2 \equiv 0, \quad \frac{\partial F}{\partial y}(P) \equiv 2 \cdot 0 \cdot 1 \equiv 0, \quad \text{and} \\ \frac{\partial F}{\partial z}(P) &\equiv 0^2 - 3 \cdot 1^2 \equiv 0 \bmod 3. \end{aligned}$$

Let $E/\mathbb{Q}$ be an elliptic curve given by a Weierstrass equation $y^2 = x^3 + Ax + B$ with integer coefficients $A, B \in \mathbb{Z}$, and let $p \geq 2$ be a prime number. If we reduce A and B modulo p, then we obtain the equation of a curve $\widetilde{E}$ given by a cubic curve and defined over the field $\mathbb{F}_p$. Even though E is smooth as a curve over $\mathbb{Q}$, the curve $\widetilde{E}$ may be singular over $\mathbb{F}_p$. In the previous example, we saw that $E/\mathbb{Q} : y^2 = x^3 + 1$ is smooth over $\mathbb{Q}$ and $\mathbb{F}_5$ but it has a singularity over $\mathbb{F}_3$. If the reduction curve $\widetilde{E}$ is smooth, then it is an elliptic curve over $\mathbb{F}_p$.

Example 16.5.2. Sometimes the reduction of a model for an elliptic curve E modulo a prime p is not smooth, but it is smooth for some other models of E. For instance, consider the curve $E : y^2 = x^3 + 15625$. Then $\widetilde{E} \equiv E \bmod 5$ is not smooth over $\mathbb{F}_5$ because the point $(0, 0) \bmod 5$ is a singular point. However, using the invertible change of variables $(x, y) \mapsto (5^2 X, 5^3 Y)$, we obtain a new model over $\mathbb{Q}$ for E given by $E' : Y^2 = X^3 + 1$, which is smooth when we reduce it modulo 5. The problem here is that the model we chose for E is not *minimal*. We describe what we mean by minimal next.

Definition 16.5.3. Let E be an elliptic curve given by $y^2 = x^3 + Ax + B$, with $A, B \in \mathbb{Q}$.

(1) We define Δ_E, the *discriminant* of E, by

$$\Delta_E = -16(4A^3 + 27B^2).$$

For a definition of the discriminant for more general Weierstrass equations, see for example [**Sil86**, p. 46].

(2) Let S be the set of all elliptic curves E' that are isomorphic to E over $\mathbb{Q}$ (see Definition 16.1.4) and such that the discriminant of E' is an integer. The

minimal discriminant of E is the integer $\Delta_{E'}$ that attains the minimum of the set $\{|\Delta_{E'}| : E' \in S\}$. In other words, the minimal discriminant is the smallest integral discriminant (in absolute value) of an elliptic curve that is isomorphic to E over $\mathbb{Q}$. If E' is the model for E with minimal discriminant, we say that E' is a *minimal model* for E.

Example 16.5.4. The curve $E : y^2 = x^3 + 5^6$ has discriminant $\Delta_E = -2^4 3^3 5^{12}$, and the curve $E' : y^2 = x^3 + 1$ has discriminant $\Delta_{E'} = -2^4 3^3$. Since E and E' are isomorphic (see Definition 16.1.4 and Example 16.5.2), then Δ_E cannot be the minimal discriminant for E and $y^2 = x^3 + 5^6$ is not a minimal model. In fact, the minimal discriminant is $\Delta_{E'} = -432$ and E' is a minimal model.

Before we go on to describe the types of reduction modulo p that one can encounter, we need a little bit of background on types of singularities. Let $\widetilde{E}$ be a cubic curve over a field K with Weierstrass equation $f(x, y) = 0$, where

$$f(x, y) = y^2 + a_1 xy + a_3 y - x^3 - a_2 x^2 - a_4 x - a_6,$$

and suppose that $\widetilde{E}$ has a singular point $P = (x_0, y_0)$; i.e., $\partial f/\partial x(P) = \partial f/\partial y(P) = 0$. Thus, we can write the Taylor expansion of $f(x, y)$ around (x_0, y_0) as follows:

$$\begin{aligned} f(x, y) &- f(x_0, y_0) \\ &= \lambda_1 (x - x_0)^2 + \lambda_2 (x - x_0)(y - y_0) + \lambda_3 (y - y_0)^2 - (x - x_0)^3 \\ &= ((y - y_0) - \alpha(x - x_0)) \cdot ((y - y_0) - \beta(x - x_0)) - (x - x_0)^3 \end{aligned}$$

for some $\lambda_i \in K$ and $\alpha, \beta \in \overline{K}$ (an algebraic closure of K).

Definition 16.5.5. The singular point $P \in \widetilde{E}$ is a *node* if $\alpha \neq \beta$. In this case there are two different tangent lines to $\widetilde{E}$ at P; namely

$$y - y_0 = \alpha(x - x_0), \quad y - y_0 = \beta(x - x_0).$$

If $\alpha = \beta$, then we say that P is a *cusp*, and there is a unique tangent line at P. See Figure 16.5.

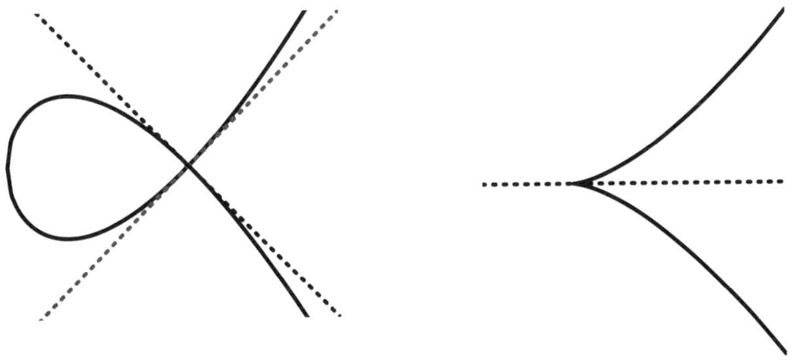

Figure 16.5. A node (left) with two tangent lines and a cusp (right) with only one tangent line.

Definition 16.5.6. Let $E/\mathbb{Q}$ be an elliptic curve given by a minimal model, let $p \geq 2$ be a prime, and let $\widetilde{E}$ be the reduction curve of E modulo p. We say that $E/\mathbb{Q}$ has *good reduction* modulo p if $\widetilde{E}$ is smooth and hence is an elliptic curve over $\mathbb{F}_p$. If $\widetilde{E}$ is singular at a point $P \in E(\mathbb{F}_p)$, then we say that $E/\mathbb{Q}$ has bad reduction at p and we distinguish two cases:

(1) If $\widetilde{E}$ has a cusp at P, then we say that E has *additive (or unstable) reduction.*

(2) If $\widetilde{E}$ has a node at P, then we say that E has *multiplicative (or semistable) reduction.* If the slopes of the tangent lines (α and β as above) are in $\mathbb{F}_p$, then the reduction is said to be *split* multiplicative (and *non-split* otherwise).

Example 16.5.7. Let us see some examples of elliptic curves with different types of reduction:

(1) $E_1\colon y^2 = x^3 + 35x + 5$ has good reduction at $p = 7$, because $y^2 \equiv x^3 + 5 \bmod 7$ is a non-singular curve over $\mathbb{F}_7$.

(2) However E_1 has bad reduction at $p = 5$, and the reduction is additive, since modulo 5 we can write the equation as $((y-0) - 0 \cdot (x-0))^2 - x^3$ and the unique slope is 0.

(3) The elliptic curve $E_2\colon y^2 = x^3 - x^2 + 35$ has bad multiplicative reduction at 5 and 7. The reduction at 5 is split, while the reduction at 7 is non-split. Indeed, modulo 5 we can write the equation as

$$((y-0) - 2(x-0)) \cdot ((y-0) + 2(x-0)) - x^3,$$

the slopes being 2 and -2. However, for $p = 7$, the slopes are not in $\mathbb{F}_7$ (because -1 is not a quadratic residue in $\mathbb{F}_7$). Indeed, when we reduce the equation modulo 7, we obtain

$$y^2 + x^2 - x^3 \bmod 7$$

and $y^2 + x^2$ can only be factored in $\mathbb{F}_7[i]$ but not in $\mathbb{F}_7$.

(4) Let E_3 be an elliptic curve given by the model $y^2 + y = x^3 - x^2 - 10x - 20$. This is a minimal model for E_3 and its (minimal) discriminant is $\Delta_{E_3} = -11^5$. The prime 11 is the unique prime of bad reduction and the reduction is split multiplicative. Indeed, the point $(5, 5) \bmod 11$ is a singular point on $E_3(\mathbb{F}_{11})$ and

$$\begin{aligned} f(x,y) &= y^2 + y + x^2 + 10x + 20 - x^3 \\ &= (y - 5 - 5(x-5)) \cdot (y - 5 + 5(x-5)) - (x-5)^3. \end{aligned}$$

Hence, the slopes at $(5,5)$ are 5 and -5, which are in $\mathbb{F}_{11}$ and distinct.

Proposition 16.5.8. *Let K be a field and let E/K be a cubic curve given by $y^2 = f(x)$, where $f(x)$ is a monic cubic polynomial in $K[x]$. Suppose that $f(x) = (x-\alpha)(x-\beta)(x-\gamma)$ with $\alpha, \beta, \gamma \in \overline{K}$ (an algebraic closure of K) and put*

$$D = (\alpha - \beta)^2(\alpha - \gamma)^2(\beta - \gamma)^2.$$

Then E is non-singular if and only if $D \neq 0$.

The proof of the proposition is left as an exercise (see Exercise 16.10.8). Notice that the quantity D that appears in the previous proposition is the *discriminant* of the polynomial $f(x)$. The discriminant of $E/\mathbb{Q}$, Δ_E as in Definition 16.5.3, is a multiple of D; in fact, $\Delta_E = 16D$. This fact together with Proposition 16.5.8 yields the following corollary:

Corollary 16.5.9. *Let $E/\mathbb{Q}$ be an elliptic curve with coefficients in $\mathbb{Z}$. Let $p \geq 2$ be a prime. If E has bad reduction at p, then $p \mid \Delta_E$. In fact, if E is given by a minimal model, then $p \mid \Delta_E$ if and only if E has bad reduction at p.*

Example 16.5.10. The discriminant of the elliptic curve $E_1 \colon y^2 = x^3 + 35x + 5$ of Example 16.5.7 is $\Delta_{E_1} = -2754800 = -2^4 \cdot 5^2 \cdot 71 \cdot 97$ (and, in fact, this is the minimal discriminant of E_1). Thus, E_1 has good reduction at 7 but it has bad reduction at 2, 5, 71, and 97. The reduction at 71 and 97 is multiplicative.

Let $\widetilde{E}$ be an elliptic curve defined over a finite field $\mathbb{F}_q$ with q elements, where $q = p^r$ and $p \geq 2$ is prime. Notice that $\widetilde{E}(\mathbb{F}_q) \subseteq \mathbb{P}^2(\mathbb{F}_q)$ and the projective plane over $\mathbb{F}_q$ has only a finite number of points (how many?). Thus, the number $N_q := |\widetilde{E}(\mathbb{F}_q)|$, i.e., the number of points on $\widetilde{E}$ over $\mathbb{F}_q$, is finite. The following theorem provides a bound for N_q. This result was conjectured by Emil Artin (in his thesis) and was proved by Helmut Hasse in the 1930s (see Figure 11.1).

Theorem 16.5.11 (Hasse). *Let $\widetilde{E}$ be an elliptic curve defined over $\mathbb{F}_q$. Then*

$$q + 1 - 2\sqrt{q} < N_q < q + 1 + 2\sqrt{q},$$

where $N_q = |\widetilde{E}(\mathbb{F}_q)|$.

See [**Sil86**, Ch. V, Theorem 1.1] for a proof of Hasse's theorem.

Remark 16.5.12. Heuristically, we expect that N_q is approximately $q + 1$, in agreement with Hasse's bound. Indeed, let $E/\mathbb{Q}$ be an elliptic curve given by $y^2 = x^3 + Ax + B$, with $A, B \in \mathbb{Z}$, and let $q = p$ be a prime for simplicity. There are p choices of x in $\mathbb{F}_p$. For each value x_0, the polynomial $f(x) = x^3 + Ax + B$ gives a value $f(x_0) \in \mathbb{F}_p$. The probability that a random element in $\mathbb{F}_p$ is a perfect square in $\mathbb{F}_p$ is $1/2$ (notice, however, that $f(x_0)$ is not random; this is just a heuristic argument). If $f(x_0)$ is a square modulo p, i.e., if there is a $y_0 \in \mathbb{F}_p$ such that $f(x_0) \equiv y_0^2 \bmod p$, then there are two points $(x_0, \pm y_0)$ in $\widetilde{E}(\mathbb{F}_p)$. If $f(x_0)$ is not a square modulo p, then there are no points in $\widetilde{E}(\mathbb{F}_p)$ with x-coordinate equal to x_0. Hence,

$$N_p \approx p \cdot \left(\frac{1}{2} \cdot 2 + \frac{1}{2} \cdot 0\right) + 1 = p + 1.$$

Notice that we have added 1 in order to account for the point at infinity.

Remark 16.5.13. Suppose that $E/\mathbb{Q}$ is an elliptic curve that has bad reduction at a prime p. How many points does the singular curve $\widetilde{E}$ have over $\mathbb{F}_p$?

Example 16.5.14. Let $E/\mathbb{Q}$ be the elliptic curve $y^2 = x^3 + 3$. Its minimal discriminant is $\Delta_E = -3888 = -2^4 \cdot 3^5$. Thus, the only primes of bad reduction are 2 and 3 and $\widetilde{E}/\mathbb{F}_p$ is smooth for all $p \geq 5$. For $p = 5$, there are precisely six points on $\widetilde{E}(\mathbb{F}_5)$; namely

$$\widetilde{E}(\mathbb{F}_5) = \{\widetilde{O}, (1, 2), (1, 3), (2, 1), (2, 4), (3, 0)\},$$

where all the coordinates should be regarded as congruences modulo 5. Thus, $N_5 = 6$, which is in the range given by Hasse's bound:

$$1.5278\ldots = 5 + 1 - 2\sqrt{5} < N_5 < 5 + 1 + 2\sqrt{5} = 10.4721\ldots.$$

Similarly, one can verify that $N_7 = 13$.

The connections between the numbers N_p and the group $E(\mathbb{Q})$ are numerous and of great interest. The most surprising relationship is captured by the Birch and Swinnerton-Dyer conjecture (see [**Loz11**, Section 5.2]) that relates the growth of N_p (as p varies) with the rank of the elliptic curve $E/\mathbb{Q}$. In the next proposition we describe a different connection between N_p and $E(\mathbb{Q})$. We shall use the following notation: if G is an abelian group and $m \geq 2$, then the points of G of order dividing m will be denoted by $G[m]$.

Proposition 16.5.15 (Ch. VII, Prop. 3.1 of [**Sil86**]). *Let $E/\mathbb{Q}$ be an elliptic curve, p a prime number, and m a natural number not divisible by p. Suppose that $E/\mathbb{Q}$ has good reduction at p. Then the reduction map modulo p,*

$$E(\mathbb{Q})[m] \longrightarrow \widetilde{E}(\mathbb{F}_p),$$

is an injective homomorphism of abelian groups. In particular, the number of elements of $E(\mathbb{Q})[m]$ divides the number of elements of $\widetilde{E}(\mathbb{F}_p)$.

The previous proposition can be very useful when calculating the torsion subgroup of an elliptic curve. Let us see an application:

Example 16.5.16. Let $E/\mathbb{Q}$: $y^2 = x^3 + 3$. In Example 16.5.14 we have seen that $N_5 = 6$ and $N_7 = 13$, and $E/\mathbb{Q}$ has bad reduction only at 2 and 3.

If $q \neq 5, 7$ is a prime number, then $E(\mathbb{Q})[q]$ is trivial. Indeed, Proposition 16.5.15 implies that $|E(\mathbb{Q})[q]|$ divides $N_5 = 6$ and also $N_7 = 13$. Thus, $|E(\mathbb{Q})[q]|$ must divide $\gcd(6, 13) = 1$.

In the case of $q = 5$, we know that $|E(\mathbb{Q})[5]|$ divides $N_7 = 13$. Moreover, by Lagrange's theorem from group theory, if $E(\mathbb{Q})[p]$ is non-trivial, then p divides $|E(\mathbb{Q})[p]|$ (it turns out that $E(\mathbb{Q})[p]$ is always a subgroup of $\mathbb{Z}/p\mathbb{Z} \times \mathbb{Z}/p\mathbb{Z}$). Since 5 does not divide 13, it follows that $E(\mathbb{Q})[5]$ must be trivial. Similarly, one can show that $E(\mathbb{Q})[7]$ is trivial, and we conclude that $E(\mathbb{Q})_{\text{torsion}}$ is trivial.

However, notice that $P = (1, 2) \in E(\mathbb{Q})$ is a point on the curve. Since we just proved that E does not have any points of finite order, it follows that P must be a point of *infinite* order, and, hence, we have shown that E has infinitely many rational points: $\pm P, \pm 2P, \pm 3P, \ldots$. In fact, $E(\mathbb{Q}) \cong \mathbb{Z}$ and $(1, 2)$ is a generator of its Mordell–Weil group.

In the previous example, the Nagell–Lutz theorem (Theorem 16.4.5) would have yielded the same result, i.e., the torsion is trivial, in an easier way. Indeed, for the curve $E : y^2 = x^3 + 3$, the quantity $4A^3 + 27B^2$ equals 3^5, so the possibilities for $y(P)^2$, where P is a torsion point of order ≥ 3, are 1, 9, or 81 (the reader can check that there are no 2-torsion points). Therefore, the possibilities for $x(P)^3 = y(P)^2 - 3$ are -2, 6, or 78, respectively. Since $x(P)$ is an integer, we reach a contradiction. In the following example, the Nagell–Lutz theorem would be a lengthier and much more tedious alternative, and Proposition 16.5.15 is much more effective.

Example 16.5.17. Let $E/\mathbb{Q} : y^2 = x^3 + 4249388$. In this case

$$4A^3 + 27B^2 = 2^4 \cdot 3^3 \cdot 11^2 \cdot 13^2 \cdot 17^2 \cdot 19^2 \cdot 23^2.$$

Therefore, $4A^3 + 27B^2$ is divisible by 192 distinct positive squares, which makes it very tedious to use the Nagell–Lutz theorem. The (minimal) discriminant of $E/\mathbb{Q}$ is $\Delta_E = -16(4A^3 + 27B^2)$ and therefore E has good reduction at 5 and 7. Moreover, $B = 4249388 \equiv 3 \bmod 35$ and therefore, by our calculations in Example 16.5.16, $N_5 = 6$ and $N_7 = 13$. Thus, Proposition 16.5.15 and the same argument we used in Example 16.5.16 show that the torsion of $E(\mathbb{Q})$ is trivial.

Incidentally, the curve $E/\mathbb{Q} : y^2 = x^3 + 4249388$ has a rational point $P = \left(\frac{25502}{169}, \frac{6090670}{2197}\right)$. Since the torsion of $E(\mathbb{Q})$ is trivial, P must be of infinite order. Here is another way to see this: since P has rational coordinates that are not integral, the Nagell–Lutz theorem implies that the order of P is infinite. In fact, $E(\mathbb{Q})$ is isomorphic to $\mathbb{Z}$ and it is generated by P.

16.6. The Rank and the Free Part of $E(\mathbb{Q})$

In the previous sections we have described simple algorithms that determine the torsion subgroup of $E(\mathbb{Q})$. Recall that the Mordell–Weil theorem (Theorem 16.3.3) says that there is a (non-canonical) isomorphism

$$E(\mathbb{Q}) \cong E(\mathbb{Q})_{\text{torsion}} \oplus \mathbb{Z}^{R_E}.$$

Our next goal is to try to find R_E generators of the free part of the Mordell–Weil group. Unfortunately, no algorithm is known that will always yield such free points. We don't even have a way to determine R_E (the rank of the curve) in general, although sometimes we can obtain upper bounds for the rank of a given curve $E/\mathbb{Q}$ (see, for instance, Theorem 16.6.4 below).

Naively, one could hope that if the coefficients of the (minimal) Weierstrass equation for $E/\mathbb{Q}$ are *small*, then the coordinates of the generators of $E(\mathbb{Q})$ should also be *small*, and perhaps a *brute force* computer search would yield these points. However, Bremner and Cassels found the following surprising example: the curve $y^2 = x^3 + 877x$ has rank equal to 1 and the x-coordinate of a generator P is

$$x(P) = (612776083187947368101/7884153586068390021 0)^2.$$

However, Serge Lang salvaged this idea and conjectured that for all $\varepsilon > 0$ there is a constant C_ε such that there is a system of generators $\{P_i : i = 1, \ldots, R_E\}$ of $E(\mathbb{Q})$ with

$$\hat{h}(P_i) \le C_\varepsilon \cdot |\Delta_E|^{1/2+\varepsilon},$$

where $\hat{h}$ is the canonical height function of $E/\mathbb{Q}$, which we define next. Lang's conjecture says that the size of the coordinates of a generator may grow exponentially with the (minimal) discriminant of a curve $E/\mathbb{Q}$.

Definition 16.6.1. We define the *height* of $\frac{m}{n} \in \mathbb{Q}$, with $\gcd(m, n) = 1$, by

$$h\left(\frac{m}{n}\right) = \log(\max\{|m|, |n|\}).$$

This can be used to define a height of a point $P = (x, y)$ on an elliptic curve $E/\mathbb{Q}$, with $x, y \in \mathbb{Q}$ by

$$H(P) = h(x).$$

Finally, we define the *canonical height* of $P \in E(\mathbb{Q})$ by

$$\hat{h}(P) = \frac{1}{2} \lim_{N\to\infty} \frac{H(2^N \cdot P)}{4^N}.$$

Note: here $2^N \cdot P$ means multiplication on the curve, using the addition law defined in Section 16.3; i.e., $2 \cdot P = P + P$, $2^2 \cdot P = 2P + 2P$, etc.

Example 16.6.2. Let $E : y^2 = x^3 + 877x$, and let P be a generator of $E(\mathbb{Q})$. Here are some values of $\frac{1}{2} \cdot \frac{H(2^N \cdot P)}{4^N}$:

$$\begin{aligned}
\frac{1}{2} \cdot H(P) &= 47.8645312628\ldots, \\
\frac{1}{2} \cdot \frac{H(2 \cdot P)}{4} &= 47.7958126219\ldots, \\
\frac{1}{2} \cdot \frac{H(2^2 \cdot P)}{4^2} &= 47.9720107996\ldots, \\
\frac{1}{2} \cdot \frac{H(2^3 \cdot P)}{4^3} &= 47.9636902383\ldots, \\
\frac{1}{2} \cdot \frac{H(2^4 \cdot P)}{4^4} &= 47.9901607777\ldots, \\
\frac{1}{2} \cdot \frac{H(2^5 \cdot P)}{4^5} &= 47.9901600133\ldots, \\
\frac{1}{2} \cdot \frac{H(2^6 \cdot P)}{4^6} &= 47.9901569227\ldots, \\
\frac{1}{2} \cdot \frac{H(2^7 \cdot P)}{4^7} &= 47.9901419861\ldots, \\
\frac{1}{2} \cdot \frac{H(2^8 \cdot P)}{4^8} &= 47.9901807594\ldots.
\end{aligned}$$

The limit is in fact equal to $\hat{h}(P) = 47.9901859939...$, well below the value $|\Delta_E|^{1/2} = 207,773.12....$

The canonical height enjoys the following properties and, in fact, the canonical height is defined so that it is (essentially) the *only* height that satisfies these properties:

Proposition 16.6.3 (Néron–Tate). *Let $E/\mathbb{Q}$ be an elliptic curve and let $\hat{h}$ be the canonical height on E.*

(1) *For all $P, Q \in E(\mathbb{Q})$, $\hat{h}(P+Q) + \hat{h}(P-Q) = 2\hat{h}(P) + 2\hat{h}(Q)$. (Note: this is called the parallelogram law.)*

(2) *For all $P \in E(\mathbb{Q})$ and $m \in \mathbb{Z}$, $\hat{h}(mP) = m^2 \cdot \hat{h}(P)$. (Note: in particular, the height of mP is much larger than the height of P, for any $m \neq 0, 1$.)*

(3) *Let $P \in E(\mathbb{Q})$. Then $\hat{h}(P) \geq 0$, and $\hat{h}(P) = 0$ if and only if P is a torsion point.*

For the proofs of these properties, see [**Sil86**, Ch. VIII, Thm. 9.3] or [**Mil06**, Ch. IV, Prop. 4.5 and Thm. 4.7].

As we mentioned at the beginning of this section, we can calculate upper bounds on the rank of a given elliptic curve (see [**Sil86**, p. 235, Exercises 8.1 and 8.2]. Here is an example:

Theorem 16.6.4 (Prop. 1.1 of [**ALP08**]). *Let $E/\mathbb{Q}$ be an elliptic curve given by a Weierstrass equation of the form*

$$E\colon y^2 = x^3 + Ax^2 + Bx, \text{ with } A, B \in \mathbb{Z}.$$

Let R_E be the rank of $E(\mathbb{Q})$. For an integer $N \geq 1$, let $\nu(N)$ be the number of distinct positive prime divisors of N. Then

$$R_E \leq \nu(A^2 - 4B) + \nu(B) - 1.$$

More generally, let $E/\mathbb{Q}$ be any elliptic curve with a non-trivial 2-torsion point and let a (resp. m) be the number of primes of additive (resp. multiplicative) bad reduction of $E/\mathbb{Q}$. Then

$$R_E \leq m + 2a - 1.$$

Example 16.6.5. Let $E/\mathbb{Q}$ be the elliptic curve $y^2 = x(x+1)(x+2)$. Since the Weierstrass equation of E is

$$y^2 = x(x+1)(x+2) = x^3 + 3x^2 + 2x,$$

it follows from Theorem 16.6.4 that the rank R_E satisfies

$$R_E \leq \nu(A^2 - 4B) + \nu(B) - 1 = \nu(1) + \nu(2) - 1 = 0 + 1 - 1 = 0,$$

and therefore the rank is 0. The reader can check that

$$E(\mathbb{Q})_{\text{torsion}} = \{\mathcal{O}, (0,0), (-1,0), (-2,0)\}.$$

Since the rank is zero, the four torsion points on $E/\mathbb{Q}$ are the only rational points on E.

Example 16.6.6. Let $E : y^2 = x^3 + 2308x^2 + 665858x$. The primes 2 and 577 are the only prime divisors of (both) B and $A^2 - 4B$. Thus,

$$R_E \leq \nu(A^2 - 4B) + \nu(B) - 1 = 2 + 2 - 1 = 3.$$

The points $P_1 = (-1681, 25543)$, $P_2 = (-338, 26)$, and $P_3 = (577/16, 332929/64)$ are of infinite order and the subgroup of $E(\mathbb{Q})$ generated by P_1, P_2, and P_3 is isomorphic to $\mathbb{Z}^3$. Therefore, the rank of E is equal to 3.

We now turn to the problem of determining if a set of points is $\mathbb{Z}$-linearly dependent or independent. Let $E/\mathbb{Q}$ be the curve defined in Example 16.6.6. We claimed that the subgroup generated by the points $P_1 = (-1681, 25543)$, $P_2 = (-338, 26)$, and $P_3 = (577/16, 332929/64)$ is isomorphic to $\mathbb{Z}^3$. But how can we show that? In particular, why is P_3 not a linear combination of P_1 and P_2? In other words, are there integers n_1 and n_2 such that $P_3 = n_1P_1 + n_2P_2$? In fact, $E/\mathbb{Q}$ has a rational torsion point $T = (0,0)$ of order 2, so could some combination of P_1, P_2, and P_3 equal T? This example motivates the need for a notion of linear dependence and independence of points over $\mathbb{Z}$.

Definition 16.6.7. Let $E/\mathbb{Q}$ be an elliptic curve. We say that the rational points $P_1, \ldots, P_m \in E(\mathbb{Q})$ are *linearly dependent over* $\mathbb{Z}$ if there are integers $n_1, \ldots, n_m \in \mathbb{Z}$ such that

$$n_1 P_1 + n_2 P_2 + \cdots + n_m P_m = T,$$

where T is a torsion point. Otherwise, if no such relation exists, we say that the points are *linearly independent* over $\mathbb{Z}$.

Example 16.6.8. Let $E/\mathbb{Q} : y^2 = x^3 + x^2 - 25x + 39$ and let

$$P_1 = \left(\frac{61}{4}, -\frac{469}{8}\right), \quad P_2 = \left(-\frac{335}{81}, -\frac{6868}{729}\right), \quad P_3 = (21, 96).$$

The points P_1, P_2, and P_3 are rational points on E and linearly dependent over $\mathbb{Z}$ because

$$-3P_1 - 2P_2 + 6P_3 = \mathcal{O}.$$

Example 16.6.9. Let $E/\mathbb{Q} : y^2 + y = x^3 - x^2 - 26790x + 1696662$ and put

$$\begin{aligned} P_1 &= \left(\frac{59584}{625}, \frac{71573}{15625}\right), \\ P_2 &= \left(\frac{101307506181}{210337009}, \frac{30548385002405573}{3050517641527}\right). \end{aligned}$$

The points P_1 and P_2 are rational points on E, and they are linearly dependent over $\mathbb{Z}$ because

$$-3P_1 + 2P_2 = (133, -685),$$

and $(133, -685)$ is a torsion point of order 5.

Now that we have defined linear independence over $\mathbb{Z}$, we need a method to prove that a number of points are linearly independent. The existence of the Néron–Tate pairing provides a way to prove independence.

Definition 16.6.10. The *Néron–Tate pairing* attached to an elliptic curve is defined by

$$\langle \cdot, \cdot \rangle : E(\mathbb{Q}) \times E(\mathbb{Q}) \to \mathbb{R}, \quad \langle P, Q \rangle = \hat{h}(P+Q) - \hat{h}(P) - \hat{h}(Q),$$

where $\hat{h}$ is the canonical height on E. Let $P_1, P_2, \ldots, P_r$ be r rational points on $E(\mathbb{Q})$. The *elliptic height matrix* associated to $\{P_i\}_{i=1}^r$ is

$$\mathcal{H} = \mathcal{H}(\{P_i\}_{i=1}^r) := (\langle P_i, P_j \rangle)_{1 \le i \le r,\ 1 \le j \le r}.$$

The determinant of $\mathcal{H}$ is called the *elliptic regulator* of the set of points $\{P_i\}_{i=1}^r$. If $\{P_i\}_{i=1}^r$ is a complete set of generators of the free part of $E(\mathbb{Q})$, then the determinant of $\mathcal{H}(\{P_i\}_{i=1}^r)$ is called the *elliptic regulator of* $E/\mathbb{Q}$.

Theorem 16.6.11. *Let $E/\mathbb{Q}$ be an elliptic curve. Then the Néron–Tate pairing $\langle \cdot, \cdot \rangle$ associated to E is a non-degenerate symmetric bilinear form on the quotient group $E(\mathbb{Q})/E(\mathbb{Q})_{\text{torsion}}$; i.e.:*

(1) *For all $P, Q \in E(\mathbb{Q})$, $\langle P, Q \rangle = \langle Q, P \rangle$.*

(2) *For all $P, Q, R \in E(\mathbb{Q})$ and all $m, n \in \mathbb{Z}$,*

$$\langle P, mQ + nR \rangle = m\langle P, Q \rangle + n\langle P, R \rangle.$$

(3) *Suppose* $P \in E(\mathbb{Q})$ *and* $\langle P, Q\rangle = 0$ *for all* $Q \in E(\mathbb{Q})$. *Then* $P \in E(\mathbb{Q})_{\text{torsion}}$. *In particular,* P *is a torsion point if and only if* $\langle P, P\rangle = 0$.

The properties of the Néron–Tate pairing follow from those of the canonical height in Proposition 16.6.3 (see Exercise 16.10.11). Theorem 16.6.11 has the following important corollary:

Corollary 16.6.12. *Let* $E/\mathbb{Q}$ *be an elliptic curve and let* $P_1, P_2, \ldots, P_r \in E(\mathbb{Q})$ *be rational points. Let* $\mathcal{H}$ *be the elliptic height matrix associated to* $\{P_i\}_{i=1}^r$. *Then:*

(1) *Suppose* $\det(\mathcal{H}) = 0$ *and* $u = (n_1, \ldots, n_r) \in \operatorname{Ker}(\mathcal{H})$, *with* $n_i \in \mathbb{Z}$. *Then the points* $\{P_i\}_{i=1}^r$ *are linearly dependent and* $\sum_{k=1}^r n_k P_k = T$, *where* T *is a torsion point on* $E(\mathbb{Q})$.

(2) *If* $\det(\mathcal{H}) \neq 0$, *then the points* $\{P_i\}_{i=1}^r$ *are linearly independent and the rank of* $E(\mathbb{Q})$ *is* $\geq r$.

Here is an example of how the Néron–Tate pairing is used in practice:

Example 16.6.13. Let $E/\mathbb{Q}$ be the elliptic curve $y^2 = x^3 + 2308x^2 + 665858x$. Put

$$\begin{aligned} P &= (-1681, 25543), \quad Q = (-338, 26), \quad \text{and} \\ R &= \left(\frac{332929}{36}, -\frac{215405063}{216}\right). \end{aligned}$$

Are P, Q, and R independent? In order to find out, we find the elliptic height matrix associated to $\{P, Q, R\}$, using a computer algebra system (such as SageMath [**Sage**]):

$$\begin{aligned} \mathcal{H} &= \begin{pmatrix} \langle P,P\rangle & \langle Q,P\rangle & \langle R,P\rangle \\ \langle P,Q\rangle & \langle Q,Q\rangle & \langle R,Q\rangle \\ \langle P,R\rangle & \langle Q,R\rangle & \langle R,R\rangle \end{pmatrix} \\ &= \begin{pmatrix} 7.397\ldots & -3.601\ldots & 3.795\ldots \\ -3.601\ldots & 6.263\ldots & 2.661\ldots \\ 3.795\ldots & 2.661\ldots & 6.457\ldots \end{pmatrix}. \end{aligned}$$

The determinant of $\mathcal{H}$ seems to be *very* close to 0 (the computer returns $3.368 \cdot 10^{-27}$). Hence Corollary 16.6.12 suggests that P, Q, and R are not independent. If we find the (approximate) kernel of $\mathcal{H}$, we discover that the (column) vector $(1, 1, -1)$ is approximately in the kernel, and therefore, $P+Q-R$ may be a torsion point. Indeed, the point $P+Q-R = (0,0)$ is a torsion point of order 2 on $E(\mathbb{Q})$. Hence, P, Q, and R are linearly *dependent* over $\mathbb{Z}$.

Instead, let $P_1 = (-1681, 25543)$, $P_2 = (-338, 26)$, $P_3 = (577/16, 332929/64)$ and let $\mathcal{H}'$ be the elliptic height matrix associated to $\{P_i\}_{i=1}^3$. Then $\det(\mathcal{H}') = 101.87727\ldots$ is non-zero and, therefore, the $\{P_i\}_{i=1}^3$ are linearly independent and the rank of $E/\mathbb{Q}$ is at least 3.

16.7. Descent and the Weak Mordell–Weil Theorem

In the previous sections we have seen methods to calculate the torsion subgroup of an elliptic curve $E/\mathbb{Q}$ and also methods to check if a collection of points are independent modulo torsion. However, we have not discussed any method to find points of infinite order. In this section, we briefly explain the *method of descent*, which facilitates the search for generators of the free part of $E(\mathbb{Q})$. Unfortunately, the method of descent is not always successful! We will try to measure the failure of the method in the following section. The method of descent (as explained here) is mostly due to Cassels. For a more detailed treatment, see [**Was08**] or [**Sil86**]. A more general descent algorithm was laid out by Birch and Swinnerton-Dyer in [**BSD63**]. The current implementation of the algorithm is more fully explained in Cremona's book [**Cre97**].

Let $E/\mathbb{Q}$ be a curve given by $y^2 = x^3 + Ax + B$, with $A, B \in \mathbb{Z}$. The most general case of the method of descent is quite involved, so we will concentrate on a particular case where the calculations are much easier: we will assume that $E(\mathbb{Q})$ has four distinct rational points of 2-torsion (including $\mathcal{O}$). As we saw before (Theorem 16.4.5 or Exercise 16.10.5), a point $P = (x, y) \in E(\mathbb{Q})$ is of 2-torsion if and only if $y = 0$ and $x^3 + Ax + B = 0$ (or $P = \mathcal{O}$). Thus, if $E(\mathbb{Q})$ has four distinct rational points of order 2, that means that $x^3 + Ax + B$ has three (integral) roots and it factors completely over $\mathbb{Z}$:

$$x^3 + Ax + B = (x - e_1)(x - e_2)(x - e_3)$$

with $e_i \in \mathbb{Z}$. Since $x^3 + Ax + B$ does not have an x^2 term, we conclude that $e_1 + e_2 + e_3 = 0$.

Suppose, then, that $E : y^2 = (x - e_1)(x - e_2)(x - e_3)$, where the roots satisfy $e_i \in \mathbb{Z}$ and $e_1 + e_2 + e_3 = 0$. We would like to find a solution $(x_0, y_0) \in E$ with $x_0, y_0 \in \mathbb{Q}$; i.e.,

$$y_0^2 = (x_0 - e_1)(x_0 - e_2)(x_0 - e_3).$$

Thus, each term $(x_0 - e_i)$ must be *almost* a square, and we can make this precise by writing

$$(x_0 - e_1) = au^2, \quad (x_0 - e_2) = bv^2, \quad (x_0 - e_3) = cw^2, \quad y_0^2 = abc(uvw)^2,$$

where $a, b, c, u, v, w \in \mathbb{Q}$, the numbers $a, b, c \in \mathbb{Q}$ are square-free, and abc is a square (in $\mathbb{Q}$).

Example 16.7.1. Let

$$E : y^2 = x^3 - 556x + 3120 = (x - 6)(x - 20)(x + 26)$$

so that $e_1 = 6$, $e_2 = 20$, and $e_3 = -26$. The point $(x_0, y_0) = (\frac{164184}{289}, \frac{66469980}{4913})$ is rational and on E. We can write

$$x_0 - e_1 = \frac{164184}{289} - 6 = 2 \cdot \left(\frac{285}{17}\right)^2$$

and, similarly, $x_0 - e_2 = (\frac{398}{17})^2$ and $x_0 - e_3 = 2 \cdot (\frac{293}{17})^2$. Thus, following the notation of the preceeding paragraphs

$$a = 2, \quad b = 1, \quad c = 2, \quad u = \frac{285}{17}, \quad v = \frac{398}{17}, \quad w = \frac{293}{17}.$$

Notice that abc is a square and $y_0^2 = (\frac{66469980}{4913})^2 = abc(uvw)^2$.

Example 16.7.2. Let $E : y^2 = x^3 - 556x + 3120$ as before, with $e_1 = 6$, $e_2 = 20$, and $e_3 = -26$. Let $P = (-8, 84)$, $Q = (24, 60)$, and $S = P + Q = (-\frac{247}{16}, -\frac{5733}{64})$. The points P, Q, and S are in $E(\mathbb{Q})$. We would like to calculate the aforementioned numbers a, b, c for each of the points P, Q, and S. For instance,

$$\begin{aligned} x(P) - e_1 &= -8 - 6 = -14 = -14 \cdot 1^2, \\ x(P) - e_2 &= -7 \cdot 4^2, \text{ and } x(P) - e_3 = 2 \cdot 3^2. \end{aligned}$$

Thus, $a_P = -14$, $b_P = -7$, and $c_P = 2$. Similarly, we calculate

$$\begin{aligned} x(Q) - 6 &= 2 \cdot 3^2, \quad x(Q) - 20 = 2^2, \quad x(Q) + 26 = 2 \cdot 5^2, \\ x(S) - 6 &= -7 \cdot \left(\frac{7}{4}\right)^2, \\ x(S) - 20 &= -7 \cdot \left(\frac{9}{4}\right)^2, \; x(S) + 26 = \left(\frac{13}{4}\right)^2. \end{aligned}$$

Thus $a_Q = 2$, $b_Q = 1$, $c_Q = 2$, and $a_S = -7$, $b_S = -7$, $c_S = 1$. Notice the following interesting fact:

$$a_P \cdot a_Q = -28 = -7 \cdot 2^2, \quad b_P \cdot b_Q = -7, \quad c_P \cdot c_Q = 4.$$

Therefore, the square-free part of $a_P \cdot a_Q$ equals $a_S = a_{P+Q} = -7$. And similarly, the square-free parts of $b_P \cdot b_Q$ and $c_P \cdot c_Q$ equal $b_S = -7$ and $c_S = 1$, respectively. Also, the reader can check that $a_{2P} = b_{2P} = c_{2P} = 1$ and $a_{2Q} = b_{2Q} = c_{2Q} = 1$.

The previous example points to the fact that there may be a homomorphism between points on $E(\mathbb{Q})$ and triples (a, b, c) of rational numbers modulo squares, or square-free parts of rational numbers; formally, we are talking about $\mathbb{Q}^\times/(\mathbb{Q}^\times)^2 \times \mathbb{Q}^\times/(\mathbb{Q}^\times)^2 \times \mathbb{Q}^\times/(\mathbb{Q}^\times)^2$. Here, the group $\mathbb{Q}^\times/(\mathbb{Q}^\times)^2$ is the multiplicative group of non-zero rational numbers, with the extra relation that two non-zero rational numbers are equivalent if their square-free parts are equal (or, equivalently, if their quotient is a perfect square). For instance, 3 and $\frac{12}{25}$ represent the same element of $\mathbb{Q}^\times/(\mathbb{Q}^\times)^2$ because $\frac{12}{25} = 3 \cdot \left(\frac{2}{5}\right)^2$. The following theorem constructs such a homomorphism. Here we have adapted the proof that appears in [**Was08**, Theorem 8.14].

Theorem 16.7.3. *Let $E/\mathbb{Q}$ be an elliptic curve*

$$y^2 = x^3 + Ax + B = (x - e_1)(x - e_2)(x - e_3)$$

with distinct $e_1, e_2, e_3 \in \mathbb{Z}$ and $e_1 + e_2 + e_3 = 0$. There is a homomorphism of groups

$$\delta : E(\mathbb{Q}) \to \mathbb{Q}^\times/(\mathbb{Q}^\times)^2 \times \mathbb{Q}^\times/(\mathbb{Q}^\times)^2 \times \mathbb{Q}^\times/(\mathbb{Q}^\times)^2$$

defined for $P = (x_0, y_0)$ by

$$\delta(P) = \begin{cases} (1, 1, 1) & \text{if } P = \mathcal{O}, \\ (x_0 - e_1, x_0 - e_2, x_0 - e_3) & \text{if } y_0 \neq 0, \\ ((e_1 - e_2)(e_1 - e_3), e_1 - e_2, e_1 - e_3) & \text{if } P = (e_1, 0), \\ (e_2 - e_1, (e_2 - e_1)(e_2 - e_3), e_2 - e_3) & \text{if } P = (e_2, 0), \\ (e_3 - e_1, e_3 - e_2, (e_3 - e_1)(e_3 - e_2)) & \text{if } P = (e_3, 0). \end{cases}$$

If $\delta(P) = (\delta_1, \delta_2, \delta_3)$, then $\delta_1 \cdot \delta_2 \cdot \delta_3 = 1$ in $\mathbb{Q}^\times/(\mathbb{Q}^\times)^2$. Moreover, the kernel of δ is precisely $2E(\mathbb{Q})$; i.e., if $\delta(Q) = (1, 1, 1)$, then $Q = 2P$ for some $P \in E(\mathbb{Q})$.

Proof. Let δ be the function defined in the statement of the theorem. Let us show that δ is a homomorphism of (abelian) groups; i.e., we want to show that $\delta(P)\cdot\delta(Q)=\delta(P+Q)$. Notice first of all that $\delta(P)=\delta(x_0,y_0)=\delta(x_0,-y_0)=\delta(-P)$, because the definition of δ does not depend on the sign of the y-coordinate of P (in fact, it only depends on whether $y(P)=0$). Thus, it suffices to prove that $\delta(P)\cdot\delta(Q)=\delta(-(P+Q))$ for all $P,Q\in E(\mathbb{Q})$.

Let $P=(x_0,y_0)$, $Q=(x_1,y_1)$, and $R=-(P+Q)=(x_2,y_2)$, and let us assume, for simplicity, that $y_i\neq 0$ for $i=1,2,3$. By the definition of the addition rule on an elliptic curve (see Figure 16.2), the points P, Q, and R are collinear. Let $\mathcal{L}=\overline{PQ}$ be the line that goes through all three points, and suppose it has equation $\mathcal{L}: y=ax+b$. Therefore, if we substitute y in the equation of $E/\mathbb{Q}$, we obtain a polynomial

$$p(x)=(ax+b)^2-(x-e_1)(x-e_2)(x-e_3).$$

The polynomial $p(x)$ is cubic, its leading term is -1, and it has precisely three rational roots, namely x_0, x_1, and x_2. Hence, it factors as

$$p(x)=(ax+b)^2-(x-e_1)(x-e_2)(x-e_3)=-(x-x_0)(x-x_1)(x-x_2).$$

If we evaluate $p(x)$ at $x=e_i$, we obtain

$$p(e_i)=(ae_i+b)^2=-(e_i-x_0)(e_i-x_1)(e_i-x_2),$$

or, equivalently, $(x_0-e_i)(x_1-e_i)(x_2-e_i)=(ae_i+b)^2$. Thus, the product $\delta(P)\cdot\delta(Q)\cdot\delta(R)$ equals

$$\begin{aligned}\delta(P)\cdot\delta(Q)\cdot\delta(R) &= (x_0-e_1,x_0-e_2,x_0-e_3)\\ &\quad\cdot(x_1-e_1,x_1-e_2,x_1-e_3)\\ &\quad\cdot(x_2-e_1,x_2-e_2,x_2-e_3)\\ &= ((x_0-e_1)(x_1-e_1)(x_2-e_1),\\ &\quad (x_0-e_2)(x_1-e_2)(x_2-e_2),\\ &\quad (x_0-e_3)(x_1-e_3)(x_2-e_3))\\ &= ((ae_1+b)^2,(ae_2+b)^2,(ae_3+b)^2)\\ &= (1,1,1)\in(\mathbb{Q}^\times/(\mathbb{Q}^\times)^2)^3.\end{aligned}$$

Hence, $\delta(P)\cdot\delta(Q)\cdot\delta(R)=1$. If we multiply both sides by $\delta(R)$ and notice that $a^2=1$ for any $a\in\mathbb{Q}^\times/(\mathbb{Q}^\times)^2$, we conclude that

$$\delta(P)\cdot\delta(Q)=\delta(R)=\delta(-(P+Q))=\delta(P+Q),$$

as desired. In order to completely prove that δ is a homomorphism, we would need to check the cases when P, Q, or R is one of the points $(e_i,0)$ or $\mathcal{O}$, but we leave those special cases for the reader to check (Exercise 16.10.14).

If $\delta(P)=(\delta_1,\delta_2,\delta_3)$, then it follows directly from the definition of δ that $\delta_1\cdot\delta_2\cdot\delta_3=1$ in $\mathbb{Q}^\times/(\mathbb{Q}^\times)^2$. Indeed, this is clear for $P=\mathcal{O}$ or $P=(e_i,0)$, and if $P=(x_0,y_0)$ with $y_0\neq 0$, then $(x_0-e_1)(x_0-e_2)(x_0-e_3)=y_0^2$, which is a square and is therefore trivial in $\mathbb{Q}^\times/(\mathbb{Q}^\times)^2$.

Next, let us show that the kernel of δ is $2E(\mathbb{Q})$. Clearly, $2E(\mathbb{Q})$ is in the kernel of δ, because δ is a homomorphism with image in $(\mathbb{Q}^\times/(\mathbb{Q}^\times)^2)^3$, as we just proved.

Indeed, if $P \in E(\mathbb{Q})$, then

$$\delta(2P) = \delta(P) \cdot \delta(P) = \delta(P)^2 = (\delta_1^2, \delta_2^2, \delta_3^2) = (1, 1, 1),$$

because squares are trivial in $\mathbb{Q}^\times/(\mathbb{Q}^\times)^2$.

Now let us show the reverse inclusion, i.e., that the kernel of δ is contained in $2E(\mathbb{Q})$. Let $Q = (x_1, y_1) \in E(\mathbb{Q})$ such that $\delta(Q) = (1, 1, 1)$. We want to find $P = (x_0, y_0)$ such that $2P = Q$. Notice that it is enough to show that $x(2P) = x_1$, because $2P$ is a point on $E(\mathbb{Q})$ and if $x(2P) = x(Q)$, then $Q = 2(\pm P)$. Hence, our goal will be to construct $(x_0, y_0) \in E(\mathbb{Q})$ such that

$$x(2P) = \frac{x_0^4 - 2Ax_0^2 - 8Bx_0 + A^2}{4y_0^2} = x_1.$$

The formula for $x(2P)$ above is given in Exercise 16.10.15.

Once again, for simplicity, let us assume $y(Q) = y_1 \neq 0$ and, as stated above, we assume $\delta(Q) = (1, 1, 1)$. Hence, $x_1 - e_i$ is a square in $\mathbb{Q}$ for $i = 1, 2, 3$. Let us write

$$x_1 - e_i = t_i^2, \quad \text{for some } t_i \in \mathbb{Q}^\times. \tag{16.8}$$

We define a new auxiliary polynomial $p(x)$ by

$$t_1\frac{(x - e_2)(x - e_3)}{(e_1 - e_2)(e_1 - e_3)} + t_2\frac{(x - e_1)(x - e_3)}{(e_2 - e_1)(e_2 - e_3)} + t_3\frac{(x - e_1)(x - e_2)}{(e_3 - e_1)(e_3 - e_2)}.$$

The polynomial $p(x)$ is an interpolating polynomial (or Lagrange polynomial) which was defined so that $p(e_i) = t_i$. Notice that $p(x)$ is a quadratic polynomial, say $p(x) = a + bx + cx^2$. Also define another polynomial $q(x) = x_1 - x - p(x)^2$ and notice that

$$q(e_i) = x_1 - e_i - p(e_i)^2 = x_1 - e_i - t_i^2 = 0$$

from the definition of t_i in (16.8). Since $q(e_i) = 0$, it follows that $(x - e_i)$ divides $q(x)$ for $i = 1, 2, 3$. Thus, $(x - e_1)(x - e_2)(x - e_3) = x^3 + Ax + B$ divides $q(x)$. In other words, $q(x) \equiv 0 \bmod x^3 + Ax + B$. Since $q(x) = x_1 - x - p(x)^2$, we can also write

$$x_1 - x \equiv p(x)^2 \equiv (a + bx + cx^2)^2 \bmod (x^3 + Ax + B).$$

We shall expand the square on the right-hand side, modulo $f(x) = x^3 + Ax + B$. Notice that $x^3 \equiv -Ax - B$ and $x^4 \equiv -Ax^2 - Bx$ modulo $f(x)$:

$$\begin{aligned} x_1 - x &\equiv p(x)^2 \equiv (a + bx + cx^2)^2 \\ &\equiv c^2x^4 + 2bcx^3 + (2ac + b^2)x^2 + 2abx + a^2 \\ &\equiv c^2(-Ax^2 - Bx) + 2bc(-Ax - B) \\ &\quad +(2ac + b^2)x^2 + 2abx + a^2 \\ &\equiv (2ac + b^2 - Ac^2)x^2 \\ &\quad +(2ab - Bc^2 - 2Abc)x + (a^2 - 2bcB), \end{aligned}$$

where all the congruences are modulo $f(x) = x^3 + Ax + B$. The congruences in the previous equation say that a polynomial of degree 1, call it $g(x) = x_1 - x$, is congruent to a polynomial of degree ≤ 2, call the last line $h(x)$, modulo a polynomial of degree 3, namely $f(x)$. Then $h(x) - g(x)$ is a polynomial of degree ≤ 2, divisible by a polynomial of degree 3. This implies that $h(x) - g(x)$ must be zero and $h(x) = g(x)$; i.e.,

$$x_1 - x = (2ac + b^2 - Ac^2)x^2 + (2ab - Bc^2 - 2Abc)x + (a^2 - 2bcB).$$

If we match coefficients, we obtain the following equalities:

$$2ac + b^2 - Ac^2 = 0, \tag{16.9}$$

$$2ab - Bc^2 - 2Abc = -1, \tag{16.10}$$

$$a^2 - 2bcB = x_1. \tag{16.11}$$

If $c = 0$, then $b = 0$ by (16.9); therefore, $p(x) = a + bx + cx^2 = a$ is a constant function, and so $t_1 = t_2 = t_3$. By (16.8), it follows that $e_1 = e_2 = e_3$, which is a contradiction to our assumptions. Hence, c must be non-zero. We multiply (16.10) by $\frac{1}{c^2}$ and (16.9) by $\frac{b}{c^3}$ to obtain

$$\frac{2ab}{c^2} - B - \frac{2Ab}{c} = -\frac{1}{c^2}, \tag{16.12}$$

$$\frac{2ab}{c^2} + \frac{b^3}{c^3} - \frac{Ab}{c} = 0. \tag{16.13}$$

We subtract (16.12) from (16.13) to get

$$\left(\frac{b}{c}\right)^3 + A\left(\frac{b}{c}\right) + B = \left(\frac{1}{c}\right)^2.$$

Hence, the point $P = (x_0, y_0) = (\frac{b}{c}, \frac{1}{c})$ is a rational point on $E(\mathbb{Q})$. It remains to show that $x(2P) = x(Q)$. From (16.13) we deduce that

$$a = \frac{\frac{Ab}{c} - \frac{b^3}{c^3}}{\frac{2b}{c^2}} = \frac{A - \left(\frac{b}{c}\right)^2}{2 \cdot \frac{1}{c}} = \frac{A - x_0^2}{2y_0},$$

and, therefore, substituting a into (16.11) yields

$$\begin{aligned} x(Q) = x_1 = a^2 - 2bcB &= \left(\frac{A - x_0^2}{2y_0}\right)^2 - 2bcB \\ &= \frac{(A^2 - 2Ax_0^2 + x_0^4) - (2bcB)(4y_0^2)}{4y_0^2} \\ &= \frac{(A^2 - 2Ax_0^2 + x_0^4) - (2bcB)(\frac{4}{c^2})}{4y_0^2} \\ &= \frac{(A^2 - 2Ax_0^2 + x_0^4) - 8Bx_0}{4y_0^2} \\ &= \frac{x_0^4 - 2Ax_0^2 - 8Bx_0 + A^2}{4y_0^2} = x(2P), \end{aligned}$$

as desired. In order to complete the proof of the fact that the kernel of δ is $2E(\mathbb{Q})$, we would need to consider the case when $y(Q) = y_1 = 0$, but we leave this special case to the reader (Exercise 16.10.17). □

Thus, the previous proposition shows that there is a homomorphism $\delta : E(\mathbb{Q}) \to (\mathbb{Q}^\times/(\mathbb{Q}^\times)^2)^3$ with kernel equal to $2E(\mathbb{Q})$. In fact, the theorem shows that there is a homomorphism from $E(\mathbb{Q})$ into

$$\Gamma = \{(\delta_1, \delta_2, \delta_3) \in (\mathbb{Q}^\times/(\mathbb{Q}^\times)^2)^3 : \delta_1 \cdot \delta_2 \cdot \delta_3 = 1 \in \mathbb{Q}^\times/(\mathbb{Q}^\times)^2\}.$$

Hence, δ induces an injection

$$E(\mathbb{Q})/2E(\mathbb{Q}) \hookrightarrow \Gamma \subset (\mathbb{Q}^\times/(\mathbb{Q}^\times)^2)^3.$$

The groups $\mathbb{Q}^\times/(\mathbb{Q}^\times)^2$ and Γ are infinite, so such an injection does not tell us much about the size of $E(\mathbb{Q})/2E(\mathbb{Q})$. However, the image of $E(\mathbb{Q})/2E(\mathbb{Q})$ is much smaller than Γ.

Example 16.7.4. Let $E : y^2 = x^3 - 556x + 3120$ as in Example 16.7.2. It turns out that $E(\mathbb{Q}) \cong \mathbb{Z}/2\mathbb{Z} \oplus \mathbb{Z}/2\mathbb{Z} \oplus \mathbb{Z}^2$. The generators of the torsion part are $T_1 = (6, 0)$ and $T_2 = (20, 0)$, and the generators of the free part are $P = (-8, 84)$ and $Q = (24, 60)$. The image of the map δ in this case is, therefore, generated by the images of T_1, T_2, P, and Q:

$$\begin{aligned} \delta(T_1) &= (-7, -14, 2), \quad \delta(T_2) = (14, 161, 46), \\ \delta(P) &= (-14, -7, 2), \quad \delta(Q) = (2, 1, 2). \end{aligned}$$

Thus, the image of δ is formed by the 16 elements that one obtains by multiplying out $\delta(T_1)$, $\delta(T_2)$, $\delta(P)$, and $\delta(Q)$, in all possible ways. Thus, $\delta(E(\mathbb{Q})/2E(\mathbb{Q}))$ is the group

$$\begin{aligned} &\{(1,1,1),\ (-7,-14,2),\ (14,161,46),\ (-2,-46,23), \\ &(-14,-7,2),\ (2,2,1),\ (-1,-23,23),\ (7,322,46), \\ &(2,1,2),\ (-14,-14,1),\ (7,161,23),\ (-1,-46,46), \\ &(-7,-7,1),\ (1,2,2),\ (-2,-23,46),\ (14,322,23)\}. \end{aligned}$$

(Exercise: check that the elements listed above form a group under multiplication.) We see that the only primes that appear in the factorization of the coordinates of elements in the image of δ are $2, 7$, and 23. Therefore, the coordinates of δ are not just in $\mathbb{Q}^\times/(\mathbb{Q}^\times)^2$ but in a much smaller subgroup of 16 elements:

$$\Gamma' = \{\pm 1,\ \pm 2,\ \pm 7,\ \pm 23,\ \pm 14,\ \pm 46,\ \pm 161,\ \pm 322\} \subset \mathbb{Q}^\times/(\mathbb{Q}^\times)^2.$$

And the image of $E(\mathbb{Q})/2E(\mathbb{Q})$ embeds into

$$\begin{aligned} \Gamma_\Delta &= \{(\delta_1, \delta_2, \delta_3) \in \Gamma' \times \Gamma' \times \Gamma' : \delta_1 \cdot \delta_2 \cdot \delta_3 = 1 \in \mathbb{Q}^\times/(\mathbb{Q}^\times)^2\} \\ &\subset \Gamma' \times \Gamma' \times \Gamma'. \end{aligned}$$

Since Γ' has 16 elements and $E(\mathbb{Q})/2E(\mathbb{Q})$ embeds into $(\Gamma')^3$, we conclude that $E(\mathbb{Q})/2E(\mathbb{Q})$ has at most $(16)^3 = 2^{12}$ elements. In fact, Γ_Δ has only 16^2 elements, so $E(\mathbb{Q})/2E(\mathbb{Q})$ has at most 2^8 elements. Notice also the following interesting "coincidence": the prime divisors that appear in Γ_Δ coincide with the prime divisors of the discriminant of E, which is $\Delta_E = 6795034624 = 2^{18} \cdot 7^2 \cdot 23^2$. In the next proposition we explain that, in fact, this is always the case.

Proposition 16.7.5. *Let $E : y^2 = (x - e_1)(x - e_2)(x - e_3)$, with $e_i \in \mathbb{Z}$. Let $P = (x_0, y_0) \in E(\mathbb{Q})$ and write*

$$(x_0 - e_1) = au^2, \quad (x_0 - e_2) = bv^2, \quad (x_0 - e_3) = cw^2, \quad y_0^2 = abc(uvw)^2,$$

where $a, b, c, u, v, w \in \mathbb{Q}$*, the numbers* $a, b, c \in \mathbb{Z}$ *are square-free, and* abc *is a square (in* $\mathbb{Z}$*). Then, if* p *divides* $a \cdot b \cdot c$*, then* p *also divides the quantity* $\Delta = (e_1 - e_2)(e_2 - e_3)(e_1 - e_3)$.

Note: the discriminant of E equals $\Delta_E = 16(e_1 - e_2)^2(e_2 - e_3)^2(e_1 - e_3)^2$, so $\Delta_E = 16\Delta^2$, where Δ is as in Proposition 16.7.5. Thus, a prime p divides Δ if and only if p divides Δ_E. (This is clear for $p > 2$; see Exercise 16.10.18 for $p = 2$.)

Proof. Suppose a prime p divides abc. Then p divides a, b, or c. Let us assume that $p \mid a$ (the same argument works if p divides b or c). Let p^k be the exact power of p that appears in the factorization of the rational number $x_0 - e_1 = au^2$. Notice that k may be positive or negative, depending on whether p divides the numerator or denominator of au^2. Notice, however, that k must be odd, because $p \mid a$ and a is square-free.

Suppose first that $k < 0$; i.e., $p^{|k|}$ is the exact power of p that divides the denominator of $x_0 - e_1$. Since $e_i \in \mathbb{Z}$, it follows that $p^{|k|}$ must divide the denominator of x_0 too, and therefore $p^{|k|}$ is the exact power that divides the denominators of $x_0 - e_2$ and $x_0 - e_3$ as well. Hence, $p^{3|k|}$ is the exact power of p dividing the denominator of $y_0^2 = \prod(x_0 - e_i)$, but this is impossible because y_0^2 is a square and $3|k|$ is odd. Thus, k must be positive.

If $k > 0$ and p divides $x_0 - e_1$, then the denominator of x_0 is not divisible by p, so it makes sense to consider $x_0 \bmod p$ and $x_0 \equiv e_1 \bmod p$. Similarly, the denominators of $x_0 - e_2$ and $x_0 - e_3$ are not divisible by p and

$$bv^2 \equiv x_0 - e_2 \equiv e_1 - e_2 \quad \text{and} \quad cw^2 \equiv x_0 - e_3 \equiv e_1 - e_3 \bmod p.$$

Since $y_0^2 = abc(uvw)^2$ and p divides a, then p must also divide one of b or c. Let us suppose it also divides b. Then $0 \equiv bv^2 \equiv x_0 - e_2 \equiv e_1 - e_2 \bmod p$ and $\Delta = (e_1 - e_2)(e_2 - e_3)(e_1 - e_3) \equiv 0 \bmod p$, as claimed. □

The definition of the map δ and the previous proposition yield the following immediate corollary:

Corollary 16.7.6. *With notation as in the previous theorem and proposition, define a subgroup* Γ' *of* $\mathbb{Q}^\times/(\mathbb{Q}^\times)^2$ *by*

$$\Gamma' = \{n \in \mathbb{Z} : 0 \neq n \text{ is square-free and if } p \mid n, \text{ then } p \mid \Delta\}/(\mathbb{Z}^\times)^2.$$

Then, δ *induces an injection of* $E(\mathbb{Q})/2E(\mathbb{Q})$ *into*

$$\begin{aligned} \Gamma_\Delta &= \{(\delta_1, \delta_2, \delta_3) \in \Gamma' \times \Gamma' \times \Gamma' : \delta_1 \cdot \delta_2 \cdot \delta_3 = 1 \in \mathbb{Q}^\times/(\mathbb{Q}^\times)^2\} \\ &\subset \Gamma' \times \Gamma' \times \Gamma'. \end{aligned}$$

We are ready to prove the weak Mordell–Weil theorem (Theorem 16.3.5), at least in our restricted case:

Corollary 16.7.7 (Weak Mordell–Weil theorem). *Let*

$$E : y^2 = (x - e_1)(x - e_2)(x - e_3)$$

be an elliptic curve, with $e_i \in \mathbb{Z}$*. Then* $E(\mathbb{Q})/2E(\mathbb{Q})$ *is finite.*

Proof. By Corollary 16.7.6, $E(\mathbb{Q})/2E(\mathbb{Q})$ injects into $\Gamma_\Delta \subset \Gamma' \times \Gamma' \times \Gamma'$. Since Γ' is finite, $E(\mathbb{Q})/2E(\mathbb{Q})$ is finite as well. □

16.8. Homogeneous Spaces

In this section we want to make the weak Mordell–Weil theorem explicit; i.e., we want

- explicit bounds on the size of $E(\mathbb{Q})/2E(\mathbb{Q})$ and
- a method to find generators of $E(\mathbb{Q})/2E(\mathbb{Q})$ (see Exercise 16.10.24, though).

Before we discuss bounds, we need to understand the structure of the quotient $E(\mathbb{Q})/2E(\mathbb{Q})$. Remember that, from the Mordell–Weil theorem (Theorem 16.3.3), $E(\mathbb{Q}) \cong T \oplus \mathbb{Z}^{R_E}$ where $T = E(\mathbb{Q})_{\text{torsion}}$ is a finite abelian group. Therefore,

$$E(\mathbb{Q})/2E(\mathbb{Q}) \cong T/2T \oplus (\mathbb{Z}/2\mathbb{Z})^{R_E}.$$

In our restricted case, we have assumed all along that $E(\mathbb{Q})$ contains four points of 2-torsion, namely $\mathcal{O}$ and $(e_i, 0)$, for $i = 1, 2, 3$. And, by Exercise 16.10.5, $E(\mathbb{Q})$ cannot have more points of order 2. Thus, $T/2T \cong \mathbb{Z}/2\mathbb{Z} \oplus \mathbb{Z}/2\mathbb{Z}$ (see Exercise 16.10.19).

Hence, the size of $E(\mathbb{Q})/2E(\mathbb{Q})$ is exactly 2^{R_E+2}, under our assumptions. Recall that we defined $\nu(N)$ to be the number of distinct prime divisors of an integer N. We prove our first bound:

Proposition 16.8.1. *Let $E : y^2 = (x - e_1)(x - e_2)(x - e_3)$ be an elliptic curve, with $e_i \in \mathbb{Z}$. Then the rank of $E(\mathbb{Q})$ is $R_E \leq 2\nu(\Delta_E)$.*

Proof. If the quantity Δ_E has $\nu = \nu(\Delta_E)$ distinct (positive) prime divisors, then we claim that the set

$$\Gamma' = \{n \in \mathbb{Z} : 0 \neq n \text{ is square-free and if } p \mid n, \text{ then } p \mid \Delta\}/(\mathbb{Z}^\times)^2$$

has precisely $2^{\nu(\Delta_E)+1}$ elements. Indeed, if $\Delta_E = p_1^{s_1} \cdots p_\nu^{s_\nu}$, then

$$\Gamma' = \{(-1)^{t_0} p_1^{t_1} \cdots p_\nu^{t_\nu} : t_i = 0 \text{ or } 1 \text{ for } i = 0, \ldots, \nu\}.$$

Thus, Γ' has as many elements as $\{(t_0, \ldots, t_\nu) : t_i = 0 \text{ or } 1\}$, which has $2^{\nu+1}$ elements. Moreover, the set Γ_Δ, as defined in Corollary 16.7.6, has as many elements as $\Gamma' \times \Gamma'$, i.e., $2^{2\nu+2}$ elements. Since $E(\mathbb{Q})/2E(\mathbb{Q})$ injects into Γ_Δ, we conclude that it also has at most $2^{2\nu+2}$ elements. Since the size of $E(\mathbb{Q})/2E(\mathbb{Q})$ is 2^{R_E+2}, we conclude that $R_E + 2 \leq 2\nu + 2$ and $R_E \leq 2\nu$, as claimed. □

Example 16.8.2. Let

$$E : y^2 = x^3 - 1156x = x(x - 34)(x + 34).$$

The discriminant of $E/\mathbb{Q}$ is $\Delta_E = 98867482624 = 2^{12} \cdot 17^6$. Hence, $\nu(\Delta_E) = 2$ and the rank of E is at most 4. (The rank is in fact 2; see Example 16.8.4 below.)

The bound $R_E \leq 2\nu(\Delta_E)$ is, in general, not very sharp (Theorem 16.6.4 is an improvement). However, the method we followed to come up with the bound yields a strategy to find generators for $E(\mathbb{Q})/2E(\mathbb{Q})$ as follows. Recall that $E(\mathbb{Q})/2E(\mathbb{Q})$ embeds into Γ_Δ via the map δ, so we want to identify which elements of Γ_Δ may belong to the image of δ. Suppose $(\delta_1, \delta_2, \delta_3) \in \Gamma_\Delta$ belongs to the image of δ and it is not the image of a torsion point. Then there exists $P = (x_0, y_0) \in E(\mathbb{Q})$ such that

$$\begin{cases} y_0^2 = (x_0 - e_1)(x_0 - e_2)(x_0 - e_3), \\ x_0 - e_1 = \delta_1 u^2, \\ x_0 - e_2 = \delta_2 v^2, \\ x_0 - e_3 = \delta_3 w^2 \end{cases}$$

for some rational numbers u, v, w. We may substitute the last equation into the previous two and obtain

$$\begin{cases} e_3 - e_1 = \delta_1 u^2 - \delta_3 w^2, \\ e_3 - e_2 = \delta_2 v^2 - \delta_3 w^2. \end{cases}$$

Recall that the elements $(\delta_1, \delta_2, \delta_3)$ that are in the image of δ satisfy $\delta_1 \cdot \delta_2 \cdot \delta_3 = 1$ modulo squares. Thus, $\delta_3 = \delta_1 \cdot \delta_2 \cdot \lambda^2$ and if we do a change of variables $(u, v, w) \mapsto (X, Y, \frac{Z}{\lambda})$, we obtain a system

$$C(\delta_1, \delta_2) : \begin{cases} e_3 - e_1 = \delta_1 X^2 - \delta_1\delta_2 Z^2, \\ e_3 - e_2 = \delta_2 Y^2 - \delta_1\delta_2 Z^2, \end{cases}$$

or, equivalently, one can subtract both equations to get

$$C(\delta_1, \delta_2) : \begin{cases} e_1 - e_2 = \delta_2 Y^2 - \delta_1 X^2, \\ e_3 - e_2 = \delta_2 Y^2 - \delta_1\delta_2 Z^2. \end{cases}$$

The space $C(\delta_1, \delta_2)$ is the intersection of two conics, and it may have rational points or not. If $(\delta_1, \delta_2, \delta_3)$ is in the image of δ, however, then the space $C(\delta_1, \delta_2)$ must have a rational point; i.e., there are $X, Y, Z \in \mathbb{Q}$ that satisfy the equations of $C(\delta_1, \delta_2)$. Moreover, if $X_0, Y_0, Z_0 \in \mathbb{Q}$ are the coordinates of a point in $C(\delta_1, \delta_2)$, then

$$P = (e_1 + \delta_1 X_0^2,\ \delta_1\delta_2 X_0 Y_0 Z_0) \tag{16.14}$$

is a rational point on $E(\mathbb{Q})$ such that $\delta(P) = (\delta_1, \delta_2, \delta_3)$. The spaces $C(\delta_1, \delta_2)$ are called *homogeneous spaces* and are extremely helpful when we try to calculate the Mordell–Weil group of an elliptic curve. We record our findings in the form of a proposition, for later use:

Proposition 16.8.3. *Let $E/\mathbb{Q}$ be an elliptic curve with Weierstrass equation $y^2 = (x-e_1)(x-e_2)(x-e_3)$, with $e_i \in \mathbb{Z}$ and $e_1+e_2+e_3 = 0$. Let $\delta : E(\mathbb{Q})/2E(\mathbb{Q}) \hookrightarrow \Gamma_\Delta$ be the injection given by Corollary* 16.7.7, *and let $\delta(E) := \delta(E(\mathbb{Q})/2E(\mathbb{Q}))$ be the image of δ in Γ_Δ. Then:*

(1) *If $(\delta_1, \delta_2, \delta_3) \in \delta(E)$, then the space $C(\delta_1, \delta_2)$ has a point (X_0, Y_0, Z_0) with rational coordinates, $X_0, Y_0, Z_0 \in \mathbb{Q}$.*
(2) *Conversely, if $C(\delta_1, \delta_2)$ has a rational point (X_0, Y_0, Z_0), then $E(\mathbb{Q})$ has a rational point*
$$P = (e_1 + \delta_1 X_0^2,\ \delta_1\delta_2 X_0 Y_0 Z_0).$$
(3) *Since δ is a homomorphism and $\delta(E)$ is the image of δ, it follows that $\delta(E)$ is a subgroup of Γ_Δ. In particular:*
    - *If $(\delta_1, \delta_2, \delta_3)$ and $(\delta_1', \delta_2', \delta_3')$ are elements of the image, then their product $(\delta_1 \cdot \delta_1',\ \delta_2 \cdot \delta_2',\ \delta_3 \cdot \delta_3')$ is also in the image.*

- *If* $(\delta_1, \delta_2, \delta_3) \in \delta(E)$ *but* $(\delta_1', \delta_2', \delta_3') \in \Gamma_\Delta$ **is not** *in the image, then their product* $(\delta_1 \cdot \delta_1',\ \delta_2 \cdot \delta_2',\ \delta_3 \cdot \delta_3')$ **is not** *in the image* $\delta(E)$.
- *If* $C(\delta_1, \delta_2)$ *and* $C(\delta_1', \delta_2')$ *have rational points, then* $C(\delta_1 \cdot \delta_1',\ \delta_2 \cdot \delta_2')$ *also has a rational point.*
- *If* $C(\delta_1, \delta_2)$ *has a rational point but* $C(\delta_1', \delta_2')$ **does not have** *a rational point, then* $C(\delta_1 \cdot \delta_1',\ \delta_2 \cdot \delta_2')$ **does not have** *a rational point.*

Example 16.8.4. Let $E : y^2 = x^3 - 1156x = x(x-34)(x+34)$. The only divisors of Δ_E are 2 and 17. Thus, $\Gamma' = \{\pm 1, \pm 2, \pm 17, \pm 34\}$. Let us choose $e_1 = 0$, $e_2 = -34$, and $e_3 = 34$. Therefore, the homogeneous spaces for this curve are all of the form

$$C(\delta_1, \delta_2) : \begin{cases} \delta_2 Y^2 - \delta_1 X^2 = 34, \\ \delta_2 Y^2 - \delta_1 \delta_2 Z^2 = 68 \end{cases}$$

with $\delta_1, \delta_2 \in \Gamma'$. We analyze these spaces, case by case. There are 64 pairs (δ_1, δ_2) to take care of:

(1) $((\delta_1, \delta_2, \delta_3) = (1, 1, 1))$. The point at infinity (i.e., the origin) is sent to $(1, 1, 1)$ via δ; i.e., $\delta(\mathcal{O}) = (1, 1, 1)$.

(2) $(\delta_1 < 0$ and $\delta_2 < 0)$. The equation $\delta_2 Y^2 - \delta_1 \delta_2 Z^2 = 68$ cannot have solutions (in $\mathbb{Q}$ or $\mathbb{R}$) because the left-hand side is always negative for any $X, Z \in \mathbb{Q}$.

(3) $(\delta_1 > 0$ and $\delta_2 < 0)$. The equation $\delta_2 Y^2 - \delta_1 X^2 = 34$ cannot have solutions (in $\mathbb{Q}$ or $\mathbb{R}$) because the left-hand side is always negative.

(4) $(\delta_1 = -1,\ \delta_2 = 34)$. The space $C(-1, 34)$ has a rational point $(X, Y, Z) = (0, 1, 1)$, which maps to $T_1 = (0, 0)$ on $E(\mathbb{Q})$ via (16.14).

(5) $(\delta_1 = -34,\ \delta_2 = 2)$. The space $C(-34, 2)$ has the rational point $(X, Y, Z) = (1, 0, 1)$, which maps to $T_2 = (-34, 0)$ on $E(\mathbb{Q})$ via (16.14).

(6) $(\delta_1 = 34,\ \delta_2 = 17)$. If $\delta(T_1) = \delta((0, 0))$ equals $(-1, 34, -34)$ and $\delta(T_2) = (-34, 2, -17)$, then

$$\delta(T_1 + T_2) = \delta(T_1) \cdot \delta(T_2) = (-1, 34, -34) \cdot (-34, 2, -17) = (34, 17, 2).$$

Thus, the space $C(34, 17)$ must have a point that maps back to $T_1 + T_2 = (34, 0)$. Indeed, $C(34, 17)$ has a point $(X, Y, Z) = (1, 2, 0)$ that maps to $(34, 0)$ via (16.14).

(7) $(\delta_1 = -1,\ \delta_2 = 2)$. The space $C(-1, 2)$ has a rational point $(X, Y, Z) = (4, 3, 5)$, which maps to $P = (-16, -120)$ on $E(\mathbb{Q})$ via (16.14). P is a point of infinite order.

(8) $((\delta_1, \delta_2) = (1, 17),\ (34, 1),$ or $(-34, 34))$. These are the pairs that correspond to $(-1, 2) \cdot \gamma$, with $\gamma = (-1, 34)$, $(-34, 2)$, or $(34, 17)$. Therefore, the corresponding spaces $C(\delta_1, \delta_2)$ must have rational points that map to $P + T_1$, $P + T_2$, and $P + T_1 + T_2$, respectively.

(9) $(\delta_1 = -2,\ \delta_2 = 2)$. The space $C(-2, 2)$ has a rational point $(X, Y, Z) = (1, 4, 3)$, which maps to $Q = (-2, -48)$ on $E(\mathbb{Q})$ via (16.14). Q is a point of infinite order.

(10) $((\delta_1,\delta_2) = (2,17), (17,1), \text{ or } (-17,34))$. These are the pairs that correspond to $(-2,2)\cdot\gamma$, with $\gamma = (-1,34)$, $(-34,2)$, or $(34,17)$. Therefore, the corresponding spaces $C(\delta_1,\delta_2)$ must have rational points that map to $Q+T_1$, $Q+T_2$, and $Q+T_1+T_2$, respectively.

(11) $((\delta_1,\delta_2) = (2,1), \text{ and } (-2,34), (-17,2), \text{ or } (17,17))$. Since $(-1,2)$ and $(-2,2)$ correspond to P and Q, respectively, then $(-1,2)\cdot(-2,2) = (2,1)$ corresponds to $P+Q$. The other pairs correspond to $(2,1)\cdot\gamma$, with $\gamma = (-1,34)$, $(-34,2)$, or $(34,17)$. Therefore, the corresponding spaces $C(\delta_1,\delta_2)$ must have rational points that map to $P+Q+T_1$, $P+Q+T_2$, and $P+Q+T_1+T_2$, respectively.

(12) $(\delta_1 = 1, \delta_2 = 2)$. The space $C(1,2)$ does not have rational points (see Exercise 16.10.20). In fact, it does not have solutions in $\mathbb{Q}_2$, the field of 2-adic numbers (see Section 11.5 for an introduction to p-adic numbers).

(13) $((\delta_1,\delta_2) = (2,2), (17,2), (34,2), (-1,1), (-2,1), (-17,1), (-34,1), (-1,17), (-2,17), (-17,17), (-34,17), (1,34), (2,34), (17,34), (34,34))$. The corresponding spaces $C(\delta_1,\delta_2)$ do not have rational points. For instance, suppose $C(2,2)$ had a point. Then $(2,2,1)$ would be in the image of δ. Since $(2,1,2)$ *is* in the image of δ (we already saw above that $C(2,1)$ has a point), then $(2,1,2)\cdot(2,2,1) = (1,2,2)$ would also be in the image of δ, but we just saw (in the previous item) that $(1,2,2)$ is *not* in the image of δ. Therefore, we have reached a contradiction and $C(2,2)$ cannot have a rational point. One can rule out all the other (δ_1,δ_2) in the list similarly.

We have analyzed all 64 possible pairs (δ_1,δ_2) and have found that the image of $E(\mathbb{Q})/2E(\mathbb{Q})$ via δ has order 2^4. Therefore, $2^{R_E+2} = 2^4$ and $R_E = 2$. The rank of the curve is exactly 2 and T_1, T_2, P, and Q (as found above) are generators of $E(\mathbb{Q})/2E(\mathbb{Q})$. (In fact, they are generators of $E(\mathbb{Q})$ as well.)

Example 16.8.5. Let $E : y^2 = x^3 - 6724x = x(x-82)(x+82)$. Let $e_1 = 0$, $e_2 = -82$, and $e_3 = 82$. The only divisors of Δ_E are 2 and 41; hence $\Gamma' = \{\pm 1, \pm 2, \pm 41, \pm 82\}$. Let us analyze the homogeneous spaces

$$C(\delta_1,\delta_2) : \begin{cases} \delta_2 Y^2 - \delta_1 X^2 = 82, \\ \delta_2 Y^2 - \delta_1\delta_2 Z^2 = 164 \end{cases}$$

as we did in the previous example. Once again, there are 64 pairs to check:

(1) $((\delta_1,\delta_2,\delta_3) = (1,1,1))$. The point at infinity (i.e., the origin) is sent to $(1,1,1)$ via δ; i.e., $\delta(\mathcal{O}) = (1,1,1)$.

(2) $(\delta_1 < 0 \text{ and } \delta_2 < 0)$. The equation $\delta_2 Y^2 - \delta_1\delta_2 Z^2 = 164$ cannot have rational solutions because the left-hand side is always negative for any $X, Z \in \mathbb{Q}$.

(3) $(\delta_1 > 0 \text{ and } \delta_2 < 0)$. The equation $\delta_2 Y^2 - \delta_1 X^2 = 82$ cannot have rational solutions because the left-hand side is always negative.

(4) $((\delta_1,\delta_2) = (-1,82), (-82,2), (82,41))$. The corresponding spaces have (trivial) rational points that map, respectively, to $T_1 = (0,0)$, $T_2 = (-82,0)$, and $T_3 = T_1 + T_2 = (82,0)$ via (16.14).

(5) $((\delta_1,\delta_2) = (1,2))$. The space $C(1,2)$ does not have rational points (same reason as for Exercise 16.10.20). In fact, it does not have any solutions over $\mathbb{Q}_2$.

(6) $((\delta_1, \delta_2) = (-1, 41), (-82, 1), (82, 82))$. The corresponding spaces cannot have rational points because these elements of Γ_Δ are the product of $(1, 2, 2)$, with no points, times $(-1, 82, -82)$, $(-82, 2, -41)$, $(82, 41, 2)$, which do have points by a previous item in this list.

How about all the other possible pairs (δ_1, δ_2)? Consider $(-1, 2, -2)$ and its homogeneous space:

$$C(-1,2) : \begin{cases} 2Y^2 + X^2 = 82, \\ 2Y^2 + 2Z^2 = 164. \end{cases}$$

Let us show that there are solutions to $C(-1, 2)$ over $\mathbb{R}$, $\mathbb{Q}_2$, and $\mathbb{Q}_{41}$:

- (Over $\mathbb{R}$). The point $(0, \sqrt{41}, \sqrt{41})$ is a point on $C(-1, 2)$ defined over $\mathbb{R}$.
- (Over $\mathbb{Q}_{41}$). Let $Y_0 = 1$ and put $f(X) = X^2 - 80$, $g(Z) = Z^2 - 81$. By Hensel's lemma (see Section 11.6 and Corollary 11.6.4), it suffices to show that there are $\alpha_0, \beta_0 \in \mathbb{F}_{41}$ such that

  $$f(\alpha_0) = g(\beta_0) \equiv 0 \bmod 41 \quad \text{and} \quad f'(\alpha_0),\ g'(\beta_0) \not\equiv 0 \bmod 41.$$

  The reader can check that the congruences $\alpha_0 \equiv 11 \bmod 41$ and $\beta_0 \equiv 9 \bmod 41$ work. Thus, there are $\alpha, \beta \in \mathbb{Q}_{41}$ such that $f(\alpha) = 0 = g(\beta)$. Hence, $(X_0, Y_0, Z_0) = (\alpha, 1, \beta)$ is a point on $C(-1, 2)$ defined over $\mathbb{Q}_{41}$, as desired.
- (Over $\mathbb{Q}_2$). Let $X_0 = 0$ and put $f(Y) = Y^2 - 41$. Let $\alpha_0 = 1$. Then $f(\alpha_0) = -40$, $f'(\alpha_0) = 82$, and

  $$3 = \nu_2(-40) > \nu_2(82^2) = \nu_2(2^2 \cdot 41^2) = 2.$$

  Thus, by Hensel's lemma (Theorem 11.6.3; see also Example 11.6.6), there is $\alpha \in \mathbb{Q}_2$ such that $f(\alpha) = 0$, or $\alpha^2 = 41$. Hence, the point $(X_0, Y_0, Z_0) = (0, \alpha, \alpha)$ is a point on $C(-1, 2)$ defined over $\mathbb{Q}_2$, as desired.

One can also show that, in fact, $C(-1, 2)$ has a point over $\mathbb{Q}_p$ *for all* $p \geq 2$. Therefore, we cannot deduce any contradictions working locally about whether $C(-1, 2)$ has a point over $\mathbb{Q}$. A computer search does not yield any $\mathbb{Q}$-points on $C(-1, 2)$. Therefore, our method breaks at this point, and we cannot determine whether there is a point on $E(\mathbb{Q})$ that comes from $C(-1, 2)$.

It turns out that $C(-1, 2)$ *does not* have rational points (but this is difficult to show). This type of space, a space that has solutions everywhere locally ($\mathbb{Q}_p$, $\mathbb{R}$) but not globally ($\mathbb{Q}$) is the main obstacle for the descent method to fully work. These ideas lead to the definition of Selmer and Shafarevich–Tate groups attached to elliptic curves. We will not define these concepts here, but the reader can learn about them in [**Loz11**] (starting with Section 2.11), which is the natural sequel of this book.

16.9. Application: The Elliptic Curve Diffie–Hellman Key Exchange

In this section we present an application of elliptic curves to cryptography. The elliptic curve Diffie–Hellman (ECDH) key exchange is a variant of the Diffie–Hellman (DH) method that we explained back in Section 8.9.1. As for the standard DH key exchange, the goal is to agree on a private key through an insecure public channel.

The ECDH protocol is currently used by popular texting apps, such as *WhatsApp*, to provide end-to-end encryption for their users' messages.

Elliptic curve Diffie–Hellman key exchange:

(1) Alice and Bob agree on a (large) prime number p, an elliptic curve E defined over $\mathbb{F}_p$, and a point $P \in E(\mathbb{F}_p)$ of large order, through a public channel.

(2) Alice chooses her secret key, an integer $a > 1$, and Bob chooses his secret key, an integer $b > 1$.

(3) Alice computes the point $A = a \cdot P$, and Bob computes $B = b \cdot P \in E(\mathbb{F}_p)$.

(4) Alice sends A to Bob, and Bob sends B to Alice, through a public channel.

(5) Alice computes $K_A = a \cdot B$, and Bob computes $K_B = b \cdot A \in E(\mathbb{F}_p)$.

(6) The secret key shared by Alice and Bob is $K = K_A = K_B \in E(\mathbb{F}_p)$.

Indeed, the keys K_A and K_B coincide as points in $E(\mathbb{F}_p)$:

$$K_A = a \cdot B = a \cdot (b \cdot P) = (ab) \cdot P = b \cdot (a \cdot P) = b \cdot A = K_B.$$

Notice that we have used the fact that addition on an elliptic curve is commutative and associative.

Remark 16.9.1. As in the security analysis of the standard Diffie–Hellman key exchange (see Remark 8.9.1), the security of the exchange relies on the fact that, given a large prime p, an elliptic curve E defined over $\mathbb{F}_p$, a point of large order $P \in E(\mathbb{F}_p)$, and some multiple A of P, it is computationally expensive (i.e., time- and memory-consuming) to find $a \geq 1$ such that $a \cdot P = A \in E(\mathbb{F}_p)$. This is called the *elliptic curve discrete logarithm problem*. The ECDH protocol is in general preferred over DH because the elliptic curve discrete logarithm problem seems to be much harder to solve than the standard discrete logarithm problem, due to the fact that the addition on an elliptic curve $E/\mathbb{F}_p$ is much more intricate than multiplication in $(\mathbb{Z}/p\mathbb{Z})^\times$. Thus, the same size keys afford greater security using an ECDH key exchange than via a DH protocol.

Example 16.9.2. Alice and Bob set up an ECDH key exchange as follows:

(1) Alice and Bob agree on a prime number $p = 103$, an elliptic curve E defined over $\mathbb{F}_{103}$,
$$E : y^2 = x^3 - 2,$$
and a point $P = (3, 5) \in E(\mathbb{F}_{103})$, through a public channel. (The order of P is 91.)

(2) Alice chooses her secret key, $a = 12$, and Bob chooses his secret key, $b = 29$.

(3) Alice computes the point $A = 12 \cdot P = (21, 35)$, and Bob computes $B = 29 \cdot P = (39, 68) \in E(\mathbb{F}_{103})$.

(4) Alice sends A to Bob, and Bob sends B to Alice, through a public channel.

(5) Alice computes $K_A = 12 \cdot B = (58, 50)$, and Bob computes $K_B = 29 \cdot A = (58, 50) \in E(\mathbb{F}_p)$.

(6) The secret key shared by Alice and Bob is $K = K_A = K_B = (58, 50) \in E(\mathbb{F}_{103})$.

Usually, we use the x-coordinate as the secret, so 58 may be used, for instance, as a key $(k_1, k_2) = (5, 8)$ for a Vigenère cipher (see Section 4.6.4). The reader can find examples of elliptic curve discrete logarithm problems and the elliptic curve Diffie–Hellman key exchange in Exercises 16.10.26, 16.10.27, and 16.10.28.

16.10. Exercises

Exercise 16.10.1. The graph of an elliptic curve E defined over $\mathbb{Q}$ is sketched in Figure 16.6. Use the geometric definition of the addition law on the points on E to draw the approximate position of the points $P + Q$, $2S = S + S$, and $P + 2S$.

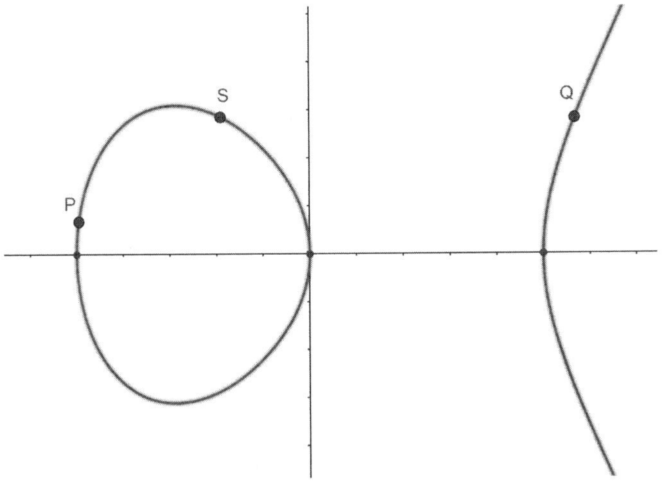

Figure 16.6. Draw the approximate position of the points $P + Q$, $2S$, and $P + 2S$.

Exercise 16.10.2. Let E be the elliptic curve $y^2 = x^3 + 3x + 5$ over the finite field $\mathbb{F}_{13}$. Is $P = (4, 4)$ a point on the curve E?

Exercise 16.10.3. Let E be the elliptic curve $y^2 = x^3 + x + 1$ defined over the field $\mathbb{F}_5$. Let $P = (0, 1)$ and $Q = (2, 4)$.

(1) Use the formulas of addition on an elliptic curve to compute the point $P + Q$.

(2) Use the formulas of addition to compute the point $2Q = Q + Q$.

(3) Use Hasse's theorem, Theorem 16.5.11, to give an upper bound for the total number of points on $E(\mathbb{F}_5)$.

(4) Find all the points on $E(\mathbb{F}_5)$, where E is $y^2 = x^3 + x + 1$ over the field $\mathbb{F}_5$.

Exercise 16.10.4. Let $E : y^2 = x^3 - 2$ be an elliptic curve defined over $\mathbb{Q}$, and let $P = (3, 5)$. Use the formulas for addition on an elliptic curve to compute $-P$, $2P$, $3P$, and $4P$.

Exercise 16.10.5. Let $E/\mathbb{Q}$ be an elliptic curve given by a Weierstrass equation of the form $y^2 = f(x)$, where $f(x) \in \mathbb{Z}[x]$ is a monic cubic polynomial with distinct roots (over $\mathbb{C}$).

(1) Show that $P = (x, y) \in E$ is a torsion point of exact order 2 if and only if $y = 0$ and $f(x) = 0$.

(2) Let $E(\mathbb{Q})[2]$ be the subgroup of $E(\mathbb{Q})$ formed by those rational points $P \in E(\mathbb{Q})$ such that $2P = \mathcal{O}$. Show that the size of $E(\mathbb{Q})[2]$ may be 1, 2, or 4.

(3) Give examples of three elliptic curves defined over $\mathbb{Q}$ where the size of $E(\mathbb{Q})[2]$ is 1, 2, and 4, respectively.

Exercise 16.10.6. Let $E_t : y^2 + (1-t)xy - ty = x^3 - tx^2$ with $t \in \mathbb{Q}$ and $\Delta_t = t^5(t^2 - 11t - 1) \neq 0$. As we saw in Example 16.4.4, every curve E_t has a subgroup isomorphic to $\mathbb{Z}/5\mathbb{Z}$. Use SageMath to find elliptic curves with torsion $\mathbb{Z}/5\mathbb{Z}$ and rank 0, 1, and 2. Also, try to find an elliptic curve E_t with rank r, as high as possible. (Note: the highest rank known, as of the writing of this book, for an elliptic curve with $\mathbb{Z}/5\mathbb{Z}$ torsion is 6, discovered by Dujella and Lecacheux in 2001; see [**Duj09**] for up-to-date records.)

Exercise 16.10.7. Let $p \geq 2$ be a prime and $E_p : y^2 = x^3 + p^2$. Show that there is no torsion point $P \in E_p(\mathbb{Q})$ with $y(P)$ equal to

$$y = \pm 1,\ \pm p^2,\ \pm 3p,\ \pm 3p^2,\ \text{or}\ \pm 3.$$

Prove that $Q = (0, p)$ is a torsion point of exact order 3. Conclude that $\{\mathcal{O}, Q, 2Q\}$ are the only torsion points on $E_p(\mathbb{Q})$. (Note: for $p = 3$, the point $(-2, 1) \in E_3(\mathbb{Q})$. Show that it is *not* a torsion point.)

Exercise 16.10.8. Prove Proposition 16.5.8, as follows:

(1) First show that if $f(x)$ is a polynomial, $f'(x)$ its derivative, and $f(\delta) = f'(\delta) = 0$, then $f(x)$ has a double root at δ.

(2) Show that if $y^2 = f(x)$ is singular, where $f(x) \in K[x]$ is a monic cubic polynomial, then the singularity must occur at $(\delta, 0)$, where δ is a root of $f(x)$.

(3) Show that $(\delta, 0)$ is singular if and only if δ is a double root of $f(x)$. Therefore $D = 0$ if and only if E is singular.

Exercise 16.10.9. Let $E/\mathbb{Q} : y^2 = x^3 + 3$. Find all the points of $\widetilde{E}(\mathbb{F}_7)$ and verify that N_7 satisfies Hasse's bound.

Exercise 16.10.10. Let $E/\mathbb{Q} : y^2 = x^3 + Ax + B$ and let $p \geq 3$ be a prime of bad reduction for $E/\mathbb{Q}$. Show that $E(\mathbb{F}_p)$ has a unique singular point.

Exercise 16.10.11. Prove parts (1) and (3) of Theorem 16.6.11. (Hint: use Definition 16.6.10 and Proposition 16.6.3.)

Exercise 16.10.12. Prove Corollary 16.6.12.

Exercise 16.10.13. Let $E : y^2 = x^3 - 10081x$. Use SageMath (or Magma [**BCP97**]) to find a minimal set of generators for the subgroup that is spanned by all these points on E:

$$(0,0),\ (-100,90),\ \left(\frac{10081}{100},\frac{90729}{1000}\right),\ (-17,408),$$
$$\left(\frac{907137}{6889},-\frac{559000596}{571787}\right),\ \left(\frac{1681}{16},\frac{20295}{64}\right),\ \left(\frac{833}{4},\frac{21063}{8}\right),$$
$$\left(-\frac{161296}{1681},\frac{19960380}{68921}\right),\ \left(-\frac{6790208}{168921},-\frac{40498852616}{69426531}\right).$$

(Hint: use Theorem 16.6.4 to determine the rank of $E/\mathbb{Q}$.)

Exercise 16.10.14. Let E and δ be defined as in Theorem 16.7.3, and suppose $P = (x_0, y_0)$ is a point on E with $y_0 \neq 0$. Show:

- $\delta(P) \cdot \delta(\mathcal{O}) = \delta(P)$.
- $\delta((e_1, 0)) \cdot \delta((e_2, 0)) = \delta((e_1, 0) + (e_2, 0))$.
- $\delta(P) \cdot \delta((e_1, 0)) = \delta(P + (e_1, 0))$.

Exercise 16.10.15. Let $E : y^2 = x^3 + Ax + B$ be an elliptic curve with $A, B \in \mathbb{Q}$, and suppose $P = (x_0, y_0)$ is a point on E, with $y_0 \neq 0$.

(1) Prove that the x-coordinate of $2P$ is given by
$$x(2P) = \frac{x_0^4 - 2Ax_0^2 - 8Bx_0 + A^2}{4y_0^2}.$$
(2) Find a formula for $y(2P)$ in terms of x_0 and y_0.

Exercise 16.10.16. The curve $E/\mathbb{Q} : y^2 = x^3 - 157^2x$ has a rational point Q with x-coordinate $x = x(Q)$ given by

$$x = \left(\frac{224403517704336969924557513090674863160948472041}{17824664537857719176051070357934327140032961660}\right)^2.$$

Show that there exists a point $P \in E(\mathbb{Q})$ such that $2P = Q$. Find the coordinates of P. (Hint: use SageMath and Exercise 16.10.15.)

Exercise 16.10.17. Let $E : y^2 = (x-e_1)(x-e_2)(x-e_3)$ with $e_i \in \mathbb{Q}$, distinct, and such that $e_1+e_2+e_3 = 0$. Additionally, suppose that $e_1 - e_2 = n^2$ and $e_1 - e_3 = m^2$ are squares. This exercise shows that, under these assumptions, there is a point $P = (x_0, y_0)$ such that $2P = (e_1, 0)$; i.e., P is a point of exact order 4.

(1) Show that $e_1 = \frac{n^2+m^2}{3}$, $e_2 = \frac{m^2-2n^2}{3}$, $e_3 = \frac{n^2-2m^2}{3}$.
(2) Find A and B, in terms of n and m, such that
$$x^3 + Ax + B = (x - e_1)(x - e_2)(x - e_3).$$
(Hint: SageMath can be of great help here.)
(3) Let $p(x) = x^4 - 2Ax^2 - 8Bx + A^2 - 4(x^3 + Ax + B)e_1$. Show that $p(x_0) = 0$ if and only if $x(2P) = e_1$ and therefore $2P = (e_1, 0)$. (Hint: use Exercise 16.10.15.)
(4) Express all the coefficients of $p(x)$ in terms of n and m. (Hint: use a computer.)
(5) Factor $p(x)$ for $(n, m) = (3, 6),\ (3, 12),\ (9, 12), \ldots$.

(6) Guess that $p(x) = (x-a)^2(x-b)^2$ for some a and b. Express all the coefficients of $p(x)$ in terms of a and b.

(7) Finally, compare the coefficients of $p(x)$ in terms of a, b and n, m and find the roots of $p(x)$ in terms of n, m. (Hint: compare first the coefficient of x^3 and then the coefficient of x^2.)

(8) Write $P = (x_0, y_0)$ in terms of n and m.

Exercise 16.10.18. Let e_1, e_2, e_3 be three distinct integers. Show that $\Delta = (e_1 - e_2)(e_2 - e_3)(e_1 - e_3)$ is always even.

Exercise 16.10.19. In this exercise we study the structure of the quotient $G/2G$, where G is a finite abelian group.

(1) Let $p \geq 2$ be a prime and let $G = \mathbb{Z}/p^e\mathbb{Z}$, with $e \geq 1$. Prove that $G/2G$ is trivial if and only if $p > 2$.

(2) Prove that if $G = \mathbb{Z}/2^e\mathbb{Z}$ and $e \geq 1$, then $G/2G \cong \mathbb{Z}/2\mathbb{Z}$.

(3) Finally, let G be an arbitrary finite abelian group. We define $G[2^\infty]$ to be the 2-primary component of G; i.e.,

$$G[2^\infty] = \{g \in G : 2^n \cdot g = 0 \text{ for some } n \geq 1\}.$$

In other words, $G[2^\infty]$ is the subgroup of G formed by those elements of G whose order is a power of 2. Prove that

$$G[2^\infty] \cong \mathbb{Z}/2^{e_1}\mathbb{Z} \oplus \mathbb{Z}/2^{e_2}\mathbb{Z} \oplus \cdots \oplus \mathbb{Z}/2^{e_r}\mathbb{Z}$$

for some $r \geq 0$ and $e_i \geq 1$ (here $r = 0$ means $G[2^\infty]$ is trivial). Also show that $G/2G \cong (\mathbb{Z}/2\mathbb{Z})^r$.

Exercise 16.10.20. Show that the space

$$C : \begin{cases} 2Y^2 - X^2 = 34, \\ Y^2 - Z^2 = 34 \end{cases}$$

does not have any rational solutions with $X, Y, Z \in \mathbb{Q}$. (Hint: modify the system so there are no powers of 2 in any of the denominators and then work modulo 8.)

Exercise 16.10.21. For the following elliptic curves, use the method of 2-descent (as in Proposition 16.8.3 and Example 16.8.4) to find the rank of $E/\mathbb{Q}$ and generators of $E(\mathbb{Q})/2E(\mathbb{Q})$. **Do not** use a computer:

(1) $E : y^2 = x^3 - 14931x + 220590$.

(2) $E : y^2 = x^3 - x^2 - 6x$.

(3) $E : y^2 = x^3 - 37636x$.

(4) $E : y^2 = x^3 - 962x^2 + 148417x$. (Hint: use Theorem 16.6.4 first to find a bound on the rank.)

Exercise 16.10.22. Find the rank and generators for the rational points on the elliptic curve $y^2 = x(x+5)(x+10)$.

Exercise 16.10.23 (Elliptic curves with non-trivial rank)**.** The goal here is a systematic way to find curves of rank at least $r \geq 0$ without using tables of elliptic curves:

(1) (Easy) Find three non-isomorphic elliptic curves over $\mathbb{Q}$ with rank ≥ 2. You must prove that the rank is at least 2. (To show linear independence, you may use SageMath or Magma to calculate the height matrix.)

(2) (Fair) Find three non-isomorphic elliptic curves over $\mathbb{Q}$ with rank ≥ 3.

(3) (Medium difficulty) Find three non-isomorphic elliptic curves over $\mathbb{Q}$ with rank ≥ 6. If so, then you can probably find three curves of rank ≥ 8 as well.

(4) (Significantly harder) Find three non-isomorphic elliptic curves over $\mathbb{Q}$ of rank ≥ 10.

(5) (You would be famous!) Find an elliptic curve over $\mathbb{Q}$ of rank ≥ 29.

Exercise 16.10.24. Let E be an elliptic curve and suppose that the images of the points $P_1, P_2, \ldots, P_n \in E(\mathbb{Q})$ in $E(\mathbb{Q})/2E(\mathbb{Q})$ generate the group $E(\mathbb{Q})/2E(\mathbb{Q})$. Let G be the subgroup of $E(\mathbb{Q})$ generated by $P_1, P_2, \ldots, P_n$.

(1) Prove that the index of G in $E(\mathbb{Q})$ is finite, i.e., the quotient group $E(\mathbb{Q})/G$ is finite.

(2) Show that, depending on the choice of generators $\{P_i\}$ of the quotient group $E(\mathbb{Q})/2E(\mathbb{Q})$, the size of $E(\mathbb{Q})/G$ may be arbitrarily large.

Exercise 16.10.25. Fermat's last theorem shows that $x^3 + y^3 = z^3$ has no integer solutions with $xyz \neq 0$. Find the first $d \geq 1$ such that $x^3 + y^3 = dz^3$ has infinitely many non-trivial solutions, find a generator for the solutions, and write down a few examples. (Hint: Example 16.1.3.)

Exercise 16.10.26. Let E be the curve $y^2 = x^3 - 2$ defined over $\mathbb{F}_{17}$, and let $P = (3, 5)$. Can you solve the elliptic curve discrete logarithm problem $x \cdot P = (13, 11)$? In other words, find an integer $x \geq 1$ such that $x \cdot P = (13, 11)$ in $E(\mathbb{F}_{17})$.

Exercise 16.10.27. Let E be the curve $y^2 = x^3 - 2$ defined over $\mathbb{F}_{103}$, and let $P = (3, 5)$.

(1) Show that the order of P in $E(\mathbb{F}_{103})$ is 91.

(2) Solve the elliptic curve discrete logarithm problem $x \cdot P = (102, 93)$ in $E(\mathbb{F}_{103})$.

(3) Show that P is a generator of $E(\mathbb{F}_{103})$; i.e., if $Q \in E(\mathbb{F}_{103})$, then the elliptic curve discrete logarithm problem $x \cdot P = Q$ always has a solution.

(Hint: use a computer and the software Magma [**BCP97**] or SageMath [**Sage**].)

Exercise 16.10.28. Let $p = 541$ (which is a prime), and let E be the elliptic curve $y^2 = x^3 + x + 1$ defined over $\mathbb{F}_{541}$. Let $P = (72, 70)$ on $E(\mathbb{F}_{541})$. The following is a list of the multiples $n \cdot P$ for $1 \leq n \leq 59$, in order:

$$\begin{gathered}
(72,70),(424,71),(9,110),(338,159),(255,123),\\
(161,528),(147,468),(168,416),(480,353),(454,92),\\
(360,174),(264,41),(152,438),(468,56),(437,44),\\
(68,447),(459,293),(115,326),(328,507),(278,318),\\
(113,117),(534,456),(307,277),(1,57,1),(491,440),\\
(107,249),(465,115),(67,517),(301,61),(301,480),\\
(67,24),(465,426),(107,292),(491,101),(1,484),\\
(307,264),(534,85),(113,424),(278,223),(328,34),\\
(115,215),(459,248),(68,94),(437,497),(468,485),\\
(152,103),(264,500),(360,367,1),(454,449),(480,188),\\
(168,125),(147,73),(161,13),(255,418),(338,382),\\
(9,431),(424,470),(72,471),(0:1:0),
\end{gathered}$$

where $(0 : 1 : 0)$ is $\mathcal{O}$, the point at infinity. In other words, $P = (72, 70)$, $2P = (424, 71)$, $3P = (9, 110)$, $\ldots$, $6P = (161, 528)$, etc.

(1) Verify that the order of P in $E(\mathbb{F}_{541})$ is 59.

(2) Rey and Finn want to set up an elliptic curve Diffie–Hellman key exchange with $p = 541$ and E and P as above. Rey chooses $a = 10$ as her secret integer. What point A should Rey send to Finn?

(3) Next, Rey receives $B = (459, 293)$ from Finn. Determine the secret point that is shared between Rey and Finn.

(4) General Hux intercepts a communication between Rey and Finn (not the one from part (2) and (3), but using the same p, E, and P). Hux now knows that Rey sent $A = (534, 456)$ to Finn and Finn sent $B = (255, 123)$ to Rey. Explain how Hux can now find the secret point that Rey and Finn share.

BIBLIOGRAPHY

[AC95] A. Adler and J. E. Coury, *Theory of numbers: A text and source book of problems*, 1st ed., Jones and Bartlett Publishers, March 1995.

[AKS04] M. Agrawal, N. Kayal, and N. Saxena, *PRIMES is in P*, Ann. of Math. (2) **160** (2004), no. 2, 781–793, DOI 10.4007/annals.2004.160.781. MR2123939

[ALP08] J. Aguirre, Á. Lozano-Robledo, and J. C. Peral, *Elliptic curves of maximal rank*, Proceedings of the "Segundas Jornadas de Teoría de Números", Bibl. Rev. Mat. Iberoamericana, Rev. Mat. Iberoamericana, Madrid, 2008, pp. 1–28. MR2603895

[Apo76] T. M. Apostol, *Introduction to analytic number theory*, Undergraduate Texts in Mathematics, Springer-Verlag, New York-Heidelberg, 1976. MR0434929

[Bak90] A. Baker, *Transcendental number theory*, 2nd ed., Cambridge Mathematical Library, Cambridge University Press, Cambridge, 1990. MR1074572

[BH62] P. T. Bateman and R. A. Horn, *A heuristic asymptotic formula concerning the distribution of prime numbers*, Math. Comp. **16** (1962), 363–367, DOI 10.2307/2004056. MR0148632

[BSD63] B. Birch and H. P. F. Swinnerton-Dyer, *Notes on elliptic curves (I) and (II)*, J. Reine Angew. Math. **212** (1963), pp. 7-25, and **218** (1965), pp. 79-108. MR0146143, MR0179168

[BCP97] W. Bosma, J. Cannon, and C. Playoust, *The Magma algebra system. I. The user language*, J. Symbolic Comput., **24** (1997), 235-265. Free Magma calculator at http://magma.maths.usyd.edu.au/calc/, doi=10.1006/jsco.1996.0125 MR1484478

[Bur10] D. Burton, *Elementary number theory*, 7th ed., McGraw-Hill Science, Engineering, Math., 2010.

[Cam01] P. J. Cameron, *The random graph revisited*, European Congress of Mathematics, Vol. I (Barcelona, 2000), Progr. Math., vol. 201, Birkhäuser, Basel, 2001, pp. 267–274. MR1905324

[Che39] J. Chernick, *On Fermat's simple theorem*, Bull. Amer. Math. Soc. **45** (1939), no. 4, 269–274, DOI 10.1090/S0002-9904-1939-06953-X. MR1563964

[Chi95] L. N. Childs, *A concrete introduction to higher algebra*, 2nd ed., Undergraduate Texts in Mathematics, Springer-Verlag, New York, 1995. MR1354141

[Coh06] H. Cohn, *A short proof of the simple continued fraction expansion of e*, Amer. Math. Monthly **113** (2006), no. 1, 57–62, DOI 10.2307/27641837. MR2202921

[Con1] K. Conrad, *Expository Papers*, http://www.math.uconn.edu/~kconrad/blurbs/

[Con2] K. Conrad, *Hensel's Lemma*, available online at http://.../~kconrad/blurbs/gradnumthy/hensel.pdf

[Con3] K. Conrad, *Sums of squares in* $\mathbf{Q}$ *and* $F(T)$, available online at http://.../~kconrad/blurbs/linmultialg/sumsquareQF(T).pdf

[Con4] K. Conrad, *Irrationality of* π *and* e, available online at http://.../~kconrad/blurbs/analysis/irrational.pdf

[Cox13] D. A. Cox, *Primes of the form* $x^2 + ny^2$*: Fermat, class field theory, and complex multiplication*, 2nd ed., Pure and Applied Mathematics (Hoboken), John Wiley & Sons, Inc., Hoboken, NJ, 2013. MR3236783

[Cre97] J. Cremona, *Algorithms for modular elliptic curves*, Cambridge University Press, 1997 (available for free online).

[Dan13] H. Daniels, *Siegel functions, modular curves, and Serre's uniformity problem*, Ph.D. Thesis (2013), University of Connecticut, available at http://alozano.clas.uconn.edu/my-students/

[Duj09] A. Dujella's website, http://web.math.hr/~duje/tors/tors.html

[DF03] D. S. Dummit and R. M. Foote, *Abstract algebra*, 3rd ed., John Wiley & Sons, Inc., Hoboken, NJ, 2004. MR2286236

[Ger08] L. J. Gerstein, *Basic quadratic forms*, Graduate Studies in Mathematics, vol. 90, American Mathematical Society, Providence, RI, 2008. MR2396246

[Gou97] F. Q. Gouvêa, *p-adic Numbers: An Introduction*, 2nd ed., Universitext, Springer-Verlag, Berlin, 1997. MR1488696

[HL23] G. H. Hardy and J. E. Littlewood, *Some problems of 'Partitio numerorum'; III: On the expression of a number as a sum of primes*, Acta Math. **44** (1923), no. 1, 1–70, DOI 10.1007/BF02403921. MR1555183

[HW38] G. H. Hardy and E. M. Wright, *An introduction to the theory of numbers*, 6th ed., revised by D. R. Heath-Brown and J. H. Silverman, with a foreword by Andrew Wiles, Oxford University Press, Oxford, 2008. MR2445243

[Hea10] T. L. Heath, *Diophantus of Alexandria: A study in the history of Greek algebra*, 2nd ed., Cambridge, England, 1910. MR1548384

[HB86] D. R. Heath-Brown, *Artin's conjecture for primitive roots*, Quart. J. Math. Oxford Ser. (2) **37** (1986), no. 145, 27–38, DOI 10.1093/qmath/37.1.27. MR830627

[HPS14] J. Hoffstein, J. Pipher, and J. H. Silverman, *An introduction to mathematical cryptography*, 2nd ed., Undergraduate Texts in Mathematics, Springer, New York, 2014. MR3289167

[IR98] K. Ireland and M. Rosen, *A classical introduction to modern number theory*, 2nd ed., Graduate Texts in Mathematics, Springer, 1998. MR1070716

[Kan91] R. Kanigel, *The man who knew infinity: A life of the genius Ramanujan*, Charles Scribner's Sons, New York, 1991. MR1113890

[Kub76] D. S. Kubert, *Universal bounds on the torsion of elliptic curves*, Proc. London Math. Soc. (3) **33** (1976), no. 2, 193–237, DOI 10.1112/plms/s3-33.2.193. MR0434947

[Lor96] D. Lorenzini, *An invitation to arithmetic geometry*, Graduate Studies in Mathematics, vol. 9, American Mathematical Society, Providence, RI, 1996. MR1376367

[Loz11] Á. Lozano-Robledo, *Elliptic curves, modular forms, and their L-functions*, Student Mathematical Library, vol. 58, American Mathematical Society, Providence, RI; Institute for Advanced Study (IAS), Princeton, NJ, 2011. IAS/Park City Mathematical Subseries. MR2757255

[Lut37] E. Lutz, *Sur l'équation $y^2 = x^3 - Ax - B$ dans les corps $\mathfrak{p}$-adiques* (French), J. Reine Angew. Math. **177** (1937), 238–247, DOI 10.1515/crll.1937.177.238. MR1581558

[Mat93] Y. V. Matiyasevich, *Hilbert's tenth problem*, MIT Press, Cambridge, Massachusetts, 1993.

[Maz78] B. Mazur, *Rational isogenies of prime degree (with an appendix by D. Goldfeld)*, Invent. Math. **44** (1978), no. 2, 129–162, DOI 10.1007/BF01390348. MR482230

[Mil06] J. S. Milne, *Elliptic curves*, Kea Books, 2006, freely available at http://www.jmilne.org/math/Books/

[Nag35] T. Nagell, *Solution de quelque problemes dans la theorie arithmetique des cubiques planes du premier genre*, Wid. Akad. Skrifter Oslo I, 1935, Nr. 1.

[Ros10] K. H. Rosen, *Elementary number theory*, 6th ed., Pearson, April 9, 2010.

[Rou91] G. Rousseau, *On the quadratic reciprocity law*, J. Austral. Math. Soc. Ser. A **51** (1991), no. 3, 423–425. MR1125443

[Row08] E. S. Rowland, *A natural prime-generating recurrence*, J. Integer Seq. **11** (2008), no. 2, Article 08.2.8, 13. MR2429995

[Sage] *SageMath, the Sage Mathematics Software System*, The Sage Developers, `http://www.sagemath.org`.

[SK52] J. G. Semple and G. T. Kneebone, *Algebraic projective geometry*, Oxford, at the Clarendon Press, 1952. MR0049579

[Ser73] J.-P. Serre, *A course in arithmetic*, translated from the French, Graduate Texts in Mathematics, No. 7, Springer-Verlag, New York-Heidelberg, 1973. MR0344216

[ShT67] I. R. Shafarevich and J. Tate, *The rank of elliptic curves*, AMS Transl. 8 (1967), 917-920.

[Sil86] J. H. Silverman, *The arithmetic of elliptic curves*, Graduate Texts in Mathematics, vol. 106, Springer-Verlag, New York, 1986. MR817210

[Sil12] J. H. Silverman, *A friendly introduction to number theory*, 4th ed., Featured Titles for Number Theory, Pearson, January 28, 2012.

[ST92] J. H. Silverman and J. Tate, *Rational points on elliptic curves*, Undergraduate Texts in Mathematics, Springer-Verlag, New York, 1992. MR1171452

[Ste08] W. Stein, *Elementary number theory: primes, congruences, and secrets. A computational approach*, Undergraduate Texts in Mathematics, Springer, New York, 2009. MR2464052

[Was08] L. C. Washington, *Elliptic curves: Number theory and cryptography*, 2nd ed., Discrete Mathematics and its Applications (Boca Raton), Chapman & Hall/CRC, Boca Raton, FL, 2008. MR2404461

[Wei17] M. H. Weissman, *An illustrated theory of numbers*, American Mathematical Society, Providence, RI, 2017. MR3677120

[Wet98] J. L. Wetherell, *Bounding the number of rational points on certain curves of high rank*, Ph.D. Thesis, University of Southern California, 1998.

[Wil95] A. Wiles, *Modular elliptic curves and Fermat's last theorem*, Ann. of Math. (2) **141** (1995), no. 3, 443–551, DOI 10.2307/2118559. MR1333035

[Zha13] Y. Zhang, *Bounded gaps between primes*, Ann. of Math. (2) **179** (2014), no. 3, 1121–1174, DOI 10.4007/annals.2014.179.3.7. MR3171761

INDEX